D1203265

Geological Society of America
Memoir 179

Regional Geology of Eastern Idaho and Western Wyoming

Edited by

Paul Karl Link
Department of Geology
Idaho State University
Pocatello, Idaho 83209

Mel A. Kuntz
U.S. Geological Survey
MS 913, Box 25046, Federal Center
Denver, Colorado 80225

Lucian B. Platt
Department of Geology
Bryn Mawr College
Bryn Mawr, Pennsylvania 19010

1992

Published by The Geological Society of America, Inc.
3300 Penrose Place, P.O. Box 9140, Boulder, Colorado 80301

GSA Books Science Editor Richard A. Hoppin

Printed in U.S.A.

Library of Congress Cataloging-in-Publication Data
Regional geology of eastern Idaho and western Wyoming / edited by Paul
Karl Link, Mel A. Kuntz, Lucian B. Platt.
p. cm. — (Memoir / Geological Society of America ; 179)
Includes bibliographical references and index.
ISBN 0-8137-1179-7
1. Geology—Idaho. 2. Geology—Wyoming. I. Link, P. K.
II. Kuntz, Mel A. III. Platt, Lucian B., 1931– . IV. Series:
Memoir (Geological Society of America) ; 179.
QE103.R44 1992
557.96—dc20 92-22017
CIP

10 9 8 7 6 5 4 3 2 1

Contents

IV. EXTENSIONAL TECTONICS

V. SNAKE RIVER PLAIN

PLATE

Dedication to Steven S. Oriel (1923–1986)

This collection of papers on the geology of eastern Idaho and western Wyoming is edited by three geologists whose careers have been permanently influenced by their contact with Steve Oriel. This dedication joins three others published previously (Reynolds, 1987; Link, 1987; Platt, 1988); and this volume (with Memoir 171) becomes the second Geological Society of America memoir dedicated to Steven S. Oriel. After Steve died, on July 6, 1986, there was a sense of unfinished business among us and other colleagues; much of the work Steve had organized and in which he had participated over the last ten years of his life was incomplete. This book is an attempt to put closure to some of that work.

Steve Oriel was a regional field geologist in the classical sense and spent most of his career in geologic mapping of the thrust belt of Idaho and Wyoming. Figure 1 shows Steve "in action" in the field over the period 1956 to 1985. During the peak of his career he was the single best source of information on all aspects of the Idaho-Wyoming thrust belt. In Steve's approach to geology, all interpretations found their source in the geologic map, an accurate depiction of things actually seen, a document of great significance. Figure 2 shows areas covered by maps for which he is an author or coauthor. Steve's bibliography lists his most important publications.

Steve was trained by pioneers in thrust belt geology: John Rodgers, his dissertation adviser at Yale University; James Gilluly, branch chief and field adviser in his early days at the U.S. Geological Survey; William W. Rubey, the chief of his first field project; Joshua I. Tracey, his colleague in the work around Kemmerer, Wyoming; and Raymond Price, then of the Geological Survey of Canada, with whom Steve spent a "sabbatical leave" from the Survey. He frequently acknowledged the contributions of these men to his career. He spent significant parts of his professional life working on publication of geologic maps made by and with these colleagues, especially Bill Rubey.

It is with a similar sense of indebtedness to Steve that we present this volume. Our interactions with Steve were in different settings: Platt as a longtime colleague during summer fieldwork, Kuntz as a coworker in the U.S. Geological Survey Snake River Plain project that Steve directed, and Link as one of the graduate students who received Steve's guidance.

From 1972 until his death, Steve was chief and coordinator of a large project of the U.S. Geological Survey to examine the geologic setting of the Snake River Plain region of eastern Idaho and western Wyoming. At the time of his death, the Snake River Plain project ended with unfinished mapping and unpublished papers, but now, six years later, a great amount of work from that and related projects has been completed.

During the 1970s and early 1980s, Steve devoted much of his field time to students working in the thrust belt. He spent more time with some graduate students than did their faculty advisers. He was a patient and demanding teacher who placed the highest value on the precision of the geologic map and cross section. His shadow fell directly on much of the thrust belt literature of those years, just as it does on many of the chapters in this volume.

Among Steve's last major accomplishments were his position as chair of the American Commission on Stratigraphic Nomenclature and his role as editor of the new Stratigraphic Code, published in the American Association of Petroleum Geologists Bulletin in 1983. This too reflects his classical and rigorous approach to geology and his lasting contribution to our science. One of Steve's primary goals was presentation of accurate and precise geologic descriptions, whether stratigraphic or structural.

This book collects several comprehensive chapters by colleagues of Steve who worked with him on the Snake River Plain project during the later years of his life and also includes chapters that were directly or indirectly influenced by Steve. Figure 3 shows the areas of coverage of chapters in this volume. The book is organized into four sections: first, a multidisciplinary overview of the evolution of the Snake River Plain and surrounding regions over the last 17 m.y. by Pierce and Morgan; second, a series of chapters on structure and stratigraphy of the thrust belts north and south of the Snake River Plain; third, several chapters on extensional tectonics of the region; and fourth, chapters on the petrologic character and tectonic development of the Snake River Plain.

The acknowledgments in many of the chapters are testimony to the authors' respect for Steven S. Oriel as a teacher, mentor, supervisor, and friend. Steve enriched our lives and left a legacy as a model of the quintessential field geologist.

Paul Karl Link
Mel A. Kuntz
Lucian B. Platt
June 1992

Figure 1. Montage of photographs of Steven S. Oriel, field geologist, over a span of thirty years. Clockwise from upper left: (A) 1956, field party in Fossil Basin, Wyoming; from left, Wesley E. Lemasurier, William W. Rubey, James Gilluly, Joshua I. Tracey, Steven S. Oriel (photograph by L. B. Platt). (B) 1979, Steve Oriel and Nick Woodward on horseback in the Snake River Range, western Wyoming (photograph by R. W. Allmendinger). The sheepherder's hat was a trademark. (C) 1967, overlooking Commissary Ridge, north of Kemmerer, Wyoming; left to right, Raymond Price, Eric Mountjoy, Steve Oriel. Steve's hand shows the flat-lying, overlapping Tertiary rocks above steeply dipping Paleozoic strata. These overlapping strata were used to date movement on the Absaroka thrust and demonstrate that thrust faults in the Idaho-Wyoming area become progressively younger to the east, toward the craton (photograph by L. B. Platt). (D) June 1985, Steve Oriel at the Absaroka thrust near Dempsey Trail, Cokeville, Wyoming, quadrangle. Steve always liked the big picture (photograph by K. S. Kellogg). (E) July 1977, Steve Oriel at Georgetown Canyon, Idaho (photograph by P. K. Link).

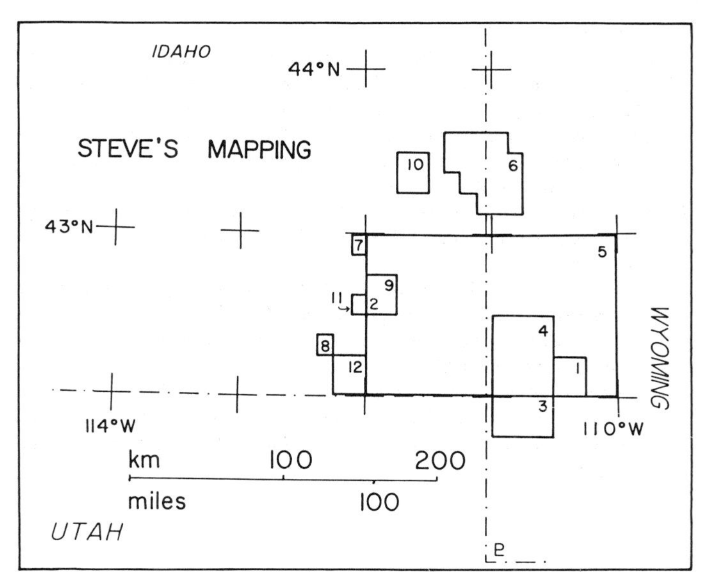

Figure 2. Map of eastern Idaho and western Wyoming showing areas mapped by Steven S. Oriel. Numbers are as follows (refer to bibliography of Oriel's publications in this dedication): 1. Oriel, 1969; 2. Oriel, 1965; 3. Rubey, Oriel, and Tracey, 1975; 4. Rubey, Oriel, and Tracey, 1980; 5. Oriel and Platt, 1968, 1980; 6. Oriel and Moore, 1985; 7. Kellogg, Oriel, Amerman, Link, and Hladky, 1989; 8. Oriel, Platt, and Allmendinger, 1991; 9. Oriel, S. S., unpublished, three-quarters of the Bancroft 15-minute quadrangle, mapped in the early 1960s; 10. Oriel, S. S., unpublished, Hell Creek 15-minute quadrangle, mapped in the mid-1970s; 11. Oriel, S. S., unpublished, Lava Hot Springs quadrangle, mapped in 1979; 12. Oriel, S. S., unpublished, Clifton, Henderson Creek, Malad City East, and Weston Canyon quadrangles, mapped in the 1970s.

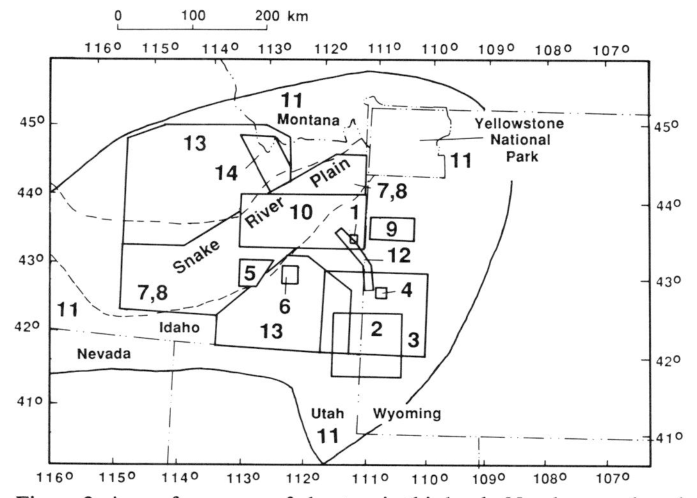

Figure 3. Area of coverage of chapters in this book. Numbers are based on alphabetical listing of authors, as follows: 1. Boyer and Hossack; 2. Coogan; 3. Craddock; 4. Fowles and Woodward; 5. Houser; 6. Kellogg; 7. Kuntz; 8. Kuntz and others; 9. Lageson; 10. Morgan; 11. Pierce and Morgan; 12. Piety and others; 13. Rodgers and Janecke; 14. Skipp and Link.

REFERENCES CITED

Link, P. K., 1987, In memoriam, Steven S. Oriel, *in* Miller, R., ed., The thrust belt revisited: Wyoming Geological Association, 38th Annual Field Conference, Guidebook, p. 6.

Platt, L. B., 1988, Dedication to Steven S. Oriel, *in* Schmidt, C., and Perry, W., eds., Interaction of the Rocky Mountain foreland and the Cordilleran thrust belt: Geological Society of America Memoir 171, p. xv.

Reynolds, M. W., 1987, Memorial to Steven S. Oriel: American Association of Petroleum Geologists Bulletin v. 71, p. 476–477.

BIBLIOGRAPHY OF IMPORTANT PUBLICATIONS OF STEVEN S. ORIEL

Armstrong, F. C., and Oriel, S. S., 1965, Tectonic development of Idaho-Wyoming thrust belt: American Association of Petroleum Geologists Bulletin, v. 49, p. 1847–1866.

Kellogg, K. S., Oriel, S. S., Amerman, R. E., Link, P. K., and Hladky, F. R., 1989, Geologic map of the Jeff Cabin Creek quadrangle, Bannock and Caribou counties, Idaho: U.S. Geological Survey Geologic Quadrangle Map GQ-1669, scale 1:24,000.

Mabey, D. R., and Oriel, S. S., 1970, Gravity and magnetic anomalies in the Soda Springs region, southeastern Idaho: U.S. Geological Survey Professional Paper 646-E, 15 p.

McKee, E. D., Oriel, S. S., and seven others, 1956, Paleotectonic maps of the Jurassic System (for the U.S.): U.S. Geological Survey Miscellaneous Geologic Investigations Map I-475, 5 folio p., 9 pls.

McKee, E. D., Oriel, S. S., and five others, 1960, Paleotectonic maps of the Triassic System (for the U.S.): U.S. Geological Survey Miscellaneous Geologic Investigations Map I-300, 33 folio p., 9 pls.

McKee, E. D., Oriel, S. S., and others, 1967, Paleotectonic maps of the Permian system (for the U.S.): U.S. Geological Survey Miscellaneous Geologic Investigations Map I-450, 164 p., 20 pls., 12 figs.

North American Commission on Stratigraphic Nomenclature (Oriel, S.S., ed.), 1983, North American Stratigraphic Code: American Association of Petroleum Geologists Bulletin, v. 67, p. 841–875.

Oriel, S. S., 1950, Geology and mineral resources of the Hot Springs Window, Madison County, North Carolina: North Carolina Division of Mineral Resources Bulletin 60, 70 p., pls., geologic map.

——, 1951, Structure of the Hot Springs Window, North Carolina: American Journal of Science, v. 249, p. 1–30, illus., geologic map.

——, 1962, Main body of the Wasatch Formation near La Barge, Wyoming: American Association of Petroleum Geologists Bulletin, v. 46, p. 2161–2173.

——, 1965, Preliminary geologic map of the southwest quarter of the Bancroft quadrangle, Bannock and Caribou Counties, Idaho: U.S. Geological Survey Miscellaneous Field Studies Map MF-299, scale 1:24,000.

——, 1968, Preliminary geologic map of the Bancroft quadrangle, Caribou and Bannock Counties, Idaho: U.S. Geological Survey Open-File Map, scale 1:48,000.

——, 1969, Geology of the Fort Hill quadrangle, Lincoln County, Wyoming: U.S. Geological Survey Professional Paper 594-M, 40 p.

Oriel, S. S., and Armstrong, F. C., 1966, Times of thrusting in the Idaho-Wyoming thrust belt, Reply *to* 'Discussion' by Eric Mountjoy: American Association of Petroleum Geologists Bulletin, v. 59, p. 2614–2621.

——, 1972, Uppermost Precambrian and lowest Cambrian rocks, southeastern Idaho: U.S. Geological Survey Professional Paper 394, 52 p.

——, 1986, Tectonic development of Idaho-Wyoming thrust belt: Authors' commentary, *in* Peterson, J. A., ed., Paleotectonics and sedimentation: American Association of Petroleum Geologists Memoir 41, p. 267–279.

Oriel, S. S., and Moore, D. W., 1985, Geologic map of the West and East Palisades RARE II Further Planning Areas, Idaho and Wyoming: U.S. Geological Survey Miscellaneous Field Studies Map MF-1619-B, scale 1:50,000.

Oriel, S. S., and Platt, L. B., 1968, Reconnaissance geologic map of the Preston

30-minute quadrangle, southeastern Idaho: U.S. Geological Survey Open-File Map, 2 sheets, scale 1:62,500.

——, 1980, Geologic map of the Preston 1° × 2° quadrangle, Idaho and Wyoming: U.S. Geological Survey Miscellaneous Investigations Series Map I-1127, scale 1:250,000.

Oriel, S. S., and Tracey, J. I., Jr., 1970, Uppermost Cretaceous and Tertiary stratigraphy of the Fossil Basin, southwestern Wyoming: U.S. Geological Survey Professional Paper 635, 53 p.

Oriel, S. S., Gazin, C. L., and Tracey, J. I., Jr., 1962, Eocene age of Almy Formation in type area, Wyoming: American Association of Petroleum Geologists Bulletin, v. 46, p. 1936–1937.

Oriel, S. S., Platt, L. B., and Allmendinger, R. W., 1991, Reconnaissance geologic map of the Elkhorn Peak quadrangle, Bannock and Oneida Counties, Idaho: U.S. Geological Survey Miscellaneous Field Investigations Map MF-2162, scale 1:24,000.

Peterson, D. D., and Oriel, S. S., 1970, Gravity anomalies in Cache Valley, Cache and Box Elder Counties, Utah, and Bannock and Franklin Counties, Idaho, *in* Geological Survey research 1970: U.S. Geological Survey Professional Paper 700-C, p. C114–C118.

Rubey, W. W. (Oriel, S. S.), 1973, Geologic map and structure sections of the Afton 30-minute quadrangle, Lincoln County, Wyoming: U.S. Geological Survey Miscellaneous Geologic Investigations Map I-686, scale 1:62,500.

Rubey, W. W., Oriel, S. S., and Tracey, J. I., 1968, Geologic map of the Kemmerer quadrangle, Lincoln County, Wyoming: U.S. Geological Survey Open-File Map, scale 1:62,500.

Rubey, W. W., Tracey, J. I., and Oriel, S. S., 1968, Geologic map of the Sage quadrangle, Lincoln County, Wyoming: U.S. Geological Survey Open-File Map, scale 1:62,500.

Rubey, W. W., Oriel, S. S., and Tracey, J. I., Jr., 1975, Geology of the Sage and Kemmerer 15-minute quadrangles, Lincoln County, Wyoming: U.S. Geological Survey Professional Paper 855, 18 p., and geological map and structure-section at 1:62,500.

——, 1980, Geologic map and structure sections of the Cokeville 30-minute quadrangle, Lincoln and Sublette Counties, Wyoming: U.S. Geological Survey Miscellaneous Investigations Series Map I-1129, 2 sheets at scale 1:62,500.

Acknowledgments

REVIEWERS OF CHAPTERS IN THIS BOOK
(IN ADDITION TO EDITORS,
P. K. LINK, M. A. KUNTZ, AND L. B. PLATT)

M. H. Anders
Bill Bonnichsen
D. L. Blackstone, Jr.
S. E. Boyer
R. M. Breckenridge
John Byrd
J. C. Coogan
R. P. Denlinger
Paul Delany
R. A. Duncan
M. A. Ellis
Eric Erslev
K. V. Evans
W. R. Hackett
M. R. Hudson
K. S. Kellogg
R. E. Klinger
P.L.K. Knuepfer
D. R. Lageson
J. D. Love
D. W. Moore
L. D. Nealey
A. R. Nelson
R. P. Nickelsen
H. T. Ore
Dean Ostenaa
K. P. Pogue
W. H. Raymond
Marith Reheis
Frank Royse, Jr.
E. T. Ruppel
T. W. Schirmer
J. G. Schmidt
W. E. Scott
Betty Skipp
J. B. Steidtmann
K. H. Wohletz
S. H. Wood
N. B. Woodward

Geological Society of America
Memoir 179
1992

Chapter 1

The track of the Yellowstone hot spot: Volcanism, faulting, and uplift

Kenneth L. Pierce and Lisa A. Morgan
U.S. Geological Survey, MS 913, Box 25046, Federal Center, Denver, Colorado 80225

ABSTRACT

The track of the Yellowstone hot spot is represented by a systematic northeast-trending linear belt of silicic, caldera-forming volcanism that arrived at Yellowstone 2 Ma, was near American Falls, Idaho about 10 Ma, and started about 16 Ma near the Nevada-Oregon-Idaho border. From 16 to 10 Ma, particularly 16 to 14 Ma, volcanism was widely dispersed around the inferred hot-spot track in a region that now forms a moderately high volcanic plateau. From 10 to 2 Ma, silicic volcanism migrated N54°E toward Yellowstone at about 3 cm/year, leaving in its wake the topographic and structural depression of the eastern Snake River Plain (SRP). This <10-Ma hot-spot track has the same rate and direction as that predicted by motion of the North American plate over a thermal plume fixed in the mantle. The eastern SRP is a linear, mountain-bounded, 90-km-wide trench almost entirely(?) floored by calderas that are thinly covered by basalt flows. The current hot-spot position at Yellowstone is spatially related to active faulting and uplift.

Basin-and-range faults in the Yellowstone-SRP region are classified into six types based on both recency of offset and height of the associated bedrock escarpment. The distribution of these fault types permits definition of three adjoining belts of faults and a pattern of waxing, culminating, and waning fault activity. The central belt, Belt II, is the most active and is characterized by faults active since 15 ka on range fronts >700 m high. Belt II has two arms forming a V that joins at Yellowstone: One arm of Belt II trends south to the Wasatch front; the other arm trends west and includes the sites of the 1959 Hebgen Lake and 1983 Borah Peak earthquakes. Fault Belt I is farthest away from the SRP and contains relatively new and reactivated faults that have not produced new bedrock escarpments higher than 200 m during the present episode of faulting. Belt III is the innermost active belt near the SRP. It contains faults that have moved since 15 to 120 ka and that have been active long enough to produce range fronts more than 500 m high. A belt with inactive faults, belt IV, occurs only south of the SRP and contains range-front faults that experienced high rates of activity coincident with hot-spot volcanism in the late Tertiary on the adjacent SRP. Comparison of these belts of fault activity with historic seismic activity reveals similarities but differences in detail.

That uplift migrated outward from the hot-spot track is suggested by (1) the Yellowstone crescent of high terrain that is about 0.5 km higher than the surrounding terrain, is about 350 km across at Yellowstone, wraps around Yellowstone like a bow wave, and has arms that extend 400 km southerly and westerly from its apex; (2) readily erodible rocks forming young, high mountains in parts of this crescent; (3) geodetic surveys and paleotopographic reconstructions that indicate young uplift near the axis of the Yellowstone crescent; (4) the fact that on the outer slope of this

Pierce, K. L., and Morgan, L. A., 1992, The track of the Yellowstone hot spot: Volcanism, faulting, and uplift, *in* Link, P. K., Kuntz, M. A., and Platt, L. B., eds., Regional Geology of Eastern Idaho and Western Wyoming: Geological Society of America Memoir 179.

crescent glaciers during the last glaciation were anomalously long compared with those of the preceding glaciation, suggesting uplift during the intervening interglaciation; (5) lateral migration of streams, apparent tilting of stream terraces away from Yellowstone, and for increasingly younger terrace pairs, migration away from Yellowstone of their divergent-convergent inflection point; and (6) a geoid dome that centers on Yellowstone and has a diameter and height similar to those of oceanic hot spots.

We conclude that the neotectonic fault belts and the Yellowstone crescent of high terrain reflect heating that is associated with the hot-spot track but has been transferred outward for distances of as much as 200 km from the eastern SRP in 10 m.y. The only practical mechanism for such heat transport would be flow of hot material within the asthenosphere, most likely by a thermal mantle plume rising to the base of the lithosphere and flowing outward horizontally for at least such 200-km distances.

The changes in the volcanic track between 16 to 10 Ma and 10 to 2 Ma is readily explained by first the head (300-km diameter) and then the chimney (10 to 20 km across) phases of a thermal mantle plume rising to the base of the southwest-moving North American plate. About 16 Ma, the bulbous plume head intercepted the base of the lithosphere and mushroomed out, resulting in widespread magmatism and tectonism centered near the common borders of Nevada, Oregon, and Idaho. Starting about 10 Ma near American Falls and progressing to Yellowstone, the chimney penetrated through its stagnating but still warm head and spread outward at the base of the lithosphere, adding basaltic magma and heat to the overriding southwest-moving lithospheric plate, leaving in its wake the eastern SRP–Yellowstone track of calderas, and forming the outward-moving belts of active faulting and uplift ahead and outward from this track.

We favor a mantle-plume explanation for the hot-spot track and associated tectonism and note the following problems with competing hypotheses: (1) for a rift origin, faulting and extension directions are at nearly right angles to that appropriate for a rift; (2) for a transform origin, geologic evidence requires neither a crustal flaw nor differential extension across the eastern SRP, and volcanic alignments on the SRP do not indicate a right-lateral shear across the SRP; and (3) for a meteorite impact origin, evidence expected to accompany such an impact near the Oregon-Nevada border has not been found. The southern Oregon rhyolite zone is not analogous to the eastern SRP and therefore does not disprove formation of the Yellowstone hot-spot track by a mantle plume.

The postulated rise of a mantle-plume head into the mantle lithosphere about 16 Ma corresponds in both time and space with the following geologic changes: (1) the start of the present pattern of basin-range extension, (2) intrusion of basalt and rhyolite along the 1,100-km-long Nevada-Oregon rift zone, (3) the main phases of flood basalt volcanism of the Columbia River and Oregon plateaus, and (4) a change from calc-alkaline volcanism of intermediate to silicic composition to basaltic *and* bimodal rhyolite/basalt volcanism.

INTRODUCTION

The track of the Yellowstone hot spot is defined by the time-transgressive centers of caldera-forming volcanism that has migrated 700 km northeastward to the Yellowstone Plateau volcanic field since 16 Ma (Plate 1; Fig. 1). We compare the progression of silicic volcanism with the timing of late Cenozoic faulting and uplift in nearby areas and suggest that a V-shaped pattern of deformation is now centered on Yellowstone. We use the term *hot spot* to describe this progression of silicic volcanism nongenetically, although we favor its formation by a mantle plume. The hot-spot track is 700 km long, and the fault belts and associated Yellowstone crescent of high terrain extend more than 200 km from the hot-spot track: We attribute these horizontal dimensions to thermal effects originating deeper than the 100(?)-km-thick lithosphere and conclude that the history of uplift, volcanism, and faulting since 10 Ma in eastern Idaho and parts of adjacent states is best explained by the west-southwest movement of the North American plate across a thermal mantle plume.

By tracking back in time from the present Yellowstone hot-spot location, we find a major change about 10 Ma. We think the Yellowstone hot-spot explanation is best appreciated by back-tracking volcanism, faulting, and uplift from the areas of present activity to older positions to the southwest. From 10 Ma to present, caldera-forming volcanism is responsible for the 90-km-wide trench of the eastern Snake River Plain; with increasing age

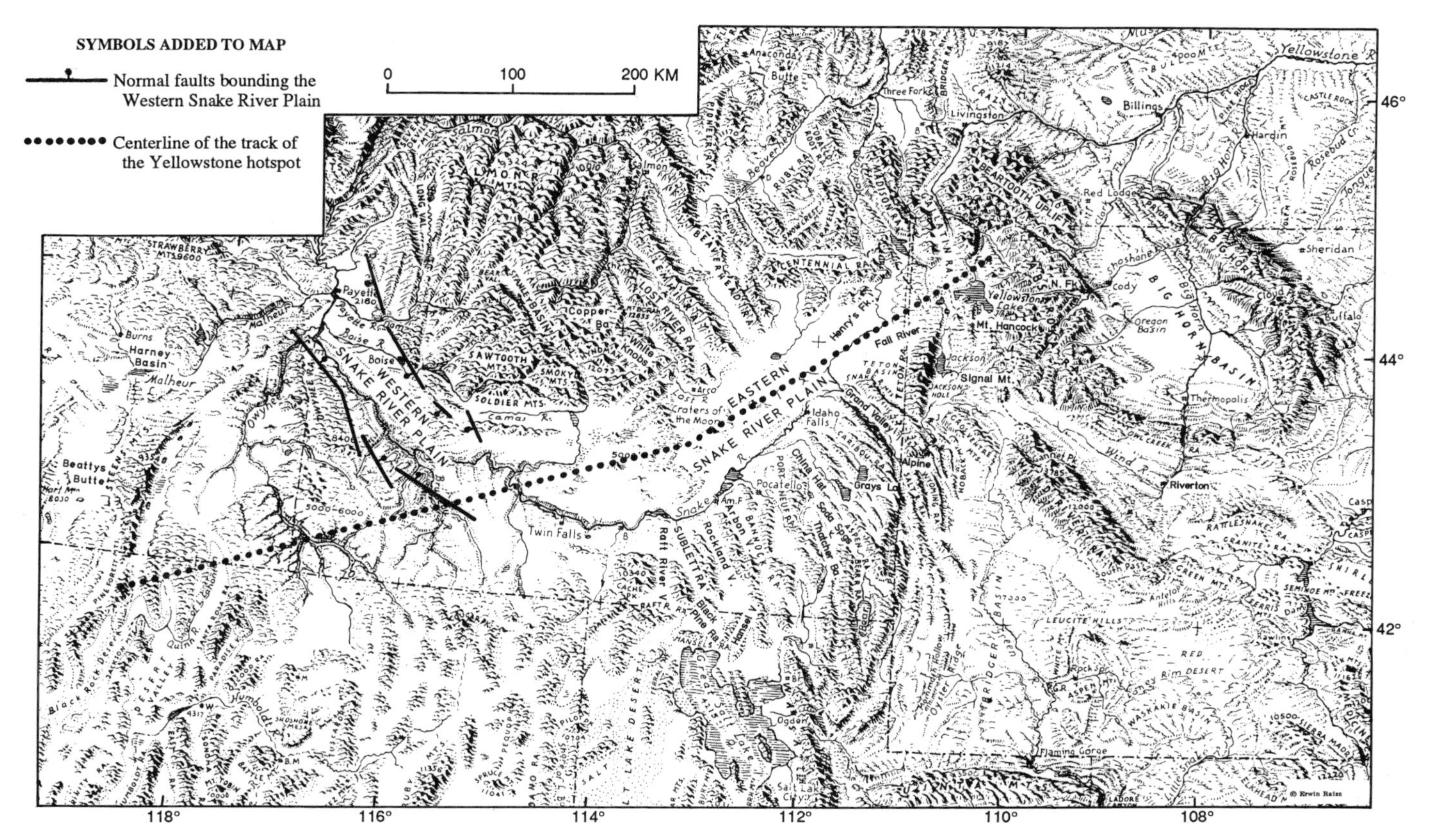

Figure 1. Location map showing geographic features in the region of the Yellowstone hot-spot track. Adapted from Raisz (1957) with minor additions.

prior to 10 Ma, volcanism was increasingly more dispersed back to the inception of the hot spot about 16 Ma. We think a reasonable explanation for this change is the transition from that of a large plume head before 10 Ma to that of a much narrower plume tail or chimney after 10 Ma (Fig. 2; Richards and others, 1989). About 16 Ma the head of a mantle plume, about 300 km in diameter, rose into the base of the southwest-moving North American plate. About 10 Ma, a narrower "tail" or chimney about 10 to 20 km across that was feeding the plume head rose through the stagnant plume head and intercepted the base of the lithosphere.

A mantle-plume hypothesis represents one side of an ongoing controversy about the origin of the eastern Snake River Plain–Yellowstone Plateau (SRP-YP) province. Plate-tectonic, global-scale studies often simply state that the province represents a hot-spot track and commonly include it in inventories of hot spots (for example, Morgan, 1972). However, several prominent researchers in the region have argued for lithospheric movements that drive asthenospheric processes such as upwelling and against an active mantle plume modifying a passive lithosphere (Christiansen and McKee, 1978; Hamilton, 1989). Our acceptance of the mantle-plume hypothesis comes after serious consideration of these models.

A hot-spot/mantle-plume mechanism, particularly the start of a hot-spot track with a large-diameter plume head, has rarely been invoked in North American geology. On a global scale, the existence and importance of hot spots and mantle plumes have gained credibility through their successful application to such topics as plate tectonics, flood basalts associated with rifts, volcanic hot-spot island chains and associated swells, and anomalies in the geoid (see Bercovici and others, 1989; Sleep, 1990; Wilson, 1990; Duncan and Richards, 1991). J. Tuzo Wilson (1963) proposed that the Hawaiian Islands, as well as other volcanic island chains, are formed by a stationary heat source located beneath the moving lithosphere—a hot spot. Morgan (1972) argued that hot spots were anchored by deep mantle plumes and showed that hot-spot tracks approximated the absolute motions of plates. Any motion of hot spots relative to each other has been recently reduced to less than 3 to 5 mm/year (Duncan and Richards, 1991) from earlier values of less than 10 mm/year (Minster and Jordan, 1978).

Hot spots and their tracks are better displayed and more common in oceanic than in continental lithosphere. They are preferentially located near divergent plate boundaries and preferentially excluded near convergent plate boundaries (Weinstein and Olson, 1989). However, during the Mesozoic–earliest Ce-

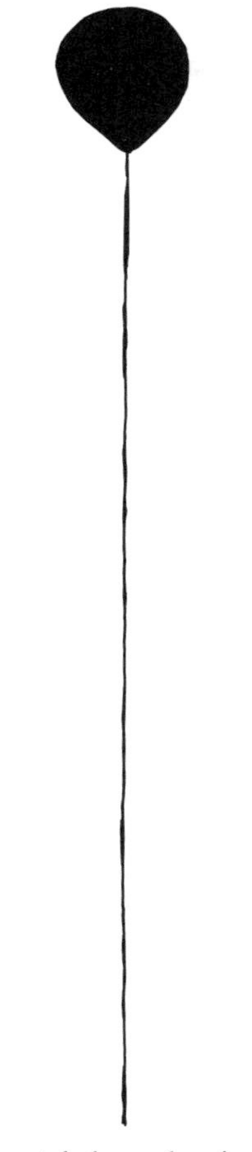

Figure 2. Sketch of experimental plume showing head and tail (chimney) parts. Drawn from photograph shown in Richards and others (1989, Fig. 2). The large bulbous plume head is fed by a much narrower tail or chimney. In this scale model, the plume head was 1.3 cm across. For the Yellowstone hot-spot track, we postulate that a plume head about 300 km across intercepted the lithosphere at 16 to 17 Ma and produced widespread volcanism and deformation, whereas from 10 Ma to present a much narrower chimney (tail) 10 to 20 km across has produced more localized volcanism, faulting, and uplift.

nozoic breakup of the supercontinent Gondwana, the continental lithosphere was greatly affected by the Reunion, Tristan, and probably the Marion hot spots (White and McKenzie, 1989). Mantle plumes feeding these hot spots rose into continental lithosphere—probably starting with mantle-plume heads—and released voluminous flood basalts. They also produced domal uplifts about 2,000 km across (Cox, 1989) and caused continental rifts, many of which evolved into the present oceanic spreading centers (Richards and others, 1989). The dispersion of continents following breakup of Gondwana included the outward movement of oceanic spreading ridges, some of which have apparently crossed hot spots; this suggests that hot spots have both a deeper origin and lesser absolute motion than the spreading centers (Duncan, 1984).

No continental analogues similar to Yellowstone–eastern Snake River Plain are known to us for a hot spot/mantle plume. The following special characteristics of the North American plate and the western United States are probably important in how the Yellowstone hot spot is manifest. (1) The North American plate moves 2 to 3 cm/year southwestward, much faster than for postulated hot spots beneath the nearly stationary African plate (Crough, 1979, 1983). The speed of the North American plate is only one-third that of the plate above the "type" Hawaiian hot spot, which penetrates oceanic crust. (2) The present location of the Yellowstone hot spot is at the northeast edge of the northeast quadrant of the active basin-range structural province bordered to the north and east by high terrain of the Rocky Mountains. (3) The Yellowstone mantle plume rose into crust thickened during the Mesozoic and earliest Tertiary orogenies (Sevier and Laramide) (Christiansen and Lipman, 1972; Wernicke and others, 1987; Molnar and Chen, 1983). (4) About 2 Ma, the Yellowstone hot spot left the thickened crust of the thrust belt and passed beneath the stable craton. (5) The plate margin southwest of the hot-spot track has been progressively changing from a subduction zone to a weak(?) transcurrent fault (Atwater, 1970) over a time span that overlaps the postulated activity of the Yellowstone hot spot since 16 Ma.

The volcanic age progression of the Yellowstone track is rather systematic along its 700-km track, whereas that for other postulated continental hot spots is less systematic, such as for the White Mountain igneous province (Duncan, 1984), the Raton (New Mexico)–Springerville (Arizona) zone (Suppe and others, 1975), and the African hot spots (Crough, 1979, 1983). However, this may in part relate to the character of the volcanic events used to trace the hot spot. We define the volcanic track of the Yellowstone hot spot using the onset of large-volume, caldera-forming ignimbrite eruptions. However, within any one volcanic field of the SRP-YP province, volcanism (in the form of basalt and rhyolitic lava flows and small-volume rhyolitic pyroclastic deposits) may have preceded the major caldera-forming event by several million years and have continued for several million years after. Thus, if it were not for the distinctive onset of the large-volume ignimbrite volcanism, Yellowstone would have a much less systematic age progression, more like the case for the above-mentioned postulated hot spots. We accept the general model (Hildreth, 1981; Leeman, 1982a, 1989; Huppert and Sparks, 1988) that the large-volume silicic magmatism along the Yellowstone hot-spot track results from partial melting of continental lithosphere by basaltic melts rising upward from the mantle.

This chapter reflects an integration of volcanology, neotectonics, geomorphology, plate tectonics, and mantle-plume dynamics. As such, this preliminary synthesis involves testable hypotheses in each of these disciplines as well as a potential framework for future studies. If our explanations are valid, studies in the Yellowstone region present unusual opportunities to study response of the continental lithosphere to such a large-scale disturbance.

Chronology of investigations

This chapter expands on an idea conceived in 1984 by Pierce relating the neotectonic deformation pattern of Idaho and adjacent States to the Yellowstone hot spot (Scott and others, 1985a, 1985b; Pierce and Scott, 1986; Pierce and others, 1988; Pierce and Morgan, 1990). Robert B. Smith and Mark H. Anders have come to similar conclusions about deformation related to the Yellowstone hot spot, based mostly on epicenter locations and undifferentiated Quaternary faulting (Smith and others, 1985; Smith and Arabasz, 1991; Anders and Geissman, 1983; Anders and Piety, 1988; Anders and others, 1989).

Myers and Hamilton (1964), in their analysis of the 1959 Hebgen Lake earthquake, suggested that active, range-front faulting on the Teton and Centennial ranges is related to the SRP-Yellowstone trend, which they considered a rift zone. Smith and Sbar (1974) suggest that radial stress distribution outward from a mantle plume beneath Yellowstone could have produced the Snake River Plain as well as the Intermountain seismic belt by rifting. Armstrong and others (1975) documented a northeast progression of rhyolitic volcanism along the eastern SRP at a rate of 3.5 cm/year. Suppe and others (1975) considered the Yellowstone hot spot responsible for both the ongoing tectonic activity between Yellowstone and the Wasatch front and for updoming 350 km wide centered on Yellowstone. Leeman (1982a, 1989) noted the Yellowstone–Snake River Plain volcanic trend extended southwest to McDermitt, found some merit in the hot-spot model, and noted that the denser and thicker lower crust in the older part of the trend might reflect basaltic underplating. High rates of faulting between 4.3 and 2 Ma followed by quiescence in the Grand Valley area were described in an abstract by Anders and Geissman (1983) and attributed to a "collapse shadow" possibly related to a northeast shift of SRP volcanic activity. Scott and others (1985a, 1985b) recognized a V-shaped pattern of the most active neotectonic faults that converged on Yellowstone like the wake of a boat about the track of the Yellowstone hot spot; they also related earlier phases of late Cenozoic deformation to older positions of the hot spot. Smith and others (1985) and Smith and Arabasz (1991) noted two belts of seismicity and late Quaternary faulting that converge on Yellowstone and a "thermal shoulder" zone of inactivity inside these belts. Piety and others (1986, p. 108–109; this volume) have concluded that the locus of faulting in the Grand Valley–Swan Valley area has moved along and outward from the track of the Yellowstone hot spot. Anders and others (1989) defined inner and outer parabolas that bound most of the seismicity in the region and present a model in which underplated basalt increases the strength of the lithosphere through time; this elegantly explains both lithospheric softening and hardening upon passage of the mantle plume. In a summary on heat flow of the Snake River Plain, Blackwell (1989) concluded that the volcanic track resulted from a mantle plume. Westaway (1989a, 1989b) argued that the V-shaped convergence of seismicity and faulting on Yellowstone could be explained by shearing interactions of an upwelling mantle plume and west-southwest motion of the North American plate. Malde (1991) has advocated the hot-spot origin of the eastern Snake River Plain and contrasted this with the graben origin of the western Snake River Plain.

VOLCANIC TRACK OF THE YELLOWSTONE HOT SPOT

The age of late Cenozoic, caldera-filled, silicic volcanic fields along the SRP increases systematically from 0 to 2 Ma on the Yellowstone Plateau to 15 to 16 Ma near the common borders of Idaho, Nevada, and Oregon (Armstrong and others, 1975) (Plate 1). Figure 3 shows the time progression of volcanism, based on the oldest caldera within a particular volcanic field. Many consider this volcanic progression to represent the trace of a mantle plume (for example, see Minster and others, 1974; Suppe and others, 1975; Crough, 1983; Anders and others, 1989; Richards and others, 1989; Blackwell, 1989; Wilson, 1990; Rodgers and others, 1990; Malde, 1991). Alternatively, this volcanic progression has been attributed to (1) a rift (Myers and Hamilton, 1964, p. 97; Hamilton, 1987), (2) volcanism localized along a crustal flaw (Eaton and others, 1975), or (3) a propagating crack along a transform fault boundary separating greater basin-range extension south of the plain from lesser extension to the north (Christiansen and McKee, 1978).

Hot-spot track 0 to 10 Ma

The onset of caldera-forming eruptions for each of the three younger volcanic fields of the SRP-YP province defines a systematic spatial and temporal progression (Plate 1; Figs. 3 and 4; Table 1). The post–10-Ma hot-spot track is also well defined by topography. From the Yellowstone Plateau to the Picabo fields, the 80 ± 20-km-wide, linear, mountain-bounded trough of the eastern SRP-YP is considered by us as floored by nearly overlapping calderas over its entire width.

The 0- to 10–Ma volcanic fields and the caldera-forming ignimbrites from them are: (1) the Yellowstone Plateau volcanic field, which produced the 2.0-Ma Huckleberry Ridge Tuff, the 1.2-Ma Mesa Falls Tuff, and the 0.6-Ma Lava Creek Tuff (Christiansen and Blank, 1972); (2) the Heise volcanic field, which produced the 6.5-Ma tuff of Blacktail Creek, the 6.0-Ma Walcott Tuff, and the 4.3-Ma tuff of Kilgore (Morgan and others, 1984; Morgan, 1988); and (3) the Picabo volcanic field, which produced the 10.3-Ma tuff of Arbon Valley (see footnote in Table 1).

Each volcanic field, commonly active for about 2 m.y., is defined by a cluster of several extremely large calderas. A systematic northeastward progression is defined by the inception age for the different volcanic fields, but no systematic age progression is apparent within a field (Plate 1). In addition, a hiatus of at least 2 m.y. probably occurs between the youngest major ignimbrite in one field and the oldest major ignimbrite in the adjacent younger

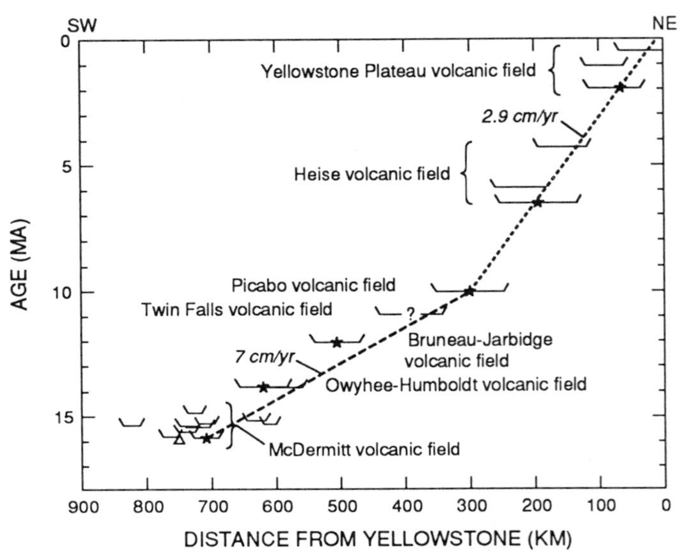

Figure 3. Plot of age of silicic volcanic centers with distance southwestward from Yellowstone. Trough-shaped lines represent caldera widths. For different volcanic fields, stars designate centers of first caldera. As defined by stars, silicic volcanism since 10 Ma has progressed N54°E at a rate of 2.9 cm/year. From 16 to 10 Ma, the apparent trend and velocity are roughly N75°E at 7 cm/year, although both the alignment of the trend and age progression are not as well defined (Plate 1 and Fig. 4). Zero distance is at northeast margin of 0.6-Ma caldera in Yellowstone. Open triangle for Steens Mountain Basalt, one of the Oregon Plateau flood basalts.

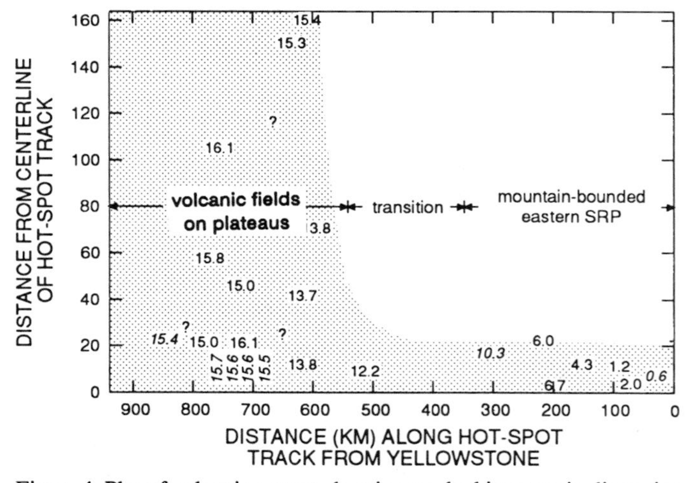

Figure 4. Plot of volcanic centers showing marked increase in dispersion (shaded area) about hot-spot centerline with distance from Yellowstone. Vertical axis is the distance of volcanic centers away from the hot-spot centerline; horizontal axis is the distance from Yellowstone. The numbers represent the age of a center, given in italics for centers south of the centerline (Plate 1). From 0 to 300 km, the centers (0.6 to 10.3 Ma) are within 20 km of the centerline. But from 600 to 800 km, the centers (mostly 13 to 16 Ma) are up to 160 km off the centerline. Although rhyolites of 13 to 16 Ma are common south of the centerline in north-central Nevada (Luedke and Smith, 1981), they are not plotted here because centers have not been located.

field (Plate 1; Table 1; Fig. 3; see Morgan and others, 1984, for further discussion).

Locations of calderas and their associated vents in the Yellowstone Plateau and Heise volcanic fields are based on analysis of many criteria, including variations in ignimbrite thickness, grain-size distribution, ignimbrite facies and flow directions, mapped field relations of the ignimbrites with associated structures and deposits, and various geophysical techniques (Morgan and others, 1984; Morgan, 1988; Christiansen, 1984; Christiansen and Blank, 1972). The location of calderas and fields older than the Yellowstone Plateau field is hampered by a thin cover of basalt; studies of the ignimbrites and their physical volcanology were used to estimate caldera locations in the Heise volcanic field (Morgan, 1988). The general location of the Picabo field is bracketed by the distribution of the 10.3-Ma tuff of Arbon Valley (Table 1; Kellogg and others, 1989), an ignimbrite readily identified by phenocrysts of biotite and bipyramidal quartz.

In addition to the known volcanic geology, the boundaries of the Picabo and Twin Falls fields (Plate 1) are drawn on the basis of similar Bouguer and isostatic gravity anomalies and of aeromagnetic and apparent-magnetic-susceptibility-contrast anomalies; these anomalies are both similar to those displayed by the Heise and Yellowstone Plateau volcanic fields based on maps provided by A. E. McCafferty (written communication, 1989) and V. Bankey (written communication, 1989). Further stratigraphic and volcanic studies are needed, however, to better define all the fields beneath the Snake River Plain, particularly the Picabo and Twin Falls volcanic fields.

Figure 3 shows that the inception of volcanic fields from 0 to 10 Ma has migrated N54 ±5°E at 2.9 ± 0.5 cm/year (errors empirically determined). For North American Plate motion at Yellowstone (lat. 44.5°N, long. 110.5°W), the HS2-NOVEL-1 model (Gripp and Gordon, 1990) defines a synthetic hot-spot track of N56±17°E at 2.2±0.8 cm/year (Alice Gipps, written communication, 1991). This calculation uses DeMets and others (1990) NUVEL-1 model for plate motions over the past 3 m.y., the hot-spot reference frame of Minster and Jordan (1978), but excludes the Yellowstone hot spot from the data set. This close correspondence of the volcanic and plate motion vectors, well within error limits, strongly supports the hypothesis that the 0- to 10-Ma hot-spot track represents a thermal plume fixed in the mantle.

Our calculated rate is 15% slower than the 3.5-cm/year rate first proposed by Armstrong and others (1975) for the last 16 m.y. based on distribution of ignimbrites rather than source calderas. Based on the volcanic track, Pollitz (1988) determined a vector of N50°E at 3.43 cm/year for the last 9 m.y., and Rodgers and others (1990) determined a vector of N56°E at 4.5 cm/year for the last 16 m.y.

Hot-spot track 10 to 16 Ma

Southwestward on the general trend of the post–10-Ma hot-spot track, about 20 mapped or inferred calderas range in age from 16 to 10 Ma (Plate 1). As noted by Malde (1991), the oldest

TABLE 1. VOLCANIC FIELDS, CALDERAS, AND IGNIMBRITES AND THEIR AGES ALONG THE TRACK OF THE YELLOWSTONE HOT SPOT

Volcanic Field/ Caldera	Ignimbrite	Age (Ma)	Reference*
Yellowstone Plateau			
Yellowstone	Lava Creek Tuff	0.6	1
Henry's Fork	Mesa Falls Tuff	1.2	1
Huckleberry Ridge	Huckleberry Ridge Tuff	2.0	1
Heise			
Kilgore	Tuff of Kilgore	4.3	2, 3
Blue Creek	Walcott Tuff	6.0	2, 3
Blacktail Creek	Tuff of Blacktail Creek	6.5	2, 3
Picabo			
Blackfoot	Tuff of Arbon Valley†	10.3	5
Picabo or Twin Falls			
?	City of Rocks Tuff	6.5	4
?	Fir Grove Tuff	?	4
?	Gwin Spring Formation	?	4
?	Older welded tuff	?	4
Twin Falls			
?	Tuff of McMullen Creek	8.6	6
?	Tuff of Wooden Shoe Butte	10.1	6
?	Tuff of Sublett Range	10.4	7
?	Tuff of Wilson Creek	11.0	8
?	Tuff of Browns Creek	11.4	8
?	Tuff of Steer Basin	12.0	6
Bruneau-Jarbidge			
?	Rhyolite of Grasmere escarpment	11.2	9
?	Cougar Point Tuff III	11.3	9
?	Cougar Point Tuff VII	12.5	9
Owyhee-Humboldt			
Juniper Mountain volcanic center	Tuff of the Badlands	12.0	8
Juniper Mountain volcanic center	Lower Lobes Tuff	13.8	8
Juniper Mountain volcanic center	Upper Lobes Tuff	13.9	8
Boulder Creek	Swisher Mountain Tuff	13.8	8, 10
McDermitt			
Whitehorse	Tuff of Whitehorse Creek	15.0	11
Hoppin Peaks	Tuff of Hoppin Peaks	15.5	11
Long Ridge	Tuff of Long Ridge 5	15.6	11
Jordan Meadow	Tuff of Long Ridge 2	15.6	11
Calavera	Tuff of Double H	15.7	11
Pueblo	Tuff of Trout Creek Mountains	15.8	11
Washburn	Tuff of Oregon Canyon	16.1	11
Unnamed volcanic fields			
Three-Finger	?	15.4	12, 13
Mahogany Mountain	Leslie Gulch Tuff	15.5	12, 13
Hog Ranch	?	15.4	12
Soldier Meadow	Soldier Meadow Tuff	15.0	14, 15

*1 = Christiansen, 1984; 2 = Morgan and others, 1984; 3 = Morgan, 1988; 4 = Leeman, 1982b; 5 = Kellogg and others, 1989; 6 = Williams and others, 1990; 7 = Williams and others, 1982; 8 = Ekren and others, 1984; 9 = Bonnichsen, 1982; 10 = Minor and others, 1986, 1987; 11 = Rytuba and McKee, 1984; 12 = Rytuba, 1989; 13 = Vander Meulen, 1989; 14 = Noble and others, 1970; 15 = Greene and Plouff, 1981; Kellogg and Marvin, 1988; Armstrong and others, 1980.

†The age of the tuff of Arbon Valley has been determined to be about 10.3 Ma at several sites on the south side of the plain over a distance of 120 km (Kellogg and Marvin, 1988; Kellogg and others, 1989; Armstrong and others, 1980). Two unpublished ages for the tuff of Arbon Valley in Rockland Valley also yield ages of about 10 Ma (Karl Kellogg and Harold Mehnert, written communication, 1989). The previously accepted age of 7.9 Ma on a similar biotite-bearing ignimbrite (Armstrong and others, 1980) was based only on one sample and is either too young or represents a more local unit. An age of about 10.3 Ma was also obtained on the north side of the SRP in the southern Lemhi Range (L. W. Snee and Falma Moye, oral communication, 1989).

set of calderas along the hot-spot track erupted the 16.1-Ma peralkaline rhyolites of the McDermitt field (Rytuba and McKee, 1984; Plate 1). In contrast with widespread silicic volcanism from 13 to 16 Ma in this area, that between 25 and 16 Ma is uncommon (Luedke and Smith, 1981, 1982; McKee and others, 1970). The 15–16 Ma ignimbrites in the northern Nevada–southwest Oregon area commonly overlie and immediately postdate Oregon Plateau basalts (Fig. 3, shown by triangle) dated generally no older than 17 Ma.

From the Picabo field to the southwest, the calderas tend to increase in age along a centerline drawn between the 10.3-Ma Picabo field and the 16.1-Ma caldera of the McDermitt field (Fig. 3; Plate 1). Silicic volcanic centers are dispersed more widely about this centerline than they are about the post–10-Ma hotspot track (Fig. 4). Silicic centers dated 13 to 16 Ma lie as much as 160 km north of this centerline. South of this centerline, rhyolitic volcanism between 13 and 16 Ma is common within 50 km and extends as much as 100 to 180 km south of this centerline (Luedke and Smith, 1981). The topography in this region is largely a plateau (Malde, 1991) rather than the mountain-bounded linear trough analogous to that of the post–10-Ma hot-spot track.

The location of the Twin Falls volcanic field is the most speculative; its approximate location is based on the distribution of 8- to 11-Ma ignimbrites exposed on both sides of the plain (Williams and others, 1982; Armstrong and others, 1980; Wood and Gardner, 1984) and on gravity and magnetic signatures that are similar to the better-known volcanic fields.

Rates of migration are difficult to calculate because of the wide geographic dispersal of silicic volcanic centers between 16.1 and 13.5 Ma. For the 350 km from the 16.1-Ma McDermitt volcanic field to the 10.3-Ma Picabo field, the apparent rate was about 7 cm/year on a trend of N70–75°E, not accounting for any basin-range extension increasing the rate and rotation to a more east-west orientation (Rodgers and others, 1990). Reasons for the contrasts in rate and other differences between the 16 to 10 Ma and 10 to 0 Ma volcanic centers are discussed later.

Western Snake River Plain and hot-spot track

The western Snake River Plain trends northwest and has a different origin than the eastern SRP (Mabey, 1982; Malde, 1991) (see Figs. 1 and 24 for locations of western plain compared to hot-spot track). The western Snake River Plain is a graben bounded by north-northwest to northwest–trending normal faults (Figs. 1 and 24) that is filled with more than 4 km of late Cenozoic deposits consisting of sedimentary and volcanic rocks, including at least 1.5 km of Columbia River basalt (Wood, 1984, 1989a; Malde, 1991; Mabey, 1982; Blackwell, 1989). This north-northwest–trending graben has generally been thought to have formed starting about 16 Ma (Malde, 1991; Zoback and Thompson, 1978; Mabey, 1982). However, Spencer Wood (1984, 1989a, written communication, 1989) suggests that available evidence indicates the western SRP graben may be as young as 11 Ma and that it transects obliquely older, more northerly trending structures that also parallel dikes associated with the Columbia River basalts. Faulted rhyolites demonstrate offset between 11 and 9 Ma, and after 9 Ma there was 1.4 km offset along the northeast margin and at least 1 km offset along the southwest margin of the western SRP (Wood, 1989a, p. 72). No such boundary faults are known for the eastern SRP.

Therefore, the physiographic Snake River Plain has two structurally contrasting parts: The eastern SRP is a northeast-trending lowland defined by post–10-Ma calderas now thinly covered by basalts, and the western SRP is a north-northwest–trending late Cenozoic graben filled with a thick sequence of primarily basalt and sediments. Major differences between the eastern SRP and the western SRP are also reflected in regional geophysical anomalies, as pointed out by Mabey (1982). The continuous physiographic province formed by linking of the volcanic eastern SRP and the graben of the western SRP does not appear to be fortuitous. Instead, graben formation of the western SRP occurred during and after passage of the eastward migrating hot spot.

A major arcuate gravity high suggests that a large, deep mafic body trends southeast along the axis of the western SRP and then changes to an easterly orientation near Twin Falls (Mabey, 1982). North of the SRP, the Idaho batholith forms a relatively massive and unextended block that interrupts basin-and-range development for about 200 km from the western SRP to the basin and range east of the batholith. If extension north of the SRP continued in the western SRP graben after hot-spot volcanism moved east of the western SRP about 12 Ma, a local right-lateral shear couple with an east-west tension gash orientation would result and might thereby produce a local transform fault similar to the regional transform of Christiansen and McKee (1978); mafic filling of this tensional shear opening could thus explain the arcuate gravity high.

NEOTECTONIC CLASSIFICATION OF FAULTING

We define six types of normal faults based on two criteria (Plate 1, Table 2): (1) recency of offset and (2) height of the associated bedrock escarpment, which includes range fronts. We use the terms *major* and *lesser* to designate the size of the bedrock escarpment, which reflects the late Cenozoic structural relief on the fault, followed by a time term such as Holocene or late Pleistocene to designate the recency of offset (Table 2). Most of the faults shown on Plate 1 are associated with sizable bedrock escarpments, primarily range fronts; faults with little or no bedrock escarpment are generally not shown unless scarps have been seen in surficial materials, a condition that also generally signifies Holocene or late Pleistocene offset. We use <15 ka for the youngest category of fault activity because 15 ka is the age of the youngest widespread alluvial fan deposits in the region near the SRP-YP province (Pierce and Scott, 1982).

TABLE 2. A CLASSIFICATION OF LATE CENOZOIC NORMAL FAULTS*

Fault Type	Escarpment	Last Offset	Characterizes Neotectonic Belt
Major Holocene	>700 m relief, high steep facets	Holocene (<15 ka)	II
Lesser and reactivated Holocene	Absent or >200 m, rejuvinated facets	Holocene (<15 ka)	I
Major late Pleistocene	>500 m relief, moderate facets	Late Pleistocene (15 to 120 ka)	III
Lesser late Pleistocene	<200 m relief, low or absent facets	Late Pleistocene (15 to 100 ka)	
Major Tertiary	>500 m, muted escarpment	Tertiary or older Quaternary	IV
Lesser Tertiary	Absent or low (<200 m) and muted	Tertiary or older Quaternary	

*For the fault-type designation, the first word applies to the escarpment height and the second to the recency of offset.

1. *Major Holocene faults* (Scott and others, 1985b) occur along precipitous mountain fronts >700 m high and have had at least 1-m offset since 15 ka. This category is intended to locate faults that have had hundreds of meters of offset in Quaternary time, as reflected by the high, steep mountain front, and that continue to be active, as attested by offsets in the last 15 ka. These faults generally have relatively high levels of activity, mostly >0.2 mm/year and locally >1.0 mm/year.

2. *Lesser and reactivated Holocene faults* are new and reactivated faults that have offsets in the last 15 ka. *Lesser Holocene faults* are not along high, precipitous range fronts, suggesting that activity on a million-year time scale has been low (<0.03 mm/year), whereas offset in the last 15 ka suggests either that the rate of activity has increased in latest Quaternary time or activity just happens to have occurred in the last 15 ka on a fault with a much longer recurrence interval. Reactivated faults may occur on sizable range fronts but are thought to have been reactivated in the Quaternary following an interval of late Cenozoic quiescence. The intent of this category is to include faults newly started or reactivated in later Quaternary time.

3. *Major late Pleistocene faults* were last active between about 100 and 15 ka. The associated range fronts are more than 500 m high but are generally neither as high nor as steep as those associated with *major Holocene* faults. They generally have low levels of activity (<0.1 mm/year). The intent of this category is to define areas that have late Quaternary rates near 0.1 mm/year, but have accumulated kilometers of structural offset.

4. *Lesser late Pleistocene faults* are new and reactivated faults that have recognizable scarps or other evidence of movement in the last 100 ka, but associated topographic escarpments are less than 200 m high. Like *lesser Holocene faults,* this category isolates relatively young faults and provides a designation separate from the *major late Pleistocene faults* that are more significant because of their much greater late Cenozoic offset.

5. *Major Tertiary faults* occur on range fronts with muted escarpments more than 500 m high. The most recent offset may be either Tertiary or older Quaternary. The escarpments suggest more than 1-km vertical offset, but absence of scarps in surficial materials indicates little or no Quaternary activity and rates probably <0.01 mm/year based on <3 m offset since 300 ka. The intent of this category is to identify faults that had moderate to high levels of late Cenozoic activity (last 15 m.y.) but have subsequently ceased movement or decelerated to much lower levels of activity.

6. *Lesser Tertiary faults* do not have scarps in surficial materials and are not known to have evidence of late Pleistocene offset. Associated bedrock escarpments are absent or less than 200 m high. The intent of this category is to separate these faults from the *major Tertiary faults.* Many faults that fit this sixth category are not shown on Plate 1.

Plate 1 shows these six fault types for the eastern SRP region: The darker the line representing a fault, the greater the ongoing activity on that fault. Four belts of faults having contrasting neotectonic character are distinguished based on the six fault types shown on Plate 1. The belts of faulting are designated by Roman numerals with Belt I forming the outside zone farthest from the SRP-YP province and Belt IV the innermost zone. The faulting in each belt is discussed next, starting with Belt II, the best-defined and most active belt.

In the study area, many faults have a history of extensional tectonism older than the current episode of extensional faulting. Montana Basin-Range structures, Idaho Basin-Range structures north of the SRP, and the Jackson Hole area of Wyoming all have Miocene normal faulting that predates the current activity (Reynolds, 1979; Fields and others, 1985; Love, 1977, p. 590; Barnosky, 1984; Burbank and Barnosky, 1990).

Belt II, Defined by major Holocene faults

This belt is characterized by *major Holocene* faults, which define the most active Quaternary structures. We use the history of the Teton fault to represent the kind of activity that illustrates *major Holocene* faulting with high rates of activity sustained over a few million years.

Teton fault history. Postglacial vertical offset on the Teton fault is largest in the middle part of the fault, reaching 20 m offset since about 15 ka near String Lake, for a rate of 1.3 mm/year (Gilbert and others, 1983; Susong and others, 1987). The range-front escarpment locally exceeds 1,500 m in height (total relief is 2,140 m) with slopes of 28° on triangular facets that extend about half way up the range front (Fig. 5). Offset on the Teton fault diminishes to no recognized scarps in surficial materials 30 km to the south and 20 km to the north of String Lake.

The available control on the history of the Teton fault sug-

Figure 5. The front of the Teton Range, showing a postglacial fault scarp 35 m high produced by a vertical offset of 20 m. Note the high, steep front of the Teton Range and high partially dissected triangular facets. Based on postglacial activity and the abruptness of the range front, this is the most active *major Holocene* fault in Belt II. For the Teton and the other *major Holocene* faults, we infer that the high, precipitous range fronts, with high, steep, relatively undissected facets, reflect high Quaternary offset rates (>0.3? mm/yr). Photograph of fault scarp just west of String Lake by R. C. Bucknam.

gests that most faulting occurred since 6 Ma and that the rate of faulting has increased into Quaternary time. On Signal Mountain, 12 km east of the Teton fault at the outlet of Jackson Lake, the 2-Ma Huckleberry Ridge Tuff dips toward the fault at 11°. Gilbert and others (1983) studied the fault geometry defined by the Huckleberry Ridge Tuff. Provided that there are no major faults between Signal Mountain and the Teton fault, they estimated a post–Huckleberry Ridge offset of 2.5 ± 0.4 km for a rate of 1.25 mm/year. These values would increase if uplift of the Teton Range, as indicated by dips of several degrees of the Huckleberry Ridge Tuff on the west side of the range, is included and would decrease if part of the dip of the Huckleberry Ridge Tuff is primary. Rather than significant primary dip for the Huckleberry Ridge Tuff at Signal Mountain, a horizontal attitude is suggested because (1) the tuff maintains a uniform thickness basinward, (2) the tuff was deposited on a quartzite-rich gravel that has a planar upper surface and not a dissected surface appropriate for a terrace tilted several degrees to the west (Gilbert and others, 1983, p. 77; Pierce, field notes), and (3) bedding features in the gravel appear conformable with the dip of the tuff.

A lithophysae-rich tuff exposed beneath the Huckleberry Ridge Tuff on Signal Mountain dips toward the Teton fault at 22°W, has a fission track age of 5.5 ± 1.0 Ma (Nancy Naeser, written communication, 1990), and probably correlates with either the 4.3-Ma tuff of Kilgore (Morgan, 1988) or the 5.99 ± 0.06-Ma Conant Creek Tuff (Christiansen and Love, 1978; Naeser and others, 1980; Gilbert and others, 1983). This ignimbrite was erupted from the Heise volcanic field (Morgan, 1988). Near Signal Mountain, dips on the Miocene Teewinot Formation are 17, 20, and 27°, for an average dip of 21.3° that is essentially the same as the 22° dip of the older tuff (Love and others, 1992). Obsidian from the upper part of the Teewinot Formation yielded a K-Ar age of 9 Ma (Love, 1977; Evernden and others, 1964), and that from the middle part yielded a K-Ar age of 10.3 ± 0.6 Ma (sample R-6826, Douglas Burbank, oral communication, 1990).

The vertical offset at String Lake, which is 13 km directly toward the fault from Signal Mountain, can be converted to a basin-tilting rate compatible with the tilting rates at Signal Mountain (Fig. 6). Assuming planar tilting of the basin into the fault out to a hinge zone 18 km from the fault, an offset rate of 20 m/15 k.y. at String Lake translates to a tilt rate at Signal Mountain of

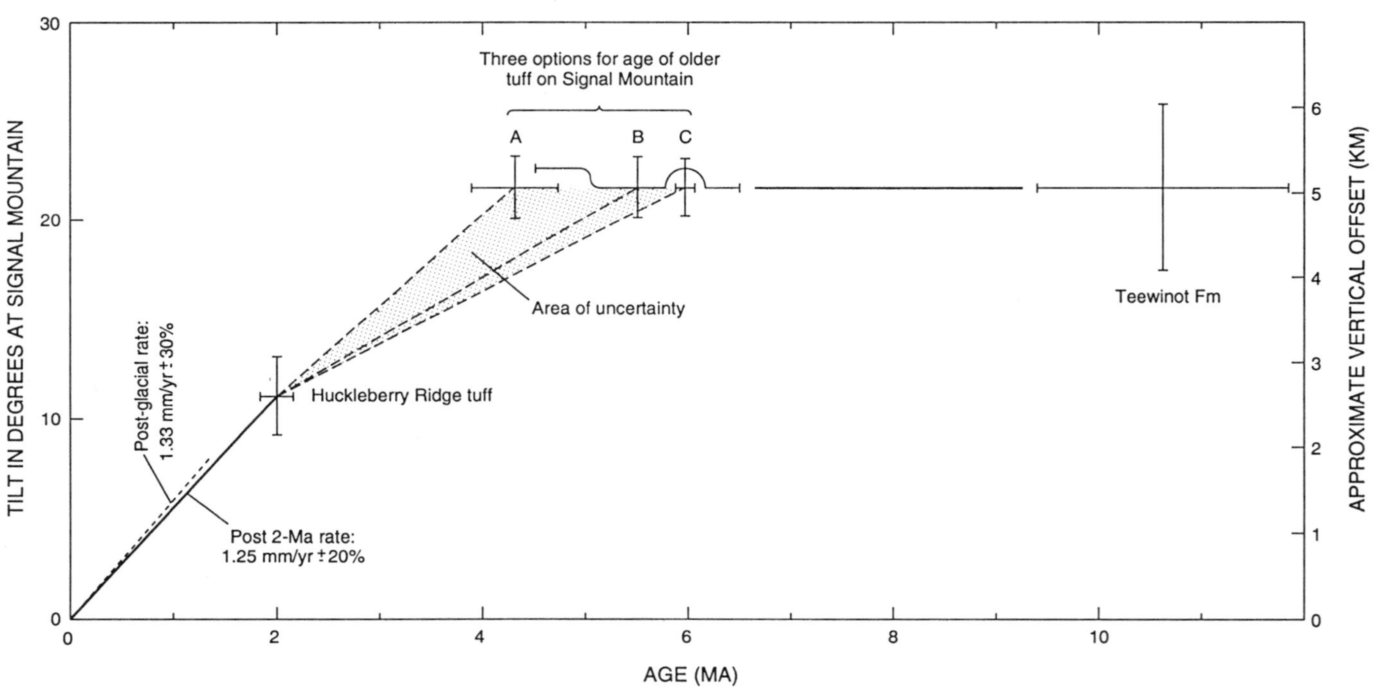

Figure 6. Inferred late Cenozoic history of the Teton fault based on tectonic rotation into the fault of beds in the Signal Mountain area. Signal Mountain is 13 km east of the central part of the Teton fault. Differences in rate between 2 and 6 Ma (shaded area) result from three options given for the age of the older tuff on Signal Mountain: A—our preferred correlation with the 4.3 tuff of Kilgore, B—a fission track age from Signal Mountain of 5.5 ± 1 Ma, and C—correlation with the 5.99 ± 0.06-Ma Conant Creek Tuff. The present episode of tilting started after 6 Ma, and high tilt rates have persisted since at least 2 Ma. Cross sections by Gilbert and others (1983) and by us indicate downfaulting of 2.5 ± 0.4 km of the 2.0-Ma tuff opposite Signal Mountain. Thus, offset rate since 2 Ma has been about 1.25 ± 0.2 mm/year (solid line). In the 15 k.y. since glacial recession near String Lake (Fig. 5), 19.2 m vertical offset (Gilbert and others, 1983) yields a rate of 1.3 ± 0.3 mm/year (dashed line).

4.2°/m.y., remarkably similar to the 5.5°/m.y. defined by the 11° dip of the 2-Ma tuff.

Assuming the dip of the units is tectonic, Figure 6 shows the rate of tilting was high from 0 to 2 Ma, either high or moderate from 2 to about 5 ± 1 Ma, and very low to virtually nonexistent prior to that. Because rates of 5.5°/m.y. from 0 to 2 Ma and 4.2°/m.y. from 0 to 15 ka are similar (Fig. 6), it is reasonable to predict that this rate will continue for at least several thousand years into the future.

Older normal faulting in and near Jackson Hole. Based on studies of the Miocene Coulter Formation, Barnosky (1984) concluded that the onset of extensional-type bimodal basalt/rhyolite volcanism sometime between 18 and 13 Ma also heralded the start of extensional faulting in northern Jackson Hole, perhaps on the Teton fault. But the Teewinot Formation, younger than the Coulter Formation, indicates no significant uplift on the Teton fault until after 9 Ma (Love, 1977). In addition, the older tuff on Signal Mountain indicates pyroclastic flow from the Heise volcanic field across the area now occupied by the Teton Range, thus indicating a much lower Teton Range about 5 Ma than now. On the Hoback fault along the eastern side of southernmost Jackson Hole, the Miocene Camp Davis Formation forms a half-graben fill that indicates more than 1.6 km of Miocene faulting (Love, 1977, p. 590; Schroeder, 1974).

Belt II, southern arm. The southern arm of Belt II extends S20°W from Yellowstone about 400 km to the Salt Lake City area, where the belt of Quaternary faulting becomes broader and trends north-south (Plate 1). The fault pattern is generally *en echelon,* right stepping with a notable gap between the Teton and Star Valley faults (Plate 1). The *major Holocene faults* are, from north to south: (1) the east Sheridan fault, which offsets more than a meter of postglacial (15 ka) talus along a 1,000-m range front (Love and Keefer, 1975); (2) the Teton fault, with up to 20-m offset since 15 ka (Gilbert and others, 1983); (3) the Star Valley fault, which has 9- to 11.6-m offset since 15 ka (Piety and others, 1986; Piety and others, this volume); (4) the Bear Lake fault, with as much as 10 m offset since 12.7 ka (McCalpin and others, 1990); (5) the east Cache fault, with 4 m offset since 15 ka (McCalpin, 1987); and (6) the Wasatch fault, where the middle six segments have Holocene activity and as much as 7 m offset since 6 ka (Machette and others, 1987, 1991). A gap in recognized Quaternary faulting occurs between Teton and Star Valley faults, although historic seismicity does occur in this gap (C. J. Langer, written communication, 1985; Wood, 1988).

Belt II, western arm. From Yellowstone, the northern belt of *major Holocene faults* extends S70°W for about 300 km to the Lost River fault but loses definition farther west in the Idaho batholith (Plate 1). Historic seismicity continues on this trend for 100 km farther west (Smith and others, 1985), but Quaternary faulting is only locally recognized (Schmidt and Mackin, 1970; Fisher and others, 1983; D. L. McIntyre, oral communication, 1988).

Faults along the western arm of Belt II are, from east to west: (1) Hebgen Lake earthquake faults, which had surface offsets during 1959 of up to 5 m and absolute subsidence of as much as 7 m, and which comprise the 12-km Hebgen Lake and 22-km Red Canyon faults (Witkind, 1964; Myers and Hamilton, 1964); (2) the southern Madison fault, which has had 5 m offset since 15 ka, including minor offset during the 1959 Hebgen Lake earthquake, and along which the 2-Ma Huckleberry Ridge Tuff exhibits roughly 800 to 900 m of downfaulting, based on cross sections we drew (Lundstrom, 1986; Mathiesen, 1983); (3) the eastern and central parts of the Centennial fault, which have had up to 20 m offset since 15 ka (rate 1.3 mm/year) (Witkind, 1975a) and a minimum of 1,500 to 1,800 m offset of the 2-Ma Huckleberry Ridge Tuff (rate >0.75 mm/year) (Sonderegger and others, 1982); (4) the 16-km Sheeps Creek and 11-km Timber Butte segments of the Red Rock fault, which have had two offsets totaling 4 m and one offset of about 2 m, respectively, since 15 ka (Stickney and Bartholomew, 1987a; Haller, 1988); (5) the 42-km Nicholia (perhaps late Pleistocene), the 23-km Leadore, and the 20-km Mollie Gulch (possibly late Pleistocene) segments of the Beaverhead fault (Plate 1, N, L, and M), which has had no more than one 1 to 2 m offset since 15 ka (Haller, 1988); (6) segments of the Lemhi fault, consisting of the 43-km Sawmill Gulch segment with two offsets since 15 ka and the 12-km Goldburg and 23-km Patterson segments with one offset each since 15 ka (Plate 1, S, G, and P; Haller, 1988; Crone and Haller, 1991); (7) the 22-km Thousand Spring and the 22-km Mackay segments of the Lost River fault (Plate 1, T and M), which have had two offsets each totaling up to 4 m since 15 ka, and the Warm Spring segment, which has had only one offset since 15 ka (Plate 1, W; Scott and others, 1985b; Crone and others, 1987); (8) the Boulder fault, 30 km east of the south end of the Sawtooth fault, with several meters offset of Pinedale moraines (<20 ka) over a distance of >5 km (Scott, 1982; A. J. Crone, oral communication, 1988); and (9) at least the northern part of the Sawtooth fault, which appears to have offset since 15 ka (Fisher and others, 1983).

The western arm, less linear than the southern arm, is divisible into two parts: from Yellowstone to the Red Rock fault, an irregular eastern part 30 to 60 km wide, and a western part about 75 km wide that parallels the SRP. The eastern part is the least systematic part of Belt II, including both north-south– and east-west–striking faults, and involves foreland uplifts exposing Archean rocks of the craton. The faults farther west in the western belt trend northwest parallel to thrust-belt structures (Rodgers and Janecke, this volume), and the active normal faults are in a row, rather than the en echelon pattern of the southern arm of Belt II (Plate 1).

Belt I, lesser and reactivated faults beyond Belt II

Belt I occurs outside Belt II and is characterized by *lesser Holocene faults* and *lesser late Pleistocene faults* as well as, for the western arm, reactivated Miocene faults. An important attribute of *lesser Holocene* and *lesser late Pleistocene faults* is the absence of high, steep range fronts. Although characterized by relatively young activity, these faults are thought to be early in their present cycle of activity, for they have not accumulated large Quaternary offsets.

Belt I, southern arm. The southern arm of Belt I is characterized by *lesser Holocene* as well as *lesser late Pleistocene faults* in a belt 30 to 70 km wide and 400 km long. These faults are discussed from north to south. The Mirror Plateau faults, occurring immediately northeast of the Yellowstone caldera (Plate 1, form an arcuate fault belt that includes both *lesser Holocene* and *lesser late Pleistocene faults,* which parallel the caldera margin at distances of 10 to 15 km. About half the faults have postglacial movement (Love, 1961; Pierce, 1974): total offset on these faults is generally less than 100 m.

Southward from the Mirror Plateau, the largest postglacial fault occurs inside the 0.6-Ma Yellowstone caldera between the East Sheridan fault and the upper Yellowstone faults (Plate 1). There, a *lesser Holocene fault* with a postglacial fault scarp nearly as high as its associated topographic escarpment offsets Holocene shorelines of Yellowstone Lake and continues southward for 20 km (Pings and Locke, 1988; Richmond, 1974). Twenty km farther south, another *lesser Holocene fault,* extending 10 km along the ridge just northeast of Bobcat Ridge, locally dams small lakes and has nearly equal scarp heights and bedrock escarpment heights from 5 to 20 m (K. L. Pierce and J. M. Good, field notes, 1987).

East of the above-listed faults, the upper Yellowstone faults (Plate 1) form a graben bounded mostly by *major late Pleistocene faults* and including two short segments of *major Holocene faults* (included here in Belt I, although they might be considered a short outlier of Belt II).

About halfway between the upper Yellowstone faults and the town of Jackson (Plate 1), two *lesser Holocene faults* offset Pinedale glacial deposits: (1) on Baldy Mountain a 5-m-high scarp occurs in Pinedale glacial deposits on Baldy Mountain (Fig. 7; K. L. Pierce and J. M. Good, unpublished data, 1985), and (2) 10 km farther east, an en echelon set of faults offsets Pinedale glacial deposits about 9 m and Bull Lake moraines about 21 m (Love and Love, 1982, p. 295; Pierce and Good, field notes, 1989, 1991).

Between Jackson and the Greys River fault, Belt II crosses into the thrust belt where a 2-km section of the Miocene Hoback fault may have been reactivated in Holocene time (Love and de la Montagne, 1956, p. 174). The Greys River fault (Plate 1) has Holocene offset with a postulated *lesser Holocene fault* length of 40 km and a total structural relief of about 500 m (Rubey, 1973; James McCalpin, written communication, 1990); the 500-m-high range front is a dip-slope thought to exaggerate late Cenozoic offset. Farther south in the Wyoming thrust belt, the Rock Creek fault has a *lesser Holocene fault* length of 44 km, along which Rubey and others (1975) note scarps 16 to 18 m high in alluvium at the base of a bedrock escarpment about 200 to 300 m high.

The Crawford Mountains fault in northeast Utah is included in Belt I (Plate 1; Gibbons and Dickey, 1983). The most recent offset is probably late Pleistocene (Susanne Hecker, oral communication, 1990). Although the range front is 500 m high and thus qualifies this fault as a *major late Pleistocene fault,* the range front is actually a dip slope, and Quaternary offset may be more like that of a *lesser late Pleistocene fault.*

Figure 7. Photograph looking south along "new" fault on Baldy Mountain 32 km east of the Teton fault, Wyoming. The 5-m-high scarp offsets Pinedale moraines whose age is probably about 20 ka. The fault trends north-south, and the east (left) side is down. The bedrock escarpment south of here is only a few times higher than the scarp, suggesting that the fault is a new late Quaternary fault.

In southwest Wyoming and still in the thrust belt, the Bear River fault zone is 37 km long and has two offsets since 5 ka (West, 1986). Surface offsets and total escarpment heights are 4 to 10 m, suggesting that total offset is late Holocene for these new or *lesser Holocene faults.* About 10 km to the east, *lesser late Pleistocene faults* offset mid-Pleistocene terrace deposits. West (1986) suggests that these *lesser late Pleistocene faults* relate to backsliding above the sub-Tertiary trace of the Darby thrust, whereas the *lesser Holocene* Bear River fault propagated to the surface above a ramp accompanying backsliding on the Darby thrust.

Belt I, western arm. The western arm of Belt I trends across basin-range structure initiated in middle Miocene time (Reynolds, 1979, p. 190). The western arm of Belt I includes a rather dispersed band 50 to 100 km wide containing *lesser late Pleistocene* and *lesser Holocene faults* in the Miocene basins and reactivated faults along the Miocene range fronts. Perhaps the contrasting, diffuse pattern results from both the differences between the orientation of Miocene faults and the present stress regimen.

In Belt I between the Emigrant fault and the Yellowstone caldera, no Holocene (postglacial) faulting is documented on important normal faults such as the east Gallatin fault (Plate 1). The Norris-Mammoth corridor (Eaton and others, 1975) occurs east of the east Gallatin fault and contains 17 volcanic vents younger than 0.6 Ma that are thought to be related to subsurface dike injection. Extension that otherwise would reach the surface on the east Gallatin fault may be accommodated by dike injection along the Norris-Mammoth corridor (Pierce and others, 1991). Active tectonism along the Norris-Mammoth corridor is indicated by historic level-line changes and by historic seismicity (Reilinger and others, 1977; Smith and others, 1985).

About 70 km north of the Yellowstone calderas, 4 to 6 m of offset on the Emigrant fault since 15 ka indicates a late Quaternary slip rate of about 0.25 mm/year (Personius, 1982). Basalt flows on the floor of the structural basin associated with the fault are tilted no more than 0.5° and are 8.4 and 5.4 Ma (Montagne and Chadwick, 1982; Burbank and Barnosky, 1990). Faulting at the late Quaternary rate must be limited to the last 0.5 Ma, preceded by quiescence to before basalt deposition at 8 Ma (Fig. 8). Prior to the 8-Ma basalt, Barstovian sediments beneath the basalts are tilted 8 to 10° toward the fault (Barnosky and Labar, 1989; Burbank and Barnosky, 1990), an attitude that requires more than 1 km of Miocene faulting (Fig. 8). This sequence indicates two pulses of activity, one between 8 and at least 15 Ma and another after 0.5 Ma, with quiescence in between (Fig. 8). A steep, fresh facet about 100 m high occurs immediately above the Emigrant fault, above which is a much higher but gentler escarpment; these landforms support the above sequence of activity on the Emigrant fault. Elsewhere in the western arm of Belt I, we suggest that Quaternary(?) reactivation of Miocene faults may be indicated by Tertiary sediments tilted toward muted range fronts with late Quaternary scarps.

In a pattern similar to that of the Emigrant fault, the Madison fault exposes Precambrian crystalline rocks in its high bedrock escarpment. Gravity surveys indicate a thick basin fill, which implies large offset during the Tertiary. The northern segment of the Madison fault is less active and appears to have experienced less Quaternary reactivation than the southern segment (Mayer and Schneider, 1985); it thus may also be a recently reactivated Miocene fault.

West of the Madison fault, definition of Belt I is problematic. Although Quaternary faulting appears to form a fringe zone outside Belt II, two problems exist: (1) Immediately west of the

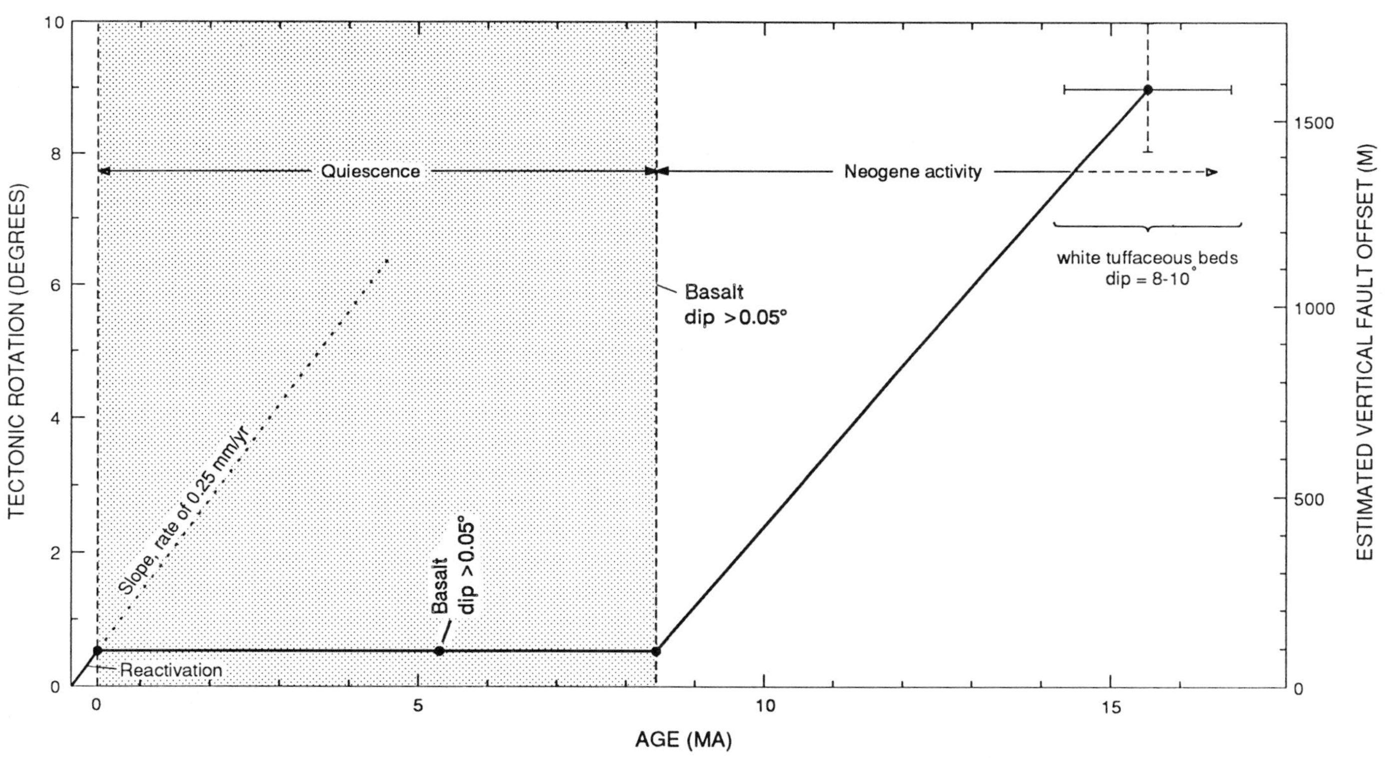

Figure 8. History of the recently reactivated Emigrant fault, a fault in Belt I about 75 km north of the Yellowstone calderas (Plate 1). Estimated fault offset is approximated from the tectonic rotation, assuming a hinge line 10 km northwest of the fault based on the present valley width. (Assuming a 20-km hinge line as for the Teton fault, offset length would be doubled.) About 4 km northwest of the fault, white tuffaceous beds about 15 Ma dip 8 to 10° into the fault, but overlying basalt flows 8.4 and 5.4 Ma (Montagne and Chadwick, 1982) dip no more than 0.5° into the fault. Late Quaternary scarps along the Emigrant fault indicate an offset rate of about 0.25 mm/year (Personius, 1982), but this rate can be extended back to no more than 0.5 Ma because of the nearly horizontal basalts. Thus, an interval of quiescence (shaded area) from about 0.5 to 8.4 Ma separates Neogene faulting from late Quaternary faulting.

Madison fault, dispersed faulting occurs in a broad north-south band and is difficult to separate from activity farther north near Helena (Stickney and Bartholomew, 1987a), and (2) farther west, later Quaternary faulting is poorly documented. In Belt I westward from the Madison to the Blacktail fault, dispersed, low-activity Quaternary faults occur in a 100-km-wide zone commonly not along range fronts. The *lesser Holocene* Sweetwater fault, 20 km northeast of the Blacktail fault, has a long history of activity for it offsets 4-Ma basalt 250 m (Stickney and Bartholomew, 1987a, b; Bartholomew and others, 1990). Farther north, the *lesser Holocene* Georgia Gulch fault and *lesser late Pleistocene* Vendome horst have muted scarps in old (150 ka) deposits that indicate low rates of late Quaternary activity on these non-range-front faults (Stickney and Bartholomew, 1987a, 1987b; Bartholomew and others, 1990). The Blacktail fault occurs on a range front 600 m high and is therefore classified as a *major(?) late Pleistocene fault* (Stickney and Bartholomew, 1987a, 1987b), although the scarp occurs at the edge of the piedmont rather than on the muted range front (Dean Ostenaa, oral communication, 1990), suggesting that it may instead be a reactivated *lesser late Pleistocene fault.* Twenty kilometers southwest of the Blacktail fault, a *lesser late Pleistocene fault* occurs across the valley and north of the Red Rock fault (Dean Ostenaa, oral communication, 1990).

West from the Blacktail fault, Quaternary faulting is poorly known, and the definition of Belt I becomes questionable. *Major late Pleistocene faults* near the northern ends of the Beaverhead, Lemhi, and Lost River ranges (Haller, 1988; Crone and Haller, 1991) are tentatively assigned to Belt I (Plate 1). The last offset on these northern segments was older than 15 ka and may have occurred in late Pleistocene time (Haller, 1988; Stickney and Bartholomew, 1987a). These faults occur on muted range fronts thought to be relict from Miocene faulting and associated basin deposition common in the Idaho and Montana basin-range structural province (Fields and others, 1985; Reynolds, 1979). Moderate levels of historic seismicity (Smith and others, 1985) are compatible with the hypothesis that these faults may be in the process of reactivation, but active at low levels.

Belt III, defined by major late Pleistocene faults

Belt III is characterized by *major late Pleistocene faults* that last moved between 15 and 125 ka and that occur on range fronts that exceed 500 m in height but are neither as high nor as steep as those associated with *major Holocene faults.* Because Belt II is the most clearly defined, the delimiting of Belt III is aided by its outer boundary's conforming with the inner boundary of Belt II. Sedimentation related to the next-to-last (Bull Lake) glaciation ended about 125 ka and commonly provides the offset datum for *major late Pleistocene faults.*

Belt III, southern arm. In the southern arm of Belt III, the two northern *major late Pleistocene faults* are the northern segment of the Bear Lake fault (McCalpin and others, 1990) and a range-front fault 20 km to the west that is associated with late Pleistocene basaltic vents (Bright, 1967; Oriel and Platt, 1980; McCoy, 1981). Farther south, the northern East Cache fault and Pocatello valley faults (McCalpin and others, 1987) and the northern three segments of the Wasatch fault (Machette and others, 1987, 1991) have late Pleistocene offsets on range fronts that are not as high, as steep, or as freshly faceted as those associated with *major Holocene faults.*

The southern arm of Belt III includes *lesser late Pleistocene faults* out in basin flats in the Grays Lake–China Hat–Soda Springs area, all associated with late Quaternary volcanism that, at China Hat is as young as 61,000 ± 6,000 years (G. B. Dalrymple, written communication, in Pierce and others, 1982). Southwest from the Soda Springs area, Belt III includes *lesser Holocene faults* active since 15 ka in the Hansel Valley–Pocatello Valley corridor (McCalpin and others, 1987), including the 1934 Hansel Valley earthquake fault. This Holocene activity in the Hansel Valley–Pocatello valley area may reflect young activity where recurrence intervals are long or may represent activity not directly related to the Yellowstone hot spot (David M. Miller, oral communication, 1991).

Belt III, western arm. The western arm of Belt III starts south of the Centennial fault; the northern boundary of this arm is located within 50 km of the eastern SRP for 150 km southwestward to the area north of Craters of the Moon National Monument. Beyond Craters of the Moon to the Boise area, *major late Pleistocene faults* may exist but are not well documented.

The most studied fault in the western arm of Belt III is the southern part of the Lost River fault, the Arco segment, which last moved about 30 ka (Fig. 9). Back tilt within 100 m of the fault results in a tectonic offset rate of about 0.07 mm/year, 30% less than the 0.1 mm/year rate based on offset datums at the fault (Fig. 9; Pierce, 1985; Scott and others, 1985b). Compared to the *major Holocene* part of the Lost River fault farther north (Plate 1), the late Quaternary slip rate is more than three times slower, the range front is lower and less precipitous, and the facets are more dissected. Neotectonic studies (Malde, 1987; Haller, 1988; Crone and Haller, 1991) of the southernmost segment of the Lemhi fault, the Howe segment, suggest late Quaternary activity similar to that seen in the Arco segment. The morphology of both scarps is similar.

Few details of the late Cenozoic history of *major late Pleistocene faults* north of the SRP are currently known. Ignimbrites older than 4.3 Ma but mostly not older than 6.5 Ma occur on the eastern slopes in the southern Lost River and Lemhi Ranges and Beaverhead Mountains and, in general, form east-facing dip slopes of 5 to 15° (McBroome, 1981). A gravel beneath a 6.5-Ma ignimbrite contains clasts of distinctive quartzite and carbonate known only 70 km to the west in the White Knob Mountains across two basin-ranges (M. H. Hait, Jr., oral communication, 1988; Scott and others, 1985b). Taken together, these relationships suggest that the present cycle of faulting started between 6.5 and 4.3 Ma (Scott and others, 1985b, p. 1058) and continued after 4.3 Ma (Rodgers and Zentner, 1988). For faults in the western arm of Belt III, structural relief is 1.5 to 2 km (Scott and others, 1985b) and is inferred to have developed between 6.5 Ma

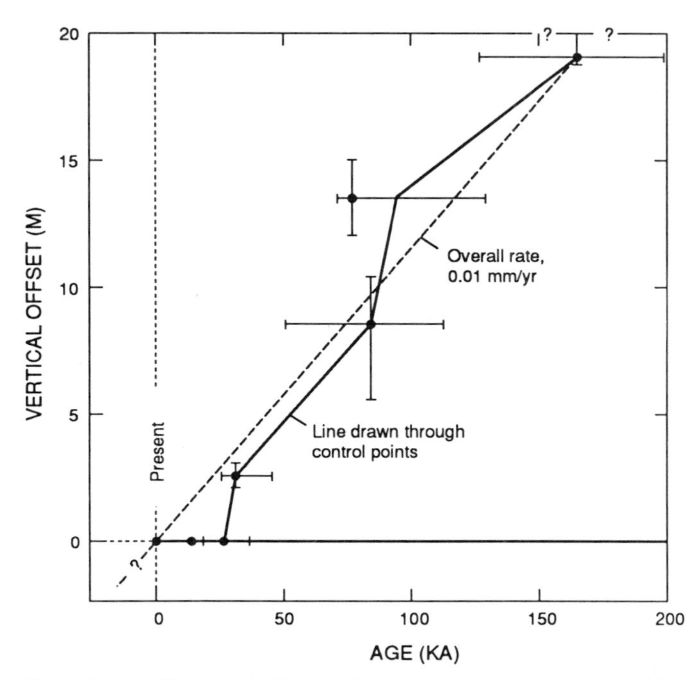

Figure 9. Late Quaternary history of the Arco segment of the Lost River fault, a fault in Belt III just north of the Snake River Plain. A 160-ka datum defines an overall rate (dotted line) of about 0.1 mm/year (Pierce, 1985). Local back tilting affects this rate, and actual vertical tectonic displacement between the mountain and basin blocks is estimated to be 0.07 mm/year. The six control points define a more irregular history, including no offsets since 30 ka (Pierce, 1985).

and present, resulting in an average rate of offset of 0.2 to 0.3 mm/year, about twice the late Quaternary rate. This suggestion that the present cycle of basin-range faulting started 6.5 Ma and decelerated to lower rates in the Quaternary needs to be further evaluated.

Belt IV, defined by major Tertiary faults

Belt IV is present only on the south side of the plain. From a vanishing point near Yellowstone, this belt widens to 50 to 80 km southwest from Yellowstone (Plate 1). Belt IV is characterized by *major Tertiary faults,* range-front faults that were active in late Tertiary time but show little or no evidence of Quaternary activity (Plate 1; Greensfelder, 1976). The ages of several of the faults in Belt IV are described in more detail in the next section. Range fronts are generally muted with no planar triangular facets and relief less than in Belts II and III, suggesting geomorphic degradation in a manner similar to scarp degradation.

NEOTECTONIC DOMAINS EMANATING FROM THE YELLOWSTONE HOT SPOT

The present cycle

Belt II is the central, most neotectonically active belt. From Yellowstone, its two arms extend more than 350 km to the south and to the west and have widths of 30 to 50 km (Plate 1). These arms diverge about the volcanic track of the Yellowstone hot spot in a V-shaped pattern analogous to the wake of a boat now at Yellowstone (Scott and others, 1985b). Belt I occurs outside Belt II, is 20 to 100? km wide, and wraps around the northeast end of the 0.6-Ma Yellowstone caldera (Plate 1). The southern arm of Belt I is characterized by *lesser Holocene* or "new" faults, whereas the western arm has reactivated Tertiary(?) and *lesser Holocene faults* as well as some *major late Pleistocene faults.* Belt III occurs inside Belt II and from a width of 50 km narrows or vanishes toward Yellowstone. This belt is characterized by *major late Pleistocene faults* that have moved between 15 to 120 ka, that have bedrock escarpments at least 500 m high, and whose associated basin fill suggests structural relief of more than 1 km. Structural relief is comparable to that of faults in Belt II, yet low offset rates of <0.1 mm/year in late Quaternary time suggest that neotectonic activity in Belt III may have waned over the last several million years.

The overall pattern of Belts I, II, and III suggests the following progression of neotectonic activity moving outward from the track of the Yellowstone hot spot. Faults in Belt I are presently active but have small amounts of structural offset in the southern arm and small amounts of offset since reactivation in the western arm. Thus, they are in a waxing state of development compared to those in Belt II. Faults in Belt II have high rates of offset that continue into the present, have produced ranges about 1 km high with large fresh triangular facets; they are in a culminating state of activity. Faults in Belt III have structural offsets similar to those in Belt II but have longer recurrence intervals between offsets and less boldly expressed facets and range heights; thus they are in a decelerating or waning state of activity. Belt IV was active in late Tertiary time but is now inactive. Taken together, the spatial arrangement of the belts suggests the following neotectonic cycle moving outward from the hot-spot track: Belt I—an initial waxing phase, Belt II—the culminating phase, Belt III—a waning phase, and Belt IV—a completed phase.

We conclude that the strong spatial and sequential tie of all four belts of faulting to the present position of the Yellowstone hot spot suggests that the same deep-seated process responsible for the volcanic track of the hot spot is also responsible for this spatial and temporal progression of faulting.

Belt IV and older cycles of faulting

If the neotectonic fault belts are genetically related to the Yellowstone hot spot, then faulting in areas adjacent to older positions of the Yellowstone hot spot should correlate in age with older hot-spot activity. Beyond the southern margin of the plain in Belt IV (Plate 1) age of faulting increases away from Yellowstone and roughly correlates with the age of silicic volcanic fields on the SRP. On the northern side of the plain, a northeast-younging progression of fault activity and cessation has not been recognized, in part because faulting there has continued into the late Pleistocene, although the effects of the hot spot may be indicated by inception of faulting roughly similar in age with the age of silicic volcanism on the SRP followed by *waning rates* of faulting. The limited data on the late Cenozoic history of faulting north of the SRP suggest higher rates of activity prior to the late Quaternary and after about 6.5 Ma (see prior discussion under Belt III). South of the SRP in Belt IV, evidence for faulting and other deformation from northeast (younger) to southwest (older) is as follows.

Grand Valley–Star Valley fault system. For Grand Valley fault (Plate 1), Anders and others (1989) used offset and tilted volcanic rocks to define an episode of deformation between 4.3 and 2 Ma that resulted in 4 km total dip-slip displacement at an average rate of 1.8 mm/year (Plate 1, Fig. 10). This 2-m.y. burst of activity was an order of magnitude more active than the interval before (0.15 mm/year) and two orders of magnitude more active than the interval after it (0.015 mm/year) (Anders and others, 1989). High rates of offset on the Grand Valley fault occurred after the 6.5 to 4.3 Ma interval of caldera-forming eruptions on the adjacent SRP in the Heise volcanic field 30 to 100 km to the northwest (Plate 1). Gravity-sliding of bedrock blocks into Grand Valley (see Boyer and Hossack, this volume) probably results from normal faulting and oversteepening of the range front. Emplacement of these slide blocks thus dates times of maximum structural activity; one block was emplaced between 6.5 and 4.4 Ma, another after 7 Ma, and another prior to 6.3 Ma (Anders, 1990; Moore and others, 1987).

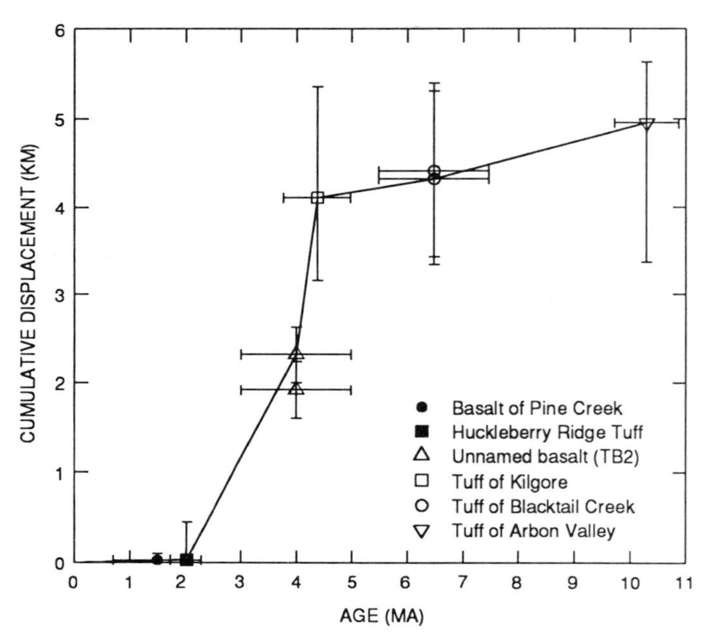

Figure 10. Late Cenozoic offset history of the Grand Valley fault, Idaho, based on tectonic rotation of volcanic layers (from Anders and others, 1989). Paleomagnetic studies of ignimbrites show that only the 2-Ma unit has a significant nontectonic dip. Cumulative displacement represents total dip-slip movement. The great majority of offset occurred between 4.3 and 2 Ma; before and after this interval, rates were one or more orders of magnitude slower. Error bars are one sigma for age and two sigma for displacement, as explained in Anders and others (1989, Fig. 14). The regional nomenclature of Morgan (1988) and Kellogg and others (1989) has been substituted for local stratigraphic nomenclature used by Anders and others (1989) as follows: tuff of Blacktail = tuff of Spring Creek; tuff of Kilgore = tuff of Heise; and tuff of Arbon Valley = tuff of Cosgrove Road.

Blackfoot Mountains. For the Gateway fault, located on the west side of the Blackfoot Mountains (Plate 1), Allmendinger (1982) concluded that most activity occurred between emplacement of ignimbrites dated 5.86 ± 0.18 Ma and 4.7 ± 0.10 Ma. These relations yield nearly 1 km vertical offset in about 1 m.y. for an offset rate of 0.8 mm/year (Allmendinger, 1982). The timing of this deformation falls in the middle of the 6.5 to 4.3 Ma interval for caldera-forming eruptions in the Heise volcanic field on the adjacent SRP (Plate 1).

Portneuf Range. For normal faulting within and on the west side of the Portneuf Range, Kellogg and Marvin (1988, p. 15) concluded that "a very large component of Basin and Range faulting occurred between 7.0 and 6.5 Ma" as indicated by bracketing ages on a boulder conglomerate at the northern end of the range. This age corresponds closely with the 6.5-Ma caldera of the Heise field on the adjacent SRP (Plate 1). Only minor subsequent faulting is recorded by 50-m offset of 2.2-Ma basalt.

Rockland Valley. Southwest of Pocatello near the >10-Ma symbol on Plate 1, extensional faulting was largely over by 8 to 10 Ma (Plate 1). The Rockland Valley fault along the west side of the Deep Creek Mountains and several other nearby basin-range normal faults have vertical displacements of more than 1 km (Trimble and Carr, 1976). Trimble and Carr (1976, p. 91) conclude that "movement on basin-and-range faults was largely completed before the outpouring of the . . . volcanic rocks that now partly fill the structural valleys." The most prominent of these volcanic rocks is the 10.3-Ma tuff of Arbon Valley (see footnote, Table 1). This ignimbrite is locally offset 100 m by younger faulting at the northern end of the Deep Creek Mountains (mountains with symbol >10 Ma, Plate 1), but Quaternary fault scarps have not been recognized in the area (Greensfelder, 1976). Thus, although the time of inception of faulting is not well defined, faulting was largely over by 10 Ma, roughly coincident with the 10.3-Ma caldera on the adjacent SRP (Plate 1).

Sublett Range. Extensional deformation for the northern part of the Sublett Range (Plate 1) was completed before emplacement of the 10.3-Ma tuff of Arbon Valley (R. L. Armstrong, oral communication, 1988). But near the southeastern end of the Sublett Range, a biotite-bearing ignimbrite tentatively correlated with the tuff of Arbon Valley is deformed by folding (M. H. Hait, Jr., oral communication, 1988). This shows that deformation 50 km south of the margin of the SRP continued somewhat after 10 Ma, consistent with a progression of deformation moving outward from the SRP (Plate 1).

Raft River Valley region. The ranges on the west side of the Raft River Valley are composed of rhyolitic Jim Sage Volcanic Member (9 to 10 Ma), the podlike form and extensive brecciation of which indicate accumulation on a wet valley floor (Armstrong, 1975; Covington, 1983; Williams and others, 1982).

Because these rocks are now 1 km above the Raft River Valley, large tectonic movements must have occurred after 10 Ma (Plate 1). Based on mapping, studies of geothermal wells, extensive seismic reflection, and other geophysical studies, Covington (1983) concluded that the rocks that now form the ranges on the east side of the Raft River Valley (Sublett and Black Pine ranges) slid eastward on a detachment fault from an original position on the flank of the Albion Range, which now forms the west side of the valley (Fig. 11). Detachment faulting resulted in formation of the proto–Raft River Valley, which was filled with volcanics of the Jim Sage Member 9 to 10 Ma. Then after 9 to 10 Ma, additional valley widening accompanied about 15 km more of eastward movement on detachment fault (Covington, 1983, Fig. 5; Williams and others, 1982). When detachment faulting ceased is poorly known, but only minor Quaternary deformation is recognized (Williams and others, 1982; K. L. Pierce, unpublished data). Cessation of detachment faulting and basin evolution to a form similar to the present by 7 to 8 Ma is indicated by the following volcanic rocks at the margins of the Raft River Valley: (1) a 7-Ma ignimbrite on the west margin and an 8- to 10-Ma ignimbrite on the northeast margin, and (2) two 8-Ma, shallow domelike intrusions on the west and southeast margins. Thus, the inferred 15 km of detachment faulting probably occurred between 10 and 7 Ma. The emplacement of 1-km-thick Jim Sage Volcanic Member of the Salt Lake Formation about 10 Ma correlates with passage of the Yellowstone hot spot on the adjacent SRP between about 12 and 10 Ma (Plate 1). The interval of deformation from >10 Ma to 7 to 10 Ma is somewhat out of sequence, being younger than that for the Sublett Range and Rockland Valley areas discussed above. The detachment faulting most likely ties to uplift of the Albion Range core complex on the west side of the Raft River Valley.

Neotectonic fault belts and historic seismicity

The two belts of faulting described here lie near the middle of the Intermountain seismic belt, an arcuate belt of historic seismicity that extends from southern Nevada northward through central Utah to Yellowstone, and thence northwestward through western Montana and Idaho (Smith and Sbar, 1974; Smith, 1978).

Our belts of faulting, based on surficial geology and geomorphology, form asymmetric V-shaped bands that converge on Yellowstone and flair outward about the track of the Yellowstone hot spot (Plate 1). This V-shaped pattern based on geologic assessment of fault activity was first pointed out by Scott and others (1985b). Smith and others (1985) independently observed that earthquake epicenters showed a similar relation to the SRP-Yellowstone hot-spot trend (Fig. 12). Anders and Geissman (1983) noted a southward progression of late Cenozoic faulting in the Grand Valley–Star Valley area that they related to migration of volcanism along the SRP. Later, Anders and others (1989) determined that two parabolas, arrayed about the path of the Yellowstone hot spot, bounded most of the earthquake activity in the SRP region (Fig. 12).

The neotectonic fault belts parallel but do not exactly correspond with concentrations of earthquake epicenters in the region (Fig. 12). These differences most likely arise from the contrasting time windows of observation; historical seismicity spans only several decades whereas the geologic record spans

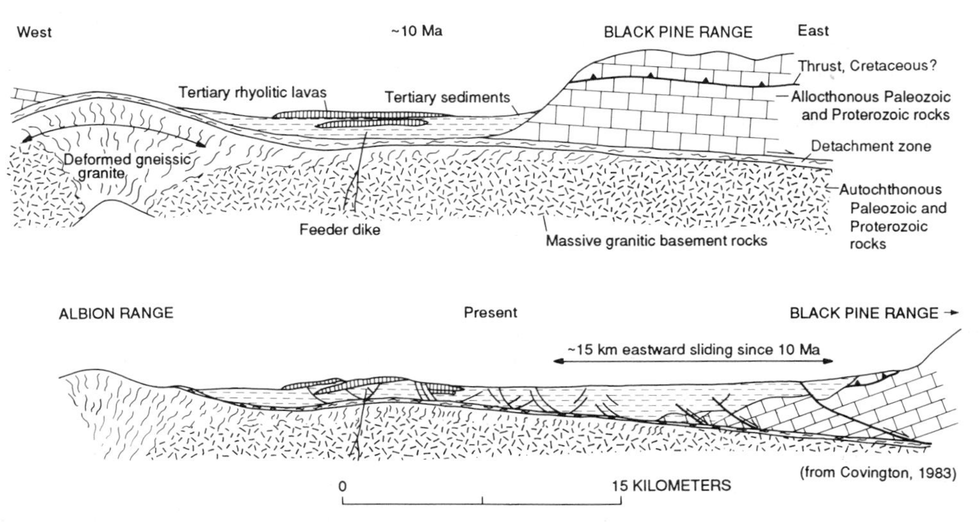

Figure 11. Deformation history of the Raft River valley and adjacent ranges showing about 15 km of subhorizontal detachment faulting since 10 Ma (from Covington, 1983). Quaternary deformation has been minor, and silicic domes and tuffs exposed around the basin margins suggest that deformation and valley filling were largely complete by 6 to 8 Ma. Silicic volcanic activity in the adjacent Snake River Plain is not well dated but probably started between 12 and 10 Ma (Plate 1).

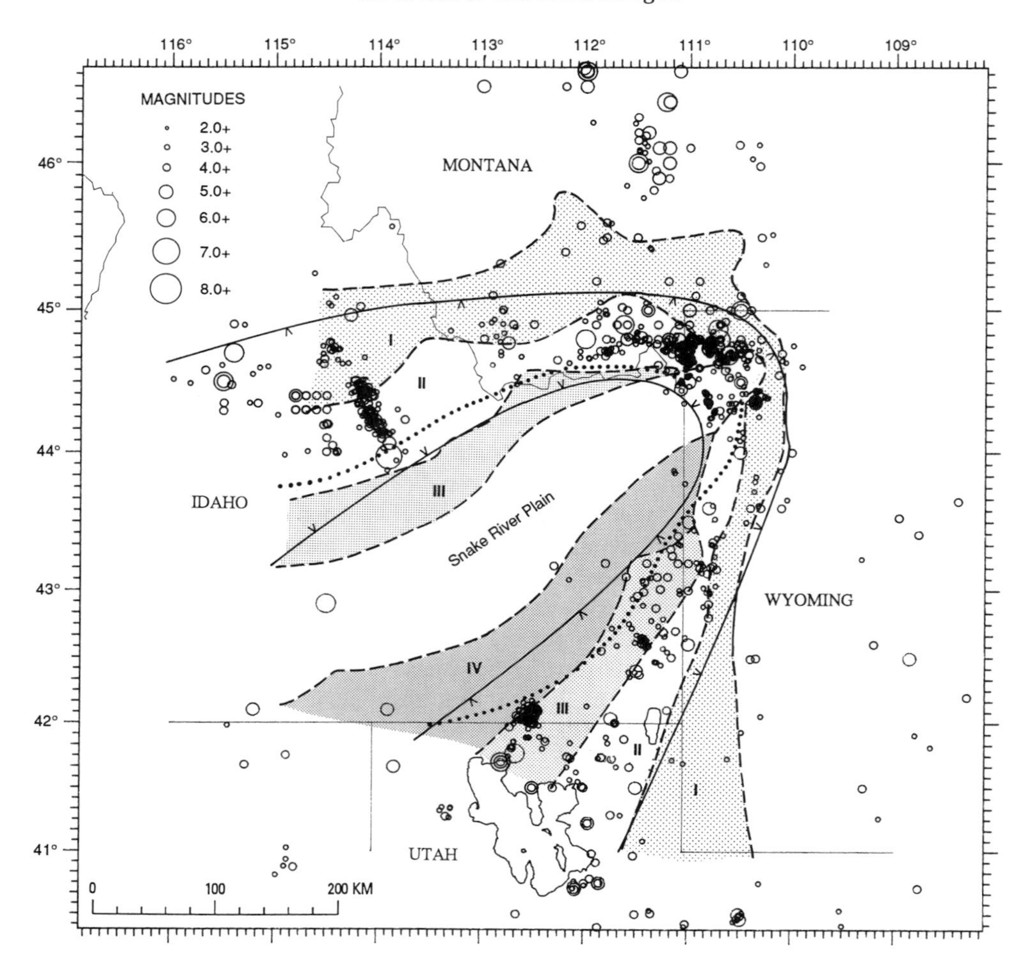

Figure 12. Correlation between belts of neotectonic faulting and historic earthquake locations in the Snake River Plain region. Belts are separated by dashed lines and variously shaded, except that Belt II, is unshaded. Epicenter map kindly provided by R. B. Smith (written communication, 1990). The earthquake record samples a much shorter time interval than our fault studies, probably explaining the differences described in the text. Dotted line is "thermal shoulder" of Smith and others (1985) inside of which earthquakes are rare and outside of which earthquakes are concentrated. The solid lines with carets are the inner and outer parabolas of Anders and others (1989), between which earthquakes have been concentrated and inside of which is their "collapse shadow."

10,000 to several million years. For example, most of the *major Holocene faults,* such as the Lemhi, Beaverhead, Red Rock, Teton, and Wasatch faults, show no historic activity (Plate 1 and Fig. 12). Prior to the 1983 Borah Peak earthquake, the western arm of Belt II had few historic earthquakes west of those in the area of the Hebgen Lake earthquake swarms.

We note the following differences and similarities between our neotectonic belts and seismic boundaries suggested by others.

1. From the SRP north about 50 km, the aseismic zone (Fig. 12) is designated a "thermal shoulder" by Smith and others (1985) and a "collapse shadow" by Anders and others (1989). Their suggestion of inactivity is based on historical seismic quiescence, but this area contains late Pleistocene fault scarps and long-term fault histories that suggest continued activity in this belt (Pierce, 1985; Malde, 1987; Haller, 1988; Crone and Haller, 1991).

2. Immediately south of the plain, the "thermal shoulder" of Smith and others (1985) and the "collapse shadow" of Anders and others (1989) lie within a zone of seismic quiescence. Our Belt IV (rundown faults) also indicates cessation of tectonic activity.

3. Unlike the symmetric inner and outer parabolas or the thermal shoulder, our neotectonic classification does not show bilateral symmetry across the plain. The late Cenozoic tectonic belts flair more outward south than north of the plain. This asymmetry results from Belt IV with *major Tertiary faults,* a belt that occurs only south of the plain.

4. North of the plain, the western arms of Belts I and II

make up the "active region" of Anders and others (1989), but south of the plain the "active region" of Anders and others (1989) includes all of the southern arms of Belts II and III and parts of Belts I and IV. South of the plain, the outer parabola bounding the southern band of historical seismicity (Anders and others, 1989) trends across Belt I and part of Belt II (Fig. 12). These differences are most likely due to the several orders of magnitude difference between the time-window of historic activity compared to that of the geologic record. If our earthquake record could be extended to an interval of several thousand years, we suggest the greatest concentration and maximum energy release from earthquakes would parallel and perhaps coincide with the trend of Belt II.

ALTITUDE AND THE HOT-SPOT TRACK

If the neotectonic belts and caldera-forming volcanism that converge at Yellowstone are the result of the North American plate moving across a relatively stationary mantle plume, then a large elevated region or swell is predicted by analogy with oceanic hot spots (Crough, 1979, 1983; Okal and Batiza, 1987). The size of three oceanic hot-spot swells is shown in Figure 13 to range between 1 to 2 km in height and 800 to 1,200 km in width.

Recognition of oceanic swells is facilitated by the relatively uniform composition and simple thermal and structural history of the oceanic crust as well as by the generally minor effects of erosion and sedimentation. The height of oceanic swells increases with the age and hence the coolness of the associated oceanic crust: The 2-km-high Cape Verde swell is in 140-Ma crust, whereas the 1.2-km high Hawaiian swell is in 90-Ma crust (Crough, 1983, Fig. 5). The Hawaiian swell has a broad arcuate front, subparallel sides, and an elevated trailing margin more than 2,000 km long.

The topographic expression of swells in continental crust is probably more difficult to recognize because of the more complicated character of the continental crust. Nevertheless, five continental hot-spot swells have been postulated to occur in the nearly static African plate; these swells are about 1,000 km across, between 700 and 2,500 m high, and have associated volcanism (Morgan, 1981; Crough, 1983, Fig. 5).

A hot spot near Yellowstone might result in additional processes that could produce patterns of uplift and subsidence with diameters smaller than the approximately 1,000-km-diameter swell. Several processes are listed here to suggest the range of diameters that might be involved in addition to the swell diameter: (1) extensive basaltic underplating or intrusion near the base of the crust and centered on the hot spot, with a width of perhaps 100 to 400 km (Leeman, 1982a, Fig. 5; Anders and others, 1989); (2) intrusion (timing uncertain) and cooling of midcrustal basaltic magmas at depths between 8 and 18 km and with width of perhaps about 100 km (Sparlin and others, 1982; Braille and others, 1982); and (3) formation and cooling of granitic magma chambers at depths of 2 to 10 km and with widths of perhaps 40 to 80 km (such magma chambers also may produce ignimbrite eruptions and rhyolite flows [Leeman, 1982a; Christiansen, 1984]).

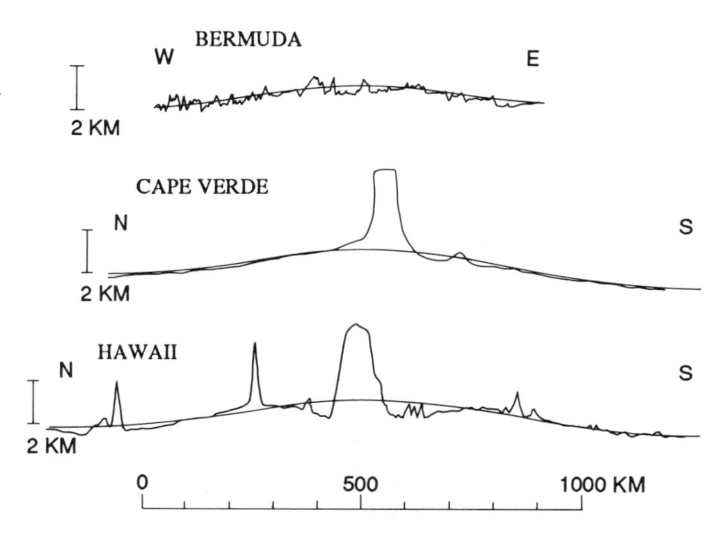

Figure 13. Profiles across three oceanic hot-spot swells showing their large-scale topographic expression with heights of 1 to 2 km and diameters of 800 to 1,200 km (from Crough, 1978). The smoothed line removes small-scale variations and the volcanic edifice and its associated isostatic depression. Although the Yellowstone hot spot is beneath continental rather than oceanic lithosphere, these profiles are given to show the scale of uplift that may be involved.

Yellowstone cresent of high terrain

In the Yellowstone region, a crescent-shaped area about 350 km across stands about 0.5 to 1 km higher than the surrounding region (Fig. 14; Plate 1). Southern and western arms extend more than 400 km from the apex of the crescent. The crest of the western arm of the crescent coincides with the belts of neotectonic faulting, whereas the crest of the southern arm is just east of these belts (Plate 1). Definition of the greater Yellowstone elevation anomaly is complicated by mountains and basins of Laramide age (formed roughly 75 to 50 Ma). Nevertheless, others have recognized such an altitude anomaly; Suppe and others (1975) described it as an "updome" about 350 km across indented on the southwest by the Snake River Plain, and Smith and others (1985) recognized a lithospheric bulge 400 km across centered on the Yellowstone Plateau. The boundary and crest of the Yellowstone crescent shown on Plate 1 were drawn by Pierce on a subjective basis, primarily using the color digital map of Godson (1981), attempting to account for Laramide uplifts.

A large part of the altitude anomaly is formed by the Absaroka Range (Fig. 1) along the eastern boundary of Yellowstone National Park, which was formed in post-Laramide time, as it consists of post-Laramide volcanic rocks. The crest of the Laramide Wind River Range (Fig. 1) seems to tilt southeasterly away from the Yellowstone hot spot, as shown by the 1-km altitude decrease from northwest to southeast of both the range crest and

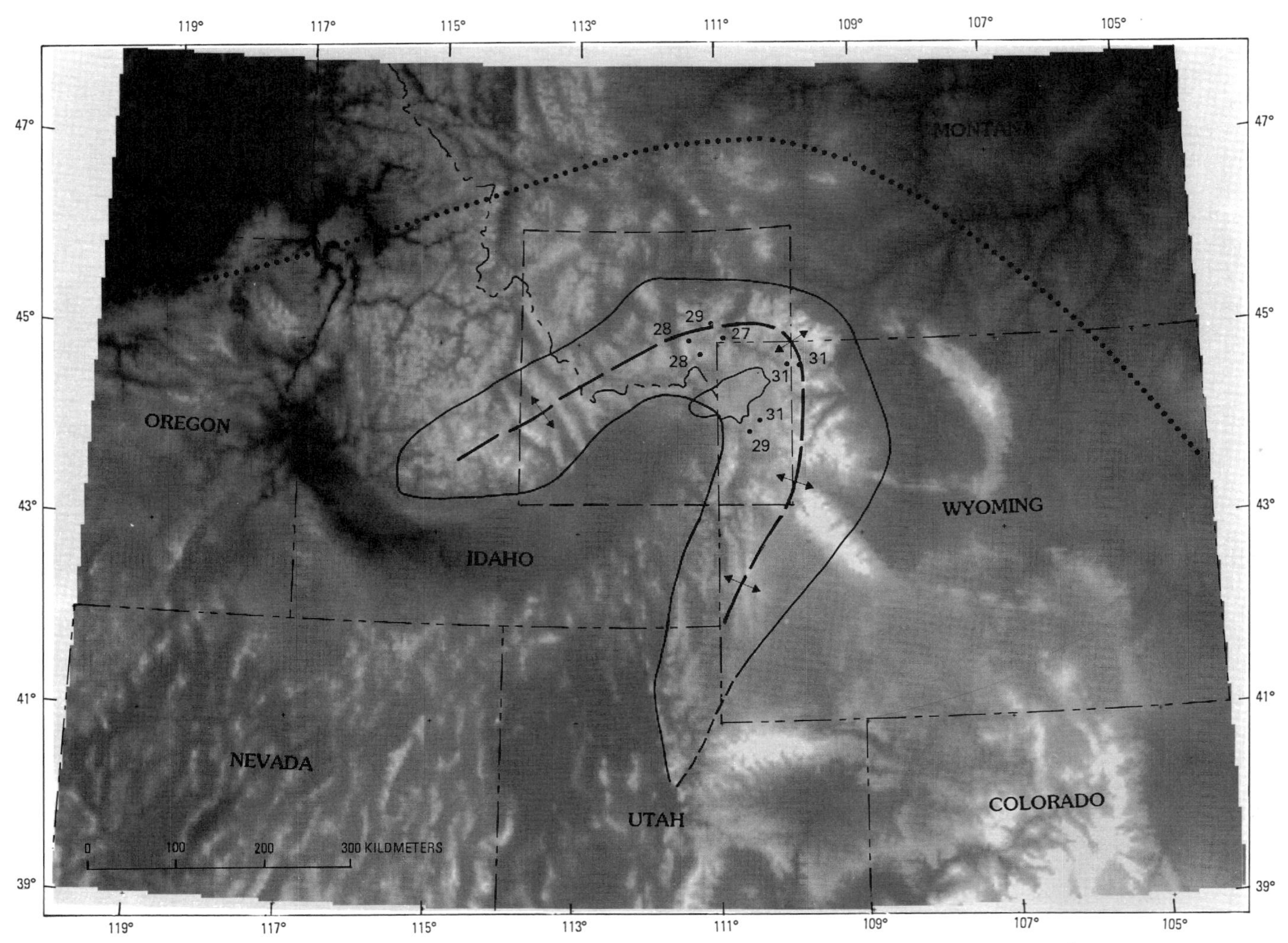

Figure 14. Gray-scale digital topography showing expression of the Yellowstone crescent of high terrain (solid line) that wraps around the Yellowstone Plateau. The higher the terrain, the lighter the shading. The Yellowstone crescent stands about 0.5 to 1 km higher than the surrounding terrain. The crest of the western arm of the crescent (long-dashed line with arrows) coincides with the belts of neotectonic faulting, whereas the crest of the southern arm is just east of these belts (Plate 1). The effects of Laramide uplifts such as the Wind River, Bighorn, and Beartooth ranges as well as the associated basins are older features that need to be discounted to more clearly define altitude anomalies associated with the Yellowstone hot spot. The numbers are the altitudes, in hundreds of meters, of selected high remnants of 2-Ma Huckleberry Ridge Tuff mapped by R. L. Christiansen (written communication, 1985), which at the margin of its caldera (oval area) is now generally between 2,000 and 2,500 m. Altitude differences exceeding 500 m between the caldera source and sites near the axis of the Yellowstone crescent indicate uplift of the Yellowstone crescent and/or subsidence of the caldera. Dotted line encloses a broader bulge of higher terrain that may also be associated with Yellowstone.

a prominent bench (erosion surface?). To the north and west, the central and western part of the head of the Yellowstone crescent is largely formed by the combined Beartooth (Snowy), Gallatin, and Madison ranges (Fig. 1), all of which stand high and have youthful topography.

Ignimbrite sheets exhibit altitude differences, that suggest uplift on the outer slope of the crescent and subsidence on its inner slope. North of the Yellowstone Plateau volcanic field near the crest of the Yellowstone crescent, the 2-Ma Huckleberry Ridge Tuff (Table 1) has altitudes of 2,800 m, 400 to 800 m higher than its altitude around its caldera (note altitudes shown on Fig. 14). Ignimbrites from the Heise volcanic field now dip as much as 20° toward the Heise center on the Snake River Plain (Plate 1); the oldest ignimbrites from the field are tilted more than the younger ones, suggesting tilting toward the Heise field when it was active (Morgan and Bonnichsen, 1989).

Stream terraces are common on the outer, leading margin of the Yellowstone crescent, suggesting incision following uplift. But on the inner, trailing margin of the crescent, including the Snake River Plain, stream terraces are much less well developed, a state that suggests subsidence rather than uplift.

Assuming the crescent is moving northeast at a plate tectonic rate of 30 km/m.y. (3 cm/year), the rate of associated uplift can be roughly estimated based on bulge height of 0.5 to 1 km and an outer slope length, measured parallel to plate motion, of 60 to 150 km. Thus, 0.5 to 1 km uplift on the outer slope of the Yellowstone crescent would last 2 to 5 m.y., yielding regional uplift rates of 0.1 to 0.5 mm/year.

Altitude changes north of the Snake River Plain

A level line across the western arm of the Yellowstone crescent shows a broad arch of uplift about 200 km across (Fig. 15B; Reilinger and others, 1977). Uplift at rates of 2 mm/year or more occurs near the axis of the Yellowstone crescent, which here also coincides with the boundary between fault Belts I and II. Contours of the rates of historic uplift define three domes (Fig. 15A; Reilinger, 1985) centered on the downfaulted basins of the *major Holocene* Red Rock, Madison-Hebgen, and Emigrant faults. These high rates of uplift (>2 mm/year) are an order of magnitude greater than the long-term regional uplift rate estimated for the arms of the Yellowstone crescent (0.2 mm/year assuming 500 m uplift in 3 m.y.). The centering of these domal uplifts on basins associated with these *major Holocene faults* may reflect a combination of local interseismic uplift and regional uplift of the Yellowstone crescent. During an earthquake, absolute basin subsidence of 1 to 2 m is likely (Barrientos and others, 1987).

For the area between the Red Rock and Madison faults, Fritz and Sears (1989) determined that a south-flowing drainage system more than 100 km long was disrupted at about the time when ignimbrites flowed into this valley system from the SRP, most likely from the 6.5- to 4.3-Ma Heise volcanic field. The

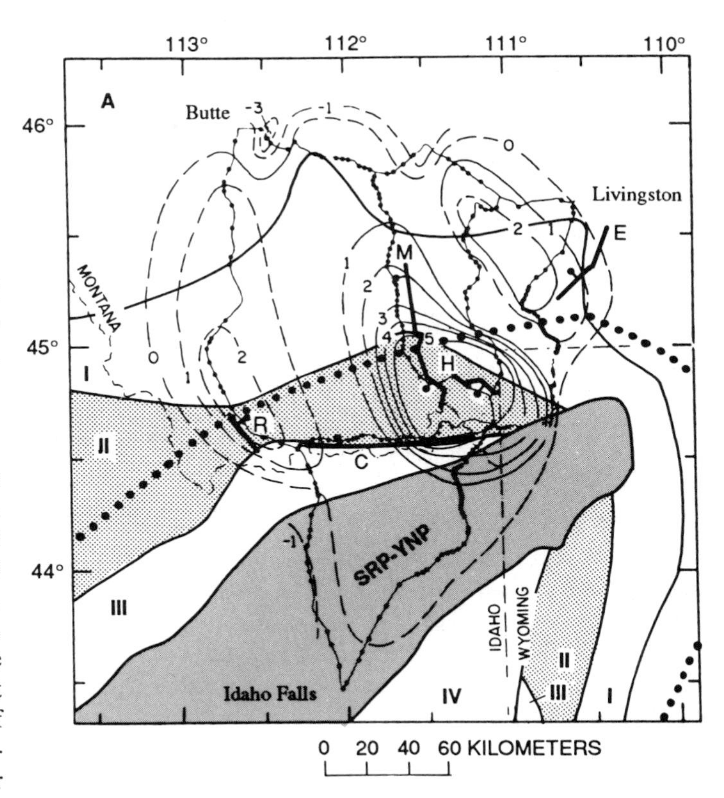

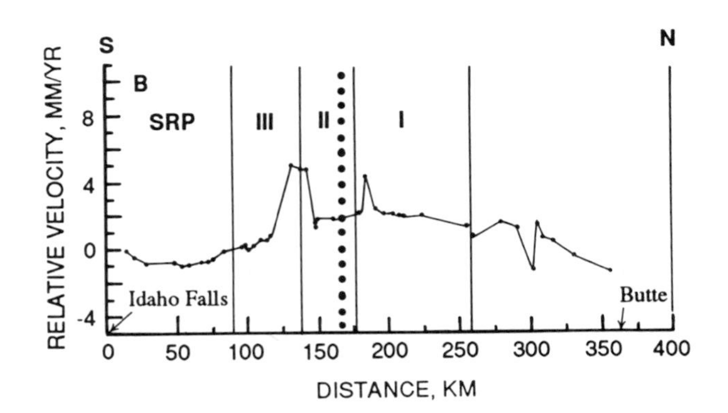

Figure 15. Historic vertical uplift across the western arm of the Yellowstone crescent of high terrain (from Reilinger, 1985). Lines with small dots indicate resurveyed benchmarks along highways. The interval between surveys was 30 to 60 years between 1903 and 1967 (Reilinger and others, 1977, Table 1). Area of map shown on Figure 14. A, Contours on historic uplift in mm/year based entirely on highway survey lines, almost all located along topographic lows. Contours based on this data, almost entirely from basins, show three domes of uplift roughly coincident both with the axis of the Yellowstone crescent and fault Belt II (shaded), excepting Emigrant fault. The highest basin uplift rates (>2 mm/year) are associated with the downthrown sides of the active Red Rock, Madison-Hebgen, and Emigrant faults (see text). B, Vertical movements derived from repeated leveling along line between Idaho Falls, Idaho, and Butte, Montana. An arch of uplift about 200 km wide centers near the boundary of Belts I and II (see part A).

drainage system was clearly reversed by the time a 4-Ma basalt flowed northward down it. This paleodrainage system was subsequently broken by basin-range faulting. Reversal of drainage by northward tilting may reflect the outer slope of the Yellowstone paleocrescent associated with hot-spot migration during the 6.5- to 4.3-Ma Heise volcanism (Plate 1), followed after 4 Ma by disruption of the drainage by basin-range faulting associated with eastward and northward migration of Belts I and II.

The three domes of historic uplift along the western arm of the Yellowstone crescent also coincide with a postulated axis of arching represented by the modern drainage divides. For the three basin-range valleys east of the Beaverhead, Lemhi, and Lost River ranges, Ruppel (1967) summarized evidence that the north half of three south-flowing drainages had been reversed from south flowing to north flowing. This reversal supports the idea of late Cenozoic arching along the present crest of the west arm of the Yellowstone crescent (Fig. 14).

Rugged mountains of readily erodible rocks

Most of the Rocky Mountains are formed of erosionally resistant rocks such as granites, gneiss, and Paleozoic limestones and sandstones. But much of the mountainous terrain forming the head of the Yellowstone crescent is underlain by relatively young, erosionally incompetent rocks. Late Cenozoic uplift is required to explain the high, steep slopes underlain by such rocks and the high rates of erosion in these areas.

The mountains that form the northern and much of the eastern margin of Jackson Hole and extend in a belt 30 to 40 km wide north to Yellowstone Lake are underlain by weakly indurated sandstones, shales, and conglomerates, mostly of Mesozoic and early Cenozoic age. Relief in this terrain is as great as 700 m over 2.4 km (16° slope) and more commonly 600 m over 1.4 to 2 km (17 to 24° slope). Landslides are common; stream valleys are choked with alluvium and have broad, active, gravel-rich channelways. Streams draining into Jackson Hole from the east, such as Pilgrim, Pacific, Lava, Spread, and Ditch creeks, have built large postglacial alluvial fans. East of a remnant of 2-Ma ignimbrite (Huckleberry Ridge Tuff) on Mt. Hancock (Fig. 1; alt. 3, 112 m), local erosion of about 800 m of Mesozoic sediments has occurred since 2 Ma. West of Mt. Hancock, structural relief on this tuff of 775 m in 6 km also demonstrates large-scale Quaternary deformation (Love and Keefer, 1975).

The Absaroka Range occurs along and east of the east boundary of Yellowstone Park and extends for about 70 km to the southeast. This range is formed largely of erodible Eocene volcaniclastic and volcanic rocks. The highest, most precipitous part of the Absaroka Range is 75 to 100 km from the center of the 0.6-Ma Yellowstone caldera. Peaks in the range reach above 3,700 m, and peak-to-valley relief is as much as 2,000 m. Mountain sides have dramatic relief, locally rising 1,000 m in 1.9 km (27°), 800 m in 1.3 km (31°), and 670 m in 1.1 km (31°). Almost yearly, snowmelt and/or flash floods result in boulder-rich deposits on alluvial fans at the base of steep slopes. The upper parts of stream courses are commonly incised in bedrock, whereas the larger streams commonly flow on partly braided floodplains with year-to-year shifts of gravel bars. Runoff is commonly turbid during snowmelt; common flash floods are accompanied by audible transport of boulders. The bedrock is dangerous to climb or even walk on because of loose and detached rock fragments. Compared to other parts of the Rocky Mountains formed of Precambrian crystalline rocks, the Absaroka Range is rapidly eroding and the presence of deep, young canyons suggests late Cenozoic uplift of a 1-km magnitude.

A widespread, low-relief erosion surface is well preserved on uplands of the Absaroka Range. In the southern Absaroka Range, basalt was erupted on this surface prior to cutting of the modern canyons (Ketner and others, 1966); one of these flows has a K-Ar age of 3.6 Ma (Blackstone, 1966).

An obsidian-bearing gravel occurs 80 km east of Jackson Lake, Wyoming, high in the Absaroka Range (alt. 3,350 m) (Fred Fisher, written communication, 1989; W. R. Keefer, oral communication, 1989; J. D. Love, oral communication, 1988). Obsidian pebbles from this gravel have a K-Ar age of 6.26 ± 0.06 Ma (Naeser and others, 1980). The possible sources of this obsidian include Jackson Hole (Love and others, 1992), the Heise volcanic field (Morgan, 1988), or two silicic volcanic deposits in the Absaroka Range (Love, 1939; Smedes and others, 1989). If the obsidian originates from either the Heise volcanic field or Jackson Hole, this gravel indicates more than a kilometer of westerly tilting since 6.3 Ma, with subsidence to the west in the Jackson Hole/Heise area on the inside of the crescent, and uplift to 3,350 m and associated deep incision near the axis of the crescent. If the source is from late Tertiary rhyolites in the Absaroka Range, the present deep canyons have been carved since 6 Ma, indicating >1 km uplift.

For the high Absaroka Range extending from the area just described for about 50 km to the north, formation of a late Cenozoic syncline with 600 m structural relief over about 40 km is described by Fisher and Ketner (1968). This deformation is on a scale an order of magnitude smaller than the Yellowstone crescent. South of the Absaroka Range, major late Cenozoic tilting along the western front of the Gros Ventre Range is described by Love and others (1988). The Pliocene Shooting Iron Formation (about 2 to 3 Ma) contains fine-grained lacustrine sediments and was deposited in a topographic low. Love and others (1988) conclude that westward tilting of the Shooting Iron Formation since 2 to 3 Ma has resulted in the 1.2-km altitude difference between remnants in Jackson Hole and those on the Gros Ventre Range.

In conclusion, the geomorphology and recent geologic history of the mountains near the head of the Yellowstone crescent of high terrain suggest late Cenozoic uplift of the magnitude of 0.5 to 1 km; and uplift is probably still continuing.

Pleistocene glacier-length ratios and altitude changes

In the Rocky Mountains, terminal moraines of the last (Pinedale) glaciation normally are found just up valley from those of the next-to-last (Bull Lake) glaciation. Pinedale terminal moraines date from between 20 and 35 ka, whereas most Bull Lake moraines date from near 140 ka, for an age difference of about 120 ± 20 k.y. (Pierce and others, 1976; Richmond, 1986). The end moraine pattern is consistent with the Oxygen-18 record from marine cores, where the estimated global ice volume of stage 2 time (Pinedale) was 95% that for stage 6 time (Bull Lake) (Shackleton, 1987). The ratio of the length of Pinedale glaciers to Bull Lake glaciers, Pd/BL ratio, is typically between about 88 and 96%. For example, in the Bighorn Mountains, 11 valleys mapped by Lon Drake and Steven Eisling (written communication, 1989) have Pd/BL ratios of 88 ± 6% (Fig. 16). In the Colorado Front Range, glacial reconstructions in five valleys yield a Pd/BL length ratio of 96 ± 3%.

Departures from the normal ratio of glacier length during the last two glaciations may indicate areas of uplift or subsidence (Fig. 16). Uplift elevates a glacier to a higher altitude during the Pinedale than it was earlier during the Bull Lake glaciation. This higher altitude would tend to increase Pd/BL ratios. Subsidence would produce a change in the opposite sense. Factors other than uplift or subsidence that may be responsible for changes in the Pd/BL ratio include (1) local responses to glacial intervals having differing values of precipitation, temperature, and duration; (2) different altitude distributions of the glaciated areas; (3) differing storm tracks between Pinedale and Bull Lake time, perhaps related to different configurations of the continental ice sheets; and (4) orographic effects of upwind altitude changes.

Figure 16 was compiled to see if high (>96%) or low (<88%) values of Pd/BL ratios define a pattern that can be explained by uplift or subsidence. In the head and southern arm of the Yellowstone crescent, the Beartooth Mountains, the Absaroka Range, and the central part of the Wind River Range have high Pd/BL ratios: Of 30 Pd/BL ratios, 24 exceed 96% and 17 exceed 100% (Fig. 16). Only one low ratio was noted: a ratio of 79% based on a moraine assigned to Bull Lake glaciation along Rock Creek southwest of Red Lodge, Montana. The belt of high values is 70 to 140 km from the center of the 0.6-Ma Yellowstone caldera and lies between the crest and outer margin of the crescent (Plate 1, Fig. 16). Uplift is expected in this area based on the northeastward migration of the Yellowstone hot spot.

For the western arm of the Yellowstone crescent of high terrain, Pd/BL ratios generally are from mountain ranges where active faulting may result in *local* rather than regional uplift. Mountains not associated with active, range-front faulting occur east of Stanley Basin (Fig. 1) and have ratios of >100%, suggesting uplift (Fig. 16).

The magnitude of the amount of uplift that might produce a 5% increase in the Pd/BL ratio can be calculated based on valley-glacier length and slope at the equilibrium line (line separating glacial accumulation area and ablation area). A vertical departure from the normal difference between Pinedale and Bull Lake equilibrium line altitudes (ELAs) would approximate the amount of uplift. The 5% increase in Pinedale glacier length from normal displaces the equilibrium line downvalley a distance equal to about half the increase in length of the glacier. For a valley glacier 15 km long, a typical slope at the ELA is about 3° (see Porter and others, 1983, Fig. 4-23). A 5% change in length is 750 m, indicating a 375-m downvalley displacement of the ELA, which for a glacier-surface slope of 3° decreases the ELA by 20 m. On the outer slope of the Yellowstone crescent, the Pd/BL ratios are about 5 to 10% above normal (Fig. 16). These glaciers are similar in size and slope to those noted above. Thus, uplift of Pinedale landscapes by several tens of meters relative to Bull Lake ones could explain this belt of high Pd/BL ratios. Uplift of 20 to 40 m in 100 to 140 k.y. (Bull Lake to Pinedale time) results in a rate of 0.15 to 0.4 mm/year, similar to 0.1 to 0.5 mm/year assuming northeast, plate-tectonic motion of the Yellowstone crescent.

A dramatic contrast in Pd/BL ratios occurs around the perimeter of the greater Yellowstone ice mass and suggests northeast-moving uplift followed by subsidence (Fig. 16). The Pd/BL ratios are >100% for glacial subsystems that terminated along the Yellowstone River and the North Fork of the Shoshone River and were *centered* to the north and east of the 0.6-Ma caldera. But Pd/BL ratios for glacial subsystems located on and south or west of the 0.6-Ma caldera are, from south to north, 59, 77, 69, and 49% (from Jackson Hole via Fall River to the Madison River and Maple Creek). These ratios are among the lowest Pd/BL ratios in the Rocky Mountains. Because the Yellowstone ice mass was an icecap commonly more than 1 km thick, any calculation of the apparent change in ELA is much more uncertain than for valley glaciers. Nevertheless, these extremely low values for the southern and western parts of the Yellowstone ice mass are compatible with perhaps 100-m *subsidence* at perhaps 1 mm/year, whereas the high Pd/BL ratios in the northern and eastern Yellowstone region suggest *uplift* at perhaps 0.1 to 0.4 mm/year.

We wish to caution that the possible uplift and subsidence pattern based on departures from the "normal" in the size of different-aged Pleistocene glaciations is built on an inadequate base of observations. Most of the Pinedale and Bull Lake age assignments and the Pd/BL ratios calculated therefrom generally are not based on measured and calibrated relative-age criteria. Additional studies are needed to verify and better quantify the pattern apparent in Figure 16.

With the above caution in mind, Figure 16 shows that high ratios, suggesting uplift, occur in an arcuate band largely coincident with the outer slope of the Yellowstone crescent of high terrain. Inside the crescent, the few ratios that were determined are mostly from ranges with active range-front faults. A concentration of low Pd/BL ratios at about the position of Twin Falls suggests subsidence near the Snake River Plain and on the inner, subsiding slope of the Yellowstone crescent. For the Yellowstone icecap, low ratios suggest subsidence in and south of the 0.6-Ma caldera; whereas to the north and east of this caldera, high values suggest uplift.

Big Horn Basin region

Unidirectional stream migrations (displacements). Abandoned terraces of many drainages east and north of Yellow-

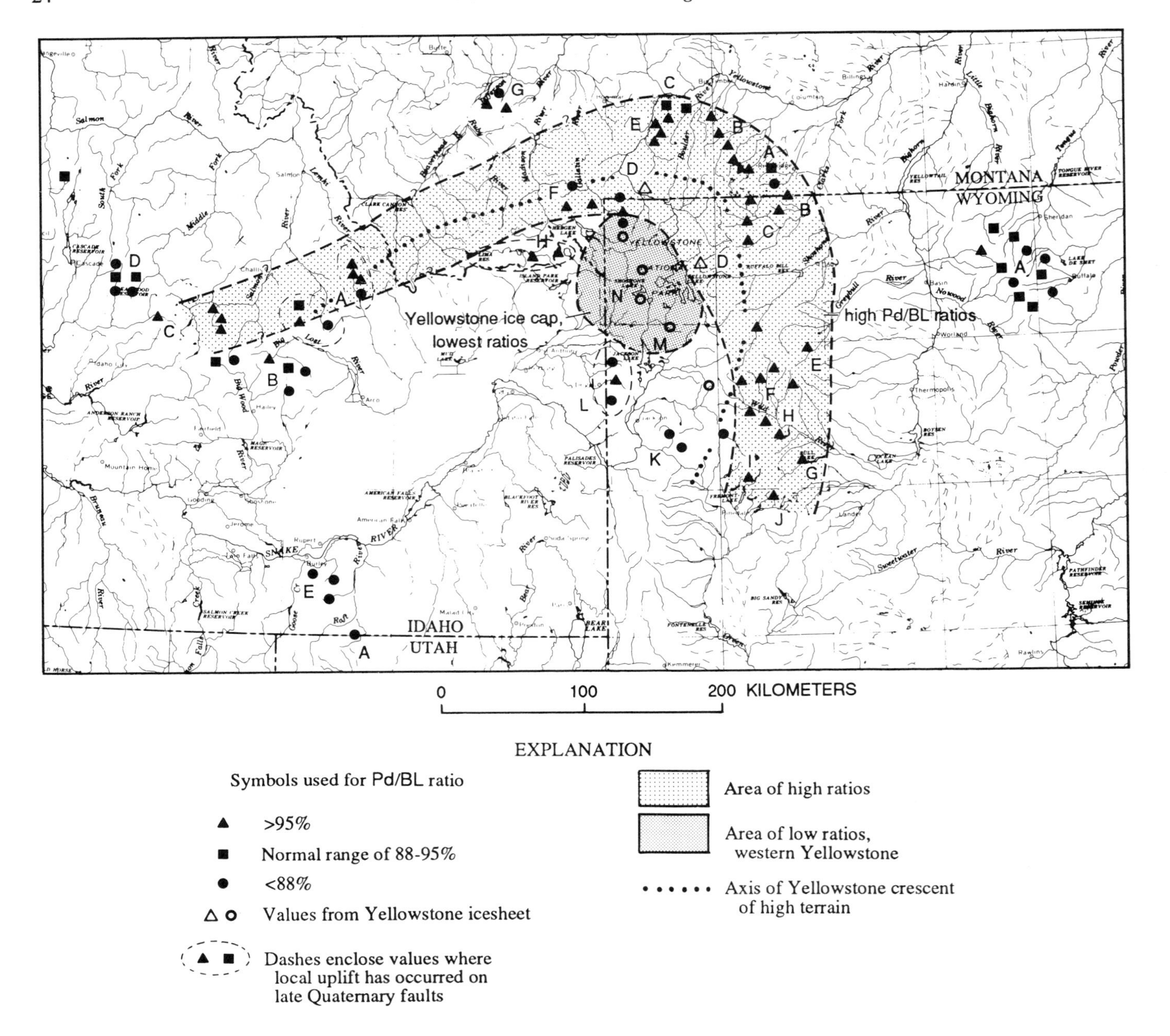

stone record a history of stream displacement, either by capture or slip off, that tends to be away from the Yellowstone crescent (Fig. 17). The main exception to this away-from-Yellowstone pattern was the capture of the 2.2-Ma Clark Fork and the subsequent migration of the 2-Ma Clark Fork (Fig. 17; Reheis and Agard, 1984). The capture appears to have been favored by the readily erodible Cretaceous and younger sediments at the northern, open end of the Big Horn Basin, but the subsequent migration is an example contrary to the overall trend of movement away from the Yellowstone crescent.

Three main processes have been evoked to explain unidirectional stream migration: (1) greater supply of sediment, particularly coarse gravel, by tributaries on one side of a stream; (2) migration down-dip, particularly with interbedded erodible and resistant strata; and (3) tectonic tilting with lateral stream migration driven either by sideways tilt of the stream or, as suggested by Karl Kellogg (written communication, 1990) by increased or decreased sediment supply to the trunk stream depending on whether tributary streams flow in or opposed to direction of tilt. Palmquist (1983) discussed reasons for the eastward migration of the Bighorn River in the Bighorn Basin and rejected all but tectonic tilting. Morris and others (1959) considered that sediment and water supply from the north were responsible for southward migration of two tributaries of the Wind River (Muddy and

Figure 16. Ratios of the length of Pinedale to Bull Lake glaciers (Pd/BL) in the Yellowstone crescent of high terrain and nearby mountains. Pinedale terminal moraines are about 20 to 35 ka, and Bull Lake ones are about 140 ka. Uplift would result in high ratios (triangles), and subsidence would result in low ratios (circles), although several nontectonic factors could also result in departures from the "normal" ratio. Area of high ratios (light shading) suggests uplift on outer slope of the Yellowstone crescent of high terrain. Area of low ratios for the Yellowstone ice sheet (darker shading) suggests subsidence of the west part of Yellowstone. Sources of data (capital letters indicate location by State on figure): *Montana:* A, East and West Rosebud, West and main fork, Rock Creek (Ritter, 1967); B, West and main fork, Stillwater River (Ten Brink, 1968); C, north side Beartooth uplift (Pierce, field notes); D, Yellowstone River, west side of Beartooth uplift, northwest Yellowstone (Pierce, 1979); E, west side of Beartooth uplift (Montagne and Chadwick, 1982); F, Taylor Fork (Walsh, 1975, with modifications by Pierce); G, Tobacco Root Mtns. (Hall and Michaud, 1988, and other reports by Hall referenced therein); H, Centennial Range (Witkind, 1975a). *Wyoming:* A, Bighorn Mtns. (Lon Drake and Steven Esling, written communication, 1989); B, east side Beartooth Mtns. (W. G. Pierce, 1965); C, Sulight Basin (K. L. Pierce, unpublished mapping); D, North Fork Shoshone River (John Good, written communication, 1980); E, Wood River (Breckenridge, 1975, reinterpreted by Pierce after consulting with Breckenridge); F, southern Absaroka Range and Wind River (Helaine Markewich, oral communication, 1989; K. L. Pierce, field notes, 1989); G, Bull Lake Creek (Richmond and Murphy, 1965; Roy and Hall, 1980); H, Dinwoody Creek, Torrey Creek, Jackeys Fork, Green River (K. L. Pierce, field notes); I, Fremont Lake, Pole Creek (Richmond, 1973); J, Big Sandy River (Richmond, 1983); K, Granite Creek and Dell Creek (Don Eschman, written communication, 1976); L, west side of Tetons (Fryxell, 1930; Scott, 1982). M, Jackson Hole (K. L. Pierce and J. D. Good, unpublished mapping); N, southwestern Yellowstone (Richmond, 1976; Colman and Pierce, 1981). *Idaho:* A, drainages adjacent to the eastern Snake River Plain not designated "B" (Scott, 1982); B, Copper Basin and area to west (Evenson and others, 1982; E. B. Evenson, written communication, 1988); C, South Fork Payette River (Stanford, 1982); D, Bear Valley and Payette River (Schmidt and Mackin, 1970; Colman and Pierce, 1986); E, Albion Range (Scott, 1982; R. L. Armstrong, written communication; K. L. Pierce, unpublished mapping). *Utah:* (A) Raft River Range (K. L. Pierce, unpublished mapping).

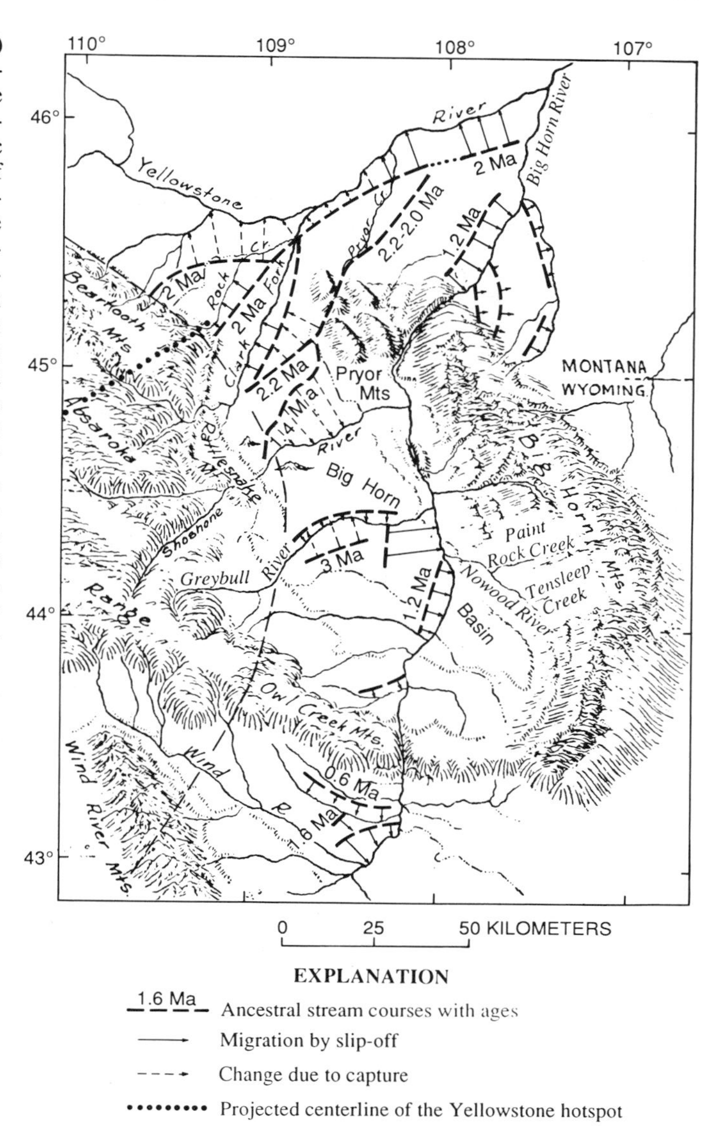

Figure 17. Direction of Quaternary unidirectional stream migrations by either slip-off or capture, Bighorn Basin and adjacent areas. Note that streams have generally migrated away from the outer margin of the Yellowstone crescent of high terrain (arcuate dashed line). Base map from Mackin (1937). Compiled from Ritter, 1967; Reheis, 1985; Reheis and Agard, 1984; Agard, 1989; Palmquist, 1983; Mackin, 1937; Andrews and others, 1947; Ritter and Kauffman, 1983; Morris and others, 1959; Hamilton and Paulson, 1968; Richards and Rogers, 1951; Agard, written communication, 1989; and Palmquist, written communication, 1989; mostly using ash-based chronology from Izett and Wilcox, 1982.

Fivemile creeks) near Riverton. However, for the southeastward migration of the Wind River, the size of tributary drainage basins flowing into the Wind River from the north (Muddy and Fivemile creeks) compared with those from the south appears to invalidate this mechanism. For drainages north of the Bighorn Mountains, migration due to unequal sediment contribution by tributaries on either side of the river is not reasonable for the northward migration of the Yellowstone River or the eastward migration of the Bighorn River and two of its tributaries (Lodge Grass and Rotten Grass creeks).

The effect of tilting on erosion and sediment supply by tributary streams seems a more potent mechanism than the tilting of the trunk stream itself. For tributaries flowing away from Yellowstone into the trunk stream, tilting would increase both their gradient and sediment delivery; for tributaries flowing toward Yellowstone, both would be decreased; the combined effect would produce migration of trunk streams away from Yellowstone.

The effect of sideways tilting of the trunk stream seems too small to produce the observed lateral migration. One km of uplift over the distance from the Yellowstone crescent axis to the east side of the Bighorn Basin, 200 km, produces a total tilt of only 0.3°. This small amount of tilting at a plate motion rate of 30 km/m.y. would occur over an interval of nearly 7 m.y., or at a rate of 0.00005° per thousand years. Although the mechanism is poorly understood, a relationship to the Yellowstone hot spot is suggested by the pattern of displacement of stream courses generally away from the axis of the Yellowstone crescent of high terrain.

Convergent/divergent terraces. Terrace profiles from some drainages in the Bighorn Basin (Fig. 17; Plate 1) suggest uplift of the western part of the basin and tilting toward the east. If tilting is in the same direction as stream flow, stream terraces will tend to diverge upstream; whereas if tilting is opposed to stream flow, terraces will tend to converge upstream. Streams flowing west into the Bighorn River and toward Yellowstone have terraces that tend to converge upstream with the modern stream, as illustrated by the terrace profiles of Paint Rock and Tensleep creeks (Figs. 17, 18). Streams flowing east into the Bighorn River have terraces that tend to diverge upstream (Mackin, 1937, p. 890), as shown by the terrace profiles for the Shoshone River, particularly the Powell and the Cody terraces, which appear to diverge at about 0.3 to 0.6 m/km (Fig. 18). For the Greybull River (Fig. 17), terraces show a lesser amount of upstream divergence (Mackin, 1937; Merrill, 1973).

Isostatic doming of the Bighorn Basin due to greater erosion of soft basin fill than the mountain rocks (Mackin, 1937; McKenna and Love, 1972) would tend to produce upstream convergence of terraces for streams flowing from the mountains toward the basin center. This effect would add to hot-spot–related upstream convergence of west-flowing streams such as Paint Rock and Tensleep creeks but would subtract from hot-spot–related upstream divergence of east-flowing streams such as the Shoshone and Greybull rivers, perhaps explaining the stronger upstream convergence than upstream divergence noted in the west- and east-flowing streams (Fig. 18).

Rock Creek (Fig. 17) flows northeast past the open, northern end of the Bighorn Basin. Terrace profiles for Rock Creek first converge and then diverge downstream. For any two terraces, the convergence/divergence point is defined by the closest vertical distance on terrace profiles. A plot of this convergence/divergence point against the mean age of the respective terrace pairs (Fig. 19) shows that this point has migrated northeastward throughout the Quaternary at an irregular rate, but overall at a rate not incompatible with hot-spot and plate-tectonic rates. The

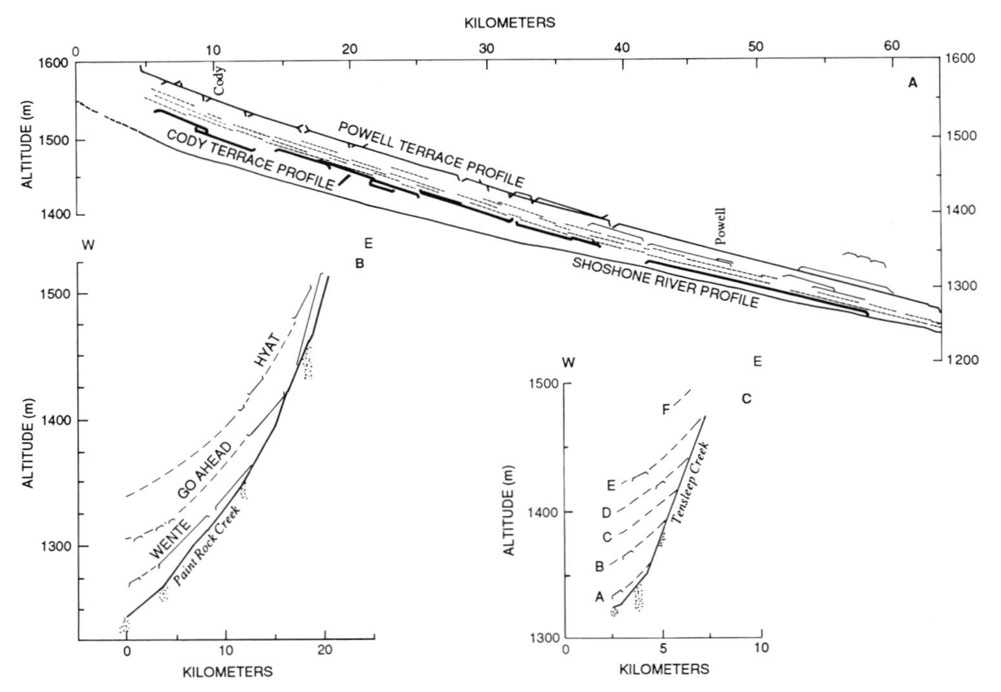

Figure 18. Terrace profiles along three selected streams, Bighorn Basin, Wyoming. A, Shoshone River; flows east away from the Yellowstone crescent, and terraces and stream profile diverge upstream. Shoshone River terraces from Mackin (1937), where solid line is altitude of top of gravel beneath variable capping of overbank and sidestream alluvium; dashes indicate top of terraces with such fine sediment. Upstream divergence of about 0.3 to 0.6 m/km based on comparison of Powell terrace with lowermost continuous Cody terrace represented by heavy line. B, Paint Rock Creek and C, Tensleep Creek. Both flow west toward the crescent; terraces converge upstream. Terrace profiles and designations from Palmquist (1983); stipples below floodplains represent depth of glaciofluvial fill.

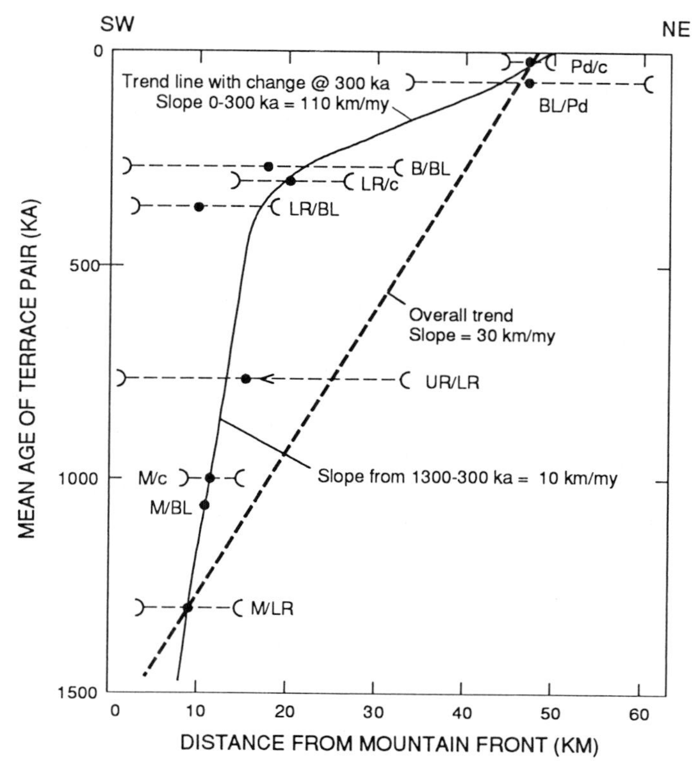

Figure 19. Northeast migration away from Yellowstone crescent and Beartooth Mountain front of the convergence/divergence points between pairs of terrace profiles, Rock Creek, Montana. Solid circle, closest place between older terrace and younger terrace; small arc, place where terrace profiles visibly diverge. Based on terrace and stream profiles drawn by Reheis (1987, Fig. 3). Terrace symbols: M = Mesa; UR = Upper Roberts; LR = Lower Roberts; B = Boyd; BL = Bull Lake; and Pd = Pinedale. Other symbols: c = channel of Rock Creek; < = convergence point lies left of dot.

convergence/divergence point might be thought of as the inflection point between slower rates of uplift to the northeast and higher rates of uplift to the southwest toward the Yellowstone crescent.

The change (Fig. 19) from an apparent rate between 1,300 and 300 ka of <10 km/m.y. to an apparent rate of 110 km/m.y. after 300 ka may suggest irregularities associated with the deep processes responsible for the inferred uplift and tilting. Whether the change in rate or even the overall trend of the plot itself has tectonic significance is open to question, for climatic, lithologic, or stream-capture factors might also relate to changes in gradient through time.

In the Bighorn Basin, the Bighorn River parallels the crest of the Yellowstone crescent, and its terraces are remarkably parallel (Palmquist, 1983). Northeastward from the Bighorn Mountains, the Bighorn flows away from the Yellowstone crescent. Here, Agard (1989) notes the lower four terraces converge with the river, whereas the upper four terraces diverge from it with an overall pattern of migration of convergence/divergence points away from the Yellowstone crescent.

The location of the maximum thickness of Quaternary gravels along the Greybull River has generally migrated about 25 to 45 km eastward in about 800 to 900 k.y. (data from R. C. Palmquist, written communication, 1989), yielding overall rates not incompatible with plate-tectonic rates.

Uplifted(?) calcic soils, Rock Creek. The upper limit of carbonate accumulation in soils is primarily controlled by (1) some maximum threshold value of precipitation, which increases with altitude, and (2) temperature, which decreases with altitude at a lapse rate of typically 5.7°C/1,000 m. In her study of the terraces of Rock Creek, Montana, Reheis (1987) determined that the upper altitude limit of calcic soils on each terrace increases with terrace age. Calcic soils occur in the 600-ka lower Roberts terrace at altitudes 300 m higher than the highest such soils in the 20-ka Pinedale terrace and 250 m higher than such soils in the 120-ka Bull Lake terrace (Fig. 20). The highest

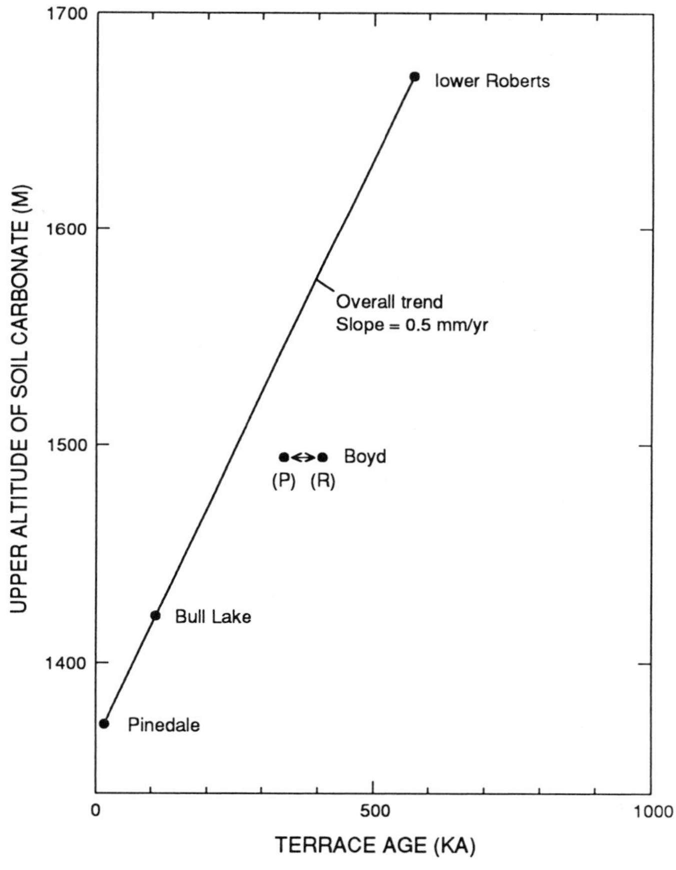

Figure 20. Decrease in the altitude of the upper limit of calcareous soils with terrace age along Rock Creek at northwest margin of Bighorn Basin, Montana (from Reheis, 1987). Regional uplift of about 0.5 mm/year could explain this change, although it may also be explained by nontectonic causes, including increase in soil fines through time and increased aridity of some increasingly older interglacial climates. Terrace names as in Reheis (1987); two age options are shown for Boyd terrace: R = Reheis, 1987; P = Palmquist, written communication, 1989. The data point for 2-Ma Mesa terrace is not included.

calcic soils forming on the 20-ka Pinedale terrace are continuing to accumulate carbonate, but the highest calcic soils forming on older terraces are relict and tending to lose their carbonate, particularly those at the highest, moister altitudes.

If the increase in altitude of relict calcic soils is due solely to uplift, uplift at 0.5 m/k.y. (0.5 mm/year) is indicated (Fig. 20). On the other hand, the increase in altitude of the upper limit of calcic soils may be due to factors other than uplift, including (1) climatic differences between interglaciations wherein some older interglacials were increasingly arid (Richmond, 1972), (2) effect of the buildup of fine material in the soil through time, discussed in Reheis (1987, p. D24), and (3) a time- and cycle-dependent process that may enhance carbonate buildup once $CaCO_3$ nucleation sites are established. Concerning factor (1), evidence for older interglacials being increasingly warmer is not supported by global oxygen-isotope records of ice volumes. As for the approximately seven interglacials that postdate the 600-ka Roberts terrace, none are clearly isotopically lighter (lesser ice volumes, warmer?) than stages 1 or 5, which are the interglacials following the 20-ka Pinedale and the 120-ka Bull Lake terraces. Only two others are isotopically similar, and the other five interglacials were isotopically heavier (greater ice volumes, colder?) than stages 1 and 5 (Shackleton, 1987).

At present, we cannot decide among these explanations for the increase in altitude of relict calcic soils. Uplift has not been previously considered, yet ongoing epeirogenic uplift is commonly thought to be occurring for much of the Rocky Mountains. The estimated rate of 0.5 mm/year assuming uplift is at the upper end of the estimated range of 0.1 to 0.5 mm/year based on plate-tectonic motion of the Yellowstone crescent (0.5 to 1 km uplift in 2 to 5 m.y.).

Problems relating regional uplift to the Yellowstone hotspot

Uplift is most clearly defined for the Yellowstone crescent on the leading margin of the Yellowstone hot spot (Fig. 14; Plate 1). The distance from the crest of the crescent to its outer margin is about 100 ± 40 km. At a migration rate of 30 km/m.y. indicated by the Yellowstone hot spot, uplift on the outer margin of the Yellowstone crescent would therefore take about 3 m.y.; uplift of 0.5 to 1 km would therefore occur at a rate of 0.1 to 0.3 mm/year.

Quantitative separation of the Yellowstone crescent from Laramide block uplifts presents a major challenge. These uplifts, presently expressed by the Gros Ventre, Beartooth, Wind River, and Blacktail-Snowcrest-Madison ranges, are mountains that expose older bedrock that has been uplifted, in part, by Laramide crustal shortening. Late Cenozoic normal faulting, however, has broken parts of these uplifts. For example, the Teton fault has broken the Gros Ventre Range (Love, 1977; Lageson, 1987, and this volume), the Madison fault has broken the west side of the Madison Range, and the Emigrant fault has broken the northwest side of the Beartooth uplift (Personius, 1982).

Beyond the Yellowstone crescent, an outer limit of moderately high terrain occurs at distances of 300 to 400 km north and east of Yellowstone, compatible with a swell 700 km across (Fig. 14, dotted line). Regrettably, we have not been able to draw a line around to the south of Yellowstone and thus define a topographic feature comparable to the 400- to 600-km radius (800- to 1,200-km diameter) of oceanic swells. This problem is greatest southeast from the Yellowstone crescent across the Green and Wind River basins and in southern Wyoming and Colorado. There is no clear boundary between high and lower topography between northwest Wyoming and southern Colorado. Suppe and others (1975) proposed the intriguing suggestion that tandem northeastward migration of the Yellowstone and Raton (New Mexico) "hot spots" has caused arching on an axis that extends between the hot spots and is responsible for the epeirogenic uplift of the Rocky Mountains–Colorado Plateau–Great Plains region. This tandem hot-spot arch is basically the same late Cenozoic uplift previously known as epeirogenic uplift of the Rocky Mountain–High Plains–Colorado Plateau area. The Raton hot spot is much less credible than the Yellowstone hot spot, having neither a systematic volcanic progression (Lipman, 1980) nor associated belts of neotectonic faulting. Consequently, the Raton hot spot, if real, is weaker and less able to affect and penetrate the continental lithosphere than is the Yellowstone hot spot.

A map of the geoid of the United States (Milbert, 1991) shows a regional geoid high that is remarkably well centered on Yellowstone (Fig. 21). That part of the geoid higher than −8 m (Fig. 21) has a diameter of about 150 km and is roughly the same as the Yellowstone crescent of high terrain. On the leading margin of the Yellowstone hot spot, the −8, −11, and −14 geoid contours (Fig. 21) sweep around the Yellowstone hot spot in a parabolic fashion consistent with plate motion to the southwest, whereas the −17, −20, and −23 m contours are semicircular about Yellowstone. We tentatively interpret this change in contour pattern to indicate that the stagnation streamline (Sleep, 1990) is near the position of the −14 m contour. Assuming this is so, the stagnation distance (Sleep, 1990) is about 250 to 400 km, the radius at 90° to the stagnation distance is 400 to 600 km, and the geoid relief inside the stagnation streamline is 6 to 9 m.

The geoid dome centered on Yellowstone is the only geophysical anomaly that we have seen that compares favorably in size and height with oceanic hot spots. Oceanic swells have geoid domes similar in dimension; for example, those shown in Figure 13 have radii of 400 to 600 km (Fig. 13), stagnation distances of 350 to 500 km and heights in the 6 to 12 m range (Sleep, 1990). For the western United States, the major topographic and bouguer gravity anomalies are quite similar and center on western Colorado, not on Yellowstone (Kane and Godson, 1989). Altitude or the bouguer gravity appear to effect the geoid map but do not dominate the signal. According to Norm Sleep (oral communication, 1992), the compensation depth for the geoid extends to the base of the lithosphere at 100 ka or so depth, whereas that for the regional topography and bouguer gravity extends to the base of the crust at 30-40 km depth.

Quaternary incision rates were examined to see if any pattern would emerge (Fig. 21). These rates are determined from

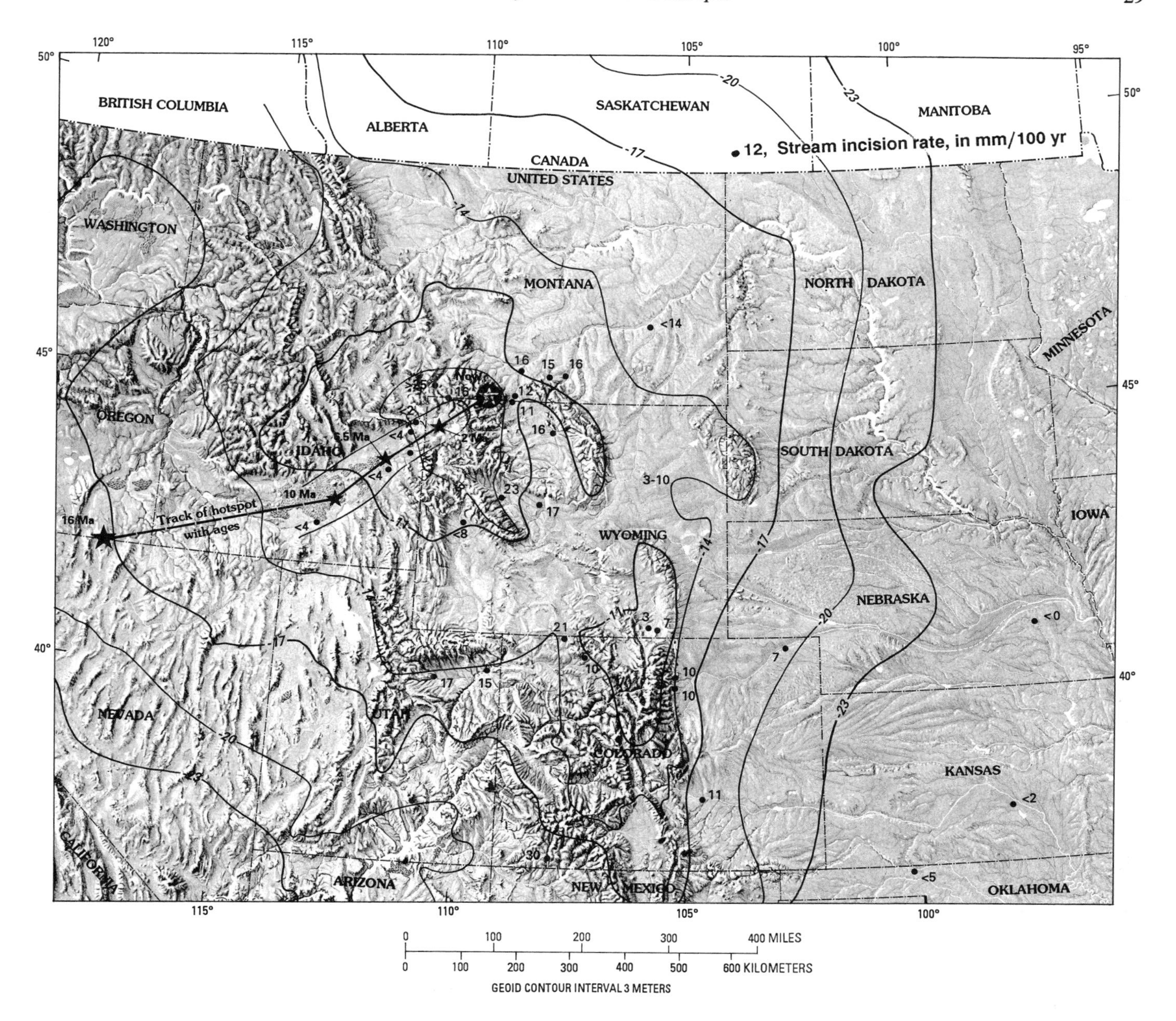

Figure 21. Contours on the geoid and some rates of stream incision in the Great Plains and part of the Rocky Mountain area. A large geoid dome centers on Yellowstone and the contours show the following pattern: (1) that part higher than –8 m has a pattern similar to the Yellowstone crescent of high terrain; (2) the –8, –11, and –14 m contours sweep in a parabolic manner around the leading edge of the Yellowstone hot spot, whereas the –17, –20, and –23 m contours are semicircular about the leading margin of the hot spot (see text). Geoid contours from Milbert (1991) are only approximately transferred and are rounded from actual values as follows (meters): 8 = 8.47, 11 = 11.43, 14 = 14.40, 17 = 17.36, 20 = 20.32, and 23 = 23.28. Contours not shown below 23 m. Incision rates generally reflect uplift rates, although other factors like the relationship between stream orientation and tilt are also involved. High incision rates of greater than 15 mm/100 year extend from just east of Yellowstone south through Colorado to New Mexico. Incision rates are based on terrace heights, the ages of which are known from volcanic ashes erupted between 0.6 and 2.0 Ma (Izett and Wilcox, 1982; Reheis and others, 1991; Carter and others, 1990). R. L. Palmquist suggested the incision-rate compilation and noted that the high incision rates correspond with the axis of uplift in the tandem Yellowstone-Raton hot-spot hypothesis of Suppe and others (1975, Fig. 7).

terraces dated by volcanic ashes that range in age from 0.6 to 2 Ma (Izett and Wilcox, 1982; Reheis and others, 1991). Relatively high incision rates extend from the Yellowstone region southeast to Colorado and New Mexico. If we assume that these incision rates might approximate uplift rates, uplift in this region has been approximately 100 to 200 m/m.y. For a hot-spot swell 1,000 m high and 1,000 km wide in a plate moving 30 km/m.y., uplift of 60 m/m.y. is predicted, a factor of 2 to 3 less than the 100 to 200 m/m.y. rates of incision. In summary, the altitude distribution and the distribution of high incision (uplift?) rates prevent the objective definition of a swell centered on Yellowstone because the "swell" is markedly enlarged on its southeast margin.

Late Cenozoic uplift and tilting of southwestern Wyoming is shown by several studies. For the Pinedale anticline area of the northern Green River Basin, studies of thermal histories of materials from a deep oil well indicate uplift started around 2 to 4 Ma and resulted in at least 20°C cooling, which is equivalent to about 1 km of uplift and erosion (Naeser, 1986; Pollastro and Barker, 1986). Farther south, Hansen (1985) noted that an eastward course of the Green River across the present Continental Divide was captured about 0.6 Ma, and gravels of this eastward course were subsequently tilted so they now slope to the west. This indicates uplift along and east of the present Continental Divide and/or subsidence to the west of it. The tandem Yellowstone-Raton hot-spot hypothesis of Suppe and others (1975) explains the Great Divide Basin (Red Desert on Fig. 1) by arching in the headwaters of a previously east-flowing drainage, thereby creating a closed basin on the Continental Divide.

DISCUSSION

Models to explain the SRP-YP volcanic province include: (1) an eastward propagating rift (Myers and Hamilton, 1964; Hamilton, 1987), (2) volcanism along the SRP-YP localized by a crustal flaw (Eaton and others, 1975), (3) a plate-interaction model with the SRP-YP trend being a "transitional transform boundary zone [at the northern margin] of the Great Basin" (Christiansen and McKee, 1978), (4) volcanism initiated by a meteorite impact (Alt and others, 1988, 1990), and (5) a deep-seated mantle plume leaving a hot-spot track in the overriding plate (Morgan, 1972, 1981; Suppe and others, 1975; Smith and Sbar, 1974; Smith and others, 1977, 1985; Zoback and Thompson, 1978; Anders and others, 1989; Blackwell, 1989; Westaway, 1989a, 1989b; Rodgers and others, 1990; Malde, 1991; Draper, 1991).

All these explanations require heat from the mantle, but in models 1, 2, and 3 the movement of the lithosphere is the operative (driving) process that causes changes in the mantle, whereas in model 5 a thermal mantle plume is the operative (driving) process that causes magmatism and deformation in the overriding plate. In model 1, asthenosphere upwelling into the space vacated by the rifting lithosphere may release melts needed to provide the heat for volcanism. Models 2 and 3 are not based primarily on lithospheric spreading across the SRP, and thus upward transport of heat by melts generated by *strongly* upwelling asthenosphere would not be available.

A mantle plume explanation may seem to have an ad hoc quality because such plumes are remote from direct observation and straightforward testing. Explanations based on response to lithospheric movements may be more appealing to geologists because such explanations relate exposed rocks and their map patterns and thus tie more closely to direct observation. We think mantle plumes are real, based on the success of hot-spot tracks in defining absolute plate motions, the more than 1,000-km dimensions of swells associated with hot spots, and the apparent need for mantle plumes in the deeper of the two convective systems operating in the mantle/lithosphere. Thus, although mantle plumes are remote from the top of the lithosphere, difficult to examine, and not well known in their effects on continental lithosphere, we think the mantle plume model merits serious consideration as an explanation for the volcanic track, neotectonic faulting, and uplift in the SRP-YP region.

Alternative models and their problems

Eastward propagating rift. Hamilton (1987, 1989) considers the lower terrain of the eastern SRP-YP province bounded by higher mountains (Plate 1) to be a rift that has propagated eastward through time. By this explanation, rifting or necking of the lithosphere has created this lower terrain; movement of asthenosphere upward into the volume vacated by lithospheric spreading results in depressurization and generation of basaltic melts. However, extension directions as well as late Cenozoic fault trends are incompatible with this mechanism.

If the SRP were a rift, extension directions of a rift should be perpendicular to its margins. But for the SRP-YP region, extension directions are subparallel to or at low angles to the margins of the SRP-YP trench (Fig. 22; S. H. Wood, 1989b; Malde, 1991). The only exceptions are a few observations near the Montana-Wyoming-Idaho border where some extension directions are nearly perpendicular to the northern margin of the SRP-YP trench. We assume the craton east of basin-range faulting is fixed, thus permitting us to give only the extension vector away from this fixed craton. South from the SRP to the Wasatch front, extension is west to west-northwest. North of the SRP, excepting the Madison-Centennial area, the extension direction is S45°W ± 15°. This extension direction continues to the northern boundary of the basin-range structural province in the Helena, Montana, area (Stickney and Bartholomew, 1987a; Reynolds, 1979, Fig. 10). In the Madison-Centennial area, extension is both S4°W ± 10° and about west, compatible with the westerly trends of the Centennial-Hebgen faults and the northerly trend of the Madison fault respectively (Stickney and Bartholomew, 1987a, Fig. 10). Fissures associated with Quaternary basalt eruptions on the SRP mostly indicate extension to the west-southwest (Kuntz and others, this volume).

Extension is approximately normal to the trend of active faults in the area of fault Belts I, II, and IV, as demonstrated

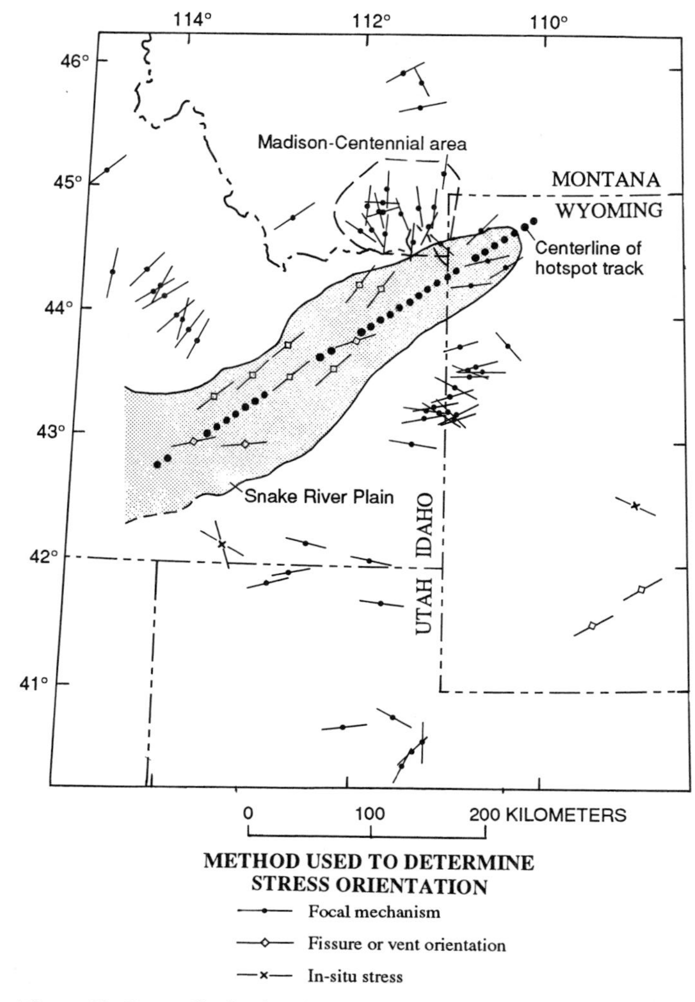

Figure 22. Stress distribution in the Snake River Plain region showing extensional (T-axis) orientations. Extension is subparallel to the length of the SRP-YP trend rather than, as expected for a rift origin, perpendicular to it. Compiled from: north of SRP (Stickney and Bartholomew 1987a); normals to fissures on SRP (Kuntz and others, this volume); Yellowstone-Teton area (Doser and Smith, 1983; Doser, 1985; Wood, 1988); Grand–Star Valley area (Piety and others, 1986); other areas south of SRP (Smith and Sbar, 1974); and general region (Zoback and Zoback, 1980).

statistically by Stickney and Bartholomew (1987a) for the area north of the plain. In addition, extension on the SRP parallels that in the nearby Basin and Ranges. Interestingly, the Great Rift fissure system (Kuntz and others, 1983) arcs through an angle of about 30°, thus joining faults of different orientation north and south of the SRP.

At the active end of the SRP-YP trend at Yellowstone, fault orientation and earthquake extension directions are inconsistent with a rift origin (Fig. 22, Plate 1). Postglacial faults trend north-south across the caldera margin and nearly perpendicular to the margins of the SRP-YP trend. On the Mirror Plateau, numerous postglacial faults trend northeasterly in a broad arc. Their orientation is 90° off the predicted trend of faults at the tip of a propagating rift on the SRP-YP trend.

Thus, actual extension directions are incompatible with a rift origin, for they generally depart more than 45° and many are near 90° from their orientation predicted by a rift origin. In addition, paleogeographic reconstructions of Paleozoic and Proterozoic facies show that opening of the SRP by rifting since the Paleozoic is not required to account for the present distribution and thicknesses of these units (Skipp and others, 1979; Skipp and Link, this volume).

From his study of late Cenozoic faulting in the Blackfoot Mountains, south of and adjacent to the Heise volcanic field, Allmendinger (1982, p. 513) concluded that "no normal faults paralleling the SRP appear to have formed prior to or synchronous with the earliest phase of magmatic activity" and that the SRP is better explained by magmatic rather than rifting processes. Elsewhere, some faults are parallel to the plain, although some of these faults may be associated with caldera margins rather than rift structures (Pankratz and Ackerman, 1982; Malde, 1991). In the southern Lemhi Range and southern Centennial Mountains, adjacent to the Blue Creek and Kilgore calderas, respectively, arcuate faults locally parallel to the plain margin apparently formed synchronously with collapse of the calderas (McBroome, 1981; Morgan, 1988). Zones of autobrecciation (poorly sorted, angular clasts of ignimbrite within a matrix of fine-grained, vapor-phase ash) are concentrated along some faults and appear restricted to the caldera margins (Morgan and others, 1984).

Hamilton (1989) considers magmatism in the SRP-YP province to be clearly rift related on the basis of its bimodal volcanic assemblage, its extensional setting, and its crustal structure. He suggests that the premagmatic crust beneath the SRP has been thinned, that in much of the province it may be absent, and that the present crust consists of mantle-derived basalt. These features and their causes are debatable as to their distinct rift signature. According to Hamilton (1989; oral communication, 1990), other models proposed for the origin of the SRP-YP province, including a hot-spot origin, require that magmatic material has been added to the crust, which would result in the plain's becoming a topographically high area rather than its present characteristic topographic low. However, several factors may explain the 500- to 700-m decrease in average altitude from the adjacent mountains down to the eastern SRP. These factors include gabbroic underplating, the plainwide extent of the calderas and volcanic fields (Morgan and others, 1989), postvolcanic thermal contraction (Brott and others, 1981), transfer of crustal material from beneath the SRP to sites outside of the plain by eruption of ignimbrite and co-ignimbrite ash (>1,000 km^3 each event), and possible changes in the mantle lithosphere. Neither the eastern SRP nor the Yellowstone Plateau appears to require a rift origin to isostatically compensate for the decrease in altitude at its margin.

In addition, the mountains at the edge of the SRP from Yellowstone to the 10.3-Ma (Picabo) caldera do not have the characteristic rift-rim topography; rather than the highest mountains being near the rift margin, the mountains near the edge are relatively low and increase in altitude for about 100 km away from the SRP margin.

Crustal flaws and the origin of the SRP-YP province. The parallelism of the northeast trend of the SRP with that of the late Cretaceous–early Tertiary Colorado Mineral Belt and the late Cenozoic Springerville-Raton lineament was noted by Lipman (1980), who suggested that these northeast trends reflect a preexisting "structural weakness" in an ancient Precambrian crust along which younger activity was concentrated. Others (Eaton and others, 1975; Mabey and others, 1978; Christiansen and McKee, 1978) suggest that the location of the SRP-YP province was controlled by a crustal flaw aligned with the northeast-trending regional aeromagnetic lineament that extends from Mono Lakes, California, into Montana (the Humboldt zone of Mabey and others, 1978). However, upon examination of regional magnetic and gravity maps, we see no strong expression of a crustal boundary along the northeast projection of the Yellowstone hot-spot track. The hot-spot track itself is well expressed in the magnetics, as the rhyolitic and basaltic rocks have relatively high magnetic susceptibilities, but this only reflects the late Cenozoic magmatism and not some preexisting crustal boundary. The projection of the Humboldt magnetic zone to northeast of Yellowstone (Eaton and others, 1975; Mabey and others, 1978) results in dimensions that seem peculiar for a crustal weakness exploited by Yellowstone-type volcanism. First, rather than a single boundary, the zone comprises commonly two lineaments nearly 100 km apart; why would two such lineaments be exploited simultaneously rather than the failing of one leading to concentration of faulting on that one? The Madison mylonite zone locally represents one of these lineaments but was *not* exploited by the Yellowstone trend. No crustal flaw is known to cross the Beartooth Mountains on the projection of the centerline of the hot-spot track (Fred Barker, oral communication, 1990). Second, these lineaments, as drawn, are absolutely straight over a distance of 1,000 km, which seems odd for a geologic boundary that is typically arcuate over such distances.

Major crustal boundaries defined by both geologic and geophysical studies do occur both north and south of the plain and appear more significant than any pre-Miocene crustal flaw coincident with the plain—for which there is no geologic evidence. The Great Falls tectonic zone, a major Archean structure active from Precambrian to Quaternary time, is subparallel to the SRP and is about 200 km north of it (O'Neill and Lopez, 1985). Similarly, the Madison mylonite zone, a major Proterozoic shear zone, is also subparallel to the SRP-YP margin but is about 30 km north of it (Erslev, 1982; Erslev and Sutter, 1990). The southern margin of the Archean Wyoming province, a major crustal boundary, is also subparallel to the plain and is 200 to 300 km south of it (J. C. Reed, written communication, 1989). A Late Proterozoic to Early Cambrian rift basin is postulated by Skipp and Link (this volume) to trend from the Portneuf Range northward across the plain to the Beaverhead Mountains. Analyses of the distribution and thicknesses of Paleozoic sedimentary facies by Skipp and others (1979) show that strike-slip offset across the SRP is not necessary, although both left-lateral (Sandberg and Mapel, 1967) and right-lateral strike-slip faulting (Sandberg and Poole, 1977) have been proposed based on similar information. Woodward (1988) presents structural and facies data to argue that no faulting is indicated across the Snake River Plain and states "There is no reason to postulate major changes [across the Snake River Plain] in Tertiary extension either." In summary, major crustal boundaries are known both north and south of the plain, whereas a major pre-Miocene boundary coincident with the plain is doubtful. If exploitation of crustal flaws is the controlling process, why were not the crustal flaws north of the plain (the Great Falls tectonic zone and the Madison mylonite zone) exploited? We conclude that the location of the Yellowstone hot-spot track probably reflects a sublithospheric process rather than a crustal flaw within the lithosphere. Preexisting crustal flaws do not have a unique relationship to the SRP, and a transcurrent fault along the SRP is not required from either stratigraphic or structural studies.

Transform boundary zone origin for the SRP. In their plate-interaction model for the basin and range structural province, Christiansen and McKee (1978) suggested the SRP-YP was a "transitional transform boundary zone [at the northern margin] of the Great Basin." They suggest that transform motion is indicated by a greater amount of basin-range extension south of the 1,000-km volcanic lowlands formed by the combined Snake River Plain and southern Oregon rhyolite belt (Christiansen and McKee, 1978; Fig. 13-8; McKee and Noble, 1986).

However, for the eastern Snake River Plain, evidence published after Christiansen and McKee (1978) does not support right-lateral offset. As discussed under "Crustal Flaws," Skipp and others (1979) and Woodward (1988) argue against any strike-slip offset based on stratigraphic facies and structural trends. The widespread basin-range extension north of the plain (Plate 1), which became *generally* known as a result of the 1983 Borah Peak earthquake, suggests that extension north of the plain is of a magnitude similar to that south of the plain and that no right-lateral offset need be occurring.

If there were right-lateral shear cross the plain, the orientation of aligned and elongated volcanic features should be in the orientation of tension gashes and oriented in a direction 30° clockwise to the trend of the SRP. As indicated by elongation and alignments of Quaternary volcanic features (Fig. 22), the volcanic rift zones trend nearly perpendicular to the margins of the SRP (Kuntz and others, this volume, Fig. 4), about 60° clockwise from the tension gash orientation. Of the nine rift zones shown by Kuntz and others (this volume, Fig. 4), only the Spencer–High Point rift zone has an orientation close to that of tension gashes that would be related to right lateral shear along the SRP.

The southern Oregon rhyolite belt has been genetically linked to the eastern SRP. Christiansen and McKee (1978) define the High Lava Plains as extending from the eastern SRP to the Brothers fault zone and parallel zones to the west. They postulate that right-lateral shear was localized in the area of the High Lava Plains because the plains are located along "the approximate northern boundary of maximum cumulative extension" of the basin and range structural province. In Christiansen and McKee's (1978) model, the eastern SRP represents an ancient crustal flaw, whereas the Brothers fault zone and parallel zones in the west represent a symmetrical tear. According to Christiansen and McKee (1978), coeval bimodal volcanism along both the eastern and western arms of the High Lava Plains originated approximately 14 to 17 Ma at the common borders of Oregon, Nevada, and Idaho and propagated symmetrically east and west from that point. Although the timing of volcanism, rates of volcanic migration, volcanic compositions, and length of the volcanic track of volcanism in southeast Oregon have been compared with those aspects of the eastern SRP (Christiansen and McKee, 1978), we think they result from different processes. Beginning in the late Miocene, silicic volcanism along the eastern SRP differs from that of the southeastern rhyolite belt in eight ways, as listed in Table 3. The earliest volcanism associated with the southeast Oregon belt is dated about 14.5 Ma and is limited to three randomly spaced events at 14.7 Ma, 14.7 Ma, and 13.5 Ma (MacLeod and others, 1976). These older volcanic events show little, if any, spatial relationship to the younger events, which began on a regular basis at about 8 Ma. Although three events occurred about 10 Ma, there is a significant lull in volcanic activity lasting 6 m.y. from the time volcanism was weakly expressed at 14.5 Ma until about 8 Ma when a volcanic progression becomes apparent in southeast Oregon. Based on the differences summarized in Table 3, we conclude that the silicic volcanism in southeast Oregon results from a different process than that on the eastern SRP.

We accept the interpretation that volcanic activity in southern Oregon is related to transform or right-lateral offset along the Brothers fault zone as advocated by Christiansen and McKee (1978), but do not extend this explanation to the eastern SRP-Yellowstone area. Draper (1991) suggests that the southern Oregon volcanic progression relates to the combined effect of extension along the Brothers fault zone and outward migration of the head of the Yellowstone plume, a process different from but related to the volcanic trend that we believe was produced by a thermal-plume chimney now at Yellowstone.

A meteorite impact origin for the SRP-YP province. Many investigators have noted that flood basalt volcanism occurs at the start of hot-spot tracks and, following this pattern, have suggested that the Columbia Plateau flood basalts relate to the Yellowstone hot spot (see, for example, Morgan, 1972, 1981; Duncan, 1982; Richards and others, 1989; White and McKenzie, 1989). However, Alt and others (1988, 1990) have gone one step farther, suggesting that meteorite impacts are responsible for the initiation of flood basalt volcanism. By analogy with the Deccan Plateau flood basalts of India and the later Reunion (Chagos-Laccadive) hot-spot track, Alt and others (1988) suggest that the initiation of the Columbia Plateau flood basalts and the Yellowstone hot-spot track are the result of a meteorite impact located in southeast Oregon that somehow initiated a deep-seated mantle plume. The sudden appearance of large lava plateaus erupting over a short interval of time without any apparent tectonic cause

TABLE 3. DIFFERENCES BETWEEN SILICIC VOLCANISM ON THE EASTERN SNAKE RIVER PLAIN AND THE SOUTHEAST OREGON RHYOLITE BELT

Factor	Eastern Snake River Plain	Southeast Oregon Rhyolite Belt	
1. Time of inception	16 to 17 Ma*	~10 Ma (majority 8 Ma)†	
2. Point of origin	Northern Nevada near McDermitt volcanic field	Southern Oregon Northern belt: SE of Harney Basin Southern belt: Beaty Butte (southern Oregon)	
3. Length of province	700 km*	250 to 300 km§	
4. Style of volcanism	Generally large-volume ignimbrites from large calderas	Generally small-volume rhyolite domes	
5. Manner in which volcanism migrated	Northeast-trending wide belt 10 to 16 Ma and narrow belt 0 to 10 Ma*	Northwest-trending wide swath along two broad belts†	
		Northern belt	Southern belt
6. Migration rate of volcanism	~7.0 cm/yr, 10 to 16 Ma* 2.9 cm/yr, 0 to 10 Ma*	4.4 cm/yr, 5 to 10 Ma† 1.5 cm/yr, 0 to 5 Ma†	2.6 cm/yr
7. Volume of erupted products	100s to >1,000 km^3	One to a few tens of km^3	
8. Faults or other structure associated with volcanic trend	Downwarp	Northwest-trending en echelon faults defining the the Brothers fault zone† §	

*This chapter
†MacLeod and others, 1976
§Walker and Nolf, 1981

is offered as a main criterion (Alt and others, 1988) to support an impact hypothesis. Alt and others (1988) also suggest the felsic lavas in the plateaus result from impact.

We question an impact hypothesis for the origin of the SRP-YP province for the following reasons. (1) The chosen impact site in southeast Oregon is not supported by any evidence in pre–Steens Basalt strata such as in the Pine Creek sequence, where evidence of impact might be expected (Scott Minor, oral communication, 1990). Established impact sites are characterized by shatter cones, shocked quartz and feldspar, high-pressure mineral phases, breccia-filled basins, and high Ni content in the associated magmatism. (2) At sites accepted as having a meteorite impact origin, such as the Manicouagan and Sudbury craters in Canada, associated magmatism has occurred for approximately a million years. However, these established sites have neither voluminous flood basalts nor protracted volcanism to produce a systematic volcanic track like that of a hot spot. In the model of Alt and others, volcanism at the surface maintains the mantle plume at depth—the hot spot continues to erupt because it continues to erupt" (Sears, Hyndman, and Alt, 1990), a mechanism we find unconvincing. (3) The argument by Alt and others (1988) for an impact producing the Deccan Plateau flood basalts may be erroneous on several accounts. First, no impact crater has been defined for the Deccan Plateau site. Second, although the Deccan Plateau basalts date from near the time of the Cretaceous/Tertiary (K-T) boundary (Courtillot and others, 1986; Alt and others, 1988), studies of the K-T boundary layer and the size of its contained shock quartz indicate the impact occurred in the Western Hemisphere (Izett, 1990), possibly in the Caribbean Basin (Hildebrand and Boynton, 1990; Bohor and Seitz, 1990). (4) Alt and others (1988) attribute an impact origin to felsic rocks that we think are associated with the southern part of the Columbia Plateau (better known as the Oregon Plateau). (5) Finally, no major faunal extinction event occurs at about 17 Ma as is thought to be associated with other postulated meteorite impacts. In conclusion, we do not accept the impact hypothesis because evidence expected to accompany such an impact in southeast Oregon has not been found.

A two-phase, mantle plume model for the SRP-YP region

We find a thermal mantle plume model best accounts for the large-scale processes operating at asthenospheric and lithospheric depths that are responsible for the observed systematic volcanic progression, regional uplift, and localization of late Cenozoic basin-range faulting. Several investigators have proposed a deep-seated mantle plume or hot-spot origin for the SRP-YP province (references listed at start of "Discussion" section) to account for the various volcanic and tectonic features present. In this section, we first explain a model and then discuss how geologic evidence supports this model, particularly as required by the large horizontal scale of the effects.

The two-phase model. Richards and others (1989) suggest that hot spots start with very large "heads" ($>10^2$ km diameter) capable of producing melting, rifting, doming, and flood basalts (White and McKenzie, 1989; Duncan and Richards, 1991). Following the path pioneered by this head is a much narrower chimney that feeds hot mantle material up into and inflates the head. Richards and others (1989) calculate that 15 to 28 m.y. are required for a plume head to rise from the core-mantle interface through the mantle to the base of the lithosphere. The head phase contains more than an order of magnitude more heat than the heat carried by 1 m.y. of chimney discharge. In addition, the central part of the head may be 50 to 100 °C hotter (White and McKenzie, 1989) than the chimney. Richards and others (1989) note that rifting may actually result from the encounter of the plume head with the lithosphere and that large amounts of continental lithospheric extension may not necessarily precede flood basalt eruptions. After the head flattens against the base of the lithosphere, the chimney eventually intercepts the lithosphere and produces a migrating sequence of volcanism and uplift in the overriding plate, a hot-spot track (Richards and others, 1989; Whitehead and Luther, 1975; Skilbeck and Whitehead, 1978). The chimney phase of the Hawaiian and other hot-spot tracks has lasted longer than 100 m.y. Although many investigators agree with a core-mantle source region for plumes (Tredoux and others, 1989), others suggest mantle plumes originate from higher in the mantle, perhaps at the 670-km seismic discontinuity (Ringwood, 1982) or possibly at even shallower depths (Anderson, 1981). Recent evidence suggests the 670-km discontinuity may be an abrupt phase change (Ito and Takahashi, 1989; B. J. Wood, 1989) and not a constraint to plume flow from the core-mantle boundary to the lithosphere.

We suggest (Morgan and Pierce, 1990; Pierce and Morgan, 1990) that a mantle-plume head about 300 km in diameter first encountered the base of the lithosphere 16 to 17 Ma near the common borders of Nevada, Oregon, and Idaho, based on volcanic and structural features in this region (Figs. 23, 24). This hypothesis was independently developed by Draper (1991). Zoback and Thompson (1978) suggested that the mantle plume associated with the Yellowstone hot spot first impinged on the lithospheric base of the North American plate around 16 to 17 Ma associated with a north-northwest–trending rift zone herein called the Nevada-Oregon rift zone (Fig. 23). The chimney, which was about 10 to 20 km in diameter and had been feeding the now stagnating head, intercepted the lithosphere about 10 Ma near American Falls, Idaho, 200 km southwest of Yellowstone (Plate 1; Figs. 23 and 24). Volcanism along the hot-spot track is more dispersed during the head phase than the chimney phase (Fig. 4).

Figure 24 is a schematic representation of the development of the thermal plume that formed the Yellowstone hot-spot track, based on other investigators' models proposed for thermal mantle plumes. The following four paragraphs explain the relations we envisage for the four parts of Figure 24.

1. The plume began deep in the mantle, perhaps at the core/mantle boundary, as a thin layer of hotter, less viscous material converging and flowing upward at a discharge rate of

perhaps 0.4 km^3/year (Richards and others, 1989) through cooler, more viscous mantle (Whitehead and Luther, 1975). To rise 3,000 km from the core/mantle boundary, Richards and others (1989) calculate the plume head took 28 m.y. to reach the lithosphere, rising at an average rate of 10 cm/year. As it rose, the plume head was constantly supplied with additional hot mantle material through its plume chimney, a conduit only about 10 or 20 km across, the narrow diameter and greater velocity of which (perhaps 1 m/year, Richards and others, 1989; or 2 to 5 m/year, Loper and Stacey, 1983) reflect the established thermal chimney where the increased conduit temperatures result in much higher strain rates toward the chimney margins. As the plume head rose, continued feeding by its chimney inflated it to a sphere nearly 300 km in diameter by the time it intercepted the lithosphere (Richards and others, 1989).

2. About 16 Ma, the plume head intercepted the base of the southwest-moving North American lithospheric plate and mushroomed out in the asthenosphere to a diameter of perhaps 600 km. Doming and lithospheric softening above this plume head resulted in east-west extension along the 1,100-km-long north-northwest–trending Nevada-Oregon rift zone (Fig. 23). Decompression melting accompanying the rise of the plume generated basaltic melt that rose into the lithosphere (White and McKenzie, 1989) and produced flood basalts in the more oceanic crust of Washington and Oregon and mafic intrusions in the northern Nevada rift(s?) (Fig. 23). Near the common boundary of Nevada, Oregon, and Idaho, basaltic melts invaded and heated more silicic crust, producing rhyolitic magmas that rose upward to form upper crustal magma chambers, from which were erupted the ignimbrites and rhyolite lava flows that now cover extensive terrain near the Nevada-Oregon-Idaho border region (Fig. 23). The plume head may have domed the overriding plate, with increased temperatures in the asthenosphere lessening drag on the overriding plate. On the west side of the dome, extension in the westward-moving plate would be favored, whereas on the east side of the dome compression seems likely.

3. Sometime between about 14 and 10 Ma, a transition from the head phase to the chimney phase occurred (Fig. 24). As the plume head spread out horizontally and thinned vertically at the base of the lithosphere, the heat per unit area beneath the lithosphere became less and the plume material became more stagnant. The chimney continued to feed upward at a relatively high velocity and carried material into the thinning, stagnating, but still hot plume head. This transition from the broad plume-head phase to the narrowly focused chimney phase is reflected in better alignment of calderas after 13 Ma (Plate 1).

4. By 10 Ma, continued upward flow of the chimney established a path through the flattened plume head and encountered the base of the lithosphere. Continued decompression melting of material rising up the chimney released basaltic melts that invaded and heated the lower crust to produce silicic magmas. These silicic magmas moved upward to form magma chambers in the upper crust, from which large-volume ignimbrites were erupted, forming the large calderas of the eastern Snake River Plain and Yellowstone Plateau. The linear, narrow volcanic track defined by the eastern SRP-YP volcanic province reflects the narrower, more focused aspect of the chimney phase of the plume (Plate 1, Fig. 4). As with the Hawaiian hot spot, flow of plume material carries heat outward for hundreds of kilometers in the asthenosphere and results in swell uplift and other lithospheric changes (Figs. 23, 24; Courtney and White, 1986; White and McKenzie, 1989). Because of the elevated temperature in the asthenosphere due to the hot spot, drag in the asthenosphere at the base of the North American plate would be lessened. Due to doming associated with the Yellowstone hot spot, perhaps including other hot spots to the south (Suppe and others, 1975), the plate on the west side may readily move westward favoring basin-range extension, whereas that on the east side continues under compression. The axis of this dome approximates the eastern margin of the basin and range structural province. In the following two sections, we discuss evidence that we think supports this two-phase, mantle plume model for the SRP-YP region.

Evidence from volcanism and rifting. The most prominent feature associated with the start of the Yellowstone hot spot is the Nevada-Oregon rift zone (Fig. 23), consistent with the observations that rifting may accompany interception of a plume head with the lithosphere (White and others, 1987) and that large amounts of continental lithospheric extension do not necessarily precede flood basalt eruptions (Richards and others, 1989).

The 1,100-km-long Nevada-Oregon rift zone is associated with 17- to 14-Ma basaltic and silicic volcanism and dike injection (Fig. 23) and is part of the 700-km-long Nevada rift zone defined by Zoback and Thompson (1978) as extending from central Nevada to southern Washington. The distinctive, linear, positive aeromagnetic anomaly associated with the central part of this rift, the northern Nevada rift, results from mafic intrusions about 16 Ma. The northern Nevada rift has recently been extended 400 km farther to southern Nevada (Blakely and others, 1989), where 16- to 14-Ma rhyolitic fields are also present (Fig. 23; Luedke and Smith, 1981). The feeder dikes for the Oregon Plateau flood basalts (including the basalts at Steens Mountain, Carlson and Hart, 1986, 1988) and the Columbia River flood basalts (Hooper, 1988; Smith and Luedke, 1984) have the same age and orientation as the northern Nevada rift and are considered by us to be a northern part of it. The eruptive volume of these flood basalts totals about 220,000 km^3 (Figs. 23, 24), based on about 170,000 km^3 for the Columbia River Basalt Group (Tolan and others, 1989) and perhaps about 50,000 km^3 for the Oregon Plateau basalts (Carlson and Hart, 1988).

White and McKenzie (1989) conclude that if mantle plumes (hot spots) coincide with active rifts, exceedingly large volumes of basalt can extend along rifts for distances of 2,000 km. The main eruption pulse of the Columbia River–Oregon Plateau flood basalts in the northern part of the 1,100-km-long Nevada-Oregon rift zone and the mid-Miocene location of the Yellowstone hot spot near the center of this rift are consistent with this pattern.

The Columbia River basalts have been referred to as the

EXPLANATION **(FIGURE 23)**

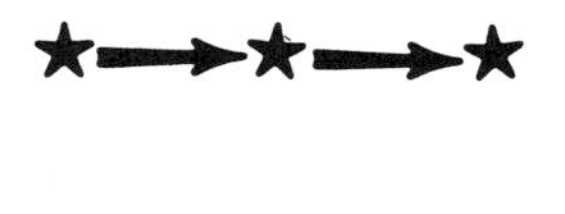

Post-10-Ma track of the Yellowstone hot spot (arrows). Arrows drawn between volcanic fields (centered at stars) with inception ages given in Ma. Margins of the Snake River Plain shown by short dashed line).

Inferred centerline of the hot-spot track from 16 to 10 Ma.

Belts characterized by following fault types: I- lesser Holocene and reactivated; II- major Holocene; III- major late Pleistocene; and IV (unshaded)- major Tertiary faults, now inactive.

Yellowstone crescent of high terrain.

Outer margin of higher terrain that may in part be related to Yellowstone, although margin cannot be drawn southeast from Yellowstone because of high terrain in southern Wyoming (? symbol) and in Colorado.

Margin of more active basin-range structural province (does not include less active Basin and Range south of Las Vegas, Nevada, area).

Area of 14- to17- Ma rhyolitic eruptions defining start of Yellowstone hotspot. ■, calderas dated between 14-17 Ma; o, rhyolite ages between 14-17 Ma (schematic for southern Nevada area).

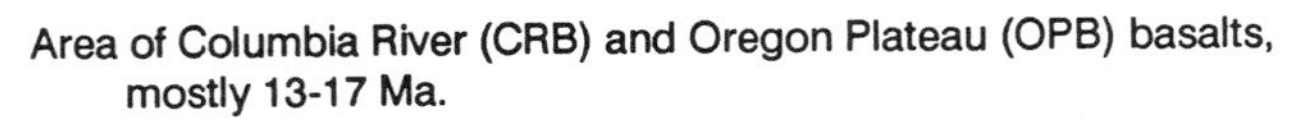

Area of Columbia River (CRB) and Oregon Plateau (OPB) basalts, mostly 13-17 Ma.

Feeder dikes for the Columbia River and Oregon Plateau basalts.

Northern Nevada rift and two similar structures to west. Heavy dashed line marks southern extension of rift.

Normal faults bounding graben of western Snake River Plain.

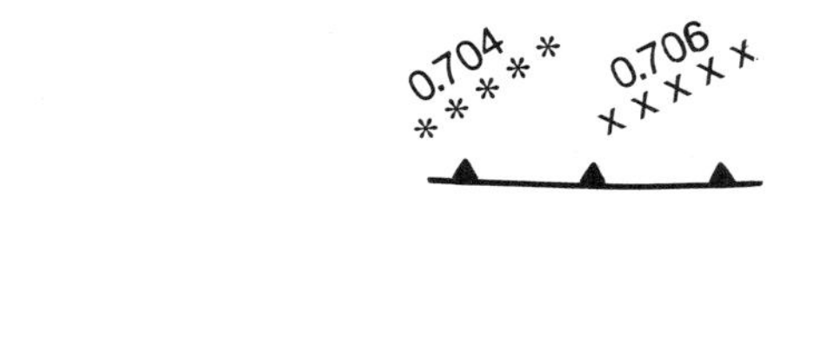

$^{87}Sr/^{86}Sr$ 0.706 and 0.704 lines. East of the 0.706 line is Precambrian sialic crust. West of 0.704 line are accreted, oceanic terrains.

Eastern limit of fold and thrust belt. East of line are continental craton and basement uplifts of Laramide age.

initial expression of the Yellowstone hot spot (Morgan, 1981; Zoback and Thompson, 1978; Duncan, 1982; Leeman, 1989; Richards and others, 1989; however, this genetic tie has been questioned by others (Leeman, 1982a, 1989; Carlson and Hart, 1988; Hooper, 1988), in part because of the lopsided position of the Columbia River basalts and feeder dikes 400 km north of the centerline of the projected hot-spot track (Fig. 23). Morgan (1981) points out that many continental flood basalts (Table 2) are associated with the initial stages of hot-spot development, and citing examples in which flood basalts are not centered on the younger hot-spot track: the Deccan Plateau flood basalts and the Reunion track, the Parana basalts and the Tristan track, and the Columbia River basalts and the Yellowstone track. The 1,100-km length of the Nevada-Oregon rift zone is actually bisected by the hot-spot track (Fig. 23), indicating that there is no asymmetry in structure but only in the volume and eruptive rates of basalts that surfaced north of the bisect line, as a result of denser, more oceanic crust to the north and less dense, continental crust to the south. The eruptive rate calculated for the Columbia River and Oregon Plateau basalts (Figs. 23, 24) is based on the fact that more than 95% of the total estimated volume of 220,000 km^3 erupted in about a 2-m.y. interval (Carlson and Hart, 1986; Baksi, 1990).

The sharp contrast in style and amount of 17- to 14-Ma volcanism along the 1,100-km long Nevada-Oregon rift zone correlates with the nature of the crust affected by the thermal plume (Figs. 23 and 24). The Columbia River and Oregon Plateau flood basalts occur in accreted oceanic terrane (Figs. 23, 24; Vallier and others, 1977; Armstrong and others, 1977) having relatively thin, dense crust (Hill, 1972). South of the Oregon Plateau basalt province into northern Nevada and southeast of the 0.704 isopleth (i.e., the initial $^{87}Sr/^{86}Sr$ = 0.704 isopleth

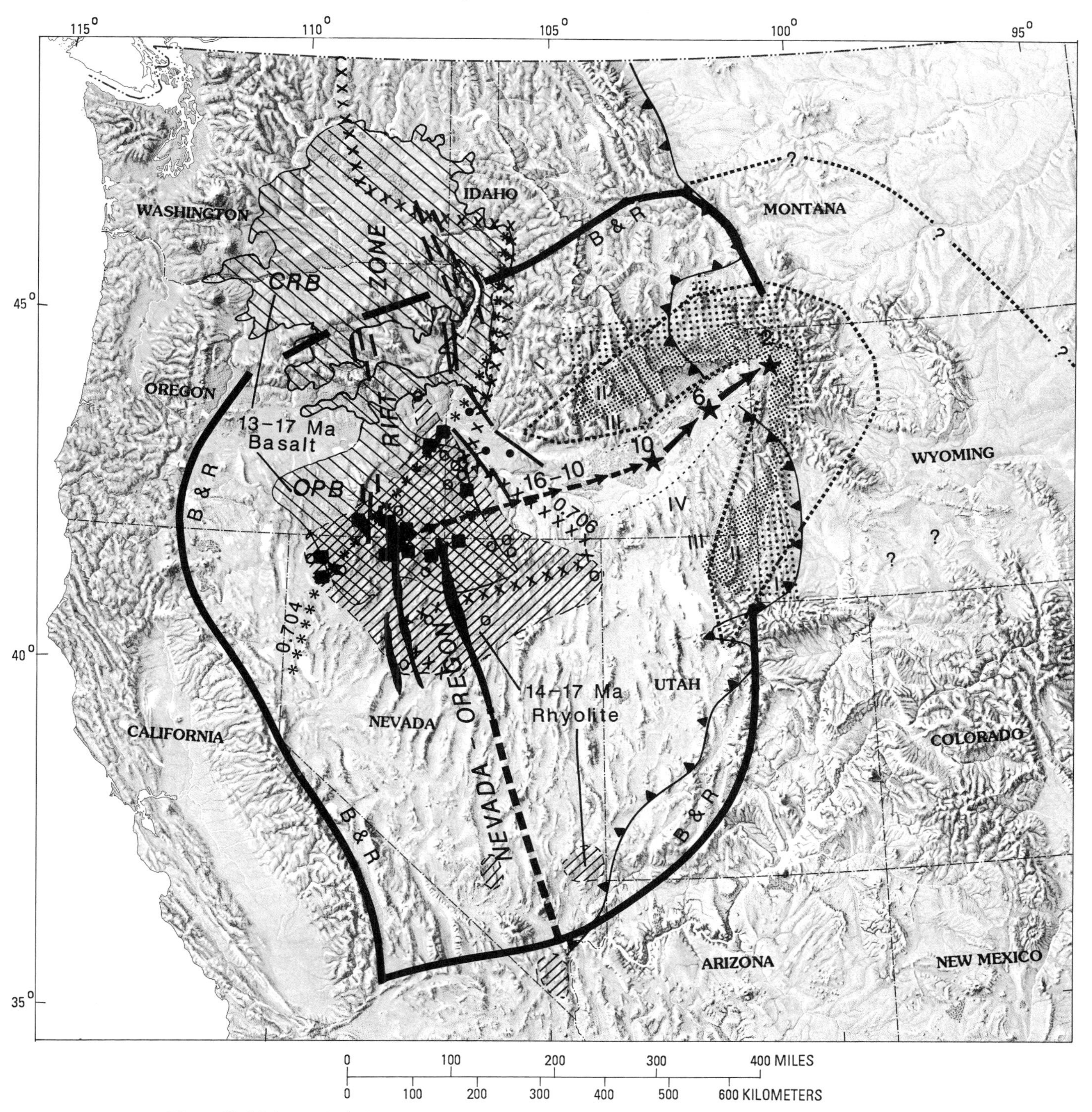

Figure 23. Major geologic features in the western United States associated with the late Cenozoic track of the Yellowstone hot spot. Compiled from: Blakely, 1988; Blakely and others, 1989; Luedke and Smith, 1981, 1982, 1983; Smith and Luedke, 1984; Hooper, 1988; Carlson and Hart, 1988; Elison and others, 1990; Kistler and Lee, 1989; Kistler and others, 1981. Base map from Harrison, 1969.

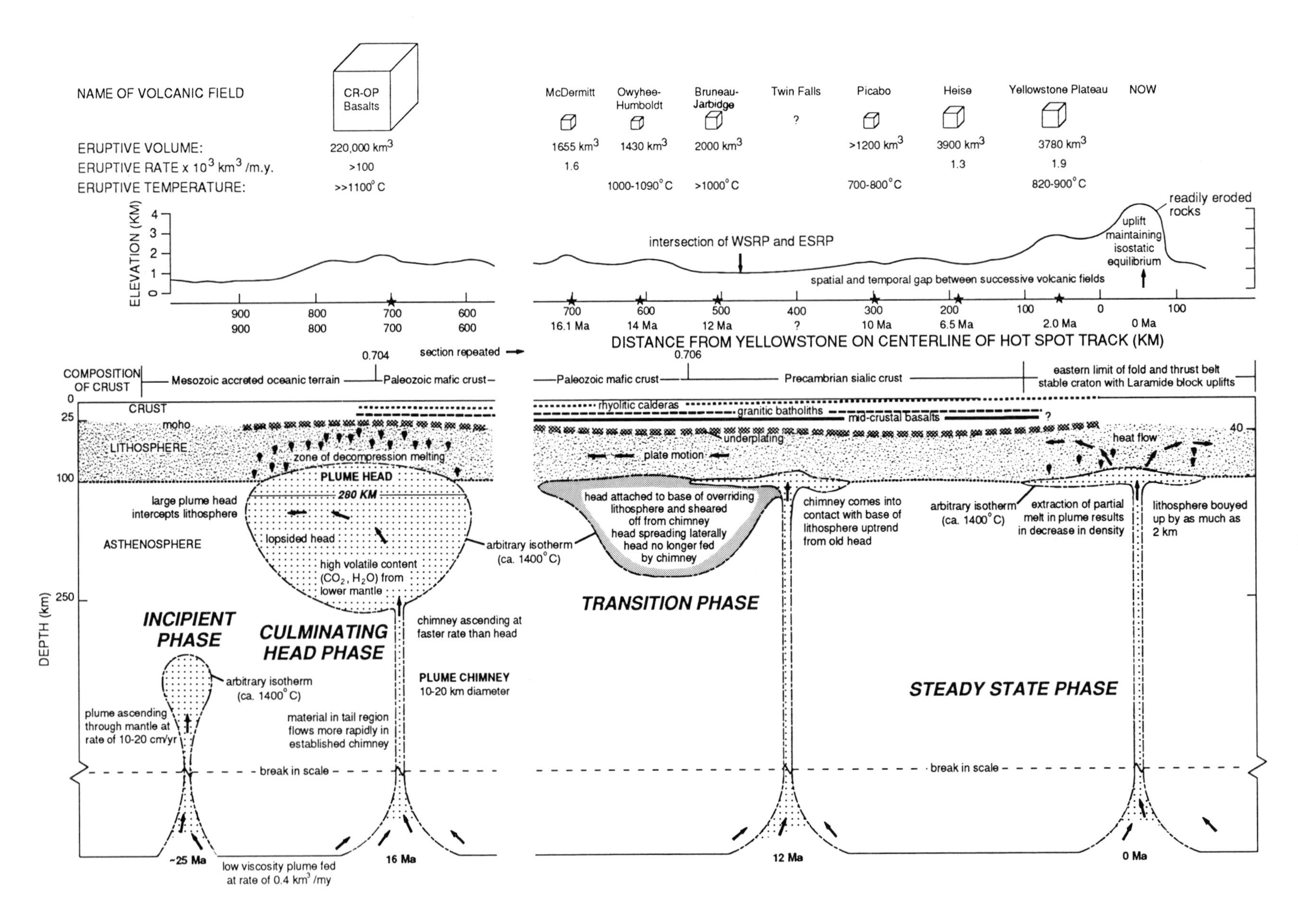

NAME OF VOLCANIC FIELD
CR-OP Basalts
McDermitt
Owyhee-Humboldt
Bruneau-Jarbidge
Twin Falls
Picabo
Heise
Yellowstone Plateau
NOW
?
ERUPTIVE VOLUME:
ERUPTIVE RATE x 10^3 km^3 /m.y.
ERUPTIVE TEMPERATURE:
220,000 km^3
>100
>>1100° C
1655 km^3
1.6
1430 km^3
1000-1090° C
2000 km^3
>1000° C
>1200 km^3
700-800° C
3900 km^3
1.3
3780 km^3
1.9
820-900° C
readily eroded rocks
uplift maintaining isostatic equilibrium
intersection of WSRP and ESRP
spatial and temporal gap between successive volcanic fields
ELEVATION (KM)
16.1 Ma
14 Ma
12 Ma
?
10 Ma
6.5 Ma
2.0 Ma
0 Ma
DISTANCE FROM YELLOWSTONE ON CENTERLINE OF HOT SPOT TRACK (KM)
0.704
section repeated
0.706
COMPOSITION OF CRUST
Mesozoic accreted oceanic terrain
Paleozoic mafic crust
Precambrian sialic crust
eastern limit of fold and thrust belt
stable craton with Laramide block uplifts
CRUST
moho
LITHOSPHERE
zone of decompression melting
PLUME HEAD
280 KM
large plume head intercepts lithosphere
ASTHENOSPHERE
lopsided head
high volatile content (CO_2, H_2O) from lower mantle
arbitrary isotherm (ca. 1400° C)
rhyolitic calderas
granitic batholiths
mid-crustal basalts
underplating
plate motion
heat flow
head attached to base of overriding lithosphere and sheared off from chimney head spreading laterally head no longer fed by chimney
chimney comes into contact with base of lithosphere uptrend from old head
extraction of partial melt in plume results in decrease in density
lithosphere bouyed up by as much as 2 km
DEPTH (km)
INCIPIENT PHASE
CULMINATING HEAD PHASE
chimney ascending at faster rate than head
TRANSITION PHASE
PLUME CHIMNEY
10-20 km diameter
STEADY STATE PHASE
plume ascending through mantle at rate of 10-20 cm/yr
material in tail region flows more rapidly in established chimney
break in scale
~25 Ma
low viscosity plume fed at rate of 0.4 km^3 /my
16 Ma
12 Ma
0 Ma

referred to by Kistler and Peterman [1978]) (Fig. 23), much of the accreted crust is somewhat thicker than to the north and is underlain by early Paleozoic or younger mafic crust having oceanic or transitional affinities (Kistler, 1983; Elison and others, 1990). This thicker (Mooney and Braile, 1989) and more brittle crust responded to the mantle plume by producing peralkaline rhyolitic volcanism in the region near the common boundaries of Nevada, Oregon, and Idaho and much less voluminous bimodal volcanism farther south along the 500-km-long Nevada sector of the rift (Blakely and others, 1989), in contrast to the voluminous flood basalts to the north (Fig. 23).

About 17 Ma, basin-range faulting became widely active. McKee and others (1970) note a hiatus in volcanic activity from about 17 to 20 Ma in the Great Basin. In addition, volcanism prior to 17 Ma was calc-alkaline of intermediate to silicic composition, whereas that after 17 Ma was basaltic *and* bimodal rhyolite/basalt volcanism (Christiansen and Lipman, 1972; Christiansen and McKee, 1978; McKee and Noble, 1986). Also at about 17 Ma, ashes erupted from this area changed from dominantly W-type (white, low iron, biotitic) to dominantly G-type (gray, high iron, nonbiotitic) (G. A. Izett, 1981, written communication, 1990). Both these changes may be associated with the head of the Yellowstone hot spot's encountering and affecting the lithosphere.

We think the rhyolite eruptions result from heat supplied to the lower crust by basalt that was formed by decompression melting in an upwelling thermal mantle plume. As outlined by Hildreth (1981), Leeman (1982a, 1989), and Huppert and Sparks (1988), hot basaltic magma from the mantle buoyantly rises to the lower crust and exchanges heat with the lower-melting-point silicic rocks, thereby forming a granitic magma that buoyantly rises to form magma chambers higher in the crust, from which the ignimbrites are erupted.

We next discuss the hot-spot track from 10 Ma to present, so we can contrast it with the more diffuse, >10-Ma track. Between 2 and 10.3 Ma, caldera-forming silicic volcanism moved at a rate of about 2.9 ± 0.5 cm/m.y. to N54 ± 5°E, resulting in the Yellowstone Plateau and Heise volcanic fields, which began at 2.0 and 6.5 Ma, respectively, and the less well understood Picabo volcanic field, which began with the 10.3-Ma tuff of Arbon Valley. The coincidence, well within error limits, for the post–10-Ma hot-spot track (N54°E at 2.9 cm/year) with the plate motion vector at Yellowstone (N56°±17E at 2.2 ± 0.8 cm/year) strongly supports the hypothesis that the hot-spot track represents a stationary mantle plume (see section titled "Hot-spot Track, 0 to 10 Ma" for basis of this absolute plate motion calculation by Alice Gipps [written communication, 1991], which does not include the Yellowstone hot spot in the data set).

A major heat source is present beneath Yellowstone, as attested by the young volcanism and active geothermal systems. Within the 0.6-Ma caldera, temperatures at only about 5 km exceeding the Curie point indicate high geothermal gradients (Smith and others, 1974). High geothermal gradients also have elevated the brittle/ductile transition from its normal depth of 15 km to about 5 km, as indicated by the absence of earthquakes below about 5 km (Smith and others, 1974). A mantle source of helium gas venting at Yellowstone is required by the high helium-3/helium-4 isotope ratios in Yellowstone thermal waters (Kennedy and others, 1987).

Beneath the Yellowstone and Heise volcanic fields, teleseismic studies also indicate high mantle temperatures. Based on P-wave delays, Iyer and others (1981) concluded that a large low-velocity body occurs beneath Yellowstone to depths of 250 to 300 km and may be best explained as a body of partial melt. The lower part of this low-velocity body extends southwestward along the SRP to the Heise volcanic field, where Evans (1982) noted a low-velocity zone to depths of 300 km. This low-velocity zone is inclined to the northwest and is offset as much as 150 km northwest from the centerline of the SRP.

The hot-spot track older than 10 Ma contrasts with the younger-than-10-Ma track in that (1) no discrete trench like the eastern SRP was formed; (2) prior to 10 Ma, silicic volcanism was spread over an area as wide as 200 km (Fig. 4) and produced ignimbrites or lava flows with higher magmatic temperatures (Ekren and others, 1984; Bonnichsen and Kauffman, 1987) than those observed in the younger ignimbrites (Hildreth, 1981) (Fig. 24); and (3) the older track has a more easterly trend and progressed at an apparent rate of 70 km/m.y., a rate at least twice that for both the North American plate motion and the younger

Figure 24. Postulated development of the Yellowstone thermal plume from its inception phase, through its huge head and transition phases, to its present chimney phase. Inspired from Richards and others (1988, 1989) and Griffiths and Richards (1989). Only the hottest part of the plume is shown—that part >1,400 °C based on the schematic isotherms of Courtney and White (1986) and Wyllie (1988). As the mantle plume rises, decompression melting produces basaltic melt that rises through the mantle lithosphere to the lower crust and may vent as flood basalts or else be emplaced in the lower crust where it exchanges its heat to produce silicic magma that rises yet higher in the crust. Scale does not permit showing the entire, approximately 1,000-km-diameter, lower-temperature plume, based on the 800- to 1,200-km diameter of oceanic hot-spot swells (Courtney and White, 1986). Volumes (minimum estimates) show flood basalts are 100 times greater in volume than ignimbrites. The schematic crustal section showing major lithologic phases (rhyolitic calderas, granitic batholiths, and mid-crustal basalts) along the hot-spot track is consistent with seismic refraction data modeled for the SRP (Sparlin and others, 1982). Rhyolites and their associated calderas are exposed at Yellowstone but beneath the SRP they are covered by a veneer of basalt and intercalated sediments. Magma and zones of partial melt presently under the Yellowstone caldera are at depths ranging from several kilometers (not shown; Clawson and others, 1989) to 250 to 300 km (Iyer and others, 1981). Stars, centers of volcanic fields. Mid-crustal basalts have not been detected seismically beneath Yellowstone (Iyer and others, 1981), but evidence of their presence may be the later post–0.6-Ma basalts that now fill much of the Henry's Fork caldera (Christiansen, 1982). Line marked 0.706 (initial $^{87}Sr/^{86}Sr$ isopleth) approximates the western edge of the Precambrian sialic crust (Armstrong and others, 1977; Kistler and Peterman, 1978); that marked 0.704 shows the western edge of Paleozoic mafic crust (Kistler, 1983; Elison and others, 1990).

track (Plate 1, Figs. 2, 23). Based on the Yellowstone hot-spot track, Pollitz (1988) noted a change in apparent plate motion at 9 Ma, but only a minor change in velocity. We are hesitant to conclude that the apparent rate and direction differences (Plate 1, Fig. 3) actually require a change in plate velocity or direction between 16 and 10 Ma and 10 and 0 Ma because: (1) crustal extension postdates the older track, (2) large imprecision exists in the location of the geographic center for the inception of volcanism within a volcanic field, and (3) the plume chimney may not center beneath the plume head both because of westward shearing of the plume head as it interacted with the North American plate (Fig. 24) and because of possible displacements between the plume head and chimney by mantle "winds" (Norm Sleep, oral communication, 1991). Factor 3 provides mechanisms that appear more able to explain the seemingly anomalous pre-10 Ma rate.

The character of the volcanic track of the Yellowstone hot spot also correlates with crustal changes. In the region of the Owyhee-Humboldt and Bruneau-Jarbidge fields, the track occurs across a relatively unfaulted plateau, the Owyhee Plateau that Malde (1991) suggests may have resisted faulting because it is underlain by a large remnant of the Idaho batholith.

The area of largest dispersion of silicic volcanic centers (Plate 1, Fig. 4) lies to the west of the 0.706 isopleth for Mesozoic and Cenozoic plutonic rocks, corresponding to the western edge of Precambrian sialic crust (Figs. 23 and 24) (i.e., the initial $^{87}Sr/^{86}Sr$ = 0.706 isopleth referred to by Kistler and Peterman, 1978; Armstrong and others, 1977; Kistler and others, 1981; Kistler and Lee, 1989). The 0.706 isopleth lies between the approximate boundaries of the Owyhee-Humboldt and Bruneau-Jarbidge volcanic fields (Figs. 23, 24; Plate 1), and east of this line the old, relatively stable cratonic crust of the Archean Wyoming province (Leeman, 1982a; J. C. Reed, written communication, 1989) may have helped restrict volcanism to the less dispersed trend noted from less than 12.5 Ma (Fig. 4).

For the volcanic track 10 to 4 Ma, the Picabo and Heise volcanic fields are flanked by Cordilleran fold-and-thrust belt terrain broken by late Cenozoic normal faults parallel to thrust belt structures, whereas the 2-Ma and younger Yellowstone Plateau field has developed in cratonic crust deformed by Laramide foreland uplifts and broken by less systematically orientated normal faults (Figs. 23, 24; Plate 1). Upon crossing from the thrust belt into the craton, the width of the volcanic belt narrows from 90 ± 10 km for the eastern SRP to 60 ± 10 km for the Yellowstone Plateau.

Eruption temperatures are higher for the pre–10-Ma part of the hot-spot track as compared with the 10-Ma and younger part. For the older part, eruption temperatures for units from the Owyhee-Humboldt field have been calculated to be in excess of 1,090°C (Ekren and others, 1984); temperatures from the Bruneau-Jarbidge field are >1,000°C (Bonnichsen, 1982) (see Fig. 24). For the 10-Ma and younger units, eruption temperatures were 820 to 900° C for the Yellowstone ignimbrites (Hildreth, 1981); we infer an eruption temperature of 700 to 800°C for the 10.3-Ma ignimbrite from the Picabo volcanic field based on that determined by Hildreth (1981) for other high silica biotite ignimbrites. This overall decrease in eruptive temperatures correlates with both hot-spot tracks traversing more continental crust as indicated by the crossing of the $^{87}Sr/^{86}Sr$ 0.706 isopleth and by the inferred plume change from head to chimney (Figs. 23, 24).

Evidence from faulting and uplift. The neotectonic belts of faulting, particularly the most active Belt II, converge on Yellowstone and define a pattern analogous to the wake of a boat that has moved up the SRP to Yellowstone (Plate 1; Scott and others, 1985b). The overall V-shaped pattern of the fault belts can be explained by outward migration of heating associated with the Yellowstone hot-spot track (Scott and others, 1985b; Smith and others, 1985; Anders and others, 1989). Heat transferred from the same source responsible for the Yellowstone hot spot has weakened the lithosphere, thereby localizing extensional faulting in Belts I, II, and III, in which nearly all extensional faulting in the northeast quadrant of the basin and range structural province is concentrated. We suggest that Belts I, II, and III represent waxing, culminating, and waning stages of fault activity respectively. Because the belts of faulting appear related to the motion of the North American plate, the belts shown in Plate 1 will move to the N55°E at 30 km/m.y. Because of this outward migration and implied sense of acceleration of activity in Belt I, culmination of activity in Belt II, and deceleration of activity in Belt III, the outer part of Belt II would have a higher rate of ongoing faulting than the inner part.

The Yellowstone crescent of high terrain has a spatial relation to the Yellowstone volcanic field similar to the fault belts. Although the following supportive evidence is incomplete and selective, ongoing uplift on the outer slope of this crescent is suggested by high Pd/BL ratios, tilting of terraces, outward migration of inflection points along terraces, uplift of ignimbrite sheets, and uplift along historic level lines. The Yellowstone crescent is also predicted to be moving to the east-northeast at a rate of 30 km/m.y.

The association of both faulting and uplift with Yellowstone cannot be explained by lateral heat *conduction* within the lithosphere because 10 to 100 m.y. are required for temperature increases to affect ductility at distances of 50 km (Anders and others, 1989). We think the transport of heat as much as 200 km outward from the SRP-Yellowstone hot-spot track (as reflected by the Yellowstone crescent of high terrain) occurs by outward flow in the asthenosphere of an upwelling hot mantle plume upon encountering the lithosphere (Crough, 1978, 1983; Sleep, 1987, 1990). However, the time available is not adequate for conductive heat transport to thermally soften the lithosphere (Houseman and England, 1986; Anders and others, 1989). Anders and others (1989) suggest that magmas rising from the outward moving plume transport heat into the lithosphere and lower the yield strength, particularly in the upper mantle part of the lithosphere, where much of the yield strength resides. The available heat from the outward motion of a mantle plume at the base of the lithosphere probably diminishes outward due to loss of heat and radial dispersal of the mantle plume.

That uplift of the Yellowstone crescent is northeast of and precedes the Yellowstone volcanism indicates that a process active in the mantle causes lithospheric deformation rather than lithospheric rifting or other deformation caused by passive mantle upwelling. For the Red Sea, Bohannon and others (1989) argue for a passive mantle model because uplift followed initial volcanism and rifting. But on the leading margin of the Yellowstone hot spot, the apex of the Yellowstone crescent of high terrain northeast of Yellowstone and geomorphic indicators of eastward tilting of the Bighorn Basin show that uplift occurs several hundred kilometers in advance of the volcanism along the hot-spot track.

Near the southern margin of the SRP, now inactive areas (Belt IV) have a late Cenozoic history of fault activity concentrated within a few million years. Extension rates at that time were similar to present ones, and extension was not oriented perpendicular to the SRP margin as a rift origin of the SRP would predict. The time-transgressive pattern of this extensional activity correlates well with the volcanic activity along the hot-spot track (Plate 1). This pattern is readily explained by the predicted east-northeast migration of the mantle plume. Faults north of the SRP are still active, and initial ages of faulting are not well constrained. Nevertheless, activation of basin-range faulting since about 10 Ma and waning of activity along the faults marginal to the SRP may reflect activity associated with passage of the hot spot.

Geophysical and petrologic studies indicate intrusions of basalt at two depths beneath the SRP (Leeman, 1982a, 1989). The deeper level represents intrusion of basaltic material near the base of the crust (crustal underplating) at a depth near 30 to 40 km (Leeman, 1982a; Anders and others, 1989). The lower crust thickens southwestward along the SRP (Leeman, 1982a; Braile and others, 1982). A mid-crustal basaltic intrusion about the same width as the SRP is indicated both by anomalously high velocities (6.5 km/second) and by high densities (2.88 g/cm^3) between depths of 10 and 20 km (Sparlin and others, 1982; Braille and others, 1982). Anders and others (1989) and Anders (1989) outline a model whereby solidification of the mid-crustal basaltic intrusions increases the strength of the crust above the crustal brittle/ductile transition.

The absence or low level of active faulting and earthquakes in Belt IV, the eastern SRP, and the 10 to 16 Ma part of the hot-spot track, combined with high heat flow both there and from the adjacent SRP, indicates that processes related to passage of the hot spot have, after first softening of the lithosphere, then resulted in both strengthening and heating of the lithosphere. This paucity of faulting and earthquake activity does not result simply from lithospheric cooling to pre–hot-spot temperatures because heat flow from these areas remains high (Brott and others, 1981; Blackwell, 1989) and basalt eruptions on the SRP have continued throughout the Quaternary. Anders and others (1989) and Anders (1989, p. 100–130) present rheological models that include injection and cooling of sub-crustal and mid-crustal and basaltic intrusions. These models predict that cooling of mid-crustal intrusions could produce mid-crustal strengthening, but below mid-crustal depths no increase in strength would occur in 5 m.y. Although mid-crustal intrusions may explain strengthening of the crust beneath the Snake River Plain, no such mid-crustal body is demonstrated *flanking* the SRP (Sparlin and others, 1982), and we doubt that the sparse volcanism in Belt IV as well as Belt III indicates the presence there of extensive mid-crustal intrusions. Thus, we find it difficult to invoke cooling of mid-crustal intrusions to strengthen the crust over this area much wider than the Snake River Plain. The hot-spot model predicts that basaltic underplating and lower crustal intrusion are likely to have occurred over this larger area, which includes Belt IV, the eastern SRP, and the 10- to 16-Ma hot-spot track (Plate 1; Fig. 12). At lower crustal depths (30 to 40 km) and temperatures, basaltic material is much stronger than silicic material (Houseman and England, 1986; see Suppe, 1985, Fig. 4-29). We suggest that the following changes associated with basaltic underplating and intrusion eventually resulted in strengthening of the lower crust after passage of the Yellowstone hot spot: (1) addition to the lower crust of basalt that, when it crystallized, would be stronger than the more silicic material that it supplanted; (2) heat exchange between basalt and silicic materials in the lower crust, resulting in partial melting and buoyant rise of both silicic magmas and heat to mid-crustal and surface levels and leaving behind more refractory residual materials with higher melting temperatures and greater strength at higher temperature than the original lower crustal material; (3) thermal purging of water from the lower crust (dehydration), resulting in higher solidus temperatures and greater strength of the remaining, less hydrous material (Karato, 1989; T. L. Grove, M.I.T., oral communication, 1991); and (4) extensional thinning of the crust bringing upper mantle materials closer to the surface and perhaps resulting in self-limiting extension (Houseman and England, 1986, p. 724).

Some extension parallel to the east-northeast trend of the SRP is indicated by fissures and fissure eruptions. These fissures are formed by near-vertical injection of basaltic dikes from near the base of the crust (Leeman, 1982a) and are locally marked by grabens (Smith and others, 1989; Parsons and Thompson, 1991). Although the surface trace of these fissures locally line up with basin-range faults marginal to the plain, they are not coplanar with the primary zone of faulting at depth; the basin-range faults dip at about 50°, whereas fissures formed by dike injection are driven vertically upward from depth perpendicular to the minimum stress, which in this extensional stress field is generally close to horizontal.

Spacing of faults becomes closer toward the SRP. For example, the Teton fault progressively splits into as many as 10 strands northward between the Teton Range and Yellowstone (Christiansen, 1984, and written communication, 1986). The same pattern occurs near the north end of the Grand Valley (Prostka and Embree, 1978), the northern part of the Portneuf Range (Kellogg, this volume), the northern part of the Blackfoot Mountains (Allmendinger, 1982), the south end of the Arco Hills (Kuntz and others, 1984), the southern part of the Lemhi Range (McBroome, 1981), and from the southern end of the Beaverhead Range to the Centennial Range (Skipp, 1988). This progression to smaller fault blocks suggests that the part of the crust involved in faulting became thinner because the depth to the brittle-ductile transition became shallower toward the SRP (R. E. Anderson, oral communication, 1988). Heating that could raise the level of brittle-ductile transition is likely from silicic and basaltic intrusions at depths of 5 to 20 km beneath the SRP.

For the southern arm of faulting, the surface topographic gradient may provide a mechanism to drive and localize faulting. There, Belts I, II, and III are on the inner, western slope of the Yellowstone crescent. But for the western arm, neither the location nor the orientation of faults appears related to the inner, southern slope of the crescent.

Further discussion of extensional faulting

The geometry of faults in the western arm produces markedly different kinematics from that in the southern arm. The *en echelon* arrangement of faults in the southern arm may produce a comparable amount of east-west extension because one fault takes over where the other dies out. But in the western arm, the faults are arranged one behind the other such that late Quaternary extension is subparallel rather than perpendicular to the length of Belt II (Plate 1). Assuming the craton east of these arms is stable, the combined effect of both arms of ongoing extensional faulting is relative transport of the Snake River Plain to the southwest.

West of the active basin-range structures north of the plain is the relatively unfaulted Idaho batholith. This area has acted like a block, but one that has moved westward relative to the craton as the basin-range faults east of it extended. West of the Idaho batholith is the western Snake River Plain, the largest and best formed graben in the entire region. The paucity of evidence for extension within the batholith appears to be compensated by strong expression of basin-range extension east and west of it.

The western Snake River Plain is a graben apparently associated with passage of the Yellowstone hot spot but it remains a fundamentally different structure from the hot-spot track in spite of the geomorphic continuity of the two lowlands. Furthermore, the gravity anomaly that appears to join these features may represent a large tension gash (Riedel-shear opening). This right-lateral, crustal-shear opening would have appropriate kinematic conditions for formation when the hot spot had moved eastward to south of the nonextending Idaho batholith block, while extension continued to be accommodated in the graben of the western Snake River Plain.

Relations between the Yellowstone hot spot and basin-range deformation

The neotectonic fault belts that converge on Yellowstone are here explained by thermal effects associated with the Yellowstone mantle plume localizing basin-range extension. This level of explanation does not attempt to explain basin-range extension. But if we backtrack from Yellowstone to the 16-Ma start of the Yellowstone hot spot, the following geologic associations between the head phase of the hot spot and the origin of the basin-range structural province suggest a strong interrelation: (1) The plume head intercepted the base of the lithosphere about 16 Ma, coinciding with the start of widespread basin-range extension and normal faulting; (2) the hot-spot track started in about the center of the active basin-range structural province (Fig. 23); (3) the change in basin-range magmatism from calc alkalic *to* basalt *and* bimodal basalt/rhyolite (Christiansen and McKee, 1978) coincided in time and space with volcanic and structural penetration of the lithosphere we relate to the plume head; (4) the 1,100-km-long Nevada-Oregon rift zone we associate with the plume head was active over a distance comparable to the diameter of the basin-range structural province (Fig. 23); (5) associated with the northern part of the Nevada-Oregon rift, voluminous flood basalt volcanism of the Columbia River and Oregon plateaus was erupted 16 ± 1 Ma through a denser, more oceanic crust; (6) a spherical plume head 300 km in diameter, if spread out at the base of the lithosphere to a layer averaging 20 km thick, would cover an area almost 1,000 km in diameter, similar in scale to the active basin-range structural province; and (7) assuming a plume takes 25 m.y. to ascend from the core mantle boundary, the plume head would contain an amount of heat approximated by that feeding the present Yellowstone thermal plume stored over an interval of 25 m.y. The hot-spot buoyancy flux of the Yellowstone hot spot is estimated to be 1.5 Mgs^{-1} (Sleep, 1990). Much of the plume's original stored heat could still reside in the asthenosphere and lower lithosphere beneath the basin-range structural province.

Thus, the active basin-range structural province may have a causal relationship with particularly the plume-head phase of the Yellowstone hot spot, as indicated by their coincidence in time and space, the key factor being the large amount of thermal energy stored in the plume head that could still be affecting the lithosphere and asthenosphere over an area perhaps as large as the active basin-range structural province (Fig. 23). The plume mechanism thus may merit integration with at least two other mechanisms considered important in the deformation of the western cordillera in late Cenozoic time. First, the Yellowstone mantle plume rose into crust thickened during the Mesozoic and earliest Tertiary orogenies (Sevier and Laramide) and subsequently softened by radiogenic heating of a thickened sialic crust (Christiansen and Lipman, 1972; Wernicke and others, 1987; Molnar and Chen, 1983). Second, in Cenozoic time the plate margin southwest of the hot-spot track has progressively changed from a subduction zone with possible back-arc spreading to a weak(?) transcurrent fault (Atwater, 1970) over a time span that overlaps the postulated activity (16 to 0 Ma) of the Yellowstone hot spot.

Thus, basin-range breakup of the continental lithosphere, which is a deformation pattern rather unusual on the Earth, may involve at least three complementary factors: (1) Sevier/Laramide orogenic thickening producing delayed radiogenic heating that resulted in thermal softening of the crust, (2) an unconfined plate margin to the west permitting westward extension faster than North American plate motion, and (3) the Yellowstone plume head providing gravitational energy through uplift as well as thermal softening of the mantle lithosphere and lower continental crust. After 10 Ma, the much thinner chimney phase of the Yellowstone hot spot penetrated to the base of the lithosphere and spread radially outward at the base of the southwest-moving North American plate; the effects of lithospheric heating from this mushrooming plume localized extension in the northeast quadrant of the active basin-range structural province.

Rise of a mantle plume would exert body forces consistent with westward pulling apart of the active basin-range structural province. Plume material rising away from the Earth's spin axis would be accelerated to the higher velocity demanded by the greater spin radius. For asthenospheric positions at the latitude of Yellowstone (about 44°N), plume rise from a spin radius of 4,300 to 4,400 km would require an eastward increase in spin velocity of 628 km/day or 7.26 m/second (an increase from 27,018 to 27,646 km/day). At a plume-head rise rate suggested earlier of 0.1 mm/year, this 100-km rise from a spin radius of 4,300 to 4,400 km would take 1 m.y. The force needed to accelerate the huge plume-head mass to this higher eastward velocity would have to be exerted through the surrounding upper mantle and lithosphere, resulting in an equal and opposite (westward) force against the surrounding mantle/lithosphere that, if weak enough, would deform by westward extension. Rise of the chimney phase of the Yellowstone plume would continue to exert westward drag on the mantle and lithosphere.

The high terrain of the western United States (largely shown on Fig. 23) can be divided into two neotectonic parts: a western part containing the active basin-range structural province and an eastern part consisting of the Rocky Mountains and High Plains. Kane and Godson (1989, Fig. 4) show that both the regional terrain and regional Bouguer gravity maps define a high 1,500 to 2,000 km across that centers in western Colorado. This high may in part relate to thermal effects associated with (1) heat remaining from the Yellowstone plume head, (2) heat from the chimney phase of the Yellowstone plume, and (3) perhaps other thermal plumes, as suggested by Wilson (1990) and Suppe and others (1975). On the other hand, the geoid (Milbert, 1991) shows a high 600 to 800 km across and centered on Yellowstone.

The western part of this high is largely occupied by the active basin-range structural province, wherein the following factors (not inclusive) favor westward extension: (1) a topographic gradient toward the west, (2) plume material rising near the center of the high that would spread outward and westward at the base of the lithosphere, (3) eastward acceleration of plume mass as it rose and thereby increased its spin radius, and (4) lithospheric softening due to thermal plume heating and previous orogenic thickening.

The Rocky Mountains–High Plains occupy the eastern part of this high, wherein the following (not inclusive) would favor compression or nonextension: (1) plate tectonic motion of the North American plate southwestward up the slope of the east side of this high, (2) outward and eastward flow of plume material from the center of this high, and (3) lack of sufficient heating to result in lithospheric softening.

A prominent difficulty relating this high, with its extension on its west and nonextension on its east side, to the postulated Yellowstone thermal plume is that the geometric center of the topographic and gravitational high is in western Colorado (Kane and Godson, 1989). Suppe and others (1975) made the ingenious suggestion that two hot spots—the Yellowstone and Raton—operating in tandem might explain the axis of high topography of the Rocky Mountains. The postulated Raton hot spot has been proposed based on domal uplift and a volcanic alignment parallel to the Yellowstone hot-spot track now beneath the Clayton volcanic field on the High Plains just south of the Colorado–New Mexico border (Suppe and others, 1975). Lipman (1980) shows that the volcanic trend of this postulated hot spot has no systematic northeastward volcanic progression and concludes that the Raton volcanic alignment is more likely controlled by a crustal flaw than a hot spot. In addition to the postulated Yellowstone and Raton hot spots, Wilson (oral communication, 1990) suggests that a third plume may be beneath the Colorado Rockies. Two or more hot spots beneath the high terrain of the western United States are not inconsistent with the observation that oceanic hot spots may occur in groups, called families (Sleep, 1990). For example, a family of three hot spots occurs off the southeast coast of Australia (Duncan and Richards, 1991). The Tasminid hot spot is about 600 km east of the east Australian hot spot, and the Lord Howe hot spot is about 600 km northeast of the Tasminid hot spot. These inter–hot-spot distances are closer than the about 1,000-km distance between the postulated Yellowstone and Raton hot spots.

CONCLUSIONS

We conclude that the temporal and spatial pattern of volcanism and faulting and the altitude changes in the Yellowstone–Snake River Plain region require a large-scale disturbance of the lithosphere that is best explained by a thermal mantle plume, starting with a head phase and followed by a chimney phase.

Temporal and spatial pattern of volcanism and faulting

After 10 Ma, inception of caldera-forming volcanism migrated east-northeast at 30 km/m.y., leaving the mountain-bounded SRP floored with a basaltic veneer on thick piles of rhyolite along its trace. Compared to after 10 Ma, migration of the Yellowstone hot spot from 16 to 10 Ma (1) produced volcanism in a less systematic pattern; (2) had an apparent rate of 7 cm/year in a more easterly orientation; (3) had loci of volcanic eruptions that were more dispersed from the axis of migration; (4) left no trenchlike analogue to the eastern SRP but rather a higher, relatively unfaulted volcanic plateau; (5) was initially accompanied by extensive north-south rifting on the 1,100-km-long Nevada-Oregon rift zone; and (6) was accompanied by extrusion of the Columbia River and Oregon Plateau flood basalts.

Neotectonic faulting in the eastern SRP region defines four belts in a nested V-shaped pattern about the post-10-Ma hot-spot track.

1. Belt II has been the most active belt in Quaternary time. Faults in this belt have had at least one offset since 15 ka, and range fronts are steep and >700 m high. From its convergence on Yellowstone, Belt II flares outward to the southwest on either side of the hot-spot track (eastern SRP).

2. Belt I contains new, small escarpments and reactivated faults. It occurs outside Belt II and appears to be waxing in activity.

3. Belt III occurs inside Belt II. Compared to faults in Belt II, those in Belt III have been less recently active and are associated with more muted, somewhat lower escarpments. Activity on these faults appears to be waning.

4. The spatial pattern of Belts, I, II, and III indicates a northeast-moving cycle of waxing, culminating, and waning fault activity, respectively, that accompanies the northeast migration of the Yellowstone hot spot. The pattern of these belts is arrayed like parts of a large wave: The frontal part of this wave is represented by waxing rates of faulting, the crestal part by the highest rates, and the backslope part by waning rates.

5. Belt IV contains quiescent late Tertiary major range-front faults and occurs only on the south side of the SRP. In this belt, the timing of faulting correlates well with the time of passage of the Yellowstone hot spot along the SRP. Southwestward from Yellowstone, the ages of major range-front faulting, or other deformation, are as follows: (1) Teton fault, <6 to 0 Ma; (2) Grand Valley fault, 4.4 to 2 Ma; (3) Blackfoot Range–front fault, 5.9 to 4.7 Ma; (4) Portneuf Range–front fault, 7 to 6.7 Ma; (5) Rockland Valley fault, 10 to >8 Ma; (6) Sublett Range folding, >10 Ma; (7) Raft River valley detachment faulting, 10 to 8(?) Ma. Although this age progression increases southwestward, it also becomes less systematic in that direction.

6. The belts of faulting flare more on the southern than on the northern side of the SRP. This asymmetry is due to the presence of Belt IV only on the south side of the eastern SRP.

Altitude changes

Another type of deformation that may mark movement of the Yellowstone hot spot is change in altitude. The pattern is similar to that for Quaternary faulting but includes a large upland area ahead of the hot-spot track. Several indicators appear to define an outward-moving, wavelike pattern, although each of the individual components suggesting altitude changes is subject to alternative interpretations. Uplift appears to be occurring on the leading slope of the wave and subsidence on the trailing slope marginal to the SRP. Indicators of regional uplift and subsidence include the following.

1. An area of high terrain defines the Yellowstone crescent of high terrain 350 km across at the position of Yellowstone; the arms of the crescent extend from the apex more than 400 km to the south and to the west. The crests of these arms parallel the neotectonic fault belts, although the southern crest is more outside the fault belts than the western crest.

2. The geoid shows a large dome that centers on Yellowstone. The highest part of this geoid anomaly is similar to the Yellowstone crescent of high terrain, excepting the geoid high includes the eastern SRP and Yellowstone Plateau. The geoid is the geophysical anomaly that compares most favorably in size and height with the geoid anomaly of oceanic swells related to hot spots.

3. From Yellowstone southwestward, the altitude of the SRP decreases. Perpendicular to the axis, the ranges adjacent to the SRP are lower than ranges farther from the SRP.

4. Three domes of historic uplift at several mm/year lie along the western arm of the Yellowstone crescent as well as the axis connecting the modern drainage divides. A 2-Ma ignimbrite also has been uplifted along the western arm and locally along the southern arm of the crest.

5. High, steep, deeply dissected mountains formed of readily erodible rocks within the Yellowstone crescent indicate neotectonic uplift. Such mountains include the Absaroka Range, the mountains of southern Yellowstone and the Bridger-Teton Wilderness area, the Mt. Leidy–northern Wind River highlands, the Gallatin Range, parts of the Madison Range, and the Centennial Range. This general pattern is complicated by mountains in the crescent, formed of resistant rocks, that were uplifted in Laramide time (about 100 to 50 Ma).

6. Departures from typical ratios for the length of glaciers during the last (Pinedale, Pd) compared to next-to-last (Bull Lake, BL) glaciation suggest uplift on the leading margin of the Yellowstone crescent and subsidence on the trailing margin. High Pd/BL ratios are common on the outer part of the Yellowstone crescent and suggest an uplift rate of perhaps 0.1 to 0.4 mm/year. Low Pd/BL ratios in the western part of the Yellowstone ice mass in the cusp of the trailing edge of the crescent suggest subsidence perhaps more than 0.1 to 0.4 mm/year.

7. Stream terraces are common on the outer, leading margin of the Yellowstone crescent, suggesting uplift. Terraces are much less well developed on the trailing edge of the crescent and on the SRP and suggest subsidence, perhaps combined with tilt directions opposed to the direction of stream flow.

8. Near the northeast margin of the Yellowstone crescent, migration of streams by both capture and by unidirectional slip-off has generally been away from Yellowstone in the direction of the outer slope of the crescent.

9. In the Bighorn Basin, tilting away from Yellowstone is suggested both by the upstream divergence of terraces of streams flowing away from Yellowstone and by the upstream convergence of terraces of streams flowing toward Yellowstone.

10. Near the Bighorn Basin, inflection points in stream profiles have migrated away from Yellowstone as shown by the downstream migration of the convergence and divergence point for pairs of terrace profiles along Rock Creek and perhaps the Bighorn River northeast of the Bighorn Mountains.

Large disturbance of the lithosphere by a thermal mantle plume

The following large-scale late Cenozoic geologic effects cover distances many times the thickness of the lithosphere and require a sub-lithospheric thermal source.

1. The total hot-spot track formed from 16 to 0 Ma is 700 km long, whereas that sector from 10 to 0 Ma is 300 km long.

2. For Belts I and II, the distance from the western belt of faulting to the southern belt of faulting ranges from less than 100 km across Yellowstone to more than 400 km across the eastern SRP at the site of the 10.3-Ma Picabo volcanic field.

3. The Yellowstone crescent of high terrain is about 350 km across near Yellowstone; each arm of the crescent is more than 400 km long.

4. The present stress pattern indicates extension generally subparallel to and nearly always within 45° of the west-southwest trend of the SRP-YP province. Such a stress pattern is not consistent with either rifting parallel to the eastern SRP-YP province or with right-lateral transform shear across the SRP-YP province. A crustal flaw origin for the SRP-YP province has the following problems: Offset across the SRP is not required by structural or stratigraphic information, and the most pronounced crustal flaws are located north or south of the SRP-YP trend and are not present on the hot-spot trend immediately northeast of Yellowstone in the Beartooth Mountains.

Thermal processes are the only reasonable explanations for the late Cenozoic volcanic, faulting, and uplift/subsidence activity. We consider that the large scale of such activity is most reasonably explained by transport of heat by outward flow of asthenosphere beneath the lithosphere. The several-million-year time scales involved are not adequate for lateral heat transport within the lithosphere over such distances. A deep-seated mantle plume best explains all these observations. A plume of hotter asthenosphere mushrooming out at the base of the lithosphere could intrude and heat the mantle lithosphere, thereby (1) weakening the mantle lithosphere (where most lithospheric strength resides) and (2) converting dense mantle lithosphere to lighter asthenosphere, resulting in isostatic uplift. Any heating and softening of the crust (upper 40 km of the lithosphere) probably is accomplished largely by upward transport of heat by magma.

We suggest that such thermal effects have localized the observed neotectonic extension pattern for the northwest quadrant of the basin-range structural province. Although this explanation seems to suggest that extension relates to a separate process, we point out that the origin of basin-range activity has three strong ties with the postulated head phase of the Yellowstone hot spot: (1) the basin-range structural province is centered on the general area of the hot-spot head about 16 Ma, (2) basin-range activity started at about 16 Ma, the same age as rhyolitic volcanism we associate with the plume head, and (3) rifting and associated volcanism we attribute to the plume head extend over a distance of 1,100 km, quite similar to the dimensions of the active basin-range structural province. Thus, we find no way to clearly separate formation of the active basin-range structural province from the plume-head origin of the Yellowstone hot spot, with both affecting a large area near the common boundary of Nevada, Idaho, and Oregon.

Future studies and predictions

In conclusion, we recognize that much more information is needed to convincingly demonstrate the histories of volcanism, faulting, and uplift/subsidence outlined in this chapter. The region within 500 km of Yellowstone is a good candidate for studies of the effects of an inferred mantle plume on the lithosphere. The space/time history of volcanism, faulting, uplift, and subsidence presented here can be readily tested. We have a model that can be evaluated in many ways and that has important implications for the lithospheric responses to a deep-seated mantle thermal plume as well as for the plume itself.

If the observed history of volcanism, faulting, uplift, and subsidence are the result of the southwest motion of the North American plate over a thermal mantle plume, then the following migration of activity should occur, based on the present patterns and inferred activity since 10 Ma.

1. Uplift of the Yellowstone crescent will migrate about N55°E at about 30 km/m.y., oblique to the trend of the western and southern wings of the crescent. On the inner side of the crescent and particularly on the eastern Snake River Plain, subsidence will occur and the Yellowstone Plateau may subside more rapidly. The broader scale of uplift east and north of the Yellowstone crescent (Fig. 14) is predicted to move east-northeast and have the aerial extent of either (1) the geoid dome (Milbert, 1991) centered on Yellowstone with a width of 600 to 800 km, or (2) the combined form of the Yellowstone hot-spot uplift in tandem with uplift related to other hot spots postulated (Suppe and others, 1975; Wilson, 1990) to be in New Mexico and perhaps Colorado.

2. Neotectonic fault Belts I, II, III, and IV will migrate about N55°E at 30 km/m.y., a direction highly oblique to the trend of both the southern and western arms.

For the next few thousand years, the highest degree of fault and earthquake activity may be concentrated in Belt II. Because Belt I is also characterized by Holocene faulting, Belt I will have activity of a magnitude similar to Belt II. The long-term seismic record will tend to fill in areas poorly represented on the short-term seismic record (Fig. 12) so the long-term pattern parallels and perhaps closely coincides with the patterns of relative activity shown by fault belts. This prediction has similarities to the "seismic gap" hypothesis, except it is based on a distribution of fault activity that is broader in both time and space.

3. Based on the spacing of volcanic fields (Plate 1, Fig. 2), injection of basalt and heat exchange to form rhyolitic magmas will occur in the lower crust beneath a region centered near Red Lodge, Montana (Fig. 1), and with a radius of several tens of kilometers. Assuming a 2-m.y. hiatus of major eruptions between adjacent fields, this new field might start 2 m.y., perhaps ± 0.5 m.y. from now. This very rough estimate depends on whether the closing large event of the 2-Ma Yellowstone Plateau volcanic field might have occurred as far back as the eruption of the 0.6-Ma Lava Creek Tuff or might still occur as much in the future as perhaps one-half million years from now. For the Red Lodge area, it is uncertain whether or not volcanic and tectonic penetration through the lithosphere to the earth's surface would occur, because the hot spot would be well beneath the continental craton where the lithosphere may be stronger than that traversed earlier by the hot spot across orogenically thickened(?) and thermally softened crust beneath the thrust belt.

ACKNOWLEDGMENTS

Major help in review as well as stimulating discussion was freely given by Karl Kellogg, Marith Reheis, Dean Ostenaa, Mel Kuntz, Paul Link, Scott Lundstrom, Lucian Platt, and Robert Duncan. We thank the following for discussions that were quite important to developing the arguments presented in this chapter: M. H. Anders, R. E. Anderson, Fred Barker, C. G. Chase, Jim Case, R. L. Christiansen, J. M. Good, M. H. Hait, W. R. Hackett, W. P. Leeman, W. W. Locke, J. D. Love, M. N. Machette, A. E. McCafferty, J. P. McCalpin, Grant Meyer, S. A. Minor, R. C. Palmquist, C. L. Pillmore, W. E. Scott, N. H. Sleep, R. B. Smith, M. W. West, and M. L. Zoback. Lorna Carter and Libby Barstow improved the prose. We particularly thank Steven S. Oriel, to whom this volume is dedicated, for his administrative leadership and scientific counsel concerning our geologic studies in the Snake River Plain region.

REFERENCES CITED

Ach, J. A., Plouff, D., and Turner, R. L., 1987, Mineral resources of the East Fork High Rock Canyon Wilderness Study Area, Washoe and Humboldt Counties, Nevada: U.S. Geological Survey Bulletin 1707B, p. B1–B14.

Agard, S. S., 1989, Map showing Quaternary and late Tertiary terraces of the lower Bighorn River, Montana: U.S. Geological Survey Miscellaneous Field Studies Map MF-2094, scale 1:100,000.

Allmendinger, R. W., 1982, Sequence of late Cenozoic deformation in the Blackfoot Mountains, southeastern Idaho, *in* Bonnichsen, B., and Breckenridge, R. M., eds., Cenozoic geology of Idaho: Idaho Bureau of Mines and Geology Bulletin 26, p. 505–516.

Alt, D., Sears, J. M., and Hyndman, D. W., 1988, Terrestrial maria: The origins of large basalt plateaus, hotspot tracks and spreading ridges: Journal of Geology, v. 96, p. 647–662.

Alt, D., Hyndman, D. W., and Sears, J. W., 1990, Impact origin of late Miocene volcanism, Pacific Northwest: Geological Society of America Abstracts with Programs, v. 22, p. 2.

Anders, M. H., 1989, Studies of the stratigraphic and structural record of large volcanic and impact events [Ph.D. thesis]: Berkeley, University of California, 165 p.

——, 1990, Late Cenozoic evolution of Grand and Swan Valleys, Idaho, *in* Roberts, S., ed., Geologic field tours of western Wyoming and parts of adjacent Idaho, Montana, and Utah: Geological Survey of Wyoming Public Information Circular 29, p. 15–25.

Anders, M. H., and Geissman, J. W., 1983, Late Cenozoic structural evolution of Swan Valley, Idaho: EOS Transactions of the American Geophysical Union, v. 64, p. 625.

Anders, M. H., and Piety, L. A., 1988, Late Cenozoic displacement history of the Grand Valley, Snake River, and Star Valley faults, southeastern Idaho: Geological Society of America Abstracts with Programs, v. 20, p. 404.

Anders, M. H., Geissman, J. W., Piety, L. A., and Sullivan, J. T., 1989, Parabolic distribution of circum-eastern Snake River Plain seismicity and latest Quaternary faulting: Migratory pattern and association with the Yellowstone hot spot: Journal of Geophysical Research, v. 94, p. 1589–1621.

Anderson, D. L., 1981, Hotspots, basalts, and the evolution of the mantle: Science, v. 213, p. 82–88.

Andrews, D. A., Pierce, W. G., and Eargle, D. H., 1947, Geologic map of the Bighorn Basin, Wyoming and Montana, showing terrace deposits and physiographic features: U.S. Geological Survey Oil and Gas Investigations, Preliminary Map 71.

Armstrong, R. L., 1975, The geochronometry of Idaho: Isochron/West, no. 14, 50 p.

Armstrong, R. L., Leeman, W. P., and Malde, H. E., 1975, K-Ar dating, Quaternary and Neogene volcanic rocks of the Snake River Plain, Idaho: America Journal of Science, v. 275, p. 225–251.

Armstrong, R. L., Taubeneck, W. H., and Hales, P. O., 1977, Rb-Sr and K-Ar geochronometry of Mesozoic granitic rocks and their Sr isotopic composition, Oregon, Washington, and Idaho: Geological Society of America Bulletin, v. 88, p. 397–411.

Armstrong, R. L., Harakal, J. E., and Neill, W. M., 1980, K-Ar dating of Snake River Plain (Idaho) volcanic rocks—new results: Isochron/West, no. 27, p. 5–10.

Atwater, T., 1970, Implications of plate tectonics for the Cenozoic tectonic evolution of western North America: Geological Society of America Bulletin, v. 81, p. 3513–3536.

Baksi, A. J., 1990, Timing and duration of Mesozoic-Tertiary flood-basalt volcanism: EOS Transactions of the American Geophysical Union, v. 71, p. 1835–1836.

Barnosky, A. D., 1984, The Coulter Formation: Evidence for Miocene volcanism in Jackson Hole, Teton County, Wyoming: Earth Science Bulletin, Wyoming Geological Association, v. 17, p. 49–97.

Barnosky, A. D., and Labar, W. J., 1989, Mid-Miocene (Barstovian) environmental and tectonic setting near Yellowstone Park, Wyoming and Montana: Geological Society of America Bulletin, v. 101, p. 1448–1456.

Barrientos, S. E., Stein, R. S., and Ward, S. N., 1987, Comparison of the 1959 Hebgen Lake, Montana, and the 1983 Borah Peak, Idaho, earthquakes from geodetic observations: Bulletin of the Seismological Society of America, v. 77, p. 784–808.

Bartholomew, M. J., Stickney, M. C., and Wilde, E. M., 1990, Late Quaternary faults and seismicity in the Jefferson Basin, *in* White, R. D., ed., Quaternary geology of the western Madison Range, Madison Valley, Tobacco Root Range, and Jefferson Valley: Rocky Mountain Friends of the Pleistocene Guidebook, p. 238–244.

Bercovici, D., Schubert, G., and Glatzmaier, G. A., 1989, Three-dimensional spherical models of convection in the earth's mantle: Science, v. 244, p. 950–955.

Blackstone, D. L., Jr., 1966, Pliocene volcanism, southern Absaroka Mountains, Wyoming: University of Wyoming, Contributions to Geology, v. 5,

p. 21–30.
Blackwell, D. D., 1989, Regional implications of heat flow of the Snake River Plain, northwestern United States: Tectonophysics, v. 164, p. 323–343.
Blakely, R. J., 1988, Curie temperature isotherm analysis and tectonic implications of aeromagnetic data from Nevada: Journal of Geophysical Research, v. 93, p. 11817–11832.
Blakely, R. J., Jachens, R. C., and McKee, E. H., 1989, The northern Nevada Rift: a 500-km-long zone that resisted subsequent deformation: EOS Transactions of the American Geophysical Union, v. 70, p. 1336.
Bohannon, R. G., Naeser, C. W., Schmidt, D. L., and Zimmermann, R. A., 1989, The timing of uplift, volcanism, and rifting peripheral to the Red Sea: A case for passive rifting? Journal of Geophysical Research, v. 94, p. 1683–1701.
Bohor, B. F., and Seitz, R., 1990, Cuban K-T catastrophe: Nature, v. 344, p. 593.
Bond, J. G., 1978, Geologic map of Idaho: Idaho Department of Lands, Bureau of Mines and Geology, Moscow, Idaho, scale 1:500,000.
Bonnichsen, B., 1982, The Bruneau-Jarbidge eruptive center, southwestern Idaho, *in* Bonnichsen, B., and Breckenridge, R. M., eds., Cenozoic geology of Idaho: Idaho Bureau of Mines and Geology Bulletin 26, p. 237–254.
Bonnichsen, B., and Kauffman, D. F., 1987, Physical features of rhyolite lava flows in the Snake River Plain volcanic province, southwestern Idaho: Geological Society of America Special Paper 212, p. 119–145.
Braile, L. W., and 9 others, 1982, The Yellowstone–Snake River Plain seismic profiling experiment: Crustal structure of the eastern Snake River Plain: Journal of Geophysical Research, v. 87, p. 2597–2609.
Breckenridge, R. M., 1975, Quaternary geology of the Wood River area, Wyoming: Wyoming Geological Association, 27th Annual Field Conference, Guidebook, p. 45–54.
Bright, R. C., 1967, Late Pleistocene stratigraphy in Thatcher Basin, southeastern Idaho: Tebiwa, The Journal of the Idaho State University Museum, v. 10, p. 1–7.
Brott, C. A., Blackwell, D. D., and Ziagos, J. P., 1981, Thermal and tectonic implications of heat flow in the eastern Snake River Plain, Idaho: Journal of Geophysical Research, v. 86, p. 11709–11734.
Burbank, D. W., and Barnosky, A. D., 1990, The magnetochronology of Barstovian mammals in southwestern Montana and implications for the initiation of Neogene crustal extension in the northern Rocky Mountains: Geological Society of America Bulletin, v. 102, p. 1093–1104.
Carlson, R. W., and Hart, W. K.,1986, Crustal genesis on the Oregon Plateau: Journal of Geophysical Research, v. 92, p. 6191–6206.
——, 1988, Flood basalt volcanism in the northwestern United States, *in* Macdougall, J. D., ed., Continental flood basalts: Dordrecht, The Netherlands, Kluwer Academic Publishers, p. 35–61.
Carter, B. J., Ward, P. A. III, and Shannon, J. T., 1990, Soil and geomorphic evolution within the Rolling Red Plains using Pleistocene volcanic ash deposits: Geomorphology, v. 3, p. 471–488.
Christiansen, R. L., 1982, Late Cenozoic volcanism of the Island Park area, eastern Idaho, *in* Bonnichsen, B., and Breckenridge, R. M., eds., Cenozoic geology of Idaho: Idaho Bureau of Mines and Geology Bulletin 26, p. 345–368.
——, 1984, Yellowstone magmatic evolution: Its bearing on understanding large-volume explosive volcanism, *in* Explosive volcanism: inception, evolution, and hazards: Washington, D.C., National Academy Press, Studies in Geophysics, p. 84–95.
Christiansen, R. L., and Blank, H. R., 1972, Volcanic stratigraphy of the Quaternary rhyolite plateau in Yellowstone National Park: U.S. Geological Survey Professional Paper 729B, 18 p.
Christiansen, R. L., and Lipman, P. W., 1972, Cenozoic volcanism and plate-tectonic evolution of the western United States. II. Late Cenozoic: Philosophical Transactions of the Royal Society of London, v. 271, p. 249–284.
Christiansen, R. L., and Love, J. D., 1978, Connant Creek Tuff: U.S. Geological Survey Bulletin 1435C, 9 p.
Christiansen, R. L., and McKee, E. H., 1978, Late Cenozoic volcanic and tectonic evolution of the Great Basin and Columbia intermontain regions, *in* Smith, R. B., and Eaton, G. P., eds., Cenozoic tectonics and regional geophysics of the western Cordillera: Geological Society of America Memoir 152, p. 283–311.
Clawson, S. R., Smith, R. B., and Benz, H. M., 1989, P-Wave attenuation of the Yellowstone caldera from three-dimensional inversion of spectral decay using explosion seismic data: Journal of Geophysical Research, v. 94, p. 7205–7222.
Colman, S. M., and Pierce, K. L., 1981, Weathering rinds on andesitic and basaltic stones as a Quaternary age indicator, western United States: U.S. Geological Survey Professional Paper 1210, 56 p.
——, 1986, The glacial record near McCall, Idaho—Weathering rinds, soil development, morphology, and other relative-age criteria: Quaternary Research, v. 25, p. 25–42.
Courtillot, V., Besse, J., Vandamme, D., Montigny, R., Jaeger, J. J., and Cappetta, H., 1986, Deccan flood basalts at the Cretaceous/Tertiary boundary?: Earth and Planetary Science Letters, v. 80, p. 361–374.
Courtney, R. C. and White, R. S., 1986, Anomalous heat flow and geoid across the Cape Verde Rise: Evidence for dynamic support from a thermal plume in the mantle: Geophysical Journal of the Royal Astronomical Society, v. 87, p. 815–867.
Covington, H. R., 1983, Structural evolution of the Raft River basin, Idaho, *in* Miller, D. M., Todd, V. R., and Howard, K. A., eds., Tectonic and stratigraphic studies in the eastern Great Basin: Geological Society of America Memoir 157, p. 229–237.
Cox, K. G., 1989, The role of mantle plumes in the development of continental drainage patterns: Nature, v. 342, p. 873–876.
Crone, A. J., and Haller, K. M., 1991, Segmentation and coseismic behavior of Basin and Range normal faults: Examples from east-central Idaho and southwestern Montana, U.S.A.: Journal of Structural Geology, v. 13, p. 151–164.
Crone, A. J., and 6 others, 1987, Surface faulting accompanying the Borah Peak earthquake, and segmentation of the Lost River fault, central Idaho: Seismological Society of America Bulletin, v. 77, p. 739–770.
Crough, S. T., 1978, Thermal origin of mid-plate hot-spot swells: Geophysical Journal of the Royal Astronomical Society, v. 55, p. 451–469.
——, 1979, Hot spot epeirogeny: Tectonophysics, v. 61, p. 321–333.
——, 1983, Hot spot swells: Annual Reviews of the Earth and Planetary Sciences, v. 11, p. 165–193.
DeMets, C., Gordon, R. G., Argus, D. F., and Stein, S., 1990, Current plate motions: Geophysical Journal International, v. 101, p. 425–478.
Doser, D. I., 1985, Source parameters and faulting processes of the 1959 Hebgen Lake, Montana, earthquake sequence: Journal of Geophysical Research, v. 90, p. 4537–4556.
Doser, D. I., and Smith, R. B., 1983, Seismicity of the Teton–southern Yellowstone region, Wyoming: Bulletin of the Seismological Society of America, v. 73, p. 1369–1394.
Draper, D. S., 1991, Late Cenozoic bimodal magmatism in the northern Basin and Range Province of southeastern Oregon: Journal of Volcanology and Geothermal Research, v. 47, p. 299–328.
Duncan, R. A., 1982, A captured island chain in the Coast Range of Oregon and Washington: Journal of Geophysical Research, v. 87, p. 10827–10837.
——, 1984, Age progressive volcanism in the New England seamounts and the opening of the central Atlantic Ocean: Journal of Geophysical Research, v. 89, p. 9980–9990.
Duncan, R. A., and Richards, M. A., 1991, Hotspots, mantle plumes, flood basalts, and true polar wander: Reviews of Geophysics, v. 29, p. 31–50.
Eaton, G. P., and 7 others, 1975, Magma beneath Yellowstone National Park: Science, v. 188, p. 787–796.
Ekren, E. B., McIntyre, D. H., and Bennett, E. H., 1984, High-temperature, large-volume, lavalike ash-flow tuffs without calderas in southwestern Idaho: U.S. Geological Survey Professional Paper 1272, 76 p.
Elison, M. W., Speed, R. C., and Kistler, R. W., 1990, Geologic and isotopic constraints on the crustal structure of the northern Great Basin: Geological Society of America Bulletin, v. 102, p. 1077–1092.
Erslev, E. A., 1982, The Madison mylonite zone: A major shear zone in the

Archean Basement of southwestern Montana: Wyoming Geological Association, 23rd Annual Field Conference, Guidebook, p. 213–221.

Erslev, E. A., and Sutter, J., 1990, Evidence for Proterozoic mylonitization in the northwestern Wyoming Province: Geological Society of America Bulletin, v. 102, p. 1681–1694.

Evans, J. R., 1982, Compressional wave velocity structure of the upper 350 km under the eastern Snake River Plain near Rexburg, Idaho: Journal of Geophysical Research, v. 87, p. 2654–2670.

Evenson, E. B., Cotter, J.F.P., and Clinch, J. M., 1982, Glaciation of the Pioneer Mountains: a proposed model for Idaho, *in* Bonnichsen, B., and Breckeridge, R. M., eds., Cenozoic geology of Idaho: Idaho Bureau of Mines and Geology Bulletin 26, p. 653–665.

Evernden, J. F., Savage, D. E., Curtis, G. H., and James, G. T., 1964, Potassium-argon dates and the Cenozoic mammalia chronology of North America: American Journal of Science, v. 265, p. 257–291.

Fields, R. W., Rasmussen, D. L., Tabrum, A. R., and Nichol, R., 1985, Cenozoic rocks of the intermontane basins of western Montana and eastern Idaho, *in* Flores, R. M., and Kaplan, S. S., eds., Cenozoic paleogeography of the West-Central United States: Denver, Society of Economic Paleontologists and Mineralogists, Rocky Mountain Section, p. 9–36.

Fisher, F. S., and Ketner, K. B., 1968, Late Tertiary syncline in the southern Absaroka Mountains, Wyoming: U.S. Geological Survey Professional Paper 600B, p. B144–B147.

Fisher, F. S., McIntyre, D. H., and Johnson, K. M., 1983, Geologic map of the Challis 1°×2° Quadrangle, Idaho: U.S. Geological Survey Open File Report 83-523, scale 1:250,000.

Fritz, W. J., and Sears, J. W., 1989, Late Cenozoic crustal deformation of SW Montana in the wake of the passing Yellowstone hot spot: Evidence from an ancient river valley: Geological Society of America Abstracts with Programs, v. 21, p. A90.

Fryxell, F. M., 1930, Glacial features of Jackson Hole, Wyoming: Augustana Library Publications, No. 13, Rock Island, Illinois, Augustana College and Theological Seminary, 129 p.

Gibbons, A. B., and Dickey, D. D., 1983, Quaternary faults in Lincoln and Uinta Counties, Wyoming, and Rich County, Utah: U.S. Geological Survey Open-File Report 83-288 (1 sheet, scale = 1:100,000).

Gilbert, J. D., Ostenaa, D., and Wood, C., 1983, Seismotectonic study, Jackson Lake Dam and Reservoir, Minidoka Project, Idaho-Wyoming: U.S. Bureau of Reclamation Seismotectonic Report 83-8, 122 p.

Godson, R. H., 1981, Digital terrain map of the United States: U.S. Geological Survey Miscellaneous Geologic Investigations Map I-1318, scale 1:7,500,000.

Greene, R. C., and Plouff, D., 1981, Location of a caldera source for the Soldier Meadow Tuff, northwestern Nevada, indicated by gravity and aeromagnetic data: Summary: Geological Society of America Bulletin, Pt. I, v. 92, p. 4–6.

Greensfelder, R. W., 1976, Maximum probable earthquake acceleration on bedrock in the State of Idaho: Boise, Idaho Department of Transportation Research Project no. 79, 69 p.

Griffiths, R. W., and Richards, M. A., 1989, The adjustment of mantle plumes to change in plate motion: Geophysical Research Letters, v. 16, p. 437–440.

Gripp, A. E., and Gordon, R. G., 1990, Current plate velocities relative to hotspots incorporating the NUVEL-1 global plate motion model: Geophysical Research Letters, v. 17, p. 1109–1112.

Hall, R. D., and Michaud, D., 1988, The use of hornblende etching, clast weathering, and soils to date alpine glacial and periglacial deposits: A study from southwestern Montana: Geological Society of America Bulletin, v. 100, p. 458–467.

Haller, K. M., 1988, Segmentation of the Lemhi and Beaverhead faults, east-central Idaho, and Red Rock fault, southwest Montana, during the late Quaternary [M.S. thesis]: Boulder, University of Colorado, 141 p.

Hamilton, L. J., and Paulson, Q. F., 1968, Geology and ground-water resources of the lower Bighorn valley, Montana: U.S. Geological Survey Water-Supply Paper 1876, 39 p.

Hamilton, W., 1987, Plate-tectonic evolution of the Western U.S.A.: Episodes, v. 10, p. 271–276.

——, 1989, Crustal geologic processes of the United States, *in* Pakiser, L. C., and Mooney, W. D., eds., Geophysical framework of the continental United States: Geological Society of America Memoir 172, p. 743–781.

Hansen, W. R., 1985, Drainage development of the Green River Basin in southwestern Wyoming and its bearing on fish biogeography, neotectonics, and paleoclimates: The Mountain Geologist, v. 22, p. 192–204.

Harrison, R. E., 1969, Shaded relief map of the United States, Albers equal area projection, U.S. Geological Survey National Atlas, sheet n. 56.

Hildebrand, A. R., and Boynton, W. V., 1990, Proximal Cretaceous-Tertiary boundary impact deposits in the Caribbean: Science, v. 24, p. 843–846.

Hildreth, W., 1981, Gradients in silicic magma chambers: Implications for lithospheric magmatism: Journal of Geophysical Research, v. 86, p. 10153–10192.

Hill, D. P., 1972, Crustal and upper-mantle structure of the Columbia Plateau from long-range seismic-refraction measurements: Geological Society of America Bulletin, v. 83, p. 1639–1648.

Hooper, P. R., 1988, The Columbia River Basalt, *in* Macdougall, J. D., ed., Continental flood basalts: Dordrecht, The Netherlands, Kluwer Academic Publishers, p. 1–33.

Houseman, G., and England, P., 1986, A dynamical model of lithospheric extension and sedimentary basins: Journal of Geophysical Research, v. 91, p. 719–729.

Huppert, H. E., and Sparks, S. J., 1988, The generation of granitic magmas by intrusion of basalt into continental crusts: Journal of Petrology, v. 29, p. 599–624.

Ito, E., and Takahashi, E. J., 1989, Postspinel transformations in the system Mg2SiO4-Fe2SiO4 and some geophysical implications: Journal of Geophysical Research, v. 94, p. 10637–10646.

Iyer, H. M., Evans, J. R., Zant, G., Stewart, R. M., Coakley, J. M., and Roloff, J. N., 1981, A deep low-velocity body under the Yellowstone caldera, Wyoming: Delineation using teleseismic P-wave residuals and tectonic interpretations—Summary: Geological Society of America Bulletin, Pt. I, v. 92, p. 792–798.

Izett, G. A., 1981, Volcanic ash beds: recorders of upper Cenozoic silicic pyroclastic volcanism in the western United States: Journal of Geophysical Research, v. 86, p. 10200–10222.

——, 1990, Cretaceous-Tertiary boundary interval, Raton Basin, Colorado and New Mexico, and its content of shock-metamorphosed minerals: Evidence relevant to the K/T boundary impact-extinction theory: Geological Society of America Special Paper 249, 100 p.

Izett, G. A., and Wilcox, R. E., 1982, Map showing localities and inferred distributions of the Huckleberry Ridge, Mesa Falls, and Lava Creek ash beds (Pearlette Family ash beds) of Pliocene and Pleistocene age in the western United States and southern Canada: U.S. Geological Survey Miscellaneous Investigations Map I-1325, scale 1:4,000,000.

Kane, M. F., and Godson, R. H., 1989: A crust/mantle structural framework of the conterminous United States based on gravity and magnetic trends, *in* Pakiser, L. C., and Mooney, W. D., eds., Geophysical framework of the continental United States, Geological Society of America Memoir 172, p. 383–403.

Karato, S-i., 1989, Defects and plastic deformation in olivine, *in* Karato, S-i., and Toriumi, M., Rheology of solids and of the earth: New York, Oxford University Press, p. 176–208.

Kellogg, K. S., and Marvin, R. F., 1988, New potassium-argon ages, geochemistry, and tectonic setting of upper Cenozoic volcanic rocks near Blackfoot, Idaho: U.S. Geological Survey Bulletin 1086, 19 p.

Kellogg, K. S., Pierce, K. L., Mehnert, H. H., Hackett, W. R., Rodgers, D. W., and Hladky, F. R., 1989, New ages on biotite-bearing tuffs of the eastern Snake River Plain, Idaho: Stratigraphic and mantle-plume implications: Geological Society of America Abstracts with Programs, v. 21, p. 101.

Kelson, K. I., and Swan, F. H., 1988, Recurrent Late Pleistocene to Holocene(?) surface faulting on the Stagner Creek Segment of the Cedar Ridge Fault, Central Wyoming: Geological Society of America Abstracts with Programs,

v. 20, p. 424.
Kennedy, B. M., Reynolds, J. H., Smith, S. P., and Truesdell, A. H., 1987, Helium isotopes: Lower Geyser Basin, Yellowstone National Park: Journal of Geophysical Research, v. 92, p. 12477–12489.
Ketner, K. B., Keefer, W. R., Fisher, F. S., Smith, D. L., and Raabe, R. G., 1966, Mineral resources of the Stratified Primitive Area, Wyoming: U.S. Geological Survey Bulletin 1230-E, 56 p.
Kistler, R. W., 1983, Isotope geochemistry of plutons in the northern Great Basin: Davis, California, Geothermal Resources Council, Special Report no. 13, p. 3–8.
Kistler, R. W. and Lee, D. E., 1989, Rubidium and strontium isotopic data for a suite of granitoid rocks from the Basin and Range Province, Arizona, California, Nevada, and Utah: U.S. Geological Survey Open-File Report 89-199, 13 p.
Kistler, R. W., and Peterman, Z. E., 1978, Reconstruction of crustal blocks of California on the basis of initial strontium isotopic compositions of Mesozoic granitic rocks: U.S. Geological Survey Professional Paper 1071, 17 p.
Kistler, R. W., Ghent, E. D., and O'Neil, J. R., 1981, Petrogenesis of garnet two-mica granites in the Ruby Mountains, Nevada: Journal of Geophysical Research, v. 86, p. 10591–10606.
Kuntz, M. A., and six others, 1983, Geological and geophysical investigations of the proposed Great Rift Wilderness area, Idaho: U.S. Geological Survey Open-File Report 80-475, 48 p.
Kuntz, M. A., Skipp, B. A., Scott, W. E., and Page, W. R., 1984, Preliminary geologic map of the Idaho National Engineering Laboratory and adjacent areas, Idaho: U.S. Geological Survey Open-File Report 84-281, 23 p., scale 1:100,000.
Lageson, D. R., 1987, Laramide controls on the Teton Range, NW Wyoming: Geological Society of America Abstracts with Programs, v. 20, p. 426.
Leeman, W. P., 1982a, Development of the Snake River Plain–Yellowstone Plateau province, Idaho and Wyoming: An overview and petrologic model, *in* Bonnichsen, B., and Breckenridge, R. M., eds., Cenozoic geology of Idaho: Idaho Bureau of Mines and Geology Bulletin 26, p. 155–178.
—— , 1982b, Geology of the Magic Reservoir area, Snake River Plain, Idaho, *in* Bonnichsen, B., and Breckenridge, R. M., eds., Cenozoic geology of Idaho: Idaho Bureau of Mines and Geology Bulletin 26, p. 369–376.
—— , 1989, Origin and development of the Snake River Plain (SRP)—An Overview, *in* Ruebelman, K. L., ed., Snake River Plain–Yellowstone Volcanic Province: 28th International Geological Congress, Field Trip Guidebook T305, p. 4–13.
Lipman, P. W., 1980, Cenozoic volcanism in the western United States: Implications for continental tectonics, *in* Continental tectonics: Washington, D.C., National Academy of Sciences, Geophysics Study Committee, p. 161–174.
Loper, D. E., and Stacey, F. D., 1983, The dynamical and thermal structure of deep mantle plumes: Physics of the Earth and Planetary Interiors, v. 33, p. 304–317.
Love, J. D., 1939, Geology along the southern margin of the Absaroka Range, Wyoming: Geological Society of America Special Paper 20, 137 p.
—— , 1961, Reconnaissance study of Quaternary faults in and south of Yellowstone National Park, Wyoming: Geological Society of America Bulletin, v. 72, p. 1749–1764.
—— , 1977, Summary of upper Cretaceous and Cenozoic stratigraphy, and of tectonic and glacial events in Jackson Hole, northwestern Wyoming: Wyoming Geological Association, 29th Annual Field Conference, Guidebook, p. 585–593.
Love, J. D., and de la Montagne, J., 1956, Pleistocene and recent tilting of Jackson Hole, Teton County, Wyoming: Wyoming Geological Association, 11th Annual Field Conference, Guidebook, p. 169–178.
Love, J. D., and Keefer, W. R., 1975, Geology of sedimentary rocks in southern Yellowstone National Park, Wyoming: U.S. Geological Survey Professional Paper 729-D, 60 p.
Love, J. D., and Love, J. M., 1982, Road log, Jackson to Dinwoody and return: Wyoming Geological Association, 34th Annual Field Conference, Guidebook, p. 283–318.
Love, J. D., Simons, F. S., Keefer, W. R., and Harwood, D. S., 1988, Geology, *in* Mineral Resources of the Gros Ventre Wilderness Study Area, Teton and Sublette Counties, Wyoming: U.S. Geological Survey Bulletin 1591, p. 6–20.
Love, J. D., Reed, J. C., Jr., and Christiansen, A. C., 1992, Geologic map of Grand Teton National Park, Teton County, Wyoming: U.S. Geological Survey Miscellaneous Geologic Investigations Map I-2031, scale 1:62,500 (in press).
Luedke, R. G., and Smith, R. L., 1981, Map showing distribution and age of late Cenozoic volcanic centers in California and Nevada: U.S. Geological Survey Miscellaneous Geologic Investigations Map I-1091-C, scale 1:1,000,000.
—— , 1982, Map showing distribution and age of late Cenozoic volcanic centers in Oregon and Washington: U.S. Geological Survey Miscellaneous Geologic Investigations Map I-1091-D, scale 1:1,000,000.
—— , 1983, Map showing distribution and age of late Cenozoic volcanic centers in Idaho, western Montana, west-central South Dakota, and northwestern Wyoming: U.S. Geological Survey Miscellaneous Geologic Investigations Map I-1091-E, scale 1:1,000,000.
Lundstrom, S. C., 1986, Soil stratigraphy and scarp morphology studies applied to the Quaternary geology of the southern Madison Valley, Montana [M.S. thesis]: Arcata, California, Humboldt State University, 53 p.
Mabey, D. R., 1982, Geophysics and tectonics of the Snake River Plain, Idaho, *in* Bonnichsen, B., and Breckenridge, R. M., eds., Cenozoic geology of Idaho: Idaho Bureau of Mines and Geology Bulletin 26, p. 139–153.
Mabey, D. R., Zietz, I., Eaton, G. P., and Kleinkopf, M. D., 1978, Regional magnetic patterns in part of the Cordillera in the western United States, *in* Smith, R. B., and Eaton, G. P., eds., Cenozoic tectonics and regional geophysics of the Western Cordillera: Geological Society of America Memoir 152, p. 93–109.
Machette, M. N., Personius, S. F., and Nelson, A. R., 1987, Quaternary geology along the Wasatch fault zone: Segmentation, recent investigations, and preliminary conclusions, *in* Gori, P. L., and Hays, W. W., eds., Assessment of regional earthquake hazards and risk along the Wasatch Front, Utah: U.S. Geological Survey Open-File Report 87-585, p. A1–A72.
—— , 1991, The Wasatch fault zone, Utah—segmentation and history of Holocene earthquakes: Journal of Structural Geology, v. 13, p. 137–149.
Mackin, J. H., 1937, Erosional history of the Big Horn Basin, Wyoming: Geological Society of America Bulletin, v. 48, p. 813–894.
MacLeod, N. S., Walker, G. W., and McKee, E. H., 1976, Geothermal significance of eastward increase in age of upper Cenozoic rhyolitic domes in southeastern Oregon, *in* Proceedings, Second United Nations Symposium on the Development and Use of Geothermal Resources, San Francisco, May 1975, Volume 1: Washington, D.C., U.S. Government Printing Office (Lawrence, Berkeley Laboratory, University of California), p. 456–474.
Madsen, D. B., and Currey, D. R., 1979, Late Quaternary glacial and vegetation changes, Little Cottonwood area, Wasatch Mountains, Utah: Quaternary Research, v. 12, p. 254–270.
Malde, H. E., 1987, Quaternary faulting near Arco and Howe, Idaho: Bulletin of the Seismological Society of America, v. 77, p. 847–867.
—— , 1991, Quaternary geology and structural history of the Snake River Plain, Idaho and Oregon, *in* Morrison, R. B., ed., Quaternary nonglacial geology: Conterminous U.S.: Boulder, Colorado, Geological Society of America, The Geology of North America, v. K-2, p. 251–281.
Mathiesen, E. L., 1983, Late Quaternary activity of the Madison fault along its 1959 rupture trace, Madison County, Montana [M.S. thesis]: Stanford, California, Stanford University, 157 p.
Mayer, L., and Schneider, N. P., 1985, Temporal and spatial relations of faulting along the Madison Range fault, Montana—Evidence for fault segmentation and its bearing on the kinematics of normal faulting: Geological Society of America Abstracts with Programs, v. 17, p. 656.
McBroome, L. A., 1981, Stratigraphy and origin of Neogene ash-flow tuffs on the north-central margin of the eastern Snake River Plain, Idaho [M.S. thesis]: Boulder, University of Colorado, 74 p.
McCalpin, J., 1987, Quaternary deformation along the East Cache fault, north-

central Utah: Geological Society of America Abstracts with Programs, v. 19, p. 320.

McCalpin, J., Robison, R. M., and Garr, J. P., 1987, Neotectonics of the Hansel Valley–Pocatello Valley Corridor, northern Utah and southern Idaho, *in* Gori, P. L., and Hays, W. W., eds., Assessment of regional earthquake hazards and risk along the Wasatch Front, Utah: U.S. Geological Survey Open-File Report 87-585, p. G1–G44.

McCalpin, J., Zhang, L., and Khromovskikh, V. S., 1990, Quaternary faulting in the Bear Lake Graben, Idaho and Utah: Geological Society of America Abstracts with Programs, v. 22, p. 38.

McCoy, W. D., 1981, Quaternary aminostratigraphy of the Bonneville and Lahontan Basins, western U.S., with paleoclimatic implications [Ph.D. thesis]: Boulder, University of Colorado, 603 p.

McKee, E. H. and Noble, D. C., 1986, Tectonic and magmatic development of the Great Basin of the western United States during late Cenozoic time: Modern Geology, v. 10, p. 39–49.

McKee, E. H., Noble, D. C., and Silberman, M. L., 1970, Middle Miocene hiatus in volcanic activity in the Great Basin area of the western United States: Earth and Planetary Sciences Letters, v. 8, p. 93–96.

McKenna, M. C., and Love, J. D., 1972, High-level strata containing Early Miocene mammals on the Bighorn Mountains, Wyoming: Novitates, no. 2490, 31 p.

Merrill, R. D., 1973, Geomorphology of terrace remnants of the Greybull River, northwestern Wyoming [Ph.D. thesis]: Austin, University of Texas, 268 p.

Milbert, D. G., 1991, GEOID90: A high-resolution geoid for the United States: EOS Transactions of the American Geophysical Union, v. 72, p. 545, 554.

Minor, S. A., Sawatzky, D., and Leszcykowski, A. M., 1986, Mineral resources of the North Fork Owyhee River Wilderness Study Area, Owyhee County, Idaho: U.S. Geological Survey Bulletin 1719-A, 10 p.

Minor, S. A., King, H. D., Kulik, D. M., Sawatzky, D., and Capstick, D. O., 1987, Mineral resources of the Upper Deep Creek Wilderness Study Area, Owyhee County, Idaho: U.S. Geological Survey Bulletin 1719-G, 10 p.

Minster, J. B., and Jordan, T. H., 1978, Present day plate motions: Journal of Geophysical Research, v. 83, p. 5331–5354.

Minster, J. B., Jordan, T. H., Molnar, P., and Haines, E., 1974, Numerical modeling of instantaneous plate tectonics: Royal Astronomical Society Geophysical Journal, v. 36, p. 541–576.

Molnar, P., and Chen, W. P., 1983, Focal depths and fault plane solutions of earthquakes under the Tibetan plateau: Journal of Geophysical Research, v. 88, p. 1180–1196.

Montagne, J., and Chadwick, R. A., 1982, Cenozoic history of the Yellowstone Valley south of Livingston Montana: Field trip Guidebook: Bozeman, Montana State University, 67 p.

Mooney, W. D., and Braile, L. W., 1989, The seismic structure of the continental crust and upper mantle of North America, *in* Bally, A. W., and Palmer, A. R., eds., The geology of North America—An overview: Geological Society of America, The Geology of North America, v. A, p. 39–52.

Moore, D. W., Oriel, S. S., and Mabey, D. R., 1987, A Neogene(?) gravity-slide block and associated slide phenomena in Swan Valley graben, Wyoming and Idaho: Geological Society of America, Centennial Field Guide, v. 2, p. 113–116.

Morgan, L. A., 1988, Explosive rhyolitic volcanism on the eastern Snake River Plain [Ph.D. thesis]: Manoa, University of Hawaii, 191 p.

Morgan, L. A., and Bonnichsen, B., 1989, Heise volcanic field, *in* Chapin, C. E., and Zidek, J., eds., Field excursions to volcanic terranes in the western United States, Vol. II: Cascades and Intermountain West: New Mexico Bureau of Mines and Mineral Resources Memoir 47, p. 153–160.

Morgan, L. A., and Pierce, K. L., 1990, Silicic volcanism along the track of the Yellowstone hotspot: Geological Society of America Abstracts with Programs, v. 22, p. 39.

Morgan, L. A., Doherty, D. J., and Leeman, W. P., 1984, Ignimbrites of the eastern Snake River Plain: Evidence for major caldera forming eruptions: Journal of Geophysical Research, v. 89, p. 8665–8678.

Morgan, L. A., Doherty, D. J., and Bonnichsen, B., 1989, Evolution of the Kilgore caldera: A model for caldera formation on the Snake River Plain–Yellowstone Plateau volcanic province: IAVCEI General Assembly Abstracts, New Mexico Bureau of Mines and Mineral Resources Bulletin 131, p. 195.

Morgan, W. J., 1972, Plate motions and deep mantle convection: Geological Society of America Memoir 132, p. 7–22.

—— , 1981, Hot spot tracks and the opening of the Atlantic and Indian Oceans, *in* Emiliani, C., ed., The sea, v. 7: New York, Wiley Interscience, p. 443–487.

Morris, D. A., Hackett, O. M., Vanlier, K. E., and Moulder, E. A., 1959, Groundwater resources of Riverton irrigation project area, Wyoming: U.S. Geological Survey Water-Supply Paper 1375, 205 p.

Myers, W. B., and Hamilton, W., 1964, Deformation accompanying the Hebgen Lake earthquake of August 17, 1959: U.S. Geological Survey Professional Paper 435, p. 55–98.

Naeser, C. W., Izett, G. A., and Obradovich, J. D., 1980, Fission-track and K-Ar ages of natural glasses: U.S. Geological Survey Bulletin 1489, 31 p.

Naeser, N. D., 1986, Neogene thermal history of the northern Green River Basin, Wyoming—Evidence from fission-track dating *in* Gautier, D. L., ed., Roles of organic matter in sediment diagenesis: Society of Economic Paleontologists and Mineralogists Special Publication 38, p. 65–72.

Noble, D. C., McKee, E. H., Smith, J. G., and Korringa, M. K., 1970, Stratigraphy and geochronology of Miocene volcanic rocks in northwestern Nevada: U.S. Geological Survey Professional Paper 700-D, p. D23–D32.

Okal, E. A., and Batiza, R., 1987, Hot spots: The first 25 years: American Geophysical Union, Geophysical Monograph 43, p. 1–11.

O'Neill, J. M., and Lopez, D. A., 1985, The Great Falls tectonic zone of east-central Idaho and west-central Montana—Its character and regional significance: American Association of Petroleum Geologists Bulletin, v. 69, p. 487–497.

Oriel, S. S., and Platt, L. A., 1980, Geologic map of the Preston 1 × 2 degree quadrangle, southeastern Idaho and western Wyoming: U.S. Geological Survey Miscellaneous Investigations Series Map I-1127, scale 1:250,000.

Osborne, G. D., 1973, Quaternary geology and geomorphology of the Uinta Basin and the south flank of the Uinta Mountains, Utah [Ph.D. thesis]: Berkeley, University of California, 266 p.

Palmquist, R. C., 1983, Terrace chronologies in the Bighorn Basin, Wyoming: Wyoming Geological Association, 34th Annual Field Conference, Guidebook, p. 217–231.

Pankratz, L. W., and Ackerman, H. D., 1982, Structure along the northwest edge of the Snake River Plain interpreted from seismic reflection: Journal of Geophysical Research, v. 87, p. 2676–2682.

Pardee, J. T., 1950, Late Cenozoic block faulting in western Montana: Geological Society of America Bulletin, v. 61, p. 359–406.

Parsons, T., and Thompson, G. A., 1991, The role of magma overpressure in suppressing earthquakes and topography: Worldwide examples: Science, v. 253, p. 1399–1402.

Perman, R. C., Swan, F. H., and Kelson, K. I., 1988, Assessment of late Quaternary faulting along the south Granite Mountains and north Granite Mountains faults of Central Wyoming: Geological Society of America Abstracts with Programs, v. 20, p. 462.

Personius, S. F., 1982, Geologic setting and geomorphic analysis of Quaternary fault scarps along the Deep Creek fault, upper Yellowstone valley, south-central Montana [M.S. thesis]: Bozeman, Montana State University, 77 p.

Pierce, K. L., 1974, Surficial geologic map of the Abiathar Peak and parts of adjacent quadrangles, Yellowstone National Park, Wyoming and Montana: U.S. Geological Survey Map I-646, scale 1:62,500.

—— , 1979, History and dynamics of glaciation in the northern Yellowstone National Park area: U.S. Geological Survey Professional Paper 729F, 90 p.

—— , 1985, Quaternary history of faulting on the Arco segment of the Lost River fault, central Idaho, *in* Stein, R. S., and Bucknam, R. C., eds., Proceedings, Workshop XXVIII on the Borah Peak, Idaho, Earthquake: U.S. Geological Survey Open-File Report 85-290, p. 195–206.

Pierce, K. L., and Morgan, L. A., 1990, The track of the Yellowstone hotspot:

Volcanism, faulting, and uplift: U.S. Geological Survey Open-File Report 90-415, 49 p.
Pierce, K. L., and Scott, W. E., 1982, Pleistocene episodes of alluvial-gravel deposition, southeastern Idaho, *in* Bonnichsen, B., and Breckenridge, R. M., eds., Cenozoic geology of Idaho: Idaho Bureau of Mines and Geology Bulletin 26, p. 658–702.
—— , 1986, Migration of faulting along and outward from the track of thermo-volcanic activity in the eastern Snake River Plain region during the last 15 m.y.: EOS Transactions of the American Geophysical Union, v. 67, p. 1225.
Pierce, K. L., Obradovich, J. D., and Friedman, I., 1976, Obsidian hydration dating and correlation of Bull Lake and Pinedale glaciations near West Yellowstone, Montana: Geological Society of America Bulletin, v. 87, p. 703–710.
Pierce, K. L., Fosberg, M. A., Scott, W. E., Lewis, G. C., and Colman, S. M., 1982, Loess deposits of southeastern Idaho: Age and correlation of the upper two loess units, *in* Bonnichsen, B., and Breckenridge, R. M., eds., Cenozoic geology of Idaho: Idaho Bureau of Mines and Geology Bulletin 26, p. 658–702.
Pierce, K. L., Scott, W. E., and Morgan, L. A., 1988, Eastern Snake River Plain neotectonics—Faulting in the last 15 Ma migrates along and outward from the Yellowstone "hot spot" track: Geological Society of America Abstracts with Programs, v. 20, p. 463.
Pierce, K. L., Adams, K. D., and Sturchio, N. C., 1991, Geologic setting of the Corwin Springs Known Geothermal Resources Area–Mammoth Hot Springs area in and adjacent to Yellowstone National Park, *in* Sorey, M. L., ed., Effects of potential geothermal development in the Corwin Springs Known Geothermal Resources Area, Montana, on the thermal features of Yellowstone National Park: U.S. Geological Survey, Water-Resources Investigation Report 91-4052, p. C-1–C-37.
Pierce, W. G., 1965, Geologic map of the Deep Lake quadrangle, Park County, Wyoming: U.S. Geological Survey Geologic Quadrangle Map GQ-478, scale 1:62,500.
Piety, L. A., Wood, C. K., Gilbert, J. D., Sullivan, J. T., and Anders, M. H., 1986, Seismotectonic study for Palisades Dam and Reservoir, Palisades Project: U.S. Bureau of Reclamation Seismotectonic Report 86-3, 198 p.
Pings, J. C., and Locke, W. W., 1988, A fault scarp across the Yellowstone caldera margin—Its morphology and implications: Geological Society of America Abstracts with Programs, v. 20, p. 463.
Pollastro, R. M., and Barker, C. E., 1986, Application of clay mineral, vitrinite reflectance, and fluid inclusion studies to the thermal and burial history of the Pinedale anticline, Green River Basin, Wyoming, *in* Gautier, D. L., ed., Roles of organic matter in sediment diagenesis: Society of Economic Paleontologists and Mineralogist Special Publication 38, p. 73–83.
Pollitz, F. F., 1988, Episodic North American and Pacific plate motions: Tectonics, v. 7, p. 711–726.
Porter, S. C., Pierce, K. L., and Hamilton, T. D., 1983, Late Pleistocene glaciation in the western United States, *in* Wright, H. E., ed., Late Quaternary environments of the United States: Minneapolis, University of Minnesota Press, p. 71–111.
Prostka, H. J., and Embree, G. F., 1978, Geology and geothermal resources of the Rexburg area, eastern Idaho: U.S. Geological Survey Open-File Report 78-1009, 15 p.
Raisz, E., 1957, Landforms of the United States: Jamaica Plains, Massachusetts, Raisz Landform Map (available from Kate Raisz, Raisz Landform Map, P.O. Box 2254, Jamaica Plains, MA, 02130), scale about 1:4,500,000.
Reheis, M. C., 1985, Evidence for Quaternary tectonism in the northern Bighorn Basin, Wyoming and Montana: Geology, v. 13, p. 364–367.
—— , 1987, Soils in granitic alluvium in humid and semiarid climates along Rock Creek, Carbon County, Montana: U.S. Geological Survey Bulletin 1590D, 71 p.
Reheis, M. C., and Agard, S. S., 1984, Timing of stream captures in the Big Horn Basin, WY and MT, determined from ash-dated gravels: Geological Society of America Abstracts with Programs, v. 16, p. 632.
Reheis, M. C., and 7 others, 1991, Quaternary history of some southern and central Rocky Mountain basins, *in* Morrison, R. B., ed., Quaternary nonglacial geology: Conterminous U.S.: Boulder, Colorado, Geological Society of America, The Geology of North America, v. K-2, p. 407–440.
Reilinger, R. E., 1985, Vertical movements associated with the 1959, M = 7.1 Hebgen Lake Montana earthquake, *in* Stein, R. S., and Bucknam, R. C., eds., Proceedings, Workshop XXVIII on the Borah Peak, Idaho, Earthquake: U.S. Geological Survey Open-File Report 85-290, p. 519–530.
Reilinger, R. E., Citron, G. P., and Brown, L. D., 1977, Recent vertical crustal movements from precise leveling data in southwestern Montana, western Yellowstone National Park and the Snake River Plain: Journal of Geophysical Research, v. 82, p. 5349–5359.
Reynolds, M. W., 1979, Character and extent of basin-range faulting, western Montana and east-central Idaho, *in* Newman, G. W., and Goode, H. D., eds., 1979 Basin and Range Symposium, Rocky Mountain Association of Geologists and Utah Geological Association, Guidebook, p. 185–193.
Richards, M. A., Hager, B. H., and Sleep, N. H., 1988, Dynamically supported geoid highs over hot spots: Observation and theory: Journal of Geophysical Research, v. 93, p. 7690–7708.
Richards, M. A., Duncan, R. A., and Courtillot, V. E., 1989, Flood basalts and hot spot tracks: Plume heads and tails: Science, v. 246, p. 103–107.
Richards, P. W., and Rogers, C. P., 1951, Geology of the Hardin area, Big Horn and Yellowstone Counties, Montana: U.S. Geological Survey Oil and Gas Investigations Map OM-111, scale 1:62,500.
Richmond, G. M., 1972, Appraisal of the future climate of the Holocene in the Rocky Mountains: Quaternary Research, v. 2, p. 315–322.
—— , 1973, Geologic map of the Fremont Lake South quadrangle, Sublette County, Wyoming: U.S. Geological Survey Geologic Quadrangle Map GQ-1138, scale 1:24,000.
—— , 1974, Surficial geology of the Frank Island quadrangle, Yellowstone National Park, Wyoming: U.S. Geological Survey Miscellaneous Geologic Investigations Map I-652, scale 1:62,500.
—— , 1976, Pleistocene stratigraphy and chronology in the mountains of western Wyoming, *in* Mahaney, W. C., ed., Quaternary stratigraphy of North America: Stroudsberg, Pennsylvania, Dowden, Hutchinson, and Ross, p. 353–379.
—— , 1983, Modification of glacial sequence along Big Sandy River, southern Wind River Range, Wyoming: Geological Society of America Abstracts with Programs, v. 15, p. 431.
—— , 1986, Stratigraphy and correlation of glacial deposits of the Rocky Mountains, the Colorado Plateau and the ranges of the Great Basin: Quaternary Science Reviews, v. 5, p. 99–127.
Richmond, G. M., and Murphy, J. F., 1965, Geologic map of the Bull Lake East quadrangle, Fremont County, Wyoming: U.S. Geological Survey Geologic Quadrangle Map GQ-431, scale 1:24,000.
Ringwood, A. E., 1982, Phase transformations and differentiation in subducted lithosphere: Implications for mantle dynamics, basalt petrogenesis, and crustal evolution: Journal of Geology, v. 90, p. 611–643.
Ritter, D. F., 1967, Terrace development along the front of the Beartooth Mountains, southern Montana: Geological Society of America Bulletin, v. 78, p. 467–484.
Ritter, D. F., and Kauffman, M. E., 1983, Terrace development in the Shoshone River Valley near Powell, Wyoming and speculations concerning the sub-Powell terrace: Wyoming Geological Association, 34th Annual Field Conference, Guidebook, p. 197–203.
Rodgers, D. W., and Zentner, N. C., 1988, Fault geometries along the northern margin of the eastern Snake River Plain, Idaho: Geological Society of America Abstracts with Programs, v. 20, p. 465.
Rodgers, D. W., Hackett, W. R., and Ore, H. T., 1990, Extension of the Yellowstone Plateau, eastern Snake River Plain, and Owyhee Plateau: Geology, v. 18, p. 1138–1141.
Roy, W. R., and Hall, R. D., 1980, Re-evaluation of the Bull Lake glaciation through re-study of the type area and studies of other localities: Geological Society of America Abstracts with Programs, v. 12, p. 302.
Rubey, W. W., 1973, Geologic map of the Afton quadrangle and part of the Big Piney quadrangle, Lincoln and Sublette Counties, Wyoming: U.S. Geologi-

cal Survey Miscellaneous Geologic Investigations Map I-686, scale 1:62,500.

Rubey, W. W., Oriel, S. S., and Tracey, J. I., Jr., 1975, Geology of the Sage and Kemmerer 15-minute quadrangles, Lincoln County, Wyoming: U.S. Geological Survey Professional Paper 855, 18 p.

Ruppel, E. T., 1967, Late Cenozoic drainage reversal, east-central Idaho, and its relation to possible undiscovered placer deposits: Economic Geology, v. 62, p. 648–663.

Rytuba, J. J., 1989, Volcanism, extensional tectonics, and epithermal mineralization in the northern Basin and Range Province, California, Nevada, Oregon, and Idaho: U.S. Geological Survey Circular 1035, p. 59–61.

Rytuba, J. J. and McKee, E. H., 1984, Peralkaline ash flow tuffs and calderas of the McDermitt volcanic field, southeast Oregon and north central Nevada: Journal of Geophysical Research, v. 89, p. 8616–8628.

Sandberg, C. A., and Mapel, W. J., 1967, Devonian of the northern Rocky Mountains and Plains, *in* Oswald, D. H., ed., International symposium on the Devonian System, Calgary, Alberta, September 1967: Calgary, Alberta Society of Petroleum Geologists, v. 1, p. 843–877.

Sandberg, C. A., and Poole, F. G., 1977, Conodont biostratigraphy and depositional complexes of Upper Devonian cratonic-platform and continental-shelf rocks, *in* Murphy, M. A., Berry, W.B.N., and Sandberg, C. A., eds., Western North America; Devonian: Riverside, California University Campus Museum Contribution 4, p. 144–182.

Schmidt, D. L., and Mackin, J. H., 1970, Quaternary geology of Long and Bear Valleys, west-central Idaho: U.S. Geological Survey Bulletin 1311-A, 66 p.

Schroeder, N. L., 1974, Geologic map of the Camp Davis quadrangle, Teton County, Wyoming: U.S. Geological Survey Geologic Quadrangle Map GQ-1160, scale 1:24,000.

Scott, W. E., 1982, Surficial geologic map of the eastern Snake River Plain and adjacent areas, 111° to 115°W, Idaho and Wyoming: U.S. Geological Survey Miscellaneous Investigations Map I-1372, scale 1:250,000.

Scott, W. E., Pierce, K. L., and Hait, M. H., Jr., 1985a, Quaternary tectonic setting of the 1983 Borah Peak earthquake, central Idaho, *in* Stein, R. S., and Bucknam, R. C., eds., Proceedings, Workshop XXVIII on the Borah Peak, Idaho, Earthquake: U.S. Geological Survey Open-File Report 85-290, p. 1–16.

——, 1985b, Quaternary tectonic setting of the 1983 Borah Peak earthquake, central Idaho: Bulletin of the Seismological Society of America, v. 75, p. 1053–1066.

Sears, J. W., Hyndman, D. W., and Alt, D., 1990, The Snake River Plain, a volcanic hotspot track: Geological Society of America Abstracts with Programs, v. 22, p. 82.

Shackleton, N. J., 1987, Oxygen isotopes, ice volume, and sea level: Quaternary Science Reviews, v. 6, p. 183–190.

Skilbeck, J. N., and Whitehead, J. A., Jr., 1978, Formation of discrete islands in linear island chains: Nature, v. 272, p. 499–501.

Skipp, B. A., 1984, Geologic map and cross-sections of the Italian Peak and Italian Peak Middle Roadless Areas, Beaverhead County, Montana, and Clark and Lemhi Counties, Idaho: U.S. Geological Survey Field Studies Map MF-1601-B, scale 1:250,000.

——, 1988, Cordilleran thrust belt and faulted foreland in the Beaverhead Mountains, Idaho and Montana, *in* Schmidt, C. J., and Perry, W. J., Jr., eds., Interaction of the Rocky Mountain Foreland and the Cordilleran Thrust Belt: Geological Society of America Memoir 171, p. 237–266.

Skipp, B. A., Sando, W. J., and Hall, W. E., 1979, Mississippian and Pennsylvanian (Carboniferous) systems in the United States—Idaho: U.S. Geological Survey Professional Paper 1110, Chap. AA, p. AA1–AA42.

Sleep, N. H., 1987, An analytical model for a mantle plume fed by a boundary layer: Geophysical Journal of the Royal Astronomical Society, v. 90, p. 119–128.

——, 1990, Hotspots and mantle plumes: Some phenomenology: Journal of Geophysical Research, v. 95, p. 6715–6736.

Smedes, H. W., M'Gonigle, M. W., and Prostka, H. J., 1989, Geologic map of the Two Ocean Pass quadrangle, Yellowstone National Park and vicinity, Wyoming: U.S. Geological Survey Geologic Quadrangle Map GQ-1667, scale 1:62,500.

Smith, R. B., 1978, Seismicity, crustal structure, and intraplate tectonics of the interior of the western Cordillera, *in* Smith, R. B., and Eaton, G. P., eds., Cenozoic tectonics and regional geophysics of the western Cordillera: Geological Society of America Memoir 152, p. 111–144.

Smith, R. B., and Arabasz, W. J., 1991, Seismicity of the Intermountain Seismic Belt, *in* Slemmons, D. B., Engdahl, E. R., Zoback, M. D., and Blackwell, D. D., eds., Neotectonics of North America: Boulder, Colorado, Geological Society of America, The Geology of North America, v. DMV, p. 185–228.

Smith, R. B., and Sbar, N. L., 1974, Contemporary tectonics and seismicity of the western United States with emphasis on the Intermountain Seismic Belt: Geological Society of America Bulletin, v. 85, p. 1205–1218.

Smith, R. B., Shuey, R. T., Freidline, R. O., Otis, R. M., and Alley, L. B., 1974, Yellowstone hot spot: New magnetic and seismic evidence: Geology, v. 2, p. 451–455.

Smith, R. B., Shuey, R. T., Pelton, J. P., and Bailey, J. P., 1977, Yellowstone hot spot: Contemporary tectonics and crustal properties from earthquake and aeromagnetic data: Journal of Geophysical Research, v. 82, p. 3665–3676.

Smith, R. B., Richins, W. D., and Doser, D. I., 1985, The 1983 Borah Peak, Idaho, earthquake: Regional seismicity, kinematics of faulting, and tectonic mechanism, *in* Stein, R. S., and Bucknam, R. C., eds., Proceedings, Workshop XXVIII on the Borah Peak, Idaho, Earthquake: U.S. Geological Survey Open-File Report 85-290, p. 236–263.

Smith, R. B., Byrd, J.O.D., Sylvester, A. G., and Susong, D. L., 1990, Neotectonics and earthquake hazards of the Teton fault: Geological Society of America Abstracts with Programs, v. 22, p. 45.

Smith, R. L., and Luedke, R. G., 1984, Potentially active volcanic lineaments and loci in western conterminous United States, *in* Boyd, F., ed., Explosive volcanism: Inception, evolution, and hazards: Washington, D.C., National Academy of Sciences, p. 47–66.

Smith, R. P., Hackett, W. R., and Rogers, D. W., 1989, Surface deformation along the Arco rift zone, eastern Snake River Plain: Geological Society of America Abstracts with Programs, v. 21, p. 146.

Sonderegger, J. L., Schofield, J. D., Berg, R. B., and Mannich, N. L., 1982, The upper Centennial valley, Beaverhead and Madison Counties, Montana: Montana Bureau of Mines and Geology Memoir 50, 54 p.

Sparlin, M. A., Braile, L. W., and Smith, R. B., 1982, Crustal structure of the eastern Snake River Plain determined from ray-trace modeling of seismic refraction data: Journal of Geophysical Research, v. 87, p. 2619–2633.

Stanford, L. R., 1982, Glacial geology of the upper South Fork Payette River, south central Idaho [M.S. thesis]: Moscow, University of Idaho, 83 p.

Stickney, M. C., and Bartholomew, M. J., 1987a, Seismicity and late Quaternary faulting of the northern Basin and Range province, Montana and Idaho: Bulletin of the Seismological Society of America, v. 77, p. 1602–1625.

——, 1987b, Preliminary map of late Quaternary faults in western Montana: Montana Bureau of Mines and Geology Open-File Report 186, scale 1:500,000.

Suppe, J., 1985, Principles of structural geology: Englewood Cliffs, New Jersey, Prentice-Hall, 537 p.

Suppe, J., Powell, C., and Berry, R., 1975, Regional topography, seismicity, Quaternary volcanism, and the present-day tectonics of the western United States: American Journal of Science, v. 275-A, p. 397–436.

Susong, D. L., Smith, R. B., and Bruhn, R. J., 1987, Quaternary faulting and segmentation of the Teton fault zone, Grand Teton National Park, Wyoming: Report submitted to the University of Wyoming–National Park Service Research Center, 13 p.

Ten Brink, N. W., 1968, Pleistocene geology of the Stillwater drainage and Beartooth Mountains near Nye, Montana [M.S. thesis]: Lancaster, Pennsylvania, Franklin and Marshall College, 172 p.

Tolan, T. L., Reidel, S. P., Beeson, M. H., Anderson, J. L., Fecht, K. R., and Swanson, D. A., 1989, Revisions of the extent and volume of the Columbia River Basalt Group, *in* Reidel, S. P., and Hooper, S. P., eds., Volcanism and tectonism in the Columbia River Flood-Basalt Province: Geological Society of America Special Paper 239, p. 1–20.

Tredoux, M., De Wit, M. J., Hart, R. J., Armstrong, R. A., Lindsay, N. M., and Sellschop, J.P.F., 1989, Platinum group elements in a 3.5 Ga nickel-iron occurrence: Possible evidence of a deep mantle origin: Journal of Geophysical Research, v. 94, p. 795–813.

Trimble, D. E., and Carr, W. J., 1976, Geology of the Rockland and Arbon quadrangles, Power County, Idaho: U.S. Geological Survey Bulletin 1399, 115 p.

U.S. Geological Survey, 1972a, Geologic map of Yellowstone National Park: U.S. Geological Survey Miscellaneous Geologic Investigations Map I-711, scale 1:125,000.

—— , 1972b, Surficial geologic map of Yellowstone National Park: U.S. Geological Survey Miscellaneous Geologic Investigations Map I-710, scale 1:125,000.

Vallier, T. L., Brooks, H. C., and Thayer, T. P., 1977, Paleozoic rocks of eastern Oregon and western Idaho, *in* Stewart, J. H., Stevens, C. H., and Fritsche, A. E., eds., Paleozoic paleogeography of the western United States: Los Angeles, California, Society of Economic Paleontologists and Mineralogists, Pacific section, Pacific Coast Paleogeography Symposium 1, p. 455–466.

Vander Meulen, D. B., 1989, Intracaldera tuffs and central-vent intrusion of the Mahogany Mountain caldera, eastern Oregon: U.S. Geological Survey Open-File Report 89-77, 69 p.

Walker, G. W., and Nolf, B., 1981, High lava plains, Brothers fault zone to Harney Basin, Oregon, *in* Johnston, D. A., and Donnelly-Nolan, J., eds., Guides to some volcanic terranes in Washington, Idaho, Oregon, and northern California: U.S. Geological Survey Circular 838, p. 105–111.

Walsh, T. H., 1975, Glaciation of the Taylor Fork Basin area, Madison Range, southwestern Montana [Ph.D. thesis]: Moscow, University of Idaho, 193 p.

Weinstein, S. A., and Olson, P. L., 1989, The proximity of hotspots to convergent and divergent plate boundaries: Geophysical Research Letters, v. 16, p. 433–436.

Wernicke, B. P., Christiansen, R. L., England, P. C., and Sonder, L. J., 1987, Tectonomagmatic evolution of Cenozoic extensions in the North American Cordillera, *in* Coward, M. P., Dewey, J. F., and Hancock, P. L., eds., Continental extensional tectonics, Geological Society of America Special Publication 28, p. 203–221.

West, M. W., 1986, Quaternary extensional reactivation of the Darby and Absaroka thrusts, southwestern Wyoming and north central Utah: Geological Society of America Abstracts with Programs, v. 18, p. 422.

Westaway, R., 1989a, Northeast Basin and Range Province active tectonics: An alternate view: Geology, v. 17, p. 779–783.

—— , 1989b, Deformation of the IVE Basin and Range Province: The response of the lithosphere to the Yellowstone plume?: Geophysical Journal International, v. 99, p. 33–62.

White, R. S., and McKenzie, D. P., 1989, Magmatism at rift zones: The generation of volcanic continental margins and flood basalts: Journal of Geophysical Research, v. 94, p. 7685–7729.

White, R. S., Spence, G. D., Fowler, S. R., McKenzie, D. P., Westbrook, G. K., and Bowen, A. D., 1987, Magmatism at rifted continental margins: Nature, v. 330, p. 439–444.

Whitehead, J. A., and Luther, D. S., 1975, Dynamics of laboratory diapir and plume models: Journal of Geophysical Research, v. 80, p. 705–717.

Williams, E. J., and Embree, G. F., 1980, Pleistocene movement on Rexburg fault, eastern Idaho: Geological Society of America Abstracts with Programs, v. 12, p. 308.

Williams, P. L., Covington, H. R., and Pierce, K. L., 1982, Cenozoic stratigraphy and tectonic evolution of the Raft River Basin, Idaho, *in* Bonnichsen, B., and Breckenridge, R. M., eds., Cenozoic geology of Idaho, Idaho Bureau of Mines and Geology Bulletin 26, p. 491–504.

Williams, P. L., Mytton, J. W., and Covington, H. R., 1990, Geologic map of the Stricker 1 quadrangle, Cassia, Twin Falls, and Jerome Counties, Idaho: U.S. Geological Survey Miscellaneous Investigations Series Map I-2078, scale 1:24,000.

Wilson, J. T., 1963, Continental drift: Scientific American, v. 208, p. 86–100.

—— , 1990, On the building and classification of mountains: Journal of Geophysical Research, v. 95, p. 6611–6628.

Witkind, I. J., 1964, Reactivated faults north of Hebgen Lake: U.S. Geological Survey Professional Paper 435-G, p. 37–50.

—— , 1975a, Geology of a strip along the Centennial fault, southwestern Montana and adjacent Idaho: U.S. Geological Survey Miscellaneous Investigations Series Map I-890, scale 1:62,500.

—— , 1975b, Preliminary map showing known and suspected active faults in western Montana: U.S. Geological Survey Open-File Report 75-285, scale 1:500,000.

—— , 1975c, Preliminary map showing known and suspected active faults in Idaho: U.S. Geological Survey Open-File Report 75-278, scale 1:500,000.

—— , 1975d, Preliminary map showing known and suspected active faults in Wyoming: U.S. Geological Survey Open-File Report 75-279, scale 1:500,000.

Wood, B. J., 1989, Mineralogical phase change at the 670-km discontinuity: Nature, v. 341, p. 278.

Wood, C., 1988, Earthquake data—1986: Jackson Lake Seismographic Network, Jackson Lake Dam, Minidoka Project, Wyoming: U.S. Bureau of Reclamation Seismotectonic Report 88-1, 33 p.

Wood, S. H., 1984, Review of late Cenozoic tectonics, volcanism, and subsurface geology of the western Snake River Plain, Idaho, *in* Beaver, P. C., ed., Geology, tectonics, and mineral resources of western and southern Idaho: Dillon, Montana, Tobacco Root Geological Society, p. 48–60.

—— , 1989a, Silicic volcanic rocks and structure of the western Mount Bennett Hills and adjacent Snake River Plain, Idaho, *in* Smith, R. P., and Downs, W. F., eds., International Geological Congress, 28th Field Trip, Guidebook T305: Washington, D.C., American Geophysical Union, p. 69–77.

—— , 1989b, Quaternary basalt vent-and-fissure alignments, principal stress directions, and extension in the Snake River Plain, Idaho: Geological Society of America Abstracts with Programs, v. 21, p. 161.

Wood, S. H., and Gardner, J. N., 1984, Silicic volcanic rocks of the Miocene Idavada Group, Bennett Mountains, southwestern Idaho: n.p., Final Contract Report to the Los Alamos National Laboratory from Boise State University, 55 p.

Woodward, N. B., 1988, Primary and secondary basement controls on thrust sheet geometries, *in* Schmidt, C. J., and Perry, W. J., Jr., eds., Interaction of the Rocky Mountain Foreland and the Cordilleran Thrust Belt: Geological Society of America Memoir 171, p. 353–366.

Wyllie, P. J., 1988, Solidus curves, mantle plumes, and magma generation beneath Hawaii: Journal of Geophysical Research, v. 93, p. 4171–4181.

Zoback, M. L., and Thompson, G. A., 1978, Basin and Range rifting in northern Nevada: Clues from a mid-Miocene rift and its subsequent offsets: Geology, v. 6, p. 111–116.

Zoback, M. L., and Zoback, M., 1980, State of stress in the conterminous United States: Journal of Geophysical Research, v. 85, p. 6113–6156.

MANUSCRIPT ACCEPTED BY THE SOCIETY JULY 19, 1991

Printed in U.S.A.

Geological Society of America
Memoir 179
1992

Chapter 2

Structural evolution of piggyback basins in the Wyoming-Idaho-Utah thrust belt

James C. Coogan*
Department of Geology and Geophysics, University of Wyoming, Laramie, Wyoming 82071-3006

ABSTRACT

The depositional histories of three piggyback basins are related to the surface and subsurface thrust geometries that controlled early Eocene basin development during the final phase of shortening in the Idaho-Wyoming-Utah thrust belt. Early Eocene strata of Bear Lake Plateau basin, Fossil Basin, and the LaBarge basin occupy similar structural positions on the Absaroka, Crawford, and Hogsback thrust sheets respectively. The basin depocenters formed above flat décollement surfaces of underlying thrust sheets. Trailing basin margins formed above the trailing footwall ramps of the thrust sheets that immediately underlie the basins. Leading basin margins formed along the leading edges of thrust sheets that are underlain by the ramps of structurally lower thrusts. Sedimentary facies distributions and sediment dispersal patterns were controlled by uplift of the basin margins rather than by subsidence of depocenters. Periods of basin margin uplift are identified by the presence of alluvial fan sediments shed from the uplifted margins, patterns of onlap and overstep along and across the margins, and cross-cutting structural relations of folded or faulted early Eocene strata along the basin margins. Uplift was accomplished through reactivation of slip along the trailing footwall ramps of the Crawford and Absaroka thrusts, late slip on the Hogsback trailing ramp, and slip on the LaBarge and Calpet thrusts in the Hogsback footwall. The late thrust slip in the interior of the thrust belt was a type of break back, or out-of-sequence thrust slip, that deviated from the previous foreland-directed sequence of thrusting.

Reactivated thrust uplift and footwall uplift are the two uplift processes common to the three basins studied in this report. Reactivated thrust uplift was generally confined to the trailing ramp areas of thrust sheets, where slip along the trailing ramp produced ramp-rooted imbricate thrusts and fault-propagation folds. Footwall uplift was confined to the leading basin margins, where the basin margins and the leading edges of the thrust sheets beneath them were translated up the ramps of structurally lower thrusts. The reactivation of thrusts beneath piggyback basins across the width of the Wyoming-Idaho-Utah thrust belt is attributed to impeded slip along the frontal thrusts of the belt. Slip impedance along the thrust front is attributed to the interaction of four processes: buttressing of the frontal thrusts by coeval foreland basement uplifts, fault deflection above preexisting foreland basement warps, decrease in regional wedge taper from the combined effects of foreland subsidence and thrust belt erosion, and rheologic changes related to stratigraphic changes toward the foreland.

---

*Present address: Mobil Exploration and Producing U.S., Denver, Colorado 80217-5444.

Coogan, J. C., 1992, Structural evolution of piggyback basins in the Wyoming-Idaho-Utah thrust belt, *in* Link, P. K., Kuntz, M. A., and Platt, L. B., eds., Regional Geology of Eastern Idaho and Western Wyoming: Geological Society of America Memoir 179.

INTRODUCTION

The southern half of the Wyoming-Idaho-Utah thrust belt is covered by early Eocene alluvial, fluvial, and lacustrine strata of the Wasatch and Green River formations that were deposited in a series of basins separated by discrete thrust-related structures during the final phase of regional thrust shortening. The basins were transported eastward throughout their development in a "piggyback" manner above the regional sole décollement. These piggyback basins (Ori and Friend, 1984) overlie individual thrust sheets that were emplaced and deeply eroded prior to basin development. The basin margins were locally sites of uplift above the trailing ramps and leading edges of thrust sheets. The early Eocene piggyback basins record a period of slip reactivation on long dormant faults in the interior of the belt that marked a change from the previous sequence of foreland-directed thrusting. The internal reactivated fault segments formed break-back thrusts behind the regional thrust front.

The purposes of this chapter are to (1) illustrate the thrust geometries and uplift processes that localized piggyback basin margins, (2) show the sedimentologic responses of individual basins to the growth of basin margin structures, and (3) provide mechanical explanations for the change in the style and sequence of thrusting that was responsible for the development of early Eocene piggyback basins.

The interpretation of break-back thrusting is a recent development in studies of the Wyoming-Idaho-Utah thrust belt. Regional studies (Armstrong and Oriel, 1965; Royse and others, 1975; Wiltschko and Dorr, 1983) present a clear record of the overall foreland-younging thrust sequence through the belt. The clearest example of a break-back thrust sequence was presented by Royse (1985) and Hunter (1988) for the Prospect thrust system in the northern Wyoming thrust belt where cross-cutting relationships and tectogenic sediments constrain thrust timing. Similar interpretations have been presented for the northern Darby (Dixon, 1982) and Absaroka (McBride and Dolberg, 1990) thrusts based on structural style. However, the lack of thrust-related sediments to date deformation permits alternative interpretations of the Darby (Jones, 1984) and Absaroka (Lageson, 1984; Woodward, 1986) thrust sequences. The piggyback basins of the central and southern thrust belt contain tectogenic sediments that are related to the subsurface structures defined by well and seismic data. The area is ideally suited for dating late thrust events, for defining the structural style of break-back thrusts, and for delineating the mechanical conditions responsible for the change in the regional thrust sequence during the last phase of thrust belt shortening.

Three basins are examined in this chapter: Bear Lake Plateau basin, which overlies the Crawford thrust sheet; Fossil Basin, which overlies the Absaroka thrust sheet; and the LaBarge basin, which overlies the Hogsback thrust sheet (Fig. 1). Both the Fossil Basin (Oriel and Tracey, 1970; Lamerson, 1982; Hurst and Steidtmann, 1986) and the LaBarge basin (Oriel, 1962, 1969; Blackstone, 1979; Dorr and Gingrich, 1980) have been studied in the past. This chapter provides the first detailed description of the structure, stratigraphy, and sedimentology of Bear Lake Plateau basin.

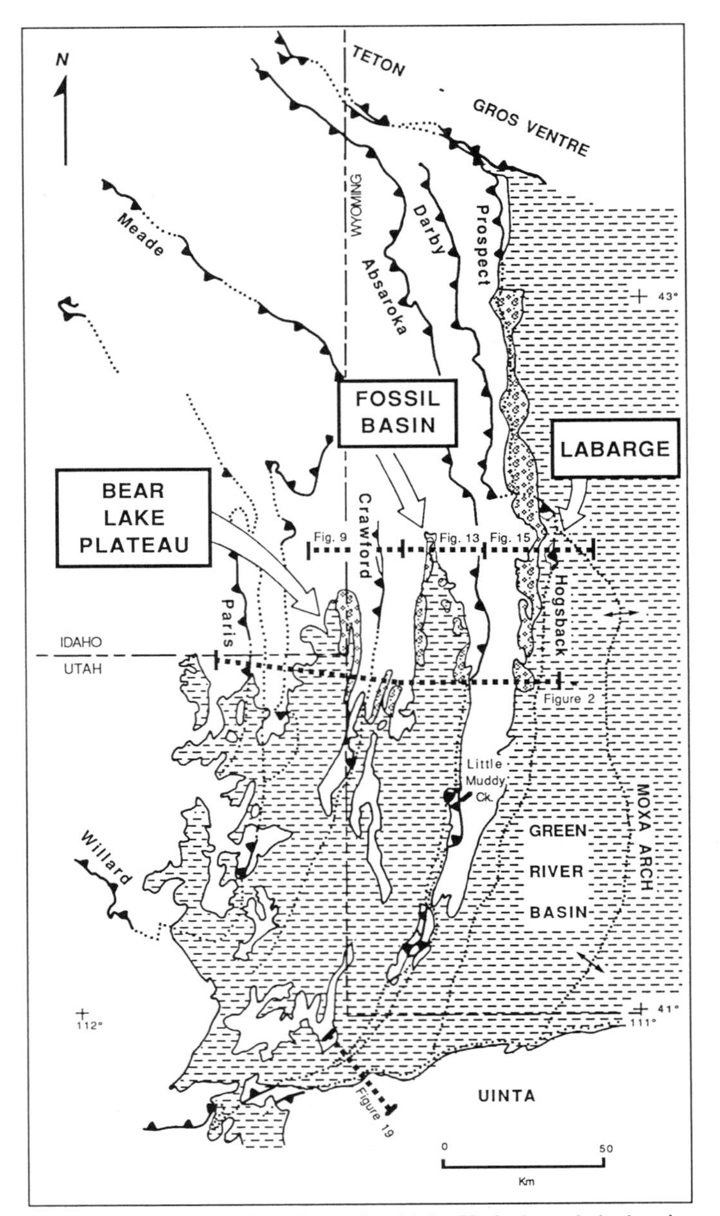

Figure 1. Index map of the Wyoming-Idaho-Utah thrust belt showing the location of principal thrust faults and foreland uplifts and their positions relative to the three early Eocene basins discussed here. The dash pattern represents fine-grained strata of the Wasatch and Green River formations. The stipple pattern represents diamictites of the Wasatch Formation (modified and generalized from Oriel and Platt, 1980; and Lamerson, 1982). Heavy dashed lines indicate cross sections shown in Figures 2, 9, 13, 15, and 19. Thrust faults, shown with solid lines where exposed and dotted lines where buried, have sawteeth on the hanging wall.

The method of investigation is to: (1) briefly describe the common structural elements of the piggyback basins and their respective thrust sheets; (2) provide geometric models to explain the structural similarities between basins; (3) relate the geometric models to the sedimentologic and structural features of Bear Lake Plateau basin, northern Fossil Basin, and the LaBarge basin; and finally (4) provide an explanation for the late deformation that caused segmentation of the top of the regional thrust wedge during the last phase of shortening across the Wyoming-Idaho-Utah thrust belt.

STRUCTURAL SETTING OF PIGGYBACK BASINS

Figure 2 illustrates several similarities between different piggyback basins. Angular unconformities between early Eocene basin sediments and Mesozoic and Paleozoic strata of underlying thrust sheets demonstrate that the basins formed well after the main-phase emplacement of their respective thrust sheets. Main-phase slip of the Crawford thrust was assigned a Coniacian-Santonian age based on the structural position and age of the Echo Canyon Conglomerate in the footwall of the southern Crawford thrust (Royse and others, 1975). This assignment is somewhat ambiguous because the precise southern trace of the Crawford thrust is concealed by Maastrichtian through early Eocene strata. Recently, DeCelles (1988) reported that the clasts in the Echo Canyon Conglomerate were probably derived from multiple sources, including the initial uplift of the basement-cored anticlinorium of the Wasatch Range carried above the Crawford thrust. By either interpretation, the Crawford thrust was emplaced by Coniacian-Santonian time. Main emplacement of the Absaroka thrust sheet is constrained as Santonian to Campanian in age by stratigraphic bracketing (Royse and others, 1975), with minor reactivation during the late Maastrichtian or early Paleocene (Lamerson, 1982). Final slip on the Hogsback thrust is stratigraphically bracketed to late Paleocene (Warner and Royse, 1987), with early Eocene footwall imbrication (Dorr and Gingrich, 1980). These timing constraints indicate that the early Eocene piggyback basins formed approximately 26 m.y. after emplacement of the Crawford thrust sheet and 16 m.y. after main-phase slip on the Absaroka thrust. The basins formed 6 m.y. after the final minor slip along the Absaroka thrust trace and immediately following slip along the Hobsback thrust trace. Piggyback basin sedimentation was concurrent with final slip on the Hogsback footwall imbricates, but it also postdates final slip on these faults.

The basins were also translated eastward during their development by slip on the underlying regional sole décollement. The basins were carried "piggyback" (Ori and Friend, 1984) above the underlying décollement during late slip on the Hogsback thrust and its footwall imbricates at the leading edge of the thrust belt.

The piggyback basins also exhibit similar patterns of deposition. The main basin depocenters lie above footwall décollements in the underlying thrust planes where thrust surfaces are parallel to bedding in low shear strength materials—such as the Jurassic salt décollement of the Crawford thrust and the Cretaceous shale décollements of the Absaroka and Hogsback thrusts. Trailing basin margins were the sites of uplift above trailing footwall ramps (position A on Figure 2), where the thrust surfaces cut upward through bedding in higher strength Paleozoic and Mesozoic units. Leading basin margins formed where the fronts of the thrust sheets were uplifted above trailing footwall ramps of structurally lower thrust sheets (position B on Figure 2).

Figure 2 also illustrates that basin deposition was not controlled by discrete subsidence of the depocenters as proposed by Oriel and Tracey (1970, p. 33) and Rubey and others (1975, p. 14) for Fossil Basin. The lack of differential subsidence between basin margins and centers is demonstrated by the homoclinal dip of the basement surface immediately beneath the basins and across the margins as constrained by seismic, magnetic, and deep well data that have become available since 1975. This study demonstrates that rather than being caused by local subsidence, these basins were formed by uplift of erosionally resistant strata along the two basin margins within the larger subsiding region of the eastern thrust belt and the Green River Basin.

GEOMETRIC MODELS FOR BASIN MARGIN UPLIFT

Basin margin uplift is controlled by two processes illustrated on Figure 3. The first process is reactivated slip on the major thrust that immediately underlies a piggyback basin. The second process is termed footwall uplift, where the leading basin margin and the leading edge of the thrust plate beneath it are translated up the trailing ramp of a structurally lower thrust. Footwall uplift previously has been referred to as *passive uplift* (Steidtmann and Schmitt, 1988) because it does not require slip across the thrust plane that bounds the thrust sheet beneath the tectogenic deposits. Schmitt and Steidtmann (1990) more recently used *interior ramp supported uplift* to describe the same phenomenon. Both terms described the case in which a mechanically locked thrust sheet in the interior of a thrust belt is uplifted above the trailing footwall ramp of a younger thrust. Because this chapter discusses uplift in the deeper, interior parts of the thrust belt as well as shallower level uplift along the leading edge of the belt, the general term *footwall uplift* is used to describe uplift of a mechanically locked thrust plane by slip on an underlying thrust.

Figure 3 presents a series of scaled geometric models for basins generated by both reactivated thrust slip and footwall uplift. The models are drawn with simple ramp-flat geometries that generalize the common structural features of the Crawford, Absaroka, and Hogsback thrust sheets. Figure 3a represents main-phase emplacement of a thrust sheet, which is shown in Figure 3b as eroded to a near-planar topographic surface prior to the development of any piggyback basin. Local topographic relief resulting from differential erosion could influence early subaerial deposition at the site of the later piggyback basin, particularly where steeply dipping, resistant strata are exposed above the trailing ramp and leading edge of the thrust sheet.

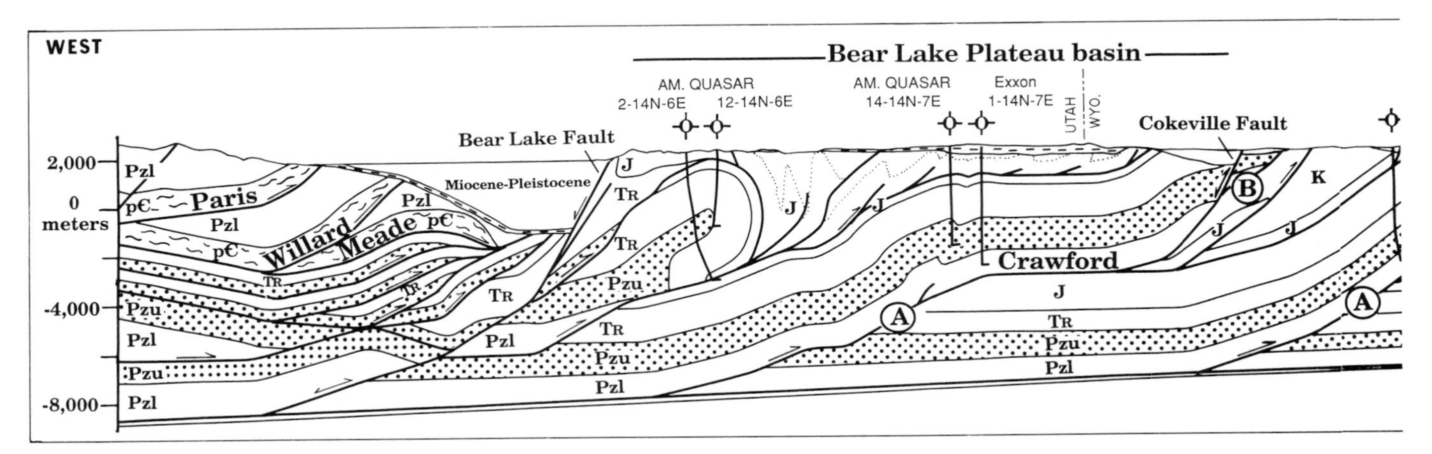

Figure 2. West-east cross section between the Paris and Hogsback thrust fronts showing the structural position of Bear Lake Plateau basin and Fossil Basin. See Figure 1 for location. Early Eocene basin fill is defined by dash pattern. The depocenters of Bear Lake Plateau basin and Fossil Basin overlie footwall décollements in the underlying Crawford and Absaroka thrust faults respectively. The trailing basin margins lie above the trailing footwall ramps (position A) of the underlying thrust sheets, and the leading basin margins lie above the leading edges of the underlying thrust sheets (position B). LaBarge basin is located 35 km (21 m) north of this line of section in the position of the western margin of the Green River Basin on the Hogsback thrust sheet. pЄ: Proterozoic, Pzl: early Paleozoic, Pzu: late Paleozoic, ₮: Triassic, J: Jurassic, K: Cretaceous rocks. Vertical scale = horizontal scale. The Absaroka and western Hogsback thrust sheets are modified from Lamerson (1982, Plate 6).

Figure 3c illustrates that reactivation of thrust slip on a previously locked thrust plane can affect either the entire thrust surface or just the trailing area of the thrust. In the first case, slip is transferred continuously from the sole décollement, up the trailing ramp, across the succeeding footwall décollement, and finally to the thrust front, where uplift is accommodated by either translation and rotation up a leading ramp or by imbrication at the thrust tip. Both the leading and trailing uplifts are source areas for clastic sediments, and the depression between them is the depocenter of the piggyback basin. Figure 3c also illustrates a second case, in which slip is confined to the trailing footwall ramp area. This slip produces a break-back thrust that bypasses the mechanically locked décollement of the thrust sheet.

Break-back thrusts and fault-propagation folds are typical late structures above trailing ramps. If such structures developed early in a foreland-directed sequence, then there should be locations along strike where the imbricate lies behind the ramp (Jones, 1984) or where it has been translated up onto the frontal décollement. Instead, these structures are consistently found immediately above the trailing ramp as a common structural motif in the Wyoming-Idaho-Utah thrust belt (Boyer, 1986, Fig. 5). Although kinematic arguments based on structural style can be ambiguous, tectogenic sediments permit independent dating of ramp-rooted structures along the trailing margins of piggyback basins.

Figure 3d illustrates a typical foreland-younging sequence of thrusting in which an early thrust sheet becomes mechanically locked, and further shortening is accommodated by propagation of a younger thrust beneath and in front of the locked thrust. The locked thrust becomes part of the hanging wall of the younger thrust and is translated toward the front of the belt by slip on the younger thrust plane. Figure 3e shows the older, locked thrust sheet translated with its footwall onto the trailing ramp of the younger thrust fault. The subsequent uplift and rotation of the older, locked thrust sheet is called footwall uplift with reference to the older, structurally higher, and mechanically locked thrust plane. Footwall uplift requires no slip on the older thrust plane. Footwall uplift can produce a leading basin margin in the hanging wall of the locked thrust, but footwall uplift does not produce the trailing basin margins seen in the Wyoming-Idaho-Utah thrust belt—where piggyback basins are bordered by uplifts on both basin margins.

Figure 3f illustrates that both reactivated thrust uplift and footwall uplift can act in unison to respectively create the trailing and leading margins of a piggyback basin. Both processes were operative in Bear Lake Plateau basin, Fossil Basin, and the LaBarge basin, although the relative timing of leading and trailing basin margin uplift differs for each basin.

Different reasons exist for the causes of uplift of trailing and leading basin margins for the simple ramp-décollement geometries illustrated in Figure 3. Uplift of the frontal basin margin can be the result of either thrust reactivation or footwall uplift, whereas uplift at the trailing basin margin provides unambiguous evidence for reactivated thrust slip.

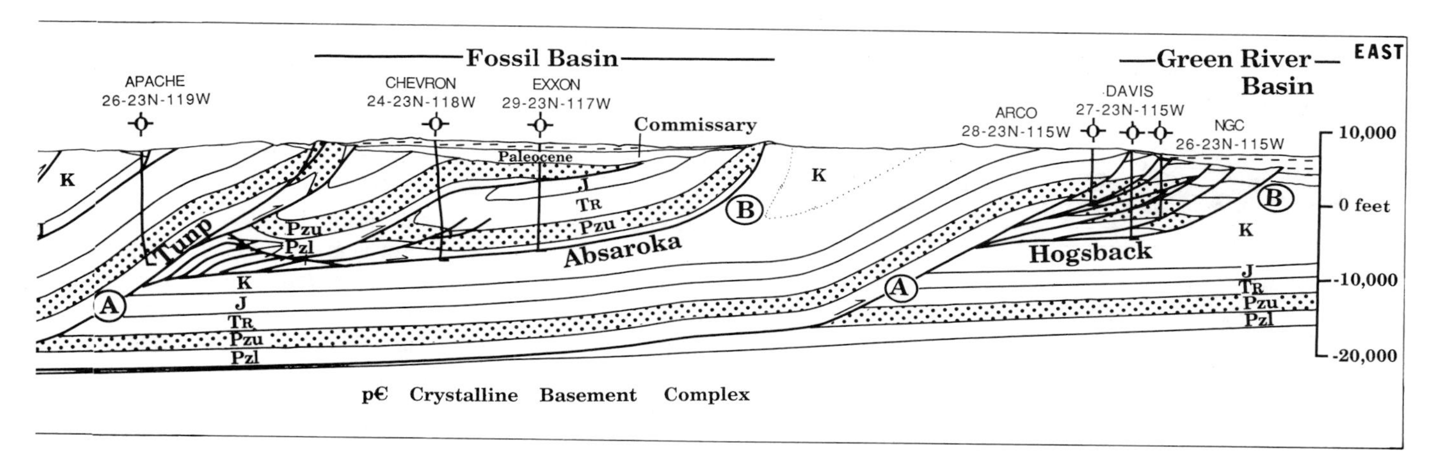

METHODS FOR DATING BASIN MARGIN UPLIFT

Periods of basin margin uplift are identified by combined observations of (1) alluvial fan sediments shed from the basin margins, (2) patterns of onlap and overstep along and across the margins, and (3) crosscutting structural relations of folded or faulted piggyback basin strata along the basin margins.

The distribution of early Eocene alluvial fan deposits in the Wyoming-Idaho-Utah thrust belt is shown in Figure 1. These deposits have been generally mapped and discussed as diamictite or conglomerate facies of the Wasatch Formation, and they include the diamictite member of the Wasatch Formation in eastern Bear Lake Plateau (this chapter); Wasatch diamictite of Oriel and Platt, 1980), the Tunp Member of the Wasatch Formation in northern Fossil Basin (Oriel and Tracey, 1970; Rubey and others, 1975; Hurst and Steidtmann, 1986), and the conglomerate member (Oriel, 1969) or Lookout Mountain Conglomerate Member (Dorr and Gingrich, 1980) of the Wasatch Formation in the La Barge basin as shown on Figure 4. The diamictites are mainly matrix-supported gravels that contain clasts derived from proximal source areas of Mesozoic and Paleozoic rocks along the structurally high margins of the piggyback basins.

Both tectonics and climate have been used to explain the origin of Wasatch diamictites. This chapter highlights the tectonic setting of the diamictites, although a favorable climate certainly contributed to diamictite deposition. Oriel and Tracey (1970, p. 29) and Tracey and others (1961) considered diamictite deposition in the Wasatch Formation to have been primarily caused by steep paleotopography generated by differential erosion combined with the humid subtropical to savanna early Eocene climate of the middle Rocky Mountains (Leopold and MacGintie, 1972). The diamictites could have been generated by periodic but intense rainfall that induced gravitational flow in areas of steep paleotopography. Oriel and Tracey (1970), Rubey and others (1975), and Tracey and others (1961) did not recognize the importance of early Eocene thrust-related uplift within the Crawford and Absaroka thrust sheets and in the western part of the Hogsback thrust sheet. Thus, the paleotopographic relief discussed by these workers would have been inherited from the topography established at the end of main emplacement of these thrust sheets. Lamerson (1982) did recognize sites of early Eocene thrust-related uplift along the margins of Fossil Basin area that Hurst and Steidtmann (1986) correlated to source areas for Wasatch diamictites. Diamictites occur only in specific structural settings in each of the three basins discussed below.

Paleotopographic relief caused by differential erosion of thrust sheets was a contributing but not sufficient condition for diamictite development because examples of erosionally resistant strata beneath fine-grained deposits of basin depocenters are found in all three basins. Instead, the diamictite source areas along the basin margins were areas of rejuvenated paleotopography that were supported by early Eocene uplift.

Patterns of onlap and overstep of early Eocene strata onto and across the basin margins can be the result of infilling of remnant paleotopography or of progressive infilling of paleotopography supported by coeval uplift. Periods of uplift are identified in this chapter by sequential angular unconformities within the Wasatch Formation and between the Wasatch Formation and older Tertiary units.

Crosscutting relations between thrust faults, thrust-related folds, and early Eocene sediments provide a third criterion for dating periods of basin margin uplift. Recent subsurface data, combined with new mapping of the Bear Lake Plateau area and field reconnaissance of the Fossil Basin and LaBarge basin, provide greater resolution of the early Eocene deformation associated with structures mapped by Oriel (1969), Rubey and others (1975, 1980), and Oriel and Platt (1980) in the three basins.

BEAR LAKE PLATEAU BASIN

Structural setting

Bear Lake Plateau basin lies on the hanging wall of the Crawford thrust sheet near the point common to Wyoming, Idaho, and Utah (Fig. 1). Two structural elements of the thrust sheet bound early Eocene depositional areas that were affected by

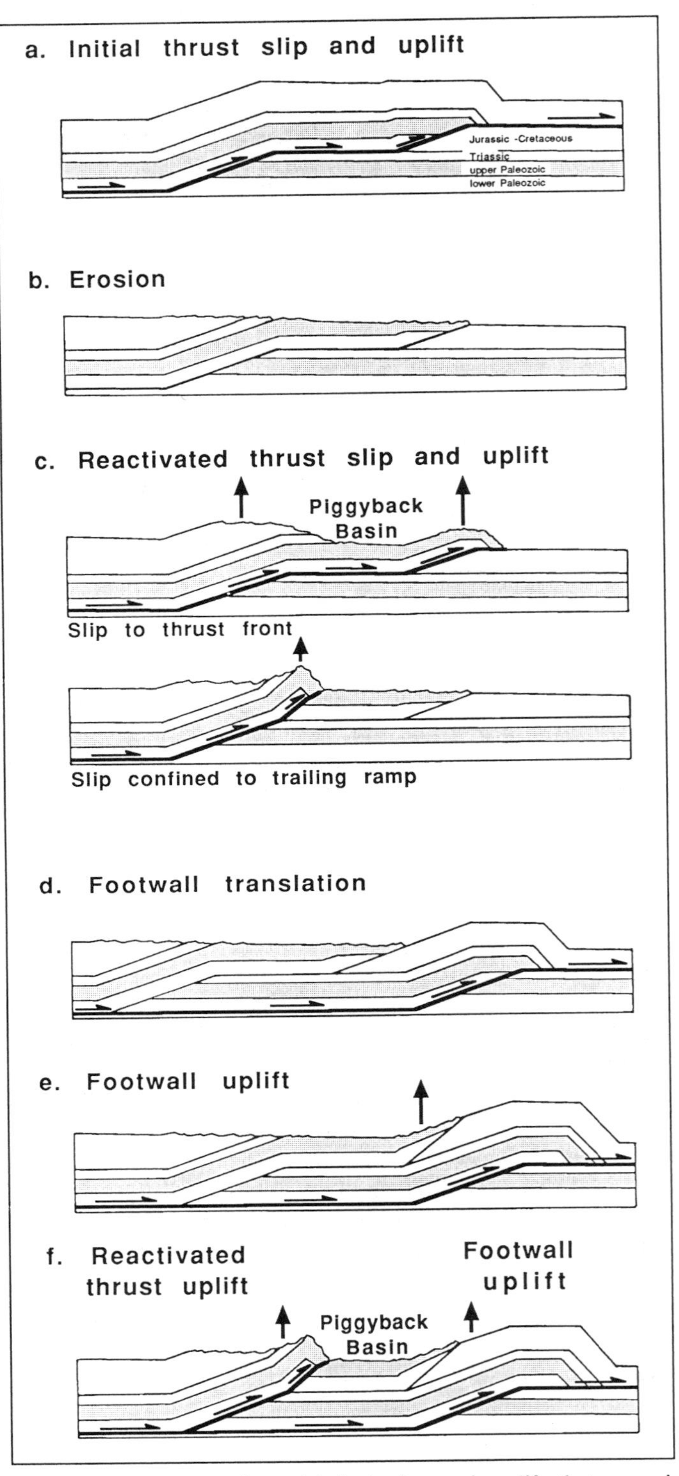

Figure 3. Scaled geometric models for basin margin uplifts that approximate the structural features common to the Crawford, Absaroka, and Hogsback thrust sheets. Thrust fault lengths undergoing slip in each diagram are shown by bold lines and arrows, whereas the other thrust faults are considered to be mechanically locked.

different sedimentation histories: (1) the leading edge of the Crawford thrust including Cokeville anticline (Fig. 5); and (2) the trailing ramp area of the Crawford thrust, which lies beneath Pegram anticline to the south and Sublette anticline to the north (Fig. 5). The thrust-related structure of Bear Lake Plateau has been modified by Miocene to Holocene normal faulting along the plateau margins. The Cokeville normal fault bounds the east side of a shallow half graben east of Bear Lake Plateau, and the Bear Lake normal fault separates the western border of the plateau from Bear Lake graben to the west (Fig. 5). Normal faults mapped by Rubey and others (1980) and Oriel and Platt (1980) on eastern Bear Lake Plateau were not found during remapping of area (Fig. 5).

Bear Lake Plateau basin is bound to the east by the leading edge of the Crawford thrust sheet. The Crawford thrust crops out east of Cokeville, Wyoming, but it is covered beneath Tertiary and Quaternary valley fill where it has been displaced by the Cokeville normal fault to the south (Fig. 5). The Stoffer Ridge thrust crops out east of Cokeville as a footwall imbricate to the Crawford thrust that branches from the frontal Crawford décollement in Jurassic salt. North of Cokeville, the leading edge of the Crawford thrust is located within a fold belt above the salt décollement (Coogan and Yonkee, 1985). Cokeville anticline is a hanging-wall ramp anticline that overlies this décollement (Coogan and Yonkee, (1985). The Wasatch Formation onlaps Mesozoic rocks of the leading edge of the Crawford sheet along the east side of Bear Lake Plateau basin where Wasatch diamictites have been rotated to westward dip above an angular unconformity.

The central part of Bear Lake Plateau basin is underlain by Pegram anticline, a fault-propagation fold that lies above the Crawford trailing ramp. The anticline is exposed along the northern margin of the plateau, and it has been drilled in the subsurface beneath the Wasatch Formation where its approximate location is dotted on Figure 5. Pegram anticline divides Bear Lake Plateau basin into two subbasins, with an early fluvial basin east of Pegram anticline and the depocenter of a late lacustrine basin to the west. Pegram anticline plunges north along the northern margin of Bear Lake Plateau, where the subsurface position of the Crawford trailing ramp shifts eastward across a lateral thrust ramp to where it underlies Sublette anticline in the northern part of Figure 5.

Mapping by Valenti (1982) and Dover (1985) along the southern edge of Bear Lake Valley demonstrates that the western boundary of Wasatch deposition extended west of Bear Lake Plateau prior to Miocene to Holocene normal faulting. The Wasatch Formation is continuous south of Bear Lake Plateau to where it onlapped Paleozoic strata of the Paris thrust hanging wall to the west (Fig. 1).

Age and stratigraphic position

Figure 4 summarizes the age and stratigraphic relatinships recognized for the Wasatch Formation and associated younger and older Tertiary deposits of Bear Lake Plateau, and Sublette

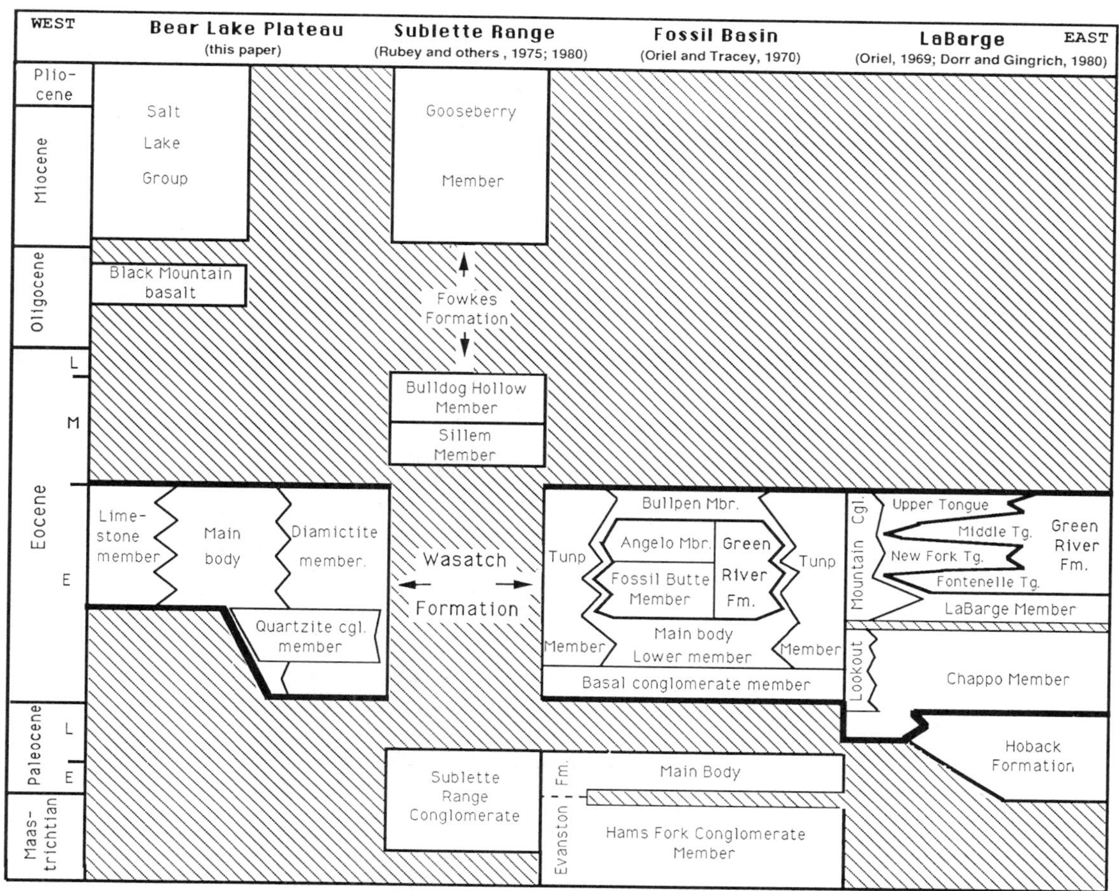

Figure 4. Latest Cretaceous and Tertiary stratigraphic relationships for the Bear Lake Plateau basin, Sublette Range, northern Fossil Basin, and LaBarge basins.

Range, Fossil Basin, LaBarge, and surrounding areas. Mansfield (1927) initially mapped most of Bear Lake Plateau as Eocene Wasatch Formation based on surface correlation to dated Wasatch deposits in other areas. Oriel and Platt (1980) subsequently mapped most of Bear Lake Plateau north of the Utah border as Pliocene and Miocene Salt Lake Formation based on a collection of *Leporidae* teeth from a white, tuffaceous conglomorate in the northwestern corner of Bear Lake Plateau (Oriel and Tracey, 1970, p. 37). Oriel and Platt (1980) also mapped thin outcrop belts of Wasatch diamictite and Maastrichtian-Paleocene Evanston Formation along the eastern margin of Bear Lake Plateau. The diamictite of Oriel and Platt (1980) is informally designated the diamictite member of the Wasatch Formation in this chapter (Figs. 4 and 5), whereas conglomerates mapped as the Evanston Formation by Oriel and Platt (1980) are here incorporated into the quartzite conglomerate member of the Wasatch Formation based on intertonguing relationships with the diamictite member.

Remapping of the area (Fig. 5) revealed three overlapping relationships for the Bear Lake Plateau basin. (1) The youngest overlapping beds are the Salt Lake strata sampled by Oriel and Tracey (1970), which are restricted to small debris flow deposits that unconformably overlie the Wasatch strata in northeastern Bear Lake Plateau (Fig. 5). (2) The Wasatch Formation is also overlain and crosscut by an olivine basalt on Black Mountain, Utah, shown in the western part of Figure 5. The basalt yields an Oligocene K-Ar whole rock age of 28.8 ± 1.7 m.y. (Mark Jensen, Utah Geological and Mineral Survey, written communication, 1989). (3) Finally, the Wasatch Formation is overlain in angular unconformity by tuffaceous beds of the middle Eocene Sillem Member of the Fowkes Formation along the eastern margin of Bear Lake Plateau (Fig. 5). Rubey and others (1980) interpreted the Fowkes-Wasatch contact as a normal fault, but the new mapping reveals that the Fowkes Formation dips 15° east above and adjacent to 15 to 25° west-dipping Wasatch Formation. This overlapping relationship provides a pre–middle Eocene upper age limit for the Wasatch Formation of Bear Lake Plateau. Within the Wasatch Formation, early Eocene gastropods occur in limestones that intertongue with the upper part of the Wasatch clastic sequence in western Bear lake Plateau basin (H. P. Buchheim, Loma Linda University, personal communication, 1989).

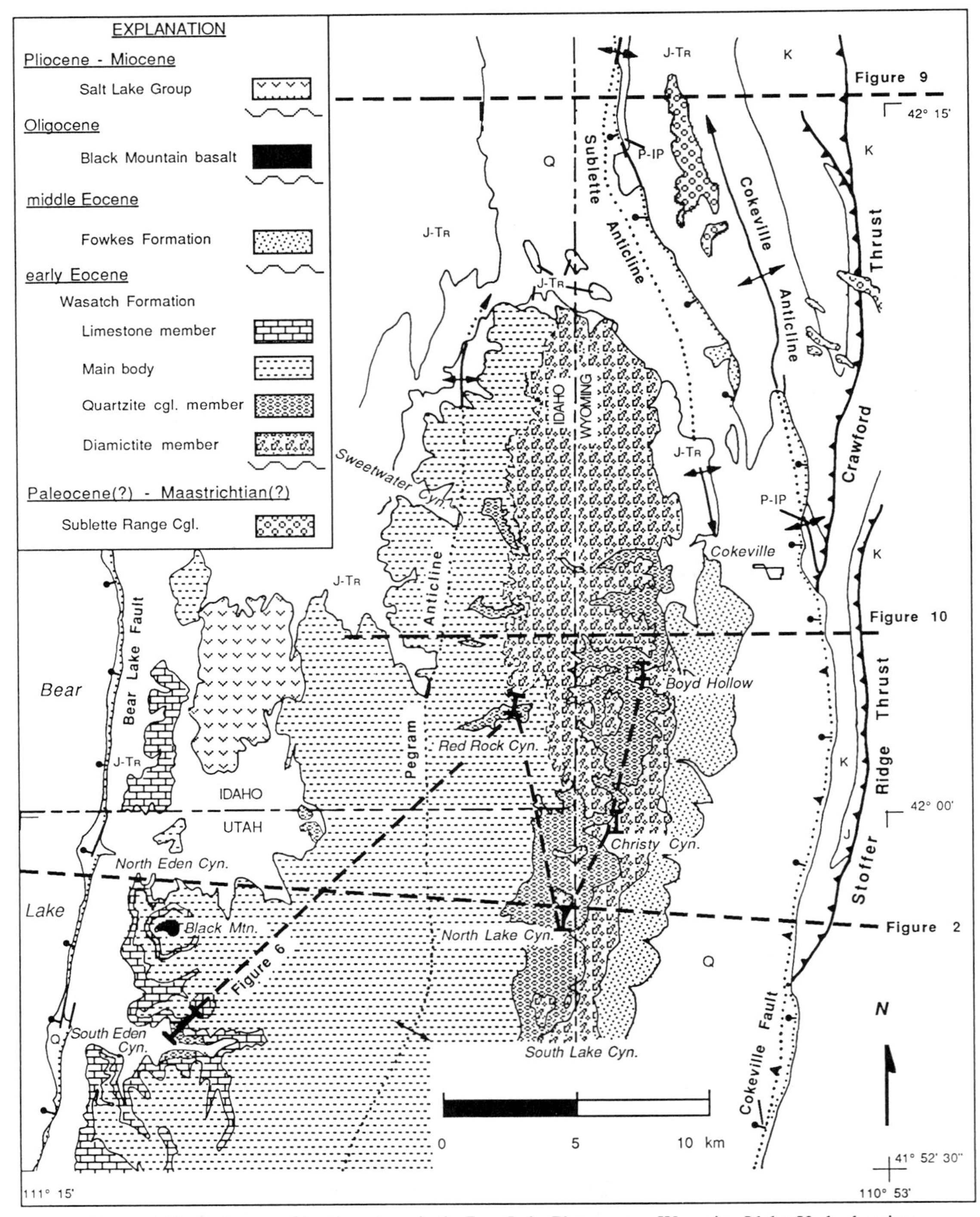

Figure 5. Geologic map of Tertiary strata in the Bear Lake Plateau area, Wyoming-Idaho-Utah, showing principal surface folds and faults of the Crawford thrust sheet. Thrust faults have sawteeth on the hanging wall; normal faults have ball on hanging wall. Faults and folds are dotted where they are buried by Tertiary and younger strata. Members of the Wasatch Formation become progressively younger westward. The Salt Lake Group, Black Mountain basalt, and Fowkes Formation unconformably overlie the Wasatch Formation. The locations of structural cross sections are shown for Figures 9, 10, and 2. The locations of measured sections are shown along the line of Figure 6. P-IP = Permian and Pennsylvanian rocks; J-TR = Jurassic and Triassic rocks; J = Jurassic rocks; K = Cretaceous rocks; Q = Quaternary rocks.

The Sublette Range Conglomerate crops out along the west flanks of Sublette and Cokeville anticlines and above the Crawford thrust trace (Fig. 5). The conglomerate is undated, but it is placed beneath the Wasatch Formation on Figure 4 based on lithologic, sedimentologic, and provenance correlations to the Paleocene-Maastrichtian Evanston Formation in adjacent areas (Salat, 1989). Like the Evanston Formation, the Sublette Range Conglomerate has abundant red quartzite boulders and cobbles that have a unique provenance in the Lake Proterozoic section of the Willard and Paris thrust sheets. Furthermore, clasts derived from the Sublette Range Conglomerate are interpreted below to have been recycled into coarse-grained Wasatch strata along the eastern margin of Bear Lake Plateau basin.

Stratigraphy and sedimentology of the Wasatch Formation

Five measured sections constrain the early Eocene stratigraphy of Bear Lake Plateau basin: Christy Canyon and Boyd Hollow on the east side of the basin, North Lake Canyon and Red Rock Canyon in the central basin, and South Eden Canyon on the west flank of Bear Lake Plateau basin (Figs. 5 and 6). Four informal members of the Wasatch Formation were measured and mapped from east to west. In ascending stratigraphic order, the members include: (1) the diamictite member, (2) the quartzite conglomerate member, (3) the main body, and (4) the limestone member. Detailed measured sections of the five members are provided in Coogan (1992). A generalized cross section through the measured sections is shown on Figure 6, in which the relative stratigraphic position of each measured section is constrained by surface and photogeologic mapping of key beds between the sections.

Diamictite member. The diamictite member is best exposed in Christy Canyon and Boyd Hollow along the eastern margin of Bear Lake Plateau basin (Figs. 5 and 6). The diamictite member consists of debris- and mud-flow deposits that were transported westward and southwestward as part of an alluvial fan system shed from the Crawford thrust front and Sublette anticline. This interpretation is based on analysis of dominant lithofacies, thickness trends, lateral relationships, and clast provenance.

The diamictite member contains three major lithofacies: matrix-supported gravel, massive clast-supported gravel, and massive mudstone. The matrix-supported gravels are moderately organized where they have sand and silt matrix and display inverse grading and faint horizontal stratification that are indicative of deposition in noncohesive debris flows (Nemec and Steel, 1984). In contrast, clay-rich matrix-supported gravels (Fig. 7) display no sorting trends or stratification and probably formed as cohesive debris flows (Nemec and Steel, 1984). Disorganized, massive, clast-supported gravels are interpreted as having formed as rapidly deposited, clay-poor, noncohesive debris flows (Nemec and Steel, 1984; DeCelles and others, 1987). Massive mudstones

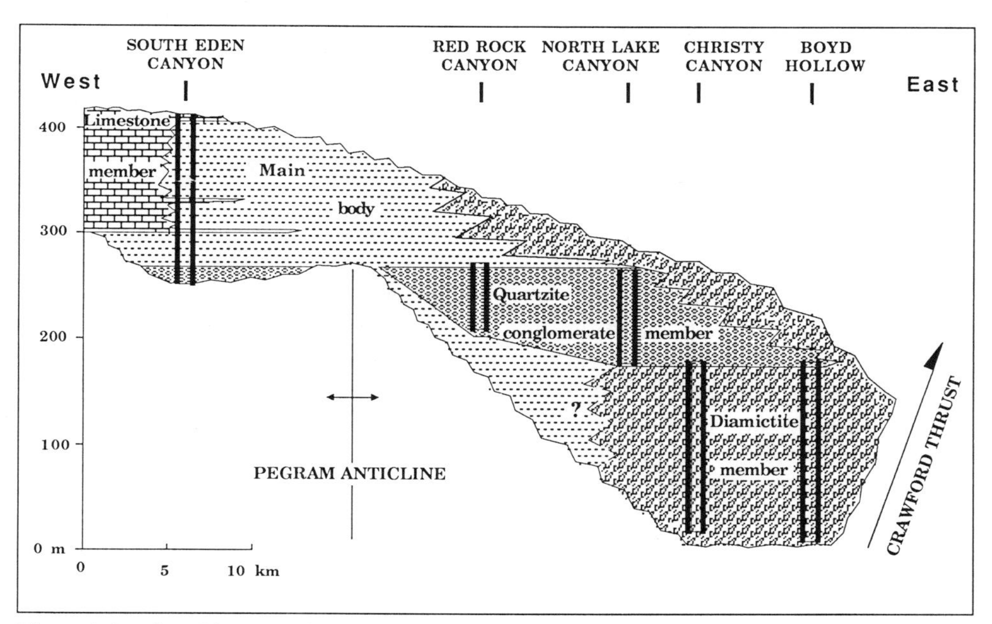

Figure 6. Stratigraphic correlation between informal members of the Wasatch Formation showing the position of underlying structures in the Crawford thrust sheet, the intertonguing relationships mapped on Figure 5, and the five measured sections (Coogan, 1992). The basin depocenter migrated through time from an early, fluvial depocenter in the quartzite conglomerate member of eastern Bear Lake Plateau basin to a late, lacustrine depocenter in western Bear Lake Plateau basin. The migration of the depocenter resulted from stratigraphic onlap and overstep across the crest of Pegram anticline.

Figure 7. Typical outcrop of disorganized matrix-supported gravel of the diamictite member of the Wasatch Formation. Notice the poor sorting of both grain size and shape. The base of the hammer rests on an angular block of Nugget Sandstone that was probably transported a short distance from the source area in the Sublette Range or along the Crawford thrust front. The point of the hammer touches a well-rounded clast of Proterozoic quartzite that probably experienced an early episode of fluvial transport prior to its deposition in the Sublette Range conglomerate as well as a second period of debris-flow transport during which it was recycled into the diamictite member during uplift of the Sublette Range and the Crawford thrust front.

locally contain widely spaced matrix-supported boulders, cobbles, and pebbles that were probably entrained in cohesive mudflows. The noncohesive debris flows and cohesive debris- and mud-flow deposits of the diamictite member are characteristic of a proximal alluvial fan depositional environment. Minor horizontally stratified and trough cross-stratified sandstone that is interbedded with the gravel and mudstone probably formed as minor sheetflood and channel deposits that are commonly found on the surface of alluvial fans (Rust and Koster, 1984).

Thickness trends and the map distribution of the diamictite member demonstrate that the alluvial fan complex was shed from a source area immediately to the east and northeast in the vicinity of the Sublette Range and the Crawford thrust front. The diamictite member thins westward and southwestward and intertongues with both the braided stream deposits of the quartzite conglomerate member and the fluvial channel and floodplain deposits of the main body of the Wasatch to the west and southwest (Figs. 5 and 6). The provenance of clasts from the diamictite member also supports the interpretation of a proximal source area to the east and northeast.

Clast counts through the vertical sections of the diamictite member at Boyd Hollow and Christy Canyon exhibit a subtle inverse stratigraphy that was likely caused by unroofing of the sublette Range and the Crawford thrust front during the early Eocene (Coogan, 1992). The stratigraphy of these potential source areas is exposed along the east flanks of Sublette and Cokeville anticlines and astride the Crawford thrust to the east where the leading edge of the Crawford thrust sheet has not been displaced by the Cokeville normal fault. The Crawford thrust and Sublette and Cokeville anticlines are overlain with angular unconformity by erosional remnants of the Sublette Range Conglomerate (Fig. 5), which contains a diagnostic assemblage of well-rounded Late Proterozoic and Paleozoic quartzite clasts as well as abundant Paleozoic chert clasts (Salat, 1989). Highly folded lower Cretaceous through Jurassic rocks lie beneath the unconformity, and Triassic and Permo-Pennsylvanian rocks have been exhumed at deeper structural levels in Sublette and Cokeville anticlines. Within the diamictite member, the well-rounded, Late Proterozoic quartzite clasts and Paleozoic quartzite and chert clasts are concentrated at the base of the unit, whereas clasts of Jurassic carbonates and the Jurassic Nugget Sandstone are rare, and Triassic clasts are absent. In contrast, the upper part of the diamictite member contains abundant angular Jurassic clasts as well as the first appearance of Triassic clasts, but well-rounded chert and quartzite clasts are less common, and Proterozoic quartzite clasts are absent in the uppermost beds. The most plausible interpretation for these clast distributions is that the stratigraphically higher diamictite beds record progressive erosion into Sublette anticline and the frontal Crawford thrust structures from the level of the Sublette Range Conglomerate that caps the anticline to the lower Triassic level in the core of the anticline.

Quartzite conglomerate member. The quartzite conglomerate member occupies a north-south trending trough in eastern Bear Lake Plateau basin, with the most complete exposures located in North Lake and Red Rock canyons (Figs. 5 and 6). The quartzite conglomerate member was deposited in a south-flowing, gravel-dominated braided stream system that reworked the toes of the diamictite member alluvial fans. This interpretation is supported by lithofacies analysis, paleocurrent data, and lateral relationships with adjacent members.

The main gravel lithofacies of the quartzite conglomerate member is massive, clast-supported gravel with up to 80% well-rounded quartzite and chert clasts. The sheetlike massive gravels grade upward into trough cross-stratified gravels and sandstones. The massive gravels contain crude horizontal stratification and imbricate cobble trains. The gravel sheets were probably deposited as individual gravel bars in a proximal, gravel-dominated, braided stream system, based on their similarity to recent and ancient examples (Boothroyd and Ashley, 1975; Rust and Koster, 1984; Miall, 1978). The trough cross-stratified gravels and sandstones probably filled mixed sand and gravel channel systems that modified the bar tops after periods of gravel-dominated deposition (Miall, 1978).

Paleocurrent data support a low-sinuosity braided stream interpretation for the quartzite conglomerate member. The individual and summed paleocurrent plots for the c-axes of imbricated pebbles in Figure 8 show an axial drainage direction subparallel to the Wasatch facies transitions and the thrust-related structural trends of Bear Lake Plateau. The paleocurrents are directionally bimodal, with a principal drainage direction to the southwest and a second minor component to the southeast.

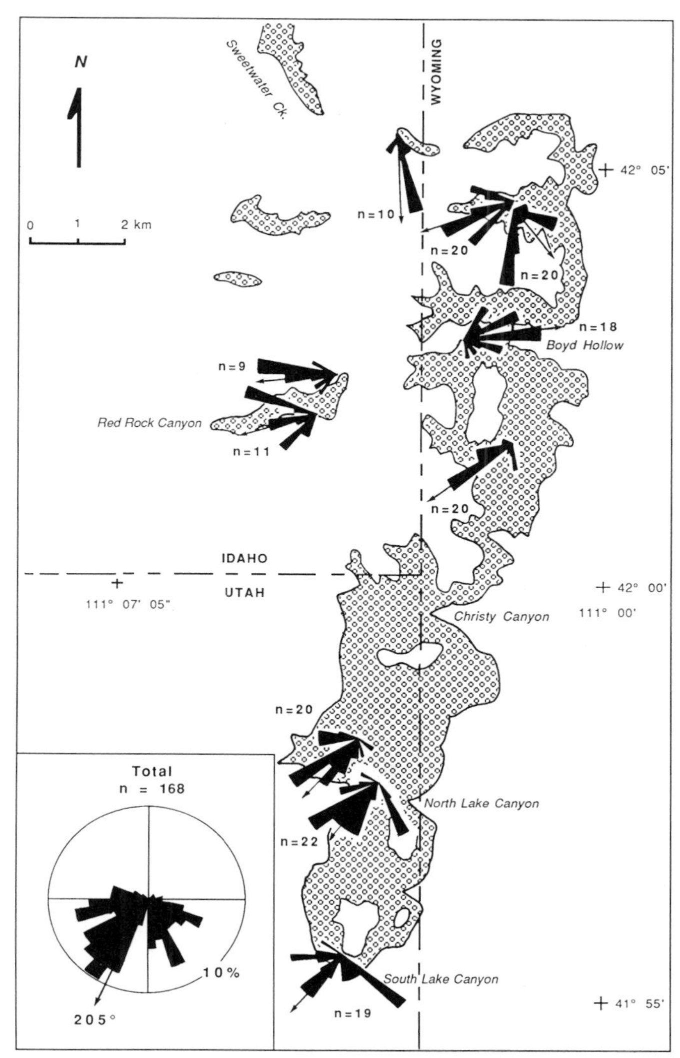

Figure 8. Rose diagrams of c-axis orientations for nonspherical cobbles in the quartzite conglomerate member of the Wasatch Formation. The vector mean for all measurements is 205°, showing a general southwestward current direction subparallel to the trend of Crawford hanging-wall structures and longitudinal to the basin axis.

The map distribution of the quartzite conglomerate member (Figs. 5 and 6) indicates that the braided alluvial plain was mainly limited to eastern Bear Lake Plateau between the diamictite member depositional front to the east and Pegram anticline to the west. Immediately west of the Red Rock Canyon section, the quartzite conglomerate member thins markedly toward the east flank of Pegram anticline, a phenomenon also observed at Sweetwater Canyon to the north (Fig. 5). Two isolated outcrops of the quartzite conglomerate member occur along the west flank of Pegram anticline overlying the basal angular unconformity above Jurassic rocks in North and South Eden canyons (Figs. 5 and 6). The conglomerate thins eastward in both locations, where it is truncated and overlapped by the main body of the Wasatch Formation. Pegram anticline was a paleotopographic high that separated the quartzite conglomerate member into eastern and western depositional areas.

Main body. The main body of the Wasatch Formation is a sand- and mud-dominated fluvial sequence that is best exposed in the South Eden section where it interfingers with the limestone member (Figs. 5 and 6). Red mudstone is volumetrically the most significant lithofacies of the main body. These mudstones are structureless wherever exposed, but they are interpreted to be floodplain muds and silts based on their relationship to laterally correlative trough cross-stratified sandstones. The sandstones are thin and laterally discontinuous with abrupt lower contacts on underlying mudstones. The bases of trough cross-stratified units often contain pebble lags that grade upward into sandstone. The lenticular shape, cross-bedding, and limited lateral extent of the trough cross-stratified sandstones lead to an interpretation of these units as channel sequences in low-sinuosity stream systems (Collinson, 1978). Horizontally laminated sandstones occur in the main body as very thin, tabular beds that occur within mudstones as well as within and above trough cross-stratified sandstones. The horizontally laminated sandstones may represent broad, shallow channel fills or possibly local crevasse splay deposits adjacent to larger channels (Collinson, 1978).

Sandstones in the main body of the Wasatch member were probably deposited in a sand- and mud-dominated fluvial system of individual short-lived stream channels bordered by mudstone deposition in interchannel areas. The main body fluvial deposits represent an intermediate transport system between the clastic source area of the diamictite member to the east and the lacustrine depocenter of the limestone member of the Wasatch Formation (Figs. 5 and 6).

Limestone member. The limestone member of the Wasatch Formation occupies a shallow lacustrine depocenter in western Bear Lake Plateau basin that is best exposed in the South Eden Canyon area (Figs. 5 and 6). The limestones contain abundant coarse clasts including oncolites, coated grains, and algal rip-up clasts. The large volume of pebbles within the limestone as oncolite nucleii and as coated grains implies that coarse material was continually supplied from the surrounding fluvial system.

The limestone member interfingers with the main body throughout the western part of Bear Lake Plateau, and it thickens dramatically to the southwest, where it is over 60 m (200 ft) thick in a continuous vertical sequence near the southeast corner of Bear lake (Figs. 5 and 6). The east to west transitions from the diamictite member alluvial fan deposits, through the main body fluvial system, and finally into the limestone member lacustrine depocenter imply a general westward or southwestward sediment transport direction during the later stages of deposition in the Bear Lake Plateau basin (Figs. 5 and 6). The locus of deposition and modes of transport contrast with the older depositional system of the quartzite conglomerate member in eastern Bear Lake Plateau. The change from the eastern to western depocenters through time was controlled by an interplay between infilling of structurally defined paleotopography and shifting sites of tectonic uplift.

Structural evolution of Bear Lake Plateau basin

Sublette anticline, Pegram anticline, and the Crawford thrust front were structural and paleotopographic highs along the margins of Bear Lake Plateau basin that served as source areas for clastic sediments. Surface and subsurface structural information is combined with stratigraphic data to establish the age and sites of uplift in the three areas.

Uplift of Sublette anticline. The clast provenance, paleocurrent directions, and stratigraphic relationships of the Wasatch Formation all indicate that Sublette anticline was a topographic high during early Eocene sedimentation in Bear Lake Plateau basin. The inverse stratigraphy of clasts in the diamictite member indicates that the Sublette Range was erosionally exhumed from Jurassic through lower Triassic levels during the early Eocene. Such downcutting requires uplift of the Sublette Range because all well and geophysical data indicate that there was no local subsidence of basement beneath the adjacent piggyback basins. Uplift is also necessary to explain the present elevation of the Sublette Range Conglomerate above and east of Sublette anticline. The Sublette Range Conglomerate crops out with approximately 10° southeast dip at elevations between 2,500 and 2,760 m (8,200 to 9,050 ft) along the east flank of Sublette anticline. Assuming that the Sublette Range Conglomerate is older than the Wasatch Formation, as indicated by Salat (1989) and this chapter, a minimum 625 m (2,050 ft) of uplift is needed to explain the elevation difference between the conglomerate and the base of the Wasatch Formation, which lies at elevations of 1,794 to 1,875 m (5,885 to 6,150 ft) in wells in the interior of eastern Bear Lake Plateau basin. The elevation difference cannot be attributed to Miocene to Holocene normal slip of the Cokeville listric normal fault because vertical throw on the fault is minimal where it soles into the footwall décollement of the Crawford thrust plane beneath the interior of the eastern basin. Similarly, a minimum of 305 m (1,000 ft) of uplift is required to explain the elevation difference between the Sublette Range Conglomerate above Sublette anticline and eastern outcrops of the conglomerate that lie at elevations between 2,195 m (7,200 ft) and 2,010 m (6,600 ft) along the east flank of Cokeville anticline and above the Crawford thrust trace (Fig. 5). A kinematic model for differential uplift of Sublette anticline relative to Bear Lake Plateau basin, Cokeville anticline, and the Crawford thrust front is developed below.

The present geometry of Sublette anticline and the structural position of the Sublette Range Conglomerate are shown in Figure 9a. Sublette anticline overlies the trailing ramp of the Crawford thrust where the ramp is constrained by the Pan Am and Cities wells as shown on Figure 9a. The positions of the Jurassic salt décollement and Afton anticline east of Sublette anticline are constrained by wells located immediately north of the line of section (see Coogan and Yonkee, 1985, Fig. 7). In the kinematic model, early Eocene uplift of Sublette anticline was accomplished by late slip on a blind imbricate thrust between Sublette and Cokeville anticlines that is rooted to the Crawford trailing ramp. Two observations support this geometric interpretation: (1) Cokeville anticline plunges beneath Sublette anticline north of Cokeville in map view (Fig. 5), and (2) tight folds in the Twin Creek Limestone between Sublette and Cokeville anticlines require a corresponding amount of shortening in older strata to achieve structural balance.

The model sequence for late imbrication beneath Sublette anticline is presented in Figures 9b and c. Figure 9b shows Sublette and Cokeville anticlines after deposition of the Sublette Range Conglomerate and prior to deposition of the Wasatch Formation. Figure 9c shows the early Eocene uplift of Sublette anticline along the ramp-rooted thrust between Sublette and Cokeville anticlines. The growth of Sublette anticline above the Crawford trailing ramp is a variation of the case in which reactivated thrust slip is confined to the trailing ramp areas of thrust faults as shown in Figure 3c.

Pegram anticline. The crest of Pegram anticline was a paleotopographic boundary that formed the western margin of the early fluvial depocenter of Bear Lake Plateau basin. Subsurface data indicate that Pegram anticline is a fault-propagation fold that overlies the Crawford trailing ramp along the length of Bear Lake Plateau basin (Figs. 5 and 10) in a structural position analogous to that of Sublette anticline to the north (Figure 9a). A lateral ramp that parallels the northern margin of Bear Lake Plateau basin accounts for the westward shift in the ramp location from north to south.

Like Sublette anticline, the structural style and position of Pegram anticline resemble break-back thrust structures that form during reactivation of trailing ramp areas of thrust sheets (Figure 3c). By analogy with Sublette anticline, the paleotopographic high along the crest of Pegram anticline could have been caused by early Eocene uplift, but it is equally plausible that the high was the result of differential erosion between the resistant Nugget Sandstone on the crest of Pegram anticline and the less resistant Twin Creek Limestone on the flanks. Most of the structural relief of Pegram anticline was clearly established prior to Wasatch deposition. Successively younger Wasatch strata onlap the east limb of Pegram anticline, and the quartzite conglomerate member pinches out immediately to the east of the anticline (Fig. 10), which indicates that any minor early Eocene uplift of Pegram anticline kept pace with infilling of the early basin. Likewise, the lack of obvious angular unconformities in Wasatch strata along the flank of Pegram anticline indicates that uplift was minimal during Wasatch deposition, and the lack of extensive eastward prograding alluvial fans along the basin margin implies that early Eocene topographic relief was modest at any one time.

The position of the early basin margin above Pegram anticline was clearly controlled by the location of the underlying thrust-related structures, but the time of uplift of Pegram anticline is uncertain. The interpretation that some component of late slip beneath Pegram anticline immediately preceded Wasatch deposition is favored here by analogy to late slip on the Crawford trailing ramp immediately to the north beneath Sublette anticline.

Crawford thrust front. The Absaroka trailing ramp immediately underlies the frontal Crawford thrust plate, an interpretation that agrees with published seismic-based structure maps of the Absaroka (Dixon, 1982) and Crawford (Valenti, 1987) thrust planes. Early Eocene deformation associated with late slip on the Absaroka trailing ramp is evident from folded Wasatch strata at the surface. The Wasatch Formation dips gently (1 to 3°) west through the interior of eastern Bear Lake Plateau basin, but the dip steepens to 15 to 25° where the Wasatch has been rotated through an axial plane along the eastern basin margin (Fig. 10).

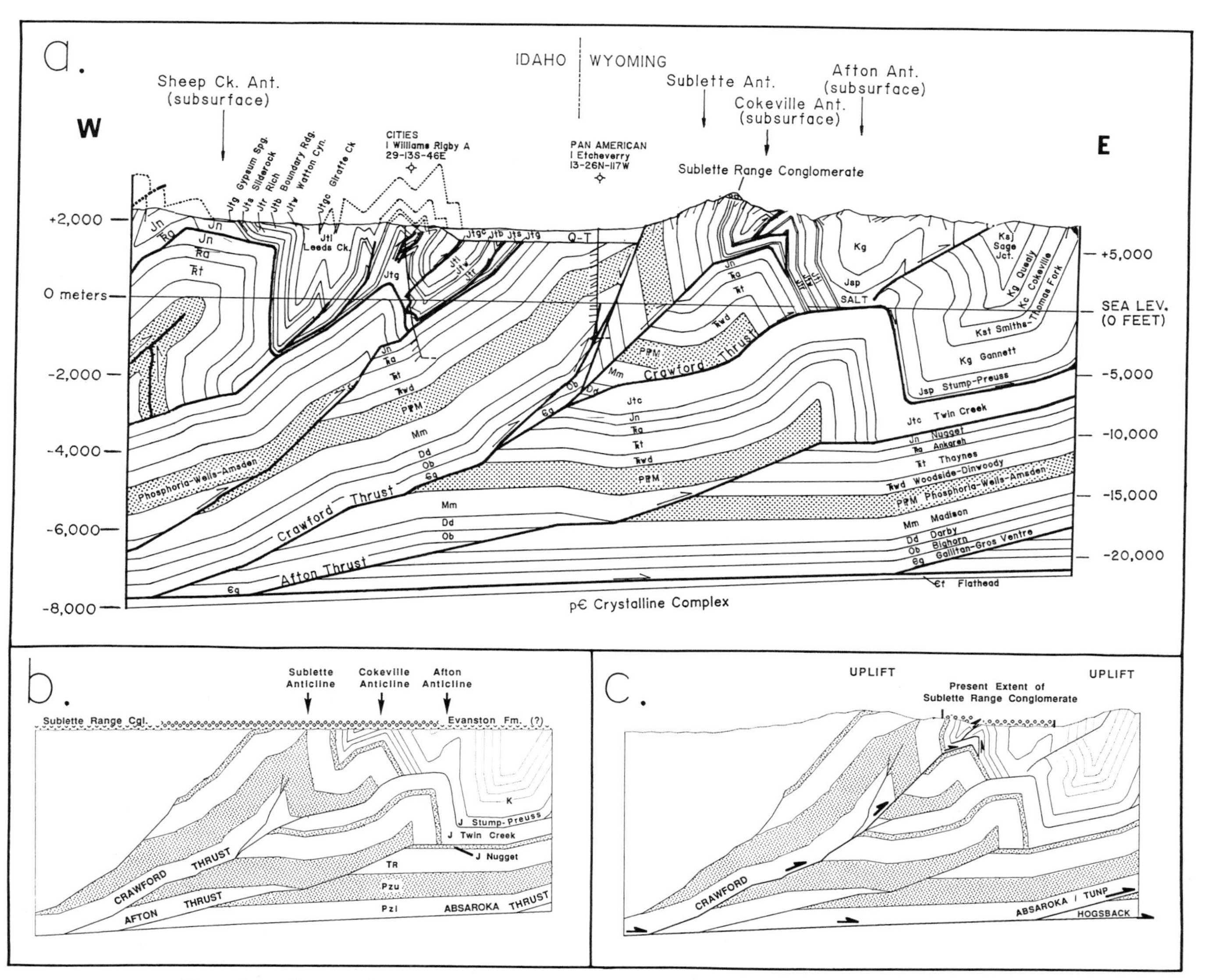

Figure 9. a, Cross section through Sublette, Cokeville, and Afton anticlines showing the present elevation and structural position of the Sublette Range Conglomerate. The trailing ramp geometry of the Crawford thrust is constrained by two wells along the west flank of Sublette anticline. Cokeville anticline plunges into the line of section from the south. Afton anticline plunges into the line of section from where it has been penetrated by two wells 9.5 km (5.1 mi) north of the section. b, Reconstruction of 9a during deposition of the Sublette Range Conglomerate. c, Reconstruction of 9a after uplift of Sublette anticline and the Sublette Range Conglomerate. Uplift was accommodated by a late ramp-rooted fault between Sublette and Cokeville anticlines immediately before or during early Eocene time. The approximate present position of the isolated outcrops of the Sublette Range Conglomerate on the east flank of Sublette anticline is shown along with the areas of uplift caused by early Eocene slip on the Crawford and Absaroka trailing ramps. Vertical scale = horizontal scale.

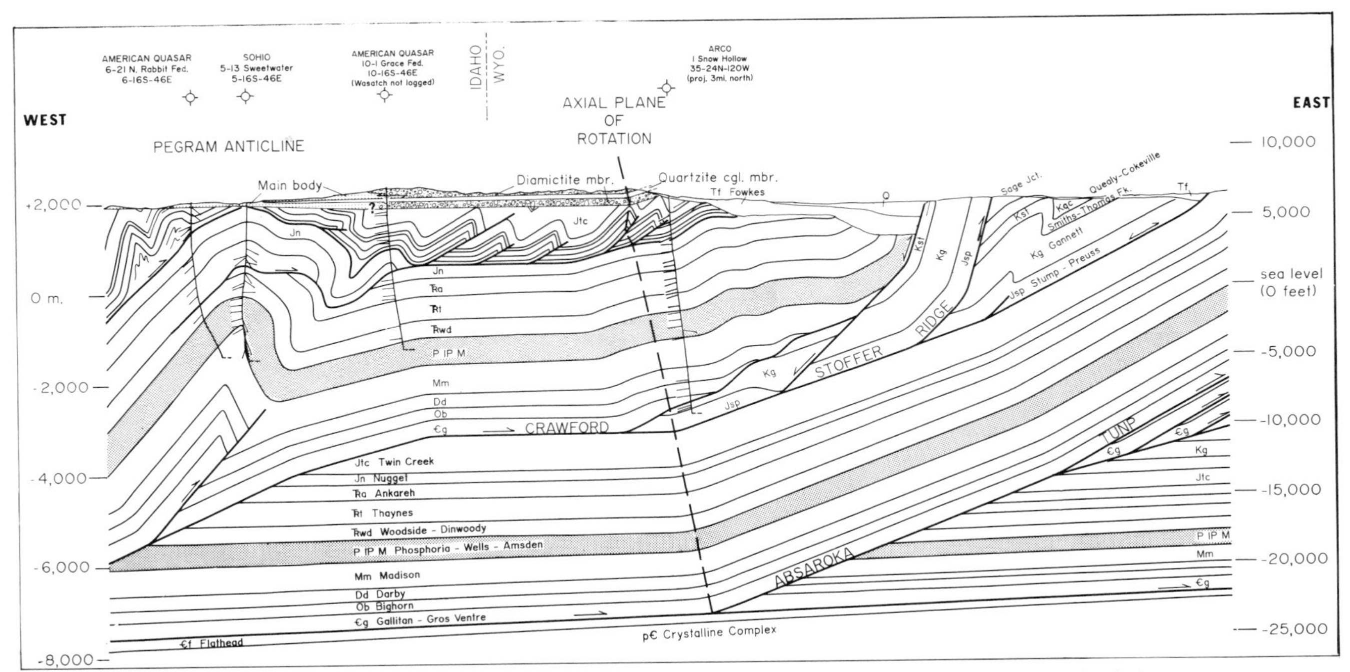

Figure 10. Cross section through eastern Bear Lake Plateau showing the geometry of the diamictite member, quartzite conglomerate member, and main body of the Wasatch Formation above Pegram anticline, Crawford thrust front, and the Absaroka trailing ramp. Pegram anticline is a fault-propagation fold above the Crawford trailing ramp that formed the western boundary of the early eastern depocenter of Bear Lake Plateau basin, as seen by the thinning of the quartzite conglomerate member along the east flank of the anticline. The leading edge of the Crawford thrust overlies the trailing ramp of the Absaroka thrust. The 20° west dip (shown by dip bars) of the Cretaceous footwall rocks at the bottom of the Arco Snow Hollow well defines the east side of an axial plane of rotation that is associated with the underlying branch point between the Absaroka thrust and the basal décollement. The Wasatch strata in the upper part of the section are also rotated approximately 20° across the axial plane. Vertical scale = horizontal scale.

The projection of the surface axial plane to depth intersects the branch point where the Absaroka thrust lifts off of the regional sole décollement in Cambrian shales at a 20° angle that is constrained by Absaroka hanging-wall dips at the base of the Arco Snow Hollow well shown in Figure 10. Deformation is clearly early Eocene in age, since the west-dipping Wasatch beds are truncated by the east-dipping middle Eocene Fowkes Formation at the surface (Figs. 5 and 10). Reactivated slip on the frontal Crawford thrust would have accomplished the same rotation of the Wasatch Formation; however, Crawford slip is precluded by the lack of substantial uplift in the Crawford trailing ramp area.

Comparison of sedimentation histories between the trailing ramp area and the leading edge of a thrust sheet is useful for determining times when through-going slip from the trailing ramp to the thrust front is possible. If thrust slip is communicated from the basal décollement, up the trailing ramp, and finally across the frontal décollement to the thrust front, a cessation of uplift in the trailing ramp region precludes any slip along the frontal length of the thrust. Sedimentation across Pegram anticline in the Crawford trailing ramp region indicates that the Crawford thrust remained locked during at least the later stages of sedimentation in Bear Lake Plateau basin. The synchronous deposition of the diamictite member along the leading edge of the Crawford thrust and the overstepping of sediments across Pegram anticline indicate that footwall uplift of the Crawford thrust front outlasted any reactivated slip and uplift on the southern Crawford trailing ramp. The duration of reactivated slip on the northern trailing ramp beneath Sublette anticline is not known. The combination of reactivated slip and uplift along the Crawford trailing ramp and footwall uplift along the Crawford thrust front represents a pattern of basin margin uplift that is also found in northern Fossil Basin and the LaBarge basin to the east.

NORTHERN FOSSIL BASIN

Structural, stratigraphic, and sedimentologic setting

Northern Fossil Basin lies within the hanging wall of the Absaroka thrust between a western basin margin adjacent to the Tunp thrust and an eastern basin margin astride the frontal trace of the Absaroka thrust (Fig. 11). Subsurface data indicate that the

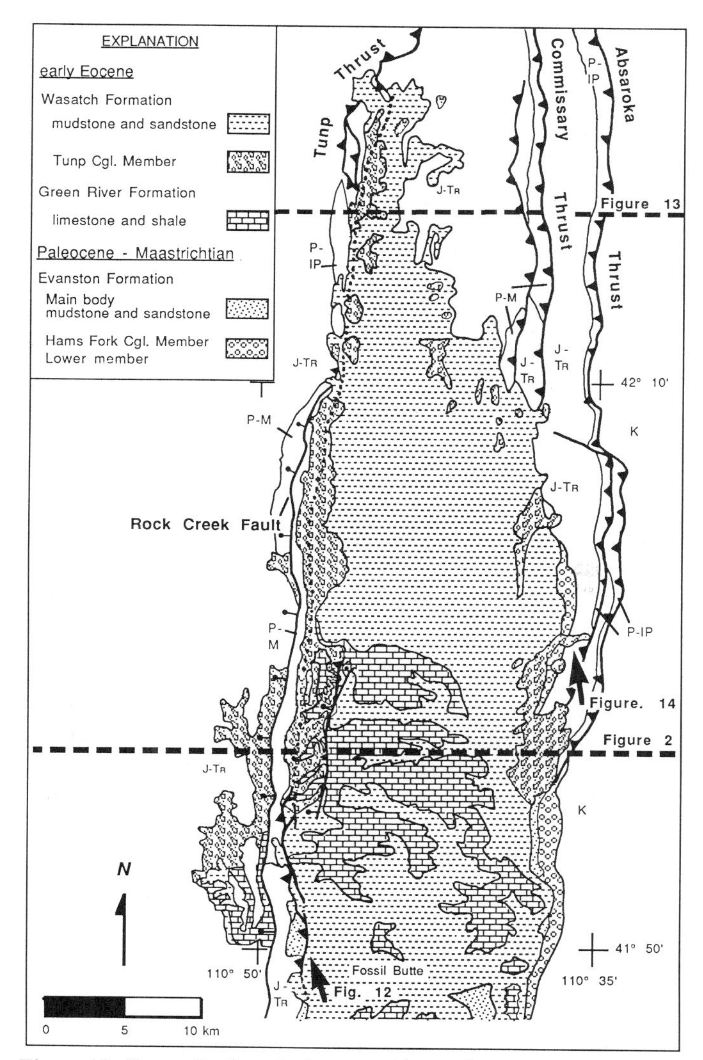

Figure 11. Generalized geologic map of latest Cretaceous and Tertiary strata in northern Fossil Basin on the Absaroka sheet. Northern Fossil basin has a symmetrical facies distribution between the coarse-grained deposits of the Tunp Member along the basin margins and the fine-grained Green River and Wasatch deposits in the basin center. The faulted western basin margin strata overlie the Tunp thrust. The trend of the eastern basin margin overlies the Hogsback thrust trailing ramp in the Absaroka footwall. P-M = Permian, Pennsylvanian, and Mississippian rocks. See Figure 5 for other stratigraphic symbols. Location of structural cross sections in Figures 2 and 13 shown by dashed lines. Location of photographs in Figures 12 and 14 shown by arrows along direction of view. From Rubey and others (1975, 1980).

Tunp thrust soles into the trailing footwall ramp of the Absaroka thrust system and that the leading edge of the Absaroka thrust plate lies above the Hogsback trailing ramp (Fig. 2).

The age and stratigraphic relationships of northern Fossil Basin (Fig. 4) were established by Oriel and Tracey (1970) and further refined by Lamerson (1982), Jacobson and Nichols (1982), and Hurst and Steidtmann (1986). Fossil Basin began as a structural and topographic depression during late Paleocene deposition of the fine-grained, fluvial main body of the Evanston Formation. Prior to that time, the Maastrichtian Hams Fork Conglomerate Member was deposited in a through-going braided stream system that crossed the area of Fossil Basin. The Hams Fork contains boulders and cobbles of Late Proterozoic quartzites that were derived from the Willard and Paris thrust sheets and transported 60 km (37 mi) across the future depocenter to their present position on the east flank of Fossil Basin (Oriel and Tracey, 1970; Salat, 1989).

Northern Fossil Basin was a well-developed north-south trending trough on the Absaroka hanging wall by early Eocene time. Both the western and eastern basin margins of northern Fossil Basin were topographically elevated at the start of Wasatch and Green River deposition, and both margins continued to be uplifted throughout infilling of the basin. The resultant early Eocene basin exhibits a symmetrical facies distribution, with the Tunp Member of the Wasatch Formation ringing the west, north, and east peripheries of the basin (Fig. 11). Hurst and Steidtmann (1986) found that the Tunp Member is dominated by matrix-supported gravels that were transported in alluvial fans toward the basin center. The Tunp Member interfingers basinward with the finer-grained fluvial lower member, main body, and Bullpen Member of the Wasatch as well as with the lacustrine limestones of the Fossil Butte and Angelo members of the Green River Formation.

Oriel and Tracey (1970) and Rubey and others (1975) considered both margins of Fossil Basin to be exclusively pre–early Eocene features. They interpreted the onlap and basinward dips of Wasatch strata along the basin margins to be the result of sedimentation onto the differentially eroded highs of the Tunp thrust hanging wall and the Absaroka thrust front, compaction of underlying sediments beneath the basin, and late block faulting. Differential erosion was important in maintaining basin margin relief throughout basin development, but it does not fully account for the early Eocene relief of the margins of Fossil Basin. For example, the lower Jurassic through Pennsylvanian ridge-forming units exposed in the Tunp hanging wall and the Absaroka thrust front are also found in subcrop beneath the center of northern Fossil Basin in the hanging wall of the Commissary thrust. As shown on Figure 2, the basal unconformity displays little relief between the resistant hanging-wall rocks and less resistant upper Jurassic footwall strata (Lamerson, 1982, Plate 6). Syndepositional uplift of erosionally resistant strata is required to explain the topographic and structural relief of both margins of northern Fossil Basin.

Structural evolution of northern Fossil Basin

Western basin margin—Tunp Thrust. The Tunp thrust has a multiphase slip history. The late Paleocene main body of the Evanston Formation (Oriel and Tracey, 1970; Lamerson, 1982; Jacobson and Nichols, 1982) overlies an angular unconformity above Jurassic and Triassic rocks of Tunp hanging wall (Rubey and others, 1975, Plate 1). Thus, initial uplift of the Tunp hanging wall occurred prior to late Paleocene time. Five features of the western margin of Fossil Basin demonstrate that the Tunp hanging wall was the site of continued uplift in the late Paleocene and early Eocene.

(1) The main body of the Evanston Formation is rotated to near vertical dips along the Tunp thrust front (Rubey and others 1975, Plate 1). The Evanston is overlain with angular unconformity by the basal conglomerate member, lower member, and main body of the Wasatch Formation (Rubey and others 1975), which indicates that uplift and east-vergent rotation occurred during latest Paleocene to earliest Eocene time.

(2) The basal conglomerate and lower members of the Wasatch Formation are also rotated to steep east dip along the southeastern edge of the Tunp hanging wall (Rubey and others, 1975, Plate 1). Intraformational unconformities between these units in the Tunp hanging wall imply that progressive tilting and uplift occurred during Wasatch deposition.

(3) The 350 m (1,150 ft) to 500 m (1,640 ft) of structural relief between equivalent late Paleocene and early Eocene stratigraphic positions in the Tunp hanging wall and footwall is adequately explained only by early Eocene uplift of the Tunp hanging wall. The unconformity between the main body of the Evanston Formation and the basal conglomerate and lower members of the Wasatch Formation lies at elevations of up to 2,200 m (7,200 ft) in the southern Tunp hanging wall (Rubey and others, 1975, Plate 1), whereas the Evanston-Wasatch contact is near 1,830 m (6,000 ft) elevation in the Tunp footwall near the basin center (Lamerson, 1982, Plate 2). Similarly, the lower member of the Wasatch Formation lies at 2,450 m (8,050 ft) elevation farther to the north in the Tunp hanging wall (Rubey and others, 1975, Plate 1), and the lower Wasatch strata are near 1,950 m (6,400 ft) elevation in the basin center (Lamerson, 1982, Plate 6). This structural relief is too large to be the result of compaction of underlying basin fill as suggested by Oriel and Tracey (1970) and must be the result of slip on the surface and buried faults along the eastern edge of the Tunp thrust hanging wall.

(4) West-dipping thrust, reverse, and normal faults along the east edge of the Tunp hanging wall truncate Wasatch strata (Figs. 11 and 12). Rubey and others (1975, p. 14) interpreted these faults as high-angle block faults related to the post-thrust regional extension seen in the Basin and Range Province to the west. The west-dipping Rock Creek fault (Fig. 11) is a late listric normal fault that is shown by well control to sole into the Tunp thrust at depth (Lamerson, 1982, Plate 2). West-dipping faults that cut Wasatch strata east of Rock Creek exhibit changes from reverse to normal offset along the strike of the same fault trace (Fig. 11; Rubey and others, 1975, Plate 1). These west-dipping faults are probably imbricates to the Tunp thrust that were locally reactivated during the later episode of normal slip on the Tunp thrust plane that was documented for the Rock Creek fault. To the north, these west-dipping faults are overlain by the Tunp Member of the Wasatch Formation, which indicates that they initiated in the early Eocene prior to deposition of the overlying strata (Rubey and others, 1975, p. 14) and prior to the inception of regional extension in late Cenozoic time (Stewart, 1978).

Figure 12. View to the north of a thrust fault on the west side of Fossil Basin that places the Triassic Ankareh Formation (Ŧ̄a) and the Jurassic Nugget Sandstone (Jn) over the lower member of the Wasatch Formation (Twl). Location is shown on Figure 11.

(5) Thin limestones of the Fossil Butte Member of the Green River Formation (Fig. 4) dip up to 58° east where they are interbedded with the Tunp Member of the Wasatch Formation above the buried Tunp thrust and its imbricates (Rubey and others 1975, Plate 1; 1980, Plate 1). These obvious tectonic dips demonstrate that uplift of the Tunp hanging wall continued into or beyond the later stages of basin deposition. The concealment of the thrust beneath the Tunp Member seems to have been one reason why earlier workers did not recognize early Eocene slip on the fault. Burial of the fault should be an expected feature along the western basin margin where sequential uplift of resistant hanging wall rocks provided elevated source areas for alluvial fans that continually buried the fault trace during sediment transport into the basin.

Figure 13 is a detailed cross section through the northern tip of Fossil Basin in the area of shallowest basin fill that illustrates the relationship between the Tunp thrust, the Absaroka thrust sheet, and the Tunp Member of the Wasatch Formation. The branching relationship between the Tunp thrust and the Absaroka trailing ramp is geometrically constrained by the intersection of the steep west-dipping panel of the Tunp hanging wall with the seismically defined position of the Absaroka trailing ramp (Dixon, 1982). The subsurface geometry of the Tunp thrust footwall is projected from the Amerada-Sunmark #1-19 Federal well (Sec. 19, T27N, R117W) 6.5 km (4 mi) north of the section and the Amerada #1-23 Federal well (Sec. 23, T25N, R118W) 14.5 km (9 mi) south of the section. Both wells penetrated the Tunp thrust, the Commissary thrust, and the underlying duplex of Cambrian and Ordovician strata. The duplex system in the Tunp footwall did not contribute to early Eocene uplift of the western margin of Fossil Basin. The Commissary and Absaroka thrusts

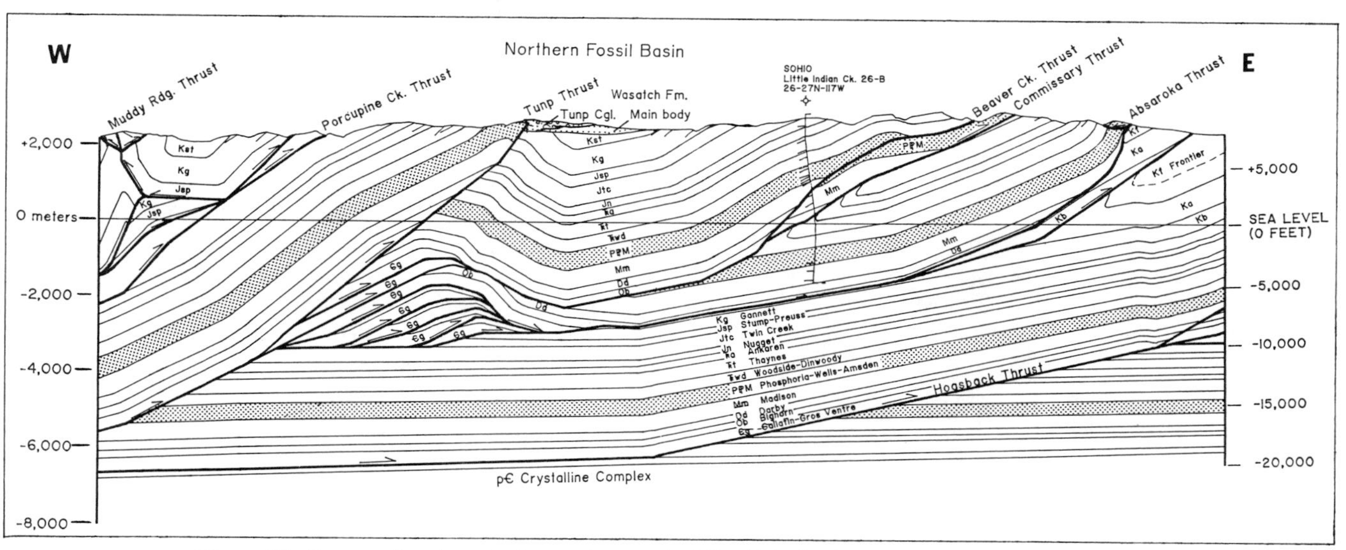

Figure 13. Cross section through the northern end of Fossil Basin showing the geometric relationship between the Tunp Member of the Wasatch Formation and its source area on the Tunp thrust hanging wall. Notice that the Tunp thrust soles into the Absaroka trailing ramp and that the eastern basin margin overlies the Hogsback trailing ramp. Vertical scale = horizontal scale.

are roof thrusts to the duplex system, and both thrusts are overlain by late Paleocene strata along strike (Figure 2; Lamerson, 1982, Plates 2 and 6), indicating that slip along the duplex imbricates predates early Eocene uplift. Consequently, reactivated slip along the Absaroka trailing ramp bypassed the mechanically locked Absaroka footwall flat. Slip on the Tunp thrust is the only possible mechanism for the early Eocene uplift along the western margin of northern Fossil Basin.

Eastern basin margin—Absaroka thrust and Hogsback ramp. The leading edge of the Absaroka thrust sheet exhibits crosscutting and overlapping relationships that firmly establish the age of slip along the Absaroka thrust itself as well as the age of footwall uplift above the Hogsback trailing ramp. The last, minor reactivated slip on the frontal Absaroka thrust during the late Maastrichtian or early Paleocene cut the Hams Fork Conglomerate, and the leading edge of the Absaroka thrust sheet was onlapped and overlain by the late Paleocene main body of the Evanston (Lamerson, 1982, Plates 2 and 9). The pre–late Paleocene uplift was the earliest expression of the eastern basin margin along the leading edge of the Absaroka thrust sheet.

The leading edge of the Absaroka thrust sheet was the site of uplift throughout early Eocene deposition of the Wasatch and Green River formations. The late Paleocene overlap of the Absaroka fault precluces reactivated slip as a mechanism for uplift, yet the main body of the Evanston Formation is rotated to 15° west dip along the eastern basin margin (Rubey and others, 1975), and it is successively overlain by the main body and Tunp Member of the Wasatch Formation (Fig. 14). Lamerson (1982, p. 298–299) first proposed that slip along the underlying Hogsback thrust ramp was responsible for uplift of the Absaroka footwall, the mechanically locked Absaroka thrust, and the overlying hanging-wall and overlap strata. Hurst and Steidtmann (1986) concluded that the alluvial fans of the Tunp Member of the Wasatch along the east side of northern Fossil Basin were shed from this area of uplift.

The relief on the eastern margin of Fossil Basin changes drastically north and south of the area shown in Figure 2 (Fig. 1). The relief is controlled by the position and amount of structural relief of the Hogsback trailing ramp beneath the Absaroka thrust sheet as well as by the erosional resistance of strata carried in the Absaroka hanging wall. There is direct correspondence between the position of the eastern basin margin and the underlying Hogsback trailing ramp. The westward deflection in the position of the Hogsback ramp from that shown in Figure 2 to that shown in Figure 13 parallels and underlies the westward deflection of the eastern margin of northern Fossil Basin (see also Dixon, 1982, Fig. 18). At the latitude of the area shown in Figure 13, footwall uplift above the Hogsback trailing ramp elevated most of the Absaroka footwall flat and was not simply limited to the leading edge of the Absaroka thrust plate. South of the area shown in Figure 2, the Hogsback trailing ramp lies immediately beneath the outcrop and subcrop trace of the Absaroka thrust (Lamerson, 1982, Plates 8, 5, 4, and 10), and the relief of the eastern margin of Fossil basin is correspondingly low. Between the area shown in Figure 2 and the Little Muddy Creek area (Fig. 1), resistant lower Jurassic through Pennsylvania rocks form the subcrop toe of the Absaroka thrust (Lamerson, 1982, Plate 8). In this case, the low basin margin relief cannot be attributed to differential erosion and must be the result of the amount of footwall uplift that was imparted by slip on the underlying Hogsback ramp. South of

Figure 14. View to the north of the eastern margin of Fossil Basin at Trail Creek. Foreground and small peak in background are capped by the flat-lying Tunp Member of the Wasatch Formation (Twt), which overlies steeply dipping Triassic (TR) and Jurassic (J) rocks of the Absaroka hanging wall as well as the steeply dipping Hams Fork Member (Keh) and lower member (Kel) of the Evanston Formation. Onlap by the shallowly west dipping lower member (Twl) and main body of the Wasatch Formation (Tw) is seen to the left. The progressive tilting and erosion of latest Cretaceous through early Eocene strata were the result of Paleocene through early Eocene footwall uplift above the Hogsback trailing ramp.

Little Muddy Creek, the low relief of the basin margin is enhanced by the erosionally soft upper Jurassic and lower Cretaceous rocks carried at the toe of the Absaroka thrust (Lamerson, 1982, Plates 5, 4, and 10), but the fact that the Absaroka thrust trace remains buried beneath late Paleocene and early Eocene basin sediments indicates that early Eocene footwall uplift was minimal along the southern length of the thrust. Finally, the early Eocene eastern margin of Fossil Basin disappears south of the Utah-Wyoming boundary where the Hogsback trailing ramp lies east of the buried Absaroka thrust trace and early Eocene deposits in southeastern Fossil Basin are continuous with those in the southwestern Green River Basin (Fig. 1).

Northern Fossil Basin provides the clearest evidence for the combined effect of reactivated thrust uplift along the trailing basin margin and footwall uplift along the leading basin margin. Unlike Bear Lake Plateau basin, Fossil Basin exhibits a symmetrical filling about a fixed depocenter, which implies that reactivated thrust uplift and erosion above the Absaroka trailing ramp kept pace with footwall uplift and erosion of the Absaroka thrust front above the Hogsback trailing ramp.

LABARGE BASIN

Structural, stratigraphic, and sedimentologic setting

A small earliest Eocene piggyback basin, which is informally referred to as the LaBarge basin, is located between the trailing ramp and the leading edge of the Hogsback thrust sheet west of LaBarge, Wyoming (Fig. 1). Later early Eocene strata that overlap the piggyback basin above an angular unconformity define the western edge of the greater Green River Basin. Figure 15 shows the structural position of the Hoback, Wasatch, and Green River formations relative to the trailing ramp, leading edge, and footwall imbricates of the Hogsback thrust at higher structural levels and to the Moxa Arch at basement level. Seismic data indicate that the Hogsback trailing ramp dips approximately 13° west from its position immediately west of LaBarge basin (Fig. 15) to its position beneath the Absaroka thrust plate (Fig. 13). The Meridian and Fort Hill thrusts are hanging-wall imbricates of the Hogsback thrust that sole into the area of transition between the trailing footwall ramp and the frontal footwall décollement of the Hogsback thrust (Fig. 15). The buried traces of the Meridian and Fort Hill thrusts lie on the western margin of LaBarge basin (Fig. 15). The LaBarge and Calpet thrusts are conjugate footwall imbricates of the Hogsback thrust that are linked to the Hogsback trailing ramp along a Cretaceous décollement in the Hogsback footwall (Fig. 15). The LaBarge and Calpet thrusts intersect in a triangle zone (Teal, 1983) east of the Hogsback thrust front where the displacement on the east-dipping Calpet thrust was transferred to the upper part of the west-dipping LaBarge thrust in the subsurface.

The age and intertonguing stratigraphic relationships between Paleocene through early Eocene strata of the LaBarge basin were first treated by Oriel (1962, 1969) and refined by Dorr and Gingrich (1980) as presented on Figure 4. The Chappo Member of the Wasatch Formation is the oldest Cenozoic unit exposed at LaBarge. The type locality of the Chappo Member (Oriel, 1962) lies east of the Hogsback thrust and contains late Paleocene (Tiffinian) strata (Dorr and Gingrich, 1980). Well correlations indicate that conglomerates at the base of the late Paleocene Chappo Member are correlative to conglomerates in the Hoback Formation east of the Hogsback thrust front (Oriel, 1969; Dorr and Gingrich, 1980). Earliest Eocene (Clarkforkian) strata were also mapped as the Chappo Member by Oriel (1969) in the hanging wall of the Hogsback thrust. The presence of two distinct outcrop belts of the Chappo Member with different ages and structural affinities is a source of confusion, and a redefinition of the Chappo Member should be considered in the future. In this study, the two stratigraphic units are discussed separately as the late Paleocene Chappo Member and the early Eocene Chappo Member, and the late Paleocene Chappo Member is grouped with the Hoback Formation in the subsurface in Figure 15.

The early Eocene LaBarge and Lookout Mountain Conglomerate members of the Wasatch Formation overlie both outcrop belts of the Chappo Member with angular unconformity. The Lookout Mountain Conglomerate consists of interbedded diamictite, conglomerate, sandstone, and mudstone that resemble the Tunp Member of Fossil Basin and the diamictite member of Bear Lake Plateau basin. The diamictites of the Lookout Mountain Conglomerate contain angular clasts of Mesozoic and Paleozoic strata derived from the trailing ramp area of the Hogsback thrust plate, including clasts up to 2 m (6 ft) in length (Oriel,

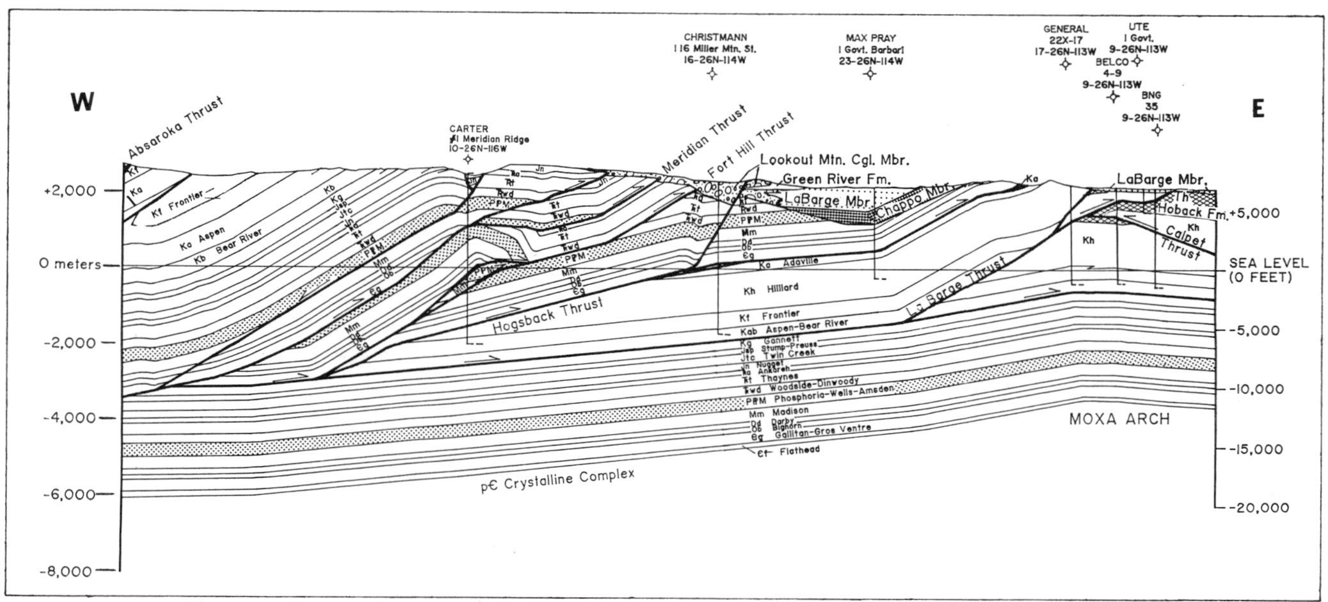

Figure 15. Cross section through the Hogsback thrust and the Moxa Arch in the LaBarge area, showing the geometry of the folded Chappo Member of the Wasatch Formation in LaBarge basin. LaBarge basin lies between an area of uplift above the Hogsback trailing ramp and the Hogsback thrust front above the LaBarge thrust ramp. The LaBarge and Calpet thrusts are conjugate footwall imbricates to the Hogsback thrust that form a triangle zone above the Moxa Arch. Later deposition along the western margin of the Green River Basin overlapped both LaBarge basin and the triangle zone with angular unconformity. The late western margin of the Green River Basin is defined by the Lookout Mountain Conglomerate alluvial fan deposits that lie above the Meridian thrust. Vertical scale = horizontal scale.

1962). These gravels are commonly matrix supported and were transported eastward on alluvial fans. The Lookout Mountain Conglomerate interfingers eastward with the fine-grained fluvial deposits of the LaBarge Member, New Fork Tongue, and upper tongue of the Wasatch Formation as well as with the lacustrine limestones of the Fontenelle tongue and the middle tongue of the Green River Formation.

Structural evolution of the LaBarge basin

LaBarge basin. LaBarge basin lies immediately west of Hogsback thrust front, the LaBarge and Calpet footwall imbricates, and the Moxa Arch basement uplift (Fig. 15). The Moxa Arch was uplifted by late Campanian time and predates all other structures in the area (Wach, 1977). The Hogsback thrust was fully emplaced by the late Paleocene (Warner and Royse, 1987). The thrust cuts the late Paleocene Hoback Formation 5 km (3 mi) north of the area shown in Figure 15 (Blackstone, 1979, Plate 6; Warner and Royse, 1987) and is in turn overlain by the earliest Eocene part of the Chappo Member (Dorr and Gingrich, 1980).

LaBarge basin developed after some 4.5 km (2.7 mi) of Cretaceous through Paleozoic strata were eroded from the Hogsback hanging wall. The early Eocene Chappo Member of the Wasatch Formation is the oldest unit exposed above the eroded Hogsback thrust sheet (Dorr and Gingrich, 1980). The early Eocene Chappo Member coarsens westward above the Hogsback thrust and contains clasts derived from Mesozoic and Paleozoic units that are exposed above the Hogsback trailing ramp to the west (Oriel, 1962). The westward onlap of Wasatch strata is evidence that considerable topographic relief existed above the Hogsback ramp in the area of known early Eocene uplift that was discussed earlier for eastern Fossil Basin. The onlap of the early Eocene Chappo Member along the eastern margin of LaBarge basin is more subtle, but the geometry of the Hogsback thrust sheet requires that deposition in the piggyback basin followed initial footwall uplift of the Hogsback thrust front.

Paleozoic rocks along the leading edge of the Hogsback thrust front did not form a high topographic boundary along the eastern basin margin prior to slip on the underlying LaBarge thrust. By restoring the Paleozoic strata to the 6° west dip that they held prior to LaBarge thrust slip, the unconformity beneath the early Eocene Chappo becomes essentially flat between its location above Mississippian rocks in the Max Pray well and its location above Cambrian rocks along the Hogsback thrust front (Fig. 15). The eastward onlap of the early Eocene Chappo Member onto the unconformity surface (Dorr and Gingrich, 1980) implies that an increment of uplift preceded Chappo deposition. The present 10 to 25° west and southwest dips of the early Eocene Chappo Member along the eastern basin margin demonstrate that uplift continued after Chappo deposition. Uplift of the eastern basin margin was caused by slip along the LaBarge thrust ramp in the footwall of the Hogsback thrust. This uplift occurred prior to deposition of the flat-lying LaBarge Member, which overlies the early Eocene Chappo Member with angular unconformity at the surface (Oriel, 1969).

Earliest Eocene slip along the LaBarge and Calpet thrusts had to be balanced by coeval slip along the Hogsback trailing ramp because the thrusts are geometrically and mechanically linked along the Cretaceous décollement in the Hogsback footwall (Fig. 15). Thus, footwall uplift of the eastern basin margin during early Eocene Chappo deposition was matched by late slip and uplift along the Hogsback trailing ramp on the western basin margin.

Western margin of Green River Basin. The present western margin of the Green River Basin developed with the onset of LaBarge Member deposition following uplift and erosion of the underlying piggyback basin strata and the LaBarge and Calpet thrust hanging walls. The depth of the contact between the LaBarge and Chappo members encountered in the Max Pray well (Fig. 15) indicates that the LaBarge Member initially infilled the piggyback basin above the Hogsback thrust sheet while onlapping the Chappo Member to the east.

Above the LaBarge Member, the western Green River Basin margin is defined by an eastward progression from alluvial fan deposits of the Lookout Mountain Conglomerate through fine-grained fluvial strata of the Wasatch Formation and into lacustrine limestones of the Green River Formation. Lookout Mountain Conglomerate clasts were clearly generated from the area of the Hogsback thrust sheet that overlies the Hogsback trailing ramp. Late slip on the Hogsback trailing ramp likely caused minor late uplift of the western basin margin based on surface and subsurface observations. Gentle warping of the LaBarge Member above the LaBarge thrust mapped by Oriel (1969) demonstrates that some minor slip reached the thrust front during Lookout Mountain deposition. However, most of the reactivated slip along the Hogsback trailing ramp probably bypassed the leading edge of the steeply dipping and mechanically locked footwall imbricates. Instead, some late shortening could have been accommodated along the Meridian thrust or its imbricates (Fig. 15). The Meridian thrust soles into the Hogsback trailing ramp area in a structural style and position analogous to that of the Tunp thrust above the Absaroka trailing ramp. Surface observations that support early Eocene slip on the Hogsback trailing ramp include the following: (1) Down-to-the-east faults mapped as block faults by Oriel (1969) both cut and overlie the Lookout Mountain Conglomerate above the Meridian thrust in a setting analogous to the Tunp Member and the Tunp thrust of western Fossil Basin, (2) gentle folds in the Lookout Mountain Conglomerate overlie the subcrop of Meridian thrust imbricates, and (3) Wasatch diamictites reported by Tracey and others (1961) were directly involved in thrusting 27 km (16 mi) north of the LaBarge area at the north end of the Hogsback thrust plate. Uplift of the eastern margin of Fossil Basin above the Hogsback trailing ramp during this period also implies continued late uplift along the western margin of the LaBarge area. In addition, Delphia and Bombolakis (1988) interpreted break-back thrust imbrication above the Hogsback trailing ramp based on palinspastic reconstructions of the Hogsback thrust sheet 50 km (30 mi) south of the LaBarge area.

DISCUSSION: CAUSE OF REGIONAL THRUST REACTIVATION

Beginning in latest Paleocene, the eastern half of the Wyoming-Idaho-Utah thrust belt experienced a fundamental change in structural style and mechanical behavior that culminated in the development of early Eocene piggyback basins. Reactivated slip along the trailing ramps of the Crawford and Absaroka thrusts and late slip along the Hogsback trailing ramp marked an episode of break-back thrusting that contrasts with the earlier main-phase, foreland-directed thrust sequence. The initiation of break-back thrusting during the early Eocene raises the question of whether the anomalous pattern of thrust slip was a response to unique mechanical conditions in the thrust belt during the final phase of regional shortening. A likely reason for break-back thrust episode is the impedance of further slip on the frontal Hogsback-Darby-Prospect thrust zone, coupled with the accommodation of further shortening by reactivation of mechanically interconnected thrusts behind the thrust front. The mechanical interconnection of the Crawford, Absaroka, and frontal thrusts is outlined below.

Mechanical organization of thrusts

Thrust faults in the Wyoming-Idaho-Utah thrust belt are divided into two separate geometric and mechanical thrust systems from west to east (Coogan, 1987). The separate thrust systems are defined by thrust surfaces that branch from a common basal décollement. The Willard, Paris, and Meade thrusts make up a western thrust system that is defined by a master sole décollement at the base of the thick Late Proterozoic clastic sequence of the Cordilleran miogeocline. In contrast, the Crawford, Absaroka, Hogsback-Darby, and Prospect thrusts are linked along a master sole décollement in Cambrian shales at the base of the thin Wyoming shelf sequence (Fig. 2).

The geometric, temporal, and mechanical separation between these two systems is displayed in the Wasatch Mountains near Ogden, Utah. There, the Willard thrust is folded above a basement anticlinorium associated with the underlying eastern thrust system (Royse and others, 1975, Plate II). The Archean Farmington Canyon Complex forms the core of a mid-crustal trailing ramp anticline in the interior of the eastern thrust system, where the basement rocks were translated onto a décollement in the Cambrian Gros Ventre shale of the Wyoming shelf sequence. The uplift onto and translation across the Cambrian décollement represent basement shortening that is equivalent to thin-skinned shortening along the successively eastward-younging Crawford,

Absaroka, and Hogsback thrusts (Coogan, 1987). The successive uplift associated with Crawford, Absaroka, and Hogsback thrusting folded the Willard thrust and mechanically isolated the western thrust system from further slip. The uplift of the western thrust system above the interior ramp of the eastern thrusts is the clearest example of the interior ramp-supported uplift mechanism discussed by Schmitt and Steidtmann (1990).

The basement anticlinorium of the Wasatch Mountains was progressively exposed as a clastic source area during main-phase Crawford (DeCelles, 1988) and Absaroka (Schmitt and Steidtmann, 1990) thrust slip, and it was the source area for conglomerates of the Wasatch Formation during the last phase of regional shortening (Crawford, 1979). The final early Eocene interior uplift of the Wasatch Mountains during deposition of the Wasatch Formation was matched to the east by uplift of piggyback basin margins above the Crawford, Absaroka, and Hogsback thrusts. Early Eocene slip along the trailing ramps of these thrusts is one expression of their mechanical interconnection along the Gros Ventre sole décollement of the eastern thrust belt. Break-back thrust slip along the Crawford and Absaroka trailing ramps accommodated shortening along the Gros Ventre décollement system during the cessation of slip on the frontal Hogsback thrust.

Causes of slip impedance along the thrust front

The cause of frontal slip impedance during the final stages of shortening in a thrust belt is part of a larger problem of why thrust belts stop moving. Recent material wedge models of thrust belts have been successful in explaining how thrust belts propagate (Chapple, 1978; Davis and others, 1983). Given sufficient push at the rear of the belt, a thrust belt with reasonable material properties should continue to maintain the shape required for it to move indefinitely (Chapple, 1978, p. 1192). Final Paleocene through early Eocene shortening in the Wyoming-Idaho-Utah thrust belt geographically and temporally overlapped shortening of basement and cover rocks in the Wyoming foreland (Perry and Schmidt, 1988). Hamilton (1988) correlates the end of foreland thin-skinned and basement shortening with a slowing of plate convergence rates in early Paleogene time. On a regional scale, the cessation of thrusting is probably the result of a diminished tectonic push at the rear of the thrust belt. On a smaller scale, the impedance of slip on the frontal thrusts during the last phase of thrust shortening can be related to mechanical interaction of the thrust front with specific and general features of the foreland. Four mechanisms for the impedance of slip along the thrust front are examined as explanations for reactivation on trailing thrusts: (1) buttressing against foreland basement uplifts, (2) thrust deflection above foreland basement warps, (3) decreased taper of the regional thrust wedge caused by foreland subsidence and thrust belt erosion, and (4) stratigraphic and rheologic changes across the eastward-thinning sedimentary section from thrust belt to foreland. These mechanisms probably interacted in different locations and at different scales during the final episode of thrust shortening.

Foreland buttressing. The term *foreland buttress* is limited in this chapter to cases in which frontal thin-skinned thrust sheets abut preexisting or coeval foreland basement uplifts. The buttress geometries consist of interconnected and crosscutting networks of west-dipping thin-skinned thrusts and east-dipping basement-rooted thrusts. The best example of a buttressed thrust front is located 100 km (62 mi) north of the LaBarge basin. There, west-dipping frontal thrusts of the Prospect thrust zone intersect coeval, east-dipping, basement-involved thrusts of the Gros Ventre uplift. Royse (1985) and Hunter (1988) demonstrated that the Prospect thrust system is composed of a break-back thrust sequence in the area where the frontal Granite Creek thrust was overridden by the basement-rooted Cache Creek blind thrust. Successive hanging-wall imbricates (Prospect, Little Granite Creek, Bull Creek, Game Creek, and Bear thrusts) to the Granite Creek thrust formed in a westward-younging sequence (Hunter, 1988). Although the northern Prospect thrust front provides a well-documented type example of foreland buttressing in the Wyoming thrust belt, the Gros Ventre buttress is too distant to have influenced Hogsback thrust slip to the south.

A foreland buttress model for the Moxa Arch was advanced by Kraig and others (1988) along the southern length of the Prospect thrust north of LaBarge, Wyoming. Dixon (1982), Royse (1985), and Kraig and others (1988) demonstrated that the Moxa Arch rapidly increases in structural relief north of the area illustrated in Figure 15. Thirty km (19 mi) to the north, the seismic expression of the Moxa Arch shows approximately 1,850 m (6,070 ft) of structural relief on the basement surface that was uplifted along a reverse fault on the west side of the arch (Kraig and others, 1988). Kraig and others (1988) interpret that pre-Paleocene slip on the basement fault propagated upward to a Triassic décollement in the cover rocks that was shared by an early frontal segment of the Prospect thrust. The pre-Paleocene basement uplift and the shared slip planes for thrust belt and foreland structures imply that the Moxa Arch was a preexisting buttress that guided slip on the frontal Prospect thrust in late Paleocene and early Eocene. It is plausible that foreland buttressing of the southern Prospect thrust could have influenced final slip on the northern Hogsback thrust immediately to the south near LaBarge (Fig. 1). At LaBarge, however, the Moxa Arch is a gentle, unfaulted basement warp where the local influence of the arch on the geometry and kinematics of the frontal thrusts is not obvious.

Foreland basement warp. Wiltschko and Eastman (1983) experimentally demonstrated that foreland basement warps can concentrate stress and deflect fault trajectories in the overlying sedimentary section during horizontal compression. Their photoelastic model of the Moxa Arch (Fig. 16) illustrates the stress distribution along a block of photoelastic material that is loaded from the left and fixed on the right above a rigid substrate (Wiltschko and Eastman, 1983, Fig. 4b). The highest stresses are concentrated along the top of the block on the fixed side of the arch. Wiltschko and Eastman (1983) indicate that shallow faults should form through this area, which generally agrees with the

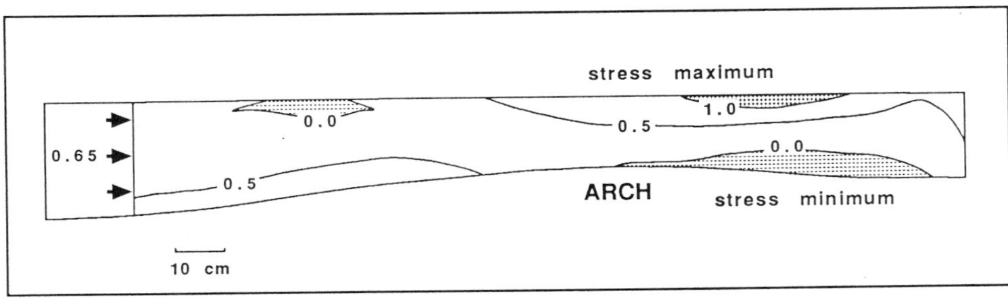

Figure 16. Stress distribution in photoelastic model of the Moxa Arch constructed by Wiltschko and Eastman (1983). A wedge of high polymer plastic was loaded laterally from the left and fixed on the right. Stress is represented by fringe orders that result from a 0.65 order load. The area of highest stress (0.5–>1.0 order) is located above the arch crest along the top of the wedge, indicating that shallow faults should form through this area. The area of lowest stress (0 order) lies in a stress shadow localized at the base of the wedge along the right flank of the arch. The model implies that foreland basement warps serve to deflect deep faults to shallower structural levels and that deep thrusts should not propagate beyond the crest of basement warps. From Wiltschko and Eastman (1983, Figure 4b).

observed position of the Calpet and LaBarge thrusts at shallow structural levels as shown in Figure 15. In contrast, the lowest stresses occur along the base of the block to the right of the arch crest. This stress shadow would impede thrust propagation near the basement-cover contact on the foreland side of the arch. Wiltschko and Eastman (1983) concluded that the Moxa Arch primarily served to deflect deep faults west of the arch to shallower structural levels above the arch and that deep thrusts should not propagate east of the arch. The model reveals that gentle basement warps can provide a tip, or sticking point, to further slip along the basal décollement near the basement cover interface in front of a thrust belt.

Foreland buttressing and thrust deflection above basement warps cannot be the only mechanisms responsible for slip impedance along the frontal thrust belt and reactivated slip in the interior. For example, Paleocene and post–early Eocene reactivated slip is documented far to the south of Moxa-Hogsback convergence for the Medicine Butte thrust along the western margin of southern Fossil Basin (Lamerson, 1982). Regional scale mechanisms for frontal slip impedance are required to explain late thrusting in the interior of the belt in areas far from foreland buttresses and basement warps.

Thrust belt taper. Late thrusting in the interior of thrust belts has been cited as a mechanism for maintaining the appropriate shape for stable sliding of thrust belts above a regional basal décollement (Woodward, 1987). Davis and others (1983) proposed that thrust belts propagate as brittlely deforming wedges above weak, through-going décollements as illustrated in Figure 17a. When subjected to a lateral load, such wedges deform internally to a critical taper, at and beyond which they slide stably. Davis and others (1983) related the topographic slope of the wedge α and the basal décollement slope β (Figs. 17a and 18) to the material properties of the wedge and décollement. The line defining the range of stable wedge shapes on Figure 18 divides potential wedge shapes into a supercritical field for stably sliding wedges with greater than critical taper and a subcritical field for wedges that experience internal deformation at less than critical taper.

Woodward (1987) proposed that late faulting in the interior of thrust belts is one expected feature of subcritical wedges. Initial shortening of a subcritical thrust belt could be accommodated by internal thrusts that elevate the rear of the thrust belt and reestablish the critical taper for stable sliding to the front of the belt. Woodward (1987) implied that break-back thrusting should be a common process throughout thrust belt evolution for critical wedge models to work. Such internal shortening and uplift would counter the taper-reducing uplift caused by the successive emplacement of thrust sheets at the front of the wedge (Woodward, 1987). The fact that break-back thrusting is only recognized for the final phase of shortening in the Wyoming-Idaho-Utah thrust belt implies that critical or supercritical taper was maintained prior to that time. Uplift along the interior ramp beneath the Wasatch basement is one mechanism that was overlooked by Woodward (1987) for maintaining an elevated rear of this thrust belt during successive emplacement of the Crawford, Absaroka, and Hogsback thrusts at the front of the belt. However, early Eocene uplift in the Wasatch Mountains area should have also contributed to maintaining wedge taper. Thus, a taper-reducing mechanism that is unique to the final phase of shortening is needed to explain the late break-back thrust events in a critical wedge model.

Differential subsidence of the Wyoming foreland is one taper-modifying mechanism that is unique to the period of final shortening in the thrust belt. Paleocene and Eocene subsidence in the foreland was the result of topographic and tectonic loads from main-phase emplacement of basement uplifts as well as sediment loads from the detritus that was shed from these uplifts into the surrounding intermontane basins. Shuster and Steidtmann (1988) demonstrated that the northern Green River Basin experienced a period of rapid tectonic subsidence in the Paleocene and early

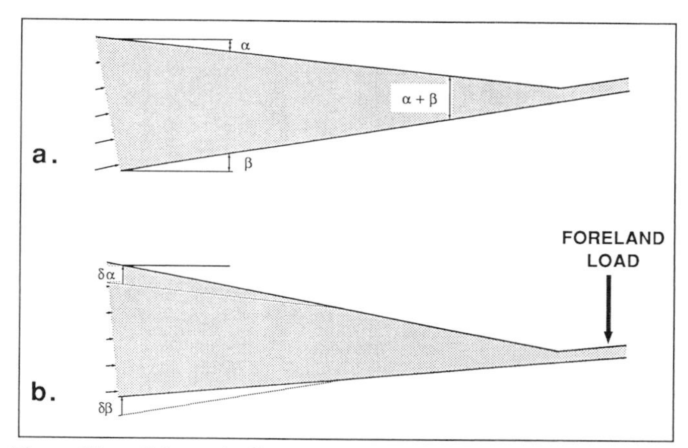

Figure 17. a. Schematic diagram that illustrates the surface slope α and décollement slope β values that define the degree of taper $\alpha + \beta$ of brittlely deforming or stably sliding wedge that is laterally loaded from the thick end and that overlies a weak basal décollement surface. b. Schematic diagram illustrating the change in topographic slope $\delta\alpha$ and décollement dip $\delta\beta$ on the wedge shape of Figure 17a subjected to a foreland flexural load (arrow) with no erosion.

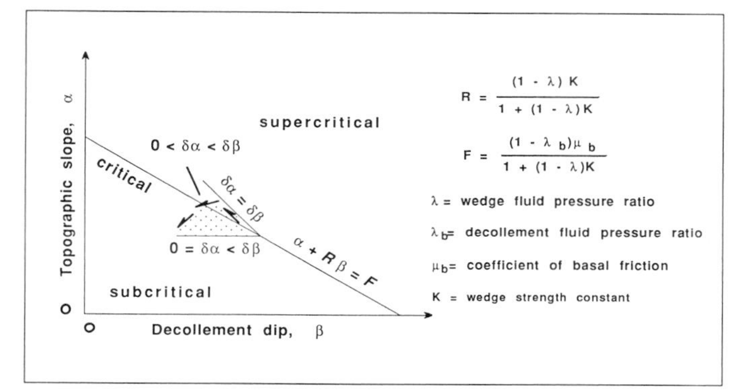

Figure 18. Wedge shape diagram divided by a line of critical wedge shapes into subcritical and supercritical fields of taper. The critical taper equation $\alpha + R\,\beta = F$ relates the topographic slope α and décollement dip β to the fluid pressure properties of the wedge and décollement as well as to the strength of the wedge and the frictional properties of the décollement (see Davis and others, [1983] for derivation). Arrows show the wedge shape path for a model wedge subjected to an instantaneous frontal load and time-dependent erosion. An initially critical wedge can pass through a continuum of supercritical, critical, and subcritical shapes by the combined effects of frontal loading and erosion.

Eocene. The Paleocene subsidence was principally a flexural response to the main phase of uplift for the Wind River and Gros Ventre–Teton basement uplifts to the east and northeast that continued through the early Eocene (Shuster and Steidtmann, 1988). Paleocene and Eocene tectonic subsidence was also demonstrated for the west-central Green River Basin near LaBarge (Shuster and Steidtmann, 1988, Fig. 7), but its relationship to distant foreland uplift loads is more speculative.

The southern margin of the thrust belt experienced similar Paleocene-Eocene subsidence and erosion during uplift of the Uinta Mountains. Figure 19 illustrates the progressive deposition, southeastward tilting, and erosion of the Maastrichtian Hams Fork Conglomerate Member of the Evanston Formation, the late Paleocene upper part of the Evanston Formation, and the early Eocene Wasatch Formation above the thrust belt. The progressive folding and erosion of the upper Evanston and Wasatch formations beneath and adjacent to the Uinta Mountain thrust demonstrate that the erosion and southward tilting of the thrust belt were synchronous with slip on the Uinta Mountain thrust (Fig. 18). The subsidence and deposition, tilting, and erosion of this part of the thrust belt are interpreted to be flexural responses to loading by the Uinta foreland uplift that is analogous to the Paleocene-Eocene flexural subsidence of the northern Green River Basin.

Flexural loads associated with Paleocene to early Eocene emplacement of the Hogsback and Prospect thrust sheets were probably insignificant because erosion of these thrust sheets largely kept pace with uplift (Warner and Royse, 1987). Such erosion is most evident in the LaBarge area where 4.5 km (2.7 mi) of Paleozoic and Mesozoic strata were removed from the Hogsback thrust sheet between late Paleocene emplacement and earliest Eocene overlap of the Wasatch Formation (Fig. 15). The general westward thinning and onlap of Paleocene and early Eocene sediements across the southern thrust belt (Fig. 1) indicate that topographic loads across the belt were insignificant during this period and that the regional subsidence responsible for this onlap was centered to the east in the foreland of the thrust belt.

Flexural subsidence centered in the foreland of a fold-and-thrust belt would reduce the basalt décollement slope (Fig. 17b). In the case in which the reduction in décollement slope $\delta\beta$ is met by an equal increase in topographic slope $\delta\alpha$, the total wedge taper remains unchanged (Fig. 17b). The path $\delta\alpha = \delta\beta$ (Fig. 18) provides one end member of wedge shape that represents instantaneous frontal loading of a wedge without erosion. $0 = \delta\alpha < \delta\beta$ represents the other end member path where uplift of the décollement slope is perfectly matched by erosion and the wedge shape lies entirely in the subcritical field (Fig. 18). Between these two paths lies a field of situations that probably represent the general case for subaerial wedges where $0 < \delta\alpha < \delta\beta$ (shaded on Fig. 18).

If the geometry of a model wedge is tracked through time (arrows on Fig. 18), it is clear that a wedge can pass through a continuum of supercritical through subcritical shapes after foreland subsidence and thrust belt erosion. Because an elastic or viscoelastic lithosphere responds to loading virtually instantaneously (Turcotte and Schubert, 1982), the initial path of a wedge subjected to a frontal load should proceed from an initial critical taper into a supercritical shape approximating the path $\delta\alpha = \delta\beta$. However, erosion rates are time dependent and proportional to relief (Pinet and Souriau, 1988), so the wedge shape would next proceed into the $0 < \delta\alpha < \delta\beta$ field, where it eventually passes through a critical, and eventually subcritical, geometry if the erosion rate approaches the uplift rate. A return to the topographic slope established prior to loading would place the wedge geometry in the subcritical field along the $0 = \delta\alpha < \delta\beta$ path. In addition, flexural or isostatic adjustment from erosion of the topographic load at the rear of the wedge would cause the décollement slope to further decrease, enhancing the subcritical character of the wedge.

The wedge shape path outlined for the frontal flexural load model is offered as a testable hypothesis for frontal slip impedance and internal deformation in the Wyoming-Idaho-Utah thrust belt during the final episode of regional shortening. Unlike most ancient thrust belts, both the décollement slope and the approximate topographic surfaces are well constrained for the late Paleocene and early Eocene of the southern Wyoming-Idaho-Utah thrust belt. In addition, the locations and magnitudes of late Paleocene and early Eocene forelands loads are reasonably contrained, as is the subsidence record of the foreland (Shuster and Steidtmann, 1988).

Foreland stratigraphic and rheologic changes. Mitra and others (1988) noted that the different structural styles of the Wyoming thrust belt and foreland basement uplifts reflect the different physical conditions at the time at which the structures formed. Deformation of the sedimentary cover in the thrust belt generally occurred at greater depths and higher initial temperatures and pressures than in the foreland, largely as the result of the thicker sedimentary section in the thrust belt. The higher temperatures permitted strain-softening mechanisms such as pressure solution slip (Wojtal and Mitra, 1986) to ease thrust slip through the sedimentary cover of the thrust belt. Thicker shale intervals in the Paleozoic rocks of the thrust belt favored the development of regional décollements, particularly where compaction of thick shales helped to maintain high fluid pressures that aid fracture propagation (Atkinson, 1984) and thrust slip (Hubbert and Rubey, 1959). Eastward thinning of important evaporite sequences, such as salt in the Jurassic Preuss Redbeds (Coogan and Yonkee, 1985), could have thwarted thrust slip at higher décollement levels where efficient slip was aided by plastic flow.

The eastward change in stratigraphically controlled rheology is perhaps the most general reason for the cessation of slip along the front of the thrust belt. As physical conditions became less favorable for fracture propagation and basal and internal slip toward the front of the belt, the amount of work required for

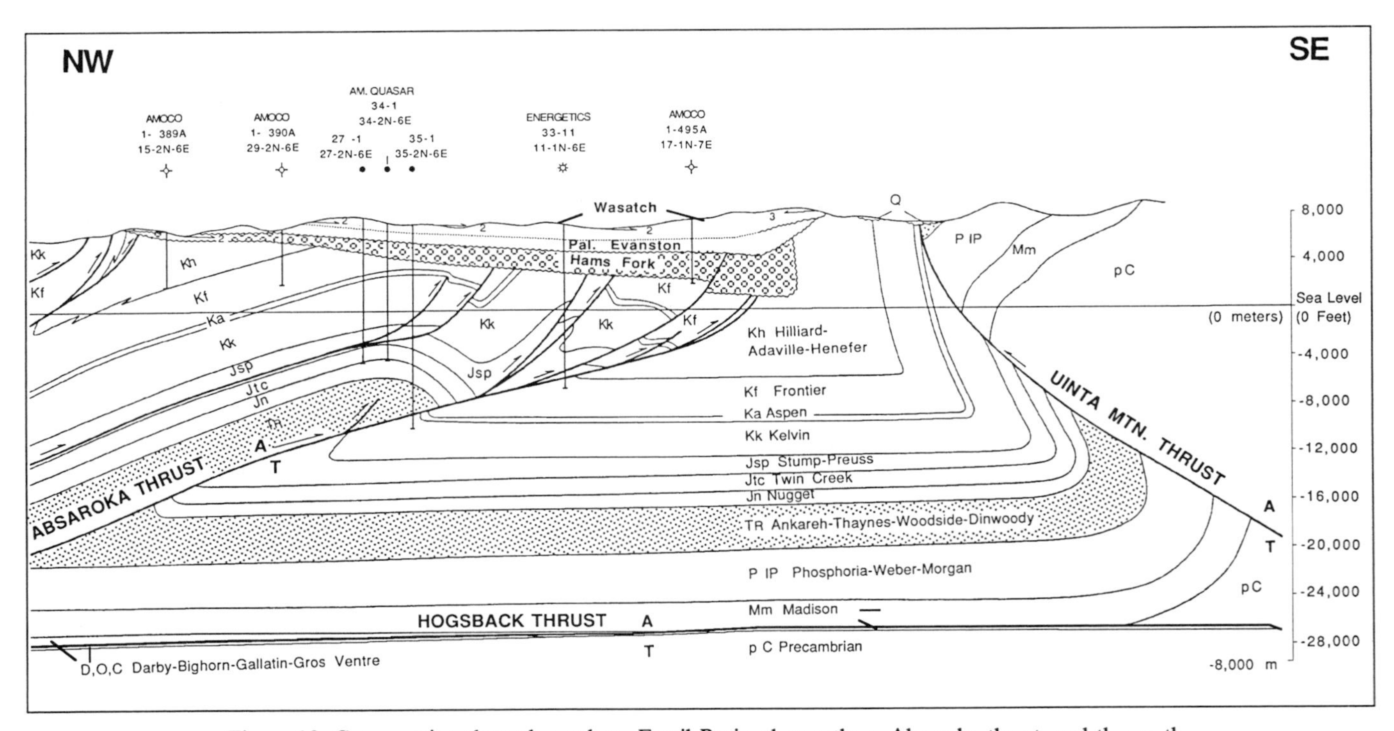

Figure 19. Cross section through southern Fossil Basin, the southern Absaroka thrust, and the north flank of the Uinta Mountains showing the Paleocene to early Eocene subsidence and erosion that accompanied emplacement of the Uinta foreland uplift. The Maastrichtian Hams Fork Conglomerate, the Paleocene Evanston Formation, and the early Eocene Wasatch Formation show southeast dip above the Absaroka thrust sheet. The southeast dip decreases up section and records progressive tilting of the thrust belt during main-phase emplacement of the Uinta uplift. The angular unconformities that bound the Tertiary sequences record continual erosion of the thrust belt during tilting. The cross section is drawn oblique to the transport direction of the Absaroka, Hogsback, and Uinta Mountain thrusts. A = slip component away from viewer; T = slip component toward viewer. Modified from Lamerson (1982, Plate 12 and Figure 22).

further frontal shortening (Mitra and Boyer, 1986) could have exceeded the energy available from the tectonic push applied to the rear of the belt. Break-back reactivation of internal thrusts might have required less work than continued slip along the frontal thrusts or propagation of new thrusts in the foreland. Foreland rheologic changes would also have affected the requirements for stable sliding of the thrust belt according to critical wedge theory, although the specific changes in wedge taper would depend on competing décollement and wedge strength effects. A less efficient basal décollement would require a greater thrust belt taper for stable sliding, whereas a foreland increase in the internal wedge strength would allow stable sliding of a wedge with lower taper (Davis and others, 1983).

CONCLUSIONS

The last phase of shortening in the Wyoming-Idaho-Utah thrust belt was marked by a period of internal thrust reactivation that segmented the top of the regional thrust wedge into a series of piggyback basins between uplifted basin margins. Surface and subsurface structural analysis combined with stratigraphic and sedimentologic study of Bear Lake Plateau basin, northern Fossil Basin, and the LaBarge basin permits the following conclusions:

1. Early Eocene deposition in piggyback basins was controlled by uplift of the basin margins rather than subsidence of basin depocenters.

2. Periods of basin margin uplift are identified by combined observations of (a) alluvial fan sediments shed from the uplifted margins; (b) patterns of onlap, overstep, and angular unconformities along and across the basin margins; and (c) crosscutting structural relations of folded or faulted early Eocene strata along the basin margins.

3. Two sites of basin margin uplift are recognized in the three piggyback basins discussed in this report: Trailing basin margin uplifts occur above the trailing footwall ramps of the thrust sheets that immediately underlie the basins, and leading basin margin uplifts occur above the leading edge of thrust sheets that are underlain by ramps of structurally lower thrusts.

4. Two uplift processes operated in all three basins: reactivated thrust uplift and footwall uplift. Reactivated thrust uplift was generally confined to the trailing ramp areas of thrust sheets as illustrated by uplift of Sublette anticline above the Crawford trailing ramp, and the last phase of uplift above the Hogsback trailing ramp. Footwall uplift was confined to the leading basin margins, where the basin margins and the leading edge of the thrust plate beneath them were translated up the ramp of a structurally lower thrust. Uplift along the leading edges of the Crawford, Absaroka, and Hoigsback thrust sheets was accommodated by uplift along the underlying Absaroka, Hogsback, and LaBarge thrust ramps respectively.

5. Correlation and comparison of times of uplift between the trailing ramp and leading edge areas of a single thrust sheet are important for distinguishing periods when through-going slip is possible. Since thrust slip is communicated from the basal décollement, up the trailing ramp, and finally across the frontal décollement to the thrust front, a cessation of uplift in the trailing ramp region precludes any slip along the frontal length of the thrust. As a corollary, thrust slip along a thrust front requires correlative slip along the trailing length of the thrust.

6. The change from a forward imbrication thrust sequence to internal thrusting during the final episode of thrust shortening was likely the result of impeded slip along the frontal thrusts of the Wyoming-Idaho-Utah thrust belt. Further shortening of the belt was accommodated by reactivated slip along mechanically interconnected thrusts in the interior of the thrust belt. The Crawford, Absaroka, Hogsback-Darby, and Prospect thrusts are mechanically interconnected along the basal décollement in the Cambrian Gros Ventre Formation.

7. Probable causes of slip impedance include foreland buttressing, fault deflection above foreland basement warps, regional wedge adjustment to subcritical taper, and foreland stratigraphic and rheologic changes. The relative contribution of each of the four slip-impeding processes proposed here probably varied in scale and location. On a subregional scale, foeland buttressing was dominant in the northern thrust belt where the frontal thrusts abut the Gros Ventre–Teton and northern Moxa Arch foreland basement uplifts. Similarly, thrust deflection above basement warps was limited to the Moxa Arch near LaBarge in the central thrust belt. On a larger scale, impedance of frontal slip could have been caused by a reduction of the regional wedge taper as a result of the combined effects of subsidence in the foreland and erosion in the thrust belt. However, stratigraphic and associated rheologic changes toward the foreland provide the broadest explanations for impedance of frontal thrust belt slip on a regional scale.

ACKNOWLEDGMENTS

Field work for this study was funded by a Steven S. Oriel Memorial Fund Grant from the Colorado Scientific Society and by a Utah Geological and Mineral Survey Graduate Program Mapping Grant. Related fieldwork, research, and data acquisition were supported by Chevron U.S.A., Sohio Petroleum Co., and Amoco Production Company. Chevron U.S.A. also provided drafting support. I thank Peter W. Huntoon and Donald L. Blackstone, Jr., for their review of early versions of this study. Subsequent review by Frank Royse, Jr., and James R. Steidtmann and comments by Lucian B. Platt improved the technical content of the paper. I also thank Paul K. Link for his helpful editing and encouragement.

My interest in the latest Cretaceous and early Eocene stratigraphy and sedimentation of the thrust belt began when I tagged along on a field trip led by Steve Oriel through Fossil Basin and the LaBarge basin in 1985. Prior to that time I considered the "cover" sequence to be a nuisance that only served to obscure the structural geology of the region. I had no idea that I would later retrace Steve's route while trying to recall his words about unconformities and diamictites. We are all fortunate that he rigorously documented his observations and ideas in print. I began this work after Steve's death, so I never had the opportunity to discuss my interpretations with him. I'm certain that he would strenuously debate some of the ideas presented here. Steve was a demanding taskmaster. He required that one formulate a hypothesis, test it in the field, and document the results in the office. He accomplished all three tasks throughout his career. During our later conversations in Denver, it was clear that Steve regarded the encouragement and guidance of young geologists as an integral part of his work. He spoke with a paternal affection of the many graduate students he had helped along the way. I will always appreciate the kind patience with which he would follow an argument, correct a misconceived notion, and then point to a straighter path toward a solution. This paper is dedicated to Steven S. Oriel.

REFERENCES CITED

Armstrong, F. C., and Oriel, S. S., 1965, Tectonic development of the Idaho-Wyoming thrust belt: American Association of Petroleum Geologists Bulletin, v. 49, p. 1847–1866.

Atkinson, B. K., 1984, Subcritical crack growth in geological materials: Journal of Geophysical Research, v. 89, p. 4077–4114.

Blackstone, D. L., Jr., 1979, Geometry of the Darby, Prospect, and LaBarge faults at their junction with the La Barge Platform, Lincoln and Sublette Counties, Wyoming: Geological Survey of Wyoming Report of Investigations 18, 34 p.

Boothroyd, J. C., and Ashley, G. M., 1975, Process, bar morphology, and sedimentary structures on braided outwash fans, northeastern Gulf of Alaska, *in* Jopling, A. V., and McDonald, B. C., eds., Glaciofluvial and glaciolacustrine sedimentation: Society of Economic Paleontologists and Mineralologists Publication 23, p. 193–222.

Boyer, S. E., 1986, Styles of folding within thrust sheets: Examples from the Appalachian and Rocky Mountains of the U.S.A. and Canada: Journal of Structural Geology, v. 8, p. 325–339.

Chapple, W. M., 1978, Mechanics of thin-skinned fold-and-thrust belts: Geological Society of America Bulletin, v. 89, p. 1189–1198.

Collinson, J. D., 1978, Alluvial sediments, *in* Reading, H. G., ed., Sedimentary environments and facies: New York, Elsevier, p. 15–59.

Coogan, J. C., 1987, Thrust systematics and displacement transfer in the Wyoming-Idaho-Utah thrust belt: Geological Society of America Abstracts with Programs, v. 19, p. 626.

—— , 1992, Thrust systems and displacement transfer in the Wyoming-Idaho-Utah thrust belt [Ph.D. thesis]: Laramie, University of Wyoming, 240 p.

Coogan, J. C., and Yonkee, W. A., 1985, Salt detachments within the Meade and Crawford thrust systems, Idaho and Wyoming, *in* Kerns, G. J., and Kerns, R. L., eds., Orogenic patterns and stratigraphy of north-central Utah and southeastern Idaho: Utah Geological Association Publication 14, p. 75–82.

Crawford, K. A., 1979, Sedimentology and tectonic significance of the Late Cretaceous–Paleocene Echo Canyon and Evanston synorogenic conglomerate of the north-central Utah thrust belt [M.S. thesis]: Madison, University of Wisconsin, 143 p.

Davis, D., Suppe, J., and Dahlen, F. A., 1983, Mechanics of fold and thrust belts and accretionary wedges: Journal of Geophysical Research, v. 88, p. 1153–1172.

DeCelles, P. G., 1988, Lithologic provenance modeling applied to the Late Cretaceous synorogenic Echo Canyon Conglomerate, Utah: A case of multiple source areas: Geology, v. 16, p. 1039–1043.

DeCelles, P. G., and 15 others, 1987, Laramide thrust-generated alluvial-fan sedimentation, Sphinx Conglomerate, southwestern Montana: American Association of Petroleum Geologists Bulletin, v. 71, p. 135–155.

Delphia, J. G., and Bombolakis, E. G., 1988, Sequential development of a frontal ramp, imbricates, and a major fold in the Kemmerer region of the Wyoming thrust belt, *in* Mitra, G., and Wojtal, S., eds., Geometries and mechanisms of thrusting, with special reference to the Appalacians: Geological Society of America Special Paper 222, p. 207–222.

Dixon, J. S., 1982, Regional structural synthesis, Wyoming salient of the western overthrust belt: American Association of Petroleum Geologists Bulletin, v. 66, p. 1560–1580.

Dorr, J. A., and Gingrich, P. D., 1980, Early Cenozoic mammalian paleontology, geologic structure, and tectonic history in overthrust belt near LaBarge, western Wyoming: Contributions to Geology, v. 18, p. 101–115.

Dover, J. H., 1985, Preliminary geologic map of the Logan 1:100,000 quadrangle, Utah: U.S. Geological Survey Open-File Report 85-216, 32 p.

Hamilton, W. B., 1988, Laramide crustal shortening, *in* Schmidt, C. J., and Perry, W. J., Jr., eds., Interaction of the Rocky Mountain foreland and the Cordilleran thrust belt: Geological Society of America Memoir 171, p. 27–39.

Hubbert, M. K., and Rubey, W. W., 1959, Role of fluid pressure in mechanics of overthrust faulting: Geological Society of America Bulletin, v. 70, p. 115–166.

Hunter, R. B., 1988, Timing and structural interaction between the thrust belt and foreland, Hoback Basin, Wyoming, *in* Schmidt, C. J., and Perry, W. J., Jr., eds., Interaction of the Rocky Mountain foreland and the Cordilleran thrust belt: Geological Society of America Memoir 171, p. 367–393.

Hurst, D. J., and Steidtmann, J. R., 1986, Stratigraphy and tectonic significance of the Tunp conglomerate in the Fossil Basin, southwest Wyoming: Mountain Geologist, v. 23, p. 6–13.

Jacobson, S. R., and Nichols, D. J., 1982, Palynological dating of syntectonic units in the Utah-Wyoming thrust belt, *in* Powers, R. B., ed., Geologic studies of the Cordilleran thrust belt: Rocky Mountain Association of Geologists, v. 2, p. 735–750.

Jones, P. B., 1984, Sequence of formation of back-limb thrusts and imbrications: Implications for development of the Idaho-Wyoming thrust belt: American Association of Petroleum Geologists Bulletin, v. 68, p. 816–818.

Kraig, D. H., Wiltschko, D. V., and Spang, J. H., 1988, The interaction of the Moxa Arch (LaBarge Platform) with the Cordilleran thrust belt, south of Snider Basin, southwestern Wyoming, *in* Schmidt, C. J., and Perry, W. J., Jr., eds., Interaction of the Rocky Mountain foreland and the Cordilleran thrust belt: Geological Society of America Memoir 171, p. 395–410.

Lageson, D. R., 1984, Structural geology of the Stewart Peak culmination, Idaho-Wyoming thrust belt: American Association of Petroleum Geologists Bulletin, v. 68, p. 401–416.

Lamerson, P. R., 1982, The Fossil Basin and its relationship to the Absaroka thrust system, Wyoming and Utah, *in* Powers, R. B., ed., Geologic studies of the Cordilleran thrust belt: Denver, Colorado, Rocky Mountain Association of Geologists, v. 1, p. 279–340.

Leopold, E. B., and MacGintie, H. D., 1972, Development and affinities of Tertiary floras in the Rocky Mountains, *in* Graham, A., ed., Floristics and paleofloristics of Asia and eastern North America: Amsterdam, Elsevier, p. 147–200.

Mansfield, G. R., 1927, Geography, geology, and mineral resources of a part of southeastern Idaho: U.S. Geological Survey Professional Paper 152, 453 p.

McBride, B. C., and Dolberg, D. M., 1990, Re-evaluation of the Stewart Peak culmination and the relationship between the St. John's and Absaroka thrust faults: Geological Society of America Abstracts with Programs, v. 22, p. 37.

Miall, A. D., 1978, Lithofacies types and vertical profile models in braided river deposits: A summary, *in* Miall, A. D., ed., Fluvial sedimentology: Canadian Society of Petroleum Geologists Memoir 10, p. 597–604.

—— , 1982, Analysis of fluvial depositional systems: American Association of Petroleum Geologists Education Course Notes Series 20, 75 p.

Mitra, G., and Boyer, S. E., 1986, Energy balance and deformation mechanisms of duplexes: Journal of Structural Geology, v. 8, p. 291–304.

Mitra, G., Hull, J. M., Yonkee, W. A., and Protzman, G. M., 1988, Comparison of mesoscopic and microscopic deformational styles in the Idaho-Wyoming thrust belt and the Rocky Mountain foreland, *in* Schmidt, C. J., and Perry, W. J., Jr., eds., Interaction of the Rocky Mountain foreland and the Cordil-

leran thrust belt: Geological Society of America Memoir 171, p. 119–141.
Nemec, W., and Steel, R. J., 1984, Alluvial and coastal conglomerates: Their significance and some comments on gravelly mass-flow deposits, *in* Koster, E. H., and Steel, R. J., eds., Sedimentology of gravels and conglomerates: Canadian Society of Petroleum Geologists Memoir 10, p. 1–31.
Ori, G. G., and Friend, P. F., 1984, Sedimentary basins formed and carried piggyback on active thrust sheets: Geology, v. 12, p. 475–478.
Oriel, S. S., 1962, Main body of Wasatch Formation near LaBarge, Wyoming: American Association of Petroleum Geologists Bulletin, v. 46, p. 3161–3173.
—— , 1969, Geology of the Fort Hill quadrangle, Lincoln County, Wyoming: U.S. Geological Survey Professional paper 594-M, 40 p.
Oriel, S. S., and Platt, L. B., 1980, Geologic map of the Preston 1° × 2° quadrangle, southeastern Idaho and western Wyoming: U.S. Geological Survey Miscellaneous Investigations Map I-1127, scale 1:250,000.
Oriel, S. S., and Tracey, J. I., Jr., 1970, Uppermost Cretaceous and Tertiary stratigraphy of Fossil basin, southwestern Wyoming: U.S. Geological Survey professional Paper 635, 53 p.
Perry, W. J., Jr., and Schmidt, C. J., 1988, Preface, *in* Schmidt, C. J., and Perry, W. J., Jr., eds., Interaction of the Rocky Mountain foreland and the Cordilleran thrust belt: Geological Society of America Memoir 171, p. ix–xi.
Pinet, P., and Souriau, M., 1988, Continental erosion and large-scale relief: Tectonophysics, v. 7, p. 563–582.
Royse, F., Jr., 1985, Geometry and timing of the Darby-Prospect-Hogsback thrust fault system, Wyoming: Geological Society of America Abstracts with Programs, v. 17, p. 263.
Royse, F., Jr., Warner, M. A., and Reese, D. L., 1975, Thrust belt structural geometry and related stratigraphic problems, Wyoming-Idaho-northern Utah, *in* Bolyard, D. W., ed., Deep drilling frontiers of the central Rocky Mountains: Rocky Mountain Association of Geologists, p. 41–54.
Rubey, W. W., Oriel, S. S., and Tracey, J. I., Jr., 1975, Geology of the Sage and Kemmerer 15-minute quadrangles: U.S. Geological Survey Professional Paper 855, 18 p.
—— , 1980, Geologic map and structure sections of the Cokeville quadrangle, Wyoming: U.S. Geological Survey Miscellaneous Investigations Map I- 1129, scale 1:62,500.
Rust, B. R., and Koster, E. H., 1984, Coarse alluvial deposits, *in* Walker, R. G., ed., Facies models: Toronto, Ontario, Geoscience Canada, Reprint Series 1, p. 53–69.
Salat, T. S., 1989, Provenance, dispersal, and tectonic significance of the Evanston Formation and Sublette Range Conglomerate, Idaho-Wyoming-Utah thrust belt [M.S. thesis]: Laramie, University of Wyoming, 100 p.
Schmitt, J. G., and Steidtmann, J. R., 1990, Interior ramp-supported uplifts: Implications for sediment provenance in foreland basins: Geological Society of America Bulletin, v. 102, p. 494–501.
Shuster, M. W., and Steidtmann, J. R., 1988, Tectonic and sedimentary evolution of the northern Green River basin, western Wyoming, *in* Schmidt, C. J., and Perry, W. J., Jr., eds., Interaction of the Rocky Mountain foreland and the Cordilleran thrust belt: Geological Society of America Memoir 171, p. 515–529.
Steidtmann, J. R., and Schmitt, J. G., 1988, Provenance and dispersal of tectogenic sediments in thin-skinned, thrusted terrains, *in* Kleinspehn, K. L., and Paola, C., eds., New perspectives in basin analysis: New York: Springer-Verlag, p. 353–356.
Stewart, J. H., 1978, Basin-range structure in western North America: A review, *in* Smith, R. B., and Eaton, G. P., eds., Cenozoic tectonics and regional geophysics of the western Cordillera: Geological Society of America Memoir 153, p. 1–31.
Teal, P. R., 1983, The triangle zone at Cabin Creek, Alberta, *in* Bally, A. W., ed., Seismic expression of structural styles: American Association of Petroleum Geologists Studies in Geology 15, v. 3, p. 3.4-1-48–3.4-1-53.
Tracey, J. I., Jr., Oriel, S. S., and Rubey, W. W., 1961, Diamictite facies of the Wasatch Formation in the Fossil Basin, southwestern Wyoming, *in* Short papers in the geologic and hydrologic sciences: U.S. Geological Survey Professional Paper 424-B, p. B149–B150.
Turcotte, D. L., and Schubert, D., 1982, Geodynamics applications of continuum physics to geological problems: New York, John Wiley & Sons, 450 p.
Valenti, G. L., 1982, Preliminary geologic map of the Laketown quadrangle, Rich County, Utah: Utah Geological and Mineral Survey Map 58, 9 p., scale 1:24,000.
—— , 1987, Review of hydrocarbon potential of the Crawford thrust plate Wyoming-Idaho-Utah thrust belt, *in* Miller, W. R., ed., The thrust belt revisited: Wyoming Geological Association, 38th Annual Field Conference, Guidebook, p. 257–266.
Wach, P. H., 1977, The Moxa Arch, an overthrust model? *in* Heisey, E. L., Norwood, E. R., Wach, P. H., and Hale, L. A., eds., Rocky Mountain thrust belt geology and resources: Wyoming Geological Association, 29th Annual Field Conference, Guidebook, p. 651–664.
Warner, M. A., and Royse, F., Jr., 1987, Thrust faulting and hydrocarbon generation: Discussion: American Association of Petroleum Geologists Bulletin, v. 71, p. 882–889.
Wiltschko, D. V., and Dorr, J. A., 1983, Timing of deformation in overthrust belt and foreland of Idaho, Wyoming, and Utah: American Association of Petroleum Geologists Bulletin, v. 67, p. 1304–1322.
Wiltschko, D. V., and Eastman, D., 1983, Role of basement warps and faults in localizing thrust fault ramps, *in* Hatcher, R. D., Jr., Williams, H., and Zeitz, I., eds., Contributions to the tectonics and geophysics of mountain chains: Geological Society of America Memoir 158, p. 177–190.
Wojtal, S., and Mitra, G., 1986, Strain hardening and strain softening in fault zones from foreland thrusts: Geological Society of America Bulletin, v. 97, p. 674–687.
Woodward, N. B., 1986, Thrust geometry of the Snake River Range, Idaho and Wyoming: Geological Society of America Bulletin, v. 97, p. 178–193.
—— , 1987, Geological applicability of critical-wedge thrust-belt models: Geological Society of America Bulletin, v. 99, p. 827–832.

Manuscript Accepted by the Society July 19, 1991

Printed in U.S.A.

Geological Society of America
Memoir 179
1992

Chapter 3

Tertiary paleogeologic maps of the western Idaho-Wyoming-Montana thrust belt

David W. Rodgers
Department of Geology, Idaho State University, Pocatello, Idaho 83209
Susanne U. Janecke*
Department of Geology and Geophysics, University of Utah, Salt Lake City, Utah 84112

ABSTRACT

Maps of the distribution of Middle Proterozoic through Cretaceous rocks beneath the Tertiary unconformities in eastern Idaho effectively remove differential uplift associated with Basin and Range tectonism and reveal preextensional structural relief associated with folds and thrusts in the western part of the Idaho-Wyoming-Montana thrust belt. North of the Snake River Plain, the paleogeologic map shows that regionally extensive Middle Proterozoic to Triassic strata and the Cretaceous Idaho batholith were variably uplifted and exposed prior to formation of the Eocene Challis volcanic field. In the Beaverhead Mountains, Ordovician and Triassic strata are juxtaposed along the Hawley Creek thrust. In the northern Lemhi and Lost River ranges, west- and south-dipping homoclines beneath the Eocene unconformity are interpreted to reflect folding above frontal and lateral ramps in the footwall of the Hawley Creek thrust system. In the southern Lemhi and Lost River ranges, the White Knob Mountains, and eastern Pioneer Mountains, Mississippian to Permian strata underlie most of the unconformity. Stratigraphic offset is evident along the exposed Pioneer and Copper Basin thrust faults, but along the exposed Glide Mountain thrust as well as the concealed White Knob, Grouse, and Lost River–Arco Hills thrusts uplift was insufficient to juxtapose rocks of different systems. To the west, the Idaho batholith and Pennsylvanian-Permian rocks are juxtaposed beneath the Eocene unconformity. Several kilometers of overburden covered both the batholith and Permian rocks in Cretaceous time and were regionally eroded prior to extrusion of Eocene volcanic rocks.

South of the Snake River Plain and west of the exposed traces of the Paris and Putnam thrust faults, Late Proterozoic to Permian strata unconformably underlie Miocene sedimentary and volcanic rocks. The Paris thrust separates Late Proterozoic through Mississippian strata in its hanging wall from Pennsylvanian through Triassic rocks in its footwall, and the Putnam thrust separates Late Proterozoic to Ordovician strata from Pennsylvanian to Triassic strata. Beneath the unconformity, stratigraphic displacement appears to diminish northwestward along the Paris thrust and southeastward along the Putnam thrust, supporting the interpretation that displacement is progressively transferred from one thrust to the other. Lower Paleozoic strata generally underlie the unconformity between the emergent thrust system and the Arbon and

*Present address: Department of Geology, Utah State University, Logan, Utah 84322-4505.

Rodgers, D. W., and Janecke, S. U., 1992, Tertiary paleogeologic maps of the western Idaho-Wyoming-Montana thrust belt, *in* Link, P. K., Kuntz, M. A., and Platt, L. B., eds., Regional Geology of Eastern Idaho and Western Wyoming: Geological Society of America Memoir 179.

Malad valleys. West of these valleys, Pennsylvanian-Permian strata everywhere underlie the unconformity. A simple flat-ramp–flat-thrust fault geometry of the Paris-Putnam thrust system may explain the outcrop pattern, with hanging-wall and footwall flats beneath the regions of low structural relief, separated by a footwall ramp located beneath the modern Malad and Arbon valleys. Paleogeologic maps show that the Paris-Putnam thrust sheet was probably not an important source of quartzose clasts in the Cretaceous-Tertiary Harebell and Pinyon formations of northwest Wyoming.

INTRODUCTION

Numerous superposed deformational events in central and southern Idaho produced a myriad of faults and folds and many kilometers of structural relief. Strata buried in excess of 10 km are exposed in the footwalls of normal faults and hanging walls of thrust faults. This structural relief reflects the cumulative uplift and subsidence of several orogenic events, in particular the late Mesozoic to Eocene Sevier orogeny and multiple episodes of Tertiary extension. The superposition of tectonic events makes it difficult to relate specific structures and structural relief to specific orogenic events.

One successful technique used to discriminate uplift and subsidence of different ages makes use of a regionally extensive unconformity to construct a paleogeologic map (e.g., Armstrong, 1968). Where a blanketing layer of volcanic or sedimentary rocks is deposited over a surface with structural relief, that relief is recorded in the rocks below the unconformity. Regardless of subsequent vertical movements, the structural relief prior to deposition of the blanketing rocks is preserved in the rocks immediately below the unconformity. In south- and east-central Idaho, a regionally extensive unconformity separates Eocene rocks from underlying Middle and Late Proterozoic, Paleozoic, and Mesozoic rocks, and in southeastern Idaho a regionally extensive unconformity separates Miocene rocks from underlying Late Proterozoic, Paleozoic, and Mesozoic rocks. Subsequent faulting and erosion has disrupted the unconformities, leaving only remnants, but these are sufficient to compile maps showing the approximate distribution of rocks beneath them. Such maps effectively remove the uplift and subsidence associated with younger Tertiary tectonism, revealing structural relief associated with older tectonic events. In this paper eastern Idaho is divided into two regions, north and south of the Neogene Snake River Plain, and for each region a paleogeologic map is presented (Figs. 1 and 2). The pre-Eocene or pre-Miocene structural relief shown on these subcrop maps is used to infer fold and fault geometries associated with the western part of the Idaho-Wyoming-Montana thrust belt.

REGIONAL STRATIGRAPHY

North of the Snake River Plain, the oldest extensive Tertiary rocks are associated with the Challis volcanic field. Early magmatism began about 51 Ma to the north (McIntyre and others, 1982; Ekren, 1985), about 49 Ma to the south (F. J. Moye and L. W. Snee, unpublished data), and 49 to 49.5 Ma in the eastern part of the field (S. U. Janecke and L. W. Snee, unpublished data). These volcanic rocks extend from the Montana border to the Idaho batholith and from the Salmon River to the Snake River Plain. Beneath the volcanic rocks is a diverse suite of sedimentary rocks ranging from Middle Proterozoic to Triassic in age (Fig. 3) (Isaacson, 1983; Skipp and Hait, 1977) and the Cretaceous Idaho batholith. In our analysis we have grouped these rocks by system, so that all rocks formed during a certain period are shown as one unit.

South of the Snake River Plain, Eocene synorogenic basins are present near the Idaho-Wyoming border (Coogan, this volume), but the oldest postthrusting rocks are the Miocene Starlight and Salt Lake formations, which formed in response to Basin and Range tectonism. Miocene sedimentary and volcanic rocks are laterally discontinuous and in most places incompletely studied, so the oldest ages are imprecisely known. Near the Snake River Plain the basal ash-flow tuffs are at least 10 Ma (Armstrong and others, 1975; Kellogg and Marvin, 1988). Beneath the Miocene unconformity, Late Proterozoic to Devonian miogeoclinal strata and Mississippian to Triassic carbonate and clastic rocks are present (Armstrong and Oriel, 1965; Oriel and Armstrong, 1986). We have grouped these rocks by system (Fig. 3).

THE PALEOGEOLOGIC MAPS

Technique

The paleogeologic maps shown in Figures 1 and 2 were constructed by compiling a map showing the ages of strata immediately beneath the lowest Tertiary rocks. Where Tertiary rocks are absent the ages of the youngest pre-Tertiary rocks were compiled, since the stratigraphic levels represented by these rocks reflect the stratigraphically lowest possible position of the pre-Tertiary unconformity. Our assumptions are as follows.

1. The oldest Tertiary rocks are approximately coeval within each region (middle Eocene north of the Snake River Plain, middle Miocene south of the Snake River Plain). Challis volcanic rocks erupted during a short time interval (Armstrong and others, 1975; McIntyre and others, 1982; S. U. Janecke, F. J. Moye, and L. W. Snee, unpublished data). This assumption is more difficult to defend in the southern area because Tertiary deposits are not well dated.

2. The topographic relief at the time of initial Tertiary deposition was small relative to the existing structural relief. In the southeastern Challis volcanic field, F. J. Moye (personal communication, 1990) recognized less than 1 km of pre-Eocene local topographic relief, and an analysis of basal Challis conglomerates

suggests that relief was less than about 500 m (Janecke, 1991 a–f; Janecke and Wilson, 1991). In southeastern Idaho, the topographic relief beneath the Miocene unconformity is locally significant but poorly characterized. Additional study of the unconformity is needed to justify our assumption.

3. The oldest Tertiary rocks predate major slip on post contractional normal or strike-slip faults. In most places the locations of abrupt structural relief coincide with known or inferred thrust faults, suggesting that the effects of normal faulting have been removed.

4. The rocks above and below the unconformity are correctly identified. Some of the area has been mapped only at a scale of 1:250,000. More detailed mapping would improve the data and interpreations.

Paleogeologic maps remove postunconformity uplift and subsidence, because rocks immediately below the unconformity are everywhere restored to a constant reference elevation. In contrast, the maps do not remove postunconformity extension or rotation. Horizontal movement associated with extension is parallel to the unconformity, and no reference marker is available to restore rocks to their preextension position. The map of the area north of the Snake River Plain (Fig. 1) shows a pre-Eocene outcrop pattern that was extended northwest-southeast in the Eocene, east-west to northeast-southwest during Eocene to Oligocene time, and northeast-southwest since the Miocene (Janecke, 1989; Janecke and Snee, 1990; Stickney and Bartholomew, 1987). The map of the area south of the Snake River Plain (Fig. 2) shows a pre-Miocene outcrop pattern extended approximately east-west. In both areas the actual amount of extension is thought to range from 30 to 100% and is the subject of continuing research.

Map patterns in fold and thrust belts

Certain map patterns are characteristic of anticlines, synclines, thrust faults, and intrustive contacts in a thrust-faulted terrain (Boyer and Elliot, 1983; Woodward and others, 1985). The type of structure, the level of erosion, and the amount of contraction will determine the map pattern. Assuming east- to northeast-directed tectonic transport, older rocks crop out to the west of thrusts and younger rocks to the east. Buried ramps in thrust faults may also be detectable with a paleogeologic map. Over a west-dipping frontal ramp breaking through footwall strata, hanging-wall rocks dip west, an anticline and syncline form over the changes from ramp to flats, and after erosion, uplifted older rocks may be exposed east of younger rocks. Buried lateral ramps also display diagnostic map patterns beneath a flat regional unconformity, with older rocks above the upper flat, younger rocks above the lower flat, and a homocline of older to younger rocks above the lateral ramp. In theory, differentiating between fault-bend folds and other folds can be difficult, but trains of short-wavelength (less than 1,000 m) décollement folds and fault propagation folds accommodate internal deformation of the major thrust sheets in eastern Idaho (Ross, 1947; Beutner, 1968; Holmes and Evans, 1989), and these folds are not likely to be confused with longer-wavelength fault-bend folds formed above major frontal ramps.

Paleogeologic map north of the Snake River Plain

Figure 1 shows the distribution of Middle Proterozoic to Cretaceous rocks beneath the Eocene unconformity in east-central Idaho. Middle Proterozoic rocks at the surface in the north give way to progressively younger strata to the west, south, and east. To the west, Pennsylvanian and Permian strata and the Cretaceous Idaho Batholith are juxtaposed. This map pattern supports the following interpretations.

Ordovician and Triassic strata are juxtaposed along the exposed and inferred northern trace of the Hawley Creek thrust (Fig. 1), whereas Ordovician and Mississippian strata are juxtaposed along its southern trace. Stratigraphic offset diminishes to the south, and displacement is transferred to the structurally lower Fritz Creek thrust (Fig. 1). The outcrop patterns on the paleogeologic map reflect a thrust geometry that was first identified by Skipp (1987, 1988a) from geologic mapping of the exposed thrusts and strata, indicating the paleogeologic map is a viable tool for deciphering structural relations.

In the central Lemhi Range, a southeast-dipping homocline is apparent beneath the unconformity (Fig. 1). Middle Proterozoic and lower Paleozoic rocks crop out in the north, and upper Paleozoic rocks crop out in the south. A similar southeast-dipping homocline is apparent in the northern Lost River Range, slightly north of the homocline in the Lemhi Range. Beneath the Neogene Pahsimeroi Valley and the Neogene Birch Creek Valley, southwest-dipping homoclines are inferred to exist (Fig. 1). The map pattern is interpreted to reflect footwall ramps of the Hawley Creek thrust system. Beneath Pahsimeroi Valley and Birch Creek Valley are frontal ramps that dip gently southwest, beneath the central Lemhi Range is a lateral ramp that dips gently southeast, and beneath the northern Lost River Range is a second lateral ramp that dips gently southeast (Fig. 4).

The inferred location of the lateral ramp beneath the Lemhi Range coincides with the southwestward projection of the Blacktail-Snowcrest foreland uplift of Scholten and others (1955). This basement uplift is exposed in southwestern Montana and based on gravity and magnetic data is interpreted to extend in the subsurface to the Beaverhead Range (Kulik and Perry, 1988). If the uplift extends to the Lemhi Range, the lateral ramp may have formed over it, or both may have followed the same preexisting structural weakness. Another related structure is a broad northwest-trending syncline in the northern Lost River Range (Mapel and others, 1965), which we interpret to be a fault-bend fold southwest of the inferred frontal ramp.

In addition to uplift over footwall ramps, the regional exposure of progressively older rocks to the north may reflect isostatic uplift of the Bitterroot lobe of the Idaho batholith and/or nondeposition. Instead of a frontal ramp beneath the Pahsimeroi Valley, an extensive, prevolcanic, northwest-striking and southwest-dipping normal fault along the Pahsimeroi Valley could account for the structural relief across the valley, although normal faults of this age have not been documented. The normal fault would end in the southeastern Pahsimeroi Valley, however, because to the south there is no structural relief between the central Lost River and Lemhi ranges (Fig. 1).

Concealed footwall ramps may influence the location of later normal faults, as shown in western Wyoming, where a concealed ramp in the Absaroka thrust is coincident with a major late Tertiary normal fault, the Grand Valley fault (Royse and others, 1975). A similar situation may characterize the Pahsimeroi Valley, where a west-dipping, Late Cenozoic normal fault along the east edge of the valley may merge with the inferred thrust ramp for part of its length. However, we do not think the northern Lemhi fault is listric and soles into a subsurface thrust because the adjacent Lost River fault is planar to a depth of 15 km (Stein and Barrientos, 1985).

Mississippian strata are flanked by Pennsylvanian-Permian strata in the south (Fig. 1). This map pattern reflects, in part, depositional patterns: thick Mississippian flysch accumulated in the White Knob, Pioneer, and White Cloud mountains east of an Antler highland (Nilsen, 1977) and was subsequently uplifted to form the Copper Basin highland between two Pennsylvanian-Permian depocenters (Skipp and Hall, 1980). Stratigraphic offset is evident along portions of the Pioneer and Copper Basin thrusts, where, respectively, Ordovician through Devonian and Mississippian strata and Mississippian and Pennsylvanian strata are juxtaposed. In contrast, Mississippian strata occupy both the footwall and hanging wall of the exposed Glide Mountain thrust, indicating that uplift was small relative to the thickness of Missis-

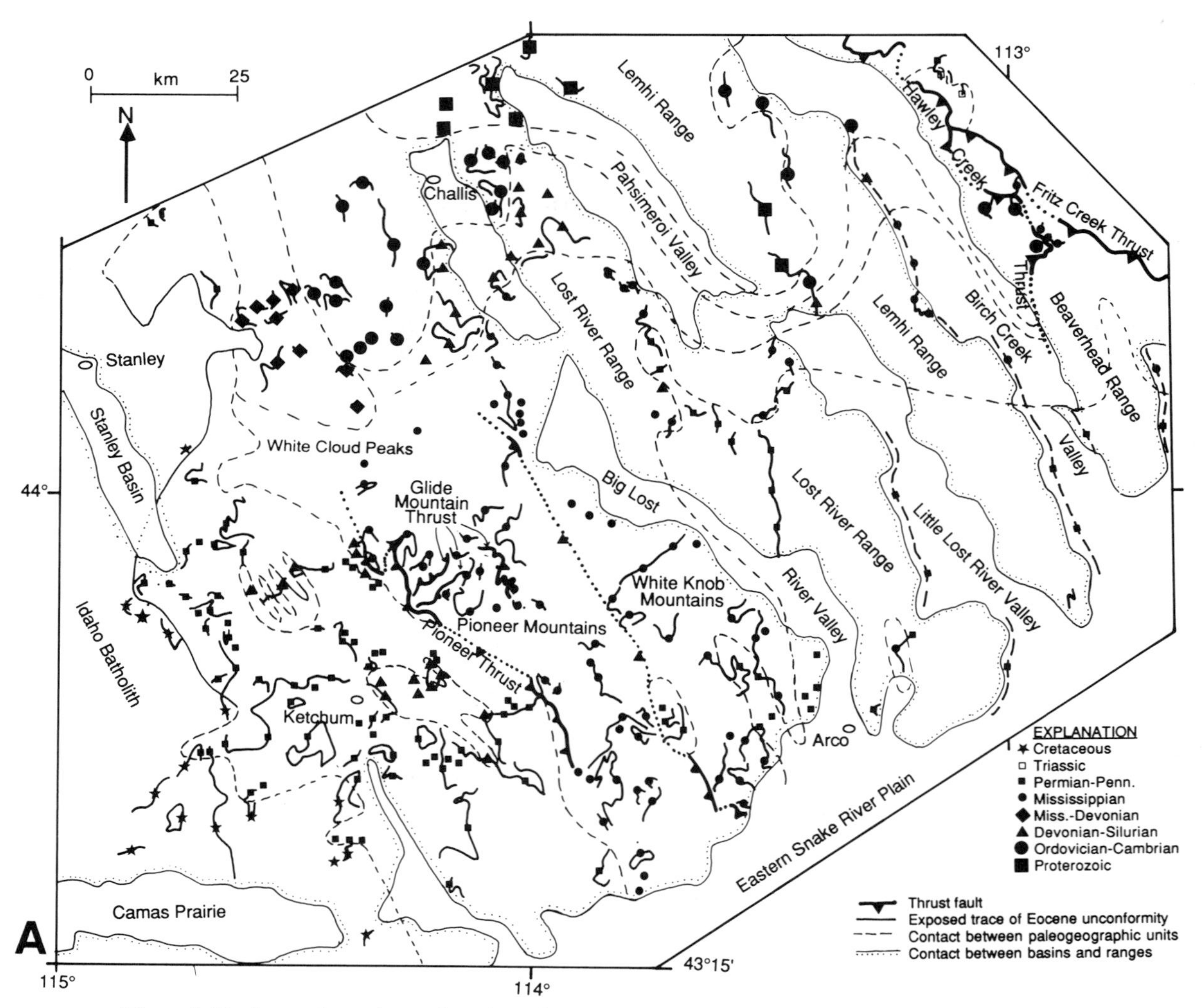

Figure 1. Tertiary paleogeology of south- and east-central Idaho. A, Map showing exposed Mesozoic thrust faults and the location and age of rocks beneath the Eocene unconformity. B, Simplified paleogeologic map and schematic cross section. Numerous small folds and local irregularities in rock distribution occur but are too small to show at the scale of the map and cross section. Ages of rock units in b are indicated by the following symbols: Coarse horizontal ruling or ZY = Middle to Late Proterozoic; coarse stipples or O€ = Ordovician-Cambrian; vertical ruling or DS = Devonian-Silurian; M =

sippian strata. Stratigraphic offset is not evident along the White Knob, Grouse, and Lost Tiver–Arco Hills thrusts, three concealed thrusts between the White Knob Mountains and the Lemhi Range. The thrusts were inferred to exist based on the telescoping of upper Paleozoic facies (Skipp and Hait, 1977; Skipp, 1987), but if they exist they are bedding-parallel with only minor uplift along them.

Devonian and Pennsylvanian-Permian rocks underlie the unconformity within and north of the western Pioneer Mountains (Fig. 1). The relatively complicated map pattern of these units in part reflects erosion through large north-northwest–trending folds, which are mapped throughout the region (Dover, 1983; Worl and others, 1991; Link, unpublished mapping). Some contacts coincide with mapped traces of northwest-striking, younger-on-older faults. These faults most recently accommodated top-to-the-northwest slip, making them right-lateral strike-slip faults (Burton and others, 1989; Burton and Link, 1989).

From east to west between Challis and Stanley, Ordovician and Cambrian rocks, poorly dated Devonian to Mississippian(?) rocks of the Salmon River assemblage, and Pennsylvanian-Permian rocks of the Sun Valley Group (Mahoney and others, 1991) progressively underlie the unconformity (Fig. 1). Numerous faults, many of which place younger rocks on older, have been identified both within and between units of different ages (Hall, 1985; Hobbs, 1985). Because of the structural complexity and uncertain ages of rocks, the paleogeologic map is of little use in resolving pre-Challis structure in this area but highlights some unresolved structural problems.

The Cretaceous Idaho batholith is overlain by Challis volcanic rocks, indicating that it was unroofed by Middle Eocene time. Metamorphic assemblages in Pennsylvanian-Permian country rock adjacent to the batholith indicate the rocks were at least 8 km deep during contact metamorphism (Whitman, 1990). The irregular distribution of satellite plutons and apophyses suggests

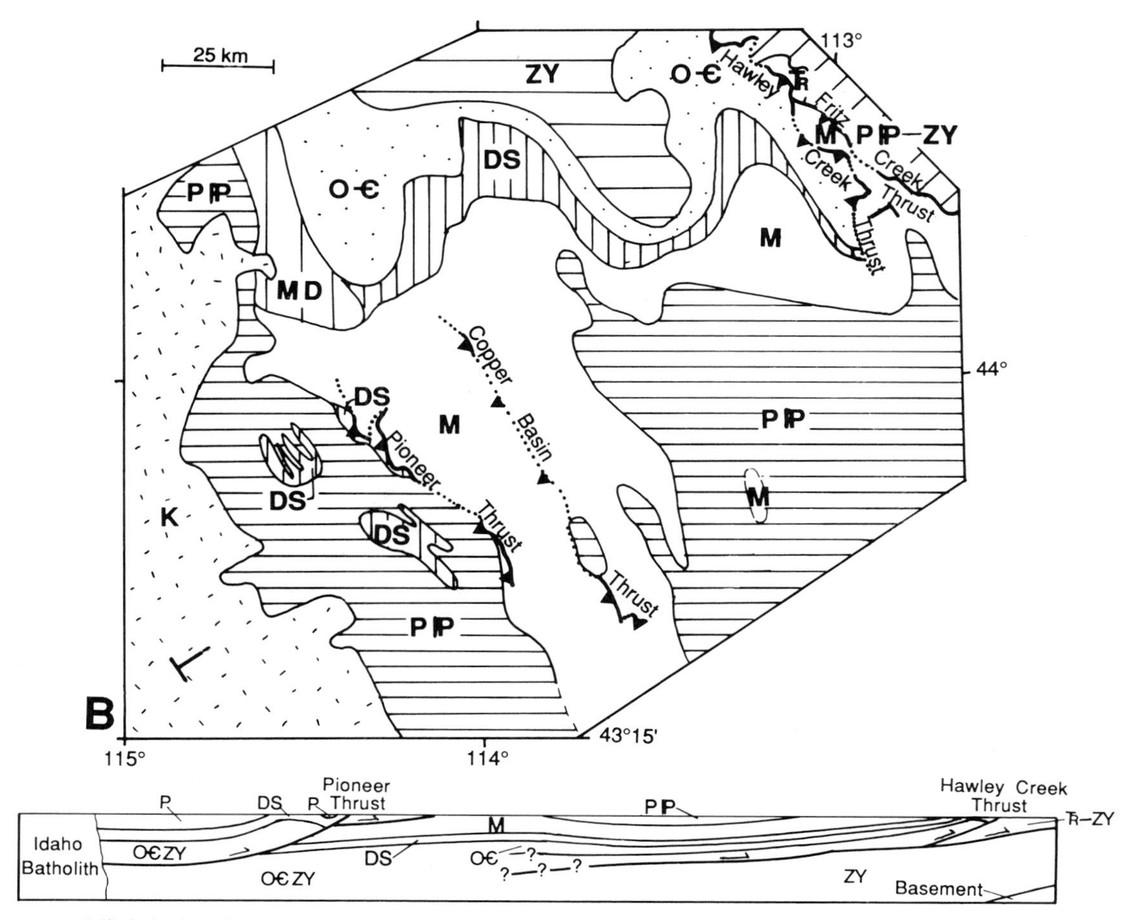

Mississippian; fine horizontal ruling or P IP = Permian-Pennsylvanian; Ʀ = Triassic; randomly oriented dashes or K = Cretaceous Idaho batholith. Sources of data include Dover (1983), Fisher and others (1983), Janecke (1992a, b, c, d, e), Janecke and Wilson (1992), Link (unpublished mapping), Link and others (1988), Mapel and others (1965), Mapel and Shropshire (1973), Nelson and Ross (1969), Rember and Bennett (1979b, c, d), Ross (1947), Ruppel (1968), Ruppel and Lopez (1981), Skipp (1985, 1988a and b, 1989), Skipp and others (1990), Worl and others (1991).

that an intrusive contact, not a fault, separates the batholith and its country rock. Evidently 8 km of overburden covered the batholith and the intruded country rocks and was completely removed by erosion during latest Cretaceous to Paleocene time. The age and extent of overburden are not known.

Paleogeologic map south of the Snake River Plain

Figure 2 shows the inferred distribution of Late Proterozoic to Mesozoic strata in southeastern Idaho prior to Miocene tectonism. From east to west, the map pattern shows stratigraphic offset across the exposed and concealed traces of the Paris and Putnam thrust faults, an expanse of Cambrian and Ordovician strata, a west-dipping homocline of Permian to Ordovician strata, and an expanse of Pennsylvanian-Permian strata.

Late Proterozoic through Mississippian and Pennsylvanian through Triassic rocks are juxtaposed along the emergent trace of the Paris thrust (Mansfield, 1927). The Paris thrust remains at approximately the same stratigraphic level through footwall strata, but to the north it cuts upsection through hanging-wall strata to form a lateral ramp (Fig. 2). Stratigraphic displacement along the thrust decreases to the north, replaced in part by folding (Armstrong, 1965).

Lower and upper Paleozoic rocks are juxtaposed along the emergent trace of the Putnam thrust in the northern Portneuf Range, and to the northwest progressively younger strata occupy the footwall and hanging wall of the thrust (Kellogg and others, 1989). Just west of the Putnam thrust are related splays of the thrust, along which uppermost Proterozoic and lowermost Cam-

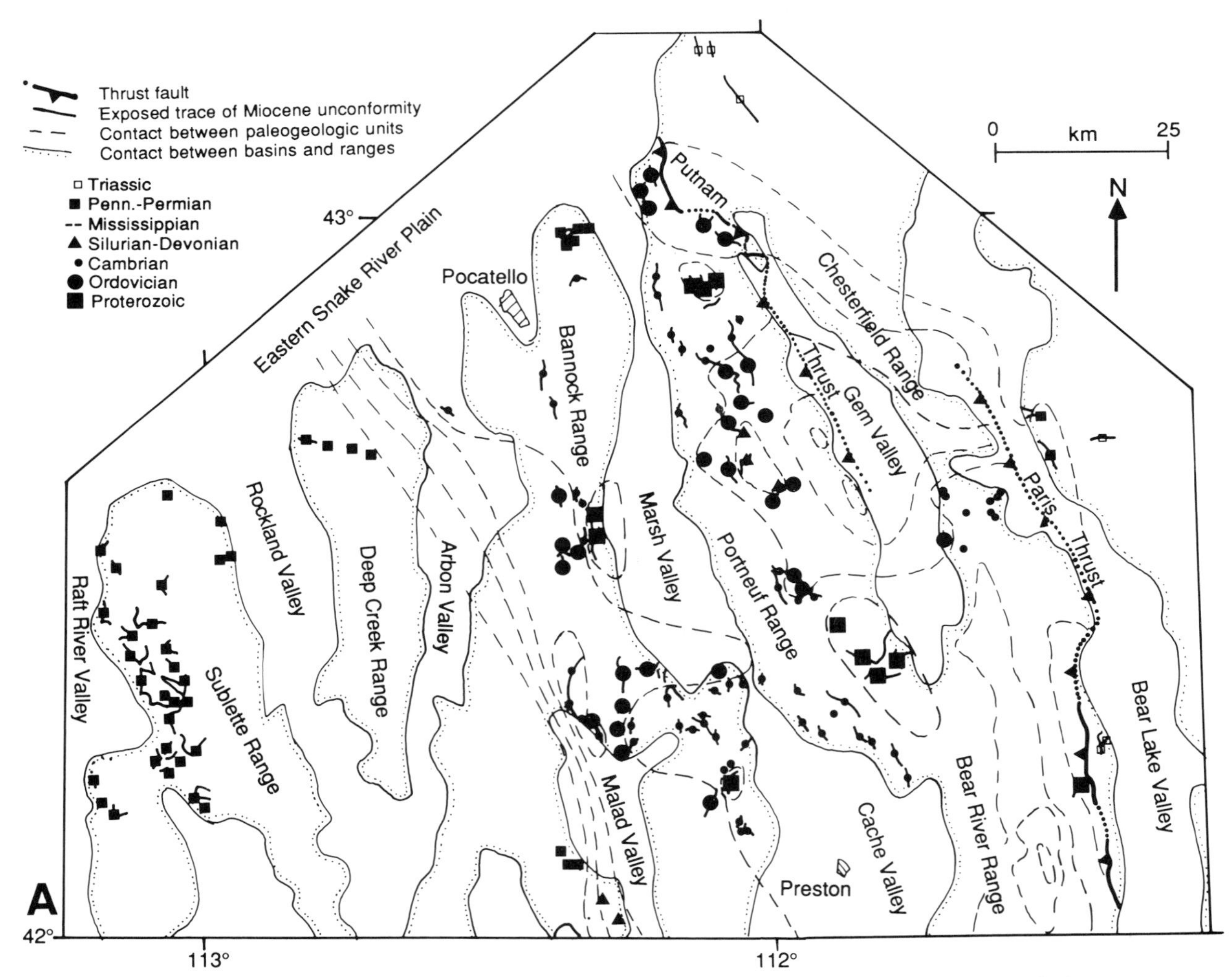

Figure 2. Tertiary paleogeology of southeastern Idaho. A, Map showing exposed Mesozoic thrust faults and the location and age of rocks beneath the Miocene unconformity. B, Simplified paleogeologic map and schematic cross section. Numerous small folds and local irregularities in rock distribution occur but are too small to show at the scale of the map and cross section. Ages of rock units in 2b are indicated by the following symbols: Fine stipples or Z = Late Proterozoic; Ꞓ = Cambrian; coarse stipples or O = Ordovician; vertical ruling or DS = Devonian-Silurian; diagonal dashes or M = Mississippian; horizontal

brian rocks overlie Cambrian and Ordovician strata (Kellogg and others, 1989; Kellogg, 1990). This region, the outcrops of Late Proterozoic rock along the Paris thrust (Mansfield, 1927), and a few isolated outcrops of Late Proterozoic rock to the west show the greatest amount of pre-Miocene uplift in southeastern Idaho. The Putnam thrust is mostly concealed beneath Neogene sedimentary rocks to the southeast, but geologic mapping (Kellogg, this volume) and the paleogeologic map (Fig. 2) indicate that progressively less stratigraphic offset occurs across the thrust. The offset traces of the Paris and Putnam thrusts, the decreasing stratigraphic offset along each thrust, and the outcrop pattern of the overlap zone support the interpretation that displacement is progressively transferred from the Paris to the Putnam thrust.

Ordovician and Cambrian strata underlie the unconformity nearly everywhere from the emergent Paris and Putnam thrusts to the Arbon and Malad valleys. The major exceptions are a large, north-trending syncline in the Bear River Range and an outcrop of upper Paleozoic rock in the northern Bannock Range that probably lies in the footwall of the Putnam thrust. The outcrop pattern of Ordovician and Cambrian strata between the emergent Paris-Putnam thrust system and Arbon and Malad valleys reflect a broad, very gentle, north-northwest–trending anticline with less than 2 km of structural relief. The angularity of the unconformity between Tertiary and older strata is typically less than 20° (Keller, 1963). We interpret the low structural relief and low-angle unconformity to indicate that the underlying, concealed Paris-Putnam thrust system has a hanging-wall–flat over footwall-flat geometry. Preextensional regional uplift and erosion

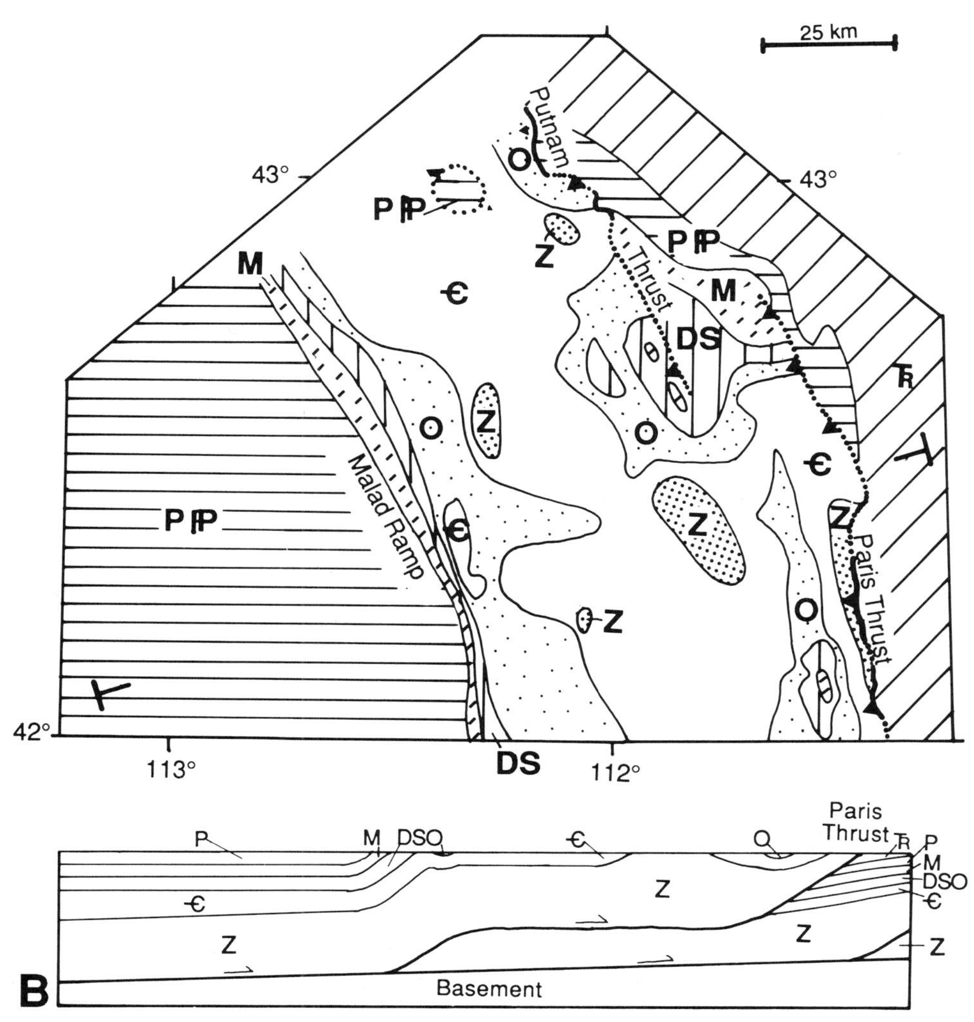

ruling or P IP = Permian-Pennsylvanian; diagonal lines or Ŧ = Triassic. Sources of data include Allmendinger (unpublished mapping, 1978), Armstrong (1969), Armstrong and others (1978), Burgel and others (1987), Corbett (1978), Hladky and Kellogg (1990), Keller (1963), Kellogg (1990), Kellogg and others (1989), Link (1982), Oriel (1965), Oriel and Platt (1980), Pierce and others (1983), Platt (1977, 1985), Pogue (1984), Rember and Bennett (1979a, e), Rodgers (unpublished mapping, 1985–1990), Trimble (1976, 1982), Trimble and Carr (1976).

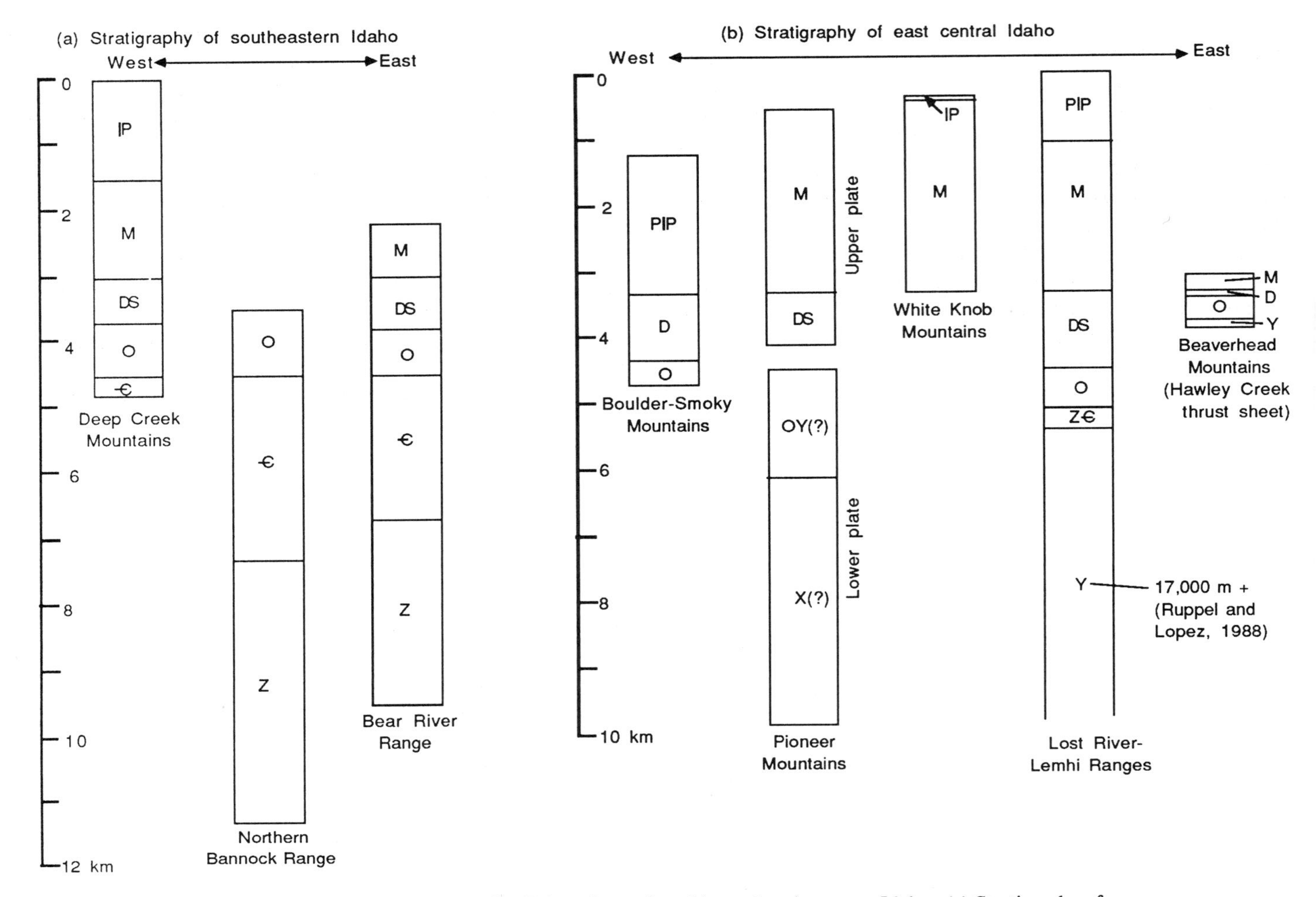

Figure 3. Composite Precambrian-Paleozoic stratigraphic sections in eastern Idaho. (a) Stratigraphy of hanging wall of Paris-Putnam thrust system, after Trimble and Carr (1976), Oriel and Platt (1980), and Link and others (1987). (b) Stratigraphic sections of several ranges in south- and east-central Idaho, after Isaacson (1983), Dover (1983), Skipp (1988a), Ruppel and Lopez (1988), Link and others (1988), and Mahoney and others (1991). Thicknesses are approximate and in places poorly known. Ages of rock units are indicated by the following symbols or combination thereof: X(?) = Early Proterozoic(?); Y = Middle Proterozoic; Z = Late Proterozoic; Є = Cambrian; O = Ordovician; S = Silurian; D = Devonian; M = Mississippian; IP = Pennsylvanian; P = Permian.

of about 4 km of middle and upper Paleozoic strata occurred throughout this area, probably due to movement of the Paris-Putnam thrust sheet over a footwall ramp to the west, described below.

Pennsylvanian and Permian strata underlie the unconformity west of Arbon and Malad valleys and show less than 2 km of structural relief. A hanging-wall–flat over footwall-flat geometry in the Paris-Putnam thrust system is similarly inferred to underlie this area.

Between the surface exposures of Pennsylvanian-Permian strata on the west and Cambrian-Ordovician strata on the east a major footwall ramp, the Malad ramp, probably exists in the Paris-Putnam thrust system. The ramp, located beneath the west-dipping homocline of Permian to Cambrian strata partially exposed on the east flank of Samaria Mountain (Platt, 1977) and concealed beneath Arbon and Malad valleys (Fig. 2), strikes north-northwest, dips west-southwest, and accommodated 4 km of differential uplift of the Paris-Putnam thrust sheet. Platt (1991) has also recognized this ramp and suggests that normal faults of the northernmost Wasatch Front may be localized by the ramp.

Thrust faults have been mapped west of the trace of the Putnam thrust but are not evident on the paleogeologic map. Trimble (1976) and Link and others (1985) identified several thrusts in the northern Bannock Range, but remapping with a revised Late Proterozoic stratigraphy (Link and others, 1987) and in greater detail (Burgel and others, 1987; Rodgers and others, unpublished mapping, 1985–1990) indicates that thrust faults are extremely rare in this area. Trimble and Carr (1976) mapped a

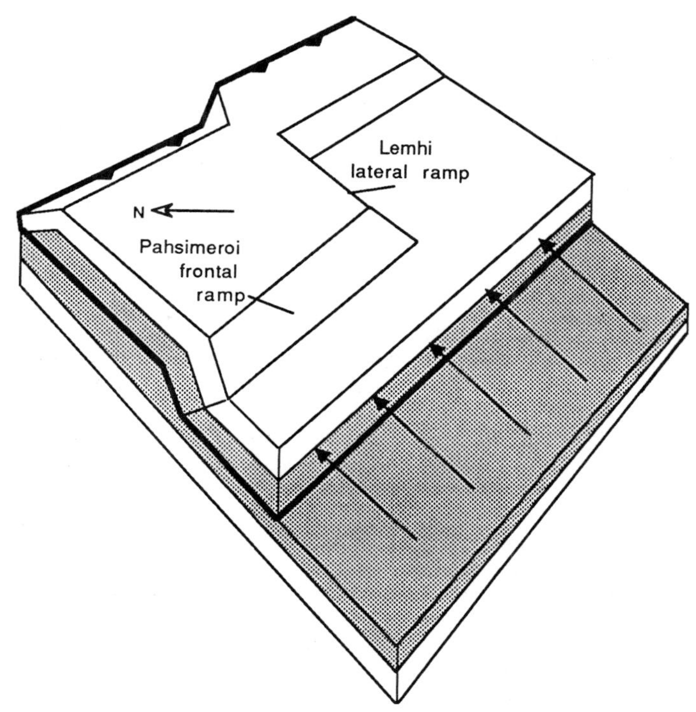

Figure 4. Schematic block diagram of concealed lateral and frontal ramps in the Hawley Creek thrust system. See text for discussion. Concept follows Ramsay and Huber (1987).

thrust fault in the northern Deep Creek Range placing Permian rocks over Ordovician rocks. We have not included this thrust on Figure 2 because of the inferred fault geometry is unusual. Based on structural styles in adjacent ranges, the absence of Tertiary rocks deposited on Orodvician rocks, and unpublished mapping by K. Kellogg (personal communication, 1991) we believe that postunconformity normal faulting better explains the juxtaposition.

Source areas for conglomerates in northwestern Wyoming

Schmitt and Steidtmann (1990) showed that uplift of an older inactive thrust sheet occurs when transported along an active underlying thrust over a buried thrust ramp. In this way older parts of the thrust belt are reestablished as sediment source areas. They proposed that quartzite clast–bearing conglomerates in northwestern Wyoming had a source in Proterozoic quartzites of the Paris–Willard thrust sheet or in lower Paleozoic to Proterozoic strata of the Medicine Lodge thrust sheet. Our paleogeologic maps show that north of the Snake River Plain, Proterozoic and lower Paleozoic quartzites were widely exposed north of 44°15′ by Eocene time. South of the plain, however, only scattered small exposures of Proterozoic quartzites are evident beneath the Miocene unconformity. This indicates the Paris-Putnam thrust sheet was not an important source of quartzose clasts in the Harebell and Pinyon Formations and that most of the detritus probably came from east-central Idaho.

It is not clear when the ramp-related uplift of the northern Lost River and Lemhi ranges brought lower Paleozoic and Proterozoic rocks to the surface. If the ramp is in the Hawley Creek thrust, as we propose above, then uplift probably occurred in Late Cretaceous time (age constraints in Skipp, 1988a). If the ramp is in a deeper thrust fault, as envisioned by Schmitt and Steidtmann (1990), then east-central Idaho may have experienced uplift into Paleocene time. We cannot discriminate these two possibilities using the paleogeologic maps presented here.

CONCLUSIONS

Map patterns on paleogeologic maps north and south of the eastern Snake River Plain show stratigraphic offset associated with some thrust faults, but not others. Stratigraphic offset is evident across several previously mapped thrust faults, including the Hawley Creek, Fritz Creek, Pioneer, Copper Basin, Putnam, and Paris thrusts. Transfer of slip between imbricate thrust faults is inferred from map patterns near the Hawley Creek–Fritz Creek and Paris-Putnam thrust systems. Near several concealed or inferred thrust faults, map patterns do not indicate stratigraphic offset, suggesting that if the faults exist they are probably of small displacement. Faults of this category include the White Knob, Grouse, Lost River–Arco Hills, and Lemhi faults north of the Snake River Plain and the Deep Creek thrust south of the Snake River Plain.

Paleogeologic maps indicate the possible location of concealed thrust ramps. A left-stepping lateral ramp through the footwall of a concealed thrust fault, probably the Hawley Creek thrust, may underlie the northern Lost River Range, Pahsimeroi Valley, and central Lemhi Range. A broad open syncline in the northern Lost River Range may be a fault-bend fold southwest of the inferred southwest-dipping frontal ramp beneath the Pahsimeroi Valley. The subsurface Paris-Putnam thrust system is interpreted to have a simple flat-ramp–flat-thrust geometry, with a major west-dipping ramp located beneath Malad and Arbon valleys. Between the emergent thrusts and the concealed ramp, subhorizontal lower Paleozoic rocks cropped out over a preextension distance of 25 to 35 km.

Regions of significant pre-Eocene uplift and erosion can be identified on the paleogeologic maps. The Idaho batholith was erosionally unroofed prior to Middle Eocene time, and about 8 km of overburden were removed. Possible source areas for thick deposits of Upper Cretaceous to Paleocene quartzite conglomerates in northwestern Wyoming were probably north of 44°15′ in east-central Idaho.

ACKNOWLEDGMENTS

The paleogeologic maps were initially compiled by Kurt Allen, Phil Bandy, Anna Baumhoff, Leigh Beem, Brad Burton, John Crocker, Dave Ettner, Alan Haslam, Don Maloney, Larry Snider, Dave Stoll, and Nick Zentner, members of the 1986 and 1988 Advanced Structural Geology classes at Idaho State University. Shelly Whitman helped synthesize the data and draft preliminary versions of the maps. Discussions with Ron Bruhn, Betty Skipp, and Dolores Kulik are appreciated. Don Blackstone, Karl Kellogg, Paul Link, Lucian Platt, Frank Royse, and Betty Skipp provided excellent reviews; if any errors remain, they are our responsibility. Funding for Janecke was provided by NSF EAR-8721100 to R. L. Bruhn and J. W. Geissman and by a Sigma Xi grant-in-aid. Although we were not personally acquainted with Steve Oriel, we admire his excellent geologic maps and terse writing style and have caught his enthusiasm for the world-class geology of southeastern Idaho.

REFERENCES CITED

Armstrong, F. C., 1965, Northwestward projection of the Paris thrust fault, southeastern Idaho [abs.]: Geological Society of America Special Paper 82, p. 317.

——, 1969, Geologic map of the Soda Springs Quadrangle, southeastern Idaho: U.S. Geological Survey Miscellaneous Geological Investigations Map I-557, scale 1:48,000.

Armstrong, F. C., and Oriel, S. S., 1965, Tectonic development of the Idaho-Wyoming thrust belt: American Association of Petroleum Geologists Bulletin, v. 49, p. 1847–1866.

Armstrong, R. L., 1968, Sevier orogenic belt in Nevada and Utah: Geological Society of America Bulletin, v. 79, p. 429–458.

Armstrong, R. L., Leeman, W. P., and Malde, H. E., 1975, K-Ar Dating, Quaternary and Neogene volcanic rocks of the Snake River Plain, Idaho: American Journal of Science, v. 275, p. 225–251.

Armstrong, R. L., Smith, J. F., Covington, H. R., and Williams, P. L., 1978, Preliminary geologic map of the west half of the Pocatello 1°×2° quadrangle, Idaho: U.S. Geological Survey Open-File Report 78-1018, scale 1:250,000.

Beutner, E. C., 1968, Structure and tectonics of the southern Lemhi Range, Idaho [Ph.D. thesis]: University Park, Pennsylvania State University, 181 p.

Boyer, S. E., and Elliott, D., 1983, Thrust systems: American Association of Petroleum Geologists Bulletin, v. 66, p. 1196–1230.

Burgel, W. D., Rodgers, D. W., and Link, P. K., 1987, Mesozoic and Cenozoic structures of the Pocatello region, southeastern Idaho, *in* Miller, W. R., ed., The thrust belt revisted: Wyoming Geological Association, 37th Annual Field Conference, Guidebook, p. 91–100.

Burton, B. R., and Link, P. K., 1989, Lake Creek mineralized area, Blaine County, Idaho, *in* Winkler, G. R., Soulliere, S. J., Worl, R. G., and Johnson, K. M., eds., Geology and mineral deposits of the Hailey and western Idaho Falls 1°×2° quadrangles, Idaho: U.S. Geological Survey Open-File Report 89-639, p. 74–85.

Burton, B. R., Link, P. K., and Rodgers, D. W., 1989, Death of the Wood River thrust; structural relations in the Pioneer and Boulder Mountains, south-central Idaho: Geological Society of America Abstracts with Programs, v. 21, p. 62.

Corbett, M. K., 1978, Preliminary geologic map of the northern Portneuf Range, Bannock and Caribou Counties, Idaho: U.S. Geological Survey Open-File Report 78-1018, scale 1:48,000.

Dover, J. H., 1983, Geologic map and sections of the central Pioneer Mountains, Blaine and Custer Counties, central Idaho: U.S. Geological Survey Miscellaneous Investigations Series Map I-1319, scale 1:48,000.

Ekren, E. B., 1985, Eocene cauldron-related volcanic events in the Challis quadrangle, *in* McIntyre, D. H., ed., Symposium on the geology and mineral deposits of the Challis 1°×2° quadrangle, Idaho: U.S. Geological Survey Bulletin 1658A-S, p. 43–58.

Fisher, F. S., McIntyre, D. H., and Johnson, K. M., 1983, Geologic map of the Challis 1°×2° quadrangle, Idaho: U.S. Geological Survey Open-File Report 83-523, scale 1:250,000.

Hall, W. E., 1985, Stratigraphy and mineralization in middle and upper Paleozoic rocks of the central Idaho black shale mineral belt: U.S. Geological Survey Bulletin 1658, chap. J, p. 117–131.

Hladky, F. R., and Kellogg, K. S., 1990, Geologic map of the Yandell Springs 7.5′ quadrangle, Bannock and Bingham counties, Idaho: U.S. Geological Survey Geologic Quadrangle Map GQ-1678, scale 1:24,000.

Hobbs, S. W., 1985, Precambrian and Paleozoic sedimentary terranes in the Bayhorse area of the Challis quadrangle, *in* McIntyre, D. H., ed., Symposium on the geology and mineral deposits of the Challis 1°×2° quadrangle, Idaho: U.S. Geological Survey Bulletin 1658D, p. 59–68.

Holmes, D. C., and Evans, J. P., 1989, Folding styles in the central Lost River Range, Idaho: Geological Society of America Abstracts with Programs, v. 21, p. 94.

Isaacson, P. E., 1983, Stratigraphic correlation chart of Paleozoic and Mesozoic rocks of Idaho: Idaho Bureau of Mines and Geology Information Circular 37, 1 p.

Janecke, S. U., 1989, Changing extension directions from Eocene to the present: Central Lost River Range and adjacent areas, Idaho: Geological Society of America Abstracts with Programs, v. 21, p. 97–98.

——, 1992a, Geologic map of the Warren Mountain, Red Hills, Short Creek, most of the Massacre Mountain, and some of the Mulkey Bar and Hawley Mountain 7.5′-quadrangles, Idaho: Idaho Geological Survey Technical Report (in press), scale 1:24,000.

——, 1992b, Geologic map of the northern two-thirds of the Methodist Creek and Mackay NE 7.5′-quadrangles, and some of the Mackay NW and Sunset Peak 7.5′-quadrangles, Idaho: Idaho Geological Survey Technical Report (in press), scale 1:24,000.

——, 1992c, Reconnaissance geologic map of the Donkey Hills and part of the Double Springs 15′-quadrangles, Idaho: Idaho Geological Survey Technical Report (in press), scale 1:62,500.

——, 1992d, Reconnaissance geologic and compilation map of the northern part of the Arco Hills and southern part of the Arco Pass 7.5′-quadrangles, Idaho: Idaho Geological Survey Technical Report (in press), scale 1:24,000.

——, 1992e, Reconnaissance geologic map of parts of the Big Windy Peak and Moffett Springs 7.5′-quadrangles, Idaho: Idaho Geological Survey Technical Report (in press), scale 1:24,000.

Janecke, S. U., and Snee, L. W., 1990, Structural, stratigraphic and geochronologic evidence for major Eocene to Oligocene extension and basin formation, east-central Idaho: Geological Society of America Abstracts with Programs, v. 22, p. 16.

Janecke, S. U., and Wilson, Eric, 1992, Geologic map of the Borah Peak, Burt Creek, Elkhorn Creek, and Leatherman Peak 7.5′-quadrangles, Idaho: Idaho Geological Survey Technical Report (in press), scale 1:24,000.

Keller, A. S., 1963, Structure and stratigraphy behind the Bannock thrust in parts of Preston and Montpelier quadrangles, Idaho [Ph.D. thesis]: New York, Columbia University, 205 p.

Kellogg, K. S., 1990, Geologic map of the south Putnam Mountain quadrangle, Bannock and Caribou counties, Idaho: U.S. Geological Survey Geologic Quadrangle Map GQ-1665, scale 1:24,000.

Kellogg, K. S., and Marvin, R. F., 1988. New potassium-argon ages, geochemistry, and tectonic setting of upper Cenozoic volcanic rocks near Blackfoot, Idaho: U.S. Geological Survey Bulletin 1806, 19 p.

Kellogg, K. S., Oriel, S. S., Amerman, R. E., Link, P. K., and Hladky, F. R., 1989, Geologic map of the Jeff Cabin Creek quadrangle, Bannock and Caribou counties, Idaho: U.S. Geological Survey Geologic Quadrangle Map GQ-1669, scale: 1:24,000.

Kulik, D. M., and Perry, W. J., Jr., 1988, The Blacktail-Snowcrest foreland uplift and its influence on structure of the Cordilleran thrust belt—Geological and geophysical evidence for the overlap province in southwestern Motana, *in* Schmidt, C. J., and Perry, W. J., Jr., eds., Interaction of the Rocky Mountain foreland and the Cordilleran thrust belt: Geological Society of America Memoir 171, p. 291–306.

Link, P. K., 1982, Geology of the upper Proterozoic Pocatello Formation, Bannock Range, southeastern Idaho [Ph.D. thesis]: Santa Barbara, University of California, 137 p.

Link, P. K., LeFebre, G. B., Pogue, K. R., and Burgel, W. D., 1985, Structural geology between the Putnam thrust and the Snake River Plain, southeastern Idaho, *in* Kerns, G. J., and Kerns, R. L., Jr., eds., Orogenic patterns and stratigraphy of north-central Utah and southeastern Idaho: Utah Geological Association Publication 14, p. 97–118.

Link, P. K., Jansen, S. T., Halimdihardja, P., Lande, A., and Zahn, P., 1987, Stratigraphy of the Brigham Group (Late Proterozoic–Cambrian), Bannock, Portneuf, and Bear River Ranges, southeastern Idaho, *in* Miller, W. R., ed., The thrust belt revisted: Wyoming Geological Association, 37th Annual Field Conference, Guidebook, p. 133–148.

Link, P. K., Skipp, B., Hait, M. H., Jr., Janecke, S., and Burton, B. R., 1988, Structural and stratigraphic transect of south-central Idaho: A field guide to the Lost River, White Knob, Pioneer, Boulder, and Smoky Mountains, *in* Link, P. K., and Hackett, W. R., eds., Guidebook to the geology of central and southern Idaho: Idaho Geological Survey Bulletin 27, p. 5–42.

Mahoney, J. B., Link, P. K., Burton, B. R., Geslin, J. K., and O'Brien, J. P., 1991, Pennsylvanian and Permian Sun Valley Group, Wood River basin, south-central Idaho, *in* Cooper, J., and Stevens, C., eds., Paleozoic paleogeography of the western United States: Los Angeles, Society of Economic Paleontologists and Mineralogists, Pacific Section, Pub. 67, v. 2, p. 551–579.

Mansfield, G. R., 1927, Geography, geology and mineral resources of part of southeastern Idaho, with descriptions of Carboniferous and Triassic fossils, by G. H. Girty: U.S. Geological Survey Professional Paper 152, 453 p.

Mapel, W. J., and Shropshire, K. L., 1973, Preliminary geologic map and sections of the Hawley Mountain quadrangle: U.S. Geological Survey Miscellaneous Field Studies Map MF-546, scale 1:62,500.

Mapel, W. J., Read, W. H., and Smith, R. K., 1965, Geologic map and sections of the Doublespring Quadrangle, Custer and Lemhi counties, Idaho: U.S. Geological Survey Geologic Quadrangle Map GQ-464, scale 1:62,500.

McIntyre, D. H., Ekren, E. D., and Hardyman, R. F., 1982, Stratigraphic and structural framework of the Challis volcanics in the eastern half of the Challis 1°×2° quadrangle, Idaho, *in* Bonnichsen, B., and Breckenridge, R. M., eds., Cenozoic geology of Idaho: Idaho Bureau of Mines and Geology Bulletin 26, p. 3–22.

Nelson, W. H., and Ross, C. P., 1969, Geologic map of the Mackay quadrangle, south central Idaho: U.S. Geological Survey Miscellaneous Geological Investigations Map I-580, scale 1:125,000.

Nilsen, T., 1977, Paleogeography of Mississippian turbidites in south central Idaho, *in* Stewart, J. H., Stevens, C. H., and Fritsche, A. E., eds., Paleozoic paleogeography of the western United States: Los Angeles, Pacific Coast Paleogeography Symposium I, Society of Economic Paleontologists and Mineralogists, Pacific Section, p. 275–299.

Oriel, S. S., 1965, Preliminary geologic map of the SW 1/4 of the Bancroft quadrangle, Bannock and Caribou Counties, Idaho: U.S. Geological Survey Mineral Investigations Field Studies Map MF-299, scale 1:24,000.

Oriel, S. S., and Armstrong, F. C., 1986, Tectonic development of the Idaho-Wyoming thrust belt: Author's commentary, *in* Peterson, J. A., ed., Paleotectonics and sedimentation: American Association of Petroleum Geologists Memoir 41, p. 267–279.

Oriel, S. S., and Platt, L. B., 1980, Geologic map of the Preston 1°×2° quadrangle, southeastern Idaho and western Wyoming: U.S. Geological Survey Miscellaneous Investigation Series Map I-1127, scale 1:250,000.

Pierce, K. L., Covington, H. R., Williams, P. L., and McIntyre, D. H., 1983, Geologic map of the Cotterel Mountains and the northern Raft River Valley, Cassia County, Idaho: U.S. Geological Survey Miscellaneous Investigation Series Map I-1450, scale 1:48,000.

Platt, L. B., 1977, Geologic map of the Ireland Springs–Samaria area, southeastern Idaho and northern Utah: U.S. Geological Survey Miscellaneous Field Studies Map MF-890, scale 1:24,000.

——, 1985, Geologic map of the Hawkins quadrangle, Bannock County, Idaho: U.S. Geological Survey Miscellaneous Field Studies Map MF-1812, scale 1:24,000.

——, 1991, Evidence for a ramp under the Paris thrust along the "Wasatch fault extended" in SE Idaho: Geological Society of America Abstracts with Programs, v. 23, p. 116.

Pogue, K. R., 1984, Geology of the Mt. Putnam area, northern Portneuf Range, Bannock and Caribou Counties, Idaho [M.S. thesis]: Pocatello, Idaho State University, 106 p.

Ramsay, J. G., and Huber, M. I., 1987, The techniques of modern structural geology, vol. 2: Folds and fractures: Orlando, Florida, Academic Press, p. 309–700.

Rember, W. C., and Bennett, E. H., 1979a, Geologic map of the Driggs quadrangle, Idaho: Idaho Bureau of Mines and Geology, Geologic Map Series, scale 1:250,000.

——, 1979b, Geologic map of the Dubois quadrangle, Idaho: Idaho Bureau of Mines and Geology, Geologic Map Series, scale 1:250,000.

——, 1979c, Geologic map of the Hailey quadrangle, Idaho: Idaho Bureau of Mines and Geology, Geologic Map Series, scale 1:250,000.

——, 1979d, Geologic map of the Idaho Falls quadrangle, Idaho: Idaho Bureau of Mines and Geology, Geologic Map Series, scale 1:250,000.

——, 1979e, Geologic map of the Pocatello quadrangle, Idaho: Idaho Bureau of Mines and Geology, Geologic Map Series, scale 1:250,000.

Ross, C. P., 1947, Geology of the Borah Peak quadrangle: Geological Society of America Bulletin, v. 58, p. 1085–1160.

Royse, F., Jr., Warner, M. A., and Reese, D. L., 1975, Thrust belt structural geometry and related stratigraphic problems, Wyoming–Idaho–northern Utah, *in* Bolyard, D. W., ed., Deep drilling frontiers of the central Rocky Mountains: Denver, Colorado, Rocky Mountains Association of Geologists Symposium, p. 41–54.

Ruppel, E. T., 1968, Geologic map of the Leadore quadrangle: U.S. Geological Survey Geologic Quadrangle Map GQ-733, scale 1:62,500.

Ruppel, E. T., and Lopez, D. A., 1981, Geologic map of the Gilmore Quadrangle, Lemhi and Custer Counties, Idaho: U.S. Geological Survey Geologic Quadrangle Map GQ-1543, scale 1:62,500.

——, 1988, Regional geology and mineral deposits in and near the central part of the Lemhi Range, Lemhi County, Idaho: U.S. Geological Survey Professional Paper 1480, 122 p.

Schmitt, J. G., and Steidtmann, J. R., 1990, Interior ramp-supported uplifts; Implications for sediment provenance in foreland basins: Geological Society of America Bulletin, v. 102, p. 494–501.

Scholten, R., Keenmon, K. A., and Kupsch, W. O., 1955, Geology of the Lima region, southwestern Montana and adjacent Idaho: Geological Society of America Bulletin, v. 66, p. 345–404.

Skipp, B., 1985, Contraction and extension faults in the southern Beaverhead Mountains, Idaho and Montana: U.S. Geological Survey Open-File Report 85-545, 170 p.

——, 1987, Basement thrust sheets in the Clearwater orogenic zone, central Idaho and western Montana: Geology, v. 15, p. 220–224.

——, 1988a, Cordilleran thrust belt and faulted foreland in the Beaverhead Mountains, Idaho and Montana, *in* Schmidt, C. J., and Perry, W. J., eds., Interaction of the Rocky Mountain foreland and the Cordilleran thrust belt: Geological Society of America Memoir 171, p. 237–265.

——, 1988b, Geologic map of Mackay 4 (Grouse) NE Quadrangle, Butte and Custer Counties, Idaho: U.S. Geological Survey Open-File Report 88-423, scale 1:24,000.

——, 1989, Geologic map of Mackay 4 (Grouse) NW Quadrangle, Butte and Custer Counties, Idaho: U.S. Geological Survey Open File Report 89-142, scale 1:24,000.

Skipp, B., and Hait, M. H., Jr., 1977, Allochthons along the northeast margin of the Snake River Plain, Idaho: Wyoming Geological Association, 29th Annual Field Conference, Guidebook, p. 499–515.

Skipp, B., and Hall, W. E., 1980, Upper Paleozoic paleotectonics and paleogeography of Idaho, *in* Fouch, T. D., and Magathan, E. R., eds., Paleozoic paleogeography of the west-central United States: Rocky Mountain Paleogeography Symposium I, Society of Economic Paleontologists and Mineralogists, Rocky Mountain section, Denver, p. 387–422.

Skipp, B., Kuntz, M. A., and Morgan, L. A., 1990, Geologic map of Mackay 4 (Grouse) SE Quadrangle, Butte County, Idaho: U.S. Geological Survey Open File Report 89-431, scale 1:24,000.

Stein, R. S., and Barrientos, S. E., 1985, The 1983 Borah Peak, Idaho, earthquake: Geodetic evidence for deep rupture on a planar fault, *in* Stein, R. S., and Bucknam, R. C., eds., Proceedings of Workshop XXVIII on the Borah Peak, Idaho, earthquake, Vol. A: U.S. Geological Survey Open-File Report 85-290, p. 459–484.

Stickney, M. C., and Bartholomew, M. J., 1987, Seismicity and late Quaternary faulting of the northern Basin and Range province, Montana and Idaho: Bulletin of the Seismological Society of America, v. 77, p. 1602–1625.

Trimble, D. E., 1976, Geology of the Michaud and Pocatello quadrangles, Bannock and Power Counties, Idaho: U.S. Geological Survey Bulletin 1400, 88 p.

——, 1982, Geologic map of the Yandell Springs quadrangle, Bannock and Bingham Counties, Idaho: U.S. Geological Survey Geologic Quadrangle Map GQ-1533, scale 1:62,500.

Trimble, D. E., and Carr, J. W., 1976, Geology of the Rockland and Arbon quadrangles, Power County, Idaho: U.S. Geological Survey Bulletin 1399, 115 p.

Whitman, S. K., 1990, Metamorphic petrology and structural geology of the Pennsylvanian-Permian Dollarhide Formation, Blaine and Camas Counties, south-central Idaho [M.S. thesis]: Pocatello, Idaho State University, 108 p.

Woodward, N. B., Boyer, S. E., and Suppe, J., 1985, An outline of balanced cross-sections: Knoxville, University of Tennessee Department of Geological Sciences Studies in Geology 11, 2d edition, 170 p.

Worl, R. G., and 7 others, 1991, Geologic map of the Hailey 1°×2° quadrangle, Idaho, U.S. Geological Survey Open-File Report 91-340, scale 1:250,000.

MANUSCRIPT ACCEPTED BY THE SOCIETY JULY 19, 1991

Printed in U.S.A.

Geological Society of America
Memoir 179
1992

Chapter 4

Cretaceous thrusting and Neogene block rotation in the northern Portneuf Range region, southeastern Idaho

Karl S. Kellogg
U.S. Geological Survey, MS 913, Box 25046, Federal Center, Denver, Colorado 80225

ABSTRACT

The Putnam thrust has long been recognized as an important Mesozoic structure in the northern Portneuf Range, southeastern Idaho. At most localities, the thrust places Ordovician rocks above Permian and Pennsylvanian rocks, although near its southeastern extent, it ramps laterally downsection to the southeast. At its southeasternmost exposures, Cambrian rocks are juxtaposed above Mississippian rocks. New work indicates that the hanging wall of the Putnam thrust contains three imbricate thrust slices or subplates, which are, from structurally lowest to highest (and generally from north to south), the Lone Pine subplate, the Narrows subplate, and the Bear Canyon–Toponce subplate.

The steeply south-dipping, east-trending Narrows thrust overlies the Lone Pine subplate, underlies the Narrows subplate, and is a lateral ramp that merges eastward into the Putnam thrust. Where exposed, the Narrows thrust places Late Proterozoic quartzite of the Brigham Group over Ordovician and Cambrian rocks. The Bear Canyon thrust overlies the Narrows subplate and underlies the Bear Canyon–Toponce subplate, dips eastward along the west side of the Portneuf Range, and places lower Brigham Group quartzite above Cambrian limestone and Cambrian and Late Proterozoic upper Brigham Group quartzite and argillite. At its northern extent, the Bear Canyon thrust curves to the east, where it merges with the Putnam thrust. On the east side of the range, the intensely folded Toponce thrust places upper Brigham Group quartzite above Ordovician rocks; the Toponce is believed to be an eastward extension of the Bear Canyon thrust.

East-dipping rocks within the Lone Pine subplate were not strongly deformed during Cretaceous thrusting, in contrast to rocks within the Narrows subplate, where east-vergent recumbent folds, cleavage directions that fan about northerly strikes, and tectonic thickening and thinning of beds indicate intense, thrust-parallel shear. The deformation and thrust geometry within the Narrows subplate suggest that the Narrows subplate actually consists of several horses within a foreland-dipping duplex.

Late Miocene and younger basin deposits occur in north-trending valleys adjacent to the northern Portneuf Range and, to the west, the Bannock and Pocatello ranges. At most places, the Neogene deposits dip to the east by as much as 35°, indicating that late Miocene and younger extension and down-to-the-east rotation occurred along mostly west-dipping listric faults that are inferred to merge on at least one regional detachment. Although range-bounding faults account for a large component of extension and rotation, an additional large component was contributed by numerous, relatively small-displacement normal faults within mountain ranges.

Kellogg, K. S., 1992, Cretaceous thrusting and Neogene block rotation in the northern Portneuf Range region, southeastern Idaho, *in* Link, P. K., Kuntz, M. A., and Platt, L. B., eds., Regional Geology of Eastern Idaho and Western Wyoming: Geological Society of America Memoir 179.

INTRODUCTION

The northern Portneuf Range lies within the Idaho-Wyoming-Utah salient of the Cordilleran fold and thrust belt, immediately south of the Snake River Plain and 20 km east of Pocatello, Idaho (Fig. 1). The range is underlain by a thick miogeoclinal sequence as old as latest Proterozoic and as young as Jurassic. It has long been recognized that the Mesozoic Putnam thrust cuts through the northern and eastern parts of the area (Mansfield, 1920, 1927; Trimble, 1982), generally placing Ordovician rocks over Permian and Pennsylvanian rocks, although the structural details have remained elusive. Much of the problem in understanding Mesozoic structures is the result of overprinting by Neogene extensional faulting in the area as well as the widespread cover of upper Tertiary and Quaternary deposits, especially at lower elevations.

The Putnam thrust was first described by Mansfield (1920), who did not attempt to correlate it with other thrusts described south of the northern Portneuf Range. Trimble and Carr (1976) suggested that the Putnam thrust may be a northward extension of the Paris thrust (the westernmost thrust of the Bannock thrust zone of Armstrong and Cressman [1963]), which is exposed on the east side of the Bear River Rnage, about 30 km southeast of the area shown in the lower right-hand corner of Figure 1. The Putnam-Paris thrust system was proposed by Trimble and Carr as the eastward boundary of an enormous klippe that extends southward into Utah and westward to the Deep Creek Mountains, although the western boundary of the proposed klippe (the Deep Creek "thrust") is probably a low-angle normal fault (K. S. Kellogg and P. K. Link, unpublished data).

The Putnam thrust may not be continuous with the Paris thrust, as Trimble and Carr (1976) proposed. Rather, the two thrusts may form a thrust transfer system (Rodgers and Janecke, this volume). Evidence presented in this chapter supports this model.

Rocks below the Putnam thrust are placed in the Meade thrust plate. The Meade thrust system (the northern end of the Bannock thrust zone of Armstrong and Cressman [1963]) is well exposed in the Blackfoot Mountains (Fig. 1), and two imbricates of the Meade thrust system, defining three subplates, have been described (Mansfield, 1952; Allmendinger, 1981).

If the Putnam thrust represents either a northern extension of the Paris-Willard thrust system or is connected to that thrust system by a transfer mechanism, then the Putnam was active probably during Early Cretaceous time. An Early Cretaceous age for the Paris thrust is based on the age of the syntectonic Ephraim Conglomerate, which had long been thought to be Late Jurassic to Early Cretaceous in age (Armstrong and Cressman, 1963). New fossil evidence, however, indicates that the age of the Ephraim Conglomerate is entirely Early Cretaceous (Heller and others, 1986).

The Portneuf Range is typical of the ranges of the Basin and Range province. It trends north-northwest and is bounded along at least part of its west side by a major range-front fault. Valleys parallel to the range are mostly filled by locally thick late Miocene and younger sedimentary and volcanic deposits (Kellogg and Marvin, 1988); northeast-dipping Eocene volcanic rocks overlie the northeast end of the range (Trimble, 1982). The range is rugged and has about 1,200 m total relief. Mount Putnam (2,678 m), in the central part of the northern Portneuf Range, is a prominent landmark along the southeastern margin of the Snake River Plain.

The geology of the northern Portneuf Range was first investigated by Mansfield (1920, 1927) during a study of the Fort Hall Indian Reservation. Subsequently, parts of the area have been reinterpreted or mapped by Corbett (1978), Trimble (1982), Pogue (1984), Hladky (1986), Amerman (1987), Hefferan (1986), Kellogg (1990), Kellogg and others (1989), Hladky and Kellogg (1990), and Hladky and others (1991). Much of the preliminary mapping in the southeastern part of the study area (the Jeff Cabin Creek 7½-minute quadrangle) was done by the late Steven S. Oriel of the U.S. Geological Survey during his last field season (1985).

Based on the work of Pogue (1984) in the northern Portneuf Range and LeFebre (1984) in the Pocatello Range to the west, Link and others (1985) proposed a model for the Mesozoic structural development of the region. In the northern Portneuf Range, the model incorporates two imbricate thrusts, the Jeff Cabin and Bear Canyon thrusts, above the Putnam thrust; rocks of the Jeff Cabin thrust are inferred to have deformed into recumbent folds before the younger and higher Bear Canyon thrust cut through the folds, locally ramping down section in the direction of transport. In the Pocatello Range, this model is invoked to account for upright, east-dipping beds that apparently overlie tightly folded and overturned beds along east-dipping thrusts (Link and others, 1985); the possible effects of regional tilting by Tertiary block faulting were not discussed by Link and his coworkers.

In this chapter, an alternate model for Mesozoic deformation in the northern Portneuf Range is proposed. This model relates folds directly to the mechanics of thrusting, shows that thrusts ramp up section in the direction of transport, and shows that the oldest thrusts are structurally highest. Two major, structurally higher imbricates of the Putnam thrust, the Narrows and Bear Canyon thrusts, define three large thrust slices within the Putnam plate. The Toponce thrust, on the east side of the range, is proposed as an eastward extension of the Bear Canyon thrust. In this chapter, the term *subplate* will be used for the separate thrust slices within the Putnam plate; the term *plate* is restricted to thrust sheets bounded above and below by "major" thrust faults (Dahlstrom, 1970).

A realistic interpretation of the structural geology of the northern Portneuf Range as well as of a large part of southeastern Idaho involves understanding the effects of Tertiary extensional faulting. The most obvious effect is down-to-the-east block fault-

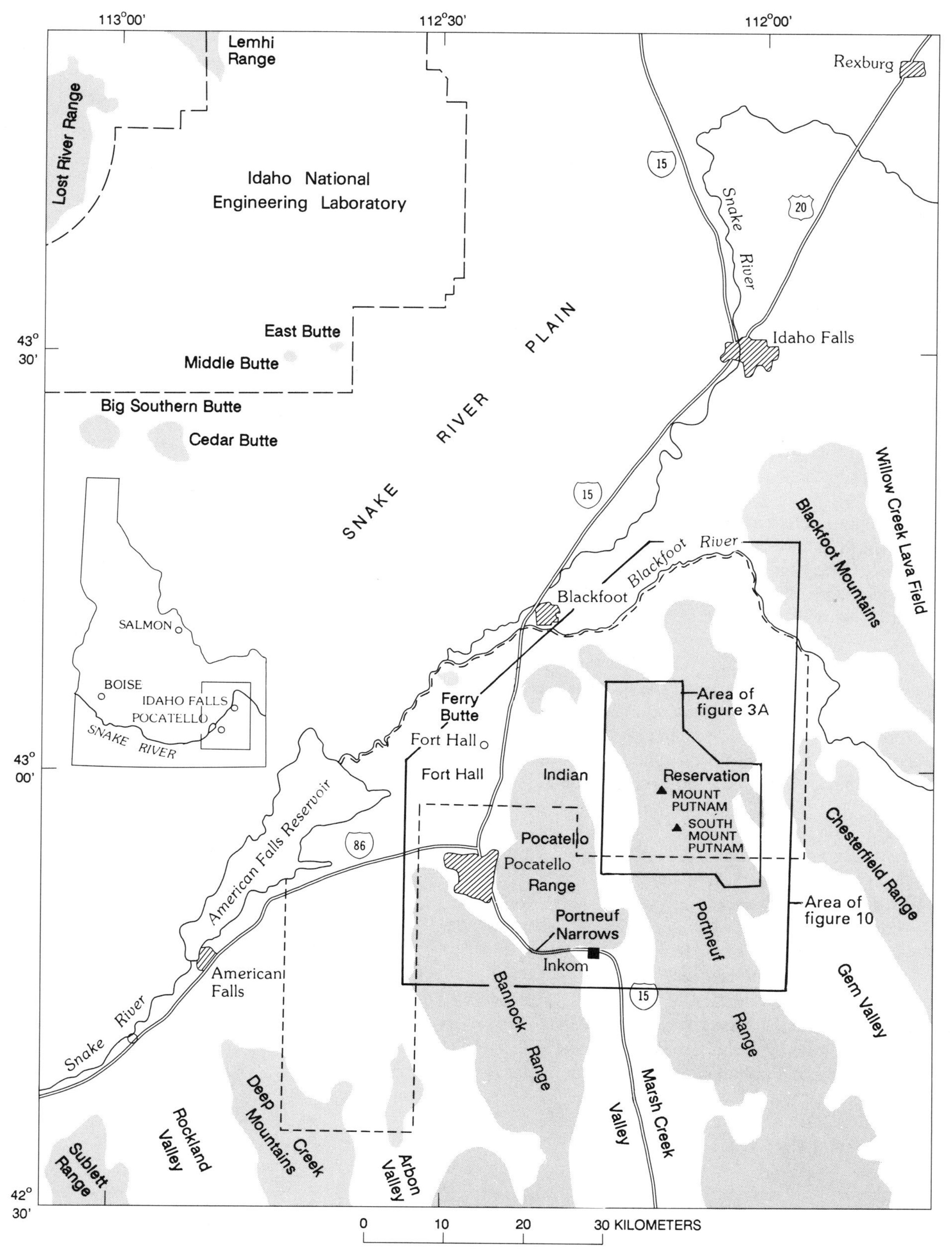

Figure 1. Index map of part of southeastern Idaho showing location of study area (Fig. 3A), area of Figure 10, and Fort Hall Indian Reservation (short-dashed line).

ing, and a model is described whereby block rotation results from movement along mostly west-facing listric faults.

STRATIGRAPHY OF THE NORTHERN PORTNEUF RANGE

A thick sequence of Precambrian to Mesozoic miogeoclinal rocks crops out in the northern Portneuf Range (Fig. 2). Detailed descriptions of these units in the study area are given by Kellogg and others (1989), Kellogg (1990), and Hladky and Kellogg (1990). The oldest and most widely exposed sequence of rocks is the Late Proterozoic and Lower Cambrian Brigham Group (Crittenden and others, 1971; Link and others, 1985, 1987), composed of about 4,000 m of quartzite and subordinate interbedded quartzose conglomerate and siltstone. The Late Proterozoic Caddy Canyon Quartzite (Crittenden and others, 1971) of the Brigham Group is the oldest exposed unit in the northern Portneuf Range, although older rocks of the lower Brigham Group and underlying Late Proterozoic Pocatello Formation crop out in the Pocatello Range immediately to the west (Trimble, 1976).

Approximately 1,000 m of section, based on estimates from Oriel and Platt (1980), have been removed structurally from the study area (Fig. 3A) by the Putnam thrust, although the total stratigraphic offset across the thrust at any one locality is considerably greater. This point will be discussed further in the next section.

Within the hanging wall of the Putnam thrust, the Brigham Group is overlain stratigraphically by several thousand meters of Cambrian to Lower Silurian carbonate and minor clastic rocks. The Upper Ordovician and Lower Silurian Fish Haven Dolomite is the youngest unit in the hanging wall, although the uppermost (Silurian) part of the unit is not believed to crop out in the study area (Fig. 3A). The Silurian Laketown Dolomite, however, which overlies the Fish Haven Dolomite, has been reported a few kilometers south of the area of Figure 3A (Corbett, 1978).

The oldest exposed unit in the footwall of the Putnam thrust, near the thrust's southeastern exposed extent, is the Mississippian Lodgepole Limestone, above which is exposed a northeast-dipping, faulted sequence of rocks as young as the Middle Jurassic Twin Creek Formation (which crops out in the northern Portneuf Range about 8 km northeast of the area of Figure 3A; Trimble, 1982; Hladky and others, 1991). This marks the westernmost exposures in the thrust belt of rocks younger than the Lower Triassic Dinwoody Formation.

STRUCTURAL GEOLOGY OF THE NORTHERN PORTNEUF RANGE

Mesozoic thrusting

The Putnam thrust and its overlying splays, including the Bear Canyon, Narrows, and Toponce thrusts, are the dominant Mesozoic structures of the northern Portneuf Range (Figs. 3 and

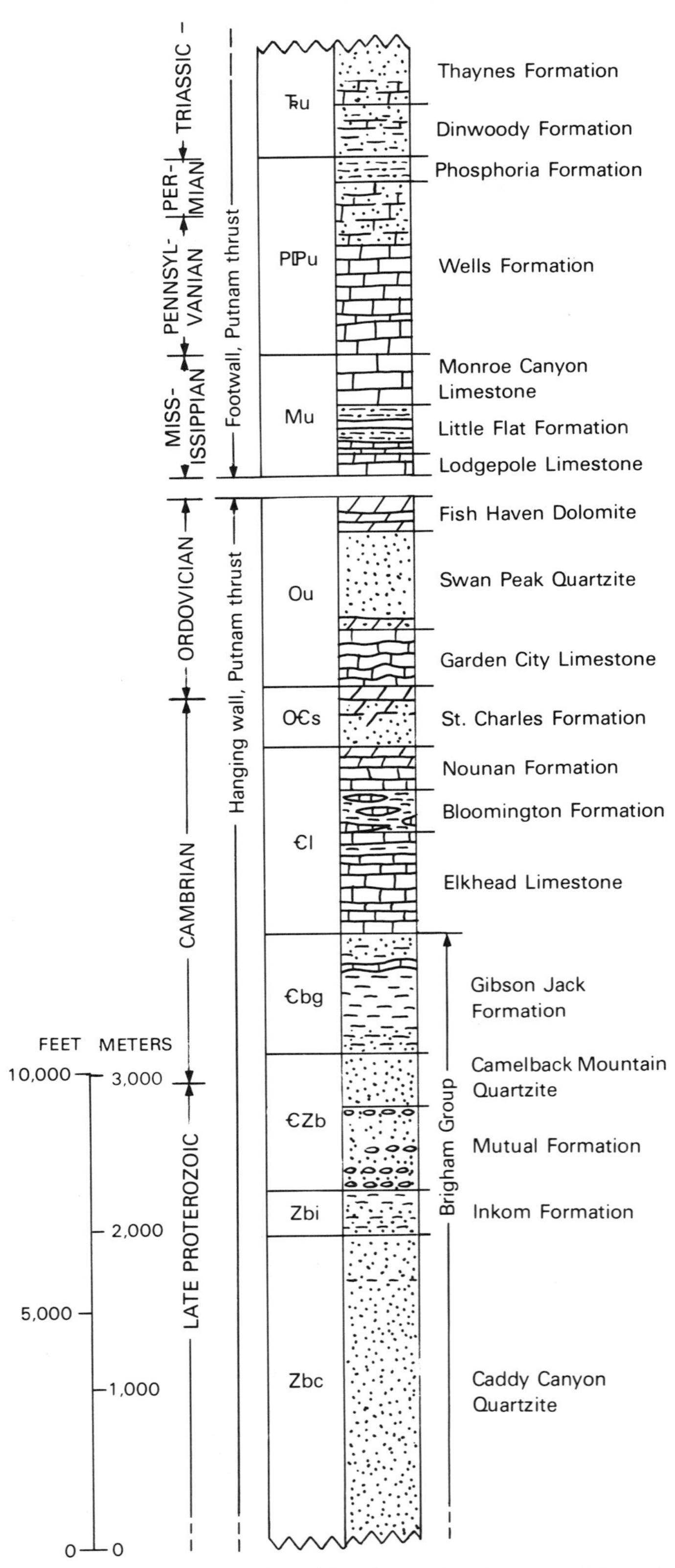

Figure 2. Generalized stratigraphic column of rocks exposed in study area (Fig. 3A). Upper part of Fish Haven Dolomite, which contains rocks as young as Silurian, is not believed to crop out in map area.

4). At its leading edge, the Putnam thrust generally places Ordovician rocks above the Pennsylvanian and Permian Wells Formation, although in its southeastern exposures, the Cambrian Nounan Formation overlies the Mississippian Little Flat Formation (Amerman, 1987). Total stratigraphic offset varies across the Putnam thrust as a result of lateral and frontal ramping (Kellogg and others, 1989; Hladky and Kellogg, 1990; Figs. 3A and 5) but does appear to decrease southeastward. In the northern part of the study area, approximate stratigraphic offset across the thrust, measured between the Ordovician Garden City Limestone and the Permian and Pennsylvanian upper Wells Formation, is about 3,700 m. At the southeastern exposures of the thrust, stratigraphic offset between the Nounan Formation and the Little Flat Formation is about 3,100 m. This apparent (though slight) decrease in stratigraphic offset to the southeast supports the idea that the Putnam and Paris thrusts form a thrust-transfer system (see Rodgers and Janecke, this volume).

A number of features contribute to a complex thrust geometry, including (1) at least three large subplates (or horses) overlying the Putnam thrust, (2) steep thrust ramps, both frontal and lateral, (3) east-trending Mesozoic tear faults (compartmental faults of Brown [1984]), which accommodate different styles of shortening on either side of the tear fault, and (4) locally developed, tight, nonparallel folds and well-developed cleavage.

Three principal subplates overlie the Putnam thrust in the northern Portneuf Range and are bounded by two thrusts (one of which, the Narrows thrust, is a steep lateral ramp) that appear to merge eastward with the Putnam thrust (Figs. 3A and 4). However, the branch lines, which define the joins between the two thrusts and the Putnam thrust, are not exposed. The three subplates are, from north to south and structurally lowest to highest, (1) the Lone Pine subplate, named for a canyon cutting through the subplate in the northern part of the area, (2) the Narrows subplate named for the east-trending Narrows thrust, which dips steeply south along The Narrows of Ross Fork Creek, and (3) the Bear Canyon–Toponce subplate, named for the Bear Canyon thrust, which bounds the subplate on the west and north, and the folded and normal-faulted Toponce thrust, exposed on the east side of the range and believed to be an eastward extension of the Bear Canyon thrust. The Bear Canyon thrust was named by Pogue (1984) for a small canyon east of Mount Putnam; the Toponce thrust, first identified by Corbett (1978) and named and described by Kellogg and others (1989), is named for nearby Toponce Creek (Fig. 3A). Several other smaller imbricate thrust slices also have either been mapped or inferred.

The Lone Pine subplate, which was extensively faulted and locally brecciated during Tertiary extension, contains exposed rocks as old as the Middle and Upper Cambrian Nounan Formation and as young sa the Upper Ordovician and Lower Silurian Fish Haven Dolomite (Hladky, 1986). Rocks of the subplate mostly dip 20 to 50° to the east, and deformation (folding and thrust-parallel shearing) during Mesozoic thrusting was apparently minimal.

In the northern part of the map area, Tertiary extensional faults have juxtaposed rocks from the upper (Putnam/Lone Pine) and lower (Meade) plates of the Putnam thrust in a complex geometry (Figs. 3A and 5). On cross section B-B′ (Fig. 5A), restoration of Tertiary movement indicates that the Putnam thrust ramps up through the Cambrian and Ordovician section in the hanging wall. In the footwall, the thrust is a flat in the lower Wells (to the west), ramps through the upper Wells Formation and the Phosphoria Formation, and appears to flat in the Dinwoody Formation (to the east). If this interpretation is correct, the two anticlines in the lower plate may either be fault-bend folds (ramp anticlines) or fault-propagation folds (Woodward and others, 1985) that overlie a deeper (Meade?) thrust.

The Narrows thrust, which defines the northern margin of the Narrows subplate, is not, strictly speaking, a thrust but a steeply south-dipping lateral ramp having a large component of strike-slip movement; Trimble (1982) called it a tear fault. The Narrows subplate contains exposed rocks as old as the Late Proterozoic Caddy Canyon Quartzite and as young as the Upper Ordovician and Lower Silurian Fish Haven Dolomite. Many of the rocks within the Narrows subplate, especially those in the Brigham Group along the west side of the range, are strongly folded, sheared, and overturned. Where the Narrows thrust is exposed along Ross Fork Creek, it places Caddy Canyon Quartzite over Lower Ordovician Garden City Limestone, although beneath the covered area to the east, the thrust is inferred to ramp steeply up section eastward in both the hanging wall and footwall. The Narrows thrust probably merges with the Putnam thrust under a thick, faulted, east-dipping (by as much as about 35°) sequence of mostly volcanogenic sediments (Hladky, 1986), correlated with the upper Miocene Starlight Formation of Carr and Trimble (1963).

The Bear Canyon–Toponce subplate contains rocks as old as the Late Proterozoic Caddy Canyon Quartzite on the west and as young as the Upper Cambrian and Lower Ordovician St. Charles Formation on the east. Most rocks within the subplate dip to the east less than 50°, although there are several gentle, north-trending folds with wave lengths of about 1 km (Fig. 3A), suggesting ramping of an underlying thrust. Internal deformation, such as shearing, is relatively slight.

Along the west side of the range south of Mill Creek, the Bear Canyon thrust dips to the east beneath mostly east-dipping Caddy Canyon Quartzite of the Bear Canyon–Toponce subplate. In the footwall (Narrows subplate), the thrust ramps down section to the south through the lower part of the Middle Cambrian Elkhead limestone, the Lower and Middle Cambrian Gibson Jack Formation, and the uppermost part of the Late Proterozoic and Lower Cambrian Camelback Mountain Quartzite (Fig. 3A). One to 2 km south of Mill Creek, excellent topographic control demonstrates that the Bear Canyon thrust dips 52° eastward, about 20° less steeply than beds both above and below the thrust (Kellogg, 1990). Offset of hanging-wall and footwall cutoffs suggests about 3 to 4 km of movement on the Bear Canyon thrust.

The clear hanging-wall–down geometry across the Bear Canyon thrust (Fig. 3B) rules out the possibility that the thrust

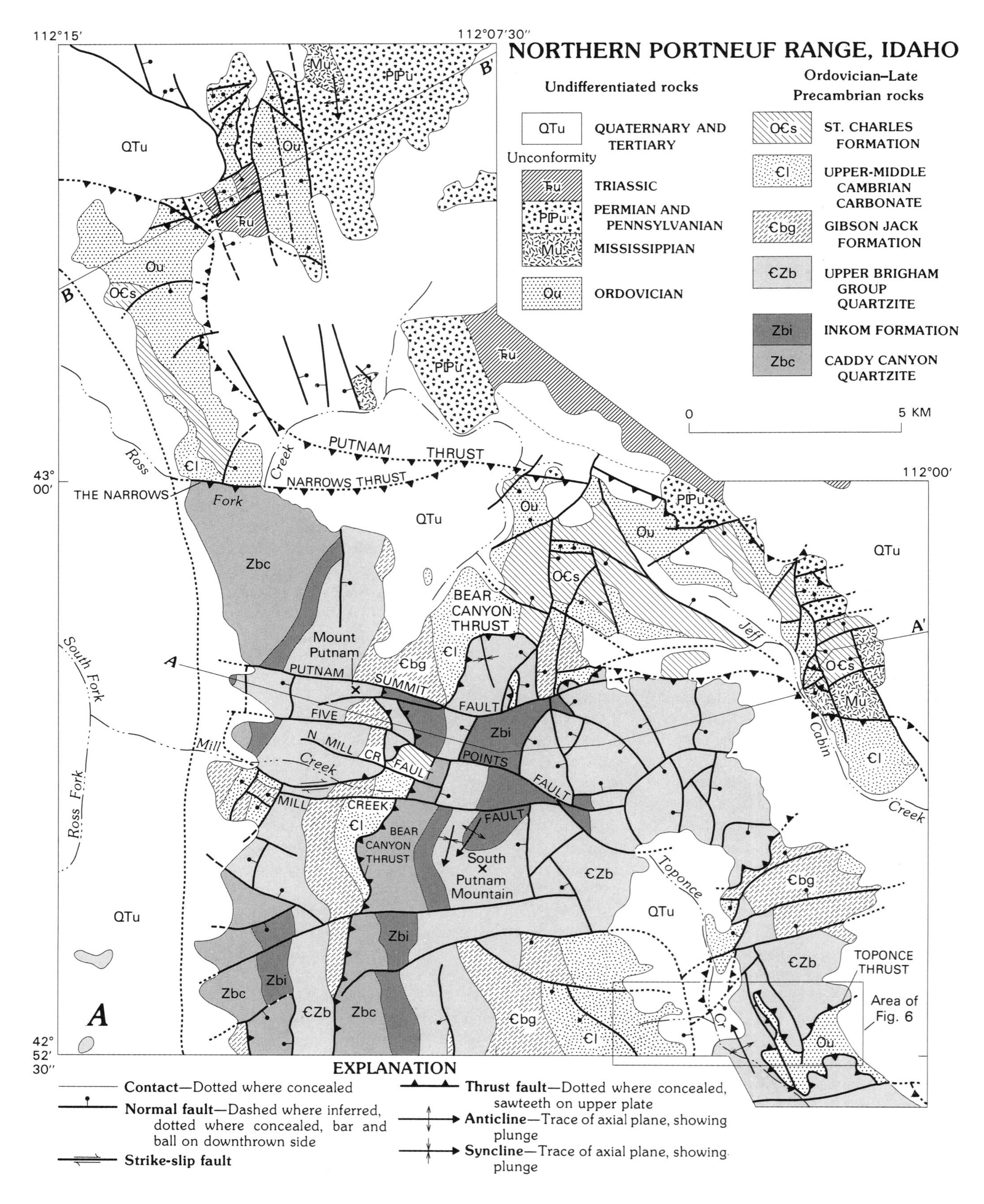

EXPLANATION

Contact—Dotted where concealed

Normal fault—Dashed where inferred, dotted where concealed, bar and ball on downthrown side

Strike-slip fault

Thrust fault—Dotted where concealed, sawteeth on upper plate

Anticline—Trace of axial plane, showing plunge

Syncline—Trace of axial plane, showing plunge

formed as a back thrust, similar to those described by Webel (1987). Prior to Neogene extension and down-to-the-east rotation (to be discussed), the Bear Canyon thrust may, in fact, have been nearly horizontal or west dipping.

The Bear Canyon thrust is offset across several east-trending faults near its northern extent where it curves to the east. It is inferred to merge with the Putnam thrust under the Quaternary and Tertiary deposits of Jeff Cabin Creek (Fig. 3A). The branch line plunges to the southeast, passing east of the area of outcrop for the Toponce thrust (Fig. 3A). This branch-line geometry is mandated if the Toponce and Bear Canyon thrusts represent the same surface, as seems reasonable; strongly deformed rocks exposed beneath the Toponce thrust would therefore be part of the Narrows subplate. Separate names are retained for the Bear Canyon and Toponce thrusts to indicate that interpretations other than a single thrust surface may be viable. For example, Kellogg and others (1989) suggest a slightly different (though not now favored) interpretation in which the Bear Canyon and Toponce thrusts represents a thrust-transfer system beneath the Bear Canyon–Toponce subplate.

Just east of the inferred intersection of the Bear Canyon and Putnam thrusts, the Putnam thrust is broken by numerous Tertiary normal faults into a zone with very complicated block-faulted geometry (Fig. 3A; Amerman, 1987). At many places, upper- and lower-plate rocks are juxtaposed across Tertiary faults and the generally southwest-dipping Putnam thrust is rotated within different fault blocks into various orientations. Several major northwest-trending Tertiary faults in this zone are downthrown on their northeast side, uncharacteristic of most normal faults in the northern Portneuf Range. Toward the south, within the zone of complex block-faulted geometry, the Putnam thrust is offset westward, and lower stratigraphic levels are exposed in both upper- and lower-plate rocks. The lower stratigraphic exposure is due to down-section frontal ramping toward the southeast. At its southernmost extent north of the inferred intersection with the Bear Canyon thrust beneath Jeff Cabin Creek, the Putnam thrust places Ordovician and Cambrian St. Charles Formation of the Narrows subplate above Mississippian Little Flat Formation of the Meade plate. South and east of the intersection, Cambrian Nounan Formation of the Bear Canyon–Toponce subplate is structurally above Mississippian Little Flat Formation.

The Toponce thrust is folded and is broken by numerous Tertiary normal faults (Fig. 6). The thrust places upright quartzite of the Brigham Group over locally highly sheared and overturned Ordovician Swan Peak Quartzite and Garden City Formation of the inferred Narrows subplate. Many white or tan quartzite beds immediately above the Toponce thrust are placed in the Mutual Formation of the Brigham Group, although their colors are uncharacteristic of this normally maroon-colored unit. The Toponce thrust is downdropped to the west under Toponce Creek along a west-dipping Tertiary normal fault having about 900 m of throw.

Pogue (1984) and Link and others (1985), following the interpretation of Corbett (1978), placed a thrust they called the Jeff Cabin thrust along upper Jeff Cabin Creek. Recent mapping (Kellogg and others, 1989) demonstrates, however, that a continuous, east-dipping section of Cambrian and Ordovician rocks exists across this area and that the Jeff Cabin thrust does not exist.

Tear or compartmental faults are common in the study area and formed concurrently with Cretaceous thrusting. Most of these faults trend east-west and are characterized by different deformational styles across them. The structural complexities are increased by recurrent movement during Tertiary extension, as shown by broad zones of brecciation along and close to the tear faults. Several of these larger tear faults are shown in Figures 3A and 4.

In Mill Creek valley, the Bear Canyon thrust is segmented

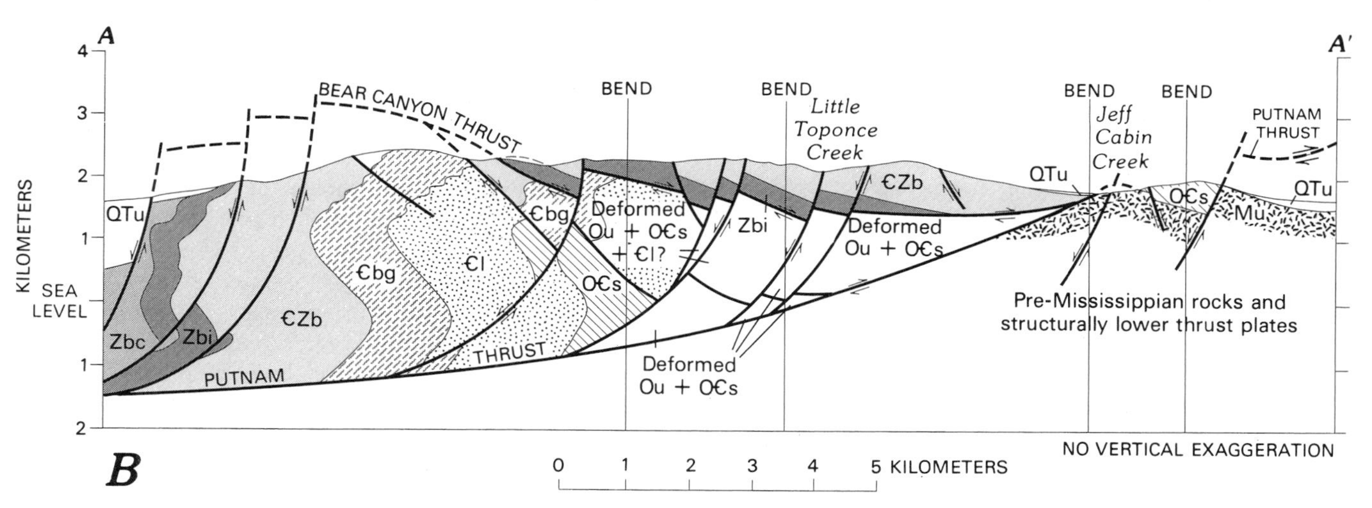

Figure 3. A, Simplified geologic map of part of the northern Portneuf Range, showing total known extent of the Putnam thrust. Geologic symbols, except for QTu (Quaternary and Tertiary deposits, undivided), are defined in Figure 2. Map based on Corbett (1978), Hladky and Kellogg (1990), Kellogg (1990), Kellogg and others (1989), and Hladky and others (1991). B, Cross section along line A-A′ on Figure 3A. Cross section B-B′ is shown in Figure 5.

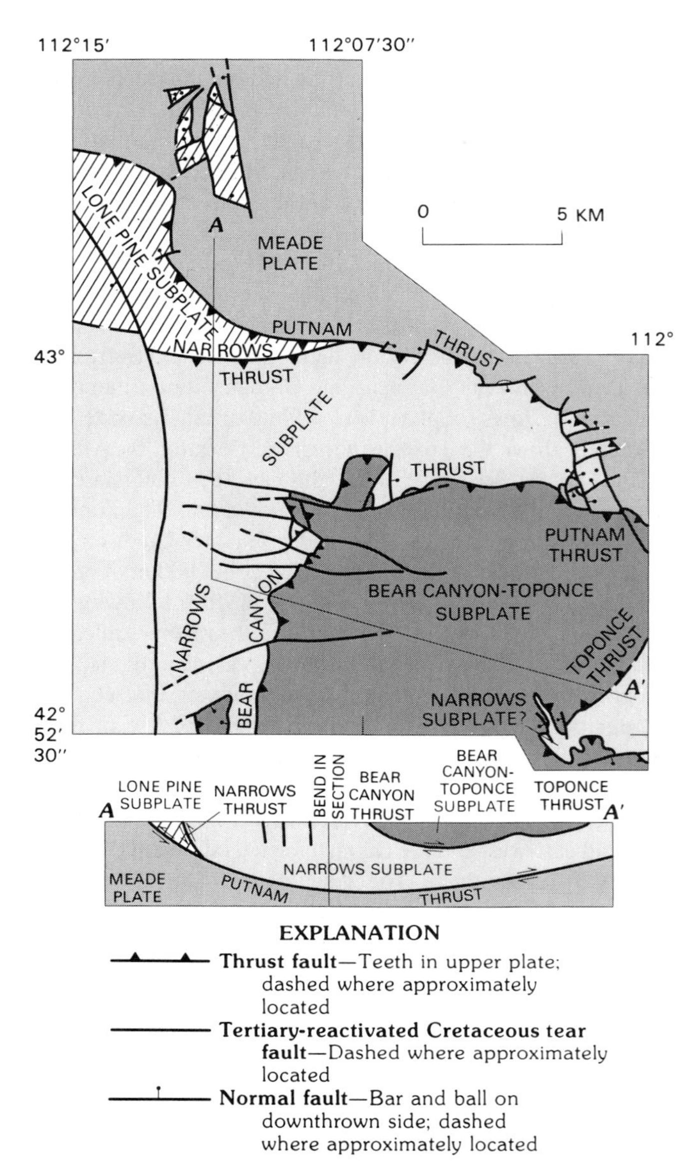

Figure 4. Diagrammatic tectonic map covering area of Figure 3A, showing the major structural elements. Cross section along A-A′ shows relationships between subplates of the Putnam thrust and the Meade plate. Scale of cross section is reduced one-third from that of map. Note that bend in cross section causes opposing sense of movement on Putnam thrust on each side of bend.

(or "compartmentalized") by several generally east-trending tear faults (Fig. 3A). Across each of these tear faults different units above and below the Bear Canyon thrust are juxtaposed, and the structural style commonly changes. For example, between the Mill Creek and North Mill Creek faults the Bear Canyon thrust places Proterozoic Caddy Canyon Quartzite over dolomitized Middle Cambrian Elkhead Limestone. Between the North Mill Creek and Five Points faults the Bear Canyon thrust places Caddy Canyon Quartzite over Ordovician and Upper Cambrian St. Charles Formation. Farther north, between the Five Points and Putnam Summit faults (the latter named by Pogue [1984]), the Bear Creek thrust divides into two thrust imbricates that may bound the west end of a small horse (Fig. 3B; see next section). This interpretation is favored because it accounts for an otherwise excessively thick section of Ordovician and Cambrian units in the Narrows subplate between the south flank of Mount Putnam and Jeff Cabin Creek. As an alternative explanation, the two thrusts may simply merge into a single Bear Canyon thrust within a kilometer or so east of their area of outcrop. In either case, it appears that the Bear Canyon thrust locally ramps down section in the direction of tectonic transport below a small anticline in Brigham Group rocks about 4 km west of Little Toponce Creek (Fig. 3B). This is the only locality in the study area where such aberrant thrust relations have been observed.

Evidence for a duplex in the Narrows subplate. A duplex is a family of imbricate subsidiary thrusts, each of which asymptotically curves downward to a sole or floor thrust and upward to a roof thrust (Fig. 7; Boyer and Elliott, 1982). In a hinterland-dipping duplex (the more commonly described type), the foreland side of a horse is uplifted as the subsequently formed horse ramps upward, causing hinterland-dipping structures. In a foreland-dipping duplex, the hinterland side of each horse is uplifted as the subsequently formed horse ramps up underneath, thereby causing both bedding and horse-bounding thrusts to rotate into a foreland-dipping orientation. Unique to a foreland-dipping duplex is that the leading-edge ramp of each horse is nose down on the floor thrust, which undergoes continued movement during thrusting.

The strong shearing and widespread east-vergent overturning that occur at many localities in the Narrows subplate, in comparison to the relatively unsheared rocks structurally above and below the Narrows subplate, strongly suggest that the Bear Canyon–Toponce thrust defines a roof thrust and that the Putnam thrust defines a floor thrust of the front of a foreland-dipping duplex. Lack of well-developed thrust-related deformation within the Lone Pine subplate, in comparison to that within the Narrows subplate, suggests that the Lone Pine subplate is not part of the duplex and that the steeply south-dipping Narrows thrust defines the northern boundary of the duplex; the Narrows thrust can be viewed as part of the floor thrust, merging with the Putnam thrust.

The Bear Canyon and Toponce thrusts both place upright, east-dipping or gently folded, essentially unsheared Brigham Group rocks above mostly overturned, west-dipping, and commonly highly sheared rocks of the Narrows subplate that are as young as Ordovician (Fig. 3A). S- or Z-shaped folds with nearly horizontal axial planes, in which west-dipping beds are overturned and east-dipping beds are upright, are characteristic of many localities in the Brigham Group quartzites (Fig. 3B) and reflect the extreme shear parallel to thrusting. Spectacular examples of nearly isoclinal folds with horizontal axial planes in Camelback Mountain Quartzite are found in the canyon of Mill Creek (Fig. 8; Pogue, 1984). Such folds have also been described in the

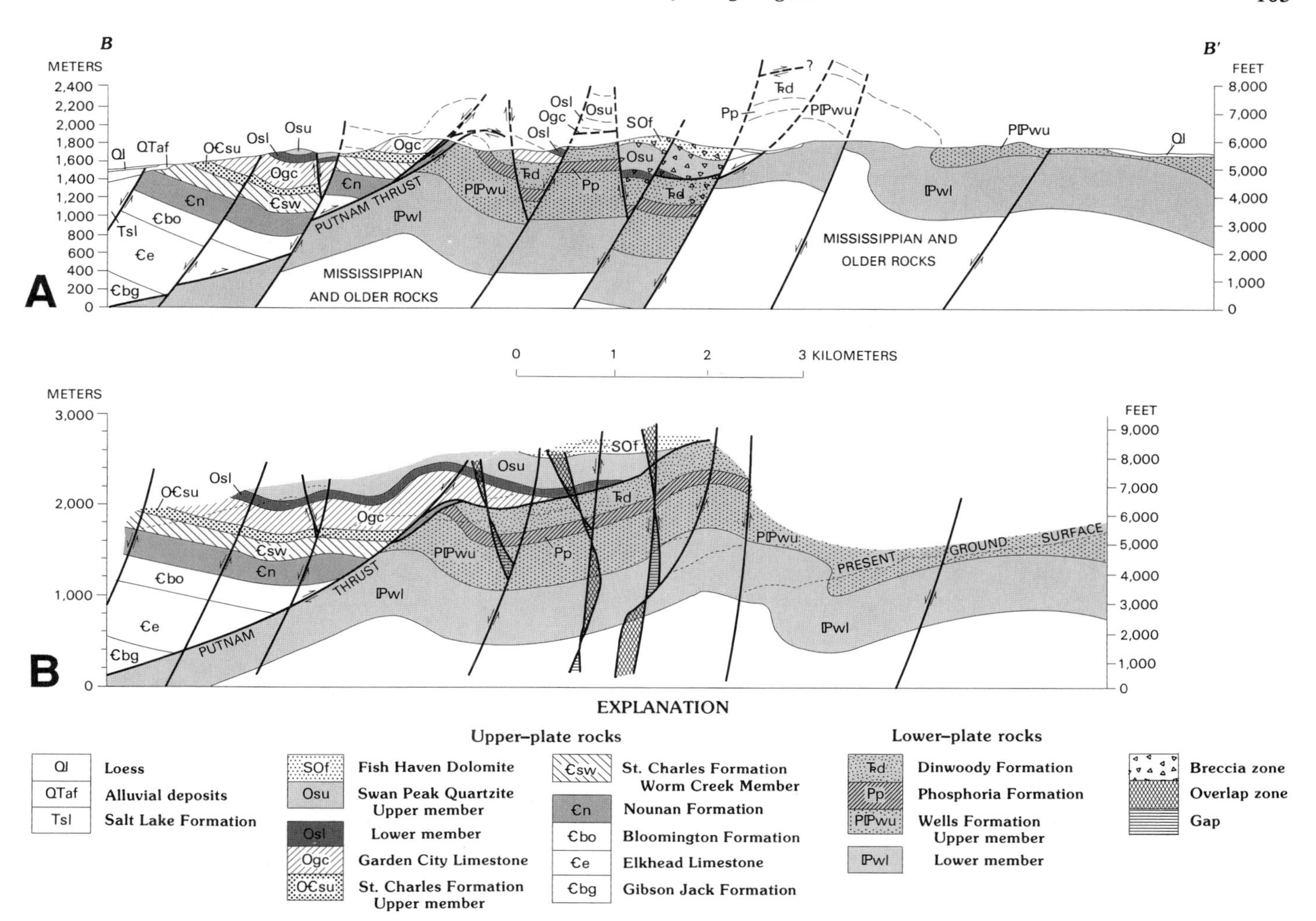

Figure 5. A, Cross section along B-B′ (shown on Fig. 3A), adapted from Hladky and Kellogg (1990). Cross section shows considerably more stratigraphic and structural detail than appears on Figure 3A. B, Cross section B-B′ after all Tertiary faults have been restored to their prefaulting positions. Cross-hatched areas are zones of overlap, and horizontally lined areas are gaps. Dotted lines represent present ground surface.

southern Pocatello Range (e.g., the Rapid Creek fold of Burgel and others, [1987]).

Postulated eastward tilting of parts of the Portneuf Range by 30 to 40° during Neogene extensional faulting (to be discussed) certainly accounts for a large component of eastward dip of structures within the Narrows subplate but leaves room for a large additional component of eastward dip owing to the formation in Cretaceous time of a foreland-dipping duplex. Prior to Neogene eastward rotation, for example, the S-shaped folds described above were probably east-vergent recumbent folds with west-dipping axial planes. Such structures are suggestive of the overturned frontal limb of a large thrust nappe; Burgel and others (1987) invoked such a model, incorporating eastward Neogene rotation, to explain the Rapid Creek fold in the southern Pocatello Range. However, the fact that a well-defined roof thrust (Bear Canyon thrust and its probable extension, the Toponce thrust) bounds the Narrows subplate makes the duplex model for the formation of the overturned structures within the Narrows subplate seem reasonable.

The extremely sheared and overturned Ordovician rocks beneath the Toponce thrust may comprise several thrust imbricates or small, foreland-dipping horses (Fig. 6). The deformational style of these footwall rocks is very similar to that observed in the canyon of Mill Creek and elsewhere in the Narrows subplate and lends support to a model for formation of a foreland-dipping duplex within the Narrows subplate.

Beds are tectonically thickened and thinned within the Narrows subplate, reflecting internal shear during thrusting. This structural style is in marked contrast to the homoclinally east-dipping beds of the Lone Pine subplate and the east-dipping to

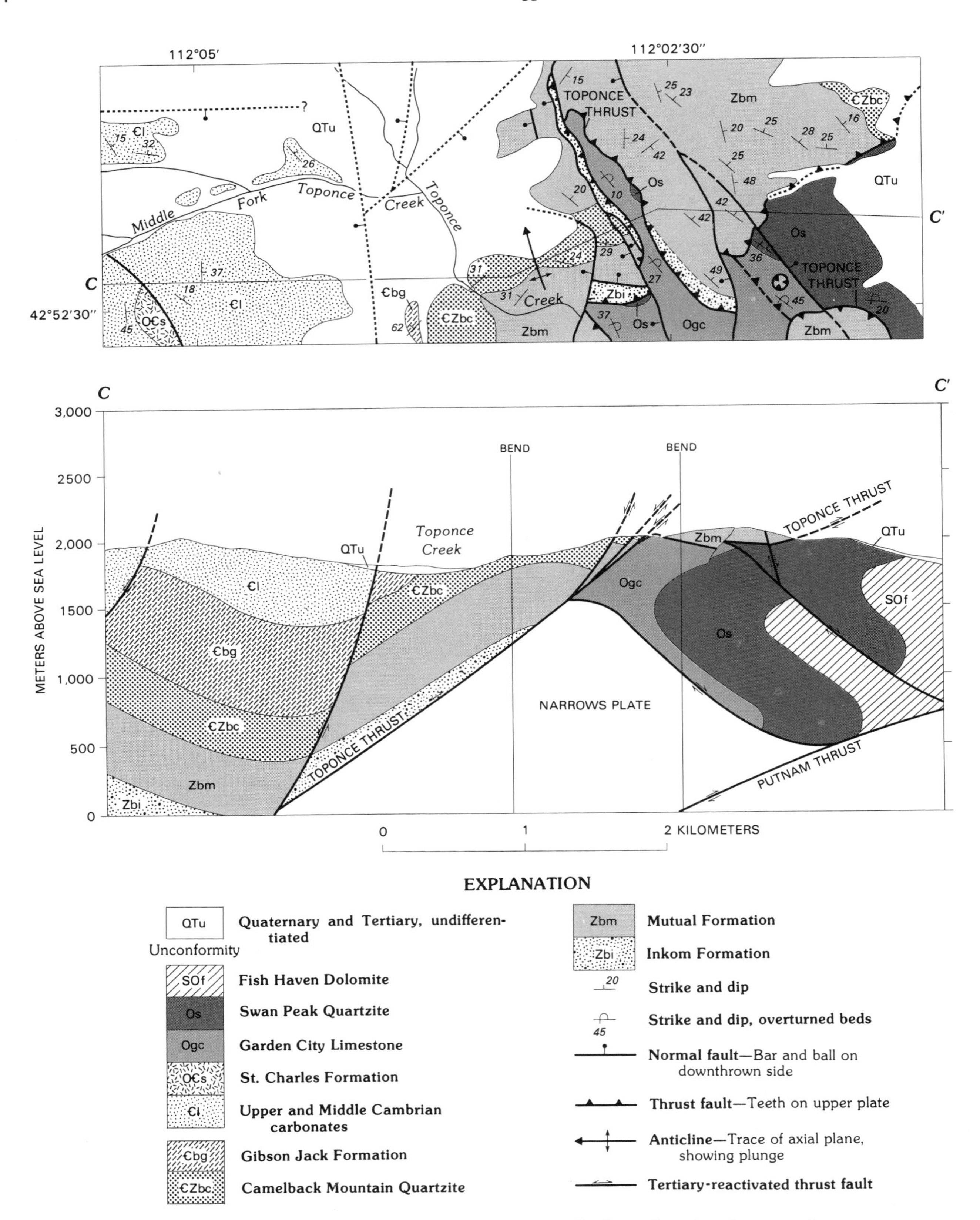

Figure 6. Enlarged geologic map and cross section of Toponce Creek area. Location of map area shown on Figure 3A.

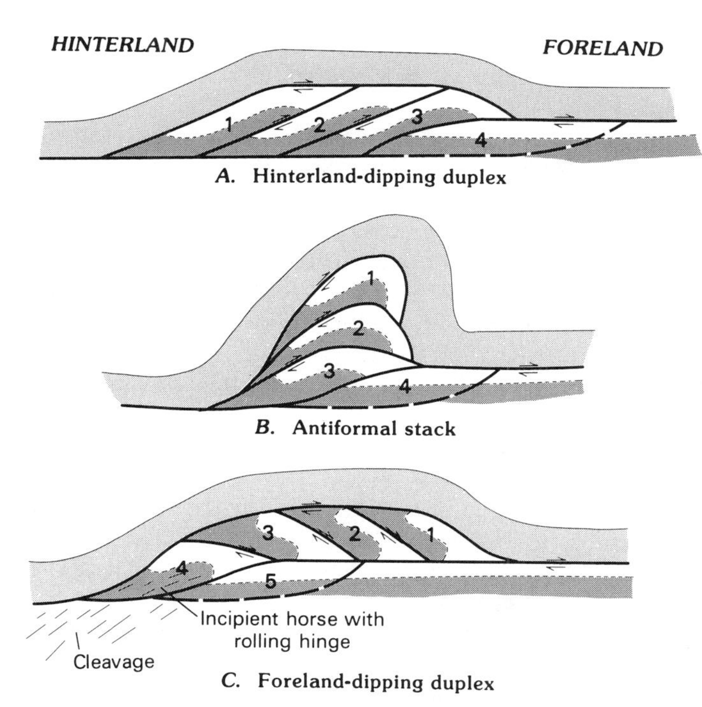

Figure 7. Schematic diagrams showing the development of (A) hinterland-dipping duplex, (B) antiformal stack (a situation intermediate to a hinterland- and foreland-dipping duplex), and (C) foreland-dipping duplex. Relative order of formation of horses is indicated by numbers, 1 being first formed; note that relative order in hinterland-dipping and foreland-dipping duplexes is reversed. Also note that any simple shear transferred to rocks above the floor thrust during both initial ramping and continued movement along the floor thrust will increase the leading-edge cutoff angle within each horse by thrust-parallel slip. Adapted from Boyer and Elliott (1982).

Figure 8. View, facing north, of recumbent syncline in Camelback Mountain Quartzite of the Narrows subplate near bottom of Mill Creek canyon. The nearly horizontal axial plane was probably west-dipping prior to Neogene rotation. Distance across outcrop is approximately 15 m. Photograph by P. K. Link.

gently folded rocks of the Bear Canyon–Toponce subplate. Thickening of beds is probably due to a combination of bedding-plane slip, small-scale duplex stacking and thrust imbrication, and ductile flow. Thinning of beds is by low-angle normal faulting and, in overturned limbs of folds, by tectonic stretching. Unrecognized map-scale faults also may have contributed to apparent changes in bedding thickness.

A good example of tectonic thickening of the Camelback Mountain Quartzite and the Mutual Formation within the Narrows subplate occurs in and north of Mill Creek, where many tight to isoclinal, parasitic folds along bedding-slip planes can be seen in the canyon walls (Fig. 8). On the west side of Mount Putnam, the overturned Mutual Formation is as thick as about 820 m, and the overturned Camelback Mountain Quartzite is as thick as about 510 m. These two formations have been tectonically thinned elsewhere, such as near the southern boundary of the map area (Fig. 3A), where the generally upright Mutual Formation is as thin as about 490 m and the upright Camelback Mountain Quartzite is as thin as about 270 m. It should be noted that the Camelback Mountain Quartzite is reported to be as thick as 1,000 m near Pocatello but is observed to be much thinner elsewhere (Trimble, 1976; Link and others, 1987); this variation suggests that tectonically induced changes in thickness of the Camelback Mountain Quartzite may be widespread. The Mutual Formation is reported to be 800 to 900 m thick near Pocatello (Trimble, 1976; Link and others, 1987).

Without additional subsurface data and detailed field studies, the lateral extent of the proposed foreland-dipping duplex in the northern Portneuf Range into areas adjacent to the range is, at best, speculative. One regional model for a foreland-dipping duplex, extending from the Portneuf Range westward at least through the northern Bannock Range, was suggested by Kellogg and Skipp (1988), although it is now apparent that their model did not adequately accommodate the role of Tertiary listric faulting and block rotation.

Cleavage in argillic rocks. Shear parallel to the direction of transport may cause initially formed cleavage to rotate, producing a fanning of cleavage orientations about a line perpendicular to the direction of tectonic transport. Fanning of cleavage directions is especially well developed during the extreme shear associated with duplex formation (Boyer and Elliott, 1982, Fig. 7). Limited fanning of "axial-plane" cleavage also commonly occurs in folds by refraction through beds of differing competency.

Well-developed slaty to phyllitic cleavage is present in all argillic rocks of the Brigham Group in the northern Portneuf Range. In addition, many of the Brigham Group quartzites, especially in the sheared, overturned limbs of recumbent folds, have developed a poorly developed shear-induced schistosity. Secondary minerals are tentatively identified as sericite, chlorite, and epidote, indicating greenschist-facies metamorphism. Metamorphic grade appears to be slightly higher in the Inkom Formation than in the stratigraphically higher Gibson Jack Formation, suggesting that burial metamorphism, predating Mesozoic thrusting, contributed at least partially to the metamorphic mineral assemblage.

Cleavage orientations were measured in argillic rocks of the Narrows and Bear Canyon–Toponce subplates. In both cases, the orientations are highly scattered (Fig. 9A), commonly changing by tens of degrees over distances of less than 100 m within rocks of apparently similar competency. Poles to cleavage generally are scattered along an east-west line, as would be expected by progressive rollover in a west-over-east simple shear regime. The east-west spread of cleavage directions might be expected to be better defined for rocks within a foreland-dipping duplex, where shear would probably be more extreme than where progressive rollover was relatively unimportant. The spread, however, appears to be as well developed in the Bear Canyon–Toponce subplate as it is in the Narrows subplate, which probably reflects the extreme ductility and resultant deformation of argillic rocks in both the Narrows and Bear Canyon–Toponce subplates. For example, abundant small-scale folds are observed in almost all argillites in the area (Kellogg, 1990; Kellogg and others, 1989).

The considerable departure of some orientations from the east-west line (Fig. 9A) is natural in any thrust system where there exist numerous, complicated, laterally varying structural domains, such as develop near lateral ramps, tear faults, and other thrust-related features (S. E. Boyer, written communication, 1989). However, except for the attempt to discriminate cleavage between the Narrows and Bear Canyon thrust subplates, no attempt was made to differentiate cleavage into different structural domains.

Cleavage also has no obviously consistent geometric relationship to bedding orientations (Fig. 9B), as would occur if cleavage developed during prefolding, layer-parallel shear (Mitra and others, 1984); cleavage orientations are clearly highly scattered with respect to the orientation of bedding planes in both the Narrows and Bear Canyon–Toponce subplates. It may be noted, however, that the method by which cleavage was rotated—about the strike of bedding by the amount of dip—is not a unique solution to the problem of restoring bedding to horizontal.

Neogene extensional faulting

Neogene extension has affected, to varying degrees, all of southeastern Idaho, and the northern Portneuf Range displays many of the extensional features typical for the region (Pierce and Morgan, this volume). A buried, linear, range-front fault is believed to bound at least part of the impressive west side of the range (Kellogg, 1990); on the east side of the range the topography is more subdued and irregular and elevations are generally lower. These features suggest that the northern Portneuf Range has the general form of a half-horst, uplifted on the west and/or downdropped on the east. The Neogene structures are complicated by an extensive network of normal faults within the range

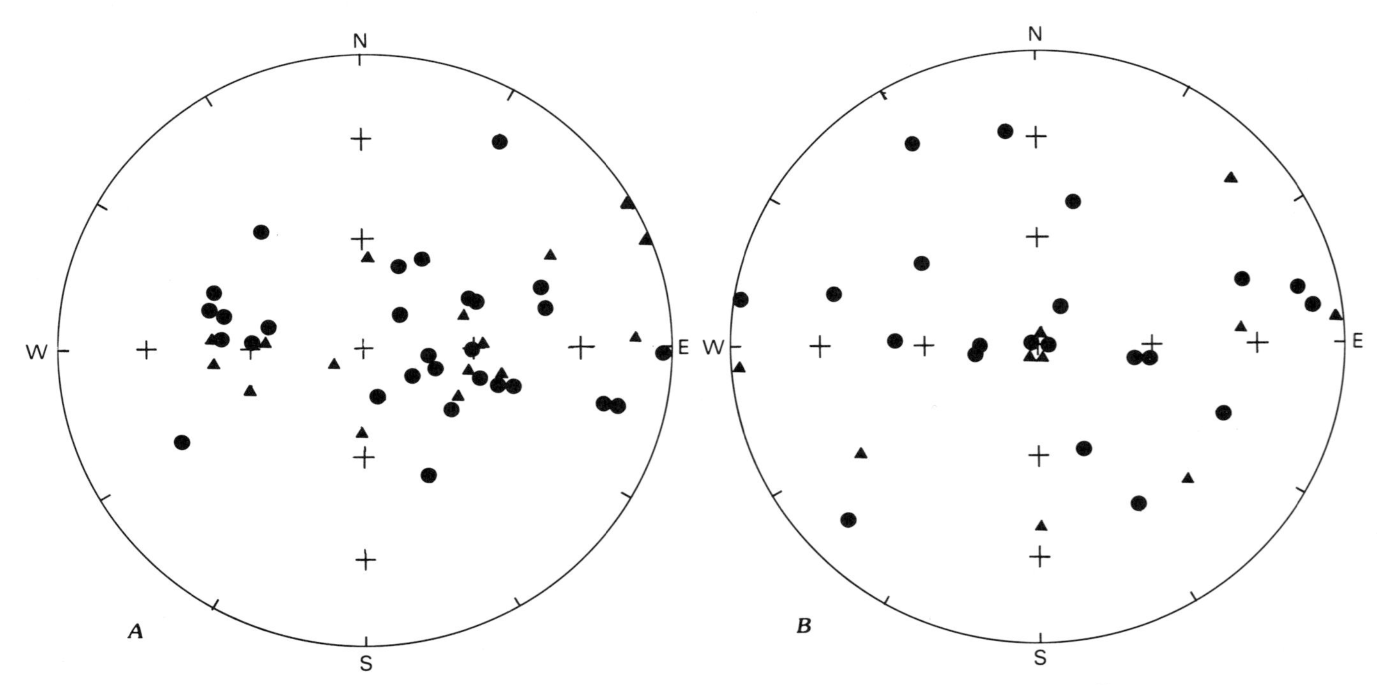

Figure 9. Equal-area projection of cleavage orientations measured in the Narrows and Bear Canyon–Toponce subplates. All poles are lower hemisphere projections. A, Poles to cleavage measured in the Narrows subplate (circles) and Bear Canyon–Toponce subplate (triangles). B, Poles to cleavage, where bedding orientations were measured in same outcrop, after rotation by an amount necessary to restore bedding to horizontal (and upward facing).

(Fig 3A). Much, if not most, of the regional extension was accommodated along these numerous, mostly west-dipping faults (Kellogg, 1987, 1990; Kellogg and others, 1989; Hladky and others, 1991).

Extensional faulting is at least as old as about 10 Ma, the maxmum-known age of valley-fill deposits in the region (Starlight Formation of Carr and Trimble [1963]), although a major pulse of Basin and Range faulting may have occurred about 7 Ma (Kellogg and Marvin, 1988).

Most intermontane Miocene and younger basin-fill deposits in the region dip to the east or northeast (Hladky, 1986; Hladky and others, 1991; Link and others, 1985; K. S. Kellogg, unpublished data and Fig. 10 of this chapter). Dips are generally less than about 35°, which is also true for Eocene volcanic rocks that bound the northeastern end of the Portneuf Range (Hladky and others, 1991). Larger eastward dips in Miocene and younger rocks, though rare, are known in the northern Portneuf Range (for example, Hladky and others, 1991). In the southern Portneuf Range, down-to-the-east rotation of Tertiary beds by as much as 90° along concave-westward listric faults has been described (Sacks and Platt, 1985); the rotation brought originally west-dipping Paleozoic and Proterozoic rocks into an east-dipping orientation.

At most places, the amount of Neogene rotation of pre-Tertiary rocks cannot be determined. However, based on known attitudes of tilted Tertiary beds in the northern Portneuf Range, it seems reasonable that the rotation at most localities was no more than about 30 to 40° eastward.

Eastward dip on Tertiary units and the existence of numerous west-dipping normal faults suggest that a regional detachment onto which concave-westward listric faults merge underlies the area. Obvious candidates for such detachment surfaces are thrust faults. Listric normal faults commonly sole into thrusts, especially near the tops of ramps (Royse and others, 1975). A good example is shown in Figure 5A (about 3 km from the area shown at the left margin), where a brecciated, west-dipping normal fault zone, about 30 m across, merges with the Putnam thrust above a footwall ramp; the normal fault is not shown on the simplified map of Fig. 3A. Another example occurs 1 km east of Toponce Creek, where several west-dipping normal faults merge with the Toponce thrust near the top of a thrust ramp (Fig. 6).

Further evidence of widespread Neogene extensional faulting in the northern Portneuf Range is the existence of breccia zones, which are not characteristic of thrust faults of the Idaho-Wyoming thrust belt (Armstrong and Oriel, 1965). Breccia zones are particularly common in Brigham Group quartzites within several tens of meters above the Bear Canyon thrust near Putnam Peak (Pogue, 1984; Kellogg, 1990) and may be due to brittle readjustment during backsliding in the zone between the curved normal fault and the more planar thrust plane. The brittle behavior reflects the lower pressures and temperatures that existed during extensional deformation as compared to those that existed during thrusting.

In the northern part of the study area, restoration of Tertiary movement suggests that about 10% extension has occurred (Fig. 5B), although the amount of extension is dependent on the original dip and curvature of the fault planes, which in most cases are poorly constrained. Topographic expression of a few fault traces in this area, however, and the fact that most Tertiary faults are west side down suggest that most Tertiary normal faults are west dipping (Hladky and Kellogg, 1990).

The existence of a thrust surface onto which the normal faults may merge is suggested by the two anticlines shown in Figure 5B—possibly fault-bend or fault-propagation folds—in the footwall of the Putnam thrust. The depth to such a surface (the Meade thrust?), which would govern the curvature of the normal faults, is conjectural. Irrespective of what curvature is chosen for the normal faults, restoring the cross section to a pre-Tertiary configuration produces large overlaps or gaps, which are largest near those Tertiary faults having the greatest offset.

The overlaps and gaps caused by Tertiary normal faulting were apparently accommodated by brecciation, which is most widespread near normal faults with the greatest throw. For example, the Fish Haven Dolomite is brecciated over an area as large as 1 by 3 km between two large north-trending faults (Fig. 5A and Hladky and others, 1991). The two faults, having vertical separations between 500 and 1,000 m, define a large graben bounded by rocks of the lower plate of the Putnam thrust.

The probable influence of Snake River Plain volcanism on the Neogene tectonics of the northern Portneuf Range cannot be ignored. Rocks are downwarped toward the plain (Sparlin and others, 1982), and the extent and complexity of normal faulting seem to increase toward the Snake River Plain margin (Kellogg, 1987; Hladky and others, 1991). There is also a striking similarity between the ages of major caldera-forming events in the plain and extensional faulting in adjacent mountain ranges (Pierce and Scott, 1986; Pierce and Morgan, this volume); both volcanism and faulting become progressively younger toward the northeast, parallel to the axis of the plain.

Faults of unequivocal Quaternary age have not been identified in the northern Portneuf Range. Immediately west of the northern part of the map area (Fig. 3A), however, northwest-trending faults, having as much as about 10 m of displacement, cut basalt flows (Kellogg and Embree, 1986) that are as young as about 2.2 Ma, or late Pliocene (Kellogg and Marvin, 1988).

STRUCTURES WEST OF THE NORTHERN PORTNEUF RANGE

The Pocatello Range lies immediately west of the northern Portneuf Range and is underlain by a mostly east-dipping, extensionally faulted section of rocks ranging from the Late Proterozoic Pocatello Formation (the oldest exposed unit in the thrust belt) to rocks as young as the Permian and Pennsylvanian Wells Formation (Trimble, 1976; Hladky and others, 1991). The northern edge of the Pocatello Range contains exposed rocks of Silurian and Devonian age (within area marked "a" in Fig. 10), unlike the northern Portneuf Range where exposed rocks of these

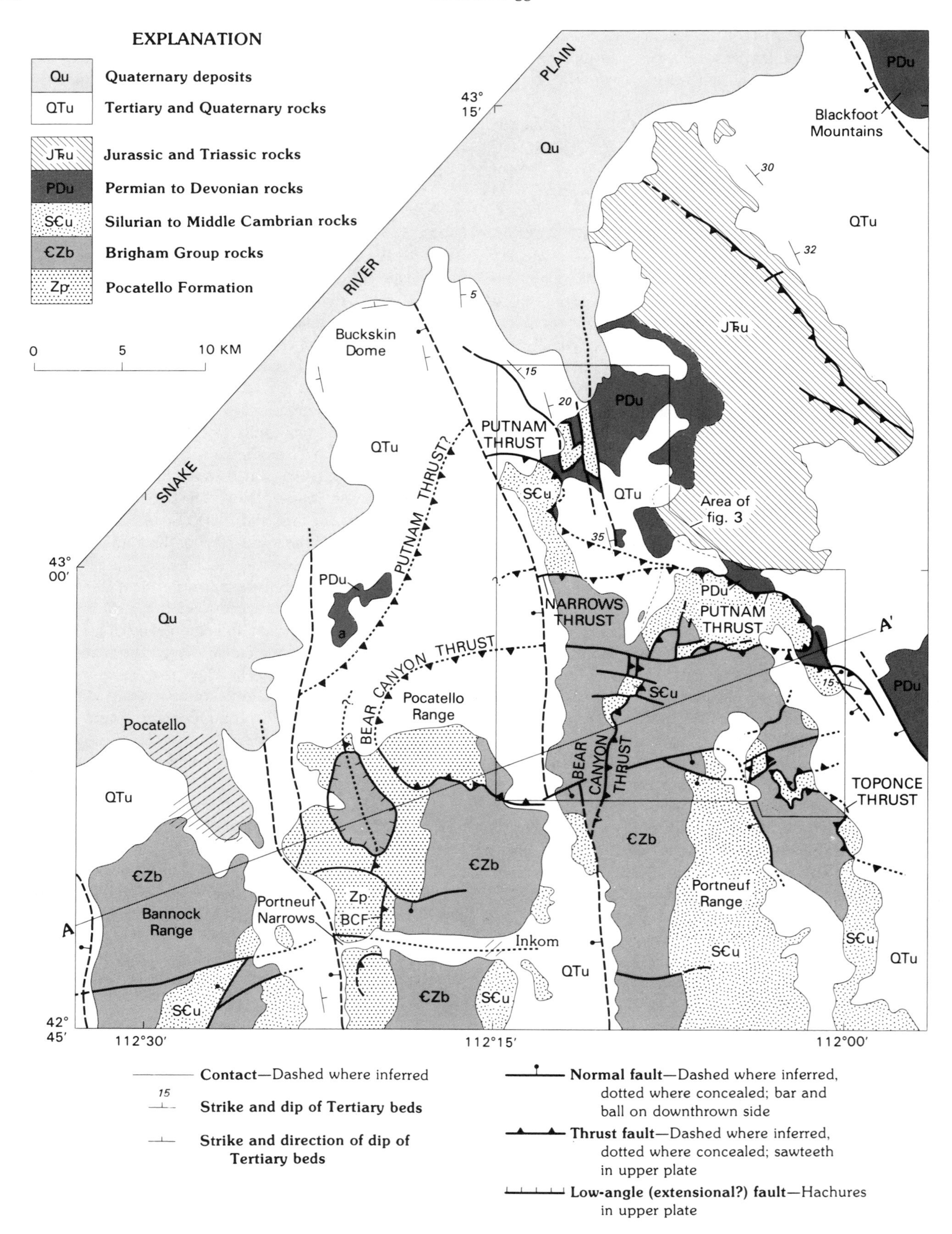
EXPLANATION
Qu Quaternary deposits
QTu Tertiary and Quaternary rocks
JTRu Jurassic and Triassic rocks
PDu Permian to Devonian rocks
S€u Silurian to Middle Cambrian rocks
€Zb Brigham Group rocks
Zp Pocatello Formation
0 5 10 KM
SNAKE RIVER PLAIN
43° 15′
43° 00′
42° 45′
112°30′
112°15′
112°00′
Blackfoot Mountains
Buckskin Dome
PUTNAM THRUST
PUTNAM THRUST?
Area of fig. 3
NARROWS THRUST
BEAR CANYON THRUST
Pocatello Range
Pocatello
TOPONCE THRUST
Portneuf Range
Bannock Range
Portneuf Narrows
BCF
Inkom
A
A′
Contact—Dashed where inferred
Strike and dip of Tertiary beds
Strike and direction of dip of Tertiary beds
Normal fault—Dashed where inferred, dotted where concealed; bar and ball on downthrown side
Thrust fault—Dashed where inferred, dotted where concealed; sawteeth in upper plate
Low-angle (extensional?) fault—Hachures in upper plate

ages have been removed structurally by the Putnam thrust. A structural interpretation of parts of the Pocatello Range was made by Ludlum (1943), refined by Trimble (1976), and later reinterpreted by LeFebre (1984), Link and others (1985), and Burgel and others (1987). West of the Pocatello Range, the northern Bannock Range contains a homoclinally east-dipping sequence of Brigham Group and Cambrian carbonates, cut by numerous Neogene extensional faults (Trimble, 1976). Figure 10 is a simplified interpretation of the geology in the region between the northern Bannock Range and the northern Portneuf Range, compiled from the above sources and from unpublished observations.

Rotation by movement along Neogene west-dipping listric faults has been suggested as the cause for eastward dip of structures in the Pocatello Range and northern Bannock Range (Burgel and others, 1987; Hersley, 1988), similar to that described for the northern Portneuf Range. Concave-westward, gently west-dipping normal faults have been described cutting homoclinally east-dipping beds of the Brigham Group, immediately south of Inkom in the northern Bannock Range (Hersley, 1988). Burgel and others (1987) also have described gently west-dipping normal faults east of the Blackrock Canyon fault (shown on Fig. 10) and immediately north of Inkom (the gently west-dipping faults are not shown on Figs. 10 and 11). A fold (Rapid Creek fold, also not shown on Figs. 10 and 11) in Brigham Group quartzites, originally with eastward vergence, is believed by Burgel and coworkers to have been rotated during normal faulting into its present position with a nearly horizontal axial plane. The Rapid Creek fold thus appears to be similar to folds observed in Mill Creek in the northern Portneuf Range (Figs. 3B and 8).

The structures underlying the area of Figure 10 are poorly known as a result of a lack of both seismic and deep-well control. The geometry of the subsurface structures, however is not purely conjectural. Based on much more extensive seismic, deep-well, and magnetic control to the east, an east-west cross section, extending down to magnetic basement and drawn as far west as the Portneuf Range, was suggested by Royse and others (1975, plate IV; the west end is approximately at 112°03′W., 42°30′N., about 25 km south of the southern border of the area shown in Fig. 10). Their cross section, although in large measure interpretive, deserves careful study and serves to make the following points: (1) depth to magnetic basement, immediately above which is an inferred basal décollement, is about 7,600 m below sea level under the Portneuf Range, (2) the basal décollement dips to the west about 7° under the Portneuf Range, (3) the Absaroka thrust soles into the basal décollement immediately east of the Portneuf Range, (4) the Paris thrust (which may be a southern extension of the Putnam thrust) and the Meade thrust also merge together immediately east of the southern Portneuf Range, (5) major down-to-the-west range-bounding faults merge with the Absaroka thrust or the basal décollement, (6) structures are inferred to be rotated into the major normal faults with a significant component of drag, such that synclines as much as 10 km wide form on the west side of the normal faults, and (7) the Meade thrust has moved about 32 km, based on offset of the Nugget Sandstone.

Figure 10. Generalized geologic map of northern Portneuf Range, Pocatello Range, and northern Bannock Range. Adapted from Mansfield (1952), Trimble (1976, 1982), Corbett (1978), Link and others (1985), Hefferan (1986), Kellogg and Marvin (1988), Kellogg (1990), Kellogg and others (1989), and Hladky and others (1991). Blackrock Canyon Limestone of Trimble (1976) is included with Pocatello Formation. Strike and dip are shown for Tertiary units only. BC is Blackrock Canyon fault. Fold symbols and numerous large normal faults are omitted from figure. Area of outcrop labeled "a" is referred to in text. Area of Figure 10 shown on Figure 1.

Based on surface outcrops and the guidelines suggested by Royse and others (1975), a schematic cross section can be drawn across the Bannock, Pocatello, and Portneuf Ranges (Fig. 11). Accepted conventions of thrust faulting (e.g., Royse and others, 1975; Boyer and Elliott, 1982) have been followed, and the cross section, as much as possible, is area balanced. A stratal wedge, tapering about 2° to the east, is assumed.

The Pocatello and Bannock ranges, similar to the Portneuf Range, are shown rotated down to the east along west-dipping Neogene listric faults that bound the ranges (including the east-bounding valleys) on their eastern sides. These major faults are inferred to sole in the basal décollement above crystalline basement and account for the regional eastward dip of pre-Tertiary beds and the generally eastward dip of Miocene and Pliocene basin-fill deposits. The dip on the Neogene beds is, however, generally less than that on the pre-Tertiary beds and structures.

In contrast to the above interpretation, Trimble (1976, p. 69) believes that the mountain block that forms the Bannock Range is not tilted. He notes the existence in the range of a nearly horizontal, "pre–basin-and-range" erosional surface that truncates east-dipping quartzites. Nonetheless, rotation of the range is favored because (1) the inferred erosional surface is not dated, so may postdate block rotation, (2) Neogene basin fill flanking the Bannock Range generally dips to the east, and (3) pre-Tertiary sedimentary rocks dip nearly homoclinally eastward.

Numerous, mostly west-side-down normal faults, smaller than the range-bounding normal faults, are found throughout the Portneuf, Pocatello, and Bannock ranges (e.g., Kellogg, 1990; Hladky and others, 1991). Most of these within-range faults (not shown on Fig. 11) are generally west-dipping listric faults, similar to the range-bounding faults, but may sole on a detachment surface (Putnam or Meade thrust?) significantly above the basal décollement; such an interpretation is made in Figure 3B. Royse and others (1975) avoid interpreting the subsurface geometry of the within-range normal faults in the Portneuf Range, which are drawn on their cross section simply as steep, straight lines that terminate at a depth of several thousand meters. The within-range normal faults may, in fact, account for a large part of the eastward dip of structures and may also explain the observed high variability in the amount of eastward rotation, which is particularly evident in the northern Portneuf Range.

Neogene extension and rotation of range-sized blocks ac-

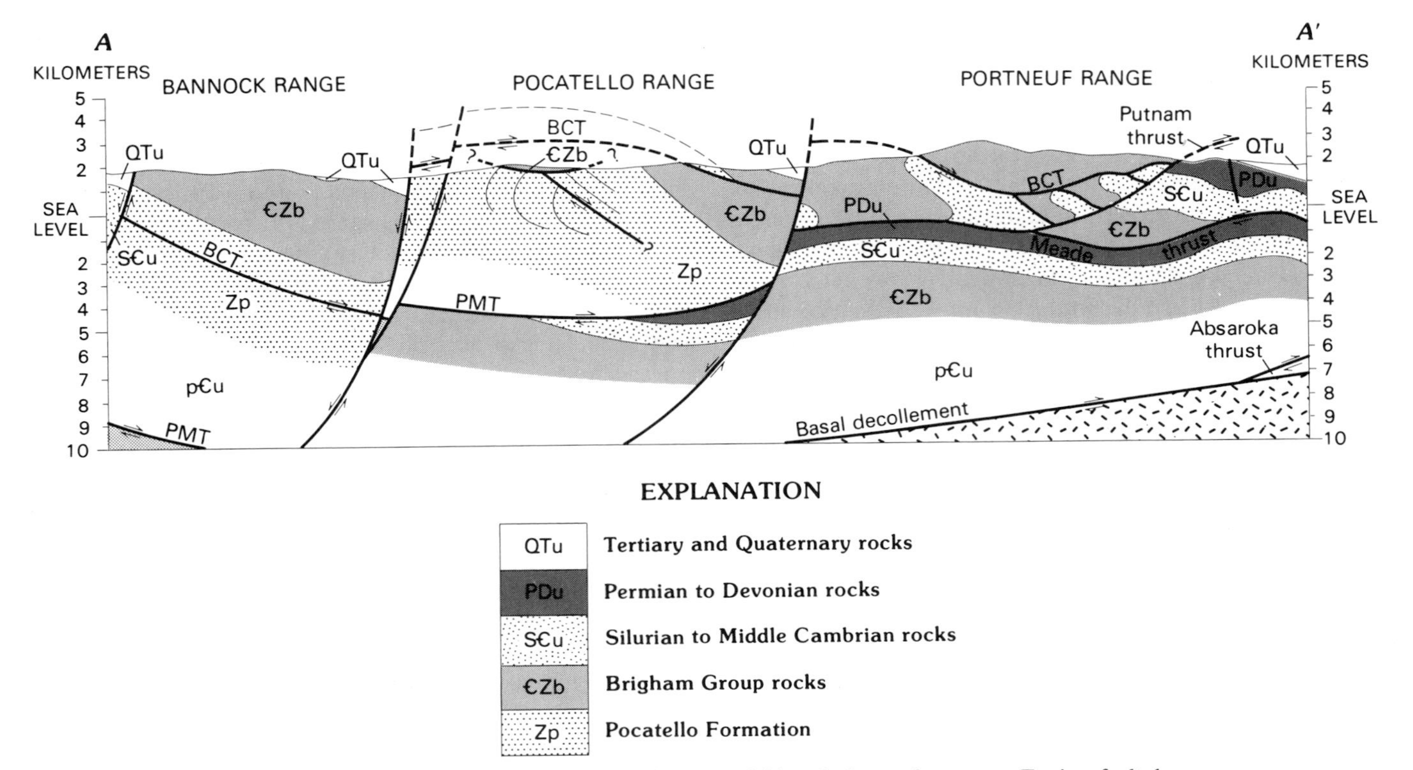

Figure 11. Cross section along line A-A′ of Figure 10. All but the largest front-range Tertiary faults have been removed. The cross section is largely schematic, due to lack of seismic and deep-well data for area, although depths to Meade thrust, Absaroka thrust, and magnetic basement are approximately known from Royse and others (1975). BCT is Bear Canyon thrust, PMT is westward extension of Putnam and Meade thrusts, p€u is Precambrian undivided, and random dash pattern is crystalline basement.

count for the surface geometry of structures in the Pocatello Range inferred to be westward projections of the Putnam and Bear Canyon thrusts. The Putnam thrust is believed to strike westward or southwestward across the northern Pocatello Range, juxtaposing Late Proterozoic rocks of the Brigham Group and Pocatello Formation to the south above middle and upper Paleozoic rocks to the north (inferred lower-plate rocks occur within the area marked "a" in Fig. 10). The precise location of the thrust is unknown because of thick, extensive deposits of Tertiary diamictite (Trimble, 1976; Hladky and others, 1991). Similar deposits in the northernmost Portneuf Range were interpreted as debris flows that formed during rapid basin-and-range uplift in late Miocene time (Kellogg and Marvin, 1988). The exposed lower-plate rocks comprise a homoclinal, east-dipping sequence as old as the Ordovician and Silurian Fish Haven Dolomite (Trimble, 1976) and as young as the Mississippian Monroe Canyon Limestone (Hladky and others, 1991; previously mapped as Wells Formation by Trimble [1976]). The generally southwestward strike and southeastward dip of the inferred Putnam thrust is due, it is believed, to tens of degrees of down-to-the-east Neogene block rotation.

The Bear Canyon thrust is believed to have been downdropped on the west by at least 2.5 km along a major Neogene fault that extends along the west side of the Portneuf Range (Figs. 10 and 11). In the Pocatello Range, the inferred westward extension of the Bear Canyon thrust is a curved east- and north-dipping surface that juxtaposes Caddy Canyon Quartzite of the lower Brigham Group and Pocatello Formation above Brigham Group rocks with about 1.2 km of stratigraphic throw (from reinterpretation of Trimble [1976]). Total movement on the Bear Canyon thrust, determined from offset of the base of the Brigham Group, is about 4.5 km (Fig. 11), in good agreement with about 3 to 4 km of movement observed across the Bear Canyon thrust in the Portneuf Range.

The above interpretation implies that rocks in the Pocatello Range and Bannock Range underlying the inferred Bear Canyon thrust are westward extensions of the Narrows subplate. Although these rocks are mostly homoclinally east dipping, many structures are similar to those found in the Narrows subplate of the northern Portneuf Range. For example, west of the town of Inkom, both north and south of the Portneuf Narrows (Fig. 10), overturned, cleaved beds within east-vergent recumbent folds in rocks of the Pocatello Formation are bounded above by at least one east-dipping, older-over-younger thrust (Blackrock Canyon fault) (LeFebre, 1984; Link and others, 1985; Fig. 10); rocks above the thrust dip homoclinally to the east (Burgel and others

[1987] interpret the Blackrock Canyon thrust to be an east-dipping normal fault). These relationships are similar, on a smaller scale, to those observed across the Bear Canyon thrust in the northern Portneuf Range. East-dipping thrusts above overturned and folded beds may be roof thrusts to small duplexes or horse-bounding thrusts within a duplex.

The inferred extension of the Bear Canyon thrust into the Pocatello Range requires in the Bannock Range that the deeply buried Bear Canyon thrust juxtapose Pocatello Formation above either Pocatello Formation or Brigham Group rocks. This in turn indicates a minimum of 5 km total west-side-down movement along the normal faults that front the west side of the Pocatello Range.

One small subplate, underlain by a low-angle fault in the west-central part of the Pocatello Range, places Brigham Group quartzite over the older Pocatello Formation (Fig. 10; Trimble, 1976). Link and others (1985) concurred with Trimble (1976) that this structure is a younger-over-older thrust and pointed out that younger-over-older flat faults have been observed in several places in the region. Alternatively, I suggest that this structure may be an extensional fault that formed considerably later than during Cretaceous thrusting but prior to Neogene high- to moderate-angle normal faulting.

CONCLUSIONS

The northern Portneuf Range and surrounding mountain ranges have undergone two major periods of deformation: Early Cretaceous compressional deformation and Neogene (mostly late Miocene) extensional faulting. Although the Putnam thrust has long been recognized as the major compressional feature in the area, the detailed structure of the complexly deformed upper plate is only now beginning to be understood.

The Putnam thrust juxtaposes Ordovician rocks above Permian and Pennsylvanian rocks along most of the thrust front. Along its southeastern exposed extent, the thrust ramps downsection toward the southeast; at its southeasternmost exposures, the thrust places Cambrian rocks above Mississippian rocks.

Two major and several minor imbricate thrusts have been identified above the Putnam thrust. From north to south and structurally lowest to highest, two of the major imbricates in the northern part of the area are the steeply south-dipping Narrows thrust and the mostly south- to east-dipping Bear Canyon thrust. The Toponce thrust, in the southeastern part of the area, is believed to be a southeastward extension of the Bear Canyon thrust.

In the northernmost part of the area, the Lone Pine subplate underlies the Narrows thrust and overlies the Putnam thrust. Where exposed, the Narrows thrust is a steep, east-striking lateral ramp, probably having a large component of strike-slip movement, that merges eastward with the Putnam thrust.

The Narrows subplate has undergone tremendous internal shear, producing recumbent folds, cleavage directions that fan about northerly strikes, east-dipping contraction faults, and tectonic thickening and thinning of beds. These features strongly suggest that the Narrows subplate forms at least part of a foreland-dipping duplex in which the Bear Canyon thrust (and its inferred extension, the Toponce thrust) forms a roof thrust and the Putnam thrust forms the main floor thrust; the east-trending Narrows thrust, which merges with the Putnam thrust, defines the northern floor-thrust boundary. Rocks within the Lone Pine subplate, structurally beneath and north of the Narrows thrust, are upright and not significantly sheared, suggesting that the Lone Pine subplate was not part of the duplex.

The extent of the proposed duplex within the Narrows subplate is unknown, and its extension into the ranges west of the northern Portneuf Range is also unknown. However, overturned, folded beds beneath inferred east-dipping thrusts in the Portneuf Narrows area near the southwest side of the Pocatello Range suggest that local duplex formation may have affected this area.

The formation of a foreland-dipping duplex cannot account for all the apparent rotation of structures within the Narrows subplate. Regional eastward dips, generally less than 35°, of Tertiary rocks throughout a large part of southeastern Idaho strongly suggest that Neogene extensional faulting caused widespread down-to-the-east rotation along concave-westward listric faults that probably merge onto a regional detachment surface or several such surfaces; these surfaces are most likely reactivated thrusts (Royse and others, 1975) along which Neogene movement was approximately opposite to that during Cretaceous thrusting.

Major range-bounding faults may sole into the basal décollement immediately above crystalline basement, which is inferred to be about 9 to 10 km beneath the Portneuf Range. Smaller, within-range normal faults probably merge on higher-level reactivated thrusts, a view supported by observed normal faults that merge with the Putnam and Toponce thrusts and by the existence of extensive zones of brecciation immediately above the Bear Canyon thrust.

Neogene block rotation also accounts for reasonable westward projections of the Putnam and Bear Canyon thrusts into the Pocatello Range.

ACKNOWLEDGMENTS

I am indebted to F. R. Hladky, R. E. Amerman, and K. R Pogue, who mapped parts of the northern Portneuf Range during the course of their graduate studies, although my structural interpretations do not necessarily agree with theirs. Of particular help were insightful reviews of the manuscript by S. E. Boyer, M. R. Hudson, W. J. Perry, Jr., L. B. Platt, T. W. Schirmer, Betty Skipp, D. L. Trimble, and one anonymous reviewer. Conversations in the field with P. K. Link helped to clarify several troublesome points. The Shoshone-Bannock Tribal Council and the Bureau of Indian Affairs were supportive during all phases of the mapping project; the Tribal Council kindly gave me permission to publish this study, which was largely undertaken on the Fort Hall Indian Reservation. However, the largest vote of thanks undoubtedly is owed to the late Steven S. Oriel, whose unsurpassed knowledge of the thrust belt provided the inspiration to get this project underway.

REFERENCES CITED

Allmendinger, R. W., 1981, Structural geometry of Meade thrust plate in northern Blackfoot Mountains, southeastern Idaho: American Association of Petroleum Geologists, v. 65, p. 509–525.

Amerman, R. E., 1987, Geology of a portion of the Jeff Cabin Creek 7½′ quadrangle, Caribou County, Idaho [M.S. thesis]: Golden, Colorado School of Mines, 138 p.

Armstrong, F. C., and Cressman, E. R., 1963, The Bannock thrust zone, southeastern Idaho: U.S. Geological Survey Professional Paper 374-J, 22 p.

Armstrong, F. C., and Oriel, S. S., 1965, Tectonic development of the Idaho-Wyoming thrust belt: American Association of Petroleum Geologists Bulletin, v. 49, p. 1847–1866.

Boyer, S. E., and Elliott, D., 1982, Thrust systems: American Association of Petroleum Geologists Bulletin, v. 66, p. 1196–1230.

Brown, W. G., 1984, Basement involved tectonics, foreland areas: American Association of Petroleum Geologists Continuing Education Course Notes No. 26, 92 p.

Burgel, W. D., Rodgers, D. W., and Link, P. K., 1987, Mesozoic and Cenozoic structures of the Pocatello region, southeastern Idaho, *in* Miller, W. R., ed., The thrust belt revisited: Wyoming Geological Association, 38th Annual Field Conference, Guidebook, p. 91–100.

Carr, W. J., and Trimble, D. E., 1963, Geology of the American Falls quadrangle, Idaho: U.S. Geological Survey Bulletin 1121-G, 44 p.

Corbett, M., 1978, Preliminary geologic map of the northern Portneuf Range, Bannock and Caribou Counties, Idaho: U.S. Geological Survey Open-File Report 78-1018, scale 1:48,000.

Crittenden, M. D., Schaeffer, F. E., Trimble, D. E., and Woodward, L. A., 1971, Nomenclature and correlation of some upper Precambrian and basal Cambrian sequences in western Utah and southeastern Idaho: Geological Society of America Bulletin, v. 82, p. 581–602.

Dahlstrom, C.D.A., 1970, Structural geology in the eastern margin of the Canadian Rocky Mountains: Bulletin of Canadian Petroleum Geology, v. 18, p. 332–406.

Hefferan, K., 1986, Geology of the Bonneville Peak quadrangle, Bannock and Caribou Counties, Idaho [M.S. thesis]: Bryn Mawr, Pennsylvania, Bryn Mawr College, 90 p.

Heller, P. L., and 7 others, 1986, Time of initial thrusting in the Sevier orogenic belt, Idaho-Wyoming, and Utah: Geology, v. 14, p. 388–391.

Hersley, C. F., 1988, Structural geology in the northern Bannock Range west of Inkom, Idaho: Geological Society of America Abstracts with Programs, v. 20, p. 420.

Hladky, F. R., 1986, Geology of an area north of The Narrows of Ross Fork canyon, northernmost Portneuf Range, Fort Hall Indian Reservation, Bannock and Bingham Counties, Idaho [M.S. thesis]: Pocatello, Idaho State University, 110 p.

Hladky, F. R., and Kellogg, K. S., 1990, Geology of the Yandell Springs, 7½′ quadrangle, Bannock and Bingham Counties, Idaho: U.S. Geological Survey Quadrangle Map GQ-1678, scale 1:24,000.

Hladky, F. R., Kellogg, K. S., Oriel, S. S., Link, P. K., Nielson, J. W., and Amerman, R. E., 1991, Geologic map of the eastern part of the Fort Hall Indian Reservation, Bannock, Bingham, and Caribou Counties, Idaho: U.S. Geological Survey Miscellaneous Investigations Map Series I-2006, scale 1:50,000.

Kellogg, K. S., 1987, Style of deformation in the northern Portneuf Range, southeastern Idaho: Geological Society of America Abstracts with Programs, v. 19, p. 286.

—— , 1990, Geologic map of the South Putnam Mountain quadrangle, Bannock and Caribou Counties, Idaho: U.S. Geological Survey Geologic Quadrangle Map GQ-1665, scale 1:24,000.

Kellogg, K. S., and Embree, G. F., 1986, Geologic map of the Stevens Peak and Buckskin basin areas, Bingham and Bannock Counties, Idaho: U.S. Geological Survey Miscellaneous Field Studies Map MF-1856, scale 1:24,000.

Kellogg, K. S., and Marvin, R. F., 1988, New potassium-argon ages, geochemistry, and tectonic setting of upper Cenozoic volcanic rocks near Blackfood, Idaho: U.S. Geological Survey Bulletin 1806, 19 p.

Kellogg, K. S., and Skipp, B., 1988, Duplex structures of the northern Portneuf Range and Pocatello Range, southeastern Idaho: Geological Society of America Abstracts with Programs, v. 20, p. 424.

Kellogg, K. S., Oriel, S. S., Amerman, R. E., Link, P. K., and Hladky, F. R., 1989, Geologic map of the Jeff Cabin Creek quadrangle, Bannock and Caribou Counties, Idaho: U.S. Geological Survey Geologic Quadrangle Map GQ-1669, scale 1:24,000.

LeFebre, G. B., 1984, Geology of the Chinks Peak area, Pocatello Range, Bannock County, Idaho [M.S. thesis]: Pocatello, Idaho State University, 61 p.

Link, P. K., LeFebre, G. B., Pogue, K. R., and Burgel, W. D., 1985, Structural geology between the Putnam thrust and the Snake River Plain, southeastern Idaho, *in* Kerns, G. J., and Kerns, R. L., Jr., eds., Orogenic patterns and stratigraphy of north-central Utah and southeastern Idaho: Utah Geological Association Publication 14, p. 97–117.

Link, P. K., Jansen, S. T., Halimdihardja, P., Lande, A., and Zahn, P., 1987, Stratigraphy of the Brigham Group (Late Proterozoic–Cambrian), Bannock, Portneuf, and Bear River Ranges, southeastern Idaho, *in* Miller, W. R., ed., The thrust belt revisited: Wyoming Geological Association, 38th Annual Field Conference, Guidebook, p. 133–148.

Ludlum, J. C., 1943, Structure and stratigraphy of part of the Bannock Range, Idaho: Geological Society of America Bulletin, v. 54, p. 973–986.

Mansfield, G. R., 1920, Geography, geology, and mineral resources of the Fort Hall Indian Reservation, Idaho: U.S. Geological Survey Bulletin 713, 152 p.

—— , 1927, Geography, geology, and mineral resources of part of the Fort Hall Indian Reservation, Idaho: U.S. Geological Survey Professional Paper 152, 453 p.

—— , 1952, Geography, geology, and mineral resources of the Ammon and Paradise Valley quadrangles, Idaho: U.S. Geological Survey Professional Paper 238, 92 p.

Mitra, G., Yonkee, W. A., and Gentry, D. J., 1984, Solution cleavage and its relationship to major structures in the Idaho-Utah-Wyoming thrust belt: Geology, v. 6, p. 354–358.

Oriel, S. S., and Platt, L. B., 1980, Geologic map of the Preston 1° × 2° quadrangle, southeastern Idaho and western Wyoming: U.S. Geological Survey Miscellaneous Investigations Series Map I-1127, scale 1:250,000.

Pierce, K. L., and Scott, W. E., 1986, Migration of faulting along and outward from the track of thermo-volcanic activity in the eastern Snake River Plain region during the past 15 m.y. [abs.]: EOS Transactions of the American Geophysical Union, v. 67, p. 1225.

Pogue, K. R., 1984, The geology of the Mt. Putnam area, northern Portneuf Range, Bannock and Caribou Counties, Idaho [M.S. thesis]: Pocatello, Idaho State University, 106 p.

Royse, F., Jr., Warner, M. A., and Reese, D. L., 1975, Thrust belt structural geometry and related stratigraphic problems, Wyoming–Idaho–northern Utah, *in* Bolyard, D. W., ed., Deep drilling frontiers in central Rocky Mountains: Denver, Rocky Mountain Association of Geologists, p. 41–54.

Sacks, P. E., and Platt, L. B., 1985, Depth and timing of décollement extension, southern Portneuf Range, southeastern Idaho: *in* Kerns, G. J., and Kerns, R. L., Jr., eds., Orogenic patterns and stratigraphy of north-central Utah and southeastern Idaho: Utah Geological Association Publication 14, p. 119–127.

Sparlin, M. A., Braile, I. W., and Smith, R. B., 1982, Crustal structure of the eastern Snake River Plain determined from ray trace modeling of seismic refraction data: Journal of Geophysical Research, v. 87, p. 2619–2633.

Trimble, D. E., 1976, Geology of the Michaud and Pocatello quadrangles, Bannock and Power Counties, Idaho: U.S. Geological Survey Bulletin 1400, 88 p.

—— , 1982, Geologic map of the Yandell Springs quadrangle, Bannock and Bingham Counties, Idaho: U.S. Geological Survey Geologic Quadrangle Map GQ-1553, scale 1:48,000.

Trimble, D. E., and Carr, W. J., 1976, Geology of the Rockland and Arbon quadrangles, Power County, Idaho: U.S. Geological Survey Bulletin 1399, 115 p.

Webel, S., 1987, Significance of back thrusting in the Rocky Mountain thrust belt, *in* Miller, W. R., ed., The thrust belt revisited: Wyoming Geological Association, 38th Annual Field Conference, Guidebook, p. 37–55.

Woodward, N. B., Boyer, S. E., and Suppe, J., 1985, An outline of balanced cross-sections: University of Tennessee, Department of Geological Sciences Studies in Geology 11, 170 p.

Manuscript Accepted by the Society July 19, 1991

Printed in U.S.A.

Geological Society of America
Memoir 179
1992

Chapter 5

LaBarge Meadows cross-strike structural discontinuity and lateral ramping in the Absaroka thrust system

Julian D. Fowles*
Department of Earth Sciences, University of Cambridgе, Cambridge, England
Nicholas B. Woodward*
Department of Geosciences and Environmental Studies, University of Tennessee, Chattanooga, Tennessee 37403

ABSTRACT

Cross-strike structural discontinuities (CSDs) involve aligned reentrants and structural trend changes oblique to major thrust and fold belt orientations. The LaBarge Meadows area of the Absaroka thrust system is the western part of an area of such structural trend changes involving the Absaroka, Darby, and Prospect thrust systems. The Absaroka system structures have not been previously discussed in the context of CSDs, but their geometry and interpretation place major constraints on more eastern parts. The tear faults (northeast-southwest and east-west trending) that directly connect the Absaroka system with the Darby-Prospect structures are of minor and dominantly dip-slip displacement in the Absaroka sheet, despite their strike-slip separation. They overprint and merely complicate the recognition of the major Absaroka system structures, namely (1) a major (1,000 m) up-to-the-south lateral ramp from Cambrian to Pennsylvanian strata, and (2) a division of the single Absaroka thrust sheet into three subsidiary thrust sheets (Tunp, Commissary, and Absaroka sheets), which begins north of the CSD but becomes obvious south of it. The increase in thrust system complexity is progressive from north to south over 30 km. It coincides with gradual thickening in the Lower Mesozoic Structural Lithic Unit and gradual thinning of the Paleozoic structural lithic units. The lateral ramp and the changes in thrust spacing are both therefore inferred to be related to variations in the relevant stratigraphic intervals and not directly related to any basement structures.

INTRODUCTION

Cross-strike structural discontinuities (CSDs; Wheeler, 1980) are areas of abrupt structural trend change, aligned at oblique angles to the dominant regional strike of structural features. They are common in fold and thrust belts and are usually inferred to be related to either basement fault control (Schmidt and Perry, 1988, and references therein) or to lateral ramping of thrust surfaces related to stratigraphic variations (Woodward and others, 1988; Woodward and Rutherford, 1989). Woodward (1987) noted that whatever the ultimate origin of lateral ramps, ones that occur in the subsurface will warp overlying thrust sheets and erosion will create reentrants in overlying sheets without those sheets showing any longitudinal structural changes themselves. In that case the alignment of "reentrants" that might define the morphology of a CSD is a result of both a structural change in a lower thrust sheet and passive folding and erosion of overlying structures.

This chapter describes in detail the small area of abrupt change of structural trend and spacing in the Idaho-Wyoming-Utah thrust belt at 42°30′ latitude (Fig. 1). The three easternmost thrusts of the belt (Prospect, Darby, and Absaroka) all change trend in the LaBarge PLatform–LaBarge Meadows area. The

*Present addresses: Fowles, Department of Earth Sciences, University of Liverpool, P.O. Box 147, Liverpool L69 3BX, England. Woodward, U.S. Department of Energy, ER-32 19901 Germantown Road, Germantown, Maryland 20874.

Fowles, J. D., and Woodward, N. B., 1992, LaBarge Meadows cross-strike structural discontinuity and lateral ramping in the Absaroka thrust system, *in* Link, P. K., Kuntz, M. A., and Platt, L. B., eds., Regional Geology of Eastern Idaho and Western Wyoming: Geological Society of America Memoir 179.

Absaroka thrust system is a dominant single thrust sheet to the north and is divided into several sheets to the south.

The Darby thrust is deflected eastward in map pattern here and is continuous with the Hogsback thrust farther south (Royse and others, 1975). The Prospect thrust ends southward in the footwall of an up-to-the-north lateral ramp ("transverse step") in the Darby-Hogsback thrust surface (Royse and others, 1975; Royse, 1985). Royse and others (1975) and Blackstone (1979) have thoroughly discussed several concepts for the geometry and possible causes of the eastern part of the CSD involving the Prospect and Darby thrusts. This discussion will be limited as much as possible to the lateral structures in the Absaroka thrust system and how their geometry casts light on previous discussions of the origin of this CSD.

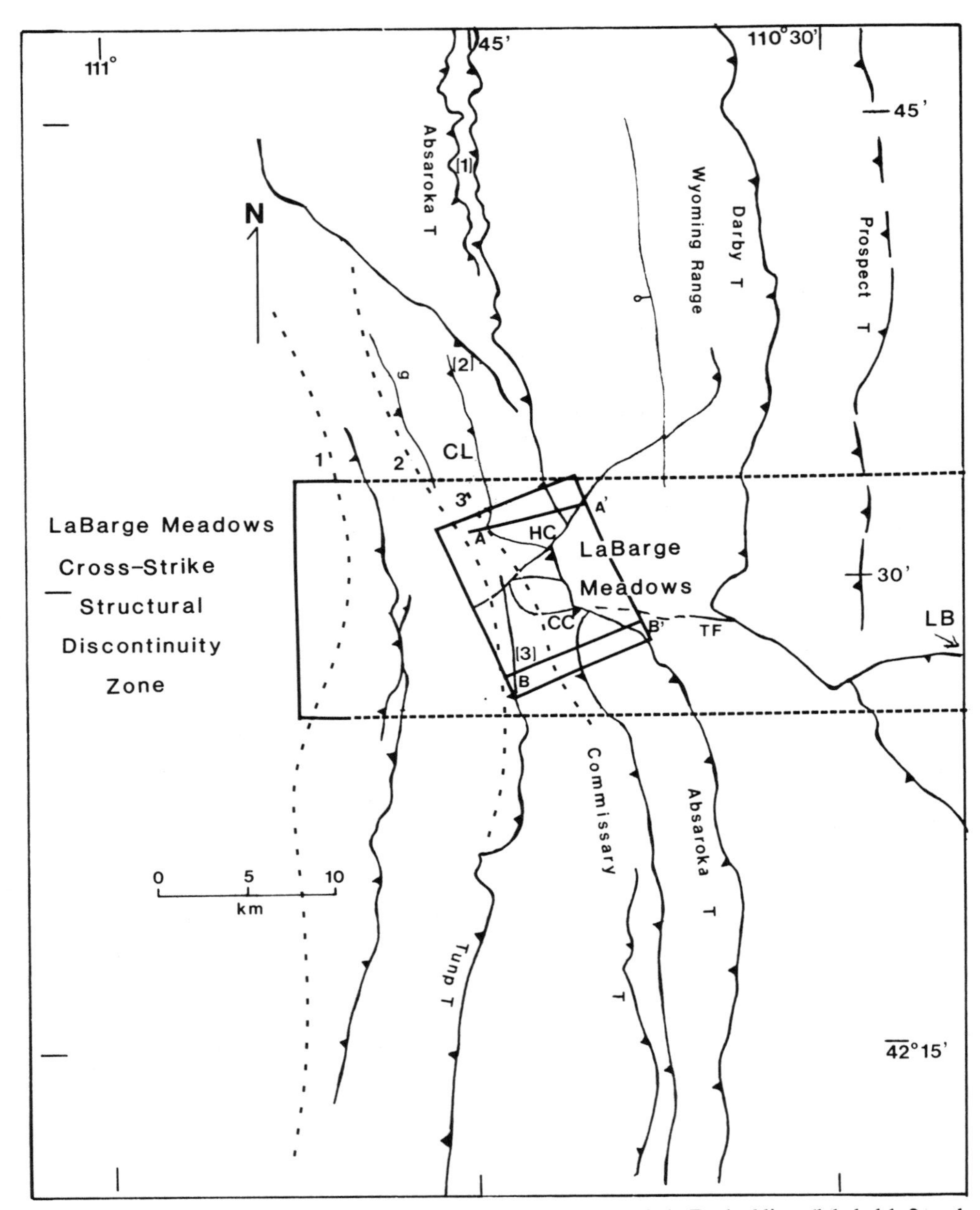

Figure 1. Regional map of the central Idaho-Wyoming-Utah thrust belt. Dashed lines (labeled 1, 2, and 3) show major fold trends. Note that the three major anticlines are continuous across the CSD within the Absaroka sheet despite the obliquely trending fault segments. Numbers in parentheses locate stratigraphic columns in Figure 7. A-A′ and B-B′ show section lines from Figure 5. Major thrusts are labeled and shown with the barbed lines; normal faults are shown with a line and ball symbol. CL—Crystal Lake thrust; TF—Thompson fault; LB—LaBarge Platform occurs just east of here; HC—Hobble Creek fault; CC—Clear Creek fault. Box locates the area shown in Figure 2A (from Oriel and Platt, 1980).

STRUCTURAL GEOMETRIES

The LaBarge Meadows and surrounding areas were mapped by Rubey (1973) and Rubey and others (1980) at 1:62,500, and they identified the major structural elements. The LaBarge Meadows area was remapped by Fowles (1985) at 1:8,000 for this project (Fig. 2). The area marks the northern extent of the Tunp and the Commissary thrusts within the regional Absaroka sheet. The thrusts are cut by the Hobble Creek, Clear Creek (called the Thompson fault farther east by Blackstone, 1979), and other minor tear faults. We have not used the Thompson fault name because we have not traced our Clear Creek fault into the Thompson fault farther east, and our interpretation of its movement history is much different than that of Blackstone. Our interpretations apply specifically to the Clear Creek fault that we mapped, and we cannot prove that they apply to any faults along the same trend farther east, although we expect they do.

There are also two major fold trends from north to south across the area, namely one within the Absaroka sheet and one within the Tunp thrust sheet.

Tear faults

Two major tear faults (Fig. 2a) cross the area from northeast to southwest (Hobble Creek Fault; HC) and east to west (Clear Creek Fault; CC). The Hobble Creek fault cuts both the Absaroka hanging wall and footwall and is cut farther east by the normal fault on the west side of the Wyoming Range (Rubey, 1973; Fig. 1). Its continuation within the Wyoming Range merges with a thrust farther east. The Hobble Creek fault displaces the Absaroka thrust trace with 2 km of dextral strike separation, but there is a progressive decrease in strike separation from east to west. There is little offset of stratigraphic contacts 4 km farther west (at locality HCJ) within the Absaroka sheet, where the HC fault can no longer be reliably mapped. Based on lower stratal dips north of the fault compared to those south of it, displacement appears to be predominantly north side down.

The Clear Creek fault also cuts through both the Absaroka hanging wall and footwall. Royse and others (1975) and Blackstone (1979) show the Clear Creek fault as the Thompson fault cutting across the entire Darby thrust sheet. The Clear Creek fault (the western part of the Thompson fault) merges with the Hobble Creek fault zone at Hobble Creek (locality HCJ in Fig. 2a), and displacement on both dies out within the Absaroka sheet. Fowles (1985) calculated about 250 m of primarily extensional dip-slip motion for the Clear Creek fault based on the offset of folds 3 and 4.

Blackstone (1979) inferred that the Thompson fault (TF in Fig. 1) accommodated much of the change in trend of the thrust belt around the LaBarge Platform area (LB in Fig. 1), which he inferred to be a previously extant basement uplift. The Thompson fault did so by allowing the Darby fault north and south of it to move separately. The sheet north of the tear apparently ramped upward west of the LaBarge Platform, whereas the sheet south of the tear ramped farther eastward.

Crowell (1959) emphasized that separation of markers in map pattern does not fairly represent fault displacement unless displacement is horizontal. Horizontal fault displacements, however, cannot simply die out laterally without significant strains in adjacent rocks. Although strike-slip faults may die out by losing discrete displacement into distributed folding and faulting, the Hobble Creek and Clear Creek faults die out to the west with little disruption of strata north, south, or west of the fault zone. Offsets of stratigraphic markers across the faults decrease gradually, suggesting to us that both faults have dominantly scissors displacements. Because the beds and faults that the Hobble Creek and Clear Creek faults cut dip moderately westward, the relative displacements in map pattern will increase systematically away from the center of rotation (that is, to the east), as observed. As scissors faults, they are not pure strike-slip faults and have a major dip-slip extensional component, with the Hobble Creek fault down-to-the-north and the Clear Creek fault down-to-the-south. The progressive increase in strike-separation across a scissors fault would also explain the difference in separation of the Absaroka thrust trace (2 to 3 km) and the Darby thrust trace (4.5 km) across the Thompson fault (Rubey and others, 1980; Blackstone, 1979).

The proposed scissors displacement for the Hobble Creek fault (modeled as a simplified up-to-the-south vertical scissors fault through homoclinal beds in Fig. 3) suggests that older strata should be exposed at the base of the thrust sheet south of the fault compared to north of the fault. The exact opposite is true, with Cambrian strata at the base of the sheet to the north and Pennsylvanian strata at the base of the sheet to the south. Thus the principal structural geometry of the Hobble Creek tear fault zone is not simply the obvious tear fault itself.

Two minor tear faults are present in the area but are restricted to the Absaroka hanging wall. The Little Corral (LC) fault is primarily a low-angle, south-dipping (strike 091°, dip 28°) extension fault (based on offset of fold 3) that ends against the Absaroka thrust to the east and the Hobble Creek fault to the west. The Springs (S) fault is a contractional oblique-slip fault that dips steeply southward (strike 090°, dip 54°; Fig. 4). It offsets a synclinal hinge of the Nugget Formation 856 m horizontally and 216 m vertically. The Springs fault is continuous farther west with the low displacement Crystal Lake thrust, and thus its south side (hanging wall) has moved upward and eastward relative to its north side (footwall). At its east end the Springs fault joins the Hobble Creek fault zone at its junction with the Absaroka thrust. Despite the relative dextral strike-slip separation of the Absaroka thrust across the Hobble Creek fault, there is no obvious correlative structure to the Springs fault between the Hobble Creek fault and the Little Creek fault. The Commissary thrust south of the Clear Creek fault occupies the same position as the Crystal Lake–Springs fault north of the Hobble Creek fault. The major folds (3 and 4) above (south of) the Springs fault appear quite continuous with a similar anticlinal-synclinal fold pair between the Little Creek and Clear Creek faults and south of the Clear Creek fault.

Overall, therefore, the minor tear faults die out within the

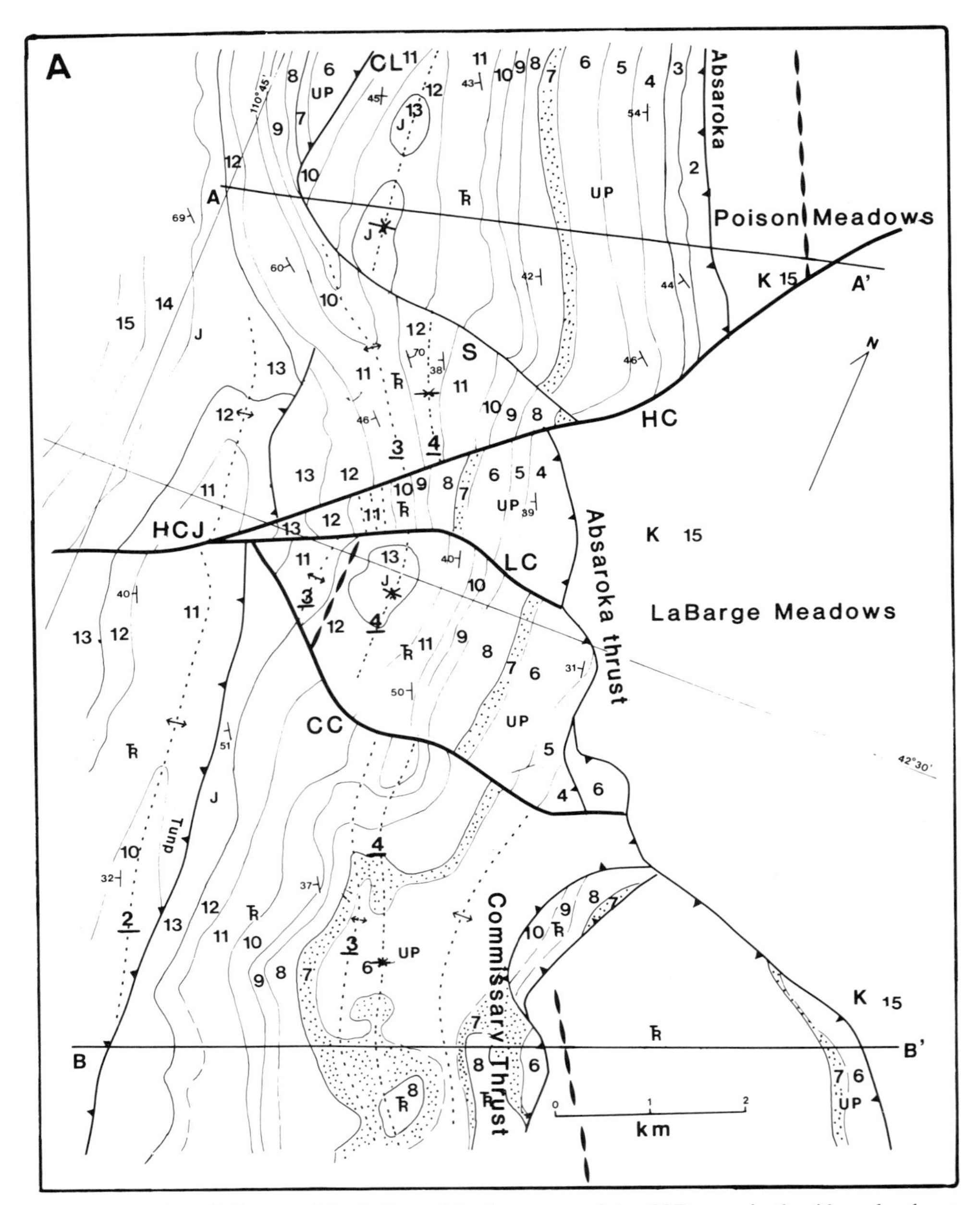

Figure 2. A, Detailed map of the LaBarge Meadows area of the CSD zone in the Absaroka thrust system. Stratigraphic units numbered as in 2B. The study area lies primarily within the Graham Meadows and Poison Creek 7½′ quadrangles. The dark dashed lines show the lower Paleozoic frontal ramp position based on the cutoff line for the Darby Formation–Lodgepole Formation (3-4) contact. The Lodgepole and Mission Canyon formations are usually grouped as the Madison Group in this area. Phosphoria Formation is speckled for map clarity. The major fold trends (fine dashed lines) are shown by underlined numbers on their axial traces: 2, 3, and 4. HC—Hobble Creek fault, CC—Clear Creek fault, LC—Little Corral fault, CL—Crystal Lake thrust, HCJ—Hobble Creek fault–Clear Creek fault junction, S—Springs fault. LP—Units 2 and 3; UP—Units 4–7; ₮—Units 8–14; K—Unit 15 (Fowles, 1985). B (on next page) Stratigraphic column for the LaBarge Meadows area, showing structural lithic unit subdivisions based on Woodward and Rutherford (1989).

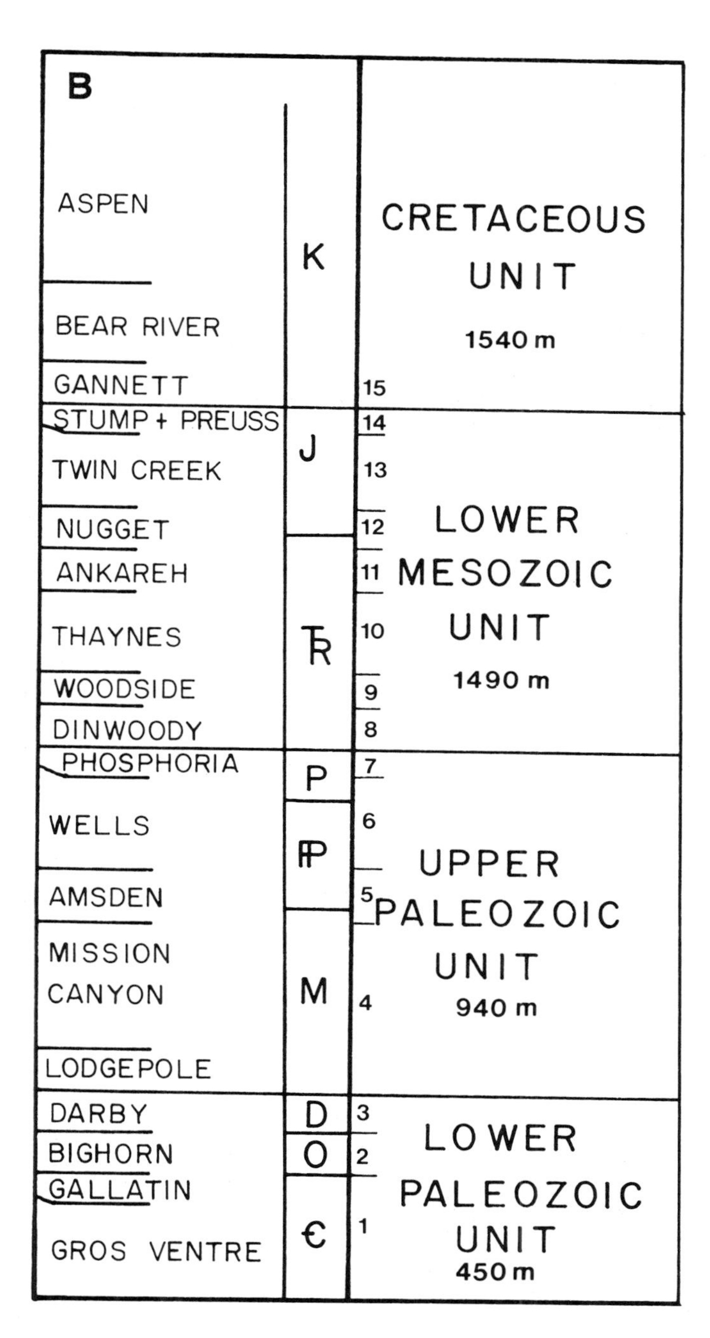

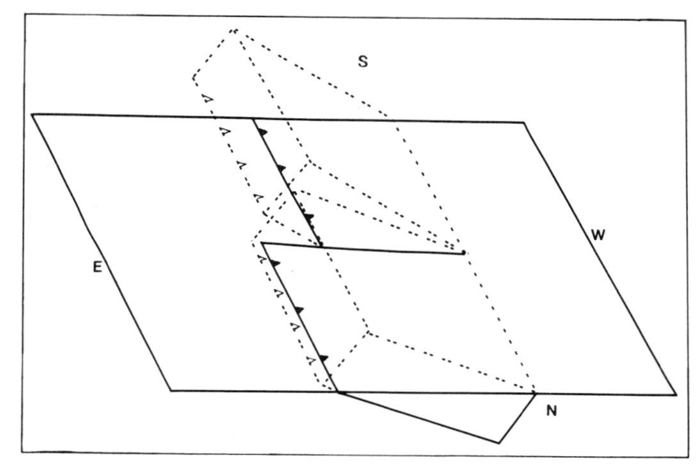

Figure 3. Idealized block diagram for a simple scissors displacement on a tear fault truncating an earlier homoclinal thrust sheet. The map pattern of the truncated fault is shown in solid lines. The restored blocks above and beneath the topographic (flat) surface are shown with double-dashed lines. If the Absaroka thrust had been offset in this manner, older strata should be exposed on the upthrown block rather than on the downthrown block.

detailed map area, and the major ones merge with other faults or extend eastward with steadily increasing strike-separation. The Springs fault is an oblique part of the Crystal Lake thrust. The Hobble Creek, the Clear Creek, and the Little Corral "tear" faults are extensional scissors faults that extend the hanging wall of the Absaroka thrust sheet in a north-south direction. The Hobble Creek and Clear Creek faults that truncate the Absaroka thrust surface are related to later deformation within the Darby thrust sheet (system), as suggested by Royse and others (1975) and Blackstone (1979). They are not simply tear faults offsetting both the Darby and Absaroka thrust surfaces equally. The Clear Creek fault is dominantly extensional within the Absaroka thrust sheet and therefore may be extensional elsewhere. The Hobble Creek fault is nearly vertical within the Absaroka sheet but may be continuous with low-angle contraction faults within the Wyoming Range (Rubey, 1973). The CSD tear-fault zone postdates the emplacement of the Absaroka thrust system and should have little to do with its structural geometry, although the Springs fault may be an Absaroka system (Crystal Lake thrust) structure.

Thrust faults

The Absaroka thrust is the major contractional fault of the area, although because of the tear faults it appears in four major segments. The Crystal Lake fault (CL) occurs within the Absaroka sheet north of the CSD, and the Tunp and Commissary thrusts occur within the Absaroka south of the CSD.

The hanging-wall strata at the base of the Absaroka thrust sheet become younger from north to south, with a major jump in stratigraphic separation across the Hobble Creek fault as mentioned before. The Absaroka hanging wall includes Cambrian Gallatin Formation (base of the Lower Paleozoic Structural Lithic Unit; Fig. 2b) north of the Hobble Creek fault (Fig. 5a) and Pennsylvanian Wells Formation (middle of the Upper Paleozoic Structural Lithic Unit) at Clear Creek south of the Clear Creek fault (Figs. 2, 5b). This is the opposite of what would be expected across a simple scissors fault cutting a homoclinally dipping thrust sheet. Given that a scissors displacement on the Hobble Creek fault explains its displacement and structural geometry most

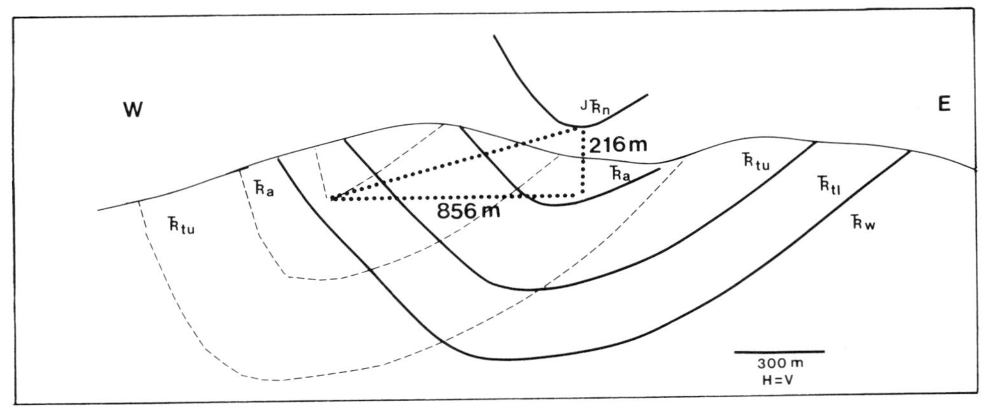

Figure 4. Projection of the Nugget Formation hinge for fold 4 across the Springs fault. Dashed lines show contacts north of the fault, and solid lines show contacts south of the fault (Fowles, 1985). The displacement is sinistral oblique.

simply, the logical further explanation is that the Absaroka was not a simple homoclinally dipping sheet prior to the formation of the Hobble Creek fault. The level of the Absaroka thrust must have changed position stratigraphically from north to south across this area. Thus a major aspect of the structure in the Absaroka thrust sheet is an up-to-the-south transverse step or lateral ramp that cuts out a thousand meters of strata from the base of the Absaroka thrust sheet.

The reconstructed minimum Devonian-Mississippian cutoff line (3-4 contact in Fig. 2a) for the Absaroka thrust illustrates the lateral ramp best. The frontal cutoffs of the Cambrian, Ordovician, and Devonian strata in the hanging wall are eroded now but would lie some distance east of the present outcrop trace of the Absaroka thrust north of Poison Meadows, as shown by the dashed line (Fig. 2a). The frontal ramp of Cambrian-Devonian strata south of the tear faults is inferred to occur at depth at least as far west as beneath the Commissary thrust trace (Fig. 5b), although it could occur farther west beneath the Tunp thrust sheet. The lateral ramp marks at least 3 km and perhaps as much as 5 to 10 km of offset of the Absaroka frontal ramp through Lower Paleozoic strata.

The Absaroka thrust system, when considered without the later, relatively small displacement tear faults (Fig. 1), shows remarkable continuity of subsidiary folds and faults across the CSD zone. The hanging-wall anticline west of the Tunp thrust (fold 2) extends 20 km north of the CSD. There are two small displacement thrusts east of this anticline north of the CSD. The western one, which extends for only a kilometer or two, is the continuation of the Tunp thrust, and the Crystal Lake thrust may represent the northern end of the Commissary thrust system. Two other imbricate thrusts within the Absaroka sheet farther north are related to the Mt. Fitzpatrick blind imbricate complex

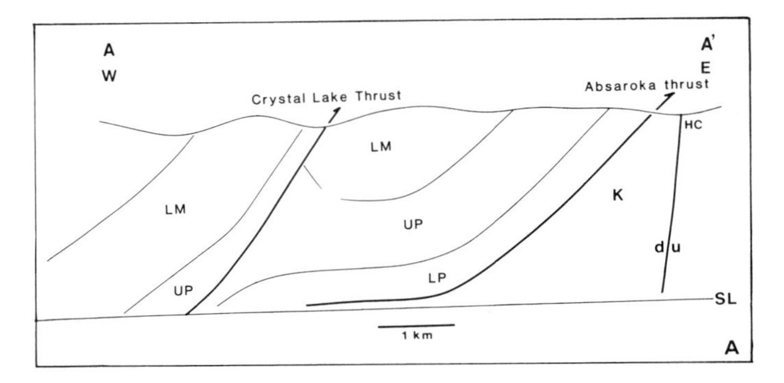

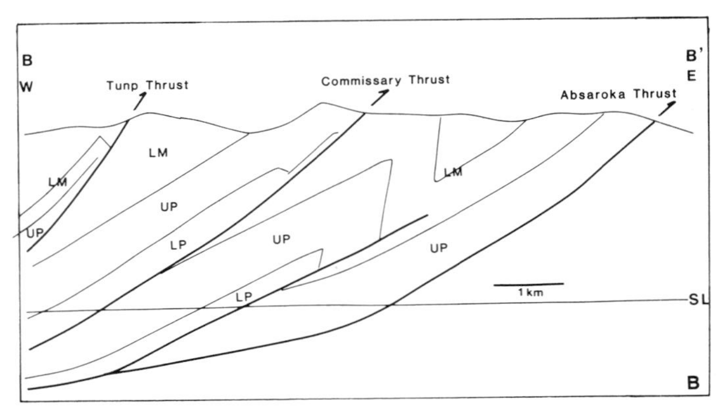

Figure 5. Cross sections across the Absaroka thrust sheet. A, Near surface section A-A′ across the sheet north of the CSD. B, Near surface section B-B′ across the sheet south of the CSD. Section locations shown in Figure 2A. Horizontal and vertical scales are equal. Other faults as designated in Figure 2A.

(Rubey, 1973; Woodward, 1986). The large-scale imbrication within the Absaroka thrust sheet becomes most obvious south of the lateral ramp and the CSD zone, but internal imbrication is common north of the CSD as well.

Extension in the hanging wall of a thrust parallel to lateral ramps is commonly observed (Butler, 1982), usually with the thinner part of the hanging wall down dropped with respect to the thicker part. Both the Little Corral and Clear Creek faults have this geometry. Because the Hobble Creek fault so precisely follows the lateral ramp in the Absaroka sheet, the western part of the fault almost certainly originally developed along it, although with uncertain displacement.

It seems most likely that when the Darby thrust system and the earlier overlying Absaroka sheet were deformed west of the LaBarge Platform, early faults associated with the Absaroka lateral ramp helped extend the tears from the Darby thrust sheet farther west into the Absaroka sheet. Both the Hobble Creek and the local Clear Creek fault were probably reactivated during Darby thrust emplacement. In any case, the present Hobble Creek fault that cuts both the Absaroka and Darby thrust sheets cuts directly through the Absaroka lateral ramp, complicating the map pattern interpretation.

The study area includes the northern tip of the Tunp thrust system (Rubey, 1973). The Tunp is cut by the Hobble Creek and Little Corral faults (Fowles, 1985; Fig. 2a). Lamerson (1982) presented evidence farther south that some motion on the Tunp is relatively late and out of sequence, but the Tunp fault surface at its north end predates the CSD tear faults. This part of the Tunp therefore predates the Darby thrusting. Because this is the northern "tip" of the Tunp, it had propagated to its full length prior to Darby motion, contrary to the assumption of most workers (Jones, 1971; Elliott, 1976) that forward motion of a thrust usually also means lateral tip migration as well. In other words, the Tunp propagated laterally to its northern tip prior to Darby thrust motion, and any later increment of forward motion farther south did not cause continued thrusting at its north end.

DISCUSSION AND CONCLUSIONS

The LaBarge Meadows area of the Absaroka thrust system is extraordinarily complex, with superimposed thrusts, folds, extensional and tear faults, and a major lateral ramp (Fig. 6). Each major structural element except the lateral ramp extends outside the area and thus can be evaluated separately without the structural superposition. On the other hand, the lateral ramp is the major structure of the area, with the two most obvious aspects of the CSD, namely the tear faults and the offset in the Absaroka trend, being relatively minor in true displacement and importance for the Absaroka system. The tear faults especially tend to hide the true continuity of structural trends within the Absaroka thrust sheet with imbricate thrusts present within the sheet both north and south of LaBarge Meadows.

Major fold trends are also continuous across the area from north to south. Spacing between folds and imbricate thrusts does increase gradually from north to south along the Absaroka thrust sheet. Woodward (1988) discussed the relative importance of different structural lithic units in the deformation style of the northern Absaroka thrust system. The three major units (Fig. 2b, 7)—the Lower Paleozoic Unit, the Upper Paleozoic Unit, and the Lower Mesozoic Unit—all change their thicknesses and facies gradually along the thrust belt from north to south. The Lower Mesozoic Unit, which forms major surficial folds within the Absaroka thrust sheet, becomes the volumetrically most important part of the stratigraphic section from the LaBarge CSD southward both by its thickening and by the thinning of the older structural lithic units. It is attractive to relate the increase in thrust spacing and fold wavelengths to the changing thicknesses of dominant members in the structural lithic units, much as Wiltschko and Dorr (1983) related overall thrust spacing to total stratigraphic thickness.

Woodward (1988) suggested that major lateral ramps through Paleozoic strata within the Absaroka thrust system farther north were related to stratigraphic facies changes, especially within the competent Mississippian carbonate section. The lateral ramps in the northern part of the thrust belt also resulted in trend changes in the thrusts. The Labarge Meadows lateral ramp may well be another example of a structure with a similar origin. Some of the tear faulting in the Absaroka sheet may have originated from extension over the lateral ramp, which has been amplified by warping during Darby thrust motion.

The up-to-the-north lateral ramp in the Darby-Hogsback thrust surface (Royse and others, 1975; Blackstone, 1979) occurs in Jurassic and Cretaceous strata and therefore cannot be related to the same stratigraphic variations as the Absaroka ramp. Both, however, step upward toward the thinner stratigraphic section (Paleozoics thin southward; Jurassic strata thin northward) and probably reflect fundamental stratigraphic control. Because the thinning is in opposite directions, it seems unlikely that a single, positive, long-lived basement uplift was involved in controlling these facies changes during sedimentation.

Possible primary basement controls (basement buttressing) on Darby thrust geometry have been suggested by Blackstone (1979), but the Absaroka thrust system developed several tens of kilometers farther west of its present position (Dixon, 1982; Woodward, 1988), and there is no evidence that any basement arching extended that far east-west beneath the thrust belt during the Absaroka's Cretaceous motion. If Blackstone's explanation of the Darby structures is accepted, the major characteristics of the CSD as a whole have no single origin, with some related to Absaroka thrust motion (and possible stratigraphic controls on its geometry) and some related to Darby thrust motion (and basement controls on its geometry). Royse (1985) maintained that the uplift of the LaBarge Platform postdated thrusting and thus could not have influenced the thrust geometry. We prefer to accept

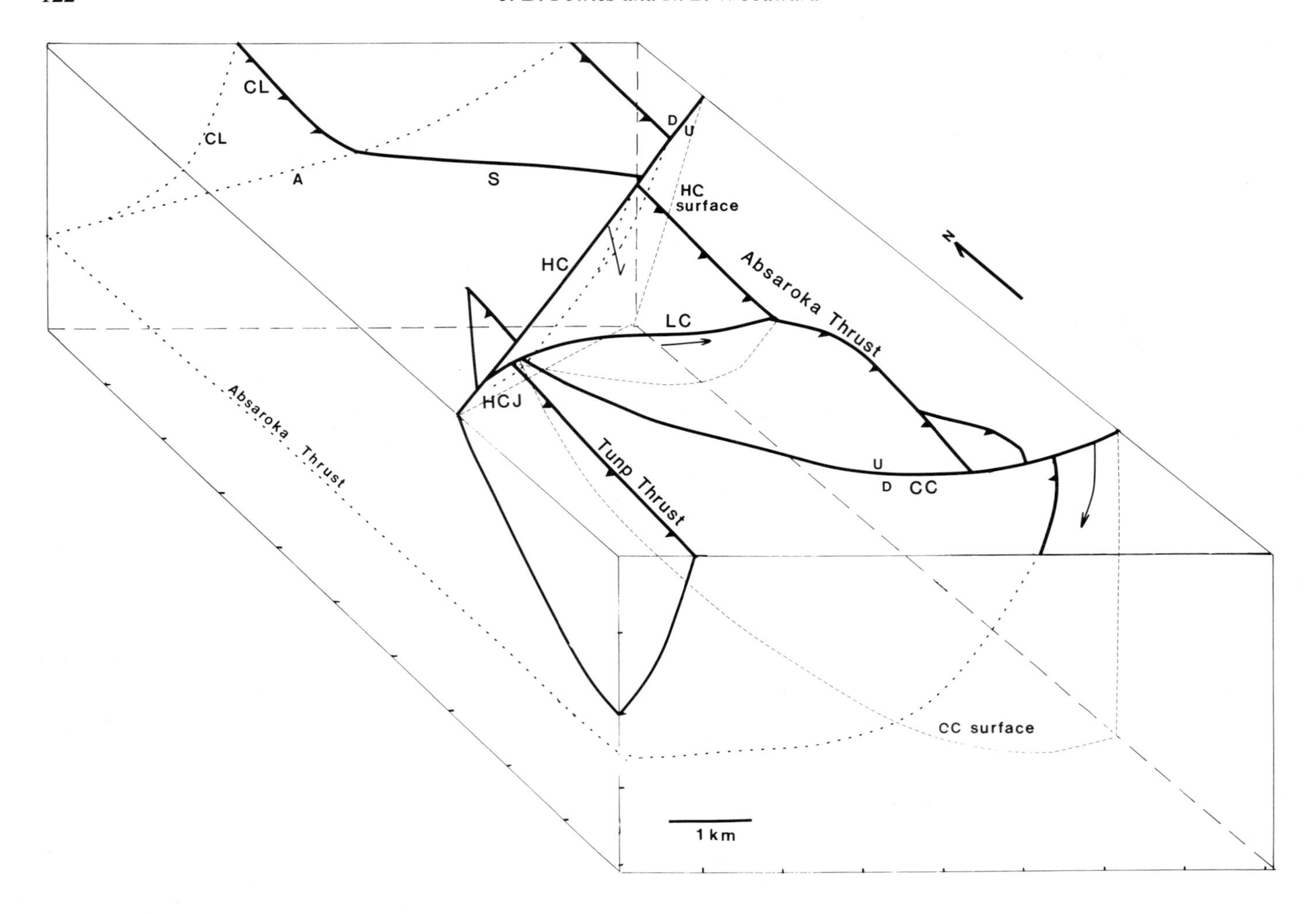

Figure 6. Block diagram of the CSD zone within the Absaroka thrust sheet. Arrows on the Hobble Creek (HC), Little Corral (LC), and Clear Creek (CC) faults show relative motions across the faults. CL—Clear Lake thrust; A—Absaroka thrust at depth; S—Springs fault.

Royse's argument, which was Blackstone's less preferred alternative, namely that the LaBarge area was uplifted after Darby motion. The result was the passive warping of that and earlier thrust systems without direct basement involvement in the thrusting. The dominant dip-slip displacement of the Hobble Creek and Clear Creek (Thompson) faults reflects extension over the warp superimposed on thrust-motion structures. This last explanation suggests that the Absaroka part of the CSD would be related to lateral ramping in the Upper Paleozoic Structural Lithic Unit, and the Darby-Prospect part of the CSD would be related to lateral ramping in the Lower Mesozoic and Cretaceous Structural Lithic Units.

ACKNOWLEDGMENTS

We wish to thank Dr. M.R.W. Johnson of the University of Edinburgh, The Weir Fund for Field Studies of the University of Edinburgh, and our many colleagues at Edinburgh and in Wyoming for their assistance. We acknowledge the support of NSF Grant #EAR-8312872 (to NBW). We also acknowledge the kind support and encouragement of Steve Oriel for work in the Wyoming thrust belt by NBW. Reviews of the manuscript by D. L. Blackstone, E. Erslev, L. B. Platt, P. K. Link, and F. Royse are gratefully acknowledged.

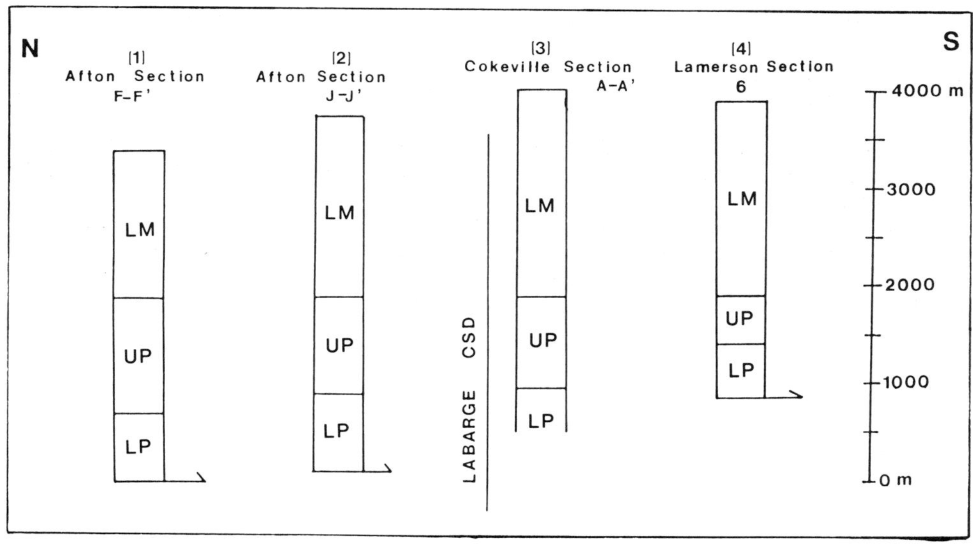

Figure 7. Comparison of stratigraphic columns from cross sections across the Absaroka thrust sheet from north to south across the LaBarge Meadows CSD. Column locations are shown on Figure 1, except for that from Lamerson (1982), which comes from 25 km farther south in the northern Fossil Basin area of the Absaroka sheet. A more extensive discussion of regional facies patterns is presented in Woodward (1992). From east to west across this part of the thrust belt the Upper Paleozoic Unit is relatively constant in thickness (Prospect, Absaroka, and Tunp sheets). The Lower Mesozoic Unit thickens 30% from the Prospect sheet westward into the Absaroka sheet as well as from north to south as shown. (Thicknesses based on Rubey, 1973; Rubey and others, 1980; and Lamerson, 1982. Thicknesses in Rubey publications were based on an extensive data catalogue, according to Oriel [personal communication].)

REFERENCES CITED

Blackstone, D. L., Jr., 1979, Geometry of the Prospect-Darby and La Barge faults at their junction with the La Barge Platform, Lincoln and Sublette Counties, Wyoming: Geological Survey of Wyoming Report of Investigations 18, 34 p.

Butler, R.W.H., 1982, The terminology of structures in thrust belts: Journal of Structural Geology, v. 4, p. 239–245.

Crowell, J. C., 1959, Problems of fault nomenclature: American Association of Petroleum Geologists Bulletin, v. 43, p. 2653–2674.

Dixon, J. S., 1982, Regional structural synthesis, Wyoming salient of Western Overthrust belt: American Association of Petroleum Geologists Bulletin, v. 66, p. 1560–1580.

Elliott, D. E., 1976, The energy balance and deformation mechanisms of thrust sheets: Philosophical Transactions Royal Society of London, Series A, v. 283, p. 289–312.

Fowles, J. D., 1985, Geology of the LaBarge Meadows area, western Wyoming, U.S.A. [Geology IV Honours Thesis]: Edinburgh, Edinburgh University, 52 p.

Jones, P. B., 1971, Folded faults and sequences of thrusting in Alberta foothills: American Association of Petroleum Geologists Bulletin, v. 55, p. 292–306.

Lamerson, P. R., 1982, The Fossil Basin area and its relationship to the Absaroka thrust fault system, *in* Powers, R. B., ed., Geologic studies of the Cordilleran thrust belt: Denver, Colorado, Rocky Mountain Association of Geologists, p. 279–340.

Oriel, S. S., and Platt, L. B., 1980, Geologic map of the Preston 1° × 2° Quadrangle, southeastern Idaho and western Wyoming: U.S. Geological Survey Miscellaneous Investigations Series, map I-1127, scale 1:250,000.

Royse, F., Jr., 1985, Geometry and timing of the Darby-Prospect-Hogsback thrust fault system, Wyoming: Geological Society of America Abstracts with Programs, v. 17, p. 263.

Royse, F., Jr., Warner, M. A., and Reese, D. L., 1975, Thrust belt structural geometry and related stratigraphic problems—Wyoming, Idaho, northern Utah, *in* Bolyard, D. W., ed., Symposium on Deep Drilling Frontiers in Central Rocky Mountains: Denver, Colorado, Rocky Mountain Association of Geologists, p. 41–54.

Rubey, W. W., 1973, Geologic map of the Afton Quadrangle and part of the Big Piney Quadrangle, Lincoln and Sublette Counties, Wyoming: U.S. Geological Survey Miscellaneous Investigations Series, Map I-686, scale 1:62,500.

Rubey, W. W., Oriel, S. S., and Tracey, J. I., 1980, Geologic map and sections of Cokeville 30′ Quadrangle, Lincoln and Sublette Counties, Wyoming: U.S. Geological Survey Miscellaneous Investigations Series, Map I-1129, scale 1:62,500.

Schmidt, C. J., and Perry, W. J., Jr., eds., 1988, Interaction of the Rocky Mountain foreland and Cordilleran thrust belt: Geological Society of America Memoir 171, 582 p.

Wheeler, R., 1980, Cross-strike structural discontinuities: Possible exploration tool for natural gas in Appalachian Overthrust belt: American Association of Petroleum Geologists Bulletin, v. 64, p. 2166–2178.

Wiltschko, D. V., and Dorr, J. A., Jr., 1983, Timing of deformation in overthrust

belt and foreland of Idaho, Wyoming and Utah: American Association of Petroleum Geologists Bulletin, v. 67, p. 1304–1322.
Woodward, N. B., 1986, Blind imbricate complexes as exploration targets in fold and thrust belts: American Association of Petroleum Geologists Bulletin, v. 70, p. 665.
—— , 1987, Stratigraphic separation diagrams and thrust belt structural analysis, *in* Miller, W. R., ed., The Overthrust Belt revisited, Wyoming Geological Association, 38th Annual Field Conference, Guidebook, p. 69–77.
—— , 1988, Primary and secondary basement controls on thrust sheet geometries, *in* Schmidt, C. J., and Perry, W. J., Jr., eds., Interaction of the Rocky Mountain foreland and Cordilleran thrust belt: Geological Society of America Memoir 171, p. 353–366.
—— , 1992, Deformation styles and geometric evolution of some Idaho-Wyoming thrust belt structures, *in* Mitra, S., and Fisher, G. W., eds., Structural geology of fold and thrust belts: Baltimore, The Johns Hopkins University Studies in Geology, The Johns Hopkins University Press (in press).
Woodward, N. B., and Rutherford, E., 1989, Structural lithic units in external orogenic zones: Tectonophysics, v. 158, p. 247–268.
Woodward, N. B., Walker, K. R., and Lutz, C. T., 1988, Relationships between lower Paleozoic facies patterns and structural trends in the Southern Appalachians: Geological Society of America Bulletin, v. 100, p. 1758–1769.

Manuscript Accepted by the Society July 19, 1991

Printed in U.S.A.

Geological Society of America
Memoir 179
1992

Chapter 6

Transpression during tectonic evolution of the Idaho-Wyoming fold-and-thrust belt

John P. Craddock
Geology Department, Macalester College, St. Paul, Minnesota 55105

ABSTRACT

A traverse across the central Idaho-Wyoming fold-and-thrust belt from the Paris thrust (west) to the Green River Basin (east) has revealed the presence of three distinct regional rock strain patterns. Within the carbonates of the six major thrust sheets, mechanical twins in calcite have recorded an early, layer-parallel and thrust transport-parallel strain and a later, synthrusting non-layer-parallel strain. Synorogenic calcite veins, also twinned, preserve a third unique strain sequence. Calcite strain magnitudes range from 2 to 16% and are highest in vein material; differential stress magnitudes, inferred for the time of twinning, are near 100 MPa for each of the three strain fabrics.

The early, prethrusting, layer-parallel twinning fabric can be used to interpret the paleo–stress and strain fields at the time of initial oblique plate convergence along the western margin of North America. Continued oblique convergence from the western dextral margin (late Jurassic to Eocene) resulted in west-to-east thrust translation and the formation of a tectonic wedge. Using modern transpressive analogues, I estimate that the ratio of dextral fault displacement to thrust shortening was approximately 7:1 for the crustal shortening associated with this fold-and-thrust belt, and calculations can be made concerning general thrust sheet piggyback rotations along the margin.

INTRODUCTION

Numerous field, analytical, and experimental efforts have supported the model of a sedimentary wedge that shortens cratonward by progressively younger and shallower thrust faults along a hinterland-dipping detachment zone that lies upon an autochthonous basement (Dahlstrom, 1970; Chapple, 1978; Boyer and Elliott, 1982; Platt, 1986). Subsequent two-dimensional theoretical modeling studies (e.g., Davis and others, 1983; Stockmal, 1983; Dahlen and others, 1984) utilized Chapple's perfectly plastic rheology as well as some of the following assumptions concerning the transformation through tens of millions of years of a sedimentary wedge into a fold-and-thrust belt: (1) plane strain (transport-parallel thrust sheet translation), (2) an autochthonous basement, and (3) an undeformed basement. I have studied the central, unbuttressed portion of the well-dated and well-exposed Idaho-Wyoming fold-and-thrust belt Cressman, 1963; Cressman and Gulbrandsen, 1964; Armstrong and Oriel, 1965; Armstrong, 1968; Rubey, 1973; Royse and others, 1975; Dorr and others, 1977; Oriel and Platt, 1980; Wiltschko and Dorr, 1983; Mitra and Yonkee, 1985; Craddock and others, 1988), which allowed for the testing of aspects of these models. In this chapter, rock strain data from a traverse across this fold-and-thrust belt are used to test the plane strain assumption and propose a tectonic model for its evolution.

Evolution of the Idaho-Wyoming fold-and-thrust belt

Laramide tectonic development of the Idaho-Wyoming portion of North America Cordillera began in the late Jurassic and ended during the Eocene epoch, spanning a time interval of 80 m.y. (Armstrong and Oriel, 1965). Heller and others (1986) have proposed, based on a basal Cretaceous unconformity, that thrusting commenced in the early Cretaceous. Crustal shortening involved the formation of seven major thrust faults, from west to east: the Paris, meade, Crawford, Absaroka, Darby, and Prospect thrusts and the Moxa Arch, which offsets the underlying Archean basement (Wiltschko and Dorr, 1983; Craddock and others,

Craddock, J. P., 1992, Transpression during tectonic evolution of the Idaho-Wyoming fold-and-thrust belt, *in* Link, P. K., Kuntz, M. A., and Platt, L. B., eds., Regional Geology of Eastern Idaho and Western Wyoming: Geological Society of America Memoir 179.

1988; Craddock, 1988; Kraig and others, 1988). The eastern portion of this belt is bounded on the north by the Archean-cored Teton–Gros Ventre Range, to the south by the Uinta Range, and on the east by the Green River Basin (Fig. 1). Northwest of the Idaho-Wyoming thrust belt lies the Snake River volcanic plain that is locally truncated by Miocene and younger listric normal faults (Blackstone, 1979, 1980). Metamorphic core complexes, such as the Albion–Raft River Range, involve deformed Precambrian crystalline rocks and were formed as Tertiary Basin and Range extensional features (Wust, 1986). They roughly define the western boundary of cratonic North America prior to Mesozoic terrane accretion (Fig. 1; Allmendinger and others, 1987).

In this chapter rock strain data are presented that were obtained for each of the major thrust sheets along a traverse across (e.g., transport-parallel) the central, unbuttessed portion of this fold-and-thrust belt. These strain data are used to interpret the kinematic development of the Idaho-Wyoming belt. In a complementary paper, the mapping and meso-scale structures, including out-of-sequence, out-of-transport, and footwall deformations, have been presented (Craddock, 1988) and have a bearing on the regional interpretation of these strain results.

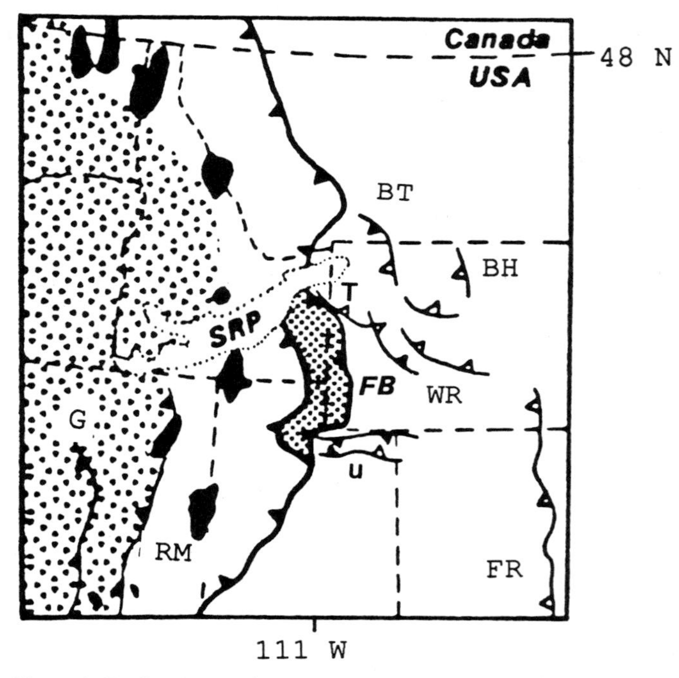

Figure 1. Regional tectonic map showing the Idaho-Wyoming fold-and thrust belt (dot shading) in relation to the Cordilleran thrust belt (darkened thrust hachures) and the adjacent foreland crystalline uplifts (open thrust hachures). Metamorphic core complexes are black (after Wust, 1986), and accreted terranes are shown in a triangular pattern. Abbreviations used are G = Golconda and RM = Roberts Mountain allochthons (Howell and others, 1985); SRP = Snake River Plain volcanic rocks; T = Teton–Gros Ventre Range; u = Uinta Mountains; FB = Green River Foreland Basin; BT = Beartooth Mountains; BH = Bighorn Mountains; WR = Wind River Range; FR = Front Ranges.

CALCITE TWIN ANALYSIS

Technique

Calcite deforms internally by mechanical twinning when differential stresses >70 to 100 bars are properly applied. Twinning is possible along three glide planes and generally causes a crystal to strain harden. Twin lamellae are 5 to 100 microns in width, making them observable with a universal stage. Analysis of twinning in calcite is possible in country rock limestones as well as in younger, cross-cutting calcite veins in these same sediments.

Mechanical twins in calcite were analyzed using the technique of Groshong (1972, 1974; Teufel, 1980; Groshong and others, 1984) to determine the orientation and magnitude of the principal strain axes. I distinguish negative expected values (NEV's), that is, grains that were not favorably oriented for twinning, from grains that recorded a positive sense of twinning (PEV's). The data were cleaned by eliminating the NEV's and/or 15 to 20% of the grains with the largest deviations (Groshong, 1974; Groshong and others, 1984). This procedure reduced the error in the strain magnitude and refined the principal strain axis orientations. In addition, compression (stress = sigma 1, 2, 3) axis contours are obtained with this method (Turner, 1953). The orientation of the odd-signed strain axis, which generally is the maximum shortening (strain = e1, e2, e3) axis, is accurate to within 6 to 8 degrees (Groshong and others, 1984). Where NEV's exceeded 40%, these data were separated and analyzed. Such an analysis yields an NEV-split that may have resulted from a second, younger strain event (Teufel, 1980). Differential stress magnitudes were calculated from twinned calcite elements using the procedure of Jamison and Spang (1976). In the following text I will refer to layer-parallel shortening (lps) fabrics in those samples where the angle between the principal shortening strain axis and bedding is less than 20 degrees (Groshong, 1972, 1974); in most samples, the contoured compression axis maxima and the maximum shortening axis are identical, within 20 degrees of bedding, and both the maximum shortening axis (e1) and the compression axis contours intersect the bedding plane great circle of the samples.

Sampling

Fifty-one oriented carbonate samples were collected for twinning strain analyses. These samples yielded 60 calcite twinning analyses from country rock limestone samples, including eight analyses given in Evans and Craddock (1985). Nineteen analyses were from twinned calcite veins. The number of samples by thrust sheet and stratigraphic unit is listed in Table 1.

Strain results

The calcite strain data from country rock and vein groups were separated for each thrust sheet. The three calcite strain fabrics (layer-parallel, non-layer-parallel, and veins) are discussed in detail below. Lower hemisphere projections of all the data are

TABLE 1. NUMBER OF ORIENTED SAMPLES PER THRUST SHEET

Thrust	Stratigraphic Unit*						
	Є	Mm	Pw	Trd	Jtc	Kp	Total
			COUNTRY ROCK SAMPLES				
Paris	7	1					8
Meade		8	1	1	1		11
Crawford					13†	1	14†
Absaroka		8	1	1	1		11
Darby		8			2	1	11
Prospect					3	2	5
Total	7	25	2	2	20†	4	60†
			VEINS IN COUNTRY ROCK				
Paris	4	1					5
Meade		4		1	1		6
Crawford					1		1
Absaroka		2			2		4
Darby					2		2
Prospect					1		1
Total	4	7	0	1	7		19

*See Figure 2 for stratigraphic abbreviations.
†Includes data from Evans, 1983; Evans and Craddock, 1985. See Figure 2 for stratigraphic names.

presented in Figure 2, and the accompanying values can be found in Table 2.

Layer-parallel fabric

Each thrust sheet contains limestones that preserve calcite strains that are parallel to bedding and generally parallel to the inferred thrust transport direction during Laramide thrusting. The average layer-parallel shortening directions in the country rock of each thrust sheet (both e1 and the trend of the Turner compression axis) show a clockwise rotation about a vertical axis from the oldest Paris thrust sheet (west) to the youngest Darby thrust (east), spanning 28 degrees (Turner compression axis) and 48 degrees (e1). The layer-parallel data are very clean, with few NEV's and moderate nominal errors. The calculated maximum shortening strain magnitudes range from 3.5 to 5.5%.

Non-layer-parallel fabric

The country rock limestones also locally preserve non-layer-parallel shortening strains with a wide range of shortening directions (Table 2). These are considered to represent local twinning strains associated with, for example, ramp bending, local folding, and "piggyback" events (see Mulugeta and Koyi, 1987). The non-layer-parallel country rock shortening strains have the highest percentage of NEV's (more than 26%), suggesting that these are secondary twinning strains (see below). Two country rock samples recorded bedding-normal shortening strains, both of which were located in locally flat-lying sediments within the folded and north-plunging Meade thrust sheet near Bear Valley, Idaho. Shortening strain magnitudes range from 16.7 to 2.7%.

Veins

The vein twinning data consistently preserve a shortening strain fabric that represents a shear strain parallel to a given vein boundary. This vein boundary-parallel strain is present regardless of vein orientation. Such fabrics largely reflect local strains that are related to vein formation and propagation and not a regional strain pattern (Craddock and van der Pluijm, 1988). Thus, the scattered e1 strain pattern was expected.

In contrast to the country rock samples, the vein data are remarkably clean, except for one sample in the Crawford sheet and samples from refolded and recemented Absaroka sheet rocks. The veins have recorded the highest shortening strain magnitudes per sample (7.13%). High strain magnitudes result from the homogeneous nature of the veins (e.g., no pore space, etc.) and record the continued high differential stress magnitudes that were present during progressive stacking of thrust sheets.

NEV-splits

Six samples contained large percentages of negative expected values (NEV's), a circumstance that allows the data for each sample to be analyzed in two sets (NEV and PEV splits); five splits in country rock carbonates and one in a vein sample were undertaken (Fig. 2). However, I did not obtain a clear regional NEV twinning fabric, a result that may in part be due to the irregular

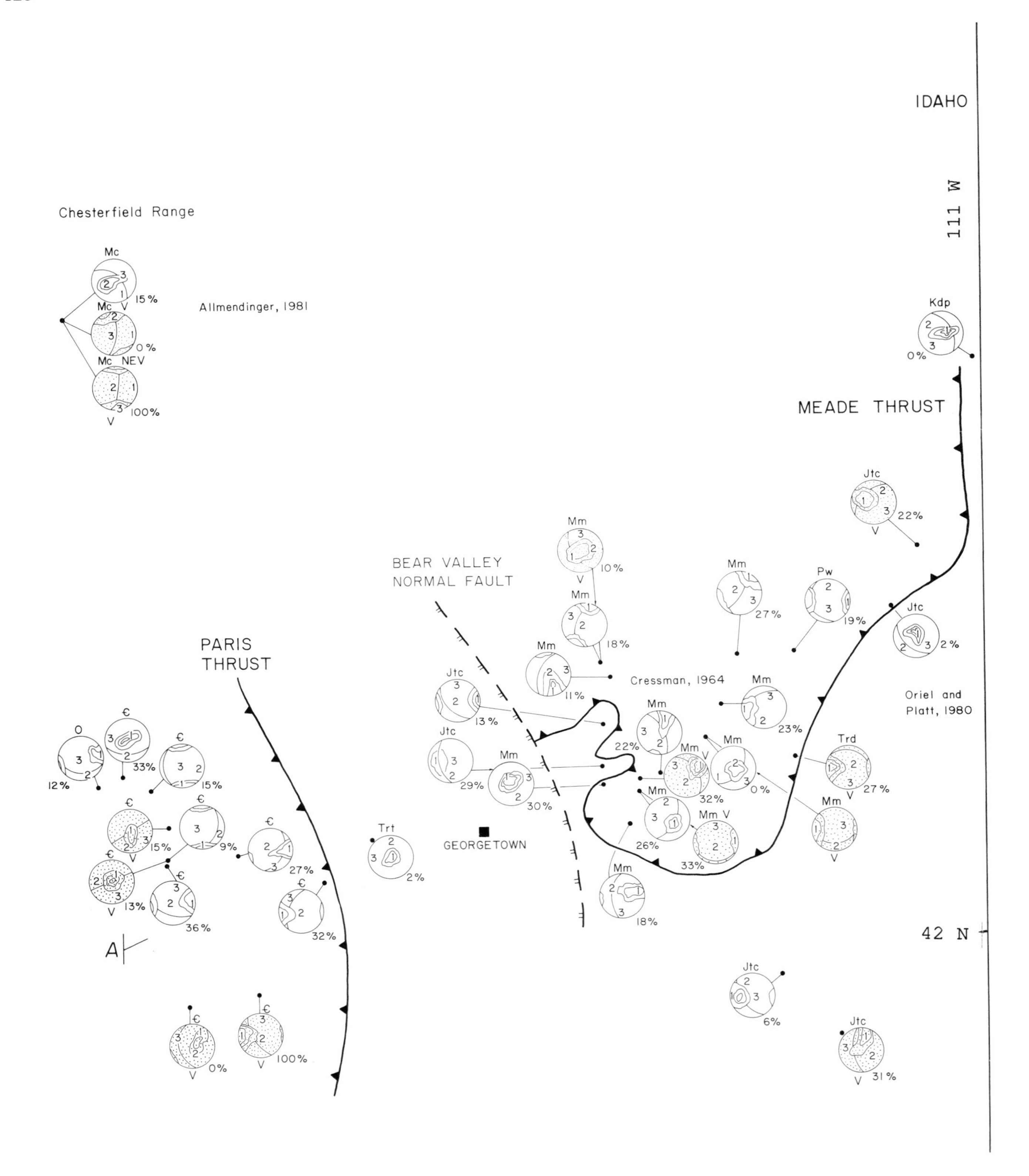

Figure 2. Map view of the Idaho-Wyoming fold-and-thrust belt with cross section location (A-A′; see Craddock, 1988, and Fig. 3) and calcite strain data. Calcite fabrics are plotted on equal area, lower hemisphere projections: 1 = e1, maximum shortening axis; 2 = e2, intermediate axis; 3 = e3, extension axis. Vein calcite results are plotted on stippled lower hemisphere projections, and open plots represent

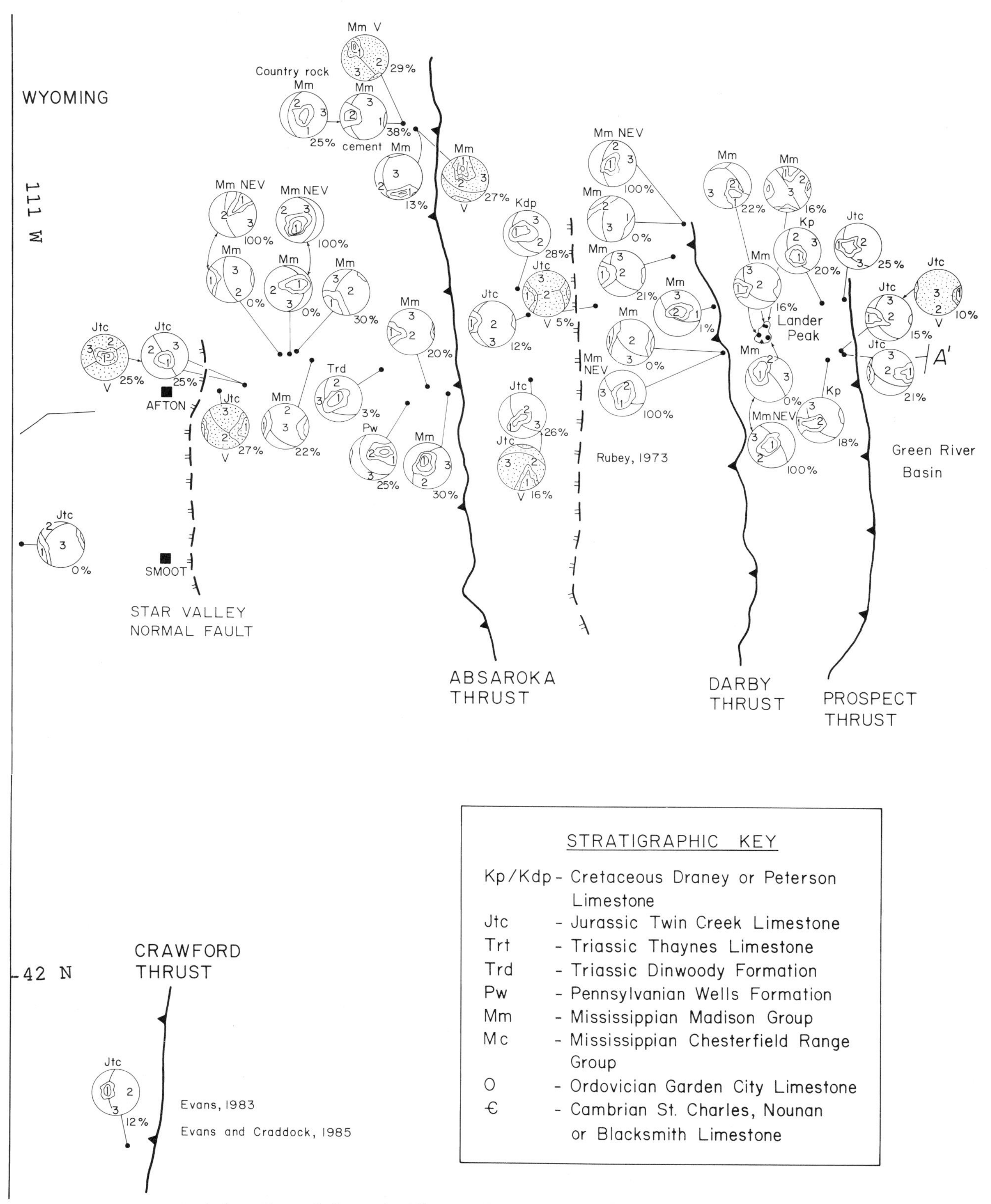

country rock data. Great circles are bedding or vein orientations. Turner (1953) compression axes are contoured starting with the third standard deviation and by increments of two. NEV values are plotted with each strain analysis as a percentage of the uncleaned data. Stratigraphic symbols are included in the key and references listed. Geologic mapping is based on Craddock (1988) and others listed on the figure (e.g., Rubey, 1973).

TABLE 2. SUMMARY OF CALCITE TWINNING DATA, BY THRUST SHEET

lps fabric*	Paris	Meade	Crawford	Absaroka	Darby	Prospect
			COUNTRY ROCK LPS FABRIC			
e1 trend	56.6	63.7	83.8	93.0	105.3	94.0
TC trend	67.5	70.7	81.0	93.0	96.0	91.6
e1 (%)	-5.5	-5.3	-3.5	-3.9	-5.5	-5.0
NEV	12.3	16.4	3.8	9.7	17.8	4.2
			COUNTRY ROCK NON-LPS FABRIC			
e1 trend	120	106.6	....	55.0	85.0	....
TC trend	80	109.0	....	101.3	107.5	....
e1 (%)	-16.7	-4.1	....	-2.7	-8.4	....
NEV	13.6	14.7	....	8.9	10.5	....
			VEIN TWINNING FABRIC			
e1 trend	57.7	62.5	10.0	98.0	127.5	68.0
TC trend	76.2	72.5	20.0	112.0	125.0	34.0
e1 (%)	-7.8	-7.8	-4.2	-6.8	-8.3	-7.9
NEV	6.6	11.7	3.2	6.7	4.7	3.5
			NEV			
Country rock			(raw percent/clean percent†)			
lps	22/0	21/1	21/2	26/2	21/1	13.5/1
Veins	7/0	23/1	31/5	23/6	11/1	0/0
Number of strain analyses	14	17	15	18	16	6

*e1 trend = orientation of maximum shortening axis in degrees; TC trend = orientation of Turner compression axis in degrees (0 - north for both); NEV = negative expected value; lps = layer-parallel shortening fabric. SE(X) + SE(Y)/2 from Groshong and others, 1984.

†Data before (raw) and after (clean) cleaning (Groshong, 1972, 1974).

distribution of these six samples. All five of the country rock samples that produced both NEV and PEV strain results were Madison Group carbonates: three from the Darby sheet, two from the Absaroka sheet. Four of the five samples produced a parent (PEV) layer-parallel strain result and a daughter (NEV) non-layer-parallel strain result. The fifth sample had a parent non-layer-parallel fabric and a layer-parallel daughter. The vein sample recorded multiple, coaxial twinning deformations (PEV and NEV splits), both of which were subhorizontal and at a high angle to bedding. The country rock samples were all from the limbs of tight, asymmetric folds, and the scatter in the NEV data presumably represents a younger twinning event associated with the folding.

Regional trends

This regional traverse is shown in cross section in Figure 3, and the A-A′ section is keyed to both Figures 2 and 4 with reference to the geology and calcite strain results. In Figure 4 the e1 shortening axis for each sample along the traverse is presented in map view. Country rock layer-parallel shortening fabrics (Fig. 4A) and country rock non-layer-parallel shortening fabrics (Fig. 4B) as well as vein data (Fig. 4C) are shown. The multiple strain results reported here complement the calcite, dolomite, and quartz microstructural strains observed by Allmendinger (1982) in the Meade sheet. The length of the e1 trend line is representative of the strain magnitude. In Figure 4A I also include the average e1 direction (large arrows) for the layer-parallel shortening fabric for each thrust sheet that presents a regional curvilinear orientation. The average e1 axis for the westernmost Paris thrust sheet trends N56E. Progressively eastward, the average e1 axis trends nearly east-west, with the easternmost value of N94E found in the Prospect thrust sheet. These summed e1 shortening directions for the layer-parallel shortening (lps) fabric are in situ values for each thrust sheet and do not represent restored or unfolded results. Fold axes are parallel to local and regional thrust fault traces (e.g., transport-normal) and are concentric folds (Dahlstrom, 1970; Dixon, 1982), such that unfolding results in a nearly identical e1 trend orientation. Similar orientations for nontwinning finite strains in the Twin Creek limestone of the Meade thrust have been reported by Protzman and Mitra (1990) for section balancing techniques.

Differential stress magnitudes

Differential stress magnitudes were calculated for each of the thrust sheets for the country rock as well as the veins (Jamison and Spang, 1976; Table 3). The values range from 47 to 102 MPa for the country rock groups, with an average value of 74 MPa. The veins recorded differential stresses that averaged 122 MPa and ranged from 159 to 81 MPa. For rocks that preserved only one twinning fabric (lps or non-lps), the layer-parallel sample was twinned by a slightly higher (74 MPa) differential stress than the non-lps specimen (67 MPa). In both the country rock and vein groups the highest calculated differential stress value was from the Paris thrust sheet in the west and the lowest from the Prospect sheet in the east, although the decrease from west to east is not linear. Furthermore, in all cases, the differential stresses associated with twinning of the synorogenic vein calcite are higher (122 MPa) than those for the country rock. This would suggest that the compressive stresses associated with country rock twinning were less than those responsible for the vein-boundary-parallel shearing that twinned the vein calcite. These differential stress magnitudes are comparable to results obtained from other low-grade thrust belt rocks in the Canadian Rockies and the Southern Appalachians (Jamison and Spang, 1976; House and Gray, 1982).

OUTCROP FABRICS

The traverse across the Idaho-Wyoming thrust belt also produced a geologic map; interpretations and descriptions of meso-scale structures; a regional, balanced cross section; and plots of outcrop fabric data for each thrust sheet (Craddock, 1988). The outcrop structural fabric data are presented in Figure 5 on lower hemisphere projections as contoured poles to planes.

Joints

Plots of systematic joints from a variety of stratigraphic units in each of the thrust sheets show a consistent pattern of one vertical joint set. The average strike of the joints from each thrust sheet rotates clockwise about a vertical axis from west to east. These joint sets are interpreted to be prethrusting extension joints that were parallel to the regional principal compressive stress direction (west-east; Craddock and others, 1988). Hence, excluding the Meade sheet data, these plots show a west-to-east clockwise rotation that is complementary to the layer-parallel calcite strain fabric.

Spaced cleavage

Solution cleavage development is similarly consistent, with one major foliation developed in each thrust sheet, except the Crawford sheet (Evans and Craddock, 1985; Mitra and Yonkee, 1985). The cleavage orientation is vertical and strikes just north of east in the Paris sheet, then strikes northwest-southeast in the Meade sheet. One cleavage in the Crawford sheet strikes northwest-southeast (shortening direction = southwest-northeast, which again correlates well with the clockwise, west-to-east thrust rotation described above.

Veins

Patterns of vein orientations increase in complexity from east to west. The increasing complexity results from the progressive piggyback stacking of the westernmost thrust sheets. The Absa-

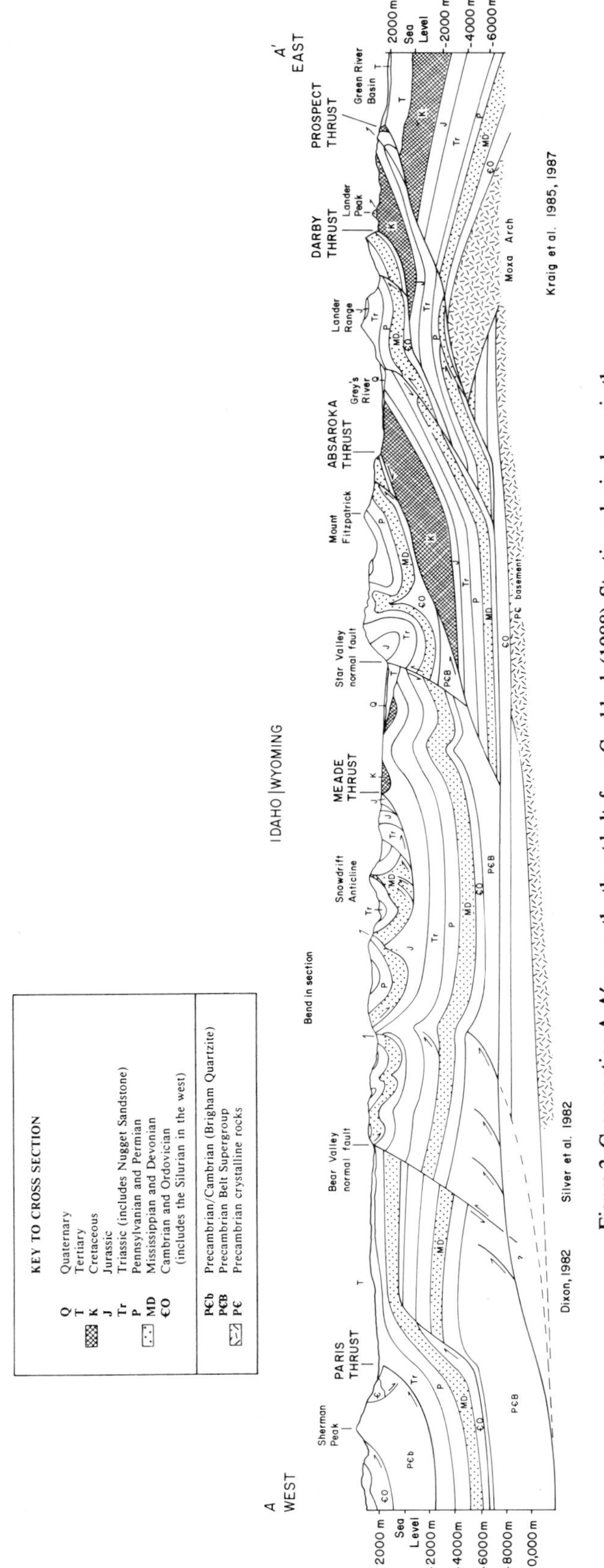

Figure 3. Cross section A-A′ across the thrust belt, from Craddock (1988). Stratigraphy is shown in the accompanying key.

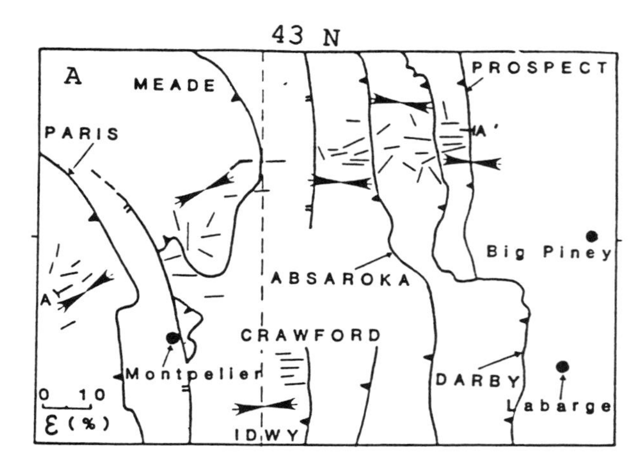

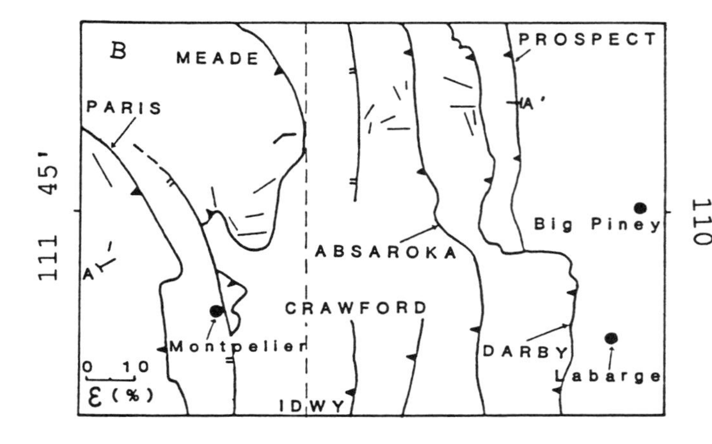

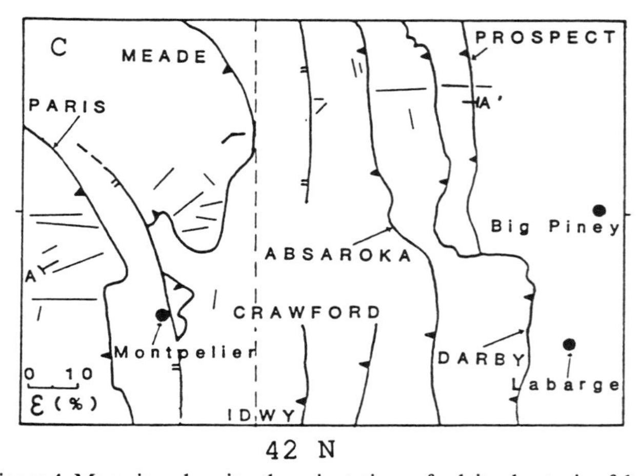

Figure 4. Map view showing the orientations of calcite shortening fabrics (e1) and their strain magnitudes. A, Layer-parallel fabric in the country rock, with averaged shortening directions (large arrows) for each thrust sheet (see text). B, Non-layer-parallel shortening axis fabrics. C, Vein shortening axis fabric. A-A′ is the general traverse line.

roka, Darby, and Prospect sheets preserve a single, vertical vein set that strikes west-east. These are probably filled extension joints (see above). The Crawford sheet also contains one vertical set, but it strikes northwest-southeast. The Meade and Paris sheets contain a number of vein sets, resulting in a diffuse girdle plot on the projection, although the Paris pattern (northwest-southeast) and the Absaroka-Prospect patterns do fit with the regional, west-to-east clockwise pattern above.

DISCUSSION

In the modeling of thrust sheet motion and sedimentary wedge mechanics, assumptions are made, including the direction and relative timing of thrust motions and the involvement of footwall and/or basement rocks (Chapple, 1978; Davis and others, 1983; Dahlen and others, 1984). The fieldwork and strain data from a "classic" belt, the Idaho-Wyoming fold-and-thrust belt, indicate that microstructural fabrics can be powerful regional and local indicators of stress and strain fields in a fold-and-thrust belt. Analysis of twinned calcite is a powerful, incremental, three-dimensional strain technique and, for these rocks, suggests that both plane strain *and* nonplane strain deformations were important during the development of the unbuttressed portion of this fold-and-thrust belt.

Margin tectonics and thrust belt shortening

Earlier studies of western North America have shown that oblique plate convergence and accretion of Cordilleran suspect terranes resulted in transpression at the late Jurassic-Cretaceous margin, with a dextral, margin-parallel, strike-slip component (Price, 1981; Monger and others, 1982; Bird, 1984; Chamberlain and St. Lambert, 1985; Price and Carmichael, 1986; Ellis, 1986; Price, 1986; Van den Driessche and Maluski, 1986; Ellis and Watkinson, 1987). Van den Driessche and Maluski (1986) have proposed that oblique convergence at a margin can be partitioned into a margin-perpendicular displacement vector and a margin-parallel displacement vector, that is, thrust and strike-slip components, respectively. Recent experimental work generally supports such a model (Van den Driessche and Brun, 1987; Lamons and others, 1987). The regional clockwise rotation of the principal shortening axis that I observed from our calcite twinning strains (Fig. 4A) may bear directly on these transpression models.

Prethrust layer-parallel shortening

It is well documented that layer-parallel shortening strains in country rocks of thrust belts predate thrust sheet translation, including those in autochthonous rocks in the foreland (Engelder and Engelder, 1977; Engelder, 1979; Spang and Brown, 1981; Spang and others, 1981; Allmendinger, 1982; Ballard and Wiltschko, 1983; Craddock and Wiltschko, 1983; Julian and Wiltschko, 1983; Evans and Craddock, 1985; Woodward and others, 1986; Craddock and others, 1988). It is possible that a younger layer-parallel strain could be imposed during folding if

TABLE 3. INFERRED DIFFERENTIAL STRESSES (MPa) BY THRUST SHEET*

Country Rock	Paris	Meade	Crawford	Absaroka	Darby	Prospect	Average
Lps†	102	81	57§	71	86	47	74
Non-lps	81	102	....	48	39	....	67
Veins	159	134	115	141	102	81	122

*Method: Jamison and Spang (1976).
†Lps = layer-parallel shortening.
§All samples from the Jurassic Twin Creek Formation.

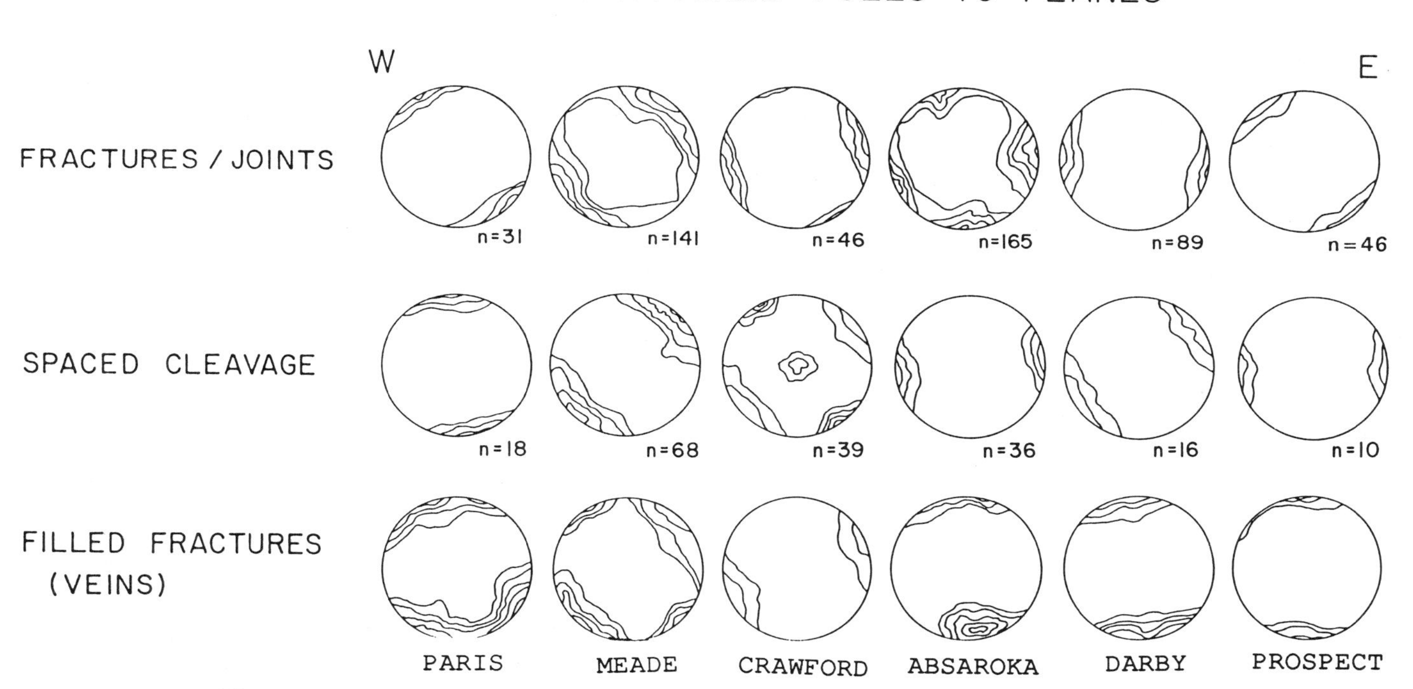

Figure 5. Lower hemisphere stereographic projections of poles to planes of various outcrop fabrics, presented by thrust sheet. Contours start with the third standard deviation and increase by increments of two.

the local rock viscosity contrasts are high. Wiltschko and others (1985) and Kilsdonk and Wiltschko (1988) have shown that where low-grade rocks in a frontal thrust sheet are well bedded, folding and thrust ramping is accomplished by bedding-parallel slip, and the older layer-parallel twinning fabric is unaltered (strain hardens) and remains parallel to the thrust transport direction.

More recently, Jackson and others (1989) and Craddock and van der Pluijm (1989) have shown that Appalacian orogenic stresses were transmitted up to 1,000 km into the foreland of an evolving thrust wedge. The stress and strain pattern is preserved by a layer-parallel, thrust transport-parallel calcite twinning strain and a magnetic anisotropy fabric. The extended footwall (e.g., foreland) sediments, in a flat-lying state, are not exposed near the Idaho-Wyoming belt. It seems unlikely that the layer-parallel calcite strain (Fig. 4A) would develop contemporaneously with the local, piggyback non-layer-parallel and vein strains (Figs. 4B and C), especially when the layer-parallel fabric is preserved in both folded and unfolded samples. The easternmost Prospect thrust preserves the most consistent layer-parallel fabric, despite the refolding of this thrust sheet (Craddock, 1988); the Prospect thrust sheet was overridden by the Darby sheet but never translated in piggyback fashion. Farther west, where piggyback deformation was more prevalent, the calcite strain data are more complicated and the early, lps fabric is overprinted. Mitra and others (1988) and Protzman and Mitra (1990) have studied finite strains in the folded Meade thrust footwall (Crawford thrust hanging wall). The micritic Jurassic Twin Creek limestone has recorded a local, prethrusting layer-parallel shortening (10 to 35%) fabric (deformed Pentacrinus, solution cleavages) that was passively rotated by later folding and thrust imbrication associated with motion by the Meade thrust sheet. These workers identified volumetric problems associated with the bulk shortening strain in the Twin Creek when balanced cross sections were constructed; including these calcite strains results would enlarge this problem.

The Idaho-Wyoming belt

In the Idaho-Wyoming thrust belt this layer-parallel shortening fabric trends approximately east-west in the eastern portion of the belt but trends progressively more northeasterly toward the west, near the former plate margin. Unfortunately, the same rocks sampled in the thrust belt are not exposed in the adjacent foreland (e.g., Green River Basin) except atop various thick-skinned mountain belts (e.g., the Wind River Range), so I can only infer the orientation of the lps fabric in the autochthonous foreland (see Craddock and van der Pluijm, 1989). A traverse from the thrust belt into the foreland (e.g., Minnesota), analyzing calcite strains and magnetic anisotropies in pre-Tertiary carbonates is underway. The present regional, curvilinear layer-parallel thrust belt fabric (Fig. 4A) is interpreted to be a result of the decreasing importance of the dextral strike-slip displacement component away from the transpressive margin (Fig. 6; see Margin Tectonics and Thrust Belt Shortening section above). The Idaho-Wyoming belt is thought by some to have formed tectonically in a fashion similar to that of the present central Andes region, where the shallow slab dip is responsible for the evolution of thick-skinned, foreland uplifts (e.g., Ramos and others, 1986) and where the subduction direction is closer to being normal, rather than oblique, to the margin (Bird, 1984, 1988; Engebretson and others, 1985)). This could have been the initial tectonic scenario in the late Jurassic along the Cordilleran margin, followed by terrane accretion and dextral motion of these terranes (Monger and others, 1982). A calcite strain study across an actively forming thrust belt has not been attempted; such a study would allow for a better interpretation of this regional, curvilinear, layer-parallel calcite strain fabric with regard to Sevier thrust belt kinematics.

It is critical to realize that during the Jurassic (Fig. 6) actual thrust motion had not yet occurred and that the sedimentary wedge had been shortened 3 to 5% parallel to bedding and perpendicular to the dextral plate margin (Mount and Suppe, 1987;

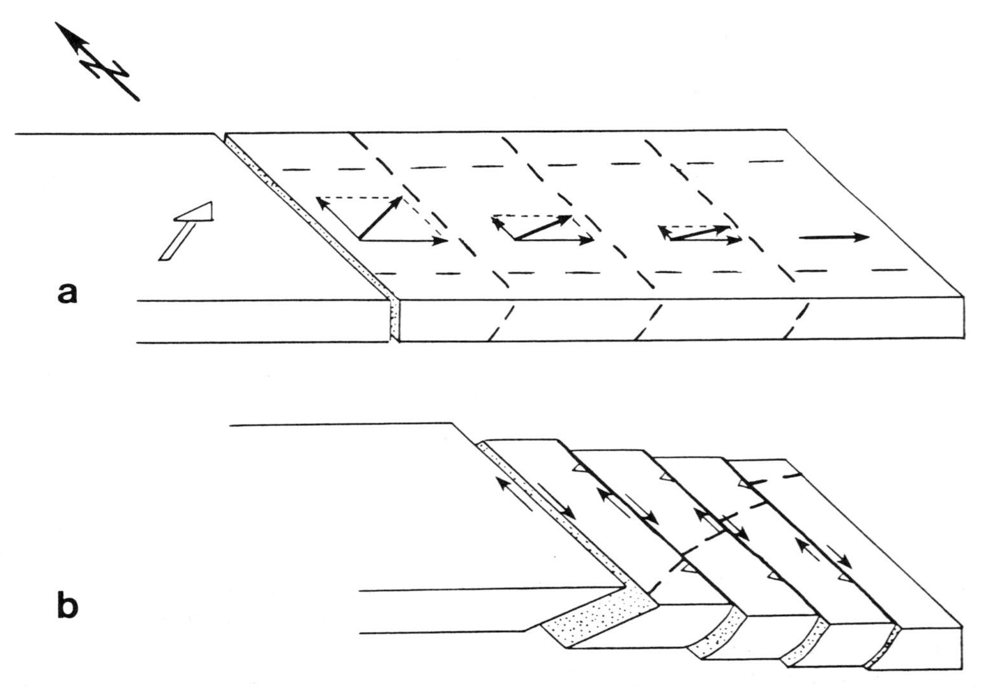

Figure 6. a, Relative dextral versus thrust fault displacement transfer (dark arrows), with the layer-parallel, transport-parallel stress and strain field that is preserved as the layer-parallel, thrust transport-parallel calcite fabric (E-W dashed lines) in the sedimentary wedge that will become the ID-WY thrust belt (future faults are N-S dashed lines). b, The decreasing importance of the dextral component of the plate motion into the foreland during regional shortening and the motion of the prethrust translation lps fabric.

Zoback and others, 1987). Not until continued convergence resulted in the localization of strain and fault propagation did the thrust sheets develop. Movement on the individual thrusts passively transported the layer-parallel fabric (Fig. 4A) to the east (Fig. 6b), resulting in non-lps calcite strains and twinned vein calcite (e.g., piggyback strains; Figs. 4B and 4C), in accordance with the regional thrust displacement kinematics (Fig. 7, Table 4). In this scenario, the thrust motion was not everywhere toward the east but rather progressively more northeasterly toward the western part of the belt where the effects of the dextral margin are greatest, and motion of the originally east-west lps fabric is rotated incrementally by thrust sheet piggyback motions. It is possible to track the motion of each thrust sheet by using the orientation of the lps fabric in each thrust sheet (Fig. 4A) and to trace the lps fabric motion throughout the orogenesis (Table 5; Fig. 7). Confirmation of these rotations is not possible by paleomagnetic methods because, as in most thrust belts and their forelands, these sediments carry a synorogenic chemical remagnetization (Jackson and others, 1989). Unfortunately, only the folded Meade thrust surface is exposed in Georgetown Canyon, with multiple striation directions present (Platt and Royse, 1989), so identification of strike-slip motions along these faults has not been possible. However, north-dipping thrust faults (e.g., south-to-north transport), east-west trending folds, and a northeast-southwest striking cleavage in the Bingham Quartzite (e.g., northwest-southeast compression) are recognized in the Paris sheet (Craddock, 1988).

Using the modern Alpine and San Andreas dextral margins as analogues, it was possible to calculate dextral and thrust fault displacements for the western margin of North America for the Sevier orogenic event (Table 6). Analysis of the Finlay-Tintina dextral fault yields a dextral to thrust displacement ratio of 7.5:1 (Table 6) and a plate convergence rate of 100 mm/yr for 30 Ma.

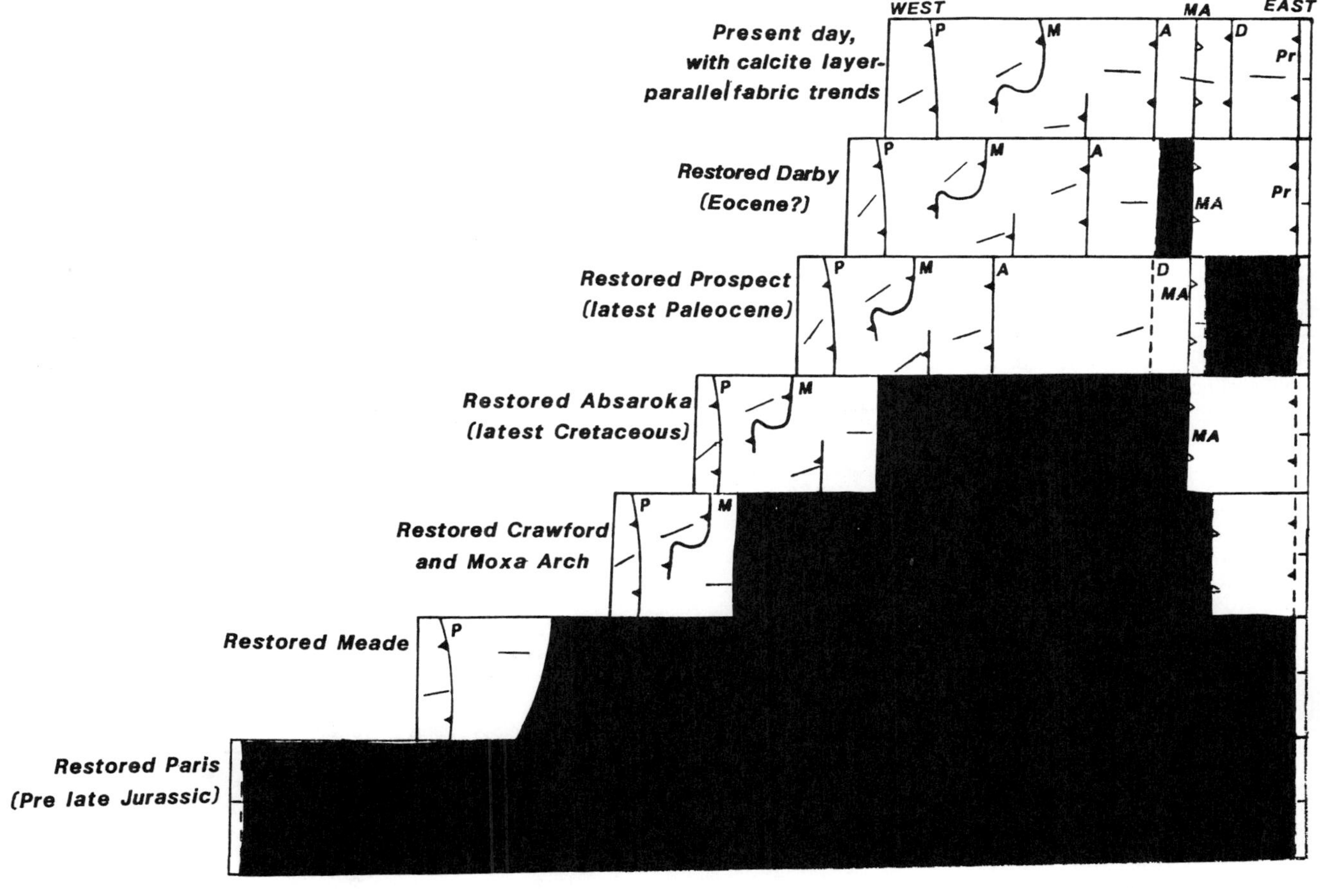

Figure 7. Map view of the kinematic development of the Idaho-Wyoming thrust belt from the Jurassic (bottom) to the Eocene (top) and motion of the prethrust translation layer-parallel calcite shortening fabric. Shaded regions are undeformed in sequence (see Craddock, 1988). Explanation: Assume that the prethrust layer-parallel, thrust transport-parallel fabric was oriented at 090 degrees, or perpendicular to the dextral margin. A restored thrust belt should preserve, or have in rough alignment, each lps fabric per thrust sheet. The Paris sheet first moved 44 km toward 082 degrees (–8 from 090). Then, the Meade sheet, carrying the Paris sheet, moved 23 km toward 070 (–20 from 090), and so on. P = Paris thrust; M = Meade; MA = Moxa Arch; A = Absaroka; Pr = Prospect; D = Darby thrust. See Tables 4 and 6 and Figure 3 for data.

TABLE 4. THRUST DISPLACEMENTS, IDAHO-WYOMING BELT

Paris (km)	Meade (km)	Absaroka (km)	Darby (km)	Prospect (km)	Moxa Arch (km)	Total (km)
44	23	35	18	18	(11)*	138 (149)

*Motion to the west.

TABLE 5. INCREMENTAL THRUST SHEET ROTATION IN THE IDAHO-WYOMING BELT

	(Late Jurassic) Paris Thrust (°)	Paris + Meade Thrusts (°)	Paris + Meade + Crawford Thrusts (°)	Paris + Meade + Crawford + Absaroka Thrusts (°)	Paris + Meade + Crawford + Absaroka + Prospect Thrusts (°)	(Eocene) Paris + Meade + Crawford + Absaroka + Prospect + Darby Thrusts (°)
			Restoration Motion			
Motion*	+8	+20	+9	+16	-4	-15
			Real, West-to-East Sequential Motion			
Motion*	-8	-20	-9	-16	+4	+15

*+ = clockwise motion; - = counterclockwise motion. Motion caused by, or related to, the subsurface Moxa Arch is unknown relative to the calcite lps fabric.

TABLE 6. DEXTRAL TRANSPRESSIVE MARGIN DYNAMICS

	Active		Inactive	
	San Andreas	Alpine	Finlay-Tintina	Idaho-Wyoming
Plate convergence	56 mm/yr	46 mm/yr	100 mm/yr	~33 mm/yr
Duration	29 Ma to 0	30 Ma to 0	30 Ma	90 Ma
Length	~1,100 km	~1,200 km	1,500 km	1,500 km
Estimated displacement	1,500 + 120 km	~480 km	1,500 km	1,650 km
Focal mechanisms	SS/T	SS/T	?	?
Dextral displacement	1,296 km	1,380 km	1,500 km	1,650 km
Associated thrust shortening	+336 km	73 km	200 km	280 km
H:V	4:1	18.9:1	7.5:1	7.5:1
References*	1	2	3	4

*1 = This chapter; Atwater, 1969; 2 = Yeats and Berryman, 1987; 3 = Price, 1981; Chamberlain and St. Lambert, 1985; 4 = This chapter; Craddock, 1988.

The Finlay-Tintina dextral system was likely related to shortening in the Idaho-Wyoming region and farther south, as the dated times of thrust sheet emplacement overlap and a multitude of non–North American terranes exist west of the Idaho-Wyoming belt (Fig. 1).

Unfortunately, no clear plate boundary, like the Finlay-Tintina fault, exists along the western fringe of the Idaho-Wyoming belt. We do know the extent of regional thin- and thick-skinned thrust shortening (280 km; see Craddock, 1988) and the duration of crustal shortening (90 m.y.). For a dextral-to-thrust desplacement ratio of 7.5:1, like the Finlay-Tintina fault, 1,650 km of dextral displacement at a rate of 33 mm/yr along a dextral boundary that is approximately 1,500 km long (Table 6) would produce the Idaho-Wyoming portion of the Sevier orogenic margin.

CONCLUSIONS

The Idaho-Wyoming fold-and-thrust belt is one of the best-dated mountain belts in the world and is representative of a typical orogenic sedimentary wedge. Deformation in this fold-and-thrust belt is characterized by local and regional out-of-sequence thrusting and footwall deformation, including the "autochthonous" basement. Non-layer-parallel country rock and vein twinning strains largely reflect local deformations, whereas the layer-parallel shortening fabric in the country rock reflects a regional, prethrust translation rock fabric.

The layer-parallel shortening strain fabric is considered to be a prethrust shortening fabric that was originally oriented normal to the right-lateral strike-slip plate margin. Subsequent west-to-east thrust translation, in response to the oblique convergence, actively transported this pretranslation fabric from west to east, causing nonplane strain twinning deformations and resulting in the curvilinear strain pattern observed along this traverse. The observation of two piggyback, nonplane strains is of concern for techniques used to restore cross sections (e.g., Dahlstrom, 1969).

ACKNOWLEDGMENTS

This paper presents a portion of my Ph.D. dissertation completed at the University of Michigan under the direction of Ben A. van der Pluijm, which included an encounter with Steve Oriel in Denver in 1983. I would like to thank the following people for their cheerful field support: Patti Craddock, Greg Miller, Drew Powers, Anne C. Campbell, and John A. Dorr, Jr. Financial support was provided in part by Atlantic Richfield Co., Champlin Petroleum Co., and the Scott F. Turner Fund and by a Rackham Faculty Grant (to Ben A. van der Pluijm) from the University of Michigan. I would also like to thank Rick Groshong, Dave Wiltschko, and Sandy Ballard for the use of various computer algorithms. Special thanks to Andy Kopania for his discussions throughout this project. Also, thanks to Marcia Bjornerud and Bob Kranz for critiques of an earlier version of this manuscript and to S. Fast for drafting Figure 2.

REFERENCES CITED

Allmendinger, R., 1981, Structural geometry of the Meade thrust plate in the northern Blackfort Mountains, southeastern Idaho: American Association of Petroleum Geologists Bulletin, v. 65, p. 509–525.

——, R., 1982, Analysis of microstructures in the Meade plate of the Idaho-Wyoming foreland thrust plate, USA: Tectonophysics, v. 85, p. 221–251.

Allmendinger, R. W., and seven others, 1987, Overview of the COCORP 40 N transect, western United States: The fabric of an orogenic belt: Geological Society of America Bulletin, v. 98, p. 308–319.

Armstron, F. C., and Cressman, E. R., 1963, The Bannock thrust zone, southeast Idaho: U.S. Geological Survey Professional Paper 374-J, 22 p.

Armstrong, F. C., and Oriel, S. S., 1965, Tectonic development of the Idaho-Wyoming Thrust belt: American Association of Petroleum Geologists Bulletin, v. 49, p. 1847–1866.

Armstrong, R. L., 1968, Sevier orogenic belt in Nevada and Utah: American Association of Petroleum Geologists Bulletin , v. 79, p. 429–458.

Atwater, T., 1969, Implications of plate tectonics for the Cenozoic tectonics of western North America: Geological Society of America Bulletin, v. 81, p. 3513–3536.

Ballard, S., and Wiltschko, D. V., 1983, Strain directions and structural style at the northeast end of the Powell Valley anticline, Pine Mountain block, southwestern Virginia: Geological Society of America Abstracts with Programs, v. 15, p. 520.

Bird, P., 1984, Laramide crustal thickening in the Rocky Mountain foreland and Great Plains: Tectonics, v. 3, p. 741–758.

——, 1988, Formation of the Rocky Mountains, western United States: A continuum computer model: Science, v. 239, p. 1501–1507.

Blackstone, D. L., Jr., 1979, Geometry of the Prospect-Darby and LaBarge faults at their junction with the LaBarge Platform, Lincoln and Sublette counties, Wyoming: Geological Survey of Wyoming Report 18, 29 p.

——, Jr., 1979, Tectonic map of the Overthrust belt, western Wyoming, southeastern Idaho, and northeastern Utah: Laramie, Geological Survey of Wyoming, tectonic map 1:316,800, economic geology map 1:316,800.

Boyer, S. E., and Elliott, D., 1982, Thrust systems: American Association of Petroleum Geologists Bulletin, v. 66, p. 1196–1230.

Chamberlain, V. E., and St. Lambert, R., 1985, Cordilleria, a newly defined Canadian microcontinent: Nature, v. 314, p. 707–713.

Chapple, W. M., 1978, Mechanics of thin-skinned fold-and-thrust belts: Geological Society of American Bulletin, v. 89, p. 1189–1198.

Craddock, J. P., 1988, Geologic map, cross section and meso-structures across the Idaho-Wyoming fold-and-thrust belt at latitude 42°45′: Geological Society of America Map and Chart Publication Series MCH-067, 1 plate and 18 pp.

Craddock, J. P., and van der Pluijm, B. A., 1988, Kinematic analysis of an en echelon–continuous vein complex: Journal of Structural Geology, v. 10, p. 445–452.

——, 1989, Late Paleozoic deformation of the cratonic carbonate cover of eastern North America: Geology, v. 17, p. 416–419.

Craddock, J. P., and Wiltschko, D. V., 1983, Strains in the Prospect thrust sheet, Overthrust belt, Wyoming: Geological Society of America Abstracts with Programs, v. 15, p. 549.

Craddock, J. P., Kopania, A. A., and Wiltschko, D. V., 1988, *in* Schmidt, C., and Perry, W., eds., Interaction between the northern Idaho-Wyoming thrust belt and bounding basement blocks, central western Wyoming: Geological Society of America Memoir 171, p. 333–352.

Cressman, E. R., 1964, Geology of Georgetown Canyon–Snowdrift Mountain Area, southeastern Idaho: U.S. Geological Survey Bulletin 1153, 105 p.

Cressman, E. R., and Gulbrandsen, R. A., 1955, Geology of the Dry Valley quadrangle, Idaho: U.S. Geological Survey Bulletin 1015-I, p. 257–270.

Dahlen, F. A., Suppe, J., and Davis, D., 1984, Mechanics of fold-and-thrust belts and accretionary wedges: Cohesive Coulomb theory: Journal of Geophysical Research, v. 89, p. 10087–10100.

Dahlstrom, C.D.A., 1969, Balanced cross sections: Canadian Journal of Earth Sciences, v. 6, p. 743–757.

—— , 1970, Structural geology in the eastern margin of the Canadian Rocky Mountains: Bulletin of Canadian Petroleum Geology, v. 18, p. 332–406.

Davis, D., Suppe, J., and Dahlen, F. A., 1983, Mechanics of fold-and-thrust belts and accretionary wedges: Journal of Geophysical Research, v. 88, p. 1153–1172.

Dixon, J. S., 1982, Regional structural synthesis, Wyoming salient of the western overthrust: American Association of Petroleum Geologists Bulletin, v. 66, p. 1560–1580.

Dorr, J. A., Jr., Spearing, D. R., and Steidtmann, J. R., 1977, Deformation and deposition between a foreland uplift and an impinging thrust belt; Hoback Basin, Wyoming: Geological Society of America Special Paper no. 177, 82 p.

Ellis, M. A., 1986, Structural morphology and associated strain in the central Cordillera (British Columbia and Washington): Evidence for oblique tectonics: Geology, v. 14, p. 647–650.

Ellis, M. A., and Watkinson, A. J., 1987, Orogen-parallel extension and oblique tectonics: The relation between stretching lineations and relative plate motions: Geology, v. 15, p. 1022–1025.

Engebretson, D. C., Cox, A., and Gordon, R. G., 1985, Relative motions between oceanic and continental plates in the Pacific basin; Geological Society of America Special Paper 206, 59 pp.

Engelder, T., 1979, The nature of deformation within the outer limits of the central Appalacian foreland fold-and-thrust belt in New York state: Tectonophysics, v. 55, p. 289–310.

Engelder, T., and Engelder, R., 1977, Fossil distortion and decollement in the Appalacian Plateau: Geology, v. 5, p. 457–460.

Evans, J. P., 1983, Structural geology of the northern termination of the Crawford thrust, western Wyoming [M.S. thesis]: College Station, Texas A & M University, 144 p.

Evans, J. P., and Craddock, J. P., 1985, Deformation history and displacement transfer between the Crawford and Meade thrust systems, Idaho-Wyoming Overthrust belt, *in* Orogenic patterns and stratigraphy of northcentral Utah and southeastern Idaho: Utah Geological Association Publication 14, p. 83–96.

Groshong, R. H., 1972, Strain calculated from twinning in calcite: Geological Society of America Bulletin, v. 83, p. 2025–2038.

—— , 1974, Experimental test of least-squares strain calculations using twinned calcite: Geological Society of America Bulletin, v. 85, p. 1855–1864.

Groshong, R. H. Teufel, L. W., and Gasteiger, C., 1984, Precision and accuracy of the calcite strain-gage technique: Geological Society of America Bulletin, v. 95, p. 357–363.

Heller, P. L., and six others, 1986, Time of initial thrusting in the Sevier orogenic belt, Idaho-Wyoming and Utah: Geology, v. 14, p. 388–391.

House, W. M., and Gray, D. R., 1982, Cataclasites along the Saltville thrust, USA, and their implications for thrust sheet emplacement: Tectonophysics, v. 3, p. 257–269.

Howell, D. G., Schermer, E. R., Jones, D. L., Ben-Avraham, Z. and Scheibner, E., 1985, Preliminary tectonostratigraphic terrane map of the Circum-Pacific region: Tulsa, Oklahoma, Circum-Pacific Map Project, American Association of Petroleum Geologists, scale 1:17,000,000, 28 p.

Jackson, M. Craddock, J. P., Ballard, M., Van Der Voo, R., and McCabe, C., 1989, Anhysteretic remanent magnetic anisotropy and calcite strains in Devonian carbonates from the Appalacian Plateau, New York: Tectonophysics, v. 161, p. 43–53.

Jamison, W. R., and Spang, J. H., 1976, Use of calcite twin lamellae to infer differential stress: Geological Society of America Bulletin, v. 87, p. 868–872.

Julian, F. E., and Wiltschko, D. V., 1983, Deformation mechanisms in a terminating thrust anticline, Sequatchie Valley anticline, Tennessee: Geological Society of America Abstracts with Programs, v. 15, p. 606.

Kilsdonk, M. W., and Wiltschko, D. V., 1988, Deformation mechanisms in the southeast ramp region of the Pine Mountain block: Geological Society of America Bulletin, v. 100, p. 653–664.

Kraig, D. H., Wiltschko, D. V., and Spang, J. H., 1988, The interaction of the Moxa Arch (LaBarge Platform) with the Cordilleran thrust belt, south of Snider Basin, Southwestern Wyoming, *in* Schmidt, C. J., and Perry, W. J., eds., Interaction of the Rocky Mountain foreland and the Cordilleran Thrust Belt: Geological Society of America Memoir 171, p. 395–410.

Lamons, R. L., Brun, J. P., and Van den Driessche, J., 1987, Sand models of oblique convergence: Geological Society of America Abstracts with Programs v. 19, p. 738.

Mansfield, G. R., 1927, Geography, geology and mineral resources of part of southeastern Idaho: U.S. Geological Survey Professional Paper 152, 453 p.

Mitra, G., and Yonkee, W. A., 1985, Relationship of spaced cleavage to folds and thrusts in the Idaho-Utah-Wyoming thrust belt: Journal of Structural Geology, v. 7, p. 361–373.

Mitra, G., Hull, J. M., Yonkee, W. A., and Protzman, G. M., 1988, Comparison of mesoscopic and microscopic deformational styles in the Idaho-Wyoming thrust belt and the Rocky Mountain forelands, *in* Schmidt, C. J., and Perry, W. J., eds., Interaction of the Rocky Mountain Foreland and the Cordilleran Thrust Belt: Geological Society of America Memoir 171, p. 119–141.

Monger, J.W.H., Price, R. A., and Templeman-Kluit, D. J., 1982, Tectonic accretion and the origin of the two major metamorphic and plutonic welts in the Canadian Cordillera: Geology, v. 10, p. 70–75.

Mount, V. S., and Suppe, J., 1987, State of stress near the San Andreas fault: Implications for wrench tectonics: Geology, v. 15, p. 1143–1146.

Mulugeta, G., and Koyi, H., 1987, Three-dimensional geometry and kinematics of experimental piggyback thrusting: Geology, v. 15, p. 1052–1056.

Oriel, S. S., and Platt, L. B., 1980, Geologic map of the Preston 1 × 2 degree quadrangle, southeastern Idaho and western Wyoming: U.S. Geological Survey Map I-1127, scale 1:250,000.

Platt, J. P., 1986, Dynamics of orogenic wedges and the uplift of high-pressure metamorphic rocks: Geological Society of America Bulletin, v. 97, p. 1037–1053.

Platt, L. B., and Royse, F., Jr., 1989, The Idaho-Wyoming thrust belt: 28th International Geological Congress, Guidebook T135, 34 pp.

Price, R. A., 1981, The Canadian foreland thrust and fold province in the southern Canadian Rocky Mountains, *in* McClay, K. R., and Price, N. J., eds., Thrust and nappe tectonics: Geological Society of London Special Publication 9, p. 427–448.

—— , 1986, The southeast Canadian Cordillera: Thrust faulting, tectonic wedging and delamination of the lithosphere: Journal of Structural Geology, v. 8, p. 239–254.

Price, R. A., and Carmichael, D. M., 1986, Geometric test for late Cretaceous–Paleogene intracontinental transform faulting in the Canadian Cordillera: Geology, v. 14, p. 468–471.

Protzman, G. M., and Mitra, G., 1990, Strain fabric associated with the Meade thrust sheet: Implications for corss-section balancing: Journal of Structural Geology, v. 12, p. 403–417.

Ramos, V. A., and six others, 1986, Paleozoic terranes of the central Argentine–Chilean Andes: Tectonics, v. 5, p. 855–880.

Royse, F., Jr., Warner, M. A., and Reese, D. L., 1975, Thrust belt structural geometry and related stratigraphic problems WY-ID-Northern UT, *in* Bolyard, D. W., ed., Symposium on deep drilling frontiers in central Rocky Mountains: Denver, Colorado, Rocky Mountain Association of Teologists, p. 41–54.

Rubey, W. W., 1973, Geologic map of the Afton quadrangle and part of the Big Piney quadrangle, Lincoln and Sublette counties, Wyoming: U.S. Geologic Survey, Map I-686, scale 1:62,500.

Rubey, W. W., and Hubbert, M. K., 1959, Role of fluid pressure in the mechanics of overthrust faulting, II: Geological Society of America Bulletin, v. 70, p. 167–206.

Spang, J. H., and Brown, S. P., 1981, Dynamic analysis of a small imbricate thrust, *in* McClay, K. R., and Price, N. J., eds., Thrust and nappe tectonics: Geological Society of London Special Publication 9, p. 143–149.

Spang, J. H., Wolcott, T. L., and Serra, S. S., 1981, Strain in the ramp regions of two minor thrusts, southern Canadian Rocky Mountains, *in* Carter, N. L. and others, eds., Mechanical behavior of crust rocks: The Handin volume: American Geophysical Union Geophysical Monograph 24, p. 243–250.

Stockmal, G. S., 1983, Modeling of large-scale accretionary wedge deformation: Journal of Geophysical Research, v. 10, p. 8271–8288.

Teufel, L. W., 1980, Strain analysis of experimental superposed deformation using calcite twin lamallae: Tectonophysics, v. 65, p. 291–309.

Turner, F. J., 1953, Nature and dynamic interpretation of deformation lamellae in calcite of marbles: American Journal of Science, v. 251, p. 276–298.

Van den Driessche, J., and Brun, J. P., 1987, Thrusting and wrenching in the Canadian Cordillera: An experimental modeling of shear partitioning in oblique collision belts: Geological Society of America Abstracts with Programs, v. 19, p. 875.

Van den Driessche, J., and Maluski, H., 1986, Mise en evidence d'un cisaillement ductile dextre d'age cretace moyen dans la region de Tete Jaune Cache (nord-est du complexe metamorphique Shuswap, Colombie-Britannique): Canadian Journal of Earth Sciences, v. 23, p. 1331–1342.

Wiltschko, D. V., and Dorr, J. A., Jr., 1983, Timing of deformation in Overthrust belt and foreland of Idaho, Wyoming and Utah: American Association of Petroleum Geologists Bulletin, v. 67, p. 1304–1322.

Wiltschko, D. V., Medwedeff, D. A., and Millson, H. E., 1985, Distribution and mechanisms of strain within rocks on the northwest ramp of Pine Mountain blocks, southern Appalacian foreland: A field test of theory: Geological Society of America Bulletin, v. 96, p. 426–435.

Woodward, N. B., Gray, D. R., and Spears, D. B., 1986, Including strain data in balanced cross-sections: Journal of Structural Geology, v. 8, p. 313–324.

Wust, S. L., 1986, Regional correlation of extension direction in Cordilleran metamorphic core complexes: Geology, v. 14, p. 828–830.

Yeats, R. S., and Berryman, K. R., 1987, South Island, New Zealand, and Transverse Ranges, California: A seismotectonic comparison: Tectonics, v. 6, p. 363–376.

Zoback, M. D., and twelve others, 1987, New evidence on the state of stress of the San Andreas fault system: Science, v. 238, p. 1105–1111.

Manuscript Accepted by the Society July 19, 1991

Printed in U.S.A.

Geological Society of America
Memoir 179
1992

Chapter 7

Middle and Late Proterozoic rocks and Late Proterozoic tectonics in the southern Beaverhead Mountains, Idaho and Montana: A preliminary report

Betty Skipp
U.S. Geological Survey, MS 913, Box 25046, Federal Center, Denver, Colorado 80225
Paul Karl Link
Department of Geology, Idaho State University, Pocatello, Idaho 83209

ABSTRACT

Two unconformity-bounded, pre-Ordovician packages of sedimentary rocks in the southern Beaverhead Mountains of southwest Montana and adjacent Idaho are recognized by differences in lithology, texture, bedding characteristics, and sedimentary structures. One package of thin-bedded arkoses and subarkoses is equivalent to rocks of the Middle Proterozoic Lemhi Group of Idaho. The other package contains thick- to thin-bedded, locally feldspathic and glauconitic, sublitharenites of shallow marine depositional facies that are assigned to the Late Proterozoic and Lower Cambrian Wilbert Formation. The Wilbert Formation resembles and correlates with the Late Proterozoic Mutual Formation and the overlying Late Proterozoic and Lower Cambrian Camelback Mountain Quartzite of the Brigham Group in Idaho south of the Snake River Plain. Distribution of rocks of the Lemhi Group and Wilbert Formation in three Mesozoic Cordilleran thrust sheets of the southern Beaverhead Mountains, the Hawley Creek, Fritz Creek, and Cabin–Medicine Lodge sheets, and the presence of only Archean(?) rocks east of the Cabin–Medicine Lodge thrust system suggest that the Wilbert Formation was deposited in a north-trending extensional basin flanked by Middle Proterozoic Lemhi Group arkoses on the northwest and by Archean(?) metamorphic rocks on the east. This basin possibly was a northern arm of a Late Proterozoic and Early Cambrian rift basin that formed at the time of the opening of the paleo-Pacific Ocean (ca. 570 Ma) and widened toward the Cordilleran miogeocline south of the Snake River Plain. Subsequent Middle to Late Cambrian uplift (Skull Canyon disturbance) in the Beaverhead Mountains could have been generated by aesthenospheric upwelling in axial parts of the Wilbert basin.

INTRODUCTION

Outcrops of Cambrian(?) and Proterozoic sandstone or quartzite of the southern Beaverhead Mountains in the southwesternmost corner of Montana and adjacent Idaho are separated from Cambrian and Late Proterozoic rocks to the south in the Pocatello area by more than 160 km (100 mi) of late Cenozoic basalt and rhyolite of the Snake River Plain (Fig. 1). Early workers assigned the sandstones in the southern Beaverhead Mountains to the Cambrian with question (Umpleby, 1913; Kirkham, 1927; Shenon, 1928). Anderson and Wagner (1944) included them with Ordovician quartzites. Later workers emphasized the resemblance of the same rocks to Middle Proterozoic sandstones of the Belt Supergroup (Scholten and others, 1955; Scholten, 1957; Scholten and Ramspott, 1968; Garmezy, 1981; Scholten, 1982; McBean, 1983; Ruppel and Lopez, 1988), though the presence of local Cambrian strata was considered probable (Scholten, 1957; Beutner and Scholten, 1967; Garmezy,

Skipp, B., and Link, P. K., 1992, Middle and Late Proterozoic rocks and Late Proterozoic tectonics in the southern Beaverhead Mountains, Idaho and Montana: A preliminary report, *in* Link, P. K., Kuntz, M. A., and Platt, L. B., eds., Regional Geology of Eastern Idaho and Western Wyoming: Geological Society of America Memoir 179.

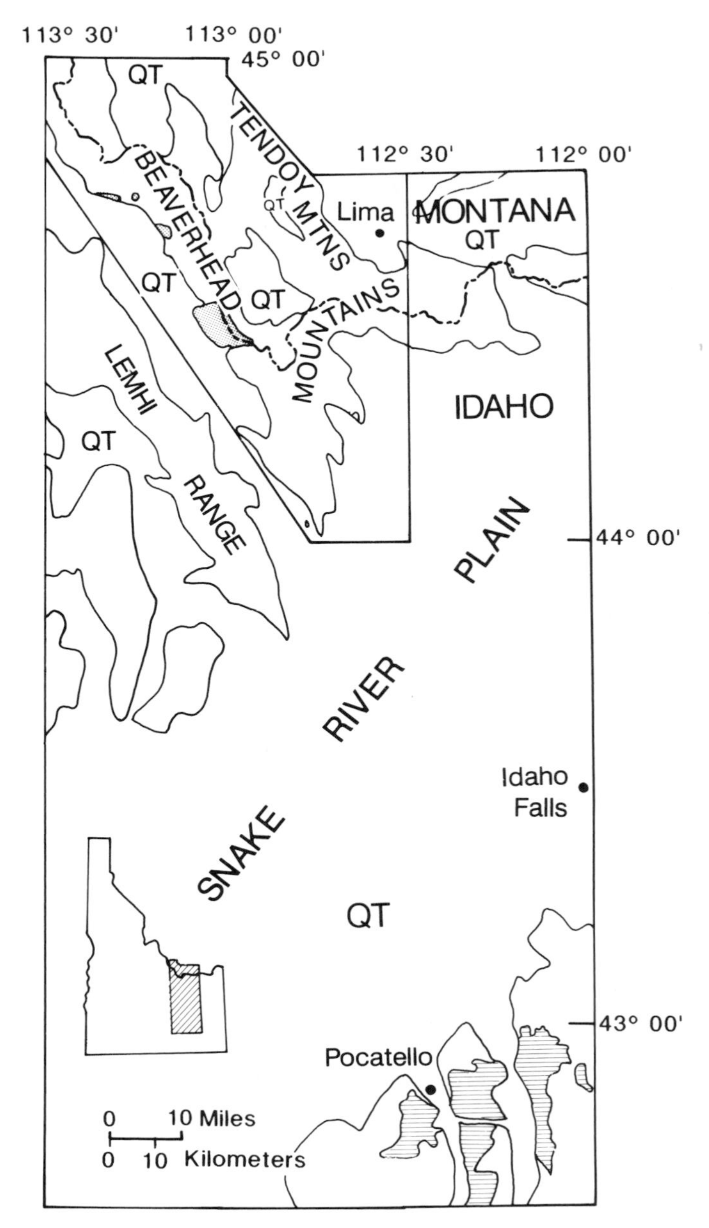

Figure 1. Index map of parts of southern Idaho and southwestern Montana showing outline of area of Figure 3 in northern part and principal geographic features. Areas labeled QT are undifferentiated Quaternary and Tertiary deposits. Unlabeled areas include all rocks older than Tertiary. Patterned areas indicate Cambrian and Late Proterozoic strata present in the Pocatello area (horizontal lines) and Ordovician granite and syenite of the Beaverhead Mountains pluton (shaded). Inset shows area of Figure 3 (diagonal lines).

1981; McBean, 1983). Field descriptions of these rocks often label them "quartzites"; they are quartz-cemented sandstones that have undergone low-grade metamorphism. We use sandstone terminology (Folk, 1974) for petrographic descriptions.

In 1974, the late Steven S. Oriel of the U.S. Geological Survey visited the senior author in the field in the southern Beaverhead Mountains and noted a lithologic similarity between some of the Proterozoic sandstones there and sandstones of the Late Proterozoic Mutual Formation of the Brigham Group in the Pocatello area. A year later, the name Wilbert Formation was proposed for Late(?) Proterozoic rocks in the southern Lemhi Range west of the Beaverhead Mountains (Ruppel and others, 1975), and in the same report, some pre-Ordovician sandstones along the western flank of the southern Beaverhead Mountains were correlated with the Wilbert. Subsequently, two maps were published in which all of the Proterozoic rocks in the southern Beaverhead Mountains were assigned to the Wilbert Formation (Skipp and Hait, 1977; Skipp and others, 1979).

In 1981, Early Cambrian fossils were reported from the upper 35 m (115 ft) of the Wilbert Formation in the Lemhi Range (Derstler and McCandless, 1981). McCandless (1982) revised the Wilbert Formation and established a new type section in the vicinity of the original type section. The revised Wilbert Formation includes about 120 m (394 ft) of light-gray to brownish-gray and pale-red, fine- to coarse-grained quartzitic sandstone, grit, and conglomerate that overlie the Middle Proterozoic Gunsight and Swauger formations with angular unconformity. The Wilbert is overlain gradationally by the Lower Cambrian Tyler Peak Formation (McCandless, 1982; Ruppel and Lopez, 1988). The revised Wilbert Formation of the Lemhi Range in Idaho contains rocks of Early Cambrian age that may grade downward into Late Proterozoic strata (Ruppel and Lopez, 1988) and that correlate with the Camelback Mountain Quartzite of the Brigham Group (Fig. 2). In the southern Beaverhead Mountains, however, where no Cambrian fossils have been recognized and where the formation is thicker, the Wilbert is considered to be of Cambrian(?) and Late Proterozoic age and is believed to correlate with the lower Camelback Mountain Quartzite and/or Mutual Formation of the Brigham Group (Fig. 2).

In 1987, the authors undertook a reexamination of most of the areas of Proterozoic sandstone shown in Figure 3 to see if there were similarities with the Cambrian and Late Proterozoic rocks south of the Snake River Plain. In the course of this fieldwork, we discovered two distinct sedimentary packages: One package resembles the Middle Proterozoic Lemhi Group, the other the Cambrian(?) and Late Proterozoic Wilbert Formation. A total of 15 sites (numbered 1 through 15; Fig. 3) were visited and sampled in the field. Sandstone compositions from eight of these sites are plotted on Figure 4. Strata at an additional six sites (numbered 16 through 21; Fig. 3) that were not studied by the authors also are discussed here; published descriptions of strata at these sites are detailed enough to permit tentative assignment to one or both packages of Proterozoic rocks based on the distinguishing criteria established in this chapter. These observations and correlations augment mapping and correlations by Skipp (1984, 1985, 1988) and Skipp and others (1988) in which the pre-Ordovician sandstones were designated as all Middle(?) Proterozoic, all Late(?) Proterozoic, or as Middle and Late Proterozoic undivided.

Recently, shatter cones and shocked rocks identified in

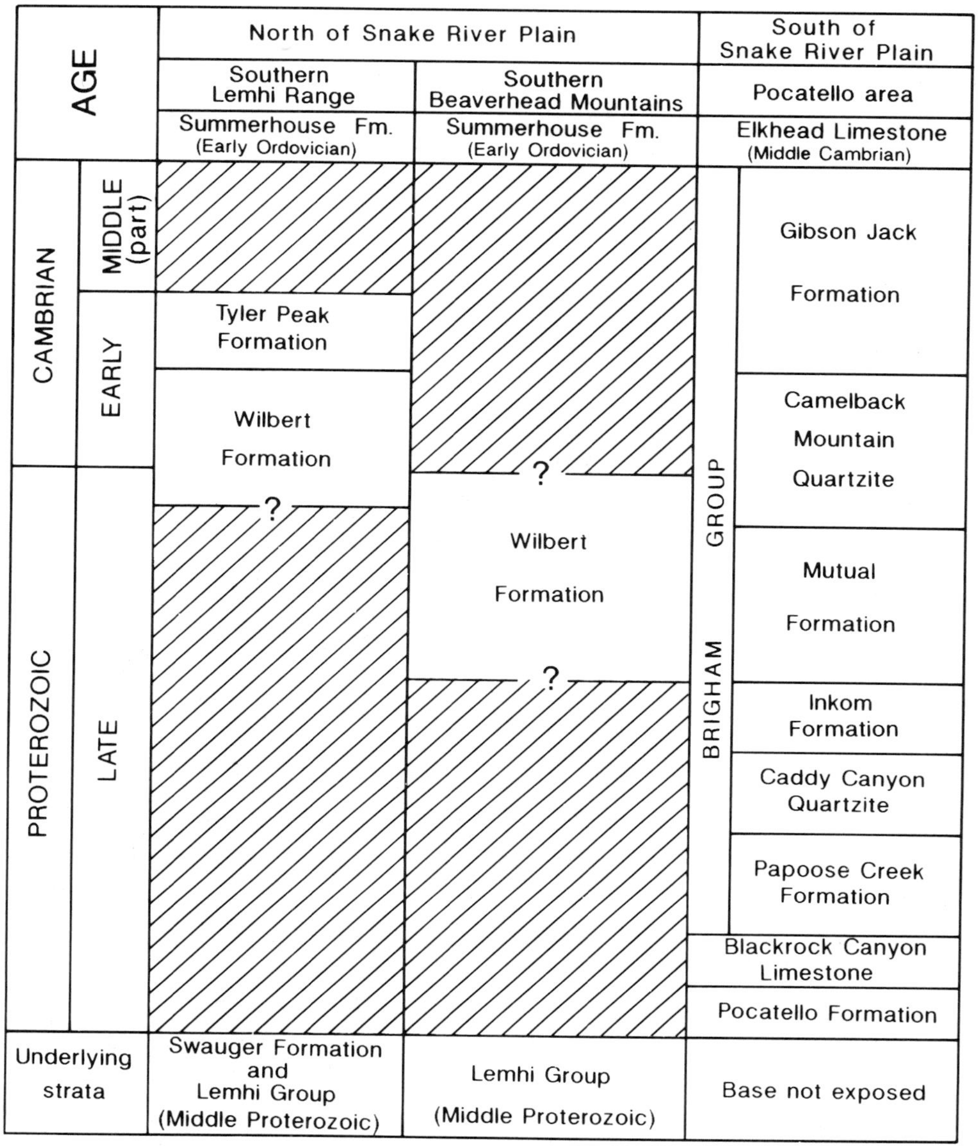

AGE		North of Snake River Plain: Southern Lemhi Range	North of Snake River Plain: Southern Beaverhead Mountains	South of Snake River Plain: Pocatello area	
		Summerhouse Fm. (Early Ordovician)	Summerhouse Fm. (Early Ordovician)	Elkhead Limestone (Middle Cambrian)	
CAMBRIAN	MIDDLE (part)			BRIGHAM GROUP	Gibson Jack Formation
CAMBRIAN	EARLY	Tyler Peak Formation		BRIGHAM GROUP	Camelback Mountain Quartzite
CAMBRIAN	EARLY	Wilbert Formation	?	BRIGHAM GROUP	Camelback Mountain Quartzite
PROTEROZOIC	LATE	?	Wilbert Formation	BRIGHAM GROUP	Mutual Formation
PROTEROZOIC	LATE		?	BRIGHAM GROUP	Inkom Formation
PROTEROZOIC	LATE			BRIGHAM GROUP	Caddy Canyon Quartzite
PROTEROZOIC	LATE			BRIGHAM GROUP	Papoose Creek Formation
PROTEROZOIC	LATE				Blackrock Canyon Limestone
PROTEROZOIC	LATE				Pocatello Formation
Underlying strata		Swauger Formation and Lemhi Group (Middle Proterozoic)	Lemhi Group (Middle Proterozoic)	Base not exposed	

Figure 2. Correlation chart of Late Proterozoic and Cambrian rocks in the southern Lemhi Range (Derstler and McCandless, 1981; McCandless, 1982; Ruppel and Lopez, 1988; Kuntz and others, 1992), southern Beaverhead Mountains (this chapter), and Pocatello area (Link, 1983; Link and others, 1987; Kellogg and others, 1989).

Proterozoic sandstones at Island Butte (locality 20, Fig. 3) and vicinity have been attributed to an ancient impact structure, possibly centered at that location (Hargraves and others, 1990). The identity and age of rocks in the vicinity of the newly recognized Beaverhead impact structure thus become important factors for determining the time of impact.

CHARACTERISTICS AND DISTRIBUTION OF MIDDLE PROTEROZOIC ROCKS

Sandstone and associated mudstone at 12 numbered localities (Fig. 3) have the characteristics listed on the left side of Table 1 and are thought to be Middle Proterozoic because they have lithologic characteristics similar to those given for rocks of the Gunsight Formation of the Lemhi Group in the Lemhi Range to the west (Ruppel, 1975; McBean, 1983; Ruppel and Lopez, 1988). In the Lemhi Range, the Gunsight Formation bears a strong resemblance to the stratigraphically lower Big Creek Formation of the Lemhi Group, with which it can be confused (Ruppel, 1975; Ruppel and Lopez, 1988). The Middle Proterozoic rocks of the Beaverhead Mountains in Idaho and Montana here are assigned to the Gunsight Formation of the Lemhi Group chiefly because that formation has the higher stratigraphic position.

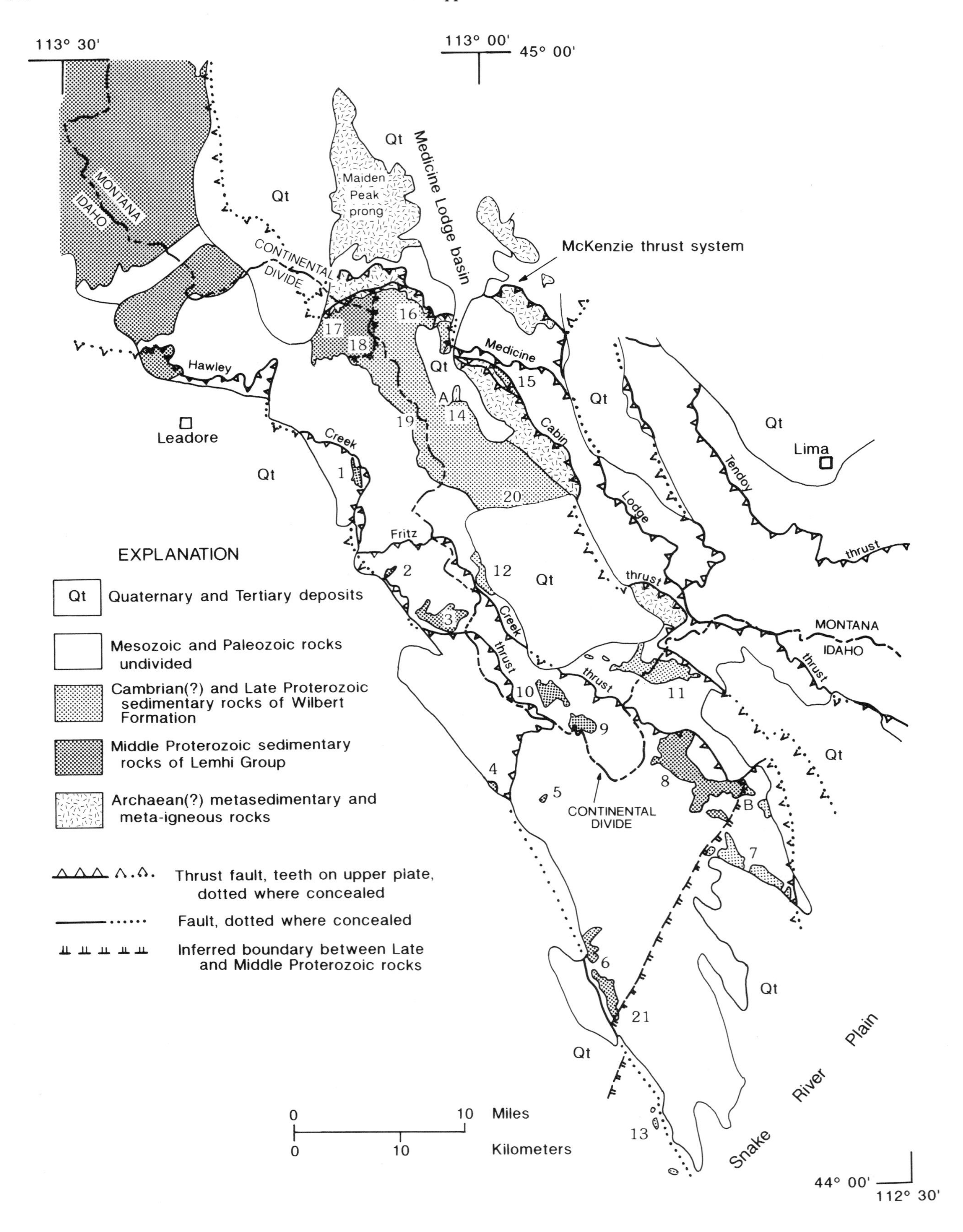
113° 30'
113° 00'
45° 00'
MONTANA
IDAHO
Qt
Maiden Peak prong
Medicine Lodge basin
CONTINENTAL DIVIDE
McKenzie thrust system
Hawley
Leadore
Creek
Medicine
Cabin
Lodge
thrust
Tendoy
Lima
Fritz
Creek
thrust
MONTANA
IDAHO
thrust
CONTINENTAL DIVIDE
A
B
1
2
3
4
5
6
7
8
9
10
11
12
13
14
15
16
17
18
19
20
21
EXPLANATION
Qt Quaternary and Tertiary deposits
Mesozoic and Paleozoic rocks undivided
Cambrian(?) and Late Proterozoic sedimentary rocks of Wilbert Formation
Middle Proterozoic sedimentary rocks of Lemhi Group
Archaean(?) metasedimentary and meta-igneous rocks
Thrust fault, teeth on upper plate, dotted where concealed
Fault, dotted where concealed
Inferred boundary between Late and Middle Proterozoic rocks
0 10 Miles
0 10 Kilometers
Snake River Plain
44° 00'
112° 30'

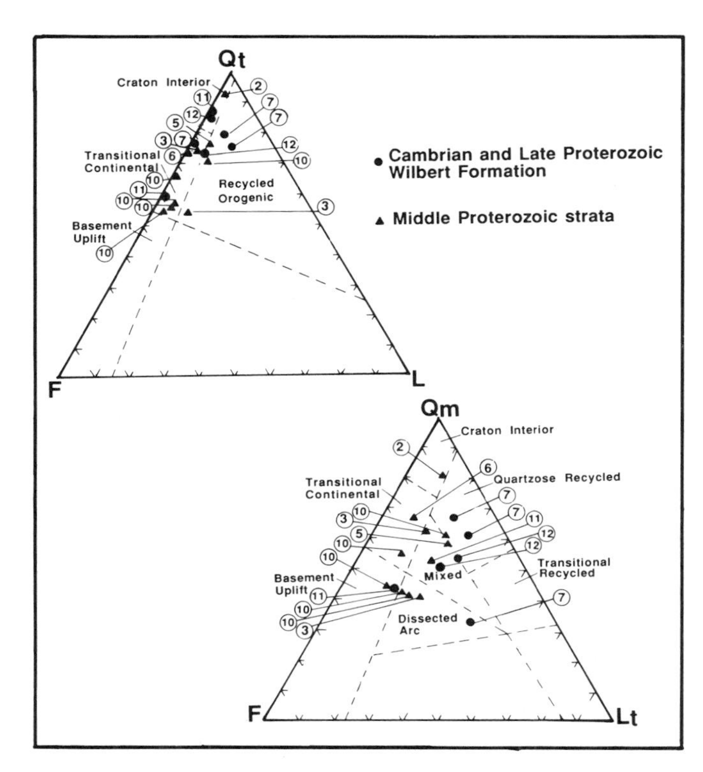

Figure 4. Ternary diagrams to illustrate composition of fine- to medium-grained sandstones of Middle Proterozoic (solid triangles) and Late Proterozoic (solid circles) rocks from the southern Beaverhead Mountains. Numbers are keyed to localities in Figure 3. Middle Proterozoic sandstones are generally more feldspathic and poorer in quartz. Late Proterozoic sandstones tend to be richer in sedimentary rock fragments. Qt is total quartz, including polycrystalline quartz and chert; Qm is monocrystalline quartz; F is feldspar; L is lithic fragments excluding polycrystalline quartz and chert; and Lt is total lithics, including polycrystalline quartz and chert. Provenance fields of Dickinson and others (1983) are shown for reference.

Figure 3. Generalized geologic map of Beaverhead Mountains south of latitude 45°00′, showing numbered localities listed below, recognized distribution of Precambrian units, and location of major thrust faults (Skipp, 1984, 1985; Skipp and others, 1988). Hachured lines are tentative boundaries between Middle Proterozoic and Cambrian(?) and Late Proterozoic rock sequences of the Hawley Creek, Fritz Creek, and Cabin–Medicine Lodge thrust plates. Locality 1 = Hawley Creek; 2 = Clear Creek; 3 = Chamberlain Creek; 4 = Wilmot Gulch and Cedar Gulch; 5 = Italian Canyon; 6 = Skull Canyon and Burnt Canyon; 7 = Myers Creek, Heart Canyon, and Warm Springs Creek; 8 = Webber Creek; 9 = Nicholia Creek; 10 = Tendoy Creek; 11 = Divide Creek; 12 = Tex Creek; 13 = Scott Butte; 14 = Erickson Creek; 15 = Head of Wilson Creek; 16 = Ayers Canyon; 17 = Cruikshank Creek; 18 = SE¼ Sec. 4, T.16N., R.28E., Idaho; 19 = Meadow Creek; 20 = Island Butte; and 21 = Long Canyon. Site A is faulted contact between Proterozoic rocks and Archean(?) basement at Erickson Creek; Site B is location of outcrops shown in Figures 5a and 5b.

On the outcrop, the Middle Proterozoic rocks are dominantly thin-bedded, pale-red, fine-grained sandstones containing local intercalated grayish-green mudstone or siltstone beds. Parallel laminae and small-scale cross-laminations of iron oxide are common, and in places the iron oxide concentrations are 5 cm (2 in) or more in thickness (Fig. 5a, b). Bedding plane concentrations of mudstone rip-up clasts locally are common, and a few isolated rounded mudstone clasts are present. Synsedimentary slump folds characterize parts of the Gunsight(?) Formation preserved beneath the overlying Wilbert Formation at Webber Creek (locality 8, site B, Figs. 3 and 5a) and beneath pebbly Ordovician(?) sandstone at Skull Canyon (locality 6). Cataclastic textures such as axial plane cleavage related to pre-Wilbert folding are present locally in Lemhi Group rocks.

The Gunsight(?) sandstones are chiefly fine- to medium-grained arkoses and subarkoses. They contain 10 to 35% feldspar, both microcline and plagioclase; 35 to 70% monocrystalline quartz; 1 to 35% lithic grains, including coarse-grained polycrystalline quartz and gneiss fragments (Fig. 4); and 0 to 4% detrital mica (McBean, 1983). Sorting generally is moderate, and grains are subangular to angular, indicating that the rocks are texturally mature to submature (Fig. 6a, b).

TABLE 1. COMPARISON OF MIDDLE AND LATE PROTEROZOIC SANDSTONES FROM THE SOUTHERN BEAVERHEAD MOUNTAINS

Middle Proterozoic Sandstones (Lemhi Group, including Gunsight Formation)	Cambrian and Late Proterozoic Wilbert Formation
LOCATION	
Northern Hawley Creek and Fritz Creek plates.	Southernmost Hawley Creek and Fritz Creek plates, Cabin–Medicine plate.
LITHOLOGY	
Arkose and subarkosic arenite, with laminae of iron oxide.	Sublitharenite, locally feldspathic. Lithic fragments include sandstone, chert, and gneiss. Locally glauconitic.
TEXTURE	
Texturally submature to mature. Fine- to coarse-grained.	Texturally immature to submature. Fine- to coarse-grained, with local granule and pebble conglomerate.
BEDDING	
Laminated to thin-bedded (0.5 to 4 cm; 0.2 to 2 in).	Thin to thick-bedded (5 to 50 cm; 2.0 to 20 in).
SEDIMENTARY STRUCTURES	
Ripple marks, cross-lamination, parallel lamination, local slump folds, rip-up clasts. No evidence of large-scale cross-bedding.	Large-scale trough, planar, and wedge cross-bedding. Pebble conglomerate lags.
COLOR	
Pale-red and grayish green.	Pale-red, grayish green, gray, brown.

Figure 5. a, Laminated beds of Lemhi Group (Gunsight(?) Formation) showing slump folds and both load and injection sedimentary structures. Length of hammer head is 16.5 cm (6.5 in). b, Small, Late Proterozoic normal faults in magnetite-rich banded feldspathic sandstone of the Middle Proterozoic Lemhi Group (Gunsight(?) Formation) overlain unconformably by unfaulted, banded, light-colored, poorly sorted, glauconitic lithic sandstone containing scattered, rounded quartz pebbles 2 cm (.78 in) in diameter of the Late Proterozoic and Lower Cambrian(?) Wilbert Formation. Arrow points to later synsedimentary fault in Wilbert Formation that lies above parallel beds in Wilbert Formation that overlap the Late Proterozoic faults. Wilbert Formation sandstones truncate magnetite-rich beds of the Gunsight Formation nearby in the same outcrop. Semiparallel strike of both sequences is approximately N.70°E., dipping 10° to the northwest. Trends of three Late Proterozoic faults that offset the Gunsight(?) Formation are N.25°E to N.65°E. View is to the northeast. Location is Webber Creek area, site B on Figure 3. Length of hammer handle showing in photo is 6.6 cm (2.6 in).

Ruppel and Lopez (1988) and McBean (1983) report the dominant sedimentary structures in the Gunsight Formation of the Lemhi Range to be parallel laminations, cross-laminations marked by concentrations of magnetite, and slump structures. Our stratigraphic reconnaissance in the Beaverhead Mountains identified these same structures in addition to ripple marks and rip-up clasts; we noted no large-scale cross-bedding typical of shoreface deposition. This suggests either a continental (lacustrine) or a sub-wave-base marine environment for the Middle Proterozoic package. A detailed study of the type locality of the Gunsight Formation in the Lemhi Range by McBean (1983) identifies large-scale trough and planar bedding in the upper part of the formation that suggests a nearshore depositional setting. The Middle Proterozoic sandstones in the southern Beaverhead

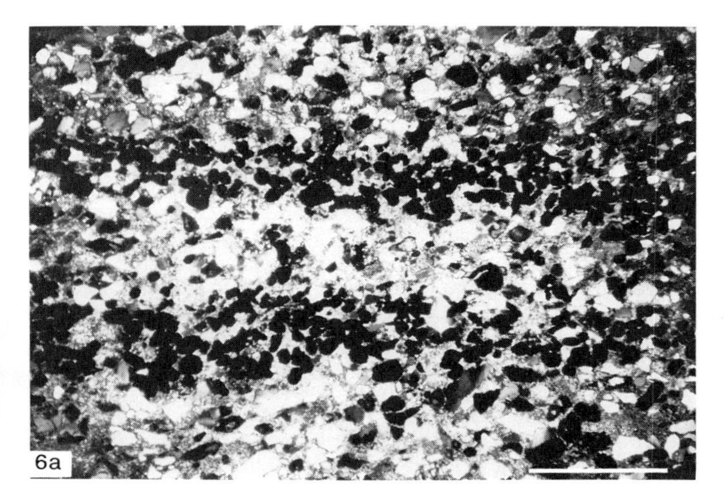

Figure 6. a, Photomicrograph of Middle Proterozoic sandstone from locality 10. Crossed polarizers. Laminae of detrital iron oxide minerals (magnetite and specular hematite) in fine-grained, well-sorted subarkose. Bar = 1 mm (.04 in) b, Photomicrograph of Middle Proterozoic sandstone from locality 3. Crossed polarizers. Well-sorted, medium-grained subarkose containing albite and microcline grains. Bar = 1 mm.

Mountains, therefore, resemble only the lower part of the type Gunsight Formation.

Middle Proterozoic rocks identified on the basis of the above characteristics were found in outcrop along the western flank of the Beaverhead Mountains from locality 6 northward to locality 1 and as far east as locality 8 (Fig. 3). Proterozoic rocks at Skull Canyon (Fig. 3, locality 6) that were assigned to the Middle Proterozoic by Beutner and Scholten (1967), Garmezy (1981), and McBean (1983) are lithologically similar to the Middle Proterozoic package defined here, rather than to the Wilbert Formation as suggested by Ruppel (1975). The description of Proterozoic rocks at Long Canyon south of Skull Canyon (Fig. 3, locality 21) given by Ruppel (1975, p. 21) suggests that Middle Proterozoic rocks may be present there below a thin Wilbert Formation. The lower, highly feldspathic beds at Long Canyon are described as medium- to thick-bedded, conspicuously cross-laminated and laminated and locally contain abundant heavy minerals. Descriptions of strata at two other localities not visited by the authors—south of the Maiden Peak prong and west of the Continental Divide (Fig. 3, localities 17 and 18)—also are suggestive of Middle Proterozoic rocks. McBean (1983, p. 49) described Proterozoic rocks in Cruikshank Creek (Fig. 3, locality 17) as "very fine-grained quartzose arkose. . . . Typical bedforms include parallel laminations, ripple cross-laminations, small-scale trough cross stratification . . . and cut-and-fill structures. . . . Dark iron oxides are common and highlight the laminations. This rock is similar to portions of the Gunsight . . . and also has some strong similarities to the Big Creek Formation." A "fine-grained quartz-mica schist" in the SE 1/4 Sec. 4, T.16N., R.28E., Idaho (Fig. 3, locality 18) described by Lucchitta (1966, p. 19 and 24) tentatively is assigned to the Middle Proterozoic package. Micaceous schistose or cleaved Lemhi Group rocks were found in Italian Canyon (locality 5), and clasts of cleaved Lemhi Group rocks are present in the overlying Wilbert Formation in the southern Lemhi Range (McCandless, 1982). Middle Proterozoic rocks north of Leadore described by Staatz (1973, 1979) and Ruppel (1968) strongly resemble Lemhi Group rocks of the Lemhi Range.

As defined above, Middle Proterozoic rocks are recognized in all three major thrust plates shown on Figure 3 (Skipp, 1984, 1985, 1988; Skipp and others, 1988). The Middle Proterozoic rocks are unconformable below Ordovician quartzites in the hanging wall of the Hawley Creek thrust plate just north of the town of Leadore and at localities 1 and 4 (Fig. 3). Middle Proterozoic rocks make up the hanging wall of the Fritz Creek thrust at localities 2, 3, 5, 6, 9, and 10 (Fig. 3), where they are overlain along an angular unconformity by Ordovician or younger Paleozoic rocks. At locality 8, site B, and possibly locality 21, Middle Proterozoic rocks are unconformable beneath the Wilbert Formation. Middle Proterozoic rocks are postulated to be present in the hanging wall of the Cabin thrust at localities 17 and 18 south of Maiden Peak prong where relationships have not been studied, but where the rocks probably are in contact with Wilbert Formation in the vicinity of the Continental Divide. Middle Proterozoic rocks north of Leadore are overlain by middle Paleozoic rocks. The Middle Proterozoic sequence is a minimum 305 m (1,000 ft) thick. Fault contacts with underlying Archean(?) rocks are north of localities 17 and 18 (M'Gonigle, 1965; Skipp, 1988) and along the northern border of the area shown in Figure 3 at Bloody Dick Creek (Hansen, 1983; Skipp, 1988).

PETROGRAPHY AND STRATIGRAPHY OF CAMBRIAN(?) AND LATE PROTEROZOIC ROCKS OF THE WILBERT FORMATION

Sandstone, conglomerate, and minor siltstone and mudstone at seven localities have the characteristics listed on the right side of Table 1 and are assigned to the Wilbert Formation. Included are localities numbered 7, 8, 11, 12, 13, 14, and 15 (Fig. 3). In addition, published descriptions of strata at localities 16, 19, 20, and 21 suggest that they are Wilbert Formation.

In general, the Wilbert Formation is thicker bedded, texturally more immature, and richer in sedimentary rock fragments and contains less feldspar than Middle Proterozoic rocks. On the outcrop, the Wilbert contains thick- to thin-bedded, pale-red, grayish-green, gray, and brown very fine-grained to coarse-grained sandstones with local interbeds of pebble conglomerate as well as minor siltstone and mudstone. Large-scale trough, planar, and wedge cross-beds are common (Fig. 7). The rocks are sublitharenites that contain 50 to 70% monocrystalline quartz, 3 to 30% feldspar (both microcline and plagioclase), and 6 to 40% lithic grains including chert, fine-grained sandstone, strained mosaics of polycrystalline quartz inferred to be gneiss fragments, and scattered detrital mica flakes and glauconite pellets (Figs. 4, 8b, and 9a). The rocks are texturally immature to submature. Grain size ranges from very fine grained sand to cobble, and grain shape varies from angular to subangular. Isolated rounded quartz pebbles about 2 cm (1 in) in diameter are common. The rocks are chiefly poorly sorted (Figs. 8a, 9a, b), though moderate sorting is present locally. The Wilbert Formation sublitharenties contain sedimentary rock fragments that may have been derived from reworked, uplifted, Middle Proterozoic rocks. Quartz cobbles in sandstone above an unconformable contact with Archean(?) metasedimentary rocks at the head of Wilson Creek suggest a nearby local source for that Weilbert detritus (locality 15, Fig. 3). Cataclastic textures, including crushed sandstone at Erickson Creek (locality 14, Fig. 3; Figs. 9a, b), may be related to the proposed Beaverhead impact structure of Hargraves and others (1990).

At all localities except 13 and 15 (Fig. 3), the Wilbert Formation is overlain unconformably by Devonian or younger Paleozoic rocks. At localities 13 and 15, the Wilbert is overlain by Ordovician quartzite. At locality 8, site B (Fig. 3), the Wilbert Formation is underlain across a folded unconformity by Middle Proterozoic rocks of the Gunsight(?) Formation (Fig. 5b). The photo (Fig. 5b) shows a portion of that unconformtiy. A similar relationship may be present at Long Canyon (locality 21), where Ruppel (1975, p. 21; Ruppel and others, 1975, p. 27) describes beds in the upper part of the Proterozoic section as "medium- to coarse-grained quartzitic sandstone and grit that contain sparse

Figure 7. Large-scale, wedge cross-bedding in Late Proterozoic or Cambrian(?) Wilbert Formation sandstone at locality 7. Cliff is approximately 8 m (26 ft) high. Photograph by P. K. Link.

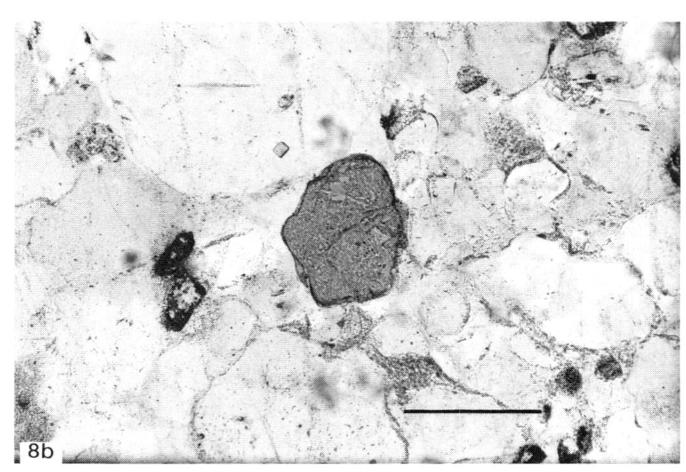

Figure 8. a, Photomicrograph of Wilbert Formation sandstone from locality 12. Crossed polarizers. Very poorly sorted, medium-grained to very coarse-grained sublitharenite containing fine-grained sedimentary rock fragments. Bar = 1 mm (.04 in). b, Photomicrograph of Wilbert Formation sandstone from locality 12. Plane light. Glauconite in medium- to coarse-grained sublitharenite. Bar = 0.25 mm (.01 in).

pebbles, [and] are glauconitic," whereas beds that he describes in the lower part of the section resemble Middle Proterozoic rocks. The section of Proterozoic rocks examined by the authors at Skull Canyon in the NW 1/4 Sec. 22, T.10N., R.30E., Idaho (locality 6, Fig 3), contains no strata beneath the angular unconformity with Ordovician(?) rocks that resemble the Wilbert Formation. However, descriptions of as much as 25 m (82 ft) of medium- to coarse-grained quartzose arkose that is locally conglomeratic (pebbles up to 2 cm [0.8 in] in diameter) in the Middle Fork of Skull Canyon beneath the unconformity (McBean, 1983, p. 56; Beutner and Scholten, 1967) fit our description of Wilbert Formation.

Rocks at three localities south of the Maiden Peak prong are assigned to the Wilbert Formation chiefly on the basis of published descriptions. Ayers Canyon (locality 16, Fig. 3) contains gneiss breccia made up of boulder-size angular blocks as much as 30 cm (12 in) in length in a matrix of arkosic pebble conglomerate (M'Gonigle, 1965; McBean, 1983). The breccia grades laterally into a pebble conglomerate that interfingers with glauconitic quartzose arkose identified in thin sections provided by J. M. M'Gonigle. The breccia is in fault contact with metasedimentary gneiss of the same composition as the breccia blocks. This breccia and the pebble conglomerate are considered to be coarse near-source facies of the Wilbert Formation as is the coarse cobble conglomerate at locality 15. At Meadow Creek (locality 19, Fig. 3), poorly sorted medium-grained to very coarse-grained feldspathic sandstone (Lucchitta, 1966, p. 20; Ruppel and others, 1975, p. 21) is assigned to the Wilbert Formation, and at Island Butte (locality 20, Fig. 3) thick-bedded fine-grained quartzose arkose containing both large-scale planar and trough cross-beds (McBean, 1983) is tentatively included with the Wilbert Formation. Shatter cones are reported from both the Island Butte and Meadow Creek localities (Hargraves and others, 1990).

Cambrian(?) and Late Proterozoic rocks have an estimated total thickness of more than 1,524 m (5,000 ft) in the southern Beaverhead Mountains (Skipp, 1988). This estimate was made by measuring minimum map thicknesses in the vicinity of the Continental Divide west of the southern part of the Medicine Lodge Basin. This estimate is emended to the more approximate "several hundred meters (thousands of feet)" because an impact may have disrupted the outcrop in the area of measurement. These rocks apparently are much thicker than the 122 m (400 ft) of Wilbert Formation in the type section at the southern end of the Lemhi Range (Ruppel and others, 1975; McCandless, 1982; Ruppel and Lopez, 1988).

The large-scale sedimentary structures of the Wilbert Forma-

Figure 9. a, Wilbert Formation sandstone deformed to flaser gneiss, locality 14. Partly crossed polarizers. Glauconite pellet in medium- to coarse-grained poorly sorted, argillaceous sublitharenite containing stained K-feldspar grains. Gneissic texture oriented diagonally from upper left to lower right. Bar = 0.50 mm (.02 in). b, Wilbert Formation sandstone from locality 14. Partly crossed polarizers. Medium- to coarse-grained, poorly sorted, argillaceous sublitharenite with possible crushed quartz grains in upper center. K-feldspar grains stained. Bar = 0.50 mm (0.02 in).

tion resemble similar structures in the upper Brigham Group south of the Snake River Plain (Link and others, 1987); the structures indicate shallow marine and shoreface depositional environments. The similarity in facies suggests a continuous north-trending latest Late Proterozoic depositional basin across what is now the Snake River Plain.

STRUCTURAL SETTING OF CAMBRIAN(?) AND LATE PROTEROZOIC WILBERT FORMATION

The Wilbert Formation is identified in the Cabin–Medicine Lodge, Fritz Creek, and Hawley Creek thrust plates of the southern and central Beaverhead Mountains (Fig. 3). Although outcrop is not continuous, relationships of the Wilbert Formation to underlying Middle Proterozoic strata or Archean metamorphic rocks can be inferred. These relationships are discussed below, starting with the Cabin–Medicine Lodge plate on the east and progressing southwestward.

Within the Cabin–Medicine Lodge thrust sheet, the Wilbert Formation is absent locally in the north and unconformably overlies either Archean basement or Middle Proterozoic strata in the central and southern parts. The northwest extension of the Cabin–Medicine Lodge thrust plate (Fig. 3; Skipp, 1987, 1988) contains a large area of thick Middle Proterozoic rocks but no younger Proterozoic rocks north of Leadore (Staatz, 1973, 1979). A contact that could be a fault, a depositional pinch out of Wilbert Formation against Middle Proterozoic rocks, or possibly a structurally deformed unconformable contact is shown as a hachured line near the Continental Divide south of Maiden Peak prong on Figure 3. The Proterozoic rocks along the Continental Divide east and south of the hachured line and north of locality 14 (Fig. 3) are assigned to the Wilbert Formation chiefly on the basis of lithologic and petrographic descriptions by Lucchitta (1966), M'Gonigle (1965), and McBean (1983). The thickness of Cambrian(?) and Late Proterozoic rocks in this area is estimated to be several hundreds of meters. A fault contact between Archean metamorphic rocks and Wilbert Formation sandstone is exposed at site A (locality 14, Fig. 3) in the southern part of the Medicine Lodge Basin east of the Continental Divide. A similar fault contact lies between gneiss breccia of the Wilbert Formation and Archean(?) gneiss at Ayers Canyon (locality 16, Fig. 3). An unfaulted unconformable contact (Landis, 1963) between metamorphic rocks and a thin Wilbert Formation is preserved northeast of site A on the eastern side of the basin at locality 15 (Fig. 3); the thin Late Proterozoic rocks at locality 15 are interpreted to be part of a thrust imbricate within the Cabin–Medicine Lodge thrust system. The contact between metamorphic basement and Wilbert Formation at all of these localities is inferred to have been originally unconformable.

In the overlying Fritz Creek thrust plate, an inferred northeast-trending contact between Middle Proterozoic rocks on the northwest and Late Proterozoic rocks on the southeast links both packages of Proterozoic sandstone at localities 8 (site B) and 21. The orientation of this line is similar to the northeast strike of extension faults at Webber Creek (locality 8) that displace Gunsight Formation down to the east. At one outcrop at this locality, the Wilbert Formation overlaps and truncates well-preserved, small, paleo–fault scarps in underlying Middle Proterozoic rocks (Fig. 5b). Separation on the northeast-striking, northwest-dipping normal faults shown in Figure 5b is small (5 to 10 cm; 2 to 4 in), and displacement is antithetic to that of the larger faults. A Late

Proterozoic age for the small faults is indicated by their preservation in Lemhi Group rocks that are truncated beneath onlapping Wilbert Formation, the absence of a basal conglomerate, and a few subsequent synsedimentary faults in the overlying Wilbert Formation (Fig. 5b). The larger faults that offset Gunsight Formation are not well exposed, but preliminary study suggests that they also may have a Late Proterozoic ancestry.

In the western Hawley Creek thrust plate, Middle Proterozoic rocks are identified at locality 4, and Wilbert Formation strata are found to the south at Scott Butte (Fig. 3, locality 13). The hachured contact separating Middle Proterozoic from Cambrian(?) and Late Proterozoic outcrops arbitarily is placed just northwest of locality 13.

The Wilbert Formation in the southern and central Beaverhead Mountains is overlain by Early Ordovician or younger Paleozoic rocks. Middle and Upper Cambrian strata are absent, probably as a result of uplift and erosion rather than nondeposition (Scholten, 1957).

EVIDENCE FOR LATE PROTEROZOIC AND CAMBRIAN RIFTING AND EARLY PALEOZOIC UPLIFT

The distribution of Precambrian and Cambrian(?) rocks in the southern Beaverhead Mountains suggests that they were deposited in a northerly trending extensional basin between Middle Proterozoic rocks on the west and Archean(?) cratonic basement on the east (Figs. 10 and 11). Evidence for the location of the Late Proterozoic basin lies in the distribution of three Precambrian stratigraphic elements (Archean metamorphic rocks, Middle Proterozoic sedimentary rocks of the Lemhi Group, and Late Proterozoic and Cambrian(?) rocks of the Wilbert Formation) in the thrust plates of east-central Idaho.

Middle Proterozoic rocks are thick and are overlain unconformably by Ordovician sandstone in the northern part of the Hawley Creek plate and northwestern part of the Fritz Creek thrust plate in the Lemhi Range and western Beaverhead Mountains. Basement is not exposed in these areas, although Archean xenoliths in Quaternary basalts as far as 160 km (100 mi) west of the Beaverhead Mountains (Leeman and others, 1985) indicate that Archean crust underlies the entire region. We suggest that Middle Proterozoic rocks of the Hawley Creek and Fritz Creek plates lay west of the Late Proterozoic Wilbert basin. The 201 m (660 ft) maximum thickness of Wilbert Formation in the southern Lemhi Range (McCandless, 1982) is thought to represent a part of the basin southwest of the main locus of subsidence in the Beaverhead Mountains.

Late Proterozoic sandstone unconformably overlies faulted Middle Proterozoic sandstone in the southern part of the Hawley

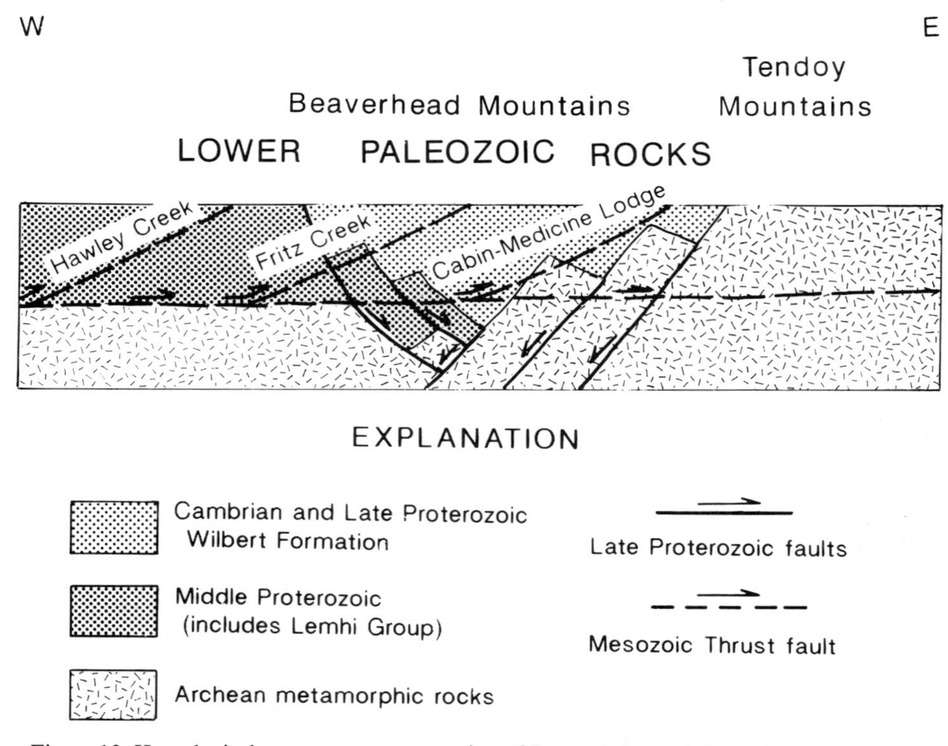

Figure 10. Hypothetical west-to-east cross section of Precambrian and Cambrian rocks in Wilbert rift basin prior to Early Ordovician time in the Beaverhead Mountains and vicinity at a latitude about 8 km (5 mi) south of Lima, Montana, based on newly recognized distribution of these rocks. Cambrian and Precambrian units are indicated by the same patterns as in Figure 3. Traces of future Mesozoic thrusts are indicated.

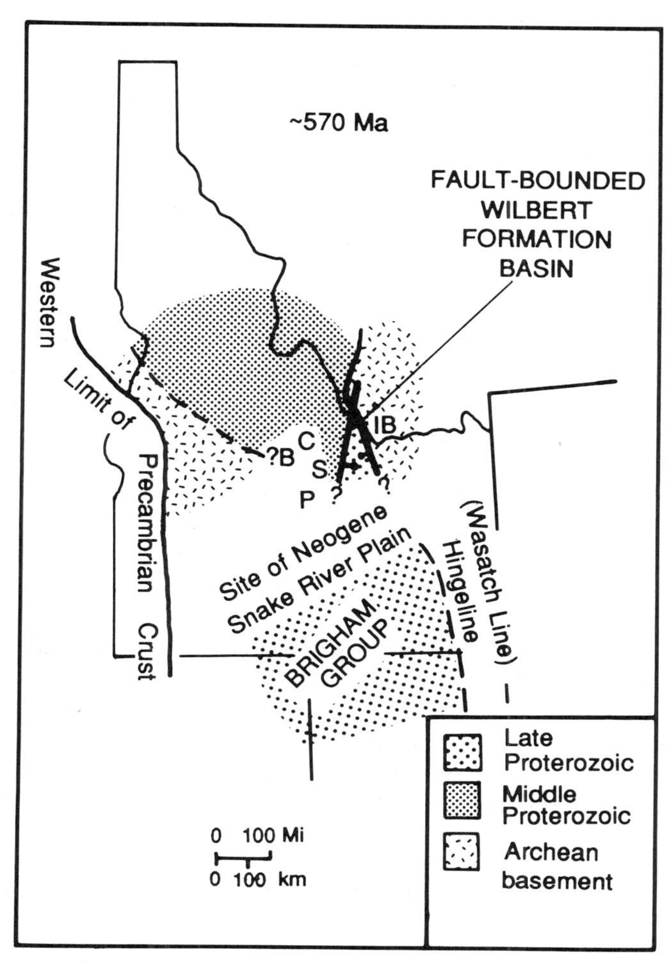

Figure 11. Speculative paleotectonic map showing position of Wilbert rift basin in latest Late Proterozoic time as a northern extension of the Brigham Group passive-margin sequence of southeastern Idaho. The western margin of the basin trends northeast as indicated on Figure 3, and the eastern margin is shown as trending north-northwest. Also shown are (1) the present unrestored location of Island Butte (IB) (locality 20, Fig. 3), the possible center of the Beaverhead impact structure, and (2) the present geographic locations of other possible Late Proterozoic and Cambrian rocks in central Idaho. S marks Sawmill Canyon near Borah Peak, where possible lithologic correlatives of the Wilbert and Tyler Peak formations have been mapped (Wilson and others, 1990). C marks a similar sequence of strata that contain some dolomite northeast of Challis (Hobbs and Hays, 1990). B marks the Bayhorse area where Middle Cambrian fossils have been recovered from a thin siltstone above the Cash Creek Quartzite, presumably also Cambrian in age. The Cash Creek Quartzite is underlain by several hundred meters of carbonate rocks, slaty shale, and pebbly quartzite. The relationship of these rocks to the Wilbert Formation remains obscure, but the succession of rock types suggests possible correlation with the Middle Cambrian cratonic sequence of southwestern Montana rather than with the Brigham Group (Hobbs and Hays, 1990). P marks metasedimentary rocks of the Hyndman Formation that underlie fossiliferous Ordovician rocks in the Pioneer gneiss dome. The Hyndman Formation tentatively has been correlated with either the Bayhorse sequence or Middle Proterozoic strata (Dover, 1981).

Creek plate and the southeastern part of the underlying Fritz Creek plate. Small-scale normal faults in Middle Proterozoic sandstones at Webber Creek (locality 8, Fig. 4b) suggest Late Proterozoic extension of Middle Proterozoic rocks prior to and during deposition of the Wilbert Formation. Marine sandstones of the Wilbert Formation overlap these scarps, which may have been part of a normal fault system along the west margin of the Wilbert basin.

In the Cabin plate, the Late Proterozoic and Lower Cambrian(?) Wilbert Formation is interpreted to lie on Archean basement and to be overlain by Devonian rocks between Divide Creek (locality 11) and localities 14 and 19 north of Island Butte. The Wilbert is estimated to be several hundred meters thick in this part of the Beaverhead Mountains. We suggest that this thickness of Wilbert Formation constitutes a near axial part of the Late Proterozoic basin that lay east of the faulted and eroded western margin of the basin that is now preserved in the Fritz Creek plate. The Wilbert Formation appears to be juxtaposed against Middle Proterozoic rocks west of the Continental Divide and east of the Cruikshank Creek (locality 17). Not far east of Cruikshank Creek, however, are the near-source gneiss breccias and conglomerates at Ayers Canyon (locality 16) (Fig. 3). This area, just south of the Maiden Peak prong, therefore appears to lie near the northern apex or termination of the Wilbert basin (Fig. 11). East and north of the Cabin–Medicine Lodge thrust system, Proterozoic rocks are absent and Ordovician(?) and Devonian rocks overlie transported Archean(?) metamorphic rocks in the McKenzie thrust system of Montana (Fig. 3; Scholten and others, 1955; M'Gonigle, 1965; Dubois, 1983; Perry and others, 1989). The Archean rocks of this thrust system lay east of the Wilbert Basin.

The diagrammatic cross section of Figure 10 illustrates an extensional basin (Wilbert basin) containing a thick section of Late Proterozoic and Cambrian(?) rocks that is bounded by Middle Proterozoic rocks on the west and Archean(?) metamorphic rocks on the east. The diagram incorporates all of the relationships observed among the three major Precambrian rock units in the southern Beaverhead Mountains. Positions of the incipient major Mesozoic thrusts along which the basin was telescoped are indicated. The basin formed west of its present geographic position, although Neogene extension has compensated for some of the Late Mesozoic eastward translation.

If transport on the Cabin–Medicine Lodge thrust system was about 40 km (25 mi) (Skipp, 1988), then the Wilbert basin probably did not exceed 48 km (30 mi) in width at present latitude 44°30′. The length of the preserved part of the Wilbert basin north of the Snake River Plain is about 80 km (50 mi). The axial part of the basin was filled with an estimated several hundred meters of detrital sediments.

The Wilbert basin probably was a northern arm of the large rift system in which Cambrian and Late Proterozoic sedimentary and volcanic rocks of the Brigham Group and Pocatello Formation accumulated south of the Snake River Plain (Fig. 11; Harper and Link, 1986; Link, 1986; Link and others, 1987; Christie-

Blick and Levy, 1989). The distinctive volcanic and diamictite-bearing strata of the Pocatello Formation (Link, 1983) at the base of the Late Proterozoic section near Pocatello (Fig. 2) are absent in the southern Beaverhead Mountains. The Wilbert Formation strongly resembles only the middle and upper parts of the Brigham Group, suggesting that onset of deposition in the southern Beaverhead Mountains was delayed until latest Proterozoic time (Fig. 11).

Although none of the Late Proterozoic rocks near Pocatello have been dated directly, regional correlation (Crittenden and others, 1983) suggests that the Pocatello Formation is at least 750 Ma. Two Late Proterozoic rifting events may have influenced eastern Idaho (Lindsey and Link, 1988; Stewart, 1991; Levy and Christie-Blick, 1991a). The first (ca. 750 Ma) resulted in deposition of the Pocatello Formation and did not affect the Beaverhead Mountains. The second (ca. 570 Ma) resulted in opening of the paleo–Pacific Ocean and deposition of sediments in the Cordilleran miogeocline (Fig. 11), producing the extensional basin of the Wilbert Formation. At the latitude of the Beaverhead Mountains, however, rifting did not result in continental separation and formation of oceanic crust, as Middle Proterozoic sandstones and evidence for an Archean crust are present far to the west.

PROTEROZOIC AND CAMBRIAN UPLIFT

Late Proterozoic and Cambrian uplift of the southern Beaverhead Mountains region has been proposed by Armstrong (1975) for the Salmon River arch, by Ruppel (1986) for the Lemhi arch, and by Scholten (1957) for the Skull Canyon disturbance. The newly recognized thick Cambrian(?) and Late Proterozoic sedimentary rocks in the Beaverhead Mountains bear on each of these concepts.

The Salmon River arch was conceived as a westward extension of the craton consisting of pre-Beltian, 1500-Ma metasedimentary rocks upraised to form an elongate arch through central and east-central Idaho (Armstrong, 1975). The arch was a convenient, hypothetical southwestern source for sediments of the Belt basin. New radiometric ages of about 1370 Ma for Middle Proterozoic granite and augen gneiss that intrude these metasedimentary rocks in east-central Idaho, however, indicate the rocks are not necessarily of pre-Beltian age (Evans and Zartman, 1990) and may be metamorphosed Belt Supergroup equivalent rocks, as suggested by Ruppel (1986). In addition, recognition of discrete thrust plates in the southern Beaverhead Mountains suggests that the area of the Salmon River arch consists of the roots of Cretaceous thrust sheets rather than a western extension of the craton (Skipp, 1987, 1988). If the transported metasedimentary rocks are equivalent to Belt Supergroup rocks, east-central Idaho was a depocenter, not a physiographic high, during Belt basin deposition. The absence of Middle Proterozoic rocks in much of the Cabin–Medicine Lodge thrust plate in the central Beaverhead Mountains suggests, however, that the central Beaverhead Mountains may have been high during Middle Proterozoic time, possibly providing a southern source for some Belt Supergroup sediments. The extremely fine grained nature of Belt-equivalent rocks of the Lemhi Group in the southern Beaverhead Mountains is not compatible with a near-source origin, but it is possible that near-source facies of the Lemhi Group were uplifted and eroded to provide some of the coarse detritus found in the younger Late Proterozoic and Lower Cambrian Wilbert Formation.

Ruppel (1986) proposed the Lemhi arch to be a large northwest-trending positive structure that was active during Late Proterozoic through early to middle Paleozoic time near the craton margin. The Beaverhead Mountains were thought to contain the decapitated crest of the Lemhi arch displaced eastward by Mesozoic thrust fault from the area of the Lemhi Range. Our study suggests that the tectonics of the craton margin were more complex than can be explained by a single recurring arch. The thick Cambrian(?) and Late Proterozoic strata in the central and southern Beaverhead Mountains indicate that the area was an Early Cambrian and Late Proterozoic depocenter surrounded by upraised Middle Proterozoic rocks on the west and Archean(?) metaigneous and metasedimentary rocks on the east.

The Beaverhead Mountains, however, were the site of Middle to Late Cambrian uplift of the Skull Canyon disturbance (Scholten, 1957). This disturbance is marked by missing strata along an angular unconformity between Proterozoic sandstone and Ordovician quartzite at Burnt Canyon, about 2.4 km (1.5 mi) north of Skull Canyon (locality 6, Fig. 3) in the southern Beaverhead Mountains (Scholten, 1957, Plate 1). The Proterozoic sandstones at this locality probably are Middle Proterozoic Lemhi Group (McCandless, 1982, p. 96). The overlying Ordovician strata have been assigned to the Middle Ordovician Kinnikinic Quartzite (Anderson and Wagner, 1944; Scholten and Ramspott, 1968; Skipp, 1988), but U-Th-Pb-zircon method ages of about 483 Ma obtained from granites and syenites of the Beaverhead Mountains pluton that intrude the quartzites (Evans and Zartman, 1988) suggest the quartzites are older, possibly Early Ordovician in age. The strong angular unconformity at locality 6, therefore, probably is between Middle Proterozoic Lemhi Group and Lower Ordovician quartzite. In the Lemhi Range to the southwest, an unconformity between fossiliferous Lower Cambrian strata of the Tyler Peak Formation that gradationally overlies the Wilbert Formation (Fig. 2) and quartzite of the Lower Ordovician Summerhouse Formation (McCandless, 1982) records Middle to Late Cambrian uplift that is roughly coincident with uplift in the Beaverhead Mountains and with the Skull Canyon disturbance of Scholten (1957). This uplift and the erosion that followed may have resulted in the removal of the upper part of the Wilbert Formation in the southern Beaverhead Mountains. The fact that Middle to Late Cambrian uplift succeeded Early Cambrian deposition closely in time and space suggests the possibility that structural and topographic inversion of the Wilbert basin may have been caused by aesthenospheric upwelling in axial portions of the basin following extension and basin filling (Vierbuchen and others, 1982).

CONCLUSIONS

Recognition of two different sandstone packages, one arkosic and one sedimentary lithic, in pre-Ordovician rocks of the southern Beaverhead Mountains and preliminary determinations of their distribution in major thrust plates of the region suggest that Lower Cambrian(?) and Late Proterozoic marine sedimentary rocks of the Wilbert Formation were deposited in an extensional basin in the southern Beaverhead Mountains. The basin was bounded on the west and north by Middle Proterozoic sedimentary rocks of the Lemhi Group and on the east by Archean metamorphic rocks of the Montana craton. All Precambrian rocks have been telescoped along Cretaceous thrust faults. Late Proterozoic and Lower Cambrian rocks of the Wilbert Formation resemble rocks of the middle and upper parts of the Brigham Group south of the Snake River Plain, suggesting that the extensional basin of the Beaverhead Mountains was a northern arm of the Cordilleran miogeocline. Correlation of the Wilbert Formation with the Brigham Group indicates rifting occurred about 600 to 570 Ma, coincident with the opening of the paleo–Pacific Ocean (Levy and Christie-Blick, 1991a; 1991b). Precambrian rocks of the area are overlain unconformably by thin lower Paleozoic sedimentary sequences that record sporadic early Paleozoic uplift in the southern Beaverhead Mountains and Lemhi Range. The Middle to Late Cambrian Skull Canyon disturbance closely followed basin filling, and this uplift may reflect aesthenospheric upwelling in axial portions of the Late Proterozoic–Early Cambrian basin.

Further detailed work in these Proterozoic sandstones and possibly related strata in other parts of central Idaho (Fig. 11) is necessary before basin configuration and stratigraphic succession can be tightly constrained. Petrographic characteristics of the two packages developed here will help in these studies.

Identification of rocks containing shatter cones and shocked rocks of the Beaverhead impact structure as Lower(?) Cambrian to Late Proterozoic Wilbert Formation eliminates Late Proterozoic impact, one of the ages originally proposed for the structure (Hargraves and others, 1990). An early Paleozoic age for the impact, post–Middle Cambrian to pre–Mississippian, remains more likely.

ACKNOWLEDGMENTS

We thank Karl V. Evans, James Schmitt, and Edward T. Ruppel for their helpful reviews. Link's research on Proterozoic strata has been supported by the Idaho State University Faculty Research Committee, the Petroleum Research Fund of the American Chemical Society, and the Kackley Endowment. Sandstone petrography was performed by Matt Barrett (Senior thesis, Idaho State University, 1988) and Travis McCling (Idaho State University).

REFERENCES CITED

Anderson, A. L., and Wagner, R. W., 1944, Lead-zinc-copper deposits of the Birch Creek district, Clark and Lemhi Counties, Idaho: Idaho Bureau of Mines and Geology Pamphlet 70, 43 p.

Armstrong, R. L., 1975, Precambrian (1,500 m.y. old) rocks of central Idaho—The Salmon River Arch and its role in Cordilleran sedimentation and tectonics: American Journal of Science, v. 275-A, p. 437–467.

Beutner, E. C., and Scholten, R., 1967, Probable Cambrian strata in east-central Idaho and their paleotectonic significance: American Association of Petroleum Geologists Bulletin, v. 51, p. 2305–2311.

Christie-Blick, N., and Levy, M., eds., 1989, Late Proterozoic and Cambrian tectonics, sedimentation, and record of metazoan radiation in the Western United States (28th International Geological Congress, Field Trip T331): Washington, D.C., American Geophysical Union, 111 p.

Crittenden, M. D., Jr., Christie-Blick, N., and Link, P. K., 1983, Evidence for two pulses of glaciation during the Late Proterozoic in northern Utah and southeastern Idaho: Geological Society of American Bulletin, v. 94, p. 437–450.

Derstler, K., and McCandless, D. O., 1981, Cambrian trilobites and trace fossils from the southern Lemhi Range, Idaho; their stratigraphic and paleotectonic significance: Geological Society of America Abstracts with Programs, v. 13, p. 194.

Dickinson, W. R., and eight others, 1983, Provenance of North American Phanerozoic sandstones in relation to tectonic setting: Geological Society of America Bulletin, v. 94, p. 222–235.

Dover, J. H., 1981, Geology of the Boulder-Pioneer Wilderness study area, Blaine and Custer counties, Idaho: U.S. Geological Survey Bulletin 1497-A, p. 21–75.

Dubois, D. P., 1983, Tectonic framework of basement thrust terrane, northern Tendoy Range, southwestern Montana, *in* Powers, R. B., ed., Geologic studies of the Cordilleran thrust belt: Denver, Rocky Mountain Association of Geologists, v. 1, p. 145–158.

Evans, K. V., and Zartman, R. E., 1988, Early Paleozoic alkalic plutonism in east-central Idaho: Geological Society of America Bulletin, v. 100, p. 1981–1987.

——, 1990, U-Th-Pb and Rb-Sr geochronology of Middle Proterozoic granite and augen gneiss, Salmon River Mountains, east-central Idaho: Geological Society of America Bulletin, v. 102, p. 63–73.

Folk, R. L., 1974, Petrology of sedimentary rocks: Austin, Texas, Hemphill Publishing Co., 182 p.

Garmezy, L., 1981, Geology and tectonic evolution of the southern Beaverhead Range, east-central Idaho [M.S. thesis]: University Park, The Pennsylvania State University, 155 p.

Hansen, P. M., 1983, Structure and stratigraphy of the Lemhi Pass area, Beaverhead Range, southwest Montana and east-central Idaho [M.S. thesis]: University Park, The Pennsylvania State University, 112 p.

Hargraves, R. B., Cullicott, C. E., Deffeyes, K. S., Hougen, S., Christiansen, P. P., and Fiske, P. S., 1990, Shatter cones and shocked rocks in southwestern Montana: The Beaverhead impact structure: Geology, v. 18, p. 832–834.

Harper, G. D., and Link, P. K., 1986, Geochemistry of Upper Proterozoic rift-related volcanics, northern Utah and southeastern Idaho: Geology, v. 14, p. 864–867.

Hobbs, S. W., and Hays, W. H., 1990, Ordovician and older rocks of the Bayhorse area, Custer County, Idaho: U.S. Geological Survey Bulletin 1891, 40 p.

Kellogg, K. S., Oriel, S. S., Amerman, R. E., Link, P. K., and Hladky, F. R., 1989, Geologic map of the Jeff Cabin Creek quadrangle, Bannock and Bingham Counties, Idaho: U.S. Geological Survey Map GQ-1669, scale 1:24,000.

Kirkham, V.R.D., 1927, A geologic reconnaissance of Clark and Jefferson, and parts of Butte, Custer, Fremont, Lemhi, and Madison Counties, Idaho: Idaho Bureau of Mines and Geology Pamphlet 19, 47 p.

Kuntz, M. A., and twelve others, 1992, Geologic map of the Idaho National Engineering Laboratory and adjoining areas, eastern Idaho: U.S. Geological Survey Miscellaneous Investigations Series Map I-2330, scale 1:100,000 (in press).

Landis, C. A., Jr., 1963, Geology of the Graphite Mountain–Teepee Mountain area, Montana-Idaho [M.S. thesis]: University Park, The Pennsylvania State University, 153 p.

Leeman, W. P., Menzies, M. A., Matty, D. J., and Embree, G. F., 1985, Strontium, neodymium, and lead isotope compositions of deep crustal xenoliths from the Snake River Plain—evidence for Archean basement: Earth and Planetary Science Letters, v. 75, p. 354–368.

Levy, M., and Christie-Blick, N., 1991a, Late Proterozoic paleogeography of the eastern Great Basin, *in* Cooper, J. D., and Stevens, C. H., eds., Paleozoic paleogeography of the western United States—II: Los Angeles, Society of Economic Paleontologists and Mineralogists, Pacific Section, v. 1, p. 371–386.

—— , 1991b, Tectonic subsidence of the early Paleozoic passive continental margin in eastern California and southern Nevada: Geological Society of America Bulletin, v. 103, p. 1590–1606.

Lindsey, K. A., and Link, P. K., 1988, Stratigraphic evidence for two episodes of continental rifting in the basal Cordilleran miogeocline from the northern Great Basin to northeastern Washington: Geological Society of America Abstracts with Programs, v. 20, p. 176.

Link, P. K., 1983, Glacial and tectonically influenced sedimentation in the upper Proterozoic Pocatello Formation, southeastern Idaho, *in* Miller, D. M., Todd, V. R., and Howard, K. A., eds., Tectonic and stratigraphic studies in the eastern Great Basin: Geological Society of America Memoir 157, p. 165–182.

—— , 1986, Tectonic model for the deposition of the Late Proterozoic Pocatello Formation, southeastern Idaho: Northwest Geology, v. 15, p. 1–7.

Link, P. K., Jansen, S. T., Halimdihardja, P., Lande, A., and Zahn, P., 1987, Stratigraphy of the Brigham Group (Late Proterozoic–Cambrian), Bannock, Portneuf, and Bear River Ranges, Southeastern Idaho, *in* Miller, W. R., ed., The thrust belt revisited: Wyoming Geological Association, 38th Annual Field Conference, Guidebook, p. 133–148.

Lucchitta, B. K., 1966, Structure of the Hawley Creek area, Idaho-Montana [Ph.D. thesis]: University Park, The Pennsylvania State University, 203 p.

McBean, A. J. II, 1983, The Proterozoic Gunsight Formation, Idaho-Montana; stratigraphy, sedimentology, and paleotectonic setting [M.S. thesis]: University Park, The Pennsylvania State University, 235 p.

McCandless, D. O., 1982, A reevaluation of Cambrian through Middle Ordovician stratigraphy of the southern Lemhi Range [M.S. thesis]: University Park, The Pennsylvania State University, 157 p.

M'Gonigle, J. W., 1965, Structure of the Maiden Peak Area, Montana-Idaho [Ph.D. thesis]: University Park, The Pennsylvania State University, 146 p.

Perry, W. J., Jr., Dyman, T. S., and Sando, W. J., 1989, Southwestern Montana recess of Cordilleran thrust belt, *in* French, D. E., and Grabb, R. F., Geologic resources of Montana: Montana Geological Society, 1989 Field Conference, Guidebook, v. 1, p. 261–270.

Ruppel, E. T., 1968, Geologic map of the Leadore Quadrangle, Lemhi County, Idaho: U.S. Geological Survey Geologic Quadrangle Map GQ-733, scale 1:62,500.

—— , 1975, Precambrian Y sedimentary rocks in east-central Idaho: U.S. Geological Survey Professional Paper 889-A, p. 1–23.

—— , 1986, The Lemhi Arch: A Late Proterozoic and Early Paleozoic landmass in Central Idaho, *in* Petersen, J. E., ed., Paleotectonics and sedimentation in the Rocky Mountain region, United States: American Association of Petroleum Geologists Memoir 41, p. 119–130.

Ruppel, E. T., and Lopez, D. A., 1988, Regional geology and mineral deposits in and near the central part of the Lemhi Range, Lemhi County, Idaho: U.S. Geological Survey Professional Paper 1480, 122 p.

Ruppel, E. T., Ross, R. J., Jr., and Schleicher, D., 1975, Precambrian Z and Lower Ordovician rocks in east central Idaho: U.S. Geological Survey Professional Paper 889-B, p. 25–34.

Scholten, R., 1957, Paleozoic evolution of the geosynclinal margin north of the Snake River Plain, Idaho-Montana: Geological Society of America Bulletin, v. 68, p. 151–170.

—— , 1982, Continental subduction in the northern Rockies—a model for back-arc thrusting in the western Cordillera, *in* Powers, R. B., ed., Geologic studies of the Cordilleran thrust belt: Denver, Rocky Mountain Association of Geologists, v. 1, p. 123–136.

Scholten, R., and Ramspott, L. D., 1968, Tectonic mechanisms indicated by structural framework of central Beaverhead Range, Idaho-Montana: Geological Society of America Special Paper 104, 71 p., map scale 1:62,500.

Scholten, R., Keenmon, K. A., and Kupsch, W. O., 1955, Geology of the Lima region, southwestern Montana and adjacent Idaho: Geological Society of America Bulletin, v. 66, p. 345–404.

Shenon, P. J., 1928, Geology and ore deposits of the Birch Creek district, Idaho: Idaho Bureau of Mines and Geology Pamphlet 27, 25 p.

Skipp, B., 1984, Geologic map and cross sections of the Italian Peak and Italian Peak Middle roadless areas, Beaverhead County, Montana, and Clark and Lemhi Counties, Idaho: U.S. Geological Survey Miscellaneous Field Studies Map MF-1601B, scale 1:62,500.

—— , 1985, Contraction and extension faults in the southern Beaverhead Mountains, Idaho and Montana: U.S. Geological Survey Open-File Report 85-545, 170 p.

—— , 1987, Basement thrust sheets in the Clearwater orogenic zone, central Idaho and western Montana: Geology, v. 15, p. 220–224.

—— , 1988, Cordilleran thrust belt and faulted foreland in the Beaverhead Mountains, Idaho and Montana, *in* Schmidt, C. J., and Perry, W. J., Jr., eds., Interaction of the Rocky Mountain foreland and Cordilleran thrust belt: Geological Society of America Memoir 171, p. 237–266, map scale 1:250,000.

Skipp, B., and Hait, M. H., Jr., 1977, Allochthons along the northeast margin of the Snake River Plain, Idaho: Wyoming Geological Association, 29th Annual Field Conference, Guidebook, p. 499–515, map scale 1:500,000.

Skipp, B., Prostka, H. J., and Schleicher, D. L., 1979, Preliminary geologic map of the Edie Ranch quadrangle, Clark County, Idaho, and Beaverhead County, Montana: U.S. Geological Survey Open-File Report 79-845, scale 1:62,500.

Skipp, B., Hassemer, J. R., Kulik, D. M., Sawatzky, D. L., Leszcykowski, A. M., and Winters, R. A., 1988, Mineral resources of the Eighteenmile Wilderness Study Area, Lemhi County, Idaho: U.S. Geological Survey Bulletin 1718-B, p. B1–B21.

Staatz, M. H., 1973, Geologic map of the Goat Mountain quadrangle, Lemhi County, Idaho, and Beaverhead County, Montana: U.S. Geological Survey Geological Quadrangle Map GQ-1097, scale 1:24,000.

—— , 1979, Geology and mineral resources of the Lemhi Pass Thorium District, Idaho and Montana: U.S. Geological Survey Professional Paper 1049-A, p. A1–A90, scale 1:31,680.

Stewart, J. H., 1991, Latest Proterozoic and Cambrian rocks of the western United States—An overview, *in* Cooper, J. D., and Stevens, C. H., eds., Paleozoic paleogeography of the western United States—II: Los Angeles, Society of Economic Paleontologists and Mineralogists, Pacific Section, v. 1, p. 13–38.

Umpleby, J. B., 1913, Geology and ore deposits of Lemhi County, Idaho: U.S. Geological Survey Bulletin 528, 182 p.

Vierbuchen, R. C., George, R. P., and Vail, P. R., 1982, A thermal-mechanical model of rifting with implications for outer highs on passive continental margins, *in* Watkins, J. S., and Drake, C. L., eds., Studies in continental margin geology: American Association of Petroleum Geologists Memoir 34, p. 765–778.

Wilson, A. B., and six others, 1990, Mineral resources of the Borah Peak wilderness study area, Custer County, Idaho: U.S. Geological Survey Bulletin 1718-E, p. E1–E15, map scale: 1:24,000.

MANUSCRIPT ACCEPTED BY THE SOCIETY JULY 19, 1991

Printed in U.S.A.

Geological Society of America
Memoir 179
1992

Chapter 8

Segmentation and paleoseismicity of the Grand Valley fault, southeastern Idaho and western Wyoming

Lucille A. Piety and J. Timothy Sullivan*
U.S. Bureau of Reclamation, Seismotectonics and Geophysics Section, D-3611, Federal Center, Denver, Colorado 80225
Mark H. Anders
Department of Geological Sciences and Lamont-Doherty Geological Observatory, Columbia University, Palisades, New York 10964

ABSTRACT

Three segments of the Grand Valley fault (from south to north the Star Valley, Grand Valley, and Swan Valley segments) in southeastern Idaho and western Wyoming were identified by combining evidence from (1) the location and extent of fault scarps on late Quaternary alluvial fans and an early Quaternary basalt flow, (2) the height of the footwall escarpment, (3) the stratigraphy of deposits on the hanging wall adjacent to the fault, and (4) the depth of dissection of these hanging-wall deposits. Multiple latest Quaternary (≤15 ka) surface ruptures are displayed on the southernmost Star Valley segment by fault scarps of different heights preserved on alluvial fans with ages estimated between 11 and 15 ka. Assuming that these scarps were formed by two to four surface ruptures, the inferred average return period for large-magnitude paleoearthquakes on this segment of the fault is 2,500 to 7,500 years. A latest Quaternary displacement rate of 0.6 mm/yr to 1.1 mm/yr is also inferred from the fault scarps. The lack of fault scarps on surfaces of similar age or older elsewhere along the Grand Valley fault readily distinguishes this segment from the other two. Height of the footwall escarpment, linearity of the range front, and shallow incision by the Salt River into deposits on the hanging wall suggest that a relatively high displacement rate persisted on the Star Valley segment throughout the Quaternary (≤2 Ma).

The lack of fault scarps on undated surfaces of colluvial, alluvial, and eolian deposits along the trace of the Grand Valley fault or on a fluvial terrace with an estimated age of 15 to 30 ka that crosses the fault suggests that no latest Quaternary surface ruptures have occurred on the central, Grand Valley segment. Marked erosion of the footwall escarpment and deep incision by the Snake River and its tributaries into lower Quaternary and older deposits on the hanging wall contrast with the youthful morphology along the Star Valley segment. Lack of tectonic tilt of sediments that could be as old as 2 Ma suggests that no major Quaternary displacement has occurred on this segment.

An average displacement rate on the northernmost Swan Valley segment of only 0.019 mm/yr is estimated from an inferred fault scarp on a basalt flow isotopically dated at 1.5 Ma. No other fault scarps were observed along this segment of the fault, but the only deposits that directly overlie the fault are loess deposits that could be as young as 11 to 30 ka. Assuming single-event surface ruptures of 2 to 6 m, the average return

*Present address: U.S. Department of Energy, Yucca Mountain Project Office, Box 98518, MS-523, Las Vegas, Nevada 89193-8518.

Piety, L. A., Sullivan, J. T., and Anders, M. H., 1992, Segmentation and paleoseismicity of the Grand Valley fault, southeastern Idaho and western Wyoming, *in* Link, P. K., Kuntz, M. A., and Platt, L. B., eds., Regional Geology of Eastern Idaho and Western Wyoming: Geological Society of America Memoir 179.

period for large-magnitude paleoearthquakes during the Quaternary is ⩾100,000 years. Tectonic tilt of late Tertiary and Quaternary volcanic units in Swan Valley indicates that the displacement rate on the Swan Valley segment has decreased since 2 Ma from a maximum rate of about 1.8 mm/yr between 2 and 4.3 Ma. This change in displacement rate on the Swan Valley segment of the Grand Valley fault, along with the timing of high displacement rates on other late Cenozoic normal faults in the region, suggests that displacement on faults immediately south of the eastern Snake River Plain is probably related to the locations of calderas on the eastern Snake River Plain. These data suggest that large-magnitude paleoearthquakes were common on the Swan Valley segment of the Grand Valley fault between about 2 and 4.3 Ma but have been infrequent since about 2 Ma.

The differences in displacement rates combined with inferred differences in the average return periods between large-magnitude paleoearthquakes during intervals of the Quaternary and the contrast in the age of the most-recent surface displacements for the various fault segments are potentially important considerations in assessing the seismic hazard posed by the Grand Valley fault.

INTRODUCTION

The Grand Valley fault is one of several north- and northwest-striking late Cenozoic normal faults that bound range blocks surrounding the northeast-trending eastern Snake River Plain (Fig. 1). The late Cenozoic normal faults in this region commonly display evidence for recurrent late Quaternary (⩽125 ka) displacement, and many have evidence for recurrent latest Quaternary (⩽15 ka) displacement along at least a portion of their extent (Anders and others, 1989; McCalpin, 1990; Pierce and Morgan, this volume). In addition, the Lost River fault, a normal fault north of the eastern Snake River Plain (Fig. 1), was the site of the 1983 Borah Peak earthquake (M_S 7.3), demonstrating that these faults should still be considered capable of generating large-magnitude, ground-rupturing earthquakes. Detailed study that would help elucidate the geologic histories of these faults has only recently begun (Lundstrom, 1986; McCalpin and others, 1987; Haller, 1988a, 1988b, 1990; Cluer, 1988; McCalpin, 1990, 1991; Rodgers and Anders, 1990; McCalpin and Forman, 1991). This chapter presents geologic evidence that suggests that the Grand Valley fault can be subdivided into three segments that have experienced different rates of Quaternary displacement and paleoseismicity.

Evidence for recurrent late Quaternary surface rupture along the Grand Valley fault was investigated by examining vertical, color aerial photographs of several scales, by flying over the area during low-sun conditions, and by mapping the fault trace, potential fault scarps, linear features, and adjacent deposits at selected localities. Soil development was used to estimate ages for deposits displaced by or overlying the fault. In addition, possible differences in Quaternary (⩽2 Ma) displacement rates along the fault are inferred from differences in morphological characteristics of the range front and valley adjacent to the fault. Furthermore, good exposures of volcanic units along the northwest portion of the fault allowed us to calculate displacement rates for several time intervals from the Miocene to the present (Anders and others, 1989). Because deposits useful in evaluating late Quaternary fault displacements along this portion of the Grand Valley fault are scarce and not well dated, we have incorporated, where possible, the older Quaternary and late Tertiary displacement rates into our assessment of displacement history on the Grand Valley fault. Central to our geologic studies of the Grand Valley fault is the concept of fault segmentation.

The concept that long faults or fault zones may be subdivided into segments is based on the observation that large, historic earthquakes on long, late Cenozoic, normal faults in the western United States do not rupture the entire length of a fault during a single earthquake (Swan and others, 1980; Schwartz and Coppersmith, 1984, 1986; Crone and others, 1987; dePolo and others, 1989). The location of these ruptures does not appear to be random but seems to be controlled by barriers along the fault plane that can often be inferred from discontinuities in the surface trace of the fault or from changes in the displacement history along the fault (Swan and others, 1980; Schwartz and Coppersmith, 1984; Bruhn and others, 1987; Crone and others, 1987; Machette and others, 1987; Fonseca, 1988; Crone and Haller, 1989). Although all historic earthquakes in the western United

Figure 1. Seismicity and major late Cenozoic normal faults surrounding the eastern Snake River Plain. Fault locations are taken or modified from Love and Montagne (1956), Royse and others (1975), Personius (1982), Gilbert and others (1983), Mathiesen (1983), Scott and others (1985), McCalpin and others (1987), Stickney and Bartholomew (1987), Haller (1988b), McCalpin (1990), Ostenaa and Wood (1990), McCalpin and Forman (1991), and Pierce and Morgan (this volume). Seismicity for earthquakes of magnitude ⩾3 or Modified Mercalli Intensity (MMI) ⩾III are compiled from the National Oceanic and Atmospheric Administration (NOAA) data file through June 1988. Earthquakes accompanied by surface rupture are indicated by the year in which they occurred: 1934 is the Hansel valley earthquake of M 6.6 (Shenon, 1936), 1959 is the Hebgen Lake earthquake of M_S 7.5 (Doser, 1985), 1975 is the Pocatello Valley earthquake of M_L 6.0 Doser and Smith, 1982), and 1983 is the Borah Peak earthquake of M_S 7.3 (National Earthquake Information Service, 1983).

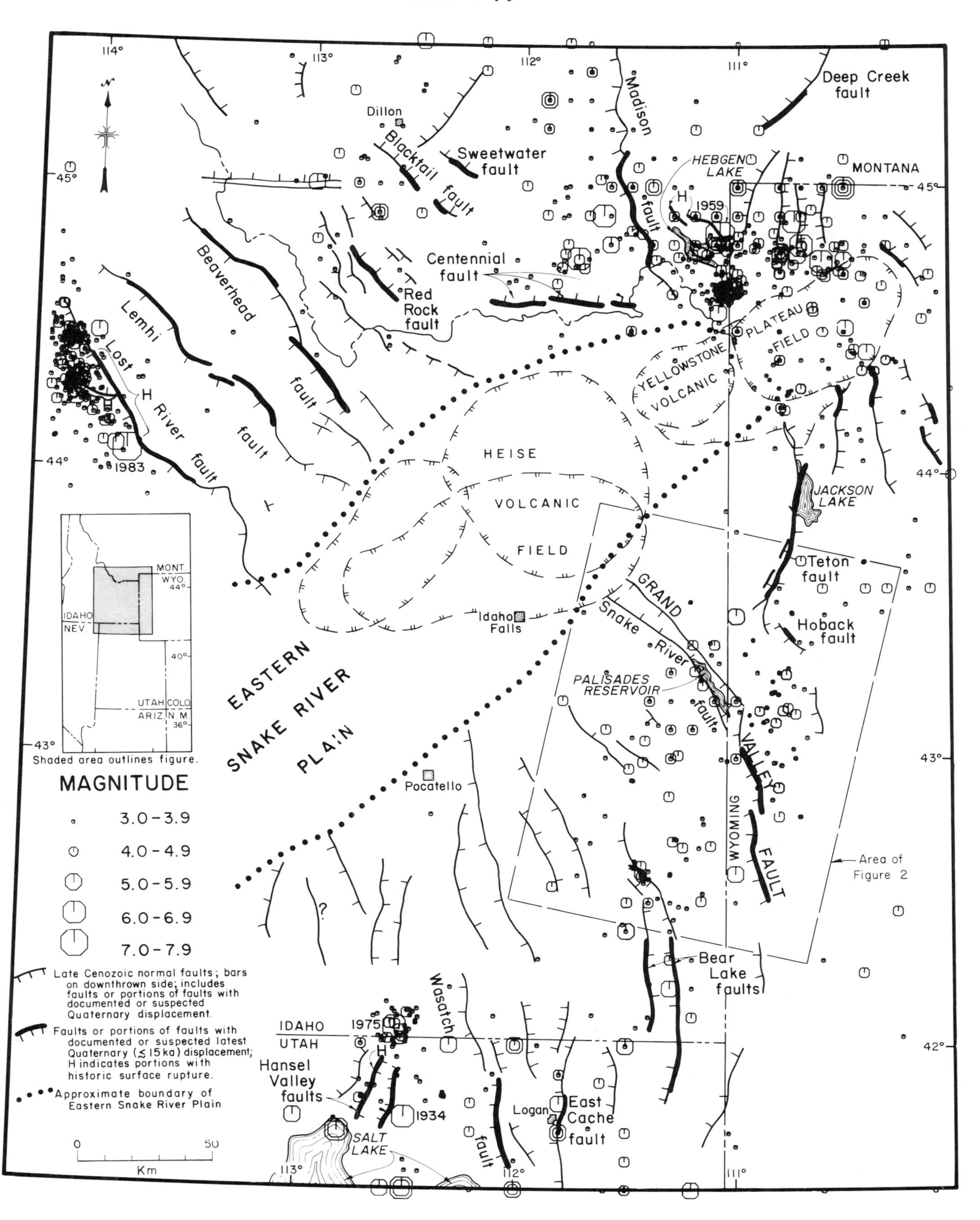
Deep Creek fault
Dillon
Blacktail fault
Sweetwater fault
Madison fault
HEBGEN LAKE
MONTANA
1959
Centennial fault
Red Rock fault
Beaverhead fault
Lemhi fault
Lost River fault
1983
YELLOWSTONE PLATEAU VOLCANIC FIELD
HEISE VOLCANIC FIELD
JACKSON LAKE
Teton fault
Hoback fault
GRAND VALLEY FAULT
Snake River fault
PALISADES RESERVOIR
Idaho Falls
EASTERN SNAKE RIVER PLAIN
Pocatello
WYOMING
Area of Figure 2
Bear Lake faults
Wasatch fault
IDAHO
UTAH
1975
Hansel Valley faults
1934
SALT LAKE
Logan
East Cache fault
MONT.
WYO.
IDAHO
NEV.
UTAH
COLO.
ARIZ.
N.M.
Shaded area outlines figure.
MAGNITUDE
3.0–3.9
4.0–4.9
5.0–5.9
6.0–6.9
7.0–7.9
Late Cenozoic normal faults; bars on downthrown side; includes faults or portions of faults with documented or suspected Quaternary displacement.
Faults or portions of faults with documented or suspected latest Quaternary (≲15 ka) displacement; H indicates portions with historic surface rupture.
Approximate boundary of Eastern Snake River Plain
0 50
Km

States of M $\geqslant$ 7 have ruptured more than one structural or geometric segment by rupturing through apparent discontinuities along the fault, about half of the endpoints in historic surface ruptures would have been predicted by identifying structural, geometric, or geologic segment boundaries (dePolo and others, 1989; Knuepfer, 1989). Thus, recognition of segment boundaries that have been persistent during past ruptures and that may continue to be so during future ruptures helps limit the length of the fault that is assumed to rupture independently (dePolo and others, 1989), an important consideration in deciding the potential seismic hazard posed by a fault.

Along the Grand Valley fault, the extent and estimated ages of fault scarps and related fluvial terraces, the height of the range front, and the age, elevation, and dissection of deposits on the hanging wall adjacent to the fault suggest that displacement rates during the latest Quaternary, and possibly throughout the Quaternary, have not been uniform along the fault. Similar criteria have been used to delineate segments of other late Cenozoic normal faults adjacent to the eastern Snake River Plain (Scott and others, 1985; Crone and others, 1987; Haller, 1987, 1988a, 1988b, 1990; Turko, 1988; Wheeler and Krystinik, 1988; Crone and Haller, 1989). Thus, the above criteria are employed to identify segments along the Grand Valley fault as described in the following sections.

GEOLOGIC SETTING

The Grand Valley fault extends 140 km from the eastern Snake River Plain in eastern Idaho into western Wyoming. It is one of five major right-stepping, *en echelon* normal faults that exhibit latest Quaternary displacements between the Yellowstone area and the Wasatch Front (Scott and others, 1985; Anders and others, 1989; McCalpin, 1990; Pierce and Morgan, this volume) (Fig. 1). This area south of the eastern Snake River Plain coincides roughly, but not exactly, with a zone of seismicity extending from southern Utah to Montana that has been termed the Intermountain seismic belt (ISB) (Sbar and others, 1972; Smith and Sbar, 1974). The southern two-thirds of the Grand Valley fault lies within this seismically active region, whereas the northernmost third of the Grand Valley fault lies within an aseismic region that is symmetric about the axis of the eastern Snake River Plain (Smith and Sbar, 1974; Pierce and Morgan, this volume) (Fig. 1). Although Smith and Sbar (1974) originally suggested that this region was undergoing extension through aseismic creep, subsequent studies showed that late Quaternary fault displacement rates in this aseismic region are significantly lower than those of faults within the surrounding more seismogenic region (Anders and Geissman, 1983; Scott and others, 1985; Piety and others, 1986; Anders and others, 1989).

The Grand Valley fault bounds the southwest and west sides of the Snake River and Salt River ranges, uplifted footwall blocks composed of Mesozoic and Paleozoic sedimentary rocks cut by late Cretaceous and early Tertiary thrust faults (Fig. 2). As defined here, this fault includes both the Grand Valley fault of Witkind (1975a) in Idaho and the Star Valley fault of Witkind (1975b) in Wyoming. The fault has a curvilinear trace as it changes in orientation from northwest-striking to north-striking, and it maintains a uniform distance from the Absaroka thrust, one of the major thrust faults composing the Idaho-Wyoming thrustbelt (Armstrong and Oriel, 1965). Relationships visible on seismic reflection profiles have been interpreted to imply that the Grand Valley fault is a listric normal fault that has reoccupied a steeply dipping subsurface ramp in the Absaroka thrust (Royse and others, 1975; Dixon, 1982). These profiles also suggest that the basin-fill deposits within each of the three basins bounded by the Grand Valley fault are between 2 and 3 km thick.

The basin-fill deposits are well exposed in Swan Valley and Grand Valley but are poorly exposed in Star Valley, where they are inferred to be buried beneath younger deposits. In Swan Valley and Grand Valley, these deposits consist of upper Tertiary–lower Quaternary clastic sediments: an upper gravelly unit of the Long Spring Formation and tuffaceous claystone and sandstone that constitute the Salt Lake Formation (Merritt, 1956, 1958). Age ranges for these clastic sediments are estimated from the ages of interbedded volcanic units in Swan Valley. Along the Snake River in the vicinity of Palisades Creek (Fig. 3), the lower Long Spring Formation is interbedded with Huckleberry Ridge Tuff, which has been isotopically dated at about 2 Ma (Christiansen and Blank, 1972; Armstrong and others, 1975, 1980; Christiansen, 1982). Thus, an age of about 2 Ma is estimated for the base of the Long Spring Formation. At Fall Creek in Swan Valley (Fig. 3), stratigraphic relationships between the Salt Lake Formation and a tuff possibly correlative with the tuff of Cosgrove Road described by Armstrong and others (1975) suggest an age of slightly greater than 10 Ma for the lower Salt Lake Formation (Anders and others, 1989). In Grand Valley, an angular unconformity between the Salt Lake Formation and the overlying Long Spring Formation indicates an age of greater than 2 Ma for the upper Salt Lake Formation there. In Swan Valley, alluvial and lacustrine deposits overlie the basalt of Irwin, which has been isotopically dated at about 4 Ma (Anders and others, 1989) (Fig. 3). The alluvial and lacustrine deposits underlie the Long Spring Formation and overlie the Salt Lake Formation. These relationships suggest that the upper Salt Lake Formation in Swan Valley is older than about 4 Ma.

Associated with the Grand Valley fault is the northwest-striking, 70-km-long Snake River fault that subparallels the Grand Valley fault along the southwest sides of Swan and Grand valleys; it has not been observed in Star Valley (Figs. 1 and 2). Exposures along Indian Creek in Swan Valley indicate that Quaternary displacement on the Snake River fault is limited to $\leqslant$2 to 3 m, much less than the Quaternary displacement on the Grand Valley fault in Swan Valley (Piety and others, 1986) (Fig. 3). In addition, the strike, relatively small displacement, down-to-northeast sense of displacement, and position at the surface and in the subsurface all suggest that the Snake River fault has had only minor displacement and is antithetic to the Grand Valley fault (Corbett, 1982; Anders and others, 1989). Consequently, the Snake River fault will not be considered further.

LATE QUATERNARY SURFACES AND THEIR AGES ESTIMATED FROM SOIL DEVELOPMENT AND GEOMORPHIC RELATIONSHIPS

Late Quaternary surfaces along the Grand Valley fault include loess-covered alluvial fans (Qla), fluvial terraces (Qt1, Qt2, and Qt3), and alluvial fans (Qa1, Qa2, and Qa3). In order to use these surfaces to discern the displacement history along the fault, their ages had to be determined or estimated. Because no datable material was recovered during this evaluation, relative ages of these surfaces were estimated using the geomorphic relationships among these surfaces and the soils developed on them. Numeric ages were estimated primarily by comparing the soils developed on these surfaces with those developed on similar surfaces in the region for which ages have been determined. Degree of soil development is commonly used in fault assessments to differentiate deposits and geomorphic surfaces of different ages (e.g., Machette, 1978, 1988; Douglas, 1980; Swan and others, 1980; McCalpin, 1982; Harden and Matti, 1989). Utilizing this technique assumes that soils began forming as soon as the surface stabilized and that degree of soil development reflects the time since stabilization. However, additional factors (e.g., climate, vegetation, topographic position, and parent material) can also influence soil properties (Jenny, 1941, 1980; Birkeland, 1984).

Along the Grand Valley fault, several of these factors do vary. For example, mean annual temperatures range from 4 to 6°C, and mean annual precipitation is between 38 and 57 cm (U.S. Weather Bureau, 1960–1978). Vegetation differences reflect local variations in climatic conditions. In addition, eolian material is variable in its distribution and thickness. Undisturbed, flat sites on surfaces away from colluvial slopes were chosen for study of soil development in order to minimize the effects of topographic position and burial by younger deposits. An insufficient number of profiles was described along the Grand Valley fault to statistically evaluate soil variability within deposits of a given age. Further study would be needed to accurately quantify how representative the descriptions presented here are of the soils developed on each surface.

Soil profiles along the Grand Valley fault were described according to Soil Survey Staff (1975, 1981), Guthrie and Witty (1982), and Birkeland (1984). Laboratory analyses were done by Colorado State University Soil Testing Laboratory, Fort Collins, following the methods of Soltanpour and Workman (1981). Soils were described to depths of 1.2 to 2.2 m in backhoe pits, in hand-dug pits, or in natural exposures. Detailed descriptions of the soil profiles are given in Piety and others (1986).

Loess-covered surfaces

Loess-covered alluvial fans (Qla surfaces) are preserved along the western front of the Snake River Range in Swan Valley and along the western front of the Salt River Range in northern Star Valley (Figs. 3 and 4). Therefore, these deposits are important to the evaluation of displacement on these portions of the Grand Valley fault. On Qla surfaces in Swan Valley, six profiles were described in a trench near Dry Canyon, and seven profiles were described in hand-dug pits near Shurtliff Canyon (Piety and others, 1986) (Fig. 3). In Star Valley, a soil was described at one locality on a Qla surface: at the mouth of Corral Canyon (Fig. 4). The loess there is about 170 cm thick. Although most of the alluvial-fan deposits in Star Valley (Qa1, Qa2, and Qa3) are fine textured in their upper parts, thick loess overlies fan alluvium chiefly in northern Star Valley (Fig. 4). Relative and numeric ages for these Qla surfaces were estimated by comparing the soils developed at these three localities with those developed on loess elsewhere in southeastern Idaho and by using the relationships between these loess-covered surfaces and a fluvial terrace of the Snake River.

Soil development. Pierce and others (1982) recognized two major loess units in southeastern Idaho. A younger loess (their loess A) covers much of the landscape and buries deposits generally older than about 15 ka. This loess accumulated between about 10 ka and about 70 ka. Where this loess is thick and underlying deposits are older than about 150 ka, an older loess unit (their loess B) is often preserved. The older loess is recognized by a buried soil that is markedly better developed than the surface soil on the younger loess unit and better developed than any buried soils within the younger loess (Lewis and Fosberg, 1982; Pierce and others, 1982). The older loess probably accumulated until shortly after about 140 to 150 ka, and the soil on it developed between about 130 ka and about 70 ka.

Several loess deposits are preserved in Swan Valley, and at least one is preserved in northern Star Valley, but correlation with the two major loess units of Pierce and others (1982) is not clear. The Qla surface soil formed on the loess near Dry Canyon in Swan Valley lacks an argillic or cambic horizon but includes accumulation of up to 27% carbonate with stage III carbonate morphology (Table 1). Total thickness of the calcic horizon is 114 cm, and carbonate has been leached to a depth of 54 cm (Table 1). Carbonate content of unaltered loess in southeastern Idaho generally ranges between 10 and 20% (Lewis and Fosberg, 1982; Pierce and others, 1982). The Qla surface soil at Shurtliff Canyon in Swan Valley, however, has an argillic horizon 80 cm thick with a 12% difference in clay content from the underlying C horizon (Table 1). This soil also includes a carbonate-rich horizon with up to 11% carbonate and stage II morphology (Table 1). The soil on the Qla surface at Corral Canyon in northern Star Valley has a strongly developed argillic horizon 144 cm thick (Table 1). A calcic horizon with stage I+ morphology and up to 23% carbonate is developed beneath the loess in pebbly alluvium.

The presence of a relatively well developed argillic horizon in the surface soils at Shurtliff Canyon and Corral Canyon may indicate that these loess deposits are older than the one near Dry Canyon. The comparison is complicated, however, because soil development in the younger loess in southeastern Idaho is highly variable depending upon factors other than surface age (Lewis and Fosberg, 1982; Pierce and others, 1982). Characteristics of soils formed on loess elsewhere in southeastern Idaho are

Figure 2. (A) LANDSAT image and (B, on facing page) index map showing physiography and faults of southeastern Idaho and western Wyoming. Major late Cenozoic normal faults are indicated with bar and ball on the downthrown side. White areas show valleys filled with upper Cenozoic sedimentary and volcanic deposits. Lightly stippled areas indicate mountainous regions composed of Paleozoic and Mesozoic sedimentary rocks. The darkly stippled area in the Teton Range indicates Precambrian crystalline rocks.

particularly dependent upon climatic conditions (Pierce and others, 1982). The surface at Shurtliff Canyon is nearly 100 m higher than the surface near Dry Canyon, and the surface at Corral Canyon is another 60 m higher than the one at Shurtliff Canyon, so that local climatic conditions may provide greater available moisture to the surfaces at Shurtliff Canyon and Corral Canyon than they do for the surface at Dry Canyon. These climatic differences, which are reflected by differences in vegetation and carbonate morphology among the sites, cannot be quantified, and it is not known if these climatic differences can account for the differences in soil properties.

Two buried soils are also preserved in the loess at both Dry Canyon and Shurtliff Canyon. Because the older loess unit of Pierce and others (1982) is buried by the younger loess unit, recognition of the older loess is based upon identification of a buried soil markedly better developed than the soil on the younger loess. Thus, development of the buried soils in Swan Valley is important to the recognition of the older loess unit of Pierce and others (1982). The buried soils at Dry Canyon, however, are not markedly better developed than the surface soil, although the first buried soil has a weakly developed argillic horizon. These buried soils have carbonate-rich horizons with 25 to 30% carbonate and stage II and III morphology (Piety and others, 1986). At Shurtliff Canyon, the first buried soil is similar

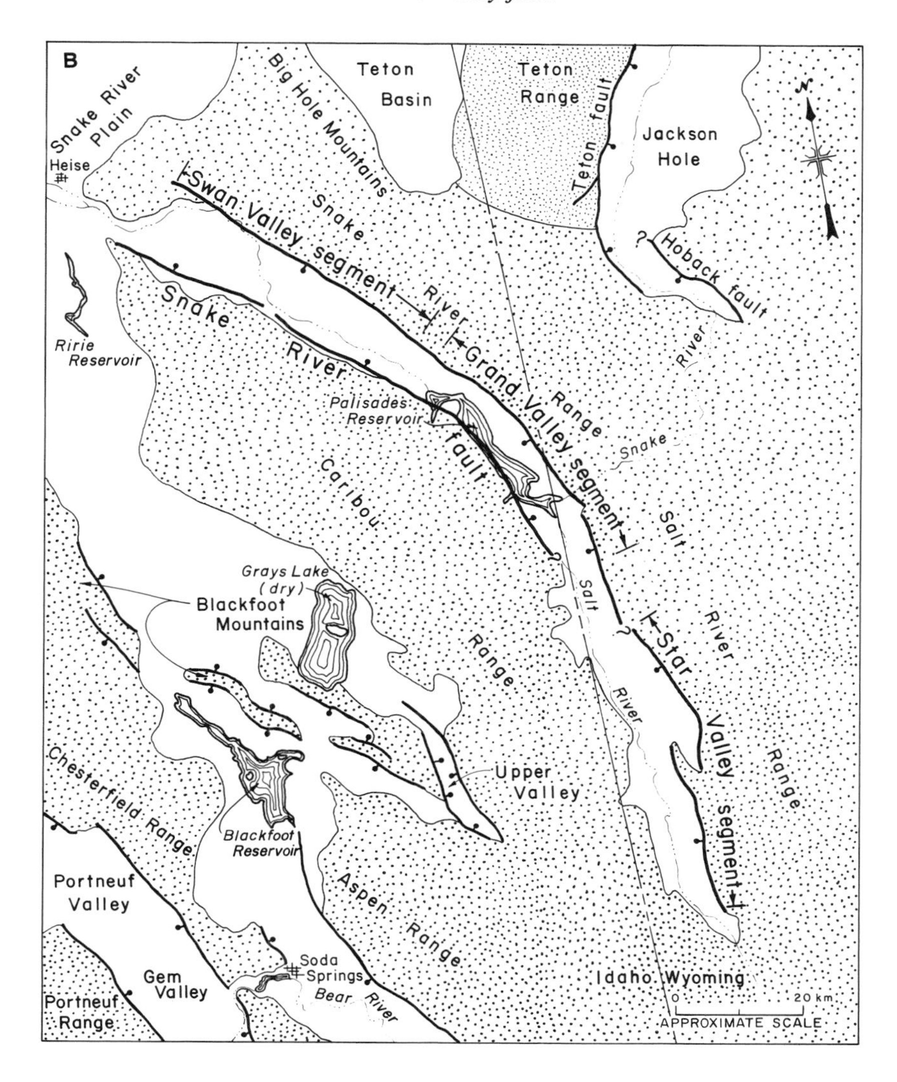

to the surface soil at this locality, whereas the second buried soil is markedly better developed (e.g., an argillic horizon with higher hue and chroma and twice the clay content of the surface soil and a carbonate horizon with three times the carbonate content of the surface soil). Differences in parent material, however, hamper the comparison of soil development. The second buried soil at Shurtliff Canyon is developed in both loess and underlying gravelly alluvium.

Geomorphic relationships. The main terrace adjacent to the Snake River near Rainey Creek (Fig. 3) is tentatively correlated with the Qt3 terrace at Alpine (Table 2). The age of this terrace is estimated to be about 11 to 15 ka (discussed in the following section). This terrace has only a thin cap of fine-grained material, either loess or overbank sediment. Although correlation of the Qla surfaces at Dry and Shurtliff canyons is complicated because the surfaces slope into Swan Valley, the Qla surface at Dry Canyon appears to be graded to a level about 15 m above the Qt3 terrace, and the Qla surface at Shurtliff Canyon grades to a level on the order of 100 m higher than the Qt3 terrace. In addition, the Qla surface at Shurtliff Canyon is much more dissected than the Qla surface at Dry Canyon. These characteristics suggest that the Qla surface at Dry Canyon is no younger than and may be slightly older than the Qt3 terrace and that the Qla surface at Shurtliff Canyon is older than the Qla surface at Dry Canyon.

Age estimates. Despite differences in soil development, the loess deposits on Qla surfaces at both Dry and Shurtliff canyons in Swan Valley and at Corral Canyon in northern Star Valley are tentatively correlated with the younger loess of Pierce and others (1982). Of the buried soils at Dry Canyon and Shurtliff Canyon in Swan Valley, only the second buried soil at Shurtliff Canyon is markedly better developed than the surface soil, and because of climatic differences and parent material changes, it is not clear if this better development reflects a time interval longer than the one during which the surface soil formed. Although the younger loess in southeastern Idaho continued to be deposited until about 10 ka (Pierce and others, 1982), suggesting a minimum age of about 10 ka for the Qla surfaces at Dry Canyon, Shurtliff Can-

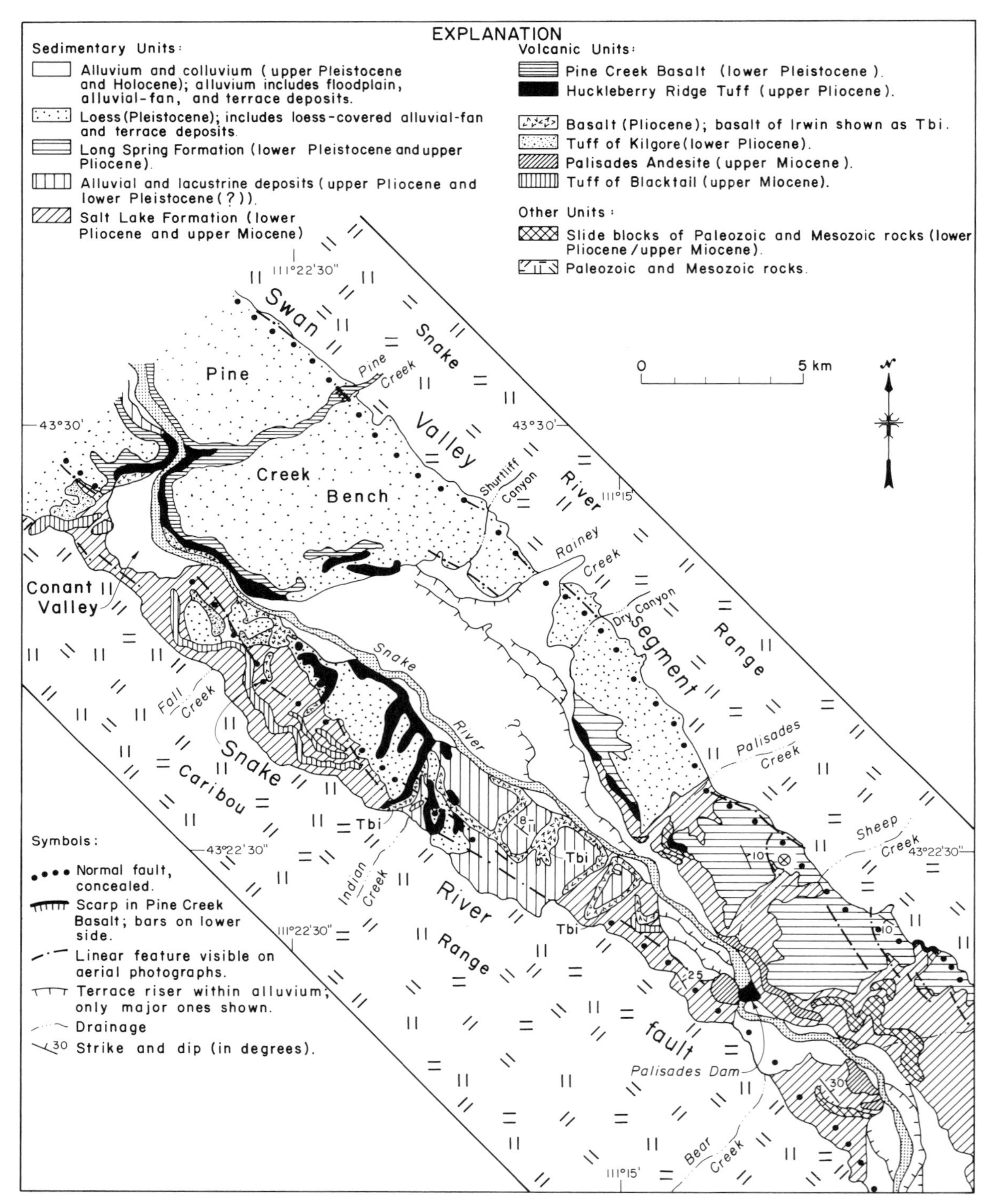

Figure 3. Generalized geologic map of Swan Valley depicting the relationships among upper Cenozoic basin fill, the Grand Valley fault, and the Snake River fault. The Swan Valley segment of the Grand Valley fault is interpreted to extend from beyond the map area on the northwest to between Palisades Creek and Sheep Creek on the southeast. Terraces Qt1, Qt2, Qt3, and Qt4 along the Snake River are not illustrated. The Qla surfaces discussed in the text are included in the areas indicated as Pleistocene loess. Outcrop patterns of upper Tertiary and Quaternary volcanic units have been modified principally from Roberts (1981) and from Jobin and Schroeder (1964a, 1964b), Jobin and Soister (1964), and Staatz and Albee (1966). Adapted from Anders and others (1989).

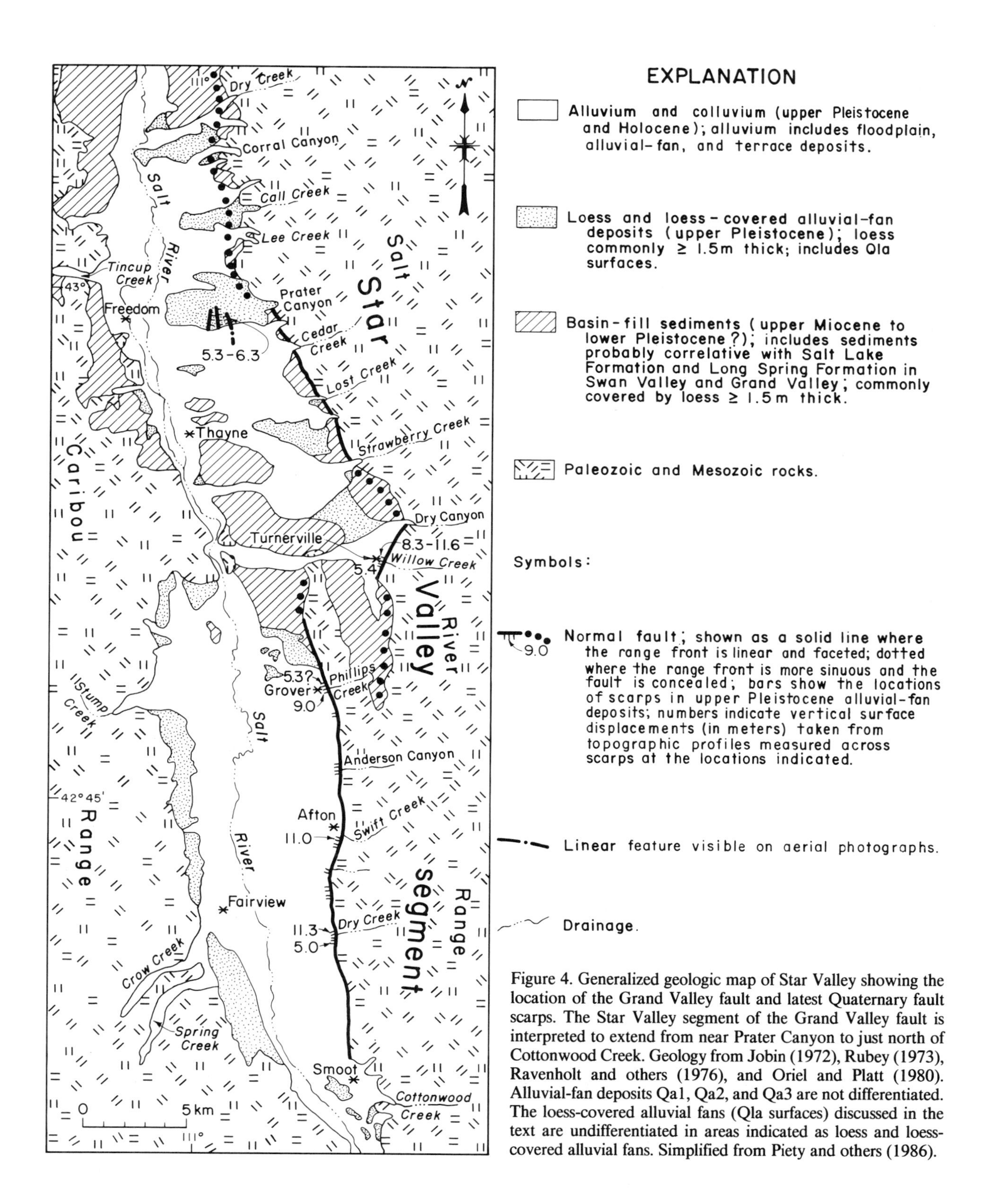

Figure 4. Generalized geologic map of Star Valley showing the location of the Grand Valley fault and latest Quaternary fault scarps. The Star Valley segment of the Grand Valley fault is interpreted to extend from near Prater Canyon to just north of Cottonwood Creek. Geology from Jobin (1972), Rubey (1973), Ravenholt and others (1976), and Oriel and Platt (1980). Alluvial-fan deposits Qa1, Qa2, and Qa3 are not differentiated. The loess-covered alluvial fans (Qla surfaces) discussed in the text are undifferentiated in areas indicated as loess and loess-covered alluvial fans. Simplified from Piety and others (1986).

TABLE 1. SUMMARY OF PROPERTIES OF SOILS DESCRIBED ON SURFACES ADJACENT TO THE GRAND VALLEY FAULT

Area	Surface* and Estimated Age (ka)	Color (dry)			Clay Accumulation			Carbonate Accumulation				Profile Development Index§	Loess Thickness (cm)
		Max. Hue	Max. Value	Max. Chroma	Max. Clay Content (%)	Clay Increase B from C (%)	Max. Horizon Thickness (cm)	Max. $CaCO_3$ Content (%)	Depth to Max. $CaCO_3$ (cm)	Max.† $CaCO_3$ Stage	Total Horizon Thickness (cm)		
Swan Valley	Qla at Dry Canyon (11 to 15)	10YR	8	3	**	...	...	27	54	III	114	0.16	475
	Qla at Shurtliff Canyon (30 to 70?)	10YR	7	3	21	12	80	11	130	II	>20	0.23	650
Grand Valley	Qt3 (11 to 15)	7.5YR	7	3	20	15	41	17	59	I+	107	0.06	4 to 33
	Qt2 (15 to 30)	7.5YR	6	4	12‡	5	40 to 44	...	71	II	24 to 32	0.09	36 to 56
Star Valley	Qa3 (2? to 5?)	7.5YR	6	3	**	...	...	11	0	I+	53	0.04	0
	Qa2 (11 to 15)	7.5YR	6	4	**	...	...	19	34 to 59	II	84 to 106	0.08	0 to 26
	Qa1 (11 to 15)	7.5YR	7	4	17†††	5§§	35	22	71 to 104	II+	107 to 113	0.08	33 to 48
	Qla at Corral Canyon (30? to 70?)	7.5YR	6	4	33	19§§	144	23***	182	I+	>26	0.36	170

*The Qla surfaces are loess-covered alluvial fans. The Qt2 and Qt3 surfaces are fluvial terraces of primarily the Snake River (Table 2). The Qa1, Qa2, and Qa3 surfaces are alluvial fans heading in tributaries of the Salt River. Properties have been summarized from the following number of detailed descriptions for each unit: four each for Qa1 and Qt3, three for Qt2, two for Qa2 and Qla at Dry Canyon, and one each for Qa3 and Qla at Shurtliff Canyon and Corral Canyon. Properties for the soils developed in loess in Swan Valley are for the surface soils only. Detailed descriptions for the soils shown here, along with those for the buried soils at Dry Canyon and Shurtliff Canyon, are in Piety and others (1986).

†Stages of carbonate morphology are from Gile and others (1966) and Birkeland (1984).

§Profile Development Index is calculated using five properties: rubification, total texture, structure, dry consistence, and clay films (Harden, 1982; Harden and Taylor, 1983; Nelson and Taylor, 1985). Properties are reported in detailed descriptions in Piety and others (1986).

**Highest clay content is in the A horizon.

‡Values for clay accumulation are for only one profile out of the three described.

§§Clay increase from the A horizon to the B horizon. The C horizon was not exposed or is a different parent material than the B horizon.

***Carbonate is accumulated in the pebbly alluvium underlying the loess.

†††This is a minimum value.

TABLE 2. CHARACTERISTICS AND ESTIMATED AGES OF FLUVIAL TERRACES ALONG THE SNAKE RIVER IN GRAND VALLEY AND SWAN VALLEY

Terrace	Height Above Snake River (m)	Soil and Geomorphic Characteristics	Estimated Age (ka)
Qt4	4.5 to 7.5	Preserved nearly continuously along and adjacent to the present Snake River; channels occasionally contain water; soils not examined	≤10
Qt3	9 to 20	Preserved as remnants inset into Qt2 terraces; channels visible on surfaces; soils contain Bw or weak Bt horizons and locally include carbonate (Bk) horizons with stage I- to I+ morphology; loess overlying the alluvium is 15 to 30 cm thick	11 to 15
Qt2	18 to 34	Preserved nearly continuously in Grand Valley; channels visible on surfaces; soils contain Bw or Bt horizons and locally include Bk horizons with stage II morphology; loess overlying the alluvium is 15 to 50 cm thick	15 to 30
Qt1	75	Preserved as a single remnant; soil is developed in loess and contains a Bt horizon and a Bk horizon with at least stage III morphology; loess overlying the alluvium is about 18 m thick and appears to contain at least one buried soil; surface partially covered by alluvial-fan and colluvial deposits	≥200

yon, and Corral Canyon, geomorphic relationships between the surfaces in Swan Valley and the Qt3 terrace adjacent to the Snake River indicate that the minimum age for the Qla surface at Dry Canyon is 11 ka to 15 ka and that the Qla surface at Shurtliff Canyon is older. How much older is unclear. If the second buried soil at Shurtliff Canyon represents the older loess deposit recognized elsewhere in southeastern Idaho, then a surface on the gravelly alluvium and overlying loess at this locality may have been stable between about 70 ka and about 130 ka, before the younger loess was deposited. Alternately, both buried soils within the loess at Shurtliff Canyon could represent relatively short breaks in deposition of the younger loess, so that the entire loess sequence could be younger than about 70 ka. The minimum age of the Qla surface at Shurtliff Canyon can only be speculated, but because this surface is higher and more dissected than the one at Dry Canyon, its minimum age is estimated to be about 30 ka.

The soil on the Qla surface at Corral Canyon in northern Star Valley is at least as well developed as the surface soil at Shurtliff Canyon. Accordingly, the Qla surface at Corral Canyon is tentatively estimated to have a minimum age of about 30 ka. Although local climatic variations and the presence of buried soils within the loess in Swan Valley complicate comparisons, the stronger development of the soil at Corral Canyon may indicate that the surface at Corral Canyon is older than the surface soil at Shurtliff Canyon. Thus, the Qla surface at Corral Canyon could be significantly older than 30 ka.

Terraces of the Snake River

At least one of the four major terraces preserved near Alpine, Wyoming, crosses the Grand Valley fault in Grand Valley and is important in establishing the absence of surface rupture on this portion of the Grand Valley fault (Fig. 5). As for the loess-covered surfaces described previously, soil development and geomorphic relationships were used to delineate relative-age differences among the terraces at Alpine and to estimate numeric ages for these surfaces. Because the area surrounding Jackson, Wyoming, located about 65 km upstream of Alpine on the Snake River (Fig. 2), was heavily glaciated several times during the late Quaternary, we assume that the major terraces at Alpine are chiefly a result of changes in climate, discharge, and load related to these glaciations. Accordingly, specific characteristics that have been utilized to correlate glacial and fluvial deposits in the Rocky Mountains (Blackwelder, 1915, 1931; Holmes and Moss, 1955; Colman and Pierce, 1977; Birkeland and others, 1979; Brookes, 1985; Colman and Pierce, 1986; Pierce, 1986) were noted for the terraces at Alpine, in addition to soil development and geomorphic relationships commonly used to estimate relative and numeric ages for fluvial terraces (Shroba and Birkeland, 1983; Birkeland, 1985). These characteristics include the preservation of the terraces, the thickness of loess overlying gravelly alluvium, the weathering of granitic clasts, and the degree of channel preservation on the terrace surfaces.

Soil development, geomorphic relationships, and terrace characteristics. The lowest terrace (Qt4) delineated at Alpine is 4.5 to 7.5 m above the Snake River; it was not studied in detail. Weak Bt or Bw development of the soils, a height of 9 to 34 m above the Snake River, extensive preservation with very little dissection, thin loess cover (15 to 50 cm), generally unweathered granitic cobbles and boulders, and readily visible channels on the terrace surfaces are characteristic of both terrace Qt2 and terrace Qt3 at Alpine (Tables 1 and 2; Piety and others, 1986). The major difference between terraces Qt2 and Qt3 is that terrace Qt2 is more extensively preserved at an elevation 9 to 14 m higher above the Snake River than terrace Qt3 (Piety and others, 1986). A higher terrace, Qt1, has a strong soil developed on a relatively thick loess deposit overlying the gravelly alluvium (Table 2). This terrace is 75 m above the Snake River and is extensively preserved at only one locality, north of the Snake River near Alpine.

Age estimates. Datable material was removed from only one of the terraces near Alpine. Amino-acid ratios obtained on snails collected by D. W. Moore (U.S. Geological Survey,

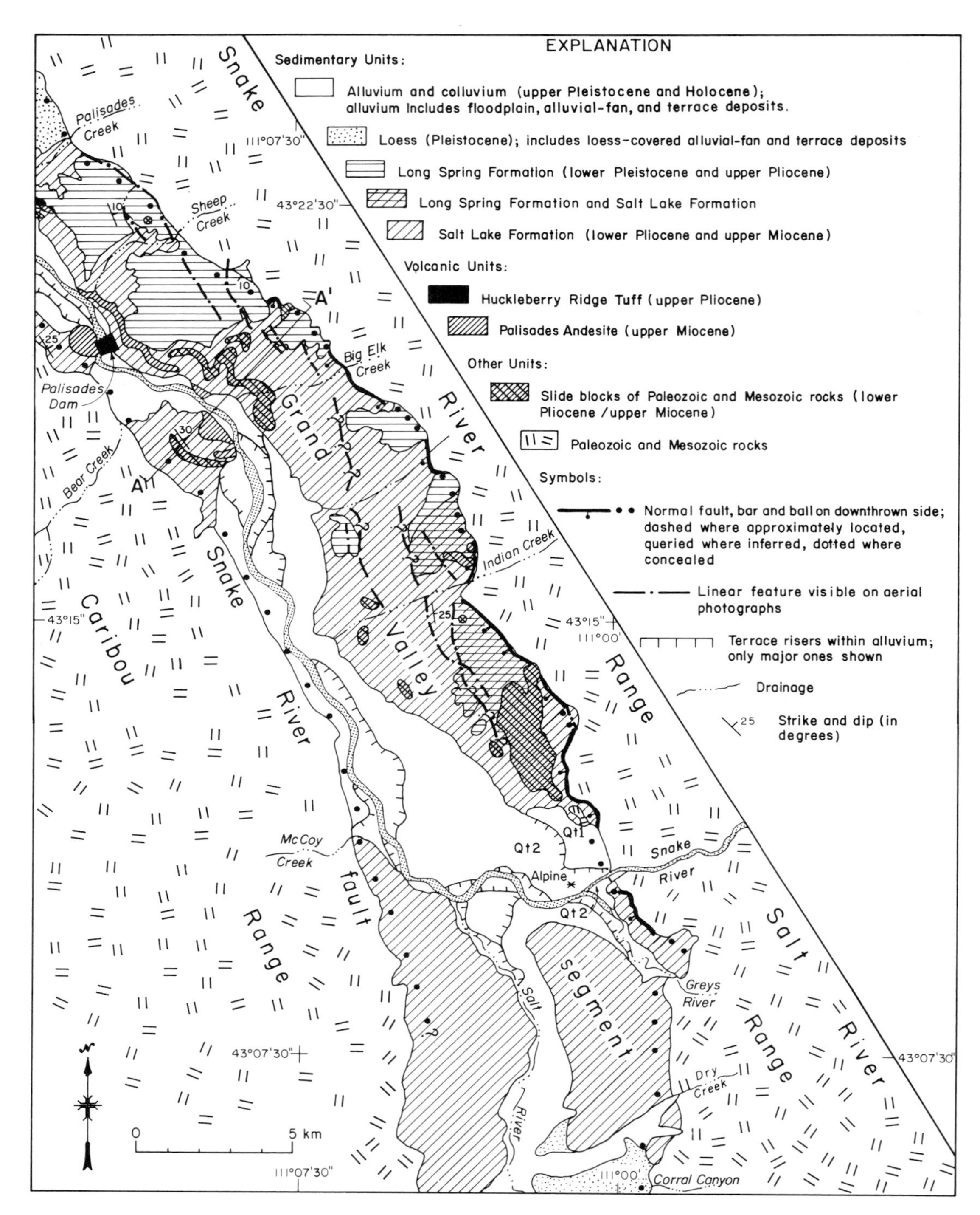

Figure 5. Generalized geologic map of Grand Valley illustrating the extent of upper Tertiary and Quaternary deposits and their relationship to the Grand Valley fault and the structurally related Snake River fault. Map has been drawn without showing Palisades Reservoir, which now covers the Pleistocene and Holocene deposits between Palisades Dam and Alpine, Wyoming. The Grand Valley segment of the Grand Valley fault is inferred to extend from the area between Palisades Creek and Sheep Creek to near Dry Creek. Terraces Qt3 and Qt4 are not shown. The Qla surfaces discussed in the text are included in the areas indicated as Pleistocene loess. Mississippian Lodgepole Limestone is included in the areas indicated as Paleozoic and Mesozoic rocks. Geology has been modified from Jobin (1972), Albee and Cullins (1975a, 1975b), and Oriel and Moore (1985). Contrast and compare the extent, ages, and dips of basin fill here with those of similar deposits in Swan Valley (Fig. 3) and in Star Valley (Fig. 4).

Denver) from the loess overlying the terrace gravel indicate an age of ⩾200 ka for terrace Qt1 (Piety and others, 1986). The apparent antiquity of this terrace does little to limit the ages of the younger terraces, Qt2 and Qt3, the ages of which must then be inferred from possible correlations to the glacial deposits upstream near Jackson. The thin cover of loess, generally unweathered nature of the gravel, and limited dissection of these two terraces suggest that they are more likely related to the last major, or Pinedale, glaciation recognized in the northern Rocky Mountains than to the penultimate, or Bull Lake, glaciation. Additionally, the soils formed on the Qt2 and Qt3 terraces are more consistent with soils developed in post-Pinedale time than with those developed in post–Bull Lake time; the soils also indicate that the terraces are pre-Holocene. More definitive ages for these terraces are difficult to determine.

Walker (1964) attempted to directly trace the fluvial terraces in Swan and Grand valleys upstream through the Snake River canyon to the glacial deposits near Jackson. He concluded that terrace Qt2 (his 80-foot terrace) correlated with the Burned Ridge moraine at Jackson and that terrace Qt3 (his 30-foot terrace) correlated with the Jackson Lake moraine at Jackson. He further concluded that these terraces were related to two ice advances during the Pinedale glaciation and speculated that only a few thousand years separated the two terraces. Recent work has better defined the sequence of glacial deposits at Jackson (Pierce and Good, 1990). Pierce and Good (1990) postulated that the age of the Jackson Lake moraine is about 15 to 30 ka but recognized three Burned Ridge moraines with ages that might range from about 15 to 70 ka.

Because of problems correlating the terraces near Alpine upstream through the Snake River canyon to Jackson, we cannot confidently correlate terraces Qt2 and Qt3 with individual glacial deposits at Jackson. However, use of the ages postulated by Pierce and Good (1990) for the glacial deposits at Jackson to limit ages for terraces Qt2 and Qt3 is considered valid. Soil development on the terraces does not seem strong enough to represent development since early Pinedale, so an age of 60 to 70 ka is discounted. Soil development and other weathering characteristics are consistent with a maximum age for the terraces of younger Pinedale and a minimum age of pre-Holocene. Consequently, we tentatively assign an age of 15 to 30 ka to terrace Qt2 and an age of 11 to 15 ka to terrace Qt3.

Alluvial fans

Late Pleistocene alluvial fans of several ages are preserved along the eastern side of Star Valley (Fig. 4). The alluvial-fan deposits are composed of generally clean, clast-supported gravel in a loose, coarse-sand matrix (Piety and others, 1986). Clast composition in these deposits is 40 to 68% limestone, and silt content at the surface is often about 50% but decreases quickly to 15 to 30% between depths of 36 and 50 cm (Piety and others, 1986). The high silt content near the ground surface suggests the addition of eolian material or overbank sediment to the upper portions of all but two of the gravelly alluvial-fan deposits studied. These silty surface horizons contain 15 to 45% gravel. Six soil profiles were described in backhoe pits, and two profiles were described in stream exposures on these surfaces. Soil development, along with geomorphic relationships, was used to estimate relative-age differences among the fan surfaces and to establish numeric ages for these surfaces.

Soil development. The soil on the youngest surface studied, Qa3, appears to lack any added fine-grained sediment (loess or overbank alluvium). A weak calcic horizon is its primary characteristic, with a maximum carbonate content of 11% that occurs near the ground surface, a maximum stage I+ $CaCO_3$ morphology, and a thickness of 53 cm (Table 1). Carbonate coats on clasts in the calcic horizon on the Qa3 surface are patchy and ⩽0.2 mm thick. The soils on the Qa1 and Qa2 surfaces have calcic horizons that are better developed than that on the Qa3 surface. These calcic horizons have a maximum carbonate content of about 20% that occurs at depth, carbonate morphology of stage II to II+, and a thickness of between 84 and 113 cm (Table 1). The carbonate is leached to a greater depth, and the calcic horizons are thicker and slightly better developed in the soils on the Qa1 surfaces than in those on the Qa2 surfaces (Table 1). Carbonate coats on clasts from the strongest calcic horizons of the soils on the Qa1 and Qa2 surfaces range in thickness from 0.2 to 1.3 mm and have an average thickness of 0.5 ± 0.2 mm (Piety and others, 1986). Some of these soils also have weak argillic or cambic horizons (Table 1).

Geomorphic relationships. Alluvial fans in Star Valley are dissected to varying degrees. In general, younger surfaces are incised into older ones (Fig. 4). This is especially true in northern Star Valley. In southern Star Valley, in contrast, Qa2 surfaces that head in drainages with basin areas of ⩽10 km^2 are usually graded to larger, older Qa1 surfaces. In addition, Qla surfaces are generally absent along the east side of southern Star Valley so that the Qa1 surfaces form a nearly continuous apron along this side of the valley. This apron is incised only 2 to 3 m, except near the Grand Valley fault where the Qa1 surfaces are incised up to 12 m in larger drainages.

Age estimates. During the Pleistocene, the Salt River Range supported only small glaciers that were limited to their source cirques (Rubey, 1973; Oriel and Platt, 1980). As a result, the alluvial fans in Star Valley cannot be traced directly to glacial deposits in the Salt River Range. Still, deposition of these alluvial fans was probably related to glacial climatic changes as was the deposition of other large late Pleistocene alluvial fans in southeastern Idaho (Pierce and Scott, 1982). Consequently, ages for the alluvial fans in Star Valley have been estimated on the basis of the similarity of their characteristics and soil development with those of other alluvial fans in the region and on the similarity of their soil development to that on the Qt3 terrace at Alpine (Table 1).

Soils on the alluvial fans in Star Valley are similar to those developed on late Pleistocene alluvial fans elsewhere in southeastern Idaho. These late Pleistocene alluvial fans have cambic or weak argillic horizons about 15 cm thick and calcic horizons 20 to 40 cm thick with stage I to II morphology (Pierce and Scott, 1982; Pierce and others, 1983; Pierce and Colman, 1986). Carbonate coats from the calcic horizons of these soils average 1.07 ± 0.16 mm thick (Pierce and Scott, 1982). In the Raft River valley, where the composition of the alluvial-fan deposits is similar to that of the alluvial-fan deposits in Star Valley, carbonate is often leached to a depth of 0 to 50 cm, and carbonate coats average 0.3 to 0.4 mm in thickness but can be as thick as 0.9 mm (Pierce and others, 1983). The age of the late Pleistocene alluvial fans in southeastern Idaho has been estimated to be 11 to 25 ka, because soil development, loess thickness, and carbonate-coat thickness are similar to those in gravel deposits elsewhere that can be directly related to deposits of the Pinedale glaciation (Pierce and Scott, 1982). Stratigraphic relationships between alluvial-fan deposits and Bonneville Flood deposits support this age estimate (Pierce and Scott, 1982).

The properties of the soils developed on the Qa1 and Qa2 alluvial fans in Star Valley suggest that the ages of these two surfaces are probably similar to each other and to those of late Pleistocene alluvial fans (about 11 to 25 ka) described by Pierce and Scott (1982), Pierce and others (1983), and Pierce and Colman (1986). Moreover, the soils on the Qa1 and Qa2 surfaces are similar to that on terrace Qt3 near Alpine, which is inferred to have an age of 11 to 15 ka (Table 1; see "Terraces of the Snake River"). On the basis of these similarities, we infer that although the Qa1 and Qa2 surfaces could be as old as about 25 ka, the best estimate of the ages of these surfaces is between 11 and 15 ka. Greater depth of carbonate leaching and the greater thickness of the calcic horizons in the soils on the Qa1 surfaces suggest that these surfaces are somewhat older than the Qa2 surfaces, but the age difference appears to be relatively small.

The soil on the youngest alluvial surfaces mapped, Qa3, is less developed than those on the Qa1 and Qa2 surfaces. Although not constrained, we infer a Holocene age (2? to 5?ka) for these surfaces on the basis of no leaching of carbonate to an appreciable depth, a maximum carbonate content of about half those in the soils on the Qa1 and Qa2 surfaces, a total thickness of the carbonate horizon that is nearly half of those in the soils on the Qa1 and Qa2 surfaces, and a lack of additional fine-grained (eolian?) sediment in the upper part of the gravelly alluvium (Table 1).

FAULT SCARPS, LINEAR FEATURES, AND TECTONIC TERRACES

Our study attempts to evaluate late Quaternary displacement on the Grand Valley fault in Swan Valley, Grand Valley, and Star Valley by focusing on the presence or absence of fault scarps on Quaternary geomorphic surfaces along the range front. Fault scarps signal those portions of the fault that have been active during the late Quaternary, if surfaces containing the scarps are this old, or over a shorter time interval, if surfaces are younger. On the other hand, surfaces without scarps can be used to demonstrate that displacements are older than the surfaces.

Potential fault scarps and linear features were located initially on black-and-white and color aerial photographs at scales of approximately 1:60,000, 1:16,000, and 1:12,000 and by aerial reconnaissance during low-sun conditions. Areas where potential fault scarps or linear features were noted were then mapped, along with adjacent geomorphic surfaces, to determine the origin of the potential fault scarps or linear features and to estimate ages for displaced or undisplaced deposits along the fault. Topographic profiles were measured across suspected fault scarps following the methods of Bucknam and Anderson (1979) and Machette (1982) to determine such scarp parameters as surface displacement and maximum slope angle.

Swan Valley

The northern 40 km of the Grand Valley fault in Swan Valley strikes about N45°W between the eastern Snake River Plain and Palisades Creek (Figs. 2 and 3). Previous investigators, on the basis of reconnaissance studies, have suggested various ages for the faults in Swan Valley. Witkind (1975a) depicted the Grand Valley fault as recurrently active since the Miocene but noted that no scarplets had been reported along the fault. He also indicated that several short, inferred northwest-striking faults within Swan Valley displace upper Quaternary deposits. Others have postulated a northeast-striking normal fault (approximately perpendicular to the Grand Valley fault) in the Rainey Creek drainage and have suggested Holocene displacement of alluvial deposits at the mouth of Rainey Creek (U.S. Bureau of Reclamation, 1976; Hait and others, 1977).

Our study attempts to evaluate these observations in Swan Valley by examining potential fault scarps and linear features at four localities from north to south: Pine Creek, Shurtliff Canyon, Rainey Creek, and Dry Canyon (Fig. 3).

Pine Creek. The only fault scarp preserved in Swan Valley is at the mouth of Pine Creek on the surface of a basalt flow, the Pine Creek Basalt, isotopically dated at 1.5 ± 0.8 Ma (Anders and others, 1989) (Fig. 3). Outcrops and water-well logs indicate that this 120-m-thick flow had a source in Conant Valley and that it fills an ancestral channel of the Snake River below the Pine Creek Bench (Piety and others, 1986) (Fig. 3). The flow is exposed along an 8-km-long section of the present Snake River Canyon and extends about 2.5 km up Pine Creek Canyon across the fault and into the Snake River Range (Fig. 3). The northwest-trending scarp records Quaternary displacement since about 1.5 Ma along this portion of the Grand Valley fault. A topographic profile surveyed on the top of the basalt indicates a net vertical displacement of 28 m, down to the southwest (Piety and others, 1986). The number of individual events is unknown. The scarp extends laterally for a distance of 0.6 km on either side of Pine Creek until it is blanketed by loess that was most likely deposited between 10 and 70 ka (see "Loess-covered surfaces").

Shurtliff Canyon. A northwest-trending linear feature, near and parallel to the inferred trace of the Grand Valley fault, is preserved on a loess-covered alluvial-fan surface (Qla) near

Shurtlilff Canyon (Fig. 3). This short, linear feature roughly correlates with a step on the upper surface of gravelly alluvium that is buried beneath loess (Piety and others, 1986). This step appears to be a loess-filled channel that has been eroded into the gravelly alluvium. However, a tectonic origin for this step cannot be completely ruled out (Piety and others, 1986). In either case, no fault scarp is preserved on the loess-covered surface above the step.

Soil development and geomorphic relationships suggest that this loess-covered surface is probably older than about 30 ka. The surface is most likely younger than about 70 ka, although an age of about 140 to 150 ka cannot be completely excluded (see "Loess-covered surfaces"). In any event, surface displacements on this portion of the Grand Valley fault are inferred to be older than about 30 ka.

Rainey Creek. Although scarps and linear features are present in Swan Valley in the vicinity of the Grand Valley fault near Rainey Creek (Fig. 3), detailed mapping shows that these features are erosional scarps rather than fault scarps (Piety and others, 1986). Scarps at the mouth of Rainey Creek are determined to be the result of erosion and downcutting along Rainey Creek and the Snake River rather than surface fault displacement because the scarps (1) are highly sinuous, (2) are short in extent, (3) are continuous as they change orientation from northeast-southwest adjacent to Rainey Creek to northwest-southeast adjacent to the Snake River, and (4) do not extend across topographically higher, loess-covered surfaces adjacent to Rainey Creek (Fig. 3).

Dry Canyon. Northwest-trending linear features, near and parallel to the inferred trace of the Grand Valley fault, are preserved on loess-covered alluvial-fan surfaces (Qla) between Dry Canyon and Rainey Creek (generalized as one linear feature southeast of Rainey Creek on Fig. 3). Similar to the linear feature preserved near Shurtliff Canyon, these short, linear features roughly correlate with steps on the surface of the gravelly alluvium. These steps are interpreted to be loess-filled channels, although a tectonic origin cannot be completely dismissed (Piety and others, 1986).

Soil development and geomorphic relationships suggest a minimum age of 11 to 15 ka for the Qla surfaces near Dry Canyon. Thus, the lack of scarps on these surfaces suggests that surface ruptures have been absent from the Grand Valley fault between Dry Canyon and Rainey Creek during at least the latest Quaternary.

Grand Valley

The middle 40 km of the Grand Valley fault strikes about N30°W from near Palisades Creek to south of the Snake River canyon near Alpine (Fig. 5). Witkind (1975a) depicted the Grand Valley fault in Grand Valley as recurrently active since the Miocene but did not report any fault scarps. Oriel and Moore (1985) showed the main trace of the Grand Valley fault as a single, highly sinuous strand at the contact between Tertiary volcaniclastic deposits and Paleozoic/Mesozoic rocks (Fig. 5). They also mapped short, down-to-the-northeast normal faults parallel to and west of the main trace of the Grand Valley fault (Fig. 5). These short faults are shown within the Tertiary sediments (Oriel and Moore, 1985). All of these short faults, plus the main trace of the Grand Valley fault, have been depicted as concealed beneath Quaternary deposits of various ages (Oriel and Moore, 1985).

Examination of aerial photographs covering the Grand Valley fault in Grand Valley indicates a potential fault scarp preserved on a late Quaternary surface, terrace Qt2, at Alpine and linear features on surfaces of older Quaternary or Tertiary sediments near Sheep and Big Elk creeks (Fig. 5).

Alpine. Only one locality along the Grand Valley fault in Grand Valley has evidence for a possible fault-related scarp on a late Quaternary surface (Piety and others, 1986). At Alpine, terrace Qt2 south of the Snake River contains a northwest-trending, 80-m-long scarp close to the concealed trace of the Grand Valley fault as mapped by Jobin (1972), Albee and Cullins (1975a), and Oriel and Moore (1985) (Fig. 5). The age of this terrace is estimated to be about 15 to 30 ka (see "Terraces of the Snake River"). The scarp varies in height from a maximum of 2 m near the Snake River to 0.5 m on its southeastern end, where it becomes obscured by a canal dug into the terrace surface. Scarp-profile measurements indicate that the maximum scarp-slope angle is 8° (Piety and others, 1986).

This scarp approximately aligns with a high-angle contact between the Mississippian Lodgepole Limestone and Qt2 terrace gravel exposed on the north side of the Snake River. This contact presumably represents the position of the Grand Valley fault. At this location, however, the contact shows no evidence of shearing (Piety and others, 1986). No scarp is present on terrace Qt2 north of the Snake River directly above this contact, although the terrace surface has been altered and is covered by road fill at this location. More important, a 240-m-long topographic profile across the scarp on the Qt2 terrace south of the Snake River indicates no net vertical displacement, strongly suggesting that the scarp is the northeast side of a channel cut into the Qt2 terrace surface (Piety and others, 1986). These observations indicate that the most-recent surface rupture along this portion of the fault predates deposition of the Qt2 terrace gravel. The Qt2 terrace gravel was, most likely, deposited over an older bedrock scarp. This conclusion is further supported by the lack of a scarp on terrace Qt1 north of the Snake River, although the location of the fault beneath this terrace is not well constrained (Piety and others, 1986) (Fig. 5).

Sheep Creek/Big Elk Creek. Short linear features are visible along the range front between Sheep Creek and Big Elk Creek on a surface underlain by Long Spring Formation (late Tertiary to early Quaternary) (Fig. 5). These features could be fault scarps. On the other hand, exposures along Sheep Creek show that the Long Spring Formation is horizontal or dipping gently to the southwest, in marked angular discordance to the underlying Salt Lake Formation, which dips 28 to 36° to the northeast into the Grand Valley fault (Berkey, 1934; Merritt,

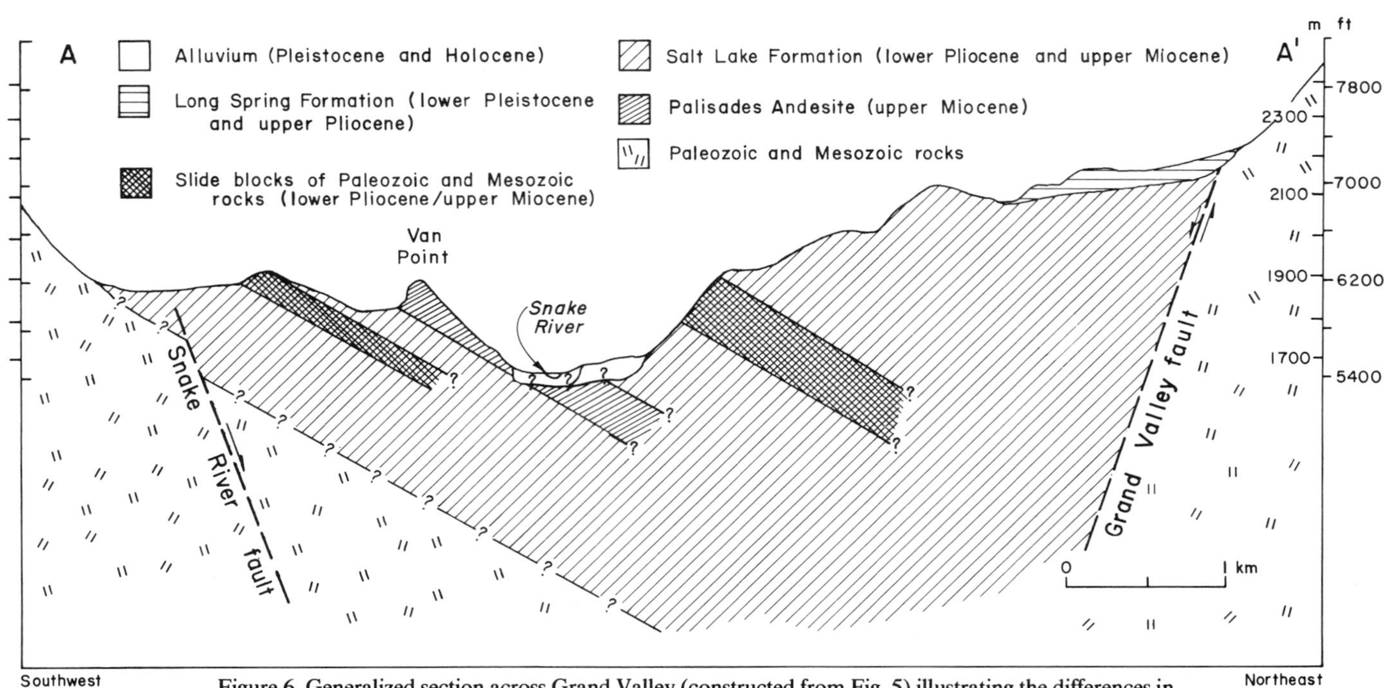

Figure 6. Generalized section across Grand Valley (constructed from Fig. 5) illustrating the differences in tilt between the Salt Lake Formation and the overlying Long Spring Formation, which partially conceals the Grand Valley fault in Grand Valley. Note that the Salt Lake Formation includes interbedded Palisades Andesite and slide blocks of Paleozoic and Mesozoic rocks.

1956, 1958; Okeson, 1958) (Figs. 5 and 6). The lack of tectonic tilt of the Long Spring Formation suggests that if these linear features are fault scarps, they represent at most only a small amount of Quaternary displacement on the Grand Valley fault at this locality.

Star Valley

The southern 58 km of the Grand Valley fault in Star Valley has been mapped at the base of the Salt River Range along the contact between Paleozoic and Mesozoic rocks and the upper Cenozoic basin-fill deposits (Rubey, 1973; Oriel and Platt, 1980) (Figs. 2 and 4). This portion of the fault strikes almost due north along most of its length. A 4-km *en echelon* right step in the trace of the fault is located at a topographic high that separates Star Valley into two structural and physiographic basins. These two basins are informally referred to as northern Star Valley and southern Star Valley.

Previous investigators, on the basis of reconnaissance studies, identified fault scarps along this southern portion of the Grand Valley fault (Star Valley fault of Rubey, 1973; Witkind, 1975b). These scarps were thought to be Holocene or Pleistocene. Linear, west-facing fault scarps are preserved discontinuously along the range front over a distance of at least 24 km along the trace of the fault in Star Valley. These scarps were evaluated at four of the larger tributaries: Willow, Phillips, Swift, and Dry creeks, from north to south (Fig. 4). These studies are discussed in Piety and others (1986) and Anders and others (1989) and are only summarized here.

Upstream of the scarps at these four large tributaries, fluvial terrace remnants are preserved in narrow valleys within the Salt River Range. Some of these remnants are directly traceable to the fault scarps, suggesting that the downcutting that resulted in terrace formation was the direct result of fault displacement. Some terraces that are not directly traceable to the fault scarps are also preserved. These terraces are used in combination with the fault scarps to help delineate the history of surface rupture on the Grand Valley fault in Star Valley.

Fault scarps. Fault scarps are easily recognizable on the Qa1 and Qa2 alluvial fans at the mouths of nine V-shaped canyons between Willow Creek in northern Star Valley and Dry Creek in southern Star Valley (Fig. 4). No fault scarps were recognized along the range front in northern Star Valley north of Willow Creek. It should be noted, however, that where alluvial-fan surfaces are preserved near the range front, these surfaces either do not cross the fault or are of such small aerial extent at the fault that scarps are difficult to locate in the thickly vegetated terrain (Piety and others, 1986). J. P. McCalpin (oral communication, 1989) noted a scarp on a surface of undetermined age at the mouth of Strawberry Creek (Fig. 4). In addition, three approximately north-trending scarps with down-to-the-west and down-to-the-east displacements are preserved about 3 km west of the range front on an alluvial surface that heads in Prater Canyon (Piety and others, 1986) (Fig. 4). The relationship between these three intrabasin scarps and the main Grand Valley fault is unknown.

Scarps on the Qa1 surfaces have vertical surface displacements between 8.3 and 11.6 m and maximum scarp-slope angles between 25 and 34° (Piety and others, 1986) (Fig. 7). Scarps on the Qa2 surfaces have vertical surface displacements between 5.0 and 5.4 m and a maximum scarp-slope angle of 27° at Dry Creek, the only site studied where a portion of a scarp on a Qa2 surface is unmodified by erosion or human activity (Piety and others, 1986). The ages of these two surfaces are estimated to be between 11 and 15 ka (see "Alluvial Fans"). At Willow Creek, a scarp that may be correlative with the scarps on the Qa2 surfaces has a vertical surface displacement of 3.3 m and a maximum angle of 17° (Piety and others, 1986). The fault scarps in Star Valley are interpreted to record at least two surface ruptures on the Grand Valley fault because the scarps on the younger Qa2 surfaces are about half as high as those on the older Qa1 surfaces.

From Prater Canyon north to Corral Canyon, Qla surfaces with an estimated minimum age of about 30 ka overlie the mapped trace of the Grand Valley fault at the mouths of large drainages, suggesting that no surface ruptures have occurred on this portion of the fault in northern Star Valley since these surfaces became stable (see "Loess-covered surfaces") (Fig. 4).

Tectonic terraces. The Qa1 and Qa2 surfaces extend upstream from the scarps several hundred meters as terraces preserved 2 to 12 m above the adjacent drainage at each of the four localities studied (Piety and others, 1986). In addition, one or two additional terraces, not directly traceable to or correlated with faulted surfaces, are preserved at three of the four localities at elevations intermediate between those of the Qa1 and Qa2 surfaces. All these terraces are thought to have a tectonic origin, at least in part, because they are preserved only upstream of the fault and because they appear to diverge downstream toward the fault, a characteristic of a tectonically formed terrace sequence (Bull, 1984). However, one or two lower terraces (Qa3 and Qa4) preserved at each of the four localities apparently do not have a chiefly tectonic origin. These terraces are 0.5 to 2.7 m above the present elevations of the drainages. Longitudinal profiles of these lower terraces are smooth across the mapped trace of the fault, so that these terraces must have formed by downcutting of the Salt River and its tributaries since the most-recent surface rupture on this portion of the Grand Valley fault.

Whereas the fault scarps indicate at least two surface ruptures on the Grand Valley fault in Star Valley since about 15 ka, one or two additional surface ruptures may be implied by the intermediate fluvial terraces preserved upstream of, but not directly at, the Grand Valley fault. A trench recently excavated by J. P. McCalpin across the 11-m scarp at Swift Creek exposed evidence for three surface ruptures since deposition of Qa1 alluvium (McCalpin, 1991). Furthermore, the presence of one or two younger terraces that cross the fault trace indicates that some time, perhaps a few thousand years, has passed since the most-recent surface rupture on the Grand Valley fault in Star Valley. Radiocarbon ages on buried soils exposed in the trench at Swift Creek indicate that the most-recent surface rupture at this site occurred about 5.5 ka (McCalpin, 1990, 1991).

Figure 7. Fault scarp in alluvium at the mouth of Swift Creek along the Grand Valley fault in southern Star Valley (Fig. 4). View is looking upvalley toward the east. Buildings are part of the town of Afton, Wyoming. Note that the scarp crosses a large, flat surface heading in the Swift Creek valley and a smaller alluvial fan to the south (right side of the photograph). On the alluvial surface of Swift Creek, the vertical surface displacement is 11 m (Piety and others, 1986). On the alluvial fan from the smaller tributary, the vertical surface displacement is 7 m (Piety and others, 1986). Note the relatively narrow, steep-walled valley of Swift Creek. Photograph taken in June 1982.

ADDITIONAL EVIDENCE FOR VARIABLE DISPLACEMENT RATES ALONG THE FAULT

Two additional types of data supplement information about displacement inferred from the locations and ages of fault scarps. First, landforms along the range front and the adjacent valley are utilized to indicate relative displacement rates along the Grand Valley fault. Certain landforms have been found to be prevalent along active range fronts, and erosion of these landforms along an inactive front probably takes 100,000 to 1,000,000 years (Bull, 1984, 1987). Because fault scarps between at least Willow Creek and Dry Creek in Star Valley indicate relatively high displacement rates along this portion of the fault during the latest Quaternary, we compare landforms along this portion of the fault with those preserved elsewhere along the Grand Valley fault. Unfortunately, several of these landforms appear to be strongly influenced by factors other than fault displacement, making a tectonic interpretation of the present landforms difficult. As a second type of data, well-exposed, datable volcanic rocks in Swan Valley are utilized to assess displacement rates on the northwest portion of the Grand Valley fault during the late Cenozoic ($\leqslant$10 Ma; Anders and others, 1989).

Range-front and valley morphology

Landforms preserved near the junction of the footwall and hanging wall can supplement information about displacement determined from fault scarps and late Quaternary surfaces (Bull, 1984, 1987). These landforms record not only uplift of the footwall and subsidence of the hanging wall but also interactions between these two processes and three other processes: channel incision across the fault and erosion and deposition in the hanging wall (Bull, 1987). Range fronts that have experienced rapid uplift along with slow channel incision and low erosion rates over the last 100,000 to 1,000,000 years typically have steep and linear profiles, narrow and V-shaped tributary valleys with steep gradients, well-defined triangular faceted spurs, and relatively young, undissected alluvial deposits on the hanging wall at the range front (Bull, 1973; Bull and McFadden, 1977; Bull, 1987). Eight landform characteristics were studied along the Grand Valley fault, following the methods of Bull and McFadden (1977) and Bull (1984, 1987) (Figs. 8 and 9), but only height of the range front, elevation and age of the oldest deposits preserved on the hanging wall adjacent to the fault, and depth of dissection of these hanging-wall deposits seem to reflect differences in displacement rates along the Grand Valley fault during the last 30,000 to 2,000,000 years (Figs. 8 and 9). These characteristics are used only as a relative measure of displacement rates along the Grand Valley fault, because of differences in climate and lithology between our study area and those of previous workers.

The highest peaks in the Snake River Range and in the Salt River Range steadily increase in elevation with distance from the eastern Snake River Plain (2,654 to 2,806 m in Swan Valley and 3,082 to 3,294 m in Star Valley; Figs. 8 and 9). However, because the elevations of the highest deposits on the hanging wall adjacent to the fault do not vary in a similar manner, the height of the range front is greatest in Star Valley between Phillips Creek and Cottonwood Creek (1,200 to 1,373 m), lowest in Grand Valley between Palisades Creek and Dry Creek (north of Corral Canyon) (650 to 800 m), and intermediate in Star Valley between Dry Creek and Strawberry Creek (1,100 to 1,190 m) and in Swan Valley northwest of Palisades Creek (800 to 975 m; Figs. 8 and 9). The depth of dissection of the deposits on the hanging wall adjacent to the fault reveals a similar pattern. Hanging-wall deposits are dissected 3 to 35 m in Star Valley between Phillips Creek and Cottonwood Creek, 299 to 595 m in Grand Valley between Palisades Creek and Dry Creek, 63 to 323 m in Swan Valley northwest of Palisades Creek, and 10 to 120 m in Star Valley between Dry Creek and about Strawberry Creek (Figs. 8 and 9). Furthermore, the ages of the oldest deposits preserved on the hanging wall adjacent to the fault also reflect this pattern. The youngest deposits (alluvial fans Qa1 and Qa2 with an estimated age of 11 to 15 ka) are preserved in Star Valley between Phillips Creek and Cottonwood Creek and at Willow Creek, Strawberry Creek, and Cedar Creek; the oldest deposits (Long Spring Formation that is inferred to be late Tertiary to early Quaternary) are preserved in Grand Valley between Palisades Creek and Dry Creek; and deposits of intermediate age (30? to 70? ka) are preserved in Swan Valley northwest of Palisades Creek and in Star Valley between Dry Creek and Prater Canyon and between Cedar Creek and Phillips Creek (Figs. 3, 4, 5, and 9).

In relating the above characteristics to displacement rates along the Grand Valley fault, we assume that these characteristics are primarily the result of interactions between the rate of downcutting along the Snake River and the rate of subsidence of each valley. We further assume that the average rate of channel downcutting has been similar along the Snake River and its tributaries in this area. Of course, fluvial terraces adjacent to the Snake River do record alternating periods of downcutting and stability, which could be the result either of changes in the climate during the Quaternary or of damming of lower Swan Valley by volcanic rocks that erupted from the eastern Snake River Plain. Despite this evidence for episodic rates of channel downcutting over time, we assume that the average downcutting rate during the Quaternary has not varied greatly for drainages along the fault.

Using the above assumptions, the greatest range-front height, the youngest deposits on the hanging wall adjacent to the fault, and the least dissection of these deposits are interpreted to support the conclusion based on the observed fault scarps that the

Figure 8. Summary of range-front and valley geomorphic parameters for the three segments of the Grand Valley fault, plus a transitional zone between the Star Valley and Grand Valley segments. Bars indicate the range of measured values; boxes indicate either single measurements for sinuosity or mean values for V_f ratios. All measurements were made following the methods of Bull and McFadden (1977) and Bull (1984, 1987), and except for sinuosity, all measurements were made using 1:24,000-scale topographic maps.

The elevations of deposits in the hanging wall are the highest elevations of deposits adjacent to the fault. Sinuosity is the ratio between the length of the range front measured along the contact between basin-fill sediments and Paleozoic/Mesozoic rocks and the straight-line length of the same portion of the range front. The closer the value is to 1.0, the more linear the range front. Sinuosity was measured using 1:250,000-scale topographic maps. The V_f ratio is the ratio of valley width to valley depth measured at a constant distance from the range front. Lower V_f ratios indicate narrower valleys, suggesting that the rate of uplift has been greater than the rate of lateral cutting of the streams. V_f ratios and depth-dissection measurements were made along drainages with basin areas of greater than 10 km^2 at distances of 1 km upstream from the fault. All parameters shown here are influenced by factors (e.g., rock type and structure, rate of channel downcutting) other than tectonic uplift (Bull and McFadden, 1977; Wells and others, 1988).

Values for Swan Valley were measured between Dry Canyon (northwest of Pine Creek just northwest of the area shown in Fig. 3) and Palisades Creek (Fig. 3). Facets, however, are present only between Dry Canyon (southwest of Rainey Creek) and Palisades Creek. Values for Grand Valley were measured between Sheep Creek and Dry Creek (near Corral Canyon in northern Star Valley, Fig. 5). Values for the transition zone were measured between Dry Creek (near Corral Canyon) and Prater Canyon (Fig. 4). Values for Star Valley were measured between Prater Canyon and Cottonwood Creek (Fig. 4).

rate of subsidence of Star Valley between Willow Creek and Cottonwood Creek along the Grand Valley fault has been faster than the rate of downcutting along the Salt River, a major tributary of the Snake River. The fault scarps indicate a relatively high rate of fault displacement during the latest Quaternary. The geomorphic parameters indicate that this rate probably persisted for several hundred thousand years and perhaps throughout the Quaternary. In contrast, the lowest range-front height, the oldest deposits on the hanging wall adjacent to the fault, and the greatest incision of these deposits are interpreted to indicate that the rate of subsidence of Grand Valley along the Grand Valley fault has been slower than the downcutting rate along the Snake River. The lack of fault scarps, even on surfaces of the Long Spring Formation, supports this conclusion.

Three other characteristics studied—range-front sinuosity, location and slope of faceted spurs, and V_f ratios—appear to be strongly influenced by factors other than range uplift and valley subsidence. Orientation of the complexly folded and faulted Paleozoic and Mesozoic beds within the Snake River Range and the Salt River Range adjacent to the Grand Valley fault appears to have been a main factor influencing these interrelated processes operating along the fault. For example, the linear, faceted portion of the fault in Swan Valley between Palisades Creek and Dry Canyon (immediately south of Rainey Creek, Figs. 2 and 3) corresponds with beds that dip 32 to 55° into Swan Valley. A similar correspondence occurs between Prater Canyon and Willow Creek in northern Star Valley (Figs. 2 and 4). In contrast, the range front is relatively sinuous (sinuosity of ⩾1.2) and lacks facets everywhere where beds dip into the range except in Star Valley between Willow Creek and Cottonwood Creek. Along this section of the fault, the range front is steep and linear with prominent facets, although beds dip 10 to 40° into the Salt River

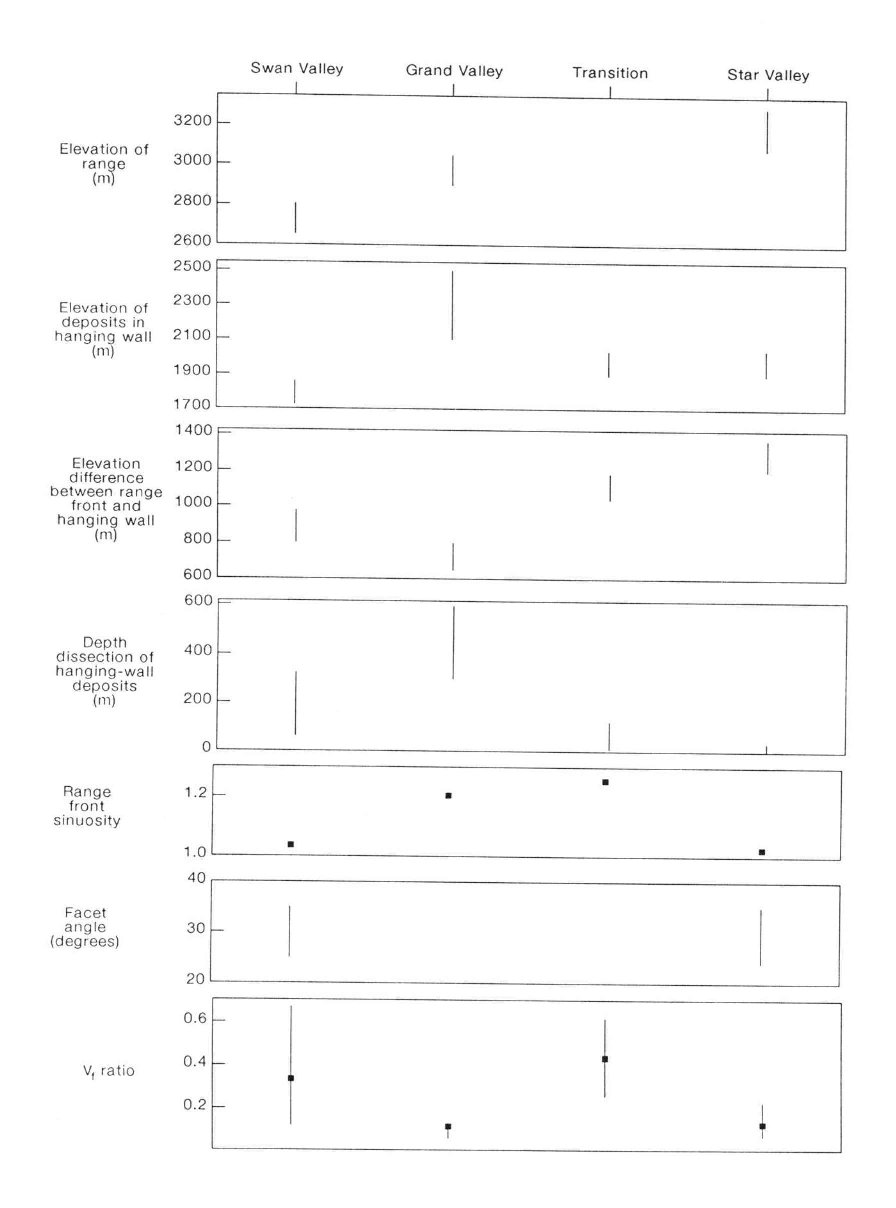

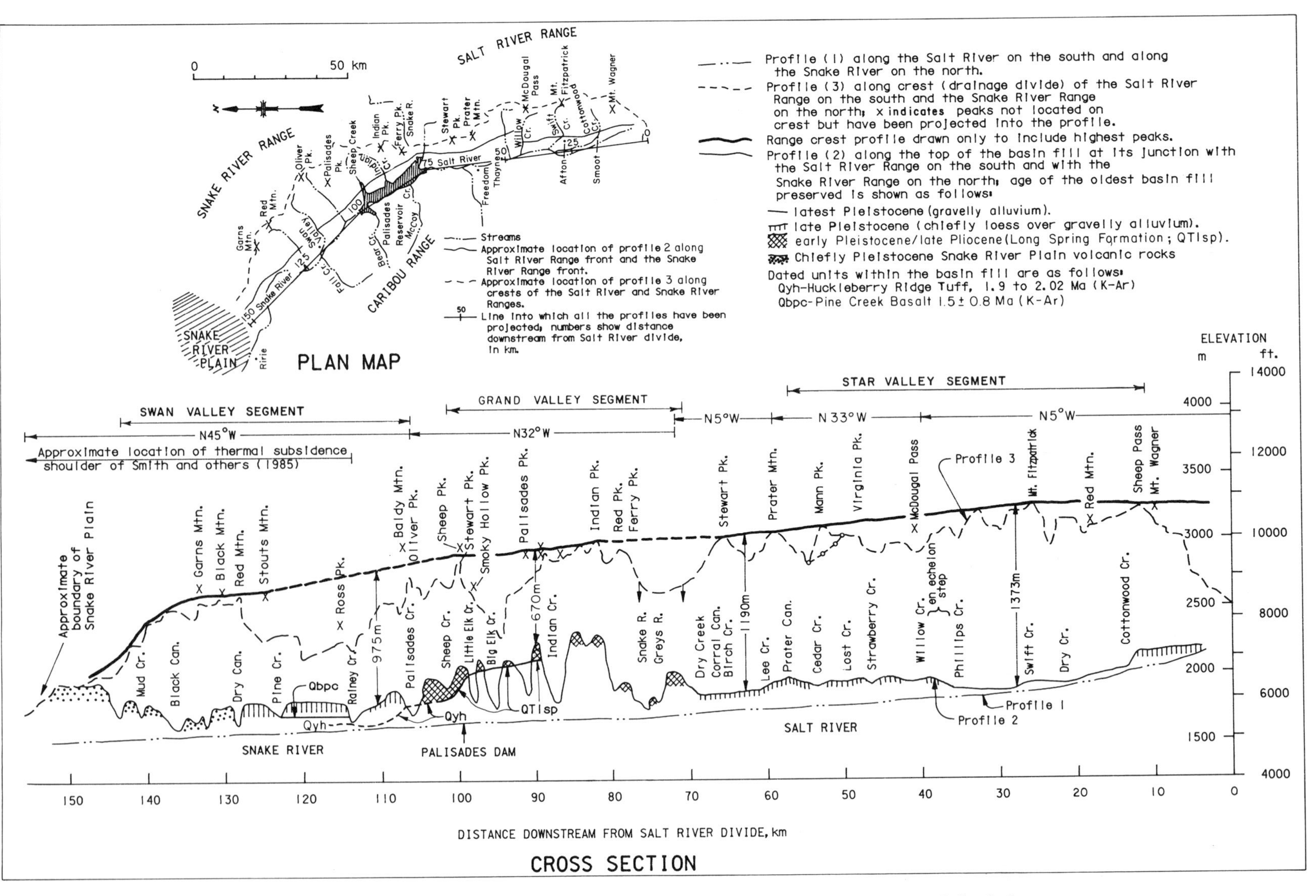

Figure 9. Topographic profiles along the footwall and hanging wall adjacent to the Grand Valley fault from southern Star Valley to the eastern Snake River Plain. Profile locations are shown on the inset map. The lines indicating the Pine Creek Basalt, the Huckleberry Ridge Tuff, and the base of the Long Spring Formation have been projected into the plane of Profile 2 from their elevations 2 to 3 km west of the Grand Valley fault (Figs. 3 and 5). These units are most likely tilted more steeply at the fault than shown here.

Range. These observations suggest that, except between Willow Creek and Cottonwood Creek, erosion along bedding-plane slopes has probably been partially, or totally, responsible for the characteristics preserved along the range front adjacent to the Grand Valley fault. Furthermore, relatively low V_f ratios both in Star Valley between Cedar Creek and Cottonwood Creek and in Grand Valley between Palisades Creek and Dry Creek (north of Corral Canyon) (Figs. 4, 5, and 8), where other data indicate differences in Quaternary displacement rates, suggest that erodibility of the footwall rocks may have influenced the shapes of the valleys within the Salt River and Snake River ranges. The wide scatter in values measured in Swan Valley (northwest of Palisades Creek) and between Dry Creek (north of Corral Canyon) and Prater Canyon supports this conclusion (Fig. 8).

Tectonic tilt of lower Quaternary and upper Tertiary units

Displacement rates during discrete temporal intervals were determined for the northwest portion of the Grand Valley fault using (1) abundant isotopic dates from Swan Valley and adjacent areas, (2) tectonic tilts or the differences between paleomagnetically determined unit mean and site mean directions from tuffs in Swan Valley and from the same units at undisturbed sites outside Swan Valley, and (3) a simple model that relates tectonic tilt and fault displacement (Fig. 12 in Anders and others, 1989). These techniques and the conclusions derived from them are presented in Anders and others (1989) and will not be repeated here.

These studies indicate that since about 10 Ma, a maximum interval displacement rate of 1.8 mm/yr occurred on the Grand Valley fault in Swan Valley between 2 and 4.3 Ma, with markedly lower rates before and after this interval: about 0.019 mm/yr since about 2 Ma and 0.12 to 0.16 mm/yr before 4.3 Ma (Table 3 and Fig. 14 in Anders and others, 1989). These studies also show that apparent northeast dips of as much as 16° on eutaxitic structures in Huckleberry Ridge Tuff on Palisades Bench (immediately northwest of Palisades Creek, Fig. 3) are due to draping of the ash-flow tuff over preexisting topography rather than to tectonic tilting.

Although similar datable units with which late Cenozoic displacement rates could be quantitatively assessed were not recovered from the volcanic and sedimentary deposits exposed in Grand Valley, similarity in tilt (28 to 36°, Fig. 6) of the Salt Lake Formation (about 10 Ma to about 4 Ma) in Grand Valley and in Swan Valley implies that the Grand Valley fault in Grand Valley may have also experienced an interval of relatively rapid displacement after about 4 Ma. In addition, the Salt Lake Formation in Grand Valley includes large, coherent masses of brecciated Cambrian to Mississippian sedimentary rocks interpreted as gravity-slide blocks (Oriel and Moore, 1985; Moore and others, 1987; Anders, 1990a; Boyer and Hossack, this volume). Sliding of thrust-sheet masses into Grand Valley during the Pliocene and Miocene has been related to displacement on the Grand Valley fault in Grand Valley between about 4 Ma and about 8 Ma (Moore and others, 1987). Sliding of these large blocks definitely occurred before deposition of the Long Spring Formation about 2 Ma (Anders, 1990a).

SEGMENTATION AND PALEOSEISMICITY OF THE GRAND VALLEY FAULT

Based on the similarity in the sizes of surface displacements associated with paleoearthquakes on the Wasatch and San Andreas faults, Schwartz and Coppersmith (1984) proposed a characteristic earthquake model that suggests that faults or fault segments tend to generate earthquakes of essentially the same size within a relatively narrow range of magnitudes near the maximum. The generally uniform behavior of faults or fault segments over time has been exhibited by historic surface ruptures in the western United States. Historic earthquakes in the ISB—the 1934 Hansel Valley earthquake of magnitude 6.6 (Shenon, 1936), the 1959 Hebgen Lake earthquake of magnitude 7.5 (Myers and Hamilton, 1964), the 1983 Borah Peak earthquake of magnitude 7.3 (Crone and Machette, 1984; Crone and others, 1987) (Fig. 1), and ten other events in the Basin and Range (Bucknam and others, 1980)—have all occurred on faults with a history of late Quaternary or younger surface displacements similar in size to the surface ruptures produced by the historic events. We assume this model is applicable to late Tertiary and Quaternary fault behavior and use it to estimate average return periods for large-magnitude, surface-rupturing paleoearthquakes on the Grand Valley fault.

The magnitudes of these large-magnitude paleoearthquakes can be estimated from fault-scarp data because both the length and size of historic surface ruptures have been related statistically to earthquake magnitude (Slemmons, 1977; Bonilla and others, 1984). The length of a fault that has ruptured repeatedly as a unit during large-magnitude ($M > 6½$) paleoearthquakes can be estimated by recognizing boundaries to rupture on a fault, that is, earthquake segments of dePolo and others (1989). The size of the surface ruptures generated by these paleoearthquakes can be estimated from relations between fault scarps and geomorphic surfaces, multiple-event fault scarp morphology, or trench data (McCalpin, 1987).

Utilizing the location and age of latest Quaternary fault scarps and the morphology of deposits in the valley adjacent to the fault, we delineate segments of the Grand Valley fault by identifying differences in displacement history and in estimated average return periods of large-magnitude paleoearthquakes along the fault. Changes in displacement history along the fault are assumed to define structural boundaries that may have served to limit the length of paleoseismic surface ruptures.

Star Valley segment

Only one portion of the Grand Valley fault, part of the fault in Star Valley, exhibits evidence for recurrent latest Quaternary surface rupture. We term this portion of the fault the Star Valley segment. This segment is characterized by scarps on late Quaternary surfaces, a linear range front with prominent facets, and relatively young, undissected deposits in Star Valley adjacent to the fault. These landform characteristics are associated with faults that have experienced rapid displacement rates relative to the downcutting and erosion rates operating at the range front.

An average latest Quaternary displacement rate of 0.6 to 1.1 mm/yr for the Star Valley segment of the Grand Valley fault is estimated from vertical displacements of 8.3 to 11.6 m on surfaces with estimated ages of 11 to 15 ka. Higher scarps on older surfaces and preservation of two to possibly four tectonic terraces in the larger drainages upstream of the fault suggest that at least two, and possibly as many as four, surface ruptures have occurred on the Star Valley segment since about 15 ka. Assuming two to four events have occurred on the fault since about 15 ka, the average return period for surface-rupturing events is 2,500 to 7,500 years (Table 3). Recent trenching across a scarp near Afton suggests three surface ruptures since 14 to 16 ka with intervals of 2,500 years to 5,900 to 6,400 years between ruptures (McCalpin, 1990, 1991). Episodic incision of the alluvial fans on which the scarps are located, which is indicated by the one or two alluvial surfaces that cross the fault undisturbed, implies that the time since the youngest event has been at least a few thousand years. McCalpin (1991) suggested that the most-recent surface rupture at Swift Creek occurred about 5.5 ka.

The time interval over which a relatively rapid displacement rate has occurred on the Star Valley segment is not known. The presence of a linear, faceted range front, which is believed to have a tectonic origin along this portion of the fault; a high range-front escarpment; and relatively young, undissected alluvial deposits that presumably bury older Quaternary deposits on the hanging wall adjacent to the fault suggest that a high displacement rate has persisted on the Star Valley segment, possibly throughout the Quaternary. The displacement rate of 0.6 to 1.1 mm/yr, which is estimated for approximately the last 15,000 years, may be representative of the longer-term displacement rate on the Star Valley segment.

The southern boundary of the Star Valley segment is placed at Cottonwood Creek (Fig. 4). Although the southernmost locality exhibiting evidence for latest Quaternary surface ruptures is Dry Creek, the segment is extended to Cottonwood Creek because fault scarps would not be readily preserved along the 5-km-long, linear, steep range front between these two localities. South of Cottonwood Creek, the physiographic expression of Star Valley dies out (Fig. 2), and the Grand Valley fault has not been mapped south of about this locality (Rubey, 1973).

The northern boundary of the Star Valley segment is more problematic. Although the northernmost locality with well-defined evidence for recurrent latest Quaternary surface ruptures is Willow Creek, such ruptures cannot be precluded between Willow Creek and Prater Canyon (Fig. 4). Several factors complicate the interpretation of fault displacement between Willow Creek and Prater Canyon. Although the range front remains steep, faceted, and linear—characteristics that suggest that the displacement rate here has been similar to that along the fault south of Willow Creek—bedrock along the range front between Willow Creek and Prater Canyon dips into Star Valley at an angle similar to that of the facets. As a result, the morphology of the range front may be in part bedrock controlled. In addition, the range front here is thickly vegetated, which makes recognition of fault scarps difficult. Furthermore, loess-covered surfaces that are inferred to be older than the surfaces containing scarps between Willow Creek and Dry Creek are preserved only between drainages and have an undiscernible relationship to the Grand Valley fault (Fig. 4). Because latest Quaternary surface ruptures cannot be precluded between Willow Creek and Prater Canyon, the northern boundary of the Star Valley segment is placed near

TABLE 3. ESTIMATED RETURN PERIODS FOR SURFACE DISPLACEMENT EVENTS AND QUATERNARY DISPLACEMENT DATA FOR SEGMENTS OF THE GRAND VALLEY FAULT

Segment	Geologic Unit Used as Datum	Age of Datum (ka)	Displacement of Datum (m)	Estimated Number of Events	Average Return Period (kyr)*
Star Valley	Alluvial fan	11 to 15	3.3 to 11.6	2 to 4	2.5 to 7.5
Swan Valley	Loess-covered alluvial fan	11 to 30†	0	0	
	Pine Creek Basalt	1,500	28	5 to 14§	107 to 300
Grand Valley	Fluvial terrace	15 to 30	0	0	

*Average return periods are calculated using the range of the age of the deposits and the range of the number of events.

†This is a minimum age estimate.

§The 28 m of displacement is assumed to have occurred in events with surface displacements ranging from 2 to 6 m.

Prater Canyon, even though displacements along the 16-km portion of the Grand Valley fault between Willow Creek and Prater Canyon may be older than those on the fault south of Willow Creek. The maximum surface rupture length of the Star Valley segment, then, is estimated to be about 45 km.

The Star Valley segment includes portions of the Grand Valley fault that are both north and south of a topographic high composed of Paleozoic-Mesozoic rocks and Tertiary-Quaternary basin fill (Fig. 4). This topographic/structural high, along with a 4-km right step in the fault, suggests a structural boundary. However, latest Quaternary fault scarps with similar total vertical displacements are preserved both north and south of this possible structural boundary, and range-front morphology is similar on both sides of the step between Cottonwood Creek and Willow Creek. Because we attempt to identify earthquake segments, portions of the fault that ruptured simultaneously during paleoearthquakes, we extend the Star Valley segment across this possible structural boundary. The interbasin high that defines the boundary could be a "leaky boundary" as described by Crone and Haller (1989). A leaky boundary is one that impedes rupture propagation but does not completely prevent rupture on an adjacent segment. Alternatively, interbasin highs, such as this one between northern and southern Star Valley, may not be rupture barriers but instead could represent regions of fault overlap resulting from distributed fault displacment. If fault rupture events occur more frequently on one of the two *en echelon* faults, then that fault will grow into the basin (or footwall) of the less active fault, causing fault overlap. These overlapping faults are called "cross-over" faults (Anders, 1990b). Because both faults have a common origin at depth, upward propagation of rupture will select only one of the overlapping faults. The surface manifestation of this is the transfer of rupture from one *en echelon* segment to the other. Such transfers do not significantly affect strain release along strike and therefore do not significantly impede rupture. Rupture barriers are common features on large normal faults, but the presence of an interbasin high such as the one between northern and southern Star Valley is not necessarily diagnostic of a rupture barrier.

The magnitude of the paleoearthquakes that caused the surface ruptures on the Star Valley segment can be estimated using two methods: the estimated maximum length of rupture and the estimated displacement that occurred during each rupture. Both of these parameters have been related statistically to earthquake magnitude using historic earthquake data (Slemmons, 1977; Bonilla and others, 1984). First, a rupture length of as much as 45 km would have been produced by paleoearthquakes of magnitude 7 to 7½, according to statistical relations between earthquake magnitude and surface rupture length of historic earthquakes (Slemmons, 1977; Bonilla and others, 1984). In addition, historic earthquakes of this magnitude in the Basin and Range and in the ISB have resulted in surface ruptures of similar length (Doser and Smith, 1989).

Second, differing scarp heights on alluvial fans of different ages suggest that individual paleoearthquakes had surface displacements of at least 2 m and perhaps as much as 6 m (Piety and others, 1986). McCalpin (1990) suggested single-event displacements of 3 to 4 m. Surface ruptures of this size have been produced by historic earthquakes of magnitude 7 to 7½ (Myers and Hamilton, 1964; Crone and Machette, 1984), and statistical relations between displacement and magnitude (Bonilla and others, 1984) support the conclusion that paleoearthquakes of this magnitude have recurred on this segment of the Grand Valley fault during the latest Quaternary.

Swan Valley segment

Fault scarps are preserved at only one locality along the Swan Valley segment of the Grand Valley fault. At this locality evidence for Quaternary surface rupture is indicated by a 28-m-high scarp on a 1.5-Ma basalt flow that crosses the fault along Pine Creek (Fig. 3). This yields an average Quaternary displacement rate of 0.019 mm/yr since 1.5 Ma. Colluvial, alluvial, and eolian deposits younger than the 1.5-Ma basalt overlie this segment of the Grand Valley fault, but these deposits could not be dated. However, loess-capped surfaces with estimated minimum ages of 11 to 30 ka overlie the mapped trace of the Grand Valley fault in at least two localities in Swan Valley. These relationships indicate that the most-recent surface rupture on the Swan Valley segment is at least older than those along the Star Valley segment to the south.

Late Tertiary and Quaternary displacement rates (since about 10 Ma) on the Swan Valley segment have been reconstructed from paleomagnetic studies of well-exposed, datable volcanic units (Anders and others, 1989). These studies indicate that since about 10 Ma, a maximum interval displacement rate of 1.8 mm/yr occurred on the Swan Valley segment between about 2 and 4.3 Ma, when caldera activity on the eastern Snake River Plain was centered just north of Swan Valley; rates before and after this interval were markedly lower (Anders and others, 1989).

The southern boundary of the Swan Valley segment is located within a 5-km zone between Palisades Creek and Sheep Creek on the basis of the lack of fault scarps on alluvial surfaces that could be as old as the basalt at Pine Creek and on the basis of geomorphic characteristics along the range front (see "Range-front and Valley Morphology"). Although not defined by our study, the physiographic expression of Swan Valley and the height of the front of the Snake River Range suggest that the Swan Valley segment extends to within about 10 km of the eastern Snake River Plain (Figs. 2 and 9). The maximum rupture length of the Swan Valley segment, then, is estimated to be about 40 km, similar to that estimated for the Star Valley segment (about 45 km). Statistical relations between rupture length and earthquake magnitude (Bonilla and others, 1984) imply that the Swan Valley segment probably experienced large-magnitude paleoearthquakes.

The morphology of the scarp at Pine Creek provides no indication of the size or timing of the individual events that formed this scarp. To estimate the average return period of scarp-forming displacements, we assume a characteristic earthquake model with individual events averaging 2 m or more (Table 3). On this basis, an average return period of about 100,000 years or longer is estimated for large-magnitude paleoearthquakes on the Swan Valley segment since about 1.5 Ma (Table 3). Higher displacement rates indicated by the paleomagnetic data for the 2-million-year interval before deposition of Pine Creek Basalt may suggest that the events recorded by the 28-m-high scarp at Pine Creek occurred more frequently early in the interval following deposition of the basalt, so that late Quaternary surface ruptures may have been less frequent than indicated by the average Quaternary rate estimated using the scarp on the Pine Creek Basalt or even nonexistent.

In short, three characteristics readily distinguish the Swan Valley segment from the Star Valley segment. First, evidence for recurrent latest Quaternary surface ruptures is lacking on the Swan Valley segment, whereas such ruptures characterize the Star Valley segment. Second, the possible Quaternary displacement rate of 0.6 to 1.1 mm/yr on the Star Valley segment contrasts with the displacement rate of 0.019 mm/yr on the Swan Valley segment since about 1.5 Ma. Third, the average return period for large-magnitude, surface-rupturing earthquakes of about 2,500 to 7,500 years on the Star Valley segment during the latest Quaternary, and possibly during the entire Quaternary, is markedly shorter than the average return period of 100,000 years or longer for similar earthquakes on the Swan Valley segment during the Quaternary.

Grand Valley segment

The Grand Valley segment of the Grand Valley fault between about Palisades Creek and Dry Creek (immediately north of Corral Canyon) lacks fault scarps or other evidence for surface ruptures, even though lower Quaternary–upper Tertiary Long Spring Formation and younger colluvial, alluvial, and eolian deposits locally overlie the fault. This lack of fault scarps is consistent with the lack of evidence for discernible tectonic tilt of the Long Spring Formation. In addition, the Long Spring Formation is dissected to depths of 299 to 595 m adjacent to the range front. This morphologic evidence indicates that the Quaternary displacement rate on the Grand Valley segment has been lower than the downcutting rate of the Snake River, and we infer, on the basis of this evidence, that the Quaternary displacement rate has been lower than the Quaternary displacement rates on either the Swan Valley segment to the northwest or the Star Valley segment to the south.

Segment boundaries are defined in part by the termination of latest Quaternary fault scarps and in part by the position of structural boundaries inferred from the morphology of deposits on the hanging wall adjacent to the fault. The southern boundary of the Grand Valley segment is placed near Dry Creek (near Corral Canyon, Fig. 4), which is located about 16 km north of the termination of the latest Quaternary fault scarps on the Star Valley segment. This locality is also the southern extent of upper Quaternary–lower Tertiary Long Spring Formation and upper Tertiary Salt Lake Formation preserved on the hanging wall adjacent to the fault (Fig. 5). Between Prater Canyon and Dry Creek in northern Star Valley (Fig. 4), loess-covered surfaces with an estimated age of at least 30 ka are preserved discontinuously over the mapped trace of the fault at the mouths of the larger drainages. This establishes that the latest Quaternary surface ruptures on the Star Valley segment did not extend north of Prater Canyon. The range front north of Prater Canyon is more sinuous than it is along the Star Valley segment, and although some facets are present, they are less distinct than those along the Star Valley segment and could be, at least in part, erosional features. Deposits preserved in northern Star Valley adjacent to the fault between Prater Canyon and Dry Creek are older than and more dissected than those preserved south of Willow Creek but are younger than and less dissected than those preserved north of Dry Creek along the Grand Valley segment (Figs. 5 and 9). This 13-km-long section of the fault could be considered a separate fault segment but here is considered a transitional zone between the Grand Valley segment to the north and the Star Valley segment to the south.

The location of the northern boundary of the Grand Valley segment at the Swan Valley segment is also defined by the extent of the lower Quaternary–upper Tertiary deposits. Although fault scarps similar to the one at Pine Creek were not recognized elsewhere in Swan Valley or in Grand Valley, characteristics of the hanging wall give some indication of the portion of the fault that has demonstrable Quaternary displacement. Southeast of about Palisades Creek, deposits preserved on the hanging wall increase in age, and these deposits are more deeply dissected than hanging-wall deposits northwest of Palisades Creek (Figs. 3 and 9). The elevation of the deposits on the hanging wall rises between Palisades Creek and Sheep Creek. In addition, surfaces underlain by Long Spring Formation (late Tertiary to early Quaternary) southeast of Palisades Creek contain no identifiable scarps similar in size to the one at Pine Creek, even though these surfaces should be old enough to record such ruptures if they extended southeast of Palisades Creek. Thus, the boundary between the Grand Valley segment and the Swan Valley segment to the north is located in a 5-km zone between Palisades Creek and Sheep Creek (Figs. 3 and 5), because the differences in the elevations and dissection of the upper Tertiary–lower Quaternary deposits infer a related difference in displacement rates along the fault. We use this difference in displacement rate to identify a structural boundary between Palisades Creek and Sheep Creek.

No quantitative data are available to constrain the Quaternary displacement rate on the Grand Valley segment, except at one locality where the Snake River crosses the fault. At Alpine, a late Quaternary terrace of the Snake River was deposited across the Grand Valley fault between 15 and 30 ka. Detailed mapping and profiling have established that no surface displacements have occurred at this location since the deposition of this terrace gravel. This conclusion is consistent with the lack of fault scarps along this segment of the fault and suggests that no large-magnitude, surface-rupturing paleoearthquakes have occurred since 15 to 30 ka. This lack of large-magnitude paleoearthquakes is in contrast to evidence for two to four such paleoearthquakes on the Star Valley segment over a similar time interval. On the other hand, the amount of tectonic tilt within the Salt Lake Formation and the lack of tectonic tilt within the Long Spring Formation suggest that the displacement history on the Grand Valley segment may have been roughly similar to that on the Swan Valley segment during the last 10,000,000 years, although no data collected during our study demonstrate such a similarity.

CONCLUSIONS

The 140-km-long Grand Valley fault is subdivided into three segments that have been distinguished on the basis of estimates of the Quaternary displacement rates, the recency of surface displacement, and the estimated average return periods of paleoearthquakes. The Star Valley segment has abundant evidence for multiple latest Quaternary surface displacements. An average return period of 2,500 to 7,500 years was estimated for large-magnitude paleoearthquakes since 11 to 15 ka. This contrasts with the Grand Valley and Swan Valley segments, where no fault scarps were identified on latest Quaternary surfaces. Average return periods of ≥100,000 years have been estimated for large-magnitude paleoearthquakes during the Quaternary on the Swan Valley segment. An average Quaternary displacement rate on this segment is estimated to be 0.019 mm/yr; morphological evidence along the Grand Valley segment suggests that its Quaternary displacement rate has been significantly lower than that on the Swan Valley segment. The height of the range front and the age, elevation, and dissection of deposits on the hanging wall combined with mapping and isotopic ages of volcanic rocks support the inference that segmentation of the Grand Valley fault persisted during the late Tertiary and Quaternary.

The segmentation of this late Cenozoic normal fault, as for other late Cenozoic normal faults in the eastern Basin and Range, is an important consideration in assessing the seismic hazard posed by the fault. The differences in interval displacement rates and in recency of displacement can be interpreted to indicate that surface displacements and associated large-magnitude paleoearthquakes have generally been restricted to individual segments. Thus, the segment length provides an estimate of the rupture length that may be associated with future earthquakes on the segment. Furthermore, the average return periods for large-magnitude paleoearthquakes during the late Quaternary and the estimated average displacement rates during the Quaternary have been shown to differ among the segments, suggesting that the likelihood of occurrence of future large-magnitude earthquakes also differs among the segments.

ACKNOWLEDGMENTS

This paper is based on an earthquake-hazard evaluation for Palisades Dam supported by the U.S. Bureau of Reclamation (Piety and others, 1986). J. D. Gilbert (U.S. Bureau of Reclamation, Boise) greatly contributed to the fieldwork and conclusions for the Grand Valley fault in Swan Valley and Grand Valley. We thank the late S. S. Oriel and D. W. Moore (U.S. Geological Survey, Denver) for sharing their extensive knowledge of the study area, C. K. Wood (U.S. Bureau of Reclamation, Denver) for providing the plot of earthquakes in Figure 1, J. S. Dixon (Champlin Petroleum Company, Denver) for allowing us access to proprietary seismic reflection data for Star Valley and Swan Valley, and G. F. Embree (Ricks College, Rexburg, Idaho) for introducing us to the volcanic stratigraphy in Swan Valley. Discussions with D. W. Moore, A. R. Nelson, and K. L. Pierce (all of the U.S. Geological Survey, Denver) helped us interpret the ages of Quaternary deposits in the study area. M. A. Ellis, P.L.K. Knuepfer, and A. R. Nelson reviewed earlier drafts of this paper and provided many helpful comments. A review by R. E. Klinger considerably improved this chapter. Results of the isotopic dating and paleomagnetic studies in Swan Valley, which included work by J. W. Geissman (University of New Mexico, Albuquerque) under contract to the U.S. Bureau of Reclamation, are published in Anders and others (1989). Numerous landowners in Star Valley and Swan Valley kindly allowed us access to their land during our study. Figures were drafted by Tim Horner, David Kellstadt, and Mike Stauffer (Drafting Section, U.S. Bureau of Reclamation, Denver).

REFERENCES CITED

Albee, H. F., and Cullins, H. L., 1975a, Geologic map of the Alpine quadrangle, Bonneville County, Idaho, and Lincoln County, Wyoming: U.S. Geological Survey Geologic Quadrangle Map GQ-1259, scale 1:24,000.

—— , 1975b, Geologic map of the Poker Peak quadrangle, Bonneville County, Idaho: U.S. Geological Survey Geologic Quadrangle Map GQ-1260, scale 1:24,000.

Anders, M. H., 1990a, Late Cenozoic evolution of Grand and Swan Valleys, Idaho, *in* Roberts, S., ed., Geologic field tours of western Wyoming and parts of adjacent Idaho, Montana, and Utah (Geological Society of America Rocky Mountain Section meeting guidebook, field trip 1, part 2): Laramie, Geological Survey of Wyoming Public Information Circular 29, p. 14–25.

—— ,1990b, Cross-over faults: Their role in fault segmentation and basin evolution: Geological Society of America Abstracts with Programs, v. 22, p. 1.

Anders, M. H., and Geissman, J. W., 1983, Late Cenozoic evolution of Swan Valley, Idaho [abs.]: EOS (Transactions of the American Geophysical Union), v. 64, p. 858.

Anders, M. H., Geissman, J. W., Piety, L. A., and Sullivan, J. T., 1989, Parabolic distribution of circum-eastern Snake River Plain seismicity and latest Quaternary faulting: Migratory pattern and association with the Yellowstone hotspot: Journal of Geophysical Research, v. 94, p. 1589–1621.

Armstrong, F. C., and Oriel, S. S., 1965, Tectonic development of the Idaho-Wyoming thrust belt: American Association of Petroleum Geologists Bulletin, v. 49, p. 1847–1866.

Armstrong, R. L., Leeman, W. P., and Malde, H. E., 1975, K-Ar dating, Quaternary and Neogene volcanic rocks of the Snake River Plain, Idaho: American Journal of Science, v. 275, p. 225–251.

Armstrong, R. L., Harakal, J. E., and Neill, W. M., 1980, K-Ar dating of Snake River Plain (Idaho) volcanic rocks—New results: Isochron/West, no. 27, p. 5–10.

Berkey, C. P., 1934, The geologic features of certain damsites on the South Fork of the Snake River at Dry Creek and Grand Valley with supplementary note on the nature of the foundation rock: Denver, Colorado, unpublished report to U.S. Bureau of Reclamation, 14 p.

Birkeland, P. W., 1984, Soils and geomorphology: New York, Oxford University Press, 372 p.

——, 1985, Quaternary soils of the western United States, *in* Boardman, J., ed., Soils and Quaternary landscape evolution: New York, John Wiley, p. 303–324.

Birkeland, P. W., Colman, S. M., Burke, R. M., Shroba, R. R., and Meierding, T. C., 1979, Nomenclature of alpine glacial deposits, or what's in a name?: Geology, v. 7, p. 401–406.

Blackwelder, E., 1915, Post-Cretaceous history of the mountains of central western Wyoming: Journal of Geology, v. 23, p. 307–340.

——, 1931, Pleistocene glaciation of the Sierra Nevada and Basin Ranges: Geological Society of America Bulletin, v. 42, p. 865–922.

Bonilla, M. G., Mark, R. K., and Lienkaemper, J. J., 1984, Statistical relations among earthquake magnitude, surface rupture length, and surface fault displacement: Bulletin of the Seismological Society of America, v. 74, p. 2379–2411.

Brookes, I. A., 1985, Weathering, *in* Rutter, N. W., ed., Dating methods of Pleistocene deposits and their problems: Geological Association of Canada Reprint Series 2, p. 61–71.

Bruhn, R. L., Gibler, P. R., and Parry, W. T., 1987, Rupture characteristics of normal faults—An example from the Wasatch fault zone, Utah, *in* Coward, M. P., Dewey, J. F., and Hancock, P. L., eds., Continental extensional tectonics: Geological Society of London Special Publication 28, p. 337–353.

Bucknam, R. C., and Anderson, R. E., 1979, Estimation of fault-scarp ages from a scarp-height-slope-angle relationship: Geology, v. 7, p. 11–14.

Bucknam, R. C., Algermissen, S. T., and Anderson, R. E., 1980, Patterns of late Quaternary faulting in western Utah and an application in earthquake hazard evaluation, *in* Proceedings, Earthquake hazards along the Wasatch and Sierra Nevada frontal fault zones, Conference X, Alta, Utah, July–August 1979: U.S. Geological Survey Open-File Report 80-801, p. 299–314.

Bull, W. B., 1973, Local base-level processes in arid fluvial systems: Geological Society of America Abstracts with Programs, v. 5, p. 562.

——, 1984, Tectonic geomorphology: Journal of Geological Education, v. 32, p. 310–324.

——, 1987, Relative rates of long-term uplift of mountain fronts, *in* Crone, A. J., and Omdahl, E. M., eds., Proceedings, Directions in paleoseismology, Conference XXXIX, Albuquerque: U.S. Geological Survey Open-File Report 87-673, p. 192–202.

Bull, W. B., and McFadden, L. D., 1977, Tectonic geomorphology north and south of the Garlock fault, California, *in* Doehring, D. O., ed., Geomorphology in arid regions, Proceedings of the Eighth Annual Geomorphology Symposium: Binghamton, State University of New York, p. 115–138.

Christiansen, R. L., 1982, Late Cenozoic volcanism of the Island Park area, eastern Idaho, *in* Bonnichsen, B., and Breckenridge, R. M., eds., Cenozoic geology of Idaho: Idaho Bureau of Mines and Geology Bulletin 26, p. 345–368.

Christiansen, R. L., and Blank, H. R., Jr., 1972, Volcanic stratigraphy of the Quaternary rhyolite plateau in Yellowstone National Park: U.S. Geological Survey Professional Paper 729-B, 18 p.

Cluer, J. K., 1988, Quaternary geology of Willow Creek and some age constraints on prehistoric faulting, Lost River Range, east-central Idaho: Bulletin of the Seismological Society of America, v. 78, p. 946–955.

Colman, S. M., and Pierce, K. L., 1977, Summary table of Quaternary dating methods: U.S. Geological Survey Miscellaneous Field Studies Map MF-904.

——, 1986, Glacial sequence near McCall, Idaho: Weathering rinds, soil development, morphology, and other relative-age criteria: Quaternary Research, v. 25, p. 25–42.

Corbett, M. K., 1982, Superposed tectonism: Northern Idaho-Wyoming thrust belt, *in* Powers, R. B., ed., Geologic studies of the Cordilleran Thrust Belt: Denver, Colorado, Rocky Mountain Association of Geologists, p. 341–356.

Crone, A. J., and Haller, K. M., 1989, Segmentation of basin-and-range normal faults: Examples from east-central Idaho and southwestern Montana, *in* Schwartz, D. P., and Sibson, R. H., eds., Proceedings, Fault segmentation and controls of rupture initiation and termination, Conference XLV, Palm Springs, California: U.S. Geological Survey Open-File Report 89-315, p. 110–130.

Crone, A. J., and Machette, M. N., 1984, Surface faulting accompanying the Borah Peak earthquake, central Idaho: Geology, v. 12, p. 664–667.

Crone, A. J., and six others, 1987, Surface faulting accompanying the Borah Peak earthquake and segmentation of the Lost River fault, central Idaho: Bulletin of the Seismological Society of America, v. 77, p. 739–770.

dePolo, C. M., Clark, D. G., Slemmons, D. B., and Aymard, W. H., 1989, Historical basin and range province surface faulting and fault segmentation, *in* Schwartz, D. P., and Sibson, R. H., eds., Proceedings, Fault segmentation and controls of rupture initiation and termination, Conference XLV, Palm Springs, California: U.S. Geological Survey Open-File Report 89-315, p. 131–162.

Dixon, J. S., 1982, Regional structural synthesis, Wyoming salient of Western Overthrust belt: American Association of Petroleum Geologists Bulletin, v. 66, p. 1560–1580.

Doser, D. I., 1985, Source parameters and faulting processes of the 1959 Hebgen Lake, Montana, earthquake sequence: Journal of Geophysical Research, v. 90, p. 4537–4555.

Doser, D. I., and Smith, R. B., 1982, Seismic moment rates in the Utah region: Bulletin of the Seismological Society of America, v. 72, p. 525–551.

——, 1989, An assessment of source parameters of earthquakes in the cordillera of the western United States: Bulletin of the Seismological Society of America, v. 79, p. 1383–1409.

Douglas, L. A., 1980, The use of soils in estimating the time of last movement of faults: Soil Science, v. 129, p. 345–352.

Fonseca, J., 1988, The Sou Hills: A barrier to faulting in the central Nevada seismic belt: Journal of Geophysical Research, v. 93, p. 475–489.

Gilbert, J. D., Ostenaa, D. A., and Wood, C. K., 1983, Seismotectonic study for Jackson Lake Dam and Reservoir, Minidoka Project, Idaho-Wyoming: Boise, Idaho, and Denver, Colorado, U.S. Bureau of Reclamation, Pacific Northwest Regional Office and Engineering and Research Center, Seismotectonic Report 83-8, 123 p.

Gile, L. H., Peterson, F. F., and Grossman, R. B., 1966, Morphological and genetic sequences of carbonate accumulation in desert soils: Soil Science, v. 101, p. 347–360.

Guthrie, R. L., and Witty, J. E., 1982, New designations for soil horizons and layers and the new soil survey manual: Soil Science Society of America Journal, v. 46, p. 443–444.

Hait, M. H., Jr., Prostka, H. J., and Oriel, S. S., 1977, Geologic relations of late Cenozoic rockslide masses near Palisades Reservoir, Idaho and Wyoming: Denver, Colorado, unpublished report to U.S. Bureau of Reclamation, 21 p.

Haller, K. M., 1987, Preliminary interpretation of segmentation of three range-front normal faults in southeastern Idaho and southwestern Montana: Geological Society of America Abstracts with Programs, v. 19, p. 280.

——, 1988a, Proposed segmentation of the Lemhi and Beaverhead faults, Idaho, and Red Rock fault, Montana—Evidence from studies of fault-scarp morphology: Geological Society of America Abstracts with Programs, v. 20, p. 418–419.

——, 1988b, Segmentation of the Lemhi and Beaverhead faults, east-central Idaho, and Red Rock fault, southwest Montana, during the late Quaternary [M.S. thesis]: Boulder, University of Colorado, 140 p.

——, 1990, Late Quaternary movement on basin-bounding normal faults north of the Snake River Plain, east-central Idaho, *in* Roberts, S., ed., Geologic field tours of western Wyoming and parts of adjacent Idaho, Montana, and Utah (Geological Society of America Rocky Mountain Section meeting guidebook, field trip 1, part 4): Laramie, Geological Survey of Wyoming Public Information Circular 29, p. 41–54.

Harden, J. W., 1982, A quantitative index of soil development from field descriptions: Examples from a chronosequence in central California: Geoderma, v. 28, p. 1–28.

Harden, J. W., and Matti, J. C., 1989, Holocene and late Pleistocene slip rates on the San Andreas fault in Yucaipa, California, using displaced alluvial-fan deposits and soil chronology: Geological Society of America Bulletin, v. 101, p. 1107–1117.

Harden, J. W., and Taylor, E. M., 1983, A quantitative comparison of soil development in four climatic regimes: Quaternary Research, v. 20, p. 342–359.

Holmes, G. W., and Moss, J. H., 1955, Pleistocene geology of the southwestern Wind River Mountains, Wyoming: Geological Society of America Bulletin, v. 66, p. 629–654.

Jenny, H., 1941, Factors of soil formation: New York, McGraw-Hill, 281 p.

—— , 1980, The soil resource: New York, Springer-Verlag, 377 p.

Jobin, D. A., 1972, Geologic map of the Ferry Peak quadrangle, Lincoln County, Wyoming: U.S. Geological Survey Quadrangle Map GQ-1027, scale 1:24,000.

Jobin, D. A., and Schroeder, M. L., 1964a, Geology of the Conant Valley quadrangle, Bonneville County, Idaho: U.S. Geological Survey Mineral Investigations Field Studies Map MF-277, scale 1:24,000.

—— , 1964b, Geology of the Irwin quadrangle, Bonneville County, Idaho: U.S. Geological Survey Mineral Investigations Field Studies Map MF-287, scale 1:24,000.

Jobin, D. A., and Soister, P. E., 1964, Geologic map of the Thompson Peak quadrangle, Bonneville County, Idaho: U.S. Geological Survey Miscellaneous Investigations Field Studies Map MF-284, scale 1:24,000.

Knuepfer, P.L.K., 1989, Implications of the characteristics of end-points of historical surface fault ruptures for the nature of fault segmentation, *in* Schwartz, D. P., and Sibson, R. H., eds., Proceedings, Fault segmentation and controls of rupture initiation and termination, Conference XLV, Palm Springs, California: U.S. Geological Survey Open-File Report 89-315, p. 193–228.

Lewis, G. C., and Fosberg, M. A., 1982, Distribution and character of loess and loess soils in southeastern Idaho, *in* Bonnichsen, B., and Breckenridge, R. M., eds., Cenozoic geology of Idaho: Idaho Bureau of Mines and Geology Bulletin 26, p. 705–716.

Love, J. D., and Montagne, J., 1956, Pleistocene and recent tilting of Jackson Hole, Teton County, Wyoming: Wyoming Geological Association, 11th Annual Field Conference, Guidebook, p. 169–178.

Lundstrom, S. C., 1986, Soil stratigraphy and scarp morphology studies applied to the Quaternary geology of the southern Madison valley, Montana [M.S. thesis]: Arcata, California, Humboldt State University, 53 p.

Machette, M. N., 1978, Dating Quaternary faults in the southwestern United States by using buried calcic paleosols: U.S. Geological Survey Journal of Research, v. 6, p. 369–381.

—— , 1982, Quaternary and Pliocene faults in the La Jencia and southern part of the Albuquerque-Belen basins, New Mexico: Evidence of fault history from fault-scarp morphology and Quaternary geology, *in* Grambling, J. A., and Wells, S. G., eds., Albuquerque County II: New Mexico Geological Society, 33rd Field Conference, Guidebook, p. 161–169.

—— , 1988, Quaternary movement along the La Jencia fault, central New Mexico: U.S. Geological Survey Professional Paper 1440, 82 p.

Machette, M. N., Personius, S. F., and Nelson, A. R., 1987, Quaternary geology along the Wasatch fault zone: Segmentation, recent investigations, and preliminary conclusions, *in* Gori, P. L., and Hays, W. W., eds., Assessment of regional earthquake hazards and risk along the Wasatch Front, Utah: U.S. Geological Survey Open-File Report 87-585, v. 1, Chapter A, p. A1–A72.

Mathiesen, E. L., 1983, Late Quaternary activity of the Madison Range fault along its 1959 rupture trace, Madison County, Montana [M.S. thesis]: Palo Alto, California, Stanford University, 157 p.

McCalpin, J., 1982, Quaternary geology and neotectonics of the west flank of the northern Sangre de Cristo Mountains, south-central Colorado: Colorado School of Mines Quarterly, v. 77, p. 1–97.

—— , 1987, Geologic criteria for recognition of individual paleoseismic events in extensional environments, *in* Crone, A. J., and Omdahl, E. M., eds., Proceedings, Directions in paleoseismology, Conference XXXIX, Albuquerque: U.S. Geological Survey Open-File Report 87-673, p. 102–114.

—— , 1990, Latest Quaternary faulting in the Northern Wasatch to Teton Corridor (NWTC): Final Technical Report to the U.S. Geological Survey, Contract No. 14-08-001-G1396, 42 p.

—— , 1991, Prehistoric surface-rupturing earthquakes in the Overthrust Belt; Did they occur on listric faults?: Geological Society of America Abstracts with Programs, v. 23, p. 47.

McCalpin, J., and Forman, S. L., 1991, Late Quaternary faulting and thermoluminescence dating of the East Cache fault zone, north-central Utah: Bulletin of the Seismological Society of America, v. 81, p. 139–161.

McCalpin, J., Robison, R. M., and Garr, J. D., 1987, Neotectonics of the Hansel Valley–Pocatello Valley corridor, northern Utah and southern Idaho, *in* Gori, P. L., and Hays, W. W., eds., Assessment of regional earthquake hazards and risk along the Wasatch Front, Utah: U.S. Geological Survey Open-File Report 87-585, v. 1, Chapter G, p. G1–G44.

Merritt, Z. S., 1956, Upper Tertiary sedimentary rocks of the Alpine-Wyoming area: Wyoming Geological Association, 11th Annual Field Conference, Guidebook, p. 117–119.

—— , 1958, Tertiary stratigraphy and general geology of the Alpine, Idaho-Wyoming area [M.A. thesis]: Laramie, University of Wyoming, 94 p.

Moore, D. W., Oriel, S. S., and Mabey, D. R., 1987, A Neogene(?) gravity-slide block and associated slide phenomena in Swan Valley graben, Wyoming and Idaho, *in* Beus, S. S., ed., Rocky Mountain Section Centennial Field Guide, vol. 2: Boulder, Colorado, Geological Society of America, p. 113–116.

Myers, W. B., and Hamilton, W., 1964, Deformation accompanying the Hebgen Lake earthquake of August 17, 1959: U.S. Geological Survey Professional Paper 435-I, p. 55–98.

National Earthquake Information Service, 1983, Preliminary determination of epicenters, monthly listing: Washington, D.C., U.S. Geological Survey, October 1983, 20 p.

Nelson, A. R., and Taylor, E. M., 1985, Automated calculation of soil profile development indices using a microcomputer and integrated spreadsheet: Geological Society of America Abstracts with Programs, v. 17, p. 258.

Okeson, C. J., 1958, Geologic report on Palisades Dam and appurtenant structures—Palisades Project, Idaho: Boise, Idaho, U.S. Bureau of Reclamation, 5 p.

Oriel, S. S., and Moore, D. W., 1985, Geologic map of the West and East Palisades Roadless Areas, Idaho and Wyoming: U.S. Geological Survey Miscellaneous Field Studies Map MF-1619-B, scale 1:50,000.

Oriel, S. S., and Platt, L. B., 1980, Geologic map of the Preston 1° × 2° quadrangle, southeastern Idaho and western Wyoming: U.S. Geological Survey Miscellaneous Investigations Map I-1127, scale 1:250,000.

Ostenaa, D. A., and Wood, C. K., 1990, Seismotectonic study for Clark Canyon Dam, East Bench Unit, Three Forks Division, Pick-Sloan Missouri Basin Program, Montana: Denver, Colorado, U.S. Bureau of Reclamation, Denver Office, Seismotectonic Report 90-4, 78 p.

Personius, S. F., 1982, Geomorphic analysis of the Deep Creek fault, upper Yellowstone Valley, south-central Montana: Wyoming Geological Association, 33rd Annual Field Conference, Guidebook, p. 203–212.

Pierce, K. L., 1986, Dating methods, *in* Active tectonics: Studies in geophysics: Washington, D.C., National Academy Press, p. 195–214.

Pierce, K. L., and Colman, S. M., 1986, Effect of height and orientation (microclimate) on geomorphic degradation rates and processes, late-glacial terrace scarps in central Idaho: Geological Society of America Bulletin, v. 97, p. 869–885.

Pierce, K. L., and Good, J. M., 1990, Quaternary geology of Jackson Hole, Wyoming, *in* Roberts, S., ed., Geologic field tours of western Wyoming and parts of adjacent Idaho, Montana, and Utah (Geological Society of America Rocky Mountain Section meeting guidebook, field trip 4): Laramie, Geological Survey of Wyoming Public Information Circular 29, p. 78–87.

Pierce, K. L., and Scott, W. E., 1982, Pleistocene episodes of alluvial-gravel deposition, southeastern Idaho, *in* Bonnichsen, B., and Breckenridge, R. M., eds., Cenozoic geology of Idaho: Idaho Bureau of Mines and Geology Bulletin 26, p. 685–702.

Pierce, K. L., Fosberg, M. A., Scott, W. E., Lewis, G. C., and Colman, S. M., 1982, Loess deposits of southeastern Idaho: Age and correlation of the upper two loess units, *in* Bonnichsen, B., and Breckenridge, R. M., eds., Cenozoic geology of Idaho: Idaho Bureau of Mines and Geology Bulletin 26, p. 717–725.

Pierce, K. L., Covington, H. R., Williams, P. L., and McIntyre, D. H., 1983, Geologic map of the Cotterel Mountains and the northern Raft River valley, Cassia County, Idaho: U.S. Geological Survey Miscellaneous Investigations Series Map I-1450, scale 1:48,000.

Piety, L. A., Wood, C. K., Gilbert, J. D., Sullivan, J. T., and Anders, M. H., 1986, Seismotectonic study for Palisades Dam and Reservoir, Palisades Project, Idaho: Denver, Colorado, and Boise, Idaho, U.S. Bureau of Reclamation, Engineering and Research Center and Pacific Northwest Regional Office, Seismotectonic Report 86-3, 198 p.

Ravenholt, H. B., Glenn, W. R., and Larson, K. N., 1976, Soil survey of Star Valley area, Wyoming-Idaho, parts of Lincoln County, Wyoming, and Bonneville and Caribou Counties, Idaho: n.p., U.S. Department of Agriculture, Soil Conservation Service and Forest Service, 74 p.

Roberts, J. C., 1981, Late Cenozoic volcanic stratigraphy of the Swan Valley graben between Palisades Dam and Pine Creek, Bonneville County, Idaho [M.S. thesis]: Pocatello, Idaho State University, 58 p.

Rodgers, D. W., and Anders, M. H., 1990, Neogene evolution of Birch Creek valley near Lone Pine, Idaho, *in* Roberts, S., ed., Geologic field tours of western Wyoming and parts of adjacent Idaho, Montana, and Utah (Geological Society of America Rocky Mountain Section meeting guidebook, field trip 1, part 3): Laramie, Geological Survey of Wyoming Public Information Circular 29, p. 26–38.

Royse, F., Jr., Warner, M. A., and Reese, D. L., 1975, Thrust belt structural geometry and related stratigraphic problems, Wyoming–Idaho–northern Utah, *in* Bolyard, D. W., ed., Deep drilling frontiers of the central Rocky Mountains: Denver, Colorado, Rocky Mountain Association of Geologists, p. 41–54.

Rubey, W. W., 1973, Geologic map of the Afton quadrangle and part of the Big Piney quadrangle, Lincoln and Sublette Counties, Wyoming: U.S. Geological Survey Miscellaneous Geologic Investigations Map I-686, scale 1:62,500.

Sbar, M. L., Barazangi, A. M., Dorman, J., Scholz, C. H., and Smith, R. B., 1972, Tectonics of the Intermountain seismic belt, western United States: Microearthquake seismicity and composite fault-plane solutions: Geological Society of America Bulletin, v. 83, p. 13–28.

Schwartz, D. P., and Coppersmith, K. J., 1984, Fault behavior and characteristic earthquakes: Examples from the Wasatch and San Andreas fault zones: Journal of Geophysical Research, v. 89, p. 5681–5698.

——, 1986, Seismic hazards: New trends in analysis using geological data, *in* Active tectonics: Studies in geophysics: Washington, D.C., National Academy Press, p. 215–230.

Scott, W. E., Pierce, K. L., and Hait, M. H., Jr., 1985, Quaternary tectonic setting of the 1983 Borah Peak earthquake, central Idaho: Bulletin of the Seismological Society of America, v. 75, p. 1053–1066.

Shenon, P. J., 1936, The Utah earthquake of March 12, 1934, excerpts from an unpublished report of the U.S. Geological Survey, *in* Neumann, F., ed., United States earthquakes 1934: U.S. Department of Commerce, Coast and Geodetic Survey Serial 593, p. 43–48.

Shroba, R. R., and Birkeland, P. W., 1983, Trends in late-Quaternary soil development in the Rocky Mountains and Sierra Nevada of the western United States, *in* Wright, H. E., Jr., ed., Late-Quaternary environments of the United States, vol. 1: The late Pleistocene: Minneapolis, University of Minnesota Press, p. 145–156.

Slemmons, D. B., 1977, Faults and earthquake magnitude: Vicksburg, Mississippi, U.S. Army Engineers Waterways Experiment Station, Miscellaneous Paper S-73-1, Report 6, 166 p.

Smith, R. B., and Sbar, M. L., 1974, Contemporary tectonics and seismicity of the western United States with emphasis on the Intermountain seismic belt: Geological Society of America Bulletin, v. 85, p. 1205–1218.

Smith, R. B., Richins, W. D., and Doser, D. I., 1985, The 1983 Borah Peak, Idaho, earthquake: Regional seismicity, kinematics of faulting, and tectonic mechanism, *in* Stein, R. S., and Bucknam, R. C., eds., Proceedings, Workshop XXVIII on the Borah Peak, Idaho, earthquake, October 1984: U.S. Geological Survey Open-File Report 85-290, Volume A, p. 236–250.

Soil Survey Staff, 1975, Soil taxonomy—A basic system of soil classification for making and interpreting soil surveys: U.S. Department of Agriculture, Soil Conservation Service, Agricultural Handbook 436, 754 p.

——, 1981, Soil survey manual—Examination and description of soils in the field (Chap. 4): U.S. Department of Agriculture, Soil Conservation Service, unpublished manuscript, p. 4-1–4-107.

Soltanpour, P. N., and Workman, S. M., 1981, Soil-testing methods used at Colorado State University Soil Testing Laboratory: Fort Collins, Colorado State University Experiment Station Technical Bulletin 142, 22 p.

Staatz, M. H., and Albee, H. F., 1966, Geology of the Garns Mountain quadrangle, Bonneville, Madison, and Teton Counties, Idaho: U.S. Geological Survey Bulletin 1205, 122 p.

Stickney, M. C., and Bartholomew, M. J., 1987, Preliminary map of late Quaternary faults in western Montana: Montana Bureau of Mines and Geology Open-File Report 186, scale 1:500,000.

Swan, F. H. III, Schwartz, D. P., and Cluff, L. S., 1980, Recurrence of moderate to large magnitude earthquakes produced by surface faulting on the Wasatch fault zone, Utah: Bulletin of the Seismological Society of America, v. 70, p. 1431–1462.

Turko, J. M., 1988, Quaternary segmentation history of the Lemhi fault, Idaho [M.A. thesis]: Binghamton, State University of New York, 91 p.

U.S. Bureau of Reclamation, 1976, Meeting notes attached to "Stability of slide masses along Palisades Reservoir, Minidoka Project, Idaho": Denver, Colorado, Memorandum from Director of Design and Construction to Regional Director, Boise, Idaho, December 6, 1976, 1 p.

U.S. Weather Bureau, 1960–1978, Climatological data, Idaho: Annual Summary, v. 63–81.

Walker, E. H., 1964, Glacial terraces along the Snake River in eastern Idaho and in Wyoming: Northwest Science, v. 38, p. 33–42.

Wells, S. G., and seven others, 1988, Regional variations in tectonic geomorphology along a segmented convergent plate boundary, Pacific coast of Costa Rica: Geomorphology, v. 1, p. 239–265.

Wheeler, R. L., and Krystinik, K. B., 1988, Segmentation of the Wasatch fault zone, Utah—Summaries, analyses, and interpretations of geological and geophysical data: U.S. Geological Survey Bulletin 1827, 47 p.

Witkind, I. J., 1975a, Preliminary map showing known and suspected active faults in Idaho: U.S. Geological Survey Open-File Report 75-278, scale 1:500,000.

——, 1975b, Preliminary map showing known and suspected active faults in Wyoming: U.S. Geological Survey Open-File Report 75-279, scale 1:500,000.

MANUSCRIPT ACCEPTED BY THE SOCIETY JULY 19, 1991

Printed in U.S.A.

Geological Society of America
Memoir 179
1992

Chapter 9

Possible Laramide influence on the Teton normal fault, western Wyoming

D. R. Lageson
Department of Earth Science, Montana State University, Bozeman, Montana 59717

ABSTRACT

The site of the modern Teton Range and Jackson Hole was formerly occupied by an extensive basement uplift, the Ancestral Teton–Gros Ventre uplift. This uplift rose during the Laramide orogeny along the northeast-dipping Cache Creek thrust fault, exposed along Teton Pass at the south end of the present Teton Range. The Cache Creek thrust fault abuts the "thin-skinned" Jackson thrust fault at Teton Pass, marking the boundary between two major tectonic provinces: the Sevier fold-thrust belt and the Laramide foreland province.

The modern Teton Range is a westward tilted block of Precambrian basement rock that is bounded along its precipitous eastern flank by the east-dipping Teton normal fault. Other normal faults at the south end of the Teton Range include the Open Canyon and Phillips Canyon normal faults in the footwall of the Teton normal fault and normal faults that bound East and West Gros Ventre Buttes in the hanging wall of the Teton normal fault. It is proposed that the Cache Creek thrust fault has been reactivated as a detachment for these late Cenozoic normal faults, resulting in rigid-body, "domino" rotation of basement blocks above the Cache Creek thrust fault. This interpretation not only explains such structural relationships at the south end of the Teton Range, as the southern termination of the Teton normal fault and adjacent normal faults at the surface trace of the Cache Creek thrust but also explains fluvial drainage patterns on the floor of Jackson Hole. Although not as common as reactivated Sevier thrust faults, the model of a reactivated Laramide thrust fault is not unique along the eastern margin of the Basin and Range province and is supported by other examples.

INTRODUCTION

The eastern margin of the Basin and Range structural province in the western United States is an ideal region in which to study the transition from unextended to extended crust. Defined by the intermountain seismic belt (Sbar and others, 1972; Smith and Sbar, 1974), the eastern margin of the Basin and Range overlaps the predominantly "thin-skinned" Sevier fold-thrust belt and the "thick-skinned" Laramide foreland province. The overlap of Basin and Range extension on the Sevier fold-thrust belt has been documented by numerous workers and is generally well known. For example, Constenius (1982) has described the Kishenen Basin, and Crosby (1984) has provided a structural-geophysical interpretation of Swan Valley in the fold-thrust belt of northwestern Montana. In the Idaho-Wyoming salient of the fold-thrust belt, Royse and others (1975) and Olson and Schmitt (1987) have described the Hoback normal fault and Hoback Basin, and Lageson (1980), McCalpin and others (1990), and Anders (1990) have described the Star-Grand-Swan fault system. Smith and Bruhn (1984) have correlated older thrust fault ramps with segments of the Wasatch normal fault in north-central Utah. In contrast, the Teton Range of western Wyoming is an example of the overlap of Basin and Range structure onto the Laramide, basement-involved, foreland province (Fig. 1). The purpose of this chapter is to discuss the possible influence of a Laramide thrust fault on the structural orientation and evolution of late Cenozoic normal faults at the south end of the Teton Range.

Lageson, D. R., 1992, Possible Laramide influence on the Teton normal fault, western Wyoming, *in* Link, P. K., Kuntz, M. A., and Platt, L. B., eds., Regional Geology of Eastern Idaho and Western Wyoming: Geological Society of America Memoir 179.

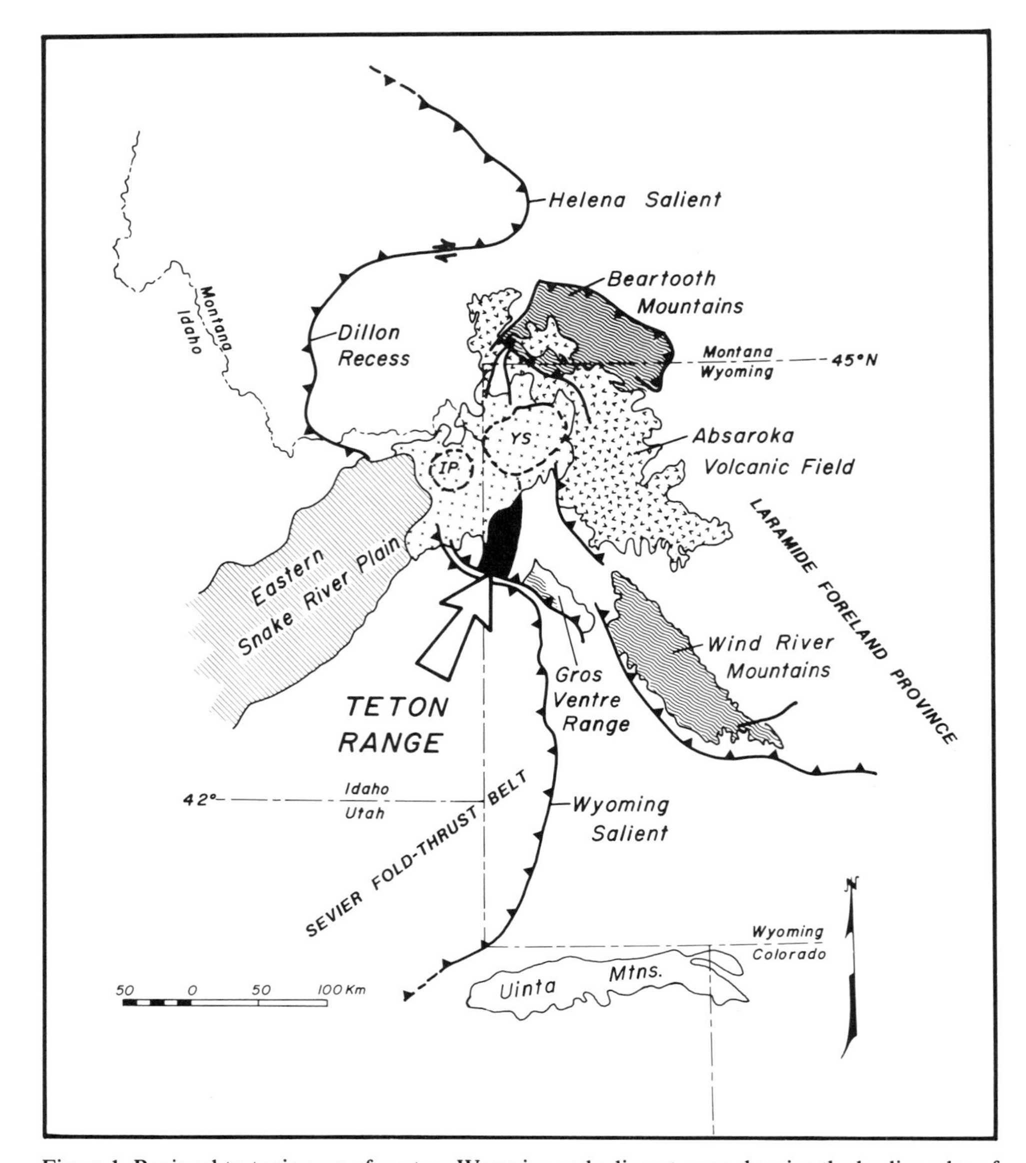

Figure 1. Regional tectonic map of western Wyoming and adjacent areas showing the leading edge of the Sevier fold-thrust belt and Laramide foreland province. Teton Range is the solid black area in the middle of the map, overlapping the Laramide foreland. Ys = Yellowstone caldera; IP = Island Park caldera; X pattern = Yellowstone volcanic field (Quaternary); V pattern = Absaroka volcanic field (Eocene); wavy pattern = Archean "basement" rocks; diagonal pattern = Snake River Plain; and dashed line = approximately located.

LARAMIDE STRUCTURE OF THE TETON AREA

Regional relationships

The modern Teton Range is the product of two spatially overlapping, yet temporally distinct, tectonic events: (1) the Late Cretaceous to early Tertiary Laramide orogeny, during which time the Ancestral Teton–Gros Ventre uplift rose as a result of regional crustal compression (Love and others, 1973), and (2) Neogene-Quaternary crustal extension and superposition of the modern Teton Range over the Ancestral Teton–Gros Ventre uplift. In this chapter, the terms *Laramide orogeny* and *Laramide province* refer to Rocky Mountain uplifts and basins of the western United States where Late Cretaceous through early Tertiary, basement-involved, dominantly compressional deformation has occurred (Dickinson and others, 1988). In general, the Laramide province lies east of the Sevier fold-thrust belt and postdates it, although the two provinces spatially and temporally overlap at many localities (Wiltschko and Dorr, 1983; Harlan and others, 1988; Kulik and Schmidt, 1988).

Two large, en echelon, Laramide uplifts dominated the present site of Grand Teton and Yellowstone national parks in

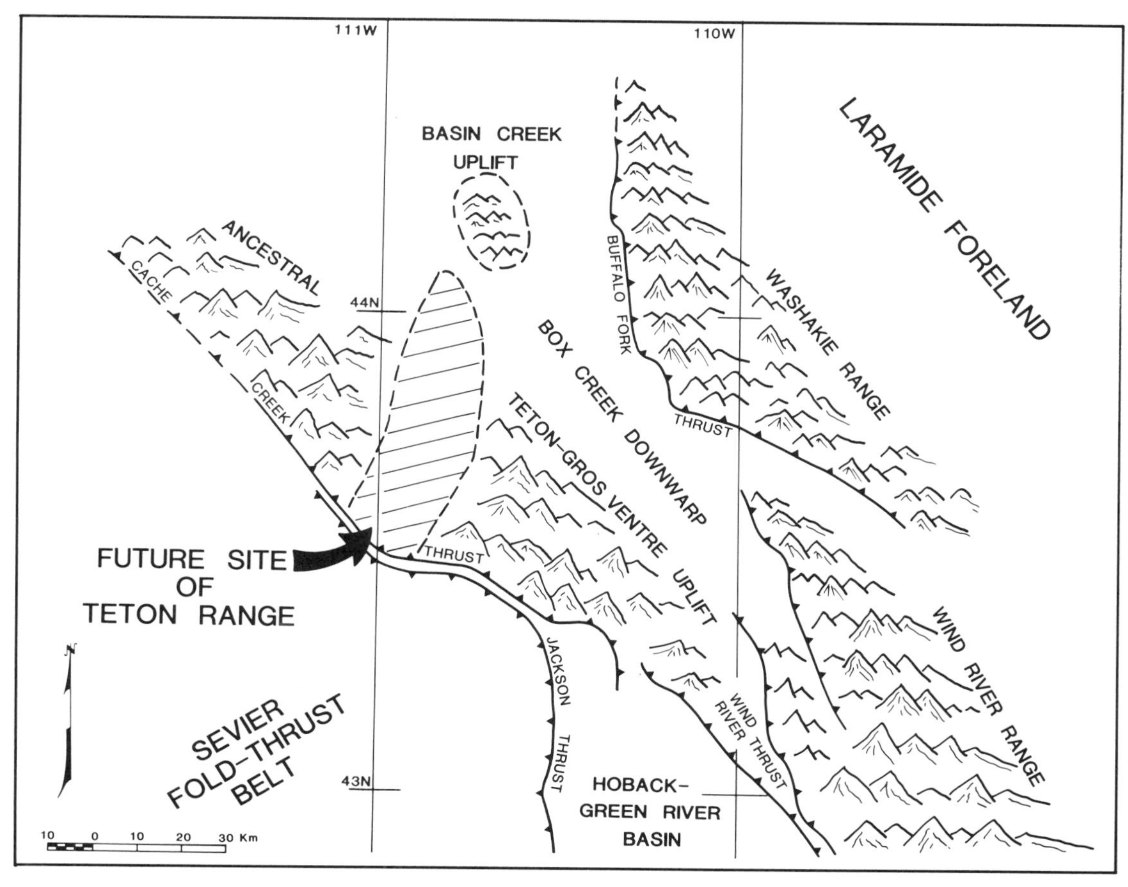

Figure 2. Diagrammatic reconstruction of major landforms and faults in northwestern Wyoming during early Eocene time. Relief implied is entirely diagrammatic. Modified from Love (1973).

northwestern Wyoming during early Tertiary time (Fig. 2). The Ancestral Teton–Gros Ventre uplift rose across the southwestern part of the region, beginning in Late Cretaceous time (Love and Reed, 1968; Love, 1977; Love and others, 1973; Roberts and Burbank, 1988) and culminating in early Eocene time (Love, 1977; Dorr and others, 1977; Wiltschko and Dorr, 1983). To the northeast, the Washakie Range, bounded by the Buffalo Fork thrust fault, experienced major uplift during the Maastrichtian (Love, 1977; Love and Keefer, 1975). The synclinal trough between these two large uplifts, the Box Creek Downwarp (Love, 1973), was locally interrupted by the smaller Basin Creek Uplift. The following discussion will focus on the Ancestral Teton–Gros Ventre uplift and its possible influence on the structural development of the modern Teton Range.

Ancestral Teton–Gros Ventre uplift

The Ancestral Teton–Gros Ventre uplift was a northwest-trending, basement-cored uplift lying west of, and en echelon with, the larger Wind River Range (Love, 1973). Before the modern Teton Range and Jackson Hole existed, this uplift spanned the distance from the present Gros Ventre Range to eastern Idaho (Fig. 2). Thus, the former topographic expression of the Ancestral Teton–Gros Ventre uplift was much larger than the modern Gros Ventre Range, which lies southeast of Jackson Hole (compare Figs. 1 and 2).

Structural continuity of the Ancestral Teton–Gros Ventre uplift has been maintained despite extension in latest Cenozoic time. The areal extent of the Ancestral Teton–Gros Ventre uplift is delimited on the south by the continuous trace of the Cache Creek thrust fault. From its southeast end in the northern Hoback Basin (Simons and others, 1988), the Cache Creek thrust fault strikes approximately N60W along the southwestern base of the modern Gros Ventre Range, crosses Jackson Hole with a strike of N80W, and continues over Teton Pass at N50W (at the south end of the Teton Range) and into Idaho, where it is ultimately covered by Quaternary deposits (Love and others, 1978). However, Oriel and Moore (1985) did not map the Cache Creek thrust northwest of Jackson, Wyoming; in its place, they mapped a down-to-the-south normal fault to explain the offset stratigraphy at Teton Pass. I do not agree with Oriel and Moore's (1985) interpretation, based on my examination of the stratigraphic and structural relationships at Teton Pass, as supported by the detailed mapping of Schroeder (1969, 1972), Zeller (1982), Vasko (1982), and Dunn (1983).

Northwest of Teton Pass, the termination of the Cache Creek thrust is equivocal because of Quaternary cover. Sales (1983) has proposed that the Cache Creek thrust turns northward to parallel the western flank of the Teton Range in a manner similar to that of the north end of the Wind River thrust. In contrast, F. Royse (personal communication, 1989) and Blackstone and De Bruin (1987) suggest that the Cache Creek thrust continues northwest from Teton Pass without significant change in strike, passing through Victor, Idaho, and crossing the southwest margin of Teton Basin (Pierre's Hole). I prefer the interpretation of Blackstone and De Bruin (1987) because the Cache Creek thrust shows no evidence of a sharp change in strike at the west end of Teton Pass, as would be necessitated by Sales's (1983) interpretation.

The large areal extent of the Ancestral Teton–Gros Ventre uplift is also outlined by the distribution of smaller, parasitic, basement-involved faults that formed on its flank. The Forellen reverse fault at the north end of the Teton Range is a good example (Fig. 3). This structure is a typical Laramide-style uplift, involving a reverse fault in the basement that is overlain and structurally balanced by folded Phanerozoic strata. The Forellen fault dips 60 to 70 degrees northeast toward the Box Creek downwarp and away from the crest of the Ancestral Teton–Gros Ventre uplift (Fig. 4). Other examples include the Buck Mountain reverse fault in the central core of the Teton Range, which strikes generally north-south and dips 60 degrees east (Bradley, 1956; Love, 1968; Smith and Lageson, 1989; Smith, 1990), and Ramshorn anticline along the eastern margin of Jackson Hole (Love, 1968).

At Teton Pass, the change in stratigraphic position of the Cache Creek fault surface reflects a possible subsurface ramp structure. Hanging-wall stratigraphy varies from Archean basement rocks at the eastern base of Teton Pass to the Triassic Chugwater Formation at the summit and upper west slope to the Lower Cretaceous Gannett Group near the western base of the pass (Love, 1956a, b; Love and Albee, 1972; Schroeder, 1969, 1972; Oriel and Moore, 1985). The change in stratigraphic position of the fault surface relative to the hanging wall indicates a ramp down the dip of the fault surface to the northeast, beyond the surface trace of the thrust. Also, Lageson (1987) argued for a lateral ramp along a segment of the Cache Creek thrust fault southeast of Jackson, based on slickenline lineations. Although ramps are more commonly associated with thin-skinned thrust faults, basement-involved thrusts can also change dip relative to their footwall and hanging-wall stratigraphy, particularly at the basement-cover interface (McBride and others, 1992). The inferred ramp geometry of the Cache Creek thrust may have played an important role in controlling the orientation of late Cenozoic normal faults at the south end of the Teton Range, although lack of detailed subsurface control does not permit a definitive solution.

Figure 3. Aerial photo looking northwest along strike of the Forellen reverse fault at the north end of the Teton Range. Basement rocks on the right (northeast) are in fault contact with up-turned Paleozoic strata in the footwall to the left (southwest). Photo by Lageson, October 1987.

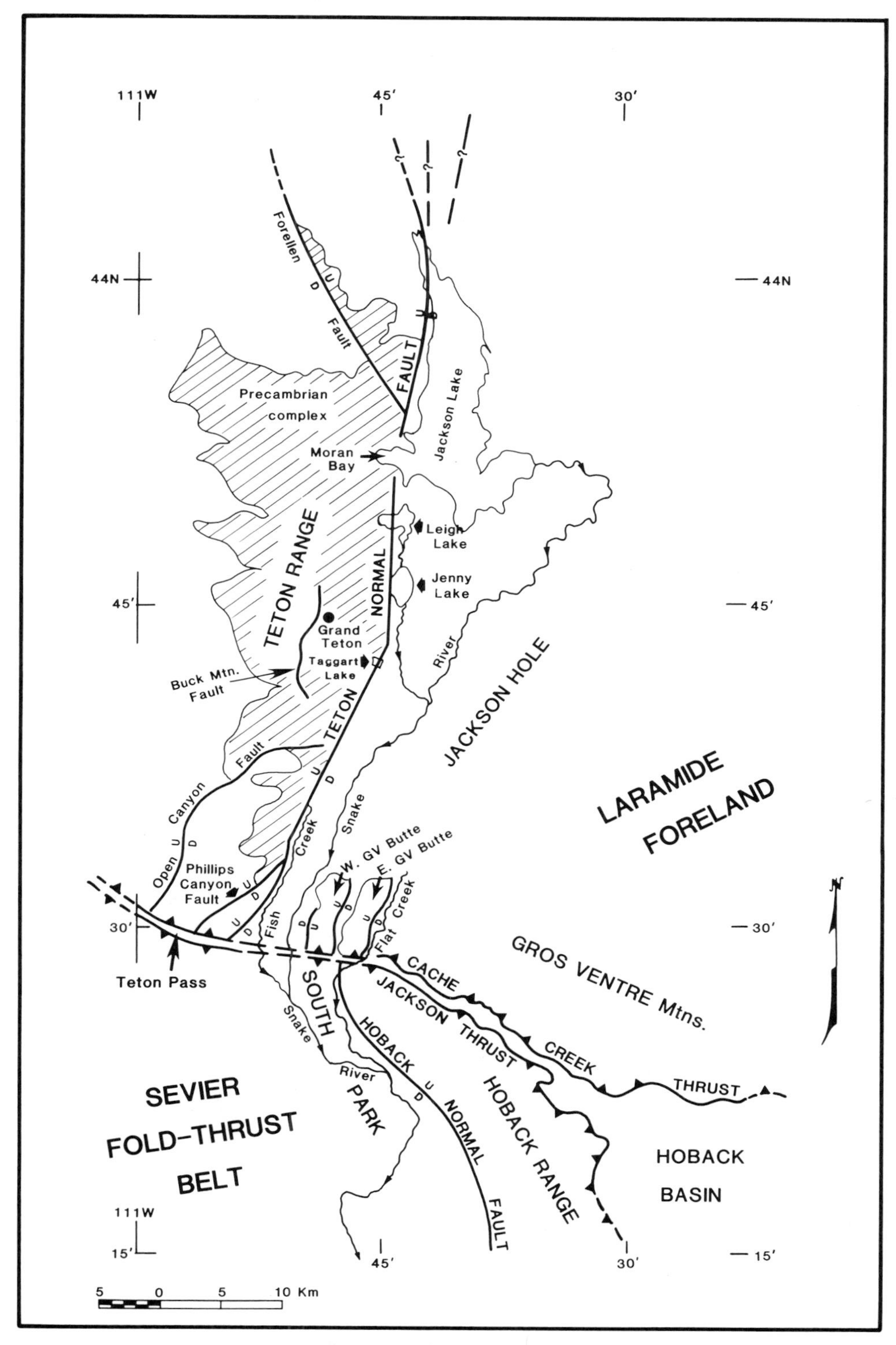

Figure 4. Structural index map of the Teton Range and adjacent areas. Modified from Behrendt and others, 1968.

TETON NORMAL FAULT

The Teton Range is one of the most spectacular topographic manifestations of late Cenozoic crustal extension in the western United States. The Teton Range trends north-south for 80 km, is approximately 20 km wide, and is distinguished by its precipitous, glaciated, east-facing escarpment. Structurally, the Teton Range is a relatively simple, block-faulted, westward-tilted uplift that has formed in response to regional east-west crustal extension (Love and Reed, 1968). The range-bounding Teton normal fault dips 45 to 55 degrees east beneath Jackson Hole and has had over 6 km of throw since the late Miocene (Behrendt and others, 1968; Love and Reed, 1968). The surface of Jackson Hole, in the hanging-wall block, is gently tilted west toward the range (Love and Montagne, 1956; Smith and others, 1990).

The central portion of the Teton Range is the highest (the summit of the Grand Teton rises to 4,197 m, approximately 2,150 m above the valley floor of Jackson Hole to the east), resulting in a broad arch in the footwall of the Teton normal fault that plunges to the north and south. The Teton Range arch is covered by Phanerozoic strata on the north, south, and west flanks of the range. The high relief of the central portion of the Teton Range is a function of three variables: (1) greater offset along the central segment of the Teton normal fault (Smith and others, 1990), (2) a massive body of quartz monzonite that is resistant to weathering and erosion (Reed, 1973), and (3) probable Laramide uplift on the Buck Mountain reverse fault along the western side of the central Teton Range (Smith, 1990).

The Teton normal fault consists of three distinct segments (Fig. 4) (Susong and others, 1987; Smith and others, 1990). The southern fault segment strikes N30E from the south end of the range (near Wilson) to Taggart Lake, where it sharply rotates about 25 degrees clockwise into the central fault segment. The central fault segment extends from Taggart Lake to Moran Bay and is characterized by the largest fault scarps at the base of the range (30+ m) with a component of left-lateral offset. The northern fault segment extends from the north side of Moran Bay to the north end of Jackson Lake where it bifurcates into multiple fault strands (Fig. 5) (Smith and others, 1990). Overall, Quaternary surface offset across the Teton normal fault increases from the north and south ends to a maximum in the central fault segment (Smith and others, 1990).

Historic earthquake data show that the Teton normal fault is a seismic gap in the intermountain seismic belt (Smith and others, 1976; Smith and others, 1990). The record of paleoseismicity as deduced from fault scarps at the base of the range does not match the current record of inactivity. Smith and others (1990) offer possible explanations for this contradiction: (1) seismicity may be migrating eastward into the Gros Ventre Range, (2) the period of historical observation may not be long enough to determine the long-term rate of seismicity, or (3) the region adjacent to the Teton normal fault may be storing elastic energy for future large-

Figure 5. Aerial photo looking south along the Teton normal fault. Moran Bay, separating the central and northern segments of the Teton normal fault, is shown in the center of the photo (white due to ice). Leigh Lake and Jenny Lake lie at the base of the range in the distance; highest peak in the distance is the Grand Teton. Abrupt break in slope shown in lower left corner marks the trace of the northern Teton normal fault segment. Photo by Lageson, October 1987.

magnitude earthquakes. Current seismicity in the Teton area is dispersed in the hanging-wall block of the Teton normal fault, east of the Teton Range. The area adjacent to the southern fault segment seems to be particularly active, with average focal depths of 4 to 5 km (R. B. Smith, personal communication, 1990); this area is underlain by the northeast-dipping Cache Creek thrust fault, which may explain the shallow focal depths.

Previous structural models

Figure 6 shows several representative structural interpretations of the Teton Range and Jackson Hole. Horberg (1938), Bradley (1956), and Love and Reed (1968) depicted the Teton Range as a westward-tilted fault block, differing only in the attitude of the Buck Mountain fault along the west side of the range and in the details of the Teton normal fault zone. Using seismic refraction and gravity data to model the Teton Range and Jackson Hole, Behrendt and others (1968) concluded that the Teton normal fault dips east at a relatively low angle (35 degrees) and that a maximum of 7 km of "vertical uplift" has occurred. Behrendt and others (1968, p. E19) state that "the fault may be a single break of relatively low dip . . . or it may consist of several closely spaced more steeply dipping faults arranged steplike to the average slope" of the gravity data. The interpretation of a series of steplike faults was favored by Behrendt and others (1968) and Love (1968) and is supported by the array of east-dipping, high-angle normal faults at the south and north ends of the Teton Range (Schroeder, 1969, 1972; U.S. Geological Survey, 1972; Smith and others, 1990); however, it remains uncertain if the central fault segment is a single fault or a series of high-angle step faults.

D. L. Reese (unpublished data, 1974) and Royse (1983) present radically different interpretations for the Teton area (Fig. 6d). In their interpretations, the Teton normal fault and the other late Cenozoic normal faults at the south end of the Teton Range sole into the older Laramide Cache Creek thrust fault at depth, indicating that the Cache Creek fault surface has been reactivated as a northeast-dipping detachment to accommodate east-west extension of the uppermost basement. Also, Sales (1983) suggested a similar, but not identical, interpretation involving reactivation of the Cache Creek thrust along a ramp. The concept of a reactivated, basement-involved, Laramide fault is not entirely new and has been proposed in other areas of the intermountain seismic belt (e.g., Royse and others, 1975; Lageson and Zim, 1985; Lageson, 1988; Guthrie and others, 1989).

Any structural model of the Teton Range and Jackson Hole must account for several structural relationships. (1) The Teton normal fault is not straight but rather has three distinct segments, as previously discussed. The southern segment strikes northeast, whereas the central and northern segments strike more northerly. (2) The Teton normal fault and other normal faults at the south end of the range end abruptly at the surface trace of the Laramide Cache Creek thrust fault; evidence for this is presented below. (3) The Teton Range and smaller, adjacent fault blocks tilt uniformly to the west (Schroeder, 1969, 1972; Love and Albee, 1972). In my opinion, reactivation of the Cache Creek thrust fault as a detachment surface for late Cenozoic normal faults at the south end of the Teton Range offers the best explanation for these structural relationships. The overall structure produced is that of several fault blocks that have been rotated like dominos across a basement-involved detachment surface (Fig. 6d). Evidence to support this interpretation is presented below.

Reactivation of the Cache Creek thrust fault

Several late Cenozoic normal faults have been mapped at the south end of the Teton Range and Jackson Hole. From west to east, these are the Open Canyon normal fault, Phillips Canyon normal fault, Teton normal fault along the eastern base of the Teton Range (the largest normal fault in the area), and faults that bound the flanks of West and East Gros Ventre Buttes (Fig. 4) (Schroeder, 1969, 1972; Love and Albee, 1972). In general, these normal faults strike N30E and dip steeply (60 to 80 degrees) southeast, separating topographically distinct basement blocks overlain by west-dipping Phanerozoic strata; the average dip of strata on the blocks is 20 degrees west. The uniform rotation of these blocks suggests that they are bounded by planar normal faults (Fig. 7), a geometric relationship demonstrated by Wernicke and Burchfiel (1982) in other parts of the Basin and Range province.

Geological maps of the area (Love and Albee, 1972; Love and Christiansen, 1985; Reed, 1973; Schroeder, 1969, 1972) clearly show that these westward-tilted, fault-bounded blocks are contained entirely within the hanging wall of the older Cache Creek thrust. The tilted blocks and associated normal faults, including the Teton normal fault, appear to terminate abruptly at the surface trace of the Cache Creek thrust, intersecting the latter at almost 90 degrees. Indeed, the enormous tilted basement block that forms the Teton Range ends with precipitous abruptness at Teton Pass, precisely at the trace of the Cache Creek thrust. The mountainous terrain south of Teton Pass is topographically much lower and is formed from northwest-striking, southwest-dipping Phanerozoic strata in the hanging wall of the Jackson thrust fault, the frontal thrust of the Idaho-Wyoming salient of the fold-thrust belt (Schroeder, 1969); this structural attitude is almost 90 degrees to the northeast-striking, northwest-dipping basement blocks north of the pass. In addition, Smith and others (1990, p. 132) state that:

> The southern end of the Teton fault appears to terminate near its intersection with older Overthrust Belt structures at the Cache Creek and Jackson thrusts. The fault may extend south of Wilson, although no Quaternary scarps are preserved there.

Therefore, I favor the hypothesis that the Teton normal fault ends at the Cache Creek thrust at Teton Pass, based on: (1) detailed mapping by previous workers (Schroeder, 1969, 1972; Vasko, 1982; Zeller, 1982), (2) the abrupt topographic termination of the

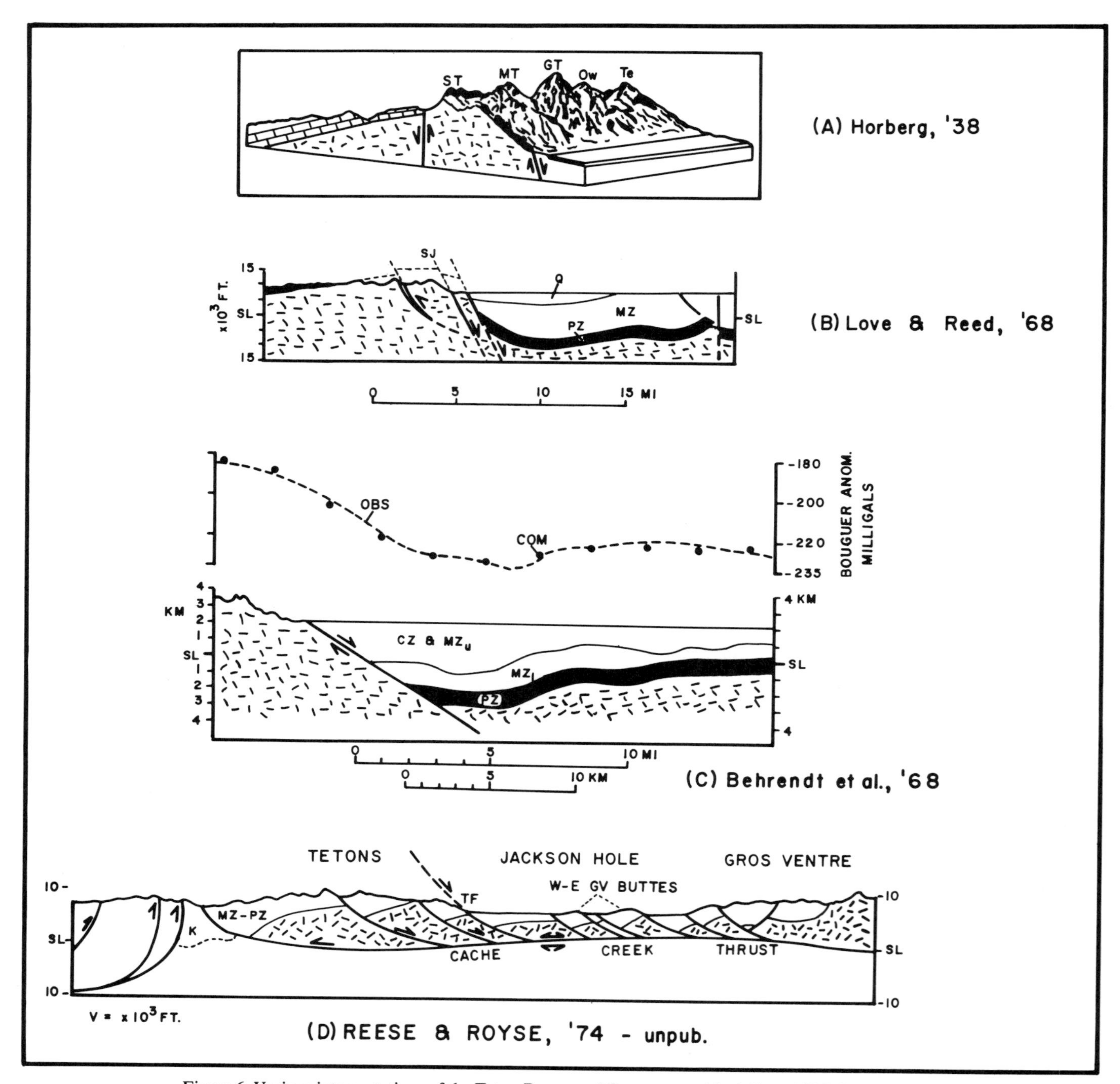

Figure 6. Various interpretations of the Teton Range and Teton normal fault from 1938 through 1974. ST = South Teton; MT = Middle Teton; GT = Grand Teton; OW = Mount Owen; Te = Mt. Teewinot; SJ = Mt. St. John; TF = Teton normal fault; GV Buttes = Gros Ventre Buttes; OBS = observed gravity; COM = computed gravity. Rock units: Q = Quaternary deposits; CZ = Cenozoic rocks; K = Cretaceous rocks; MZu = Upper Mesozoic rocks; MZ = Mesozoic rocks; MZl = Lower Mesozoic rocks; MZ-PZ = Mesozoic and Paleozoic rocks, unidivided; PZ = Paleozoic rocks; and Random hachures = Precambrian basement rocks.

Teton Range at Teton Pass, (3) 90-degree change in strike from the Teton block to the Jackson thrust sheet, (4) lack of evidence for fault scarps south of Teton Pass (Smith and others, 1990), and (5) increase in Quaternary surface offset across the Teton normal fault northward from Teton Pass (Smith and others, 1990).

An alternate interpretation for the southern termination of the Teton normal fault has been offered by Love and Albee (1972). They have placed the Teton normal fault more than 2 km east of the mountain front at the eastern base of Teton Pass and have extended it several kilometers to the south as a "postulated fault" through Quaternary floodplain alluvium of the Snake River, undifferentiated terrace deposits, glacial outwash, and unstable (slumped) outcrops of the Cretaceous Aspen Shale. Their postulated Teton fault coincides with the braided channel of the Snake River across most of the Jackson 7.5-minute quadrangle. Farther south, Albee (1968) mapped the "concealed" Teton normal fault through Quaternary colluvium that overlies folded Cretaceous shales in the northeast corner of the area on the Munger Mountain 7.5-minute quadrangle. Therefore, neither Love and Albee (1972) nor Albee (1968) presents convincing evidence for continuing the Teton normal fault south of the surface trace of the Cache Creek thrust.

Given the hypothesis that the Teton fault and other late Cenozoic normal faults at the south end of the Teton Range terminate against the surface trace of the Cache Creek thrust, it is likely that the two fault systems merge at depth. Ramsay and Huber (1987, p. 507) discuss the possible solutions presented by the termination of one fault system against another:

> A fault may also terminate abruptly against another fault, either because it has been truncated by a late fault, or because the two differently oriented systems acted together in a way which is geometrically related to give an overall compatible movement plan.

The Cache Creek thrust could not have truncated the normal faults because it is a Laramide fault (Dorr and others, 1977; Wiltschko and Dorr, 1983). Therefore, the Cache Creek thrust and younger normal faults are interpreted to interact in "an overall compatible movement plan," accomplished by transferring normal fault displacement to the Cache Creek fault surface at depth. In this manner, the Cache Creek fault has accommodated the extension of superjacent, tilted fault blocks by providing a detachment surface on which the normal faults have soled, similar in gross respects to Wernicke and Burchfiel's (1982) model of simple-shear domino extension.

Reactivation of the Cache Creek thrust is further supported by the observation that displacement on the Teton normal fault, and to a lesser degree the Phillips Canyon and Open Canyon normal faults, increases to the northeast away from Teton Pass in the dip-direction of the Cache Creek thrust. Throw on the Teton fault increases to approximately 6 km in the vicinity of Jenny Lake, where the Teton Range rises to its highest elevations (Behrendt and others, 1968). Also, Quaternary surface offsets across the Teton normal fault increase from an average of 10 m

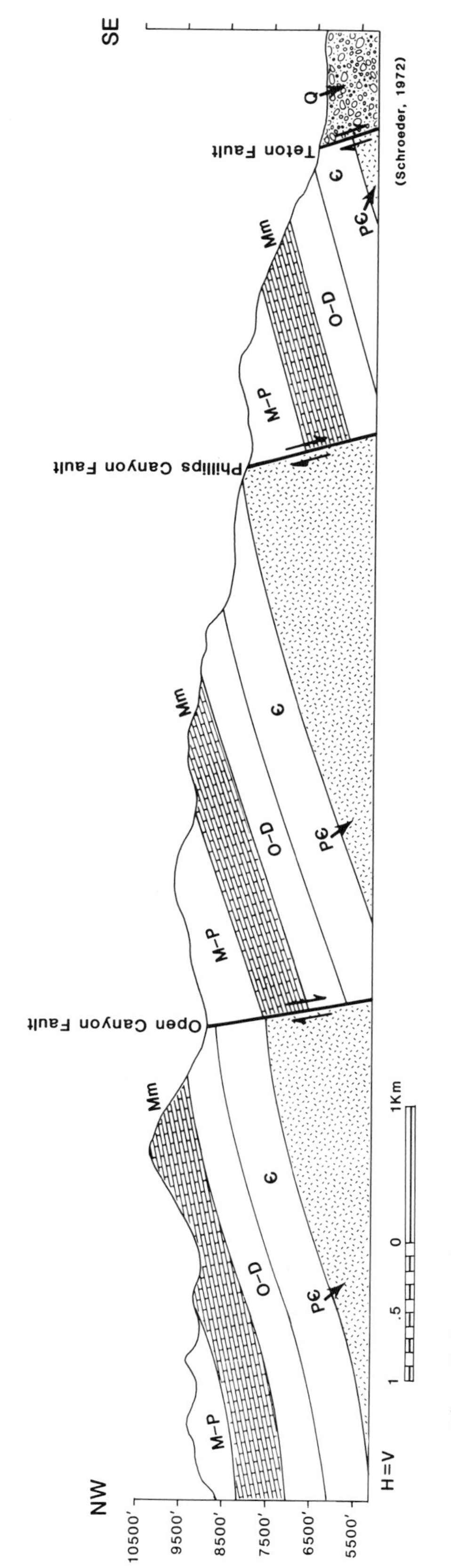

Figure 7. Cross section at the south end of the Teton Range showing uniform tilt of strata and basement blocks in the footwall of the Teton, Phillips Canyon, and Open Canyon normal faults (from Schroeder, 1972). Rock units: Q = Quaternary deposits; M-P = Mississippian through Permian rocks, undivided; Mm = Mississippian Madison Limestone; O-D = Ordovician through Devonian rocks, unidivided; P€ = Precambrian rocks.

just north of Teton Pass to a maximum of 30+ m along the central fault segment (Smith and others, 1990); similar relationships have been documented at the north end of the Teton normal fault. Therefore, the Teton fault appears to be hinged at its south end, resulting in left-lateral, oblique-slip displacement northward along the fault as well as greater throw or vertical separation to the north. Left-lateral offset of Pinedale moraine has been documented at Taggart Lake (Fig. 8) and Jenny Lake (Susong and others, 1987; Smith and others, 1990). Throw on the Open Canyon and Phillips Canyon normal faults also increases northward, as reflected by offset stratigraphy along the strike of the faults (Schroeder, 1972). These observations support the conclusion that the Teton and other normal faults sole into the northeast-dipping Cache Creek thrust, parallel to the direction of increasing normal fault displacement.

Surface tilting and drainage patterns

Other evidence for detachment normal faulting involving the Cache Creek thrust includes surface tilt directions and drainage patterns. Love and Montagne (1956) used altimeter surveys to establish that the surface of Jackson Hole, north of the Cache Creek thrust, is tilted west toward the Teton normal fault; this fact has recently been substantiated by a highly precise first-order level line (Smith and others, 1990). In contrast, the surface of South Park, directly south of the Cache Creek thrust and the town of Jackson, appears to be tilted east toward the Hoback normal fault (Fig. 9) (Love and Montagne, 1956; Love and Albee, 1972; Royse and others, 1975). Streams that flow across Jackson Hole and South Park directly reflect these opposite senses of surface rotation. For example, the Snake River flows southwest across Jackson Hole in an entrenched channel from its outlet at Jackson Lake. The westward component of the flow direction of the river is a result of the west tilt of the surface of Jackson Hole, forming a topographic depression at the base of the range (Love and Montagne, 1956). Cottonwood Creek and Fish Creek flow south in this range-front depression for several kilometers before eventually joining the Snake River. Between the village of Moose and the trace of the Cache Creek thrust, the Snake River flows at its closest position to the range. Quoting Love and Montagne (1956, p. 175):

> It is thought that the Teton fault dies out in the vicinity of Trail Creek [Cache Creek thrust], and, within two miles to the south of this locality where westward tilting has been eliminated, Fish Creek leaves the mountain front, turns southeast, and joins the Snake River.

South of the Cache Creek thrust, the Snake River slowly turns southeast toward the Hoback normal fault, reflecting the east-tilt of the surface of South Park. Flat Creek, on the east side of South Park, flows several kilometers south in a range-front depression before joining the Snake River. Quoting Love and Montagne (1956, p. 171):

> South and west of the town of Jackson, in the area where the Teton fault on the west side of the valley has died out and the Hoback normal fault has developed on the east side of the valley, the drainage relationships are exactly the reverse of those to the north.

Figure 8. Aerial photo of Taggart Lake (left) and Bradley Lake (right) showing left-lateral offset of Pinedale lateral moraine between the two lakes. Photo by Lageson, October 1987.

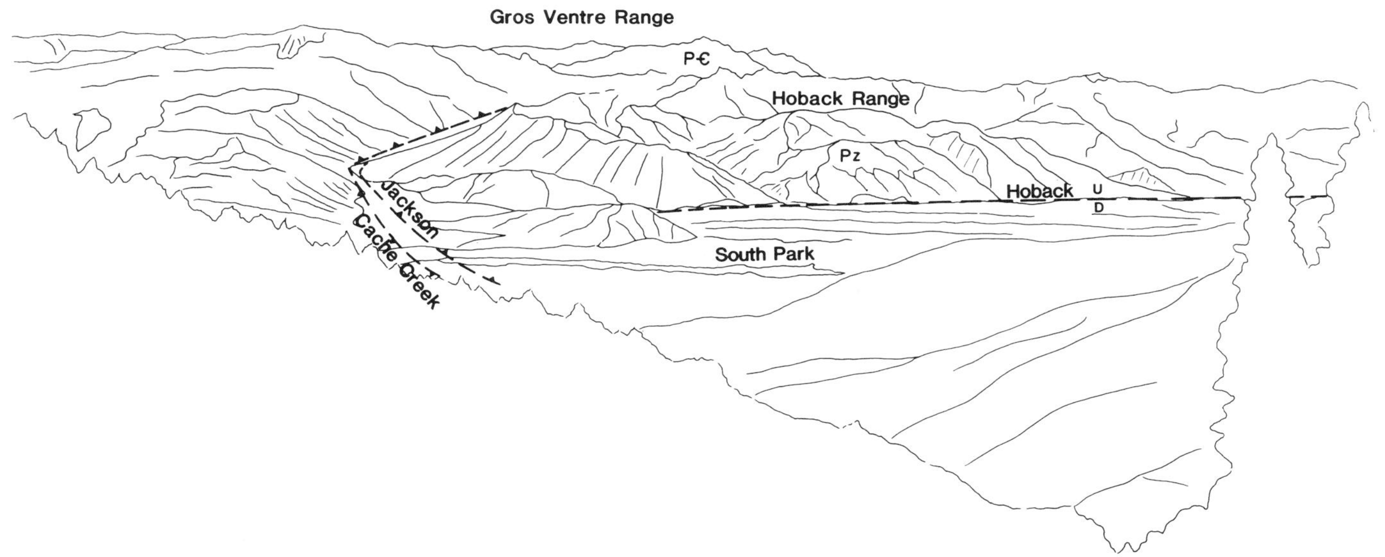

Figure 9. Sketch looking east from Teton Pass at the Hoback Range and Hoback normal fault in the hanging wall of the Jackson thrust fault. The surface of South Park is tilted east into the Hoback normal fault.

These observations lead to the conclusion that Jackson Hole and South Park are exhibiting independent structural behavior. The surface of Jackson Hole is tilted west whereas the surface of South Park is tilted east, and the boundary between the two is the Cache Creek thrust fault. This is further evidence that the Teton normal fault terminates at the Cache Creek thrust and that the two faults are structurally coupled at depth.

DISCUSSION

As previously mentioned, soling of normal faults into preexisting Laramide thrust faults is not without precedent as a structural style along the eastern margin of the Basin and Range province. Moreover, the process of hanging-wall rotations above an extensional detachment has been vividly demonstrated by the clay cake experiments of Cloos (1968) and later developed into the models of listric normal faulting (Dahlstrom, 1970; Royse and others, 1975; Proffett, 1977; Gibbs, 1983) and domino block faulting above low-angle detachments (Morton and Black, 1975; Davis, 1979; Wernicke and Burchfiel, 1982).

The model presented here does not imply that preexisting, reactivated thrust faults are a necessary prerequisite for extension of the eastern Basin and Range. There are several examples of extended fault blocks and detachments that have ignored preexisting thrust faults (Allmendinger and Platt, 1983; Burgel and others, 1987). However, along the eastern margin of the Basin and Range province, where the crust is thicker (Smith, 1978) and where there has been less total extension, it is possible that favorably oriented, preexisting zones of weakness in the uppermost crust and sedimentary cover may be important as detachments for late Cenozoic normal faults. For example, the Hoback normal fault and Grand Valley normal fault are well-known examples of thrust fault ramps that have been reactivated by late Cenozoic normal faults in the Wyoming salient of the fold-thrust belt (Royse and others, 1975). McDonald (1976) and Allmendinger and others (1983) suggested the possibility of a low-angle, west-dipping detachment beneath west-central Utah (Sevier Desert detachment) that may in part be a reactivated Mesozoic thrust fault. Using industry seismic reflection data, Smith and Bruhn (1984) observed that listric and shallow-dipping planar normal faults sole into preexisting thrust faults; they also suggested that segments of the Wasatch normal fault zone are collocated with a large, Mesozoic thrust ramp in basement rocks and that segment boundaries are controlled by lateral terminations in older thrust faults.

In northwestern Montana, fault-bounded, tilted blocks of Proterozoic strata form mountain ranges (Flathead, Swan, and Mission mountains), similar in scale to the Teton Range, that are distributed across a reactivated thrust belt décollement (Lageson and Sheriff, 1986). Similarly, late Cenozoic back sliding on basement-involving Laramide thrust faults has been suggested in the Bridger Range and Snowcrest Range of southwestern Montana (Lageson and Zim, 1985; Lageson, 1989; McBride and others, 1992). Therefore, where Basin and Range normal faults overlap the Sevier fold-thrust belt or the Laramide foreland province, preexisting zones of weakness have often played an important role in distributing the resulting extension. Structural interaction of the Teton normal fault and the Cache Creek thrust at the south end of the Teton Range is a hypothesis that should receive consideration and further study.

CONCLUSIONS

The Laramide paleogeography of western Wyoming and southeastern Idaho was dominated by the Ancestral Teton–Gros Ventre uplift. This northwest-trending, basement-cored uplift was bounded along its southwestern flank by the northeast-dipping Cache Creek thrust fault.

The modern Teton Range and Jackson Hole have been superimposed across the Ancestral Teton–Gros Ventre uplift. Structural elements of the former Laramide uplift, such as the Buck Mountain and Forellen reverse faults and the Cache Creek thrust fault, are evident within and adjacent to the modern Teton Range. Of these, the Cache Creek thrust is interpreted to have played the most important role in determining the orientation of the Teton normal fault at the south end of the Teton Range.

It is proposed that the Cache Creek thrust was reactivated as a sole fault for the extension and rotation of overlying basement blocks. Evidence for this includes the termination of fault blocks and normal faults against the trace of the Cache Creek thrust, surface indications of a thrust ramp coincident with the Teton fault zone, increasing displacement of the Teton fault zone to the northeast, and evidence for oppositely tilted crustal blocks north and south of the Cache Creek thrust that have affected drainage patterns.

Reactivation of a basement-involved thrust fault by a normal fault is not a unique occurrence in the eastern Basin and Range province. Several examples are cited where older Sevier thrust faults and Laramide basement-involved faults have been reactivated by late Cenozoic normal faults, typically along ramp interfaces. This style of extension may be more common within the transitional zone of thicker crust, as compared to more highly extended parts of the Basin and Range.

ACKNOWLEDGMENTS

I wish to thank Paul K. Link, J. David Love, Lucian Platt, and Frank Royse, Jr., for their rigorous reviews of this manuscript. I acknowledge D. L. Reese, Frank Royse, Jr., and J. Sales for providing the original insight to the detachment model for the Teton Range presented herein. I accept responsibility for the accuracy of observations and interpretations presented in this manuscript. Acknowledgment is made to the donors of The Petroleum Research Fund, administered by the American Chemical Society, for partial support of this research. Additional funding was generously provided by Montana State University (Faculty Research Development Grant), National Science Foundation (EPSCoR Program), Chevron USA, and Marathon Oil Company. The University of Wyoming–National Park Service Research Center kindly provided housing during the 1987 and 1988 field seasons.

REFERENCES CITED

Albee, H. F., 1968, Geologic map of the Munger Montain quadrangle, Teton and Lincoln Counties, Wyoming: U.S. Geological Survey, GQ-705, scale 1:24,000.

Allmendinger, R. W., and Platt, L. B., 1983, Stratigraphic variation and low-angle faulting in the North Hansel Mountains and Samaria Mountain, southern Idaho, *in* Miller, D. M., Todd, V. R., and Howard, K. A., eds., Tectonics and stratigraphic studies in the eastern Great Basin: Geological Society of America Memoir 157, p. 149–163.

Allmendinger, R. W., and six others, 1983, Cenozoic and Mesozoic structure of the eastern Basin and Range province, Utah, from COCORP seismic-reflection data: Geology, v. 11, p. 532–536.

Anders, M. H., 1990, Late Cenozoic evolution of Grand and Swan Valleys, Idaho, *in* Roberts, S., ed., Geologic field tours of western Wyoming: Geological Survey of Wyoming Public Information Circular 29, p. 15–25.

Behrendt, J. C., Tibbetts, B. L., Bonini, W. E., Lavin, P. M., Love, J. D., and Reed, J. C., 1968, A geophysical study in Grand Teton National Park and vicinity, Teton County, Wyoming: U.S. Geological Survey Professional Paper 516-E, 23 p.

Blackstone, D. L., Jr., and DeBruin, R. H., 1987, Tectonic map of the overthrust belt, western Wyoming, northwestern Utah, and southeastern Idaho, showing oil and gas fields and exploratory wells in the overthrust belt and adjacent Green River Basin: The Geological Survey of Wyoming, Map Series 23, scale 1:316,800.

Bradley, C.C., 1956, The Precambrian complex of Grand Teton National Park, Wyoming, *in* Berg, R. R., ed., Jackson Hole field conference guidebook: Wyoming Geological Association, 11th Annual Field Conference, Guidebook, p. 34–42.

Burgel, W. D. Rodgers, D. W., and Link, P. K., 1987, Mesozoic and Cenozoic structures of the Pocatello region, southeastern Idaho, *in* Miller, W. R., ed., The thrust belt revisited: Wyoming Geological Association, 38th Annual Field Conference, Guidebook, p. 91–100.

Cloos, E., 1968, Experimental analysis of Gulf Coast fracture patterns: American Association of Petroleum Geologists Bulletin, v. 52, p. 420–444.

Constenius, K. T., 1982, Relationship between the Kishenehn Basin and the Flathead listric normal fault system and Lewis thrust salient, *in* Powers, R. B., ed., Geologic studies of the Cordilleran thrust belt, v. 2: Denver, Rocky Mountain Association of Geologists, p. 817–830.

Crosby, G. W., 1984, Structural-geophysical interpretation of Swan Valley, Montana, *in* McBane, J. D., and Garrison, P. B., eds., Northwest Montana and adjacent Canada: Montana Geological Society, Field Conference, Guidebook, p. 245–251.

Dahlstrom, C.D.A., 1970, Structural geology in the eastern margin of the Canadian Rocky Mountains: Bulletin of Canadian Petroleum Geology, v. 18, p. 332–406.

Davis, G. A., 1979, Problems of intraplate extensional tectonics, western United States, with special emphasis on the Great Basin, *in* Newman, G. W., and Goode, H. D., eds., Basin and Range Symposium and Great Basin Field Conference: Denver, Rocky Mountain Association of Geologists and Utah Geological Association, p. 41–54.

Dickinson, W. R., and six others, 1988, Paleogeographic and paleotectonic setting of Laramide sedimentary basins in the central Rocky Mountain region: Geological Society of America Bulletin, v. 100, p. 1023–1039.

Dorr, J. A., Spearing, D. R., and Steidtmann, J. R., 1977, Deformation and deposition between a foreland uplift and an impinging thrust belt: Hoback Basin, Wyoming: Geological Society of America Special Paper 177, 82 p.

Dunn, S.L.D., 1983, Timing of foreland and thrust belt deformation in an overlap area, Teton Pass, Idaho and Wyoming, *in* Lowell, J. D., ed., Rocky Mountain foreland basins and uplifts: Denver, Rocky Mountain Association of Geologists, p. 263–269.

Gibbs, A. D. 1983, Balanced cross-section construction from seismic sections in areas of extensional tectonics: Journal of Structural Geology, v. 5, p. 153–160.

Guthrie, G. E., and McBride, B. C., 1989, Hydrocarbon generation and Late Cretaceous Laramide deformation in the central Snowcrest Range, southwestern Montana, *in* French, D. E, and Grabb, R. F., eds., Geologic resources of Montana, v. 1: Montana Geological Society, Field Conference, Guidebook, p. 311–324.

Harlan, S. S., Geissman, J. W., Lageson, D. R., and Snee, L. W., 1988, Paleomagnetic and isotopic dating of thrust-belt deformation along the eastern edge of the Helena salient, northern Crazy Mountains Basin, Montana: Geological Society of America Bulletin, v. 100, p. 492–499.

Horberg, L., 1938, The structural geology and physiography of the Teton Pass area, Wyoming: Rock Island, Illinois, Augustana College, Augustana Library Publication 16, 86 p.

Kulik, D. M., and Schmidt, C. J., 1988, Region of overlap and styles of interaction of Cordilleran thrust belt and Rocky Mountain foreland, *in* Schmidt, C. J., and Perry, W. J., Jr., eds., Interaction of the Rocky Mountain foreland and the Cordilleran thrust belt: Geological Society of America Memoir 171, p. 75–98.

Lageson, D. R., 1980, Structural geology of the Stewart Peak quadrangle, Lincoln County, Wyoming, and adjacent parts of the Idaho-Wyoming thrust belt [Ph.D. thesis]: Laramie, University of Wyoming, 293 p.

—— , 1987, Laramide uplift of the Gros Ventre Range and implications for the origin of the Teton fault, Wyoming, *in* Miller, W. R., ed., The thrust belt revisited: Wyoming Geological Association, 38th Annual Field Conference, Guidebook, p. 79–89.

—— , 1988, Laramide controls on the Teton Range, northwest Wyoming: Geological Society of America Abstracts with Programs, v. 20, p. 426.

—— , 1989, Reactivation of a Proterozoic continental margin, Bridger Range, southwestern Montana, *in* French, D. E., and Grabb, R. F., eds., Geological resources of Montana, v. 1: Montana Geological Society, Field Conference, Guidebook, p. 279–298.

Lageson, D. R., Sheriff, S., 1986, Kinematic development of Neogene extension, northwest Montana: Geological Society of America Abstacts with Programs, v. 18, p. 389.

Lageson, D. R., and Zim, J. C., 1985, Uplifted basement wedges in the northern Rocky Mountains: Geological Society of America Abstracts with Programs, v. 17, p. 250.

Love, J. D., 1956a, Cretaceous and Tertiary stratigraphy of the Jackson Hole area, northwestern Wyoming, *in* Berg, R. R., ed., Jackson Hole field conference guidebook: Wyoming Geological Association, 11th Annual Field Conference, Guidebook, p. 79–94.

—— , 1956b, Summary of geologic history of Teton County, Wyoming, during late Cretaceous, Tertiary, and Quaternary times, *in* Berg, R. R., ed., Jackson Hole field conference guidebook: Wyoming Geological Association, 11th Annual Field Conference, Guidebook, p. 140–150.

—— , 1968, Geologic cross section of the Teton Range: U.S. Geological Survey topographic map of Grand Teton National Park, Teton County, Wyoming, scale 1:62,500.

—— , 1973, Harebell Formation (Upper Cretaceous) and Pinyon Conglomerate (Uppermost Cretaceous and Paleocene), northwestern Wyoming: U.S. Geological Survey Professional Paper 734-A, 54 p.

—— , 1977, Summary of Upper Cretaceous and Cenozoic stratigraphy, and of tectonic and glacial events in Jackson Hole, northwestern Wyoming, *in* Heisey, E. L., and others, eds., Rocky Mountain thrust belt geology and resources: Wyoming Geological Association, 29th Annual Field Conference, Guidebook, p. 585–593.

Love, J. D., and Albee, H. F., 1972, Geologic map of the Jackson Quadrangle, Teton County, Wyoming: U.S. Geological Survey Map I-769-A, scale 1:24,000.

Love, J. D., and Christiansen, A. C., 1985, Geologic map of Wyoming: U.S. Geological Survey, scale 1:500,000.

Love, J. D., and Keefer, W. R., 1975, Geology of sedimentary rocks in southern Yellowstone National Park, Wyoming: U.S. Geological Survey Professional Paper 729-D, 60 p.

Love, J. D., and Montagne, J., 1956, Pleistocene and Recent tilting of Jackson Hole, Teton County, Wyoming, *in* Berg, R. R., Jackson Hole field conference guidebook: Wyoming Geological Association, 11th Annual Field Conference, Guidebook, p. 169–178.

Love, J. D., and Reed, J. C., Jr., 1968, Creation of the Teton landscape—the geologic story of Grand Teton National Park: Moose, Wyoming, Grand Teton Natural History Association, 120 p.

Love, J. D., Reed, J. C., Jr., Christiansen, R. L., and Stacy, J. R., 1973, Geologic block diagram and tectonic history of the Teton region, Wyoming-Idaho: U.S. Geological Survey Map I-730.

Love, J. D., Leopold, E. B., and Love, D. W., 1978, Eocene rocks, fossils, and geologic history, Teton Range, northwestern Wyoming: U.S. Geological Survey Professional Paper 932-B, 40 p.

McBride, B. C., Schmidt, C. J., Guthrie, G. E., and Sheedlo, M. K., 1992, Multiple reactivation of a collisional boundary: An example from southwestern Montana, *in* Bartholomew, M. J., Hyndman, D. W., Mogk, D. W., and Mason, R., eds., Characterization and comparison of ancient and Mesozoic continental margins—Proceedings of the 8th International Conference on Basement Tectonics, Butte, Montana, USA: Dordrecht, The Netherlands, Kluwer Academic Publishing, p. 343–358.

McCalpin, J. P., Piety, L. A., and Anders, M. H., 1990, Latest Quaternary faulting and structural evolution of Star Valley, Wyoming, *in* Roberts, S., ed., Geologic field tours of western Wyoming: Geological Survey of Wyoming Public Information Circular 29, p. 5–12.

McDonald, R. E., 1976, Tertiary tectonics and sedimentary rocks along the transition, Basin and Range province to plateau and thrust belt province, *in* Hill, J. G., ed., Symposium on geology of the Cordilleran hingeline: Denver, Rocky Mountain Association of Geologists, p. 281–318.

Morton, W. H., and Black, R., 1975, Crustal attenuation in Afar, *in* Pilger, A., and Rosler, A., eds., Afar depression of Ethiopia (v. 1): Stuttgart, Germany, Schweizerbart'sche Verlagsbuchhaandlung, p. 55–65.

Olson, T. J., and Schmitt, J. G., 1987, Sedimentary evolution of the Miocene-Pliocene Camp Davis Basin, northwestern Wyoming, *in* Miller, W. R., ed., The thrust belt revisited: Wyoming Geological Association, 38th Annual Field Conference, Guidebook, p. 225–243.

Oriel, S. S., and Moore, D. W., 1985, Geologic map of the West and East Palisades roadless areas, Idaho and Wyoming: U.S. Geological Survey Map MF-1619-B, scale 1:50,000.

Proffett, J. M., Jr., 1977, Cenozoic geology of the Yerington district, Nevada, and implications for the nature and origin of Basin and Range faulting: Geological Society of America Bulletin, v. 88, p. 247–266.

Ramsay, J. G., and Huber, M. I., 1987, The techniques of modern structural geology, v. 2: Folds and fractures: London, Academic Press, 700 p.

Reed, J. C., Jr., 1973, Geologic map of the Precambrian rocks of the Teton Range, Wyoming: U.S. Geological Survey Open File Report 73-230, scale 1:62,500.

Roberts, S. M., and Burbank, D. W., 1988, Late Cretaceous to Recent differential uplift in the Teton Range, N.W. WY, by apatite fission-track dating: Geological Society of America Abstracts with Programs, v. 20, p. 465.

Royse, F., Jr., 1983, Extensional faults and folds in the foreland thrust belt, Utah, Wyoming, Idaho: Geological Society of America Abstracts with Programs, v. 15, p. 295.

Royse, F., Jr., Warner, M. A., and Reese, D. L., 1975, Thrust belt structural geometry and related stratigraphic problems, Wyoming–Idaho–northern Utah, *in* Bolyard, D. W., ed., Symposium on deep drilling frontiers in the central Rocky Mountains: Denver, Rocky Mountain Association of Geologists, p. 41–54.

Sales, J. K., 1983, Collapse of Rocky Mountain basement uplifts, *in* Lowell, J. D., ed., Rocky Mountain foreland basins and uplifts: Denver, Rocky Mountain Association of Geologists, p. 79–97.

Sbar, M. L., Barazangi, M., Dorman, J., Scholz, C. H., and Smith, R. B., 1972, Tectonics of the intermountain seismic belt, western United States: Microearthquake seismicity and composite fault plane solutions: Geological Society of America Bulletin, v. 83, p. 13–28.

Schroeder, M. L., 1969, Geologic map of the Teton Pass quadrangle, Teton

County, Wyoming: U.S. Geological Survey Map GQ-793, scale 1:24,000.
——, 1972, Geologic map of the Rendezvous Peak quadrangle, Teton County, Wyoming: U.S. Geological Survey Map GQ-980, scale 1:24,000.
Simons, F. S., Love, J. D., Keefer, W. R., Harwood, D. S., Kulik, D. M., and Bieniewski, C. L., 1988, Mineral resources of the Gros Ventre Wilderness study area, Teton and Sublette Counties, Wyoming: U.S. Geological Survey Bulletin, 1591, 65 p.
Smith, D. J., 1990, Structural analysis of the Buck Mountain fault and related intra-range faults, Teton Range, Wyoming [M.S. thesis]: Bozeman, Montana State University, 92 p.
Smith, D. J., and Lageson, D. R., 1989, Structural geology and history of the Buck Mountain fault, Teton RAnge, western Wyoming: Geological Society of America Abstracts with Programs, v. 21, p. 145.
Smith, R. B., 1978, Seismicity, crustal structure, and intraplate tectonics of the interior of the western Cordillera, *in* Smith, R. B., and Eaton, G. P., eds., Cenozoic tectonics and regional geophysics of the western Cordillera: Geological Society of America Memoir 152, 111–144.
Smith, R. B., and Bruhn, R. L., 1984, Intraplate extensional tectonics of the eastern Basin-Range: Inferences on structural style from seismic reflection data, regional tectonics, and thermal-mechanical models of brittle-ductile deformation: Journal of Geophysical Research, v. 89, p. 5733–5762.
Smith, R. B., and Sbar, M. L., 1974, Contemporary tectonics and seismicity of the western United States, with emphasis on the intermountain seismic belt: Geological Society of America Bulletin, v. 85, p. 1205–1218.
Smith, R. B., Pelton, J. R., and Love, J. D., 1976, Seismicity and the possibility of earthquake related landslides in the Teton–Gros Ventre–Jackson Hole area, Wyoming: University of Wyoming Contributions to Geology v. 14, p. 57–64.
Smith, R. B., Byrd, J.O.D., and Susong, D. D., 1990, Neotectonics and structural evolution of the Teton fault, *in* Roberts, S., ed., Geologic field tours of western Wyoming: Geological Survey of Wyoming Public Information Circular 29, p. 127–138.
Susong, D. D., Smith, R. B., and Bruhn, R. L., 1987, Quaternary faulting and segmentation of the Teton fault zone, Grand Teton National Park, Wyoming: EOS Transactions of the American Geophysical Union, v. 68, p. 1452.
U.S. Geological Survey, 1972, Geologic map of Yellowstone National Park: U.S. Geological Survey, Map I-711, scale 1:125,000.
Vasko, A. M., 1982, Structural setting of Teton Pass with emphasis on fault breccia associated with the Jackson thrust fault, Wyoming [M.S. thesis]: Bozeman, Montana State University, 64 p.
Wernicke, B., and Burchfiel, B. C., 1982, Modes of extensional tectonics: Journal of Structural Geology, v. 4, p. 105–115.
Wiltschko, D. J., and Dorr, J. A., Jr., 1983, Timing of deformation in overthrust belt and foreland of Idaho, Wyoming, and Utah: American Association of Petroleum Geologists Bulletin, v. 67, p. 1304–1322.
Zeller, P., 1982, Structural geology along Teton Pass, Wyoming [M.S. thesis]: Columbia, University of Missouri.

Manuscript Accepted by the Society July 19, 1991

Printed in U.S.A.

Geological Society of America
Memoir 179
1992

Chapter 10

Structural features and emplacement of surficial gravity-slide sheets, northern Idaho–Wyoming thrust belt

Steven E. Boyer
Department of Geological Sciences, University of Washington, Seattle, Washington 98195
John R. Hossack
BP Exploration, Inc., 5151 San Felipe, Houston, Texas 77210

ABSTRACT

Gravity-slide sheets are common in areas of sufficient topographic relief provided by extensional and compressional tectonism. They occur throughout the Basin and Range Province of Nevada and Utah but are perhaps best exposed in the thrust belt of southeast Idaho and northwest Wyoming. The Idaho-Wyoming slide sheets developed adjacent to extensional basin-bounding faults, which in turn were initiated at a ramp in the Absaroka thrust. Initial coherent sheets may have been up to 12 km^2 in aerial extent, but as a result of differential movement rates, these larger sheets were broken into smaller blocks by strike-parallel extension during emplacement. Folds on the down-slope edges of the sheets suggest movement analogous to a tank tread: Through interbed shear the limestone beds rolled forward and were dragged beneath the leading edge of each sheet. It is not known whether initial emplacement was catastrophic or by creep, but undisrupted cleavages within footwall Tertiary mudstones and coherent bedding within the sheets suggest creep for at least the later phases of movement.

INTRODUCTION

Gravity-slide blocks associated with Tertiary extensional faulting occur throughout the Basin and Range Province of Nevada, Utah, and Idaho (Longwell, 1951; Drewes, 1959, 1963, 1967; Cook, 1960; Mackin, 1960; Young, 1960; Kurie, 1966; Seager, 1970; Moores, 1968; Shackelford, 1975; Krieger, 1977; Bally and others, 1981; Todd, 1983; Faugére and others, 1986; Sable and Anderson, 1986; 1987, p. 444; Boyer & Allison, 1987) and adjacent portions of the Idaho-Utah-Wyoming thrust belt (Beutner, 1972; Hait and others, 1977; Moore and others, 1984; Oriel and Moore, 1985; Moore and others, 1987; Hait, 1987; Skipp, 1988; Janecke, 1989; Anders, 1990; McCalpin and others, 1990). As gravity-slide surfaces are difficult to distinguish from thrusts or extensional faults and may have been misinterpreted as such (Boyer and Allison, 1987), gravity-slide surfaces are probably more extensive than has been previously recognized. Indeed, many of these slide masses have previously been interpreted as Tertiary thrust sheets (Reber, 1952). Although the identification of gravity-slide sheets, and thus their extent and significance, have often been disrupted (Proffett, 1977), their existence in much of the Basin and Range Province cannot be questioned, as they have been encountered in wells drilled for petroleum and minerals exploration and have been imaged by reflection seismic methods (Bortz and Murray, 1979; Boyer and Allison, 1987).

Gravity-slide sheets are especially common within half grabens of southeast Idaho and northwest Wyoming (Fig. 1). As the slides in the Idaho-Wyoming thrust belt placed older (lower Paleozoic) upon younger (Miocene and Pliocene) strata, they were initially interpreted as west-vergent thrusts (Love, 1956a, b; Merritt, 1956). Subsequently Eardley (1960) and Sehnke (1969) suggested that these features were gravity-induced slide sheets related to Tertiary extensional faulting, a conclusion supported by subsequent workers (Beutner, 1972; Schroeder, 1974; Albee and others, 1977; Dorr and others, 1977; Hait and others, 1977; Oriel and Platt, 1980; Moore and others, 1984; Moore and others, 1987; Oriel and Moore, 1985; Hait, 1987; Olson and Schmitt, 1987; Webel, 1987; McCalpin and others, 1990; Anders, 1990).

Boyer, S. E., and Hossack, J. R., 1992, Structural features and emplacement of surficial gravity-slide sheets, northern Idaho–Wyoming thrust belt, *in* Link, P. K., Kuntz, M. A., and Platt, L. B., eds., Regional Geology of Eastern Idaho and Western Wyoming: Geological Society of America Memoir 179.

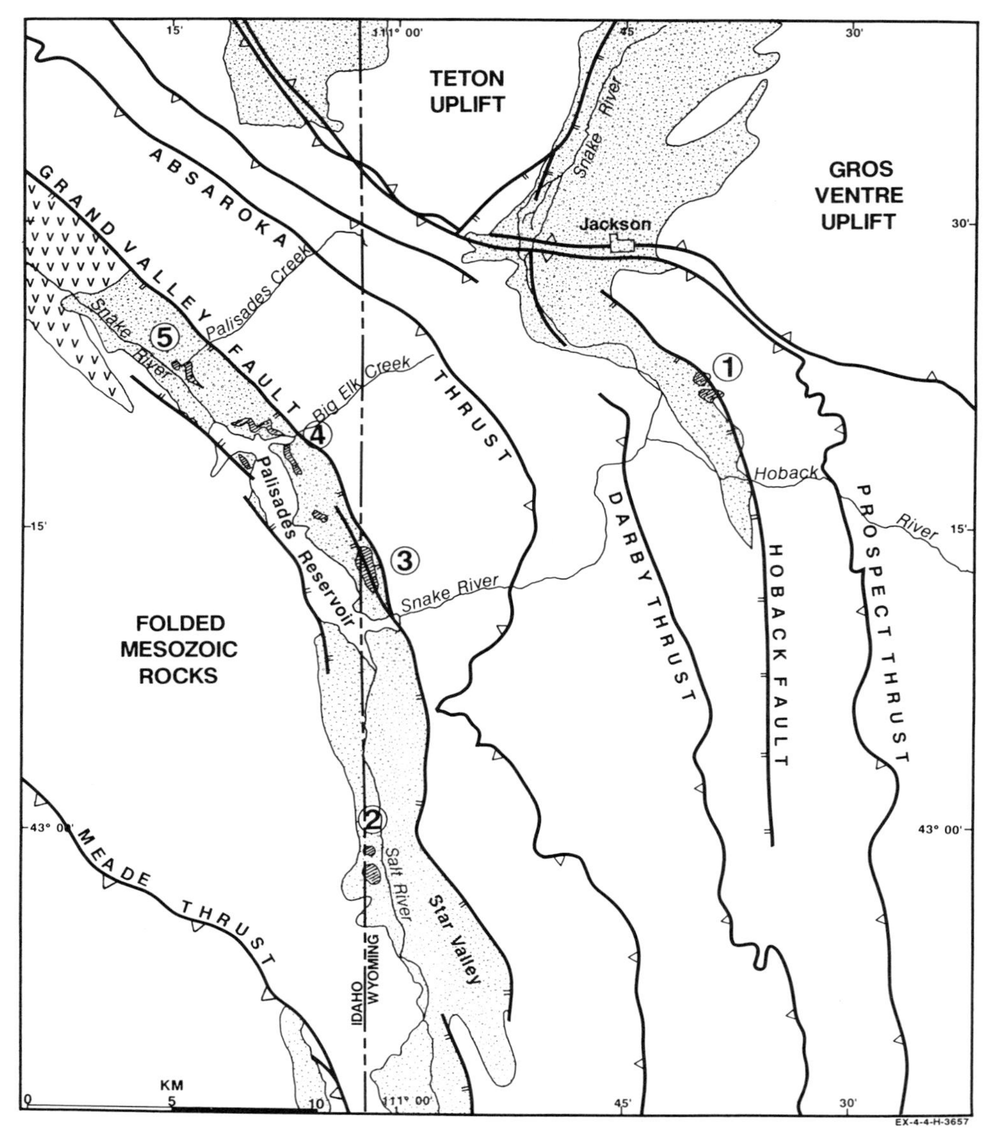

Figure 1. Distribution of gravity-slide sheets in the northern part of the Idaho-Wyoming thrust belt. The following references are maps and/or other publications that display or describe the various slides at the indicated locations: (1) along trace of Hoback normal fault: Sehnke, 1969; Schroeder, 1974; Dorr and others, 1977; Olson and Schmitt, 1987; (2) west side of Star Valley: Hait and others, 1977; Oriel and Platt, 1980; McCalpin and others, 1990; (3) Alpine Junction: Albee and Cullins, 1975; Moore and others, 1987; Oriel and Moore, 1985; (4) Big Elk Creek: Hait and others, 1977; Oriel and Moore, 1985; Anders, 1990; and (5) Palisades Creek: Albee and others, 1977; Oriel and Moores, 1985. A map and section of the Alpine slide sheet (3) appear in Figures 4 and 5. Refer to Figures 6 through 13 for photographs and sketches of the Palisades Creek slides and associated deformational features.

It has been known for some time that extensional faulting and resultant isostatic uplift of footwalls provide the topographic relief and gravitational instability necessary for the initiation of large-scale gravity sliding (Fig. 2), but to the best of our knowledge no studies have been initiated to understand the mechanisms of movement or the rate of emplacement of the gravity-slide sheets exposed in Grand Valley. Did the slides move catastrophically, being emplaced in a matter of seconds or a few minutes, or did they creep down alluvial fans and into adjacent valleys over hundreds of thousands of years? The purpose of this chapter is to

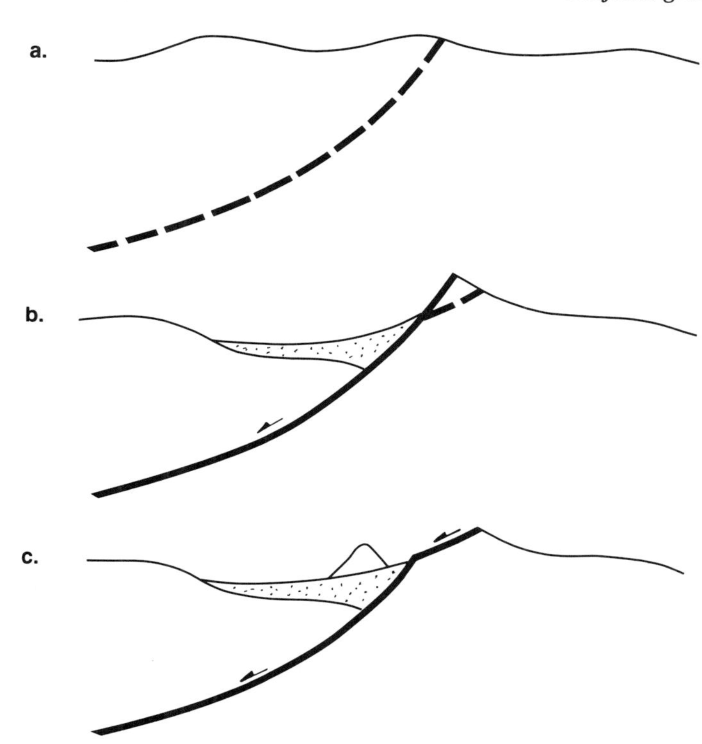

Figure 2. Model of gravity-slide emplacement (after Drewes, 1959, 1963). a, Dashed line is trace of future extensional fault. b, During fault displacement half graben is filled with alluvial deposits derived from the adjacent range. Footwall peak is gravitationally unstable. c, Exposed segment of footwall detaches along bedding or fault planes and slides onto surface of an alluvial fan.

describe some of the features associated with the emplacement of the gravity-slide blocks and to utilize these observations to place constraints on the mode and rate of emplacement. We begin with a discussion of the Alpine slide sheet, located at the mouth of the Snake River Canyon near Alpine Junction (3 in Fig. 1), which demonstrates the relationship of the gravity-slide sheets to extensional faulting and preexisting thrust geometries, and then describe the Palisades Creek slides, 23 km to the north, illustrating the internal deformation that accompanied and facilitated their emplacement.

RELATIONSHIP OF SLIDES TO PREEXISTING STRUCTURES

Gravity slides in southeast Idaho and adjacent Wyoming formed adjacent to Miocene/Pliocene extensional faults (Hait and others, 1977; Moore and others, 1987; Anders, 1990), the locations of which were controlled by the older thrust geometry. As is the case in the Canadian Rocky Mountains (Bally and others, 1966; Dahlstrom, 1970; Labrecque and Shaw, 1973), available seismic data indicate that most Tertiary extensional faults in the northern part of the Idaho-Wyoming thrust belt appear to merge at depth with preexisting thrust planes (Royse and others, 1975, p. 47). The thrust faults, being planes of weakness with an orientation favoring reactivation, controlled the loci of later extensional faults (Fig. 3). The relationship of extensional faults to underlying thrust faults is illustrated by a large gravity slide located at the mouth of the Snake River canyon near Alpine Junction, Wyoming (Fig. 4; location 3 in Fig. 1), the exposed parts of which cover approximately 7 km^2. A map of the slide was first published by Albee and Cullins (1975), and the geometry and extent of the slide have become better understood as a result of mapping by Oriel and Moore (1985). With the aid of a cross section, Moore and others (1987) discussed the general geometry of the block and described deformational features associated with its emplacement. The following description of the slide sheet is drawn from the map of Oriel and Moore (1985) and the discussion by Moore and others (1987).

The Alpine slide consists of two distinct segments: a western block of Cambrian through Devonian rocks upon Miocene and Pliocene volcaniclastics and conglomerates and an eastern block of Cambrian through Devonian strata overlying Ordovician dolomite (Figs. 4 and 5). The western block includes a truncated thrust that brought the hanging-wall Cambrian Gros Ventre Formation and Ordovician Bighorn Dolomite in contact with a footwall syncline of Ordovician dolomite, the Devonian Darby Formation, and the lower Mississippian Lodgepole Limestone. The Cambrian strata along the northwestern side of the western block are overturned and dip 25 to 55° toward the east-southeast.

The transported thrust may correlate to the Ferry Peak thrust (Moore and others, 1987), which is exposed at an elevation of 2,010 to 2,870 m (6,600–9,400 ft) and approximately 2.4 km

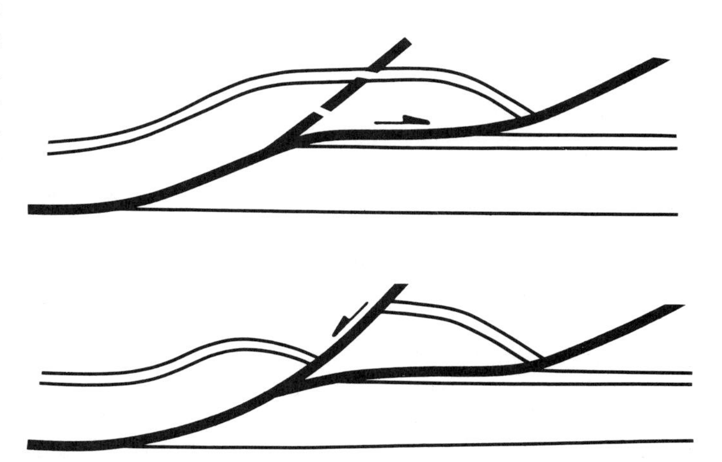

Figure 3. In the northern (Idaho/Wyoming) part of the thrust belt extensional faults propagated from the top of thrust ramps (Royse and others, 1975).

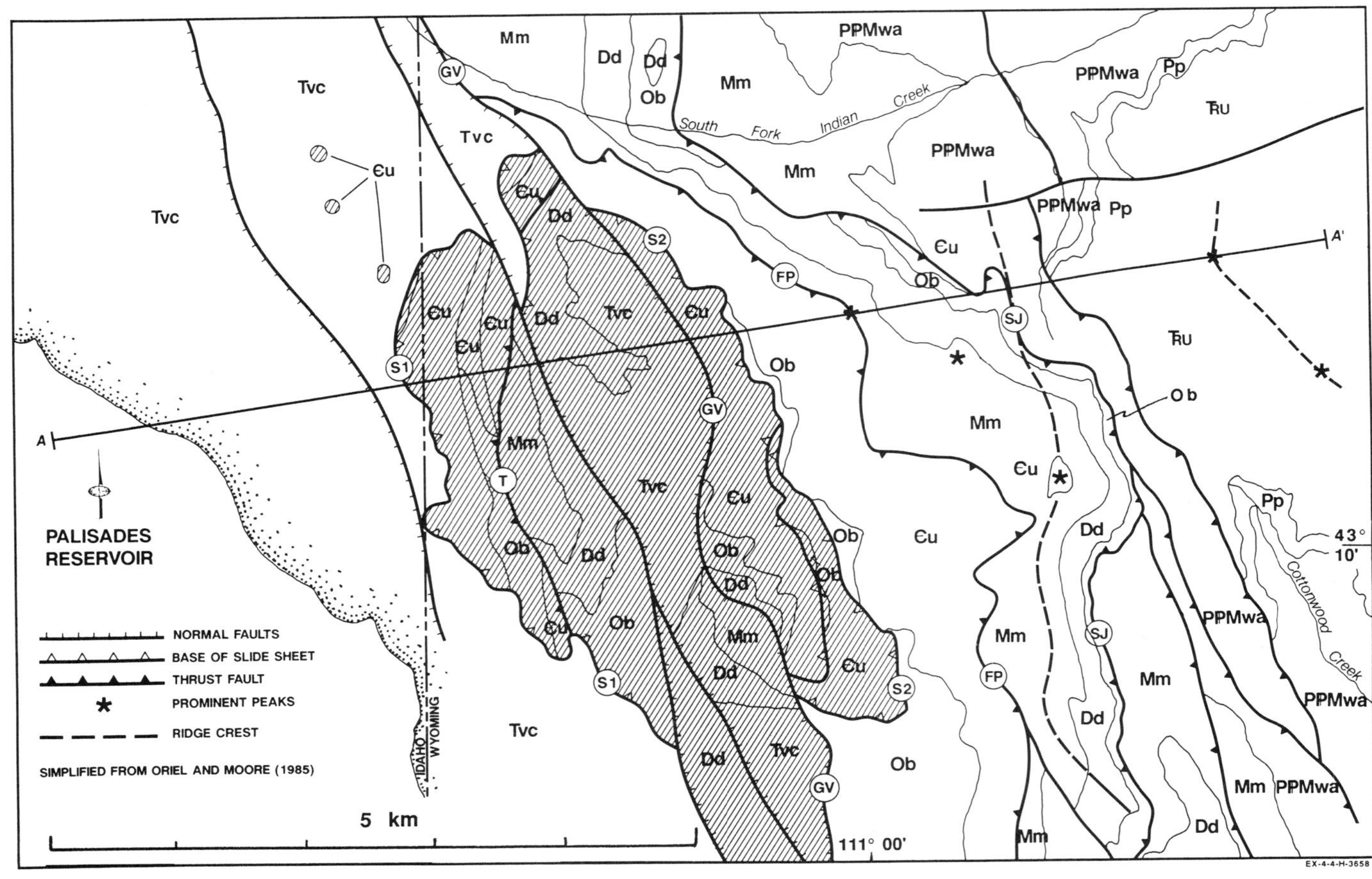

Figure 4. The Alpine Junction slide sheet, based on the mapping of Oriel and Moore (1985) and Moore and others (1987). Slide mass and overlying Tertiary sediments are indicated by hachures. GV = Grand Valley normal fault; SJ = St. John thrust. FP = Ferry Peak thrust; S1 = base of slide surface to west of Grand Valley normal fault, where the slide block overlies Tertiary sedimentary rocks; S2 = slide surface to east of Grand Valley normal fault; T = thrust within slide sheet. Abbreviations of map units correspond to abbreviations shown in Fig. 5. Cross section A = A′ is shown in Figure 5.

to the east of the center of the slide sheet (Jobin, 1972; Albee and Cullins, 1975; Oriel and Moore, 1985). However, if the gravity slide truncated and translated the Ferry Peak thrust, rocks in the slide sheet should contain only rocks younger than the corresponding hanging-wall and footwall rocks exposed within the range. Instead, within the range Mississippian strata comprise the footwall of the Ferry Peak thrust, whereas within the slide block the footwall rocks of the transported thrust include rocks as old as Ordovician. Therefore, the thrust embedded within the slide sheet may actually correlate to a structurally higher imbricate, perhaps the Needle Peak thrust (Albee and Cullins, 1975), which has been partly removed by erosion and gravity sliding. Regardless of these correlations, the minimum displacement on the slide sheet is 2.4 km, the distance between the downhill side of the sheet and the Grand Valley normal fault.

The slide sheets at Alpine Junction are of interest because they provide examples of the relationship of gravity slides to preexisting structure and of the large physical dimensions that these sheets may attain. Additional slide sheets are exposed in the intervening region between Palisaes Creek and Alpine Junction, being best exposed in a number of drainages and ridges on the northeast side of Palisades Reservoir and a "peninsula" that juts into the southwest corner of the reservoir 3 km southeast of Palisades Dam (see the map of Oriel and Moore, 1985, for detail, and Fig. 1 for approximate location).

We now turn our attention to the exposures near Palisades Creek (Fig. 6), 22 km north of the Alpine Junction slide blocks, where a number of well-exposed segmented slide sheets display deformational features associated with slide emplacement.

INTERNAL DEFORMATION FEATURES

Immediately south of the Palisades Creek (location 5 in Fig. 1) is a very large limestone sheet of Cambrian Gros Ventre(?)

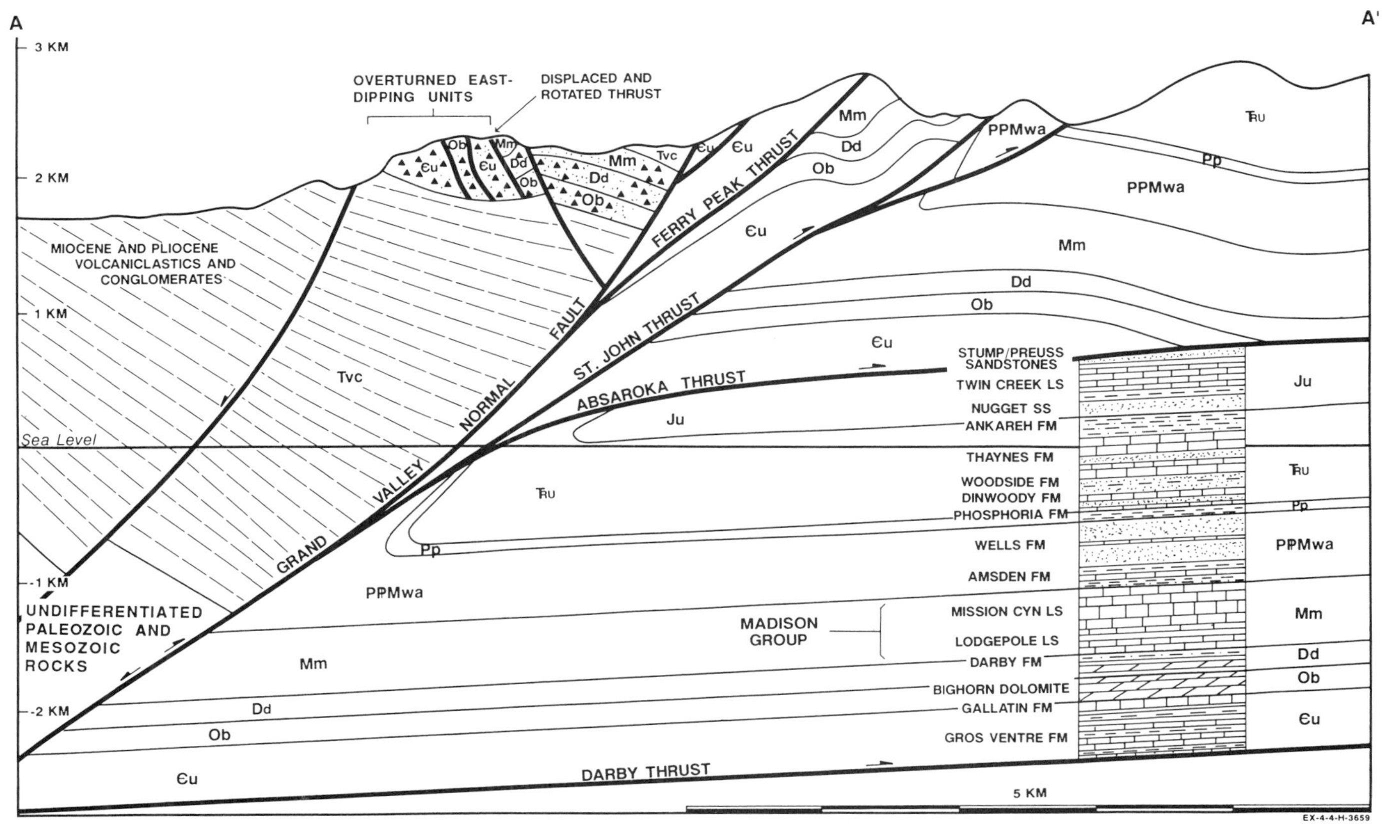

Figure 5. Cross section A-A′ of Figure 4, modified from the section of Moore and others (1987). No vertical exaggeration. Tvc = Miocene and Pliocene volcaniclastics.

Figure 6. Slide sheets and isolated blocks to the north of Palisades Dam and immediately south of Palisades Creek. View from the southwest. The north end of the large slide sheet at left is shown in greater detail in Figure 7. The largest block has an exposed face of 75 m high by 400 m long.

Figure 7. View east into the face of the largest slide block of Figure 6. Cliff is approximately 75 m high. The slide sheet has been segmented and extended by a south-dipping fault that separates the larger part of the sheet from a smaller block at left. This segmentation, fault, dashed line in photo, flattens at the top of the prominent overhang (in shadow, slightly left of center) and then steepens again toward the base of the outcrop. The fault is exposed on the base of the overhang, where it is decorated with south-oriented slickenlines. A series of isolated blocks trails to the south of the larger block (Fig. 6), and we presume that they were also broken from the largest sheet by a number of additional cross-strike-oriented segmentation faults.

Formation, the exposed face of which is approximately 50 to 75 m high and 400 m long (Fig. 7). Extending southward from this large sheet are a number of smaller blocks that are strung out at approximately the same stratigraphic level within the Tertiary sequence. These slide sheets overlie conglomerates containing cobbles and boulders of what appear to be Pennsylvanian quartzites, red sandstones of either the Mississippian Pennsylvanian Amsden or Triassic Woodside Formations, and Mississippian(?) carbonates. The slides are in turn overlain by limestone (Cambrian Gros Ventre Formation(?) conglomerates. Thus, the conglomerates constitute an inverse stratigraph;y reflecting unroofing of the adjacent Snake River Range.

Strike-parallel segmentation of slide sheets

A down-to-the-south fault defines the northern termination of the largest block from a much smaller, partially detached block to the north (Fig. 7). Slickenlines on the fault plane trend north-northwest, parallel to the strike of Grand Valley. We interpret this fault, which is internal to the slide sheet, to be related to along-strike segmentation of the slide sheets during gravity emplacement. It is not related to regional extension. In an order to distinguish such gravity-related faults from basin-bounding faults accompanying crustal extension, we will subsequently refer to the former as "segmentation faults."

Near the south end of the master sheet are exposures of extensional shear bands (Fig. 8) and an east-west (cross-strike) clastic dike injected from the base of the sheet (Fig. 9). These features, which strike parallel to the internal extensional fault (Fig. 7), also indicate north-south (strike-parallel) segmentation of the slide sheets during emplacement.

We infer from these relationships that the numerous small slide sheets of the Palisades Creek area (Fig. 1) were derived from large sheets by a "calving" process. Other slide masses exposed between Palisades Creek and Alpine Junction (Fig. 1) are geometrically similar. They appear as coherent continuous sheets from a distance, but upon closer examination, they are found to be broken into smaller discontinuous blocks. Thus, it is unlikely that all slide masses in Grand valley moved simultaneously as coherent sheets. As the larger sheets moved down alluvial fans, various parts of the slide may have moved more rapidly than adjacent parts, leading to strike-parallel segmentation. This model would explain the trails of smaller blocks that extend north and south from some of the larger masses (Fig. 6)

Rate of emplacement

Did the sheets move catastrophically from the high ridges of the adjacent Snake River Range, or did they creep slowly down the slopes of alluvial fans? What features could be evaluated to determine the rates of emplacement? In the following discussion we use *creep* not in the rock mechanic sense but as a geomorphic term: "The slow, more or less continuous downslope movement of . . . rock . . . under gravitational stresses" (Bates and Jackson, 1987, p. 153). Various criteria to distinguish catastrophic emplacement from creep include the nature of the boundary zone and the internal coherence of the rock mass.

If the rock mass moved catastrophically onto unconsolidated alluvial fan deposits, the boundary zone should consist of a chatoic mixture of hanging-wall limestone breccia and footwall conglomerate clasts and boulders. Instead, the boundary between alluvial fan deposits and the basal breccias of the overlying sheets is knife-sharp (Fig. 10). Since there is no mixing of hanging-wall and footwall units, we infer that the sheets moved by creep.

Furthermore, immediately south of Palisades Creek, cleavage is developed within red silty mudstones separating the footwall Tertiary conglomerates from the hanging-wall slide sheets of Cambrian (?) limestone (Fig. 10B). The mudstone may have been a weathering horizon or colluvial deposit developed upon the alluvial fan, as it contains small clasts of footwall and hanging-wall units. The lower boundary of the cleaved mudstone is a knife-sharp plane, a number of clasts of the footwall having been severed by the slide surface. Cleavage within the mudstones of the slide zone indicates that the mudstone was sheared between the overriding limestone slide sheet and the footwall conglomerates. These fabrics could not have been preserved if the slide block had been emplaced catastrophically.

Internal coherence of slide blocks may provide some clue to the rate of emplacement. Except along a basal boundary zone, bedding is relatively well preserved within the slide masses, especially at Palisades Creek (Fig. 8). One would not expect this to be the case with catastrophically emplaced landslide deposits, such as the 1903 slide at Frank, Alberta (Sharpe, 1938). Therefore, we feel that the nature of the slide surface and the internal consistency of the blocks is evidence of periodic creep.

Evidence from the study of other slide masses, which one might use as analogues, is contradictory. In many instances the rocks are entirely brecciated, but internal stratigraphy is maintained and underlying material is relatively undisturbed: slide breccias in the plains of Alès, southern France (Denizot, 1931, 1937; Goguel, 1936; Koop, 1952); areas of southern Nevada and western Arizona (Longwell, 1951), and Death Valley (Topping, 1990).

In the case of the Grand Valley slides, Anders (1990) argued for single-event slides rather than prolonged, periodic motion (termed *chronic* by Koop, 1952), and in the case of slide blocks near the southeastern end of northern Star Valley, McCalpin and others (1990, Fig. 1 and p. 9) argued that the random orientation of blocks within those masses indicated emplacement as a fluidized rockslide. Therefore, we cannot rule out the possibility that initial catastrophic emplacement was followed by chronic creep, at which time the observed characteristics of the boundary zone developed. It is also possible that the larger continuous sheets of Grand Valley moved largely by creep, whereas some of the smaller blocks of Star Valley and Grand Valley are remnants of catastrophic landslides. We intend to examine additional basal surfaces in the hope of constraining rates of emplacement.

Frontal (downslope) folds

Folds that envelop the frontal (downslope) side of the slide blocks are observed and also appear compatible with creep. Such folds are well exposed on the west side of a small block north of Palisades Creek (Fig. 11; see Fig. 1 for location). Beds on the downslope side dip steeply to the west and are folded beneath the block. As the beds rolled under the front of the block they become intensely brecciated (Figs. 12 and 13). Despite intense brecciation, a hint of bedding remains in the overturned beds (Fig. 11).

The overturned brecciated carbonates overlie a recessive deposit containing grayish, fine-grained detritus and small angular clasts of hanging-wall limestone and occasional cobble and boulders derived from the footwall conglomerates (Fig. 12). Separated from this deposit by a thin hematitic zone is an underlying fine-grained recessive unit, which is reddish in color and is similar in appearance to the cleaved mudstone observed to the south of Palisades Creek (Fig. 12b). This latter recessive unit contains only clasts and brecciated fragments of cobble and boulders derived from the footwall.

Initially, we interpreted the hematitic horizon to be the slide surface. However, exotic boulders (for example, the Pennsylvanian quartzite in Figure 12b) sandwiched between the limestone block and the upper grayish recessive unit are slickensided, leading us to the conclusion that the slide surface corresponds to this contact. The hematitic zone is possibly a weathering horizon at the top of colluvial or alluvial deposit that covered the alluvial-fan conglomerates.

MODEL OF EMPLACEMENT

Based on the above observations of the relationships of the slide blocks to adjacent thrust and extensional structures and of the geometry and distribution of internal and basal deformational features, we propose the following model for the emplacement of these gravity-drive slide sheets.

During thrust emplacement and the formation of imbricate thrust faults at the top of the Absaroka ramp, the location of the present Grand Valley was probably the site of a piggyback basin, rimmed on the east by the Snake Range. During Miocene/Pliocene Basin and Range extension, the Absaroka ramp served to nucleate extensional faults, which broke toward the earth's surface within the complex of thrust imbricates (Fig. 14a). Some of these imbricates may have been reactivated as normal fault splays.

Extensional faulting produced relief on the order of 750 to 900 m. During extension, the downdip segments of many of the thrust imbricates were truncated by the basin-bounding normal faults. The resulting gravitational head, combined with creation of a stress-free toe at the downdip ends of the basinward-dipping imbricate thrust sheets, provided the instability required to drive the emplacement of the slide masses (Fig. 14b).

Earthquakes accompanying continuing extensional faulting may have triggered the gravity slides. Many of the slides were initially up to 5 km in length (measured parallel to the axis of the valley). As a result of variations of dip and coefficient of friction on alluvial fans, parts of the slide masses moved more rapidly than others (Fig. 14c), resulting in numerous cross-strike normal faults and extensional fractures and kink bands (Figs. 7, 8, and 9). As the larger masses moved downslope they "calved" and abandoned smaller fault-bounded blocks (Fig. 14d).

With the data obtained to date there is no way of determining whether the initial emplacement was by rapid catastrophic landsliding or by creep. However, the presence of calcite-filled extension fractures and delicate cleavages (Fig. 10) preserved along the base of some of the larger sheets indicates that the majority of the movement was by creep.

Folds enveloping the downdip faces of the slide sheets suggest that they crept forward in a manner analogous to a tank or Caterpillar tractor trend (Fig. 15). Gouge-filled dikes along the base of the sheet (Fig. 9) would lead one to surmise that movement was aided by fluid-pressure, but coefficients of friction along the base must have been sufficient from time to time to permit the development of the drag-folds at the leading edges of the slide sheets.

With continued movement on the normal faults, the slide surfaces were rotated toward the range and the lowered dips on the alluvial fans terminated movement on the slide sheets. Deactivated sheets were covered by continuing deposition of alluvial fans, and with recurrent extension and earthquake activity, additional slides developed.

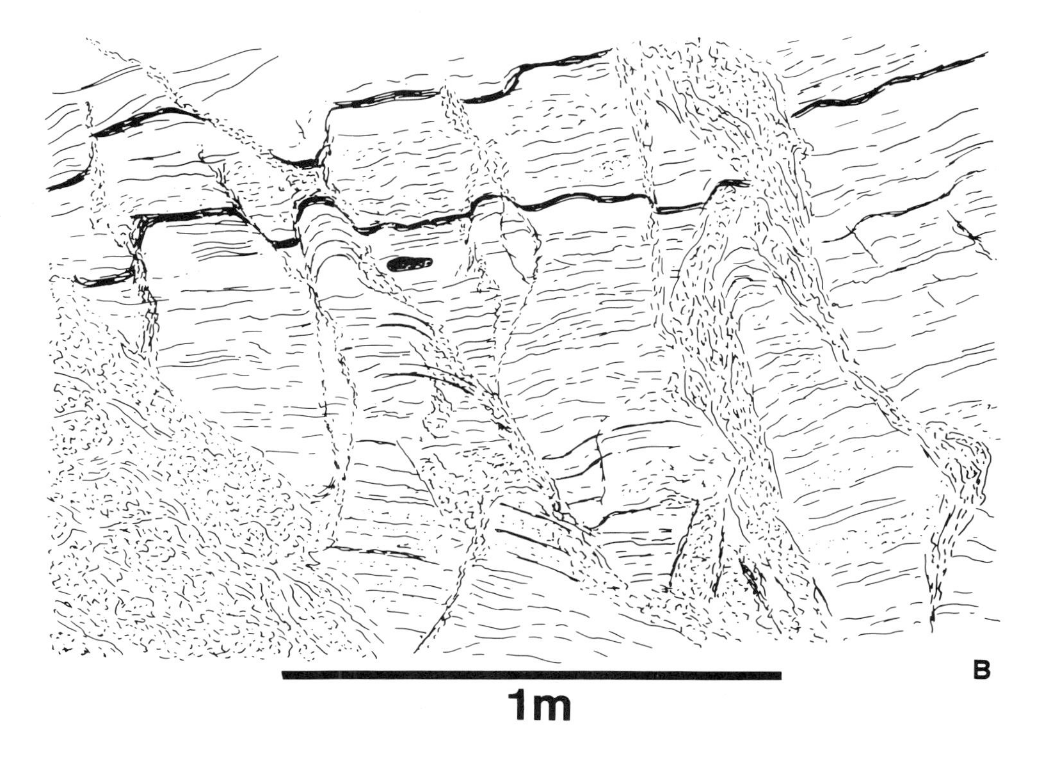

Figure 8. Extensional shear bands disrupting bedding near the south end and at the base of the largest slide block in Figures 6 and 7. A (on facing page), Bedding dips at approximately 15° toward the left. Shear bands form conjugate sets; one set dips 80 to 90° and the other dips 45 to 60° to the right. The larger outline indicates the location of Figure 8B, and the smaller rectangle is the area shown in Figure 8C. Note 15-cm scale. B, Sketch from photo. Shear bands are marked by gouge and are bordered by zones of intense fracturing. C, A sharp shear band boundary. Bedding dipping gently to the left is visible to the right, whereas bedding within the shear band is totally destroyed by brecciation. Knife is 9 cm long.

Figure 9. A, Clastic dike in base of the largest sheet (Figs. 7 and 8). B, Close-up of gouge that fills base of dike. Scale is 15 cm long.

Figure 10. Cleavage developed in mudstones, which are sandwiched between footwall conglomerates and the overriding slide block. A, At the base of the mudstones, the slide surface (black lines) dips approximately 50° to the east (right) and the dip of cleavage is 60°. Slide block originally dipped 10 to 25° toward the west, parallel to the underlying alluvial fan surface. Eastward dip was attained during continued movement on the basin-bounding Grand Valley normal fault 2.5 km to the east. Above and to left of knife, note dolomite cobble that has been ground flat on its upper surface by movement on the slide surface. B, Close-up of cleavage in mudstones. Heavy black line indicates trend of cleavage. Note truncated pebbles below slide surface.

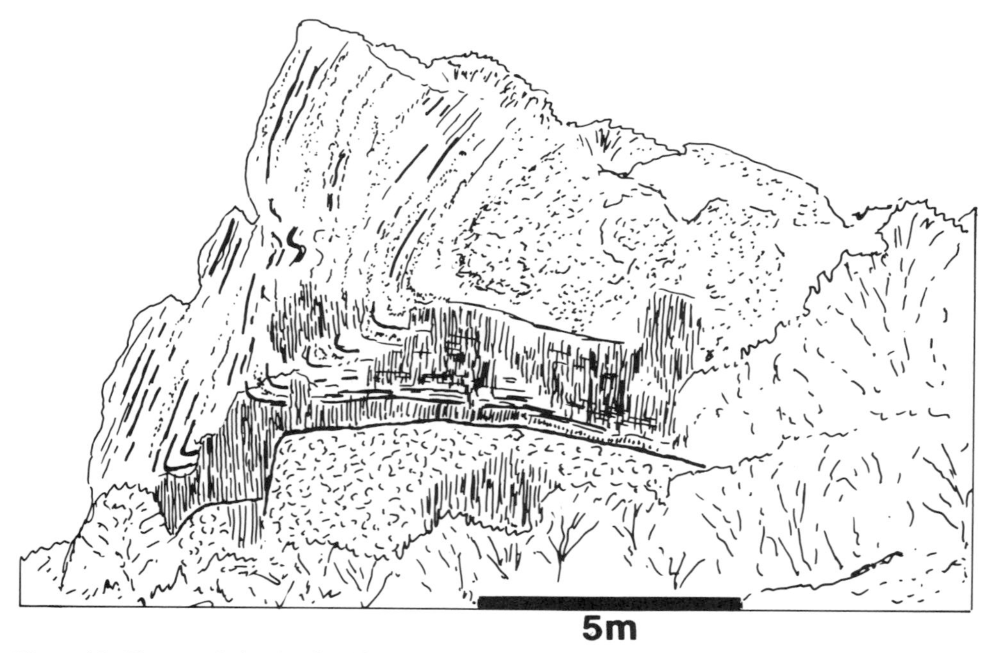

Figure 11. Photo and sketch of a slide block (upper two-thirds of the photo) immediately north of Palisades Creek (location 5, Fig. 1). View to the north. The lower third of the exposure is composed of footwall fanglomerates. Bedding at the downdip side (to the left) of the outcrop is vertical, with stratigraphic tops presumed to be toward the left. The beds rolled beneath the block. The overturned beds along the base of the slide sheet are completely brecciated (see Fig. 12).

Figure 12. a, Glide zone between slide block of Cambrian limestones and footwall Miocene conglomerates. b, Close-up photo of glide zone. Small boulder at upper right is Pennsylvanian quartzite. The same boulder is visible above and to right of scale in the upper photo (a). Sub-horizontal dark band in middle of photograph is probably hematitic, but we have not sampled it. Beneath the dark band is a muddy siltstone containing pebbles of upper Pennsylvanian and Triassic reddish siltstones, fine-grained sandstone, and Mississippian(?) carbonates. The composition of the clasts is comparable to that of the conglomerates in the lower part of the outcrop (see upper photo). Above the "hematitic" zone is a breccia composed of Cambrian limestone of the slide block. The few exotic blocks, such as the Pennsylvanian quartzite boulder, are also entrained in the breccia. We interpret the slide surface to lie at the top of the breccia.

Figure 13. Breccia within slide block, approximately 10 cm above base of block and 25 cm to the left of the scale in Fig. 12a.

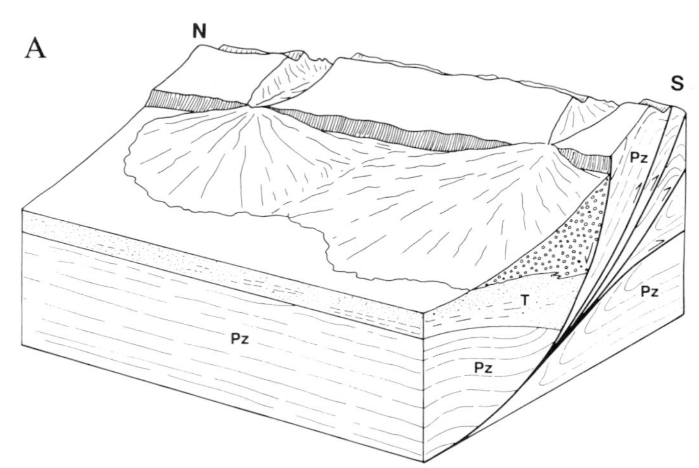

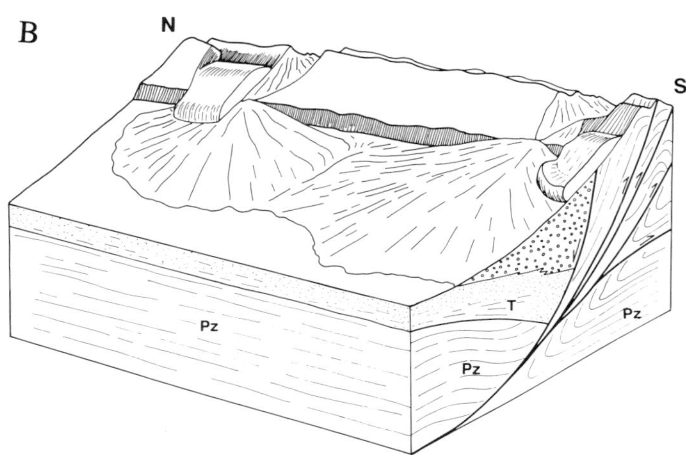

Figure 14. Schematic model for evolution of gravity slides. The relative positions and sizes of the blocks are not meant to be related to particular slide sheets within the valley. Open circles = Tertiary alluvial fan deposits. Stippled and short dashes = finer grained Tertiary clastics and volcanic deposits. T = Tertiary. Pz = Paleozoic strata. A, The first event was movement on the Absaroka thrust at approximately 80 to 70 Ma (Wiltschko and Dorr, 1983), followed or accompanied by formation of thrust imbricates at the top of a ramp in the Absaroka thrust. These imbricates enhance the topographic relief of what is known today as the Snake River Range. Late Miocene extensional faults, such as the Grand Valley fault, merged at depth with the Absaroka ramp. Drainages cut across the fault scarp and into the range, their alluvial fans radiating outward from the mouths of the canyons. B, The topographic relief, created by thrust ramping, imbrication, and extension introduced gravitational instabilities within the steeply dipping Paleozoic rocks of the range. As a consequence huge slabs of Paleozoic carbonate rocks slid down the flanks of the range, crossing the trace of the Grand Valley fault

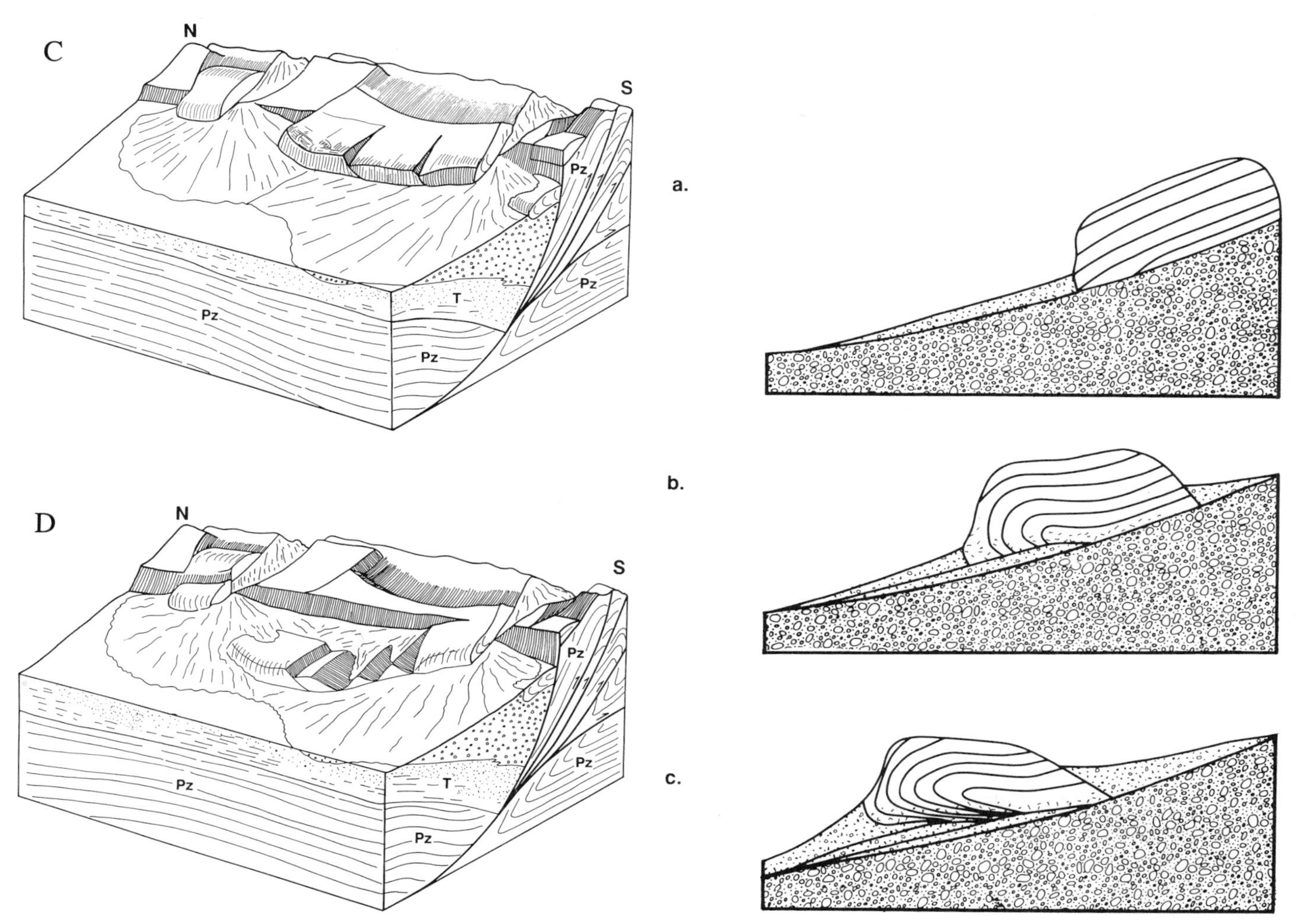

(Hait and others, 1977; Moore and others, 1987; McCalpin and others, 1990). C, Movement on the faults was probably episodic and irregularly distributed along the range front. The southern segment of the fault is shown being reactivated while the northern portion remains inactive. This movement truncated the southern slide sheet, which had come to rest upon the trace of the fault. A third large sheet broke from the range. The central segment is depicted as having moved more rapidly than the north and south ends, leading to strike-parallel extension within the block. d, As normal faulting resumed in the north, the northern slide block and central slide sheets were cut by the basin-bounding normal fault, and the central sheet continued to move down the surface of the alluvial fans. The southern slide and portions of the central sheet became buried beneath alluvial-fan conglomerates. With continued movement on the normal faults, the slide sheets were rotated toward the range, and the dip on the slide surfaces decreased until gravitational sliding could no longer continue. All the slide masses were eventually covered by sediments and were exhumed during Quaternary erosion.

Figure 15. Overturned beds (Figs. 4 and 5) and folds (Fig. 11) enveloping the downdip snouts of the slides suggest that the slides moved like the track of a Caterpillar tractor. a, Initially, beds were approximately parallel to the slide surface at the top of the alluvial fan. During periods of inactivity alluvial detritus accumulated around and in front of the slide block. b. During renewed movement, the slide sheet overrode the younger alluvial deposits. As beds at the leading edge were dragged beneath the sheet, they were intensely brecciated. The brecciated zone is indicated by the random dash pattern at the leading edge and along the base. Erosion of the brecciated carbonates provided angular fragments to the detritus deposited in front of the advancing sheet. c, The slide sheet continued to move, overriding its own breccia and alluvial detritus. The beds continued to roll beneath the sheet as it advanced. Downdip interbed shear would have been required to accommodate the drag-folding at the nose of the slide sheets.

CONCLUSIONS

Gravity slide sheets in the Idaho-Wyoming thrust belt were emplaced during Miocene/Pliocene Basin and Range extension. Extensional faults initiated at the top of a ramp in the Absaroka thrust and climbed to the earth's surface through a complex of thrust imbricates that had developed at the top of the ramp during Late Cretaceous to early Tertiary thrusting. Extensional faults truncated the downdip portions of the thrust imbricates, thus freeing the slabs of Paleozoic carbonates to slide during continued extensional faulting and related seismicity. Initial emplacement of the slide sheets, some measuring up to 250 m high by 5 km long and 2 km wide, may have been catastrophically induced by seismicity, but subsequent movement was by creep, during which time cleavages developed along the base of the sheets and footwall gouge was injected into basal fractures. As the sheets crept down the surface of alluvial fans, variations in basal sliding friction or differential dip on the fans led to extensional breakup; smaller blocks, bounded by the extension faults, were calved and abandoned as the larger "mother" sheet continued downslope. The slide sheets moved in the manner of a tank or Caterpillar tractor tread. As the limestone beds rolled forward, they were dragged beneath the leading face of the slide sheet and brecciated. During continued movement the slides overrode their own breccia.

ACKNOWLEDGMENTS

We thank the following individuals for reviewing the manuscript: David Moore, Mark Anders, Darrel Cowan, and Nicholas Woodward and the editors of this volume, Paul Link, Mel Kuntz, and Lucian Platt. We are especially grateful to David Moore and Mark Anders, who suggested numerous changes to improve the clarity of the text and the rigor of our arguments and provided several published and unpublished references of which we were unaware. The readability of the text was much improved by the comments of Mel Kuntz. We hope that we have succeeded in implementing their suggestions.

We dedicate this chapter to the memory of Steve Oriel, whose maps and published observations have stood the test of time and will serve as a valuable resource for many future generations of earth scientists as they attempt to unravel the geological mysteries of the Rocky Mountains.

REFERENCES CITED

Albee, H. F., and Cullins, H. L., 1975, Geologic map of the Alpine Quadrangle, Bonneville County, Idaho, and Lincoln County, Wyoming: U.S. Geological Survey Map GQ-1259, scale 1:24,000.

Albee, H. F., Lingley, W. S., Jr., and Love, J. D., 1977, Trip Two, Geology of the Snake River Range and adjacent areas, *in* Heisey, E. L., and others, eds., Rocky Mountain Thrust Belt geology and resources: Wyoming Geological Association, 29th Annual Field Conference, Guidebook, p. 769–783.

Anders, M. H., 1990, Late Cenozoic evolution of Grand and Swan Valleys, Idaho, *in* Roberts, S., ed., Geologic field tours of western Wyoming and parts of adjacent Idaho: Geological Survey of Wyoming, Public Information Circular 29, p. 15–25.

Bally, A. W., Gordy, P. L., and Stewart, G. A., 1966, Structure, seismic data and orogenic evolution of southern Canadian Rocky Mountains: Bulletin of Canadian Petroleum Geology, v. 14, p. 337–381.

Bally, A. W., Bernoulli, D., Davis, G. A., and Montadert, L., 1981, Listric normal faults: Oceanologica Acta, Proceedings, 26th International Geological Congress, Geology of continental margins symposium, Paris, July 7–17, 1980: n.p., p. 87–101.

Bates, R. L., and Jackson, J. A., 1987, Glossary of geology (third edition): Alexandria, Virginia, American Geological Institute, 788 p.

Beutner, E. C., 1972, Reverse gravitative movement on earlier overthrusts, Lemhi Range, Idaho: Geological Society of America Bulletin, v. 83, p. 839–846.

Bortz, L. C., and Murray, D. K., 1979, Eagle Springs oil field, Nye County, Nevada, *in* Newman, G. W., and Goode, H. D., eds., 1979, Basin and Range symposium: Denver, Colorado, Rocky Mountain Association of Geologists and Utah Geological Association, p. 441–453.

Boyer, S. E., and Allison, M. L., 1987, Estimates of extension in the Basin and Range Province: Geological Society of America Abstracts with Programs, v. 19, p. 597.

Cook, E. F., 1960, Breccia blocks (Mississippian) of the Welcome Spring area, southwest Utah: Geological Society of America Bulletin, v. 71, p. 1709–1712.

Dahlstrom, C.D.A., 1970, Structural geology in the eastern margin of the Canadian Rocky Mountains: Bulletin of Canadian Petroleum Geology, v. 18, p. 332–406.

Denizot, G., 1931, Les affleurements crétaciques dans la plaine tertiaire d'Alès: Bulletin Société Géologique de France, sér 5, tome 1, p. 397–428.

—— , 1937, Affleurements crétaciques brèches tectoniques et brèches sédimentaires de la plaine d'Alès: Bulletin Société Géologique de France, sér. 5, tome 7, p. 187–202.

Dorr, J. A., Jr., Spearing, D. R., and Steidtmann, J. R., 1977, Deformation and deposition between a foreland uplift and an impinging thrust belt: Hoback Basin, Wyoming: Geological Society of America Special Paper 177, 82 p.

Drewes, H., 1959, Turtleback faults of Death Valley, California: A reinterpretation: Geological Society of America Bulletin, v. 70, p. 1497–1508.

—— , 1963, Geology of the Funeral Peak quadrangle, California, on the east flank of Death Valley: U.S. Geological Survey Professional Paper 413, 78 p.

—— , 1967, Geology of the Connors Pass quadrangle, Schell Creek Range, east-central Nevada: U.S. Geological Survey Professional Paper 557, 93 p.

Eardley, A. J., 1960, Phases of orogeny in the fold belt of western Wyoming and southeastern Idaho, *in* Overthrust Belt of southwestern Wyoming and adjacent areas: Wyoming Geological Association, 15th Annual Field Conference, Guidebook, p. 37–40.

Faugére, E., Choukroune, P., and Angelier, J., 1986, Presence of olistolites in the syndistensive Miocene basin of the Las Vegas area, Nevada: Structural implications: Bulletin Société Géologique de France, sér. 8, tome 2, p. 879–884.

Goguel, J., 1936, Les brèches Urgoniennes d'Alès (Gard): Bulletin Société Géologique de France, sér. 5, tome 6, p. 219–235.

Hait, H. H., Jr., 1987, Southern Lemhi Range, east-central Idaho, *in* Beus, S. S., ed., Centennial Field Guide, v. 2, Rocky Mountain Section of the Geological Society of America, p. 99–102.

Hait, M. H., Jr., Prostka, H. J., and Oriel, S. S., 1977, Geologic relations of late Cenozoic rockslide masses near Palisades Reservoir, Idaho and Wyoming: U.S. Geological Survey report for the U.S. Bureau of Reclamation (unpublished), 21 p.

Janecke, S.U., 1989, Changing extension directions from Eocene to the present: Central Lost River Range and adjacent areas, Idaho: Geological Society of America Abstracts with Programs, v. 21, no. 5, p. 97–98.

Jobin, D. A., 1972, Geologic map of the Ferry Peak quadrangle, Lincoln County, Wyoming: U.S. Geological Survey Map GQ-1027, scale 1:24,000.

Koop, O. J., 1952, A megabreccia formed by sliding in southern France: American Journal of Science, v. 250, p. 822–828.

Krieger, M. H., 1977, Large landslides, composed of megabreccia, interbedded in Miocene basin deposits, southeastern Arizona: U.S. Geological Survey Professional Paper 1008, 25 p.

Kurie, A. E., 1966, Recurrent structural disturbance of the Colorado Plateau margin near Zion National Park, Utah: Geological Society of America Bulletin, v. 77, p. 867–872.

Labrecque, J. E., and Shaw, E. W., 1973, Restoration of Basin and Range faulting across the Howell Creek Window and Flathead Valley of southeastern British Columbia: Bulletin of Canadian Petroleum Geology, v. 21, p. 117–122.

Longwell, C. R., 1951, Megabreccia developed downslope from large faults: American Journal of Science, v. 245, p. 343–355.

Love, J. D., 1956a, Cretaceous and Tertiary stratigraphy of the Jackson Hole area, northwestern Wyoming: Wyoming Geological Association, 11th Annual Field Conference, Guidebook, p. 76–94.

—— , 1956b, Summary of geologic history of Teton County, Wyoming, during late Cretaceous, Tertiary and Quaternary times: Wyoming Geological Association, 11th Annual Field Conference, Guidebook, p. 140–150.

Mackin, J. H., 1960, Structural significance of Tertiary volcanic rocks in southwestern Utah: American Journal of Science, v. 258, p. 81–131.

McCalpin, J. P., Piety, L. A., and Anders, M. H., 1990, Latest Quaternary faulting and structural evolution of Star Valley, Wyoming, *in* Roberts, S., ed., Geologic field tours of western Wyoming and parts of adjacent Idaho: Geological Survey of Wyoming Public Information Circular 29, p. 5–12.

Merritt, Z. S., 1956, Upper Tertiary sedimentary rocks of the Alpine, Idaho-Wyoming area: Wyoming Geological Association, 11th Annual Field Conference, Guidebook, p. 117–119.

Moore, D. W., Woodward, N. B., and Oriel, S. S., 1984, Preliminary geologic map of the Mount Baird Quadrangle, Bonneville County, Idaho, and Lincoln and Teton Counties, Wyoming: U.S. Geological Survey Open-File Report 84-776, scale 1:24,000.

Moore, D. W., Oriel, S. S., and Mabey, D. R., 1987, A Neogene (?) gravity-slide block and associated slide phenomena in Swan Valley graben, Wyoming and Idaho, *in* Beus, S. S., ed., Centennial Field Guide, v. 2, Rocky Mountain Section of the Geological Society of America, p. 113–116.

Moores, E. M., 1968, Mio-Pliocene sediments, gravity slides, and their tectonic significance, east-central Nevada: Journal of Geology, v. 76, p. 88–98.

Olson, T. J., and Schmitt, J. G., 1987, Sedimentary evolution of the Miocene-Pliocene Camp Davis Basin, northwestern Wyoming, *in* Miller, W., ed., The Thrust Belt revisited, Wyoming Geological Association, 38th Field Conference, Guidebook, p. 225–243.

Oriel, S. S., and Moore, D. W., 1985, Geologic map of the West and East Palisades RARE II further planning areas, Idaho and Wyoming: U.S. Geological Survey Map MF-1619-B, scale 1:50,000.

Oriel, S. S., and Platt, L. B., 1980, Geologic map of the Preston 1° × 2° quadrangle, southeastern Idaho and western Wyoming: U.S. Geological Survey Map I-1127, scale 1:24,000.

Proffett, J. M., Jr., 1977, Cenozoic geology of the Yerington district, Nevada, and implications for nature and origin of basin and range faulting: Geological Society of America Bulletin, v. 88, p. 247–266.

Reber, S. J., 1952, Stratigraphy and structure of the south-central and northern Beaver Dam Mountains, Utah: Guidebook to the Geology of Utah, v. 7: Utah Geological Society, p. 101–108.

Royse, F., Jr., Warner, M. A., and Reese, D. L., 1975, Thrust belt structural geometry and related stratigraphic problems, Wyoming–Idaho–Northern Utah, *in* Bolyard, D. W., ed., Deep drilling frontiers of the Central Rocky Mountains: Denver, Colorado, Rocky Mountain Association of Geologists, p. 41–54.

Sable, E. G., and Anderson, J. J., 1986, Tertiary tectonic slide megabreccias, Markagunt Plateau, southwestern Utah: Geological Society of America Abstracts with Programs, v. 17, p. 263.

Schroeder, M. L., 1974, Geologic map of the Camp Davis quadrangle, Teton County, Wyoming: U.S. Geological Survey Map GQ-1160, scale 1:24,000.

Seager, W. R., 1970, Low-angle gravity glide structures in the northern Virgin Mountains, Nevada and Arizona: Geological Society of America Bulletin, v. 18, p. 1517–1538.

Sehnke, E. D., 1969, Gravitational gliding structures, Horse Creek area, Teton County, Wyoming [M.S. thesis]: Ann Arbor, University of Michigan, 41 p.

Shackelford, T. J., 1975, Late Tertiary gravity sliding in the Rawhide Mountains, western Arizona: Geological Society of America Abstracts with Programs, v. 7, p. 372–373.

Sharpe, C.F.S., 1938, Landslides and related phenomena: A study of mass-movements of soil and rock: New York, Columbia University Press.

Skipp, B., 1988, Cordilleran thrust belt and faulted foreland in the Beaverhead Mountains, Idaho and Montana, *in* Schmidt, C. J., and Perry, W. J., Jr., eds., Interaction of the Rocky Mountain Foreland and the Cordilleran Thrust Belt: Geological Society of America Memoir 171, p. 237–266.

Todd, V.R., 1983, Late Miocene displacement of Pre-Tertiary and Tertiary rocks in the Matlin Mountains, northwestern Utah, *in* Miller, D. M., Todd, V. R., and Howard, K. A., eds., Tectonic and stratigraphic studies in the eastern Great Basin: Geological Society of America Memoir 157, p. 239–270.

Topping, D. J., 1990, Large landslides and Miocene extension in the Amargosa Chaos Basin, southern Death Valley, California: EOS Transactions of the American Geophysical Union, v. 71, p. 1612.

Webel, S., 1987, Significance of backthrusting in the Rocky Mountain thrust belt, *in* Miller, W. R., ed., The Thrust Belt revisited, Wyoming Geological Association 38th Annual Field Conference, Guidebook, p. 37–53.

Wiltschko, D. V., and Dorr, J. A., Jr., 1983, Timing of deformation in Overthrust Belt and foreland of Idaho, Wyoming, and Utah: American Association of Petroleum Geologists Bulletin, v. 67, p. 1304–1322.

Woodward, N. B., 1986, Thrust fault geometry of the Snake River Range, Idaho and Wyoming: Geological Society of America Bulletin, v. 97, p. 178–193.

Young, J. C., 1960, Structure and stratigraphy in north central Schell Creek Range, *in* Boettcher, J. W., and Sloan, W. W., Jr., eds., Guidebook to the geology of East Central Nevada: Intermountain Association of Petroleum Geologists, p. 158–172.

Manuscript Accepted by the Society July 19, 1991

Printed in U.S.A.

Geological Society of America
Memoir 179
1992

Chapter 11

Stratigraphic relations and paleomagnetic and geochemical correlations of ignimbrites of the Heise volcanic field, eastern Snake River Plain, eastern Idaho and western Wyoming

Lisa A. Morgan
U.S. Geological Survey, MS 964, Box 25046, Federal Center, Denver, Colorado 80225-0046

ABSTRACT

Paleomagnetic studies resolve stratigraphic and correlational problems of several ignimbrites exposed along the margins of the eastern Snake River Plain (ESRP) in eastern Idaho and western Wyoming. Each ignimbrite exhibits a consistent magnetic remanence direction from its base to top. All units studied are high-silica, densely welded, metaluminous ignimbrites, the radiometric ages of which range from 7 Ma to 4 Ma. This study investigates whether ignimbrites of the Heise volcanic field—the tuffs of Blacktail Creek, Blue Creek, Elkhorn Spring, and Kilgore—correlate with the lithologically and chronologically similar Walcott and Conant Creek tuffs, two ignimbrite sheets marginal to the Heise field.

Correlation of the tuff of Elkhorn Spring with the tuff of Blue Creek appears to be erroneous. Rather, paleomagnetic and chemical data indicate that the tuff of Blue Creek is equivalent to the Walcott Tuff. Results also suggest that the Conant Creek Tuff may actually consist of two separate ignimbrites: (1) an older unit that is tentatively correlated with the tuff of Elkhorn Spring and (2) a younger unit that is correlative with the tuff of Kilgore. The older unit appears to be restricted to the Heise cliffs and to the west side of the Teton Range. The tuff of Blacktail Creek remains a separate older unit.

The Blue Creek and Kilgore calderas are now recognized as sources for the Walcott Tuff and the younger Conant Creek Tuff. The source of the older Conant Creek Tuff remains unidentified. Minimal areal distribution estimates are 15,000 km^2 for the Walcott Tuff and 18,000 km^2 for the tuff of Kilgore.

INTRODUCTION

A paleomagnetic study has been conducted on the major ignimbrites of the Heise volcanic field (the tuffs of Blacktail Creek, Blue Creek, and Kilgore) and on the Walcott and Conant Creek tuffs in an attempt to resolve correlational problems and to further understand the stratigraphic framework of regional ignimbrites in the eastern Snake River Plain (ESRP). This study investigates whether the Walcott and Conant Creek tuffs, two ignimbrite sheets marginal to the Heise volcanic field, correlate with similar ignimbrites of the Heise field (Fig. 1). The study also provides evidence as to whether the tuff of Elkhorn Spring, exposed on the southern margin of the ESRP at Heise, is correlative with the tuff of Blue Creek as previously proposed (McBroome and others, 1981; Morgan and others, 1984). The 2.0-Ma Huckleberry Ridge Tuff from the Yellowstone Plateau volcanic field to the northeast (Christiansen, 1982) is also present in the study area but is not discussed in detail in this chapter; its magnetic remanence direction (Reynolds, 1977) is distinct from the units described herein. Other ignimbrite sheets are also present on the ESRP, but they are demonstrably older (such as the 10.3-Ma biotite-bearing tuff of Arbon Valley; Kellogg and others [1989]) or are less extensive (such as the 9.9-Ma tuff of Kyle Canyon exposed in the southern Lemhi Range; McBroome [1981]).

Morgan, L. A., 1992, Stratigraphic relations and paleomagnetic and geochemical correlations of ignimbrites of the Heise volcanic field, eastern Snake River Plain, eastern Idaho and western Wyoming, *in* Link, P. K., Kuntz, M. A., and Platt, L. B., eds., Regional Geology of Eastern Idaho and Western Wyoming: Geological Society of America Memoir 179.

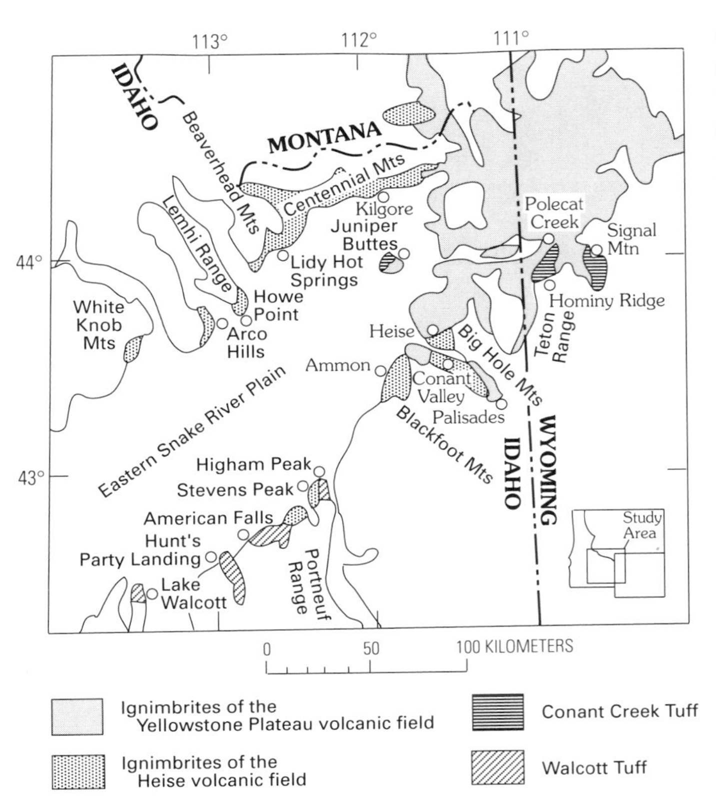

Figure 1. Index map showing the areal distribution of the Yellowstone Plateau volcanic field, the three major ignimbrites in the Heise volcanic field, the Conant Creek Tuff, and the Walcott Tuff.

Paleomagnetic and geochemical data show that the tuff of Blue Creek, exposed extensively on the northern margin of the ESRP, does not correlate with the tuff of Elkhorn Spring (Morgan and others, 1984) but is equivalent to the Walcott Tuff. The Conant Creek Tuff may consist of two separate ignimbrite sheets: (1) an older unit that is from the type section of the Conant Creek Tuff and may be correlative with the tuff of Elkhorn Spring and (2) a younger unit that is equivalent to the 4.3-Ma tuff of Kilgore. The tuff of Blacktail Creek remains an older separate unit.

MAJOR NEOGENE IGNIMBRITES OF THE ESRP

The Heise volcanic field

The Heise volcanic field consists of volcanic units exposed along and within the ESRP that are: (1) *younger* than the approximately 9.0-Ma and older sedimentary and volcaniclastic units of the Medicine Lodge Volcanics along the northern margin of the ESRP (Skipp and others, 1979), (2) *within* the upper sections of the Salt Lake Formation along the southern margin of the ESRP (Mansfield, 1927; Oriel and Moore, 1985), and (3) *older* than the 2.0- to 2.5-Ma volcanic rocks from the Yellowstone Plateau volcanic field (Christiansen and Blank, 1972). Three large-volume, densely welded, high-silica, rhyolitic ignimbrites form a framework for the Heise field (Morgan and others 1984). In ascending stratigraphic order, these are the 6.5-Ma tuff of Blacktail Creek, the 6.0-Ma tuff of Blue Creek, and the 4.3-Ma tuff of Kilgore. The Heise field also contains other smaller volume ignimbrites such as the tuff of Elkhorn Spring (undated), which is limited to the southern margin of the ESRP near Heise, Idaho (Fig. 1).

The tuff of Blacktail Creek. The 6.5-Ma tuff of Blacktail Creek is a rhyolitic, relatively crystal-rich, densely welded ignimbrite exposed extensively along the northern margin of the ESRP from the southern Beaverhead Range to the White Knob Mountains and along the southern margin from the Big Hole Mountains to the Portneuf Range (Fig. 1). It has a 10 to 20% total phenocryst content, consisting of plagioclase, quartz, sanidine, augite, magnetite, and zircon. Lithologic characteristics of the tuff of Blacktail Creek are (1) white-haloed, orange lithic clasts in its basal parts and (2) a high phenocryst content. It is a simple cooling unit (cf., Smith, 1960) and the oldest major ignimbrite in the Heise field.

The tuff of Blue Creek. The 6.0-Ma tuff of Blue Creek is a rhyolitic, crystal-poor, densely welded ignimbrite exposed along the northern ESRP margin from the southern Beaverhead Mountains to the White Knob Mountains and along the southern ESRP margin in the Conant Valley area (Fig. 1). It has a 0.5 to 3% total phenocryst content, consisting of plagioclase, quartz, sanidine, augite, orthopyroxene, magnetite, zircon, and trace amounts of biotite. Lithologic characteristics of the tuff of Blue Creek are (1) a basal welded fall deposit of orange pumice clasts in a matrix of black shards in deposits along the northern margin, (2) a nearly aphyric basal black glass, and (3) a low phenocryst content.

The tuff of Elkhorn Spring. The tuff of Elkhorn Spring is a rhyolitic, crystal-poor, densely welded ignimbrite exposed in prominent cliffs above the town of Heise along the southern margin of the ESRP. It has <2% total phenocryst content of plagioclase, quartz, sanidine, green pyroxene, and zircon. The tuff of Elkhorn Spring is similar to the tuff of Blue Creek in that it is phenocryst poor and is stratigraphically below the tuff of Kilgore and above the tuff of Blacktail Creek. However, the tuff of Elkhorn Spring is distinct from other Heise field ignimbrites because it contains a distinctive green pyroxene not recognized in other ignimbrites (D. J. Doherty, written communication, 1983; Morgan and others, 1984). It is a simple cooling unit (cf., Smith, 1960).

The tuff of Kilgore. The 4.3-Ma tuff of Kilgore is a rhyolitic, relatively crystal-poor, densely welded ignimbrite exposed along the northern ESRP margin from the Kilgore area to the southern Beaverhead Range and along the southern ESRP margin from the Palisades area to the Blackfoot Mountains. It has 2 to 10% total phenocryst content, consisting of plagioclase, quartz, sanidine, augite, magnetite, and zircon. Drillhole data indicate that the tuff of Kilgore is present within the ESRP south of the Lemhi Range (Doherty and others, 1979; McBroome, 1981) and north of Heise (Embree and others, 1978; Morgan and others, 1984). It is exposed at Juniper Buttes (Kuntz, 1979; Morgan, 1988). Its lithologic characteristics include (1) abundant, small (1 to 2 cm diameter), round-to-oval lithophysae common in upper parts, (2) maroon-colored pumice that occurs locally, and (3) phenocryst content intermediate to those of the two older major units. It is a compound cooling unit (cf., Smith, 1960) in exposures along the southern ESRP margin and is the youngest major ignimbrite in the Heise field.

Walcott and Conant Creek tuffs

In general, exposures recognized as the Walcott Tuff and the Conant Creek Tuff are located southwest and east, respectively, of the Heise volcanic field (Fig. 1). Because of their separate areas of exposure, stratigraphic relationships among the Walcott Tuff, the Conant Creek Tuff, and ignimbrites of the Heise field cannot be established from field relations.

The Walcott Tuff. As described by Carr and Trimble (1963), the Walcott Tuff contains two distinctive members: (1) a lower, well-bedded, friable, medium- to fine-grained ash and (2) an upper, welded ignimbrite. The upper member is a rhyolitic, crystal-poor, densely welded ignimbrite that has an average thickness of 15 m. Locally, a black basal vitrophyre and a capping red vitrophyre are exposed; a central zone containing small (<2 cm), spherical lithophysae is also typical. Published K-Ar whole-rock ages for the Walcott Tuff range from 6.3 Ma (Table 1) (Armstrong and others, 1975) to 6.9 Ma (Trimble and Carr, 1976).

The Conant Creek Tuff. The Conant Creek Tuff is stratigraphically below the 2.0-Ma Huckleberry Ridge Tuff at its type section on the west side of the Teton Range at Hominy Ridge (Fig. 1). Christiansen and Love (1978) described the Conant Creek Tuff as a rhyolitic, crystal-poor ignimbrite that is exposed in isolated outcrops on the east, west, and north sides of the Teton

TABLE 1. COMPILATION OF PUBLISHED RADIOMETRIC AGES OF UNITS DISCUSSED IN TEXT*

Unit	Method	Age (Ma)	Reference†	Site Locations
Tuff of Kilgore	K-Ar (wr)	4.1 ± 0.1	1	Ammon
Tuff of Kilgore	Fission-track (zircon)	4.2 ± 0.3	2	
Tuff of Kilgore	Fission-track (zircon)	4.4 ± 0.4	3	
Tuff of Kilgore	K-Ar (san)	4.3 ± 0.3	3	
Tuff of Kilgore	K-Ar (wr)	4.3 ± 0.2	4	
Tuff of Kilgore	Fission-track (zircon)	4.4 ± 0.6	3	
Tuff of Kilgore	K-Ar	4.7 ± 0.1	1	
Tuff of Kilgore	K-Ar (san)	4.8 ± 0.3	5	
Conant Creek Tuff	Fission-track (glass)	4.3 ± 0.5	6	Hominy Ridge, basal vitrophyre
Conant Creek Tuff	K-Ar (wr)	5.99 ± 0.06	6, 7	Hominy Ridge, basal vitrophyre
Tuff of Blue Creek	Fission-track (zircon)	5.6 ± 0.6	2	Howe Point
Tuff of Blue Creek	K-Ar (wr)	6.0 ± 0.2	8	Howe Point
Walcott Tuff	K-Ar (wr)	6.3 ± 0.3	5	
Walcott Tuff	K-Ar (wr)	6.3 ± 0.3	9	American Falls
Walcott Tuff	K-Ar (wr)	6.5 ± 0.1	1	Blackfoot
Walcott Tuff	K-Ar (wr)	6.5 ± 0.1	1	American Falls
Walcott Tuff??	K-Ar (wr)	6.9 ± 0.4	9	American Falls
Tuff of Blacktail Ck.	K-Ar (wr)	6.3 ± 0.3	10	
Tuff of Blacktail Ck.	K-Ar (wr)	6.5 ± 0.1	1	
Tuff of Blacktail Ck.	K-Ar (wr)	6.5 ± 0.3	10	
Tuff of Blacktail Ck.	Fission-track (zircon)	6.5 ± 0.3	3	
Tuff of Blacktail Ck.	K-Ar (wr)	6.5 ± 0.3	10	
Tuff of Blacktail Ck.	Fission-track (zircon)	6.6 ± 0.7	2	
Tuff of Blacktail Ck.	K-Ar (plag)	6.6 ± 0.3	11	Blackfoot
Tuff of Blacktail Ck.	K-Ar (san)	6.7 ± 0.2	11	Blackfoot
Tuff of Blacktail Ck.	K-Ar (plag)	6.7 ± 0.3	11	Blackfoot

*K-Ar (wr), K-Ar determinations of whole-rock samples; K-Ar (plag), K-Ar determinations of plagioclase phenocrysts; K-Ar (san), K-Ar determinations of K-feldspar phenocrysts.

†1 = Armstrong and others, 1975; 2 = McBroome, 1981; 3 = Morgan and others, 1984; 4 = Armstrong and others, 1980; 5 = Marvin and others, 1970; 6 = Christiansen and Love, 1978; 7 = Naeser and others, 1980; 8 = Marvin, written communication, 1986; 9 = Trimble and Carr, 1976; 10 = Anders and others, 1989; 11 = Kellogg and Marvin, 1988.

Range. The ignimbrite locally contains small (1 to 3 cm), spherical, lithophysal cavities and a black, basal vitrophyre. Ages for the Conant Creek Tuff are 5.99 Ma (K-Ar whole-rock age) and 4.3 Ma (fission-track age) (Table 1) (Christiansen and Love, 1978; Naeser and others, 1980). On the basis of lithologic similarities, Christiansen and Love (1978) suggest that the Conant Creek Tuff might correlate with welded ignimbrites in the Heise volcanic field that underlie the Huckleberry Ridge Tuff on and near the ESRP. They proposed a source for the Conant Creek Tuff on the ESRP.

RADIOMETRIC AGES

Available K-Ar and fission-track ages show broad ranges for the tuff of Blacktail Creek, the tuff of Blue Creek, the tuff of Kilgore, and the Walcott and Conant Creek tuffs (Table 1). K-Ar and fission-track ages for the tuff of Blacktail Creek range from 6.3 Ma to 6.7 Ma; the average age of the tuff of Blacktail Creek is 6.5 Ma. Two ages for the tuff of Blue Creek are a zircon fission-track age of 5.6 Ma and a whole-rock K-Ar age of 6.0 Ma; both ages are indistinguishable within analytical uncertainties (Table 1). Radiometric ages are unavailable for the tuff of Elkhorn Spring, but stratigraphic relations at Heise indicate that the age of the tuff of Elkhorn Spring must be between 4.3 and 5.7 Ma. Ages for the tuff of Kilgore range from 4.1 Ma to 4.8 Ma; however, most ages for the tuff of Kilgore fall around 4.3 Ma. A broad range of K-Ar whole-rock ages exists for the Walcott Tuff for samples obtained on the southern ESRP margin; ages range from 6.3 Ma (Marvin and others, 1970) to 6.9 Ma (Trimble and Carr, 1976). Two ages for the Conant Creek Tuff were obtained from the basal vitrophyre at the type section at Hominy Ridge: a fission-track age of 4.3 Ma (Naeser and others, 1980) and a K-Ar whole-rock age of 5.99 Ma (Christiansen and Love, 1978; Naeser and others, 1980).

Data in this chapter indicate that the Walcott Tuff on the southern ESRP margin and the tuff of Blue Creek on the northern ESRP margin are correlative units. Furthermore, most of the radiometric ages for these ignimbrites are indistinguishable within analytical uncertainty. A sample from the reference section of the tuff of Blue Creek at Howe Point gives a K-Ar whole-rock age of 6.0 Ma (R. F. Marvin, U.S. Geological Survey, written communication, 1986), which agrees well with most ages for the Walcott Tuff (Table 1). The age of one sample from the American Falls area (dated at 6.9 Ma; Trimble and Carr, 1976) is much older than ages obtained from the Walcott Tuff elsewhere; this sample may have been collected from misidentified 6.5-Ma tuff of Blacktail Creek. Farther east in the Stevens Peak area, Armstrong and others (1975) suggested that the youngest widespread ignimbrite in the vicinity had an age of 6.5 Ma and was possibly correlative with the Walcott Tuff. However, mapping and remanence directions suggest that, although the Walcott Tuff and tuff of Blacktail Creek are present in the area, the youngest widespread ignimbrite is the 4.3-Ma tuff of Kilgore. The 6.5-Ma unit dated by Armstrong and others (1975) is most likely the tuff of Blacktail Creek, for which the reported age ranges from 6.3 Ma to 6.7 Ma (Table 1).

The two available ages for the Conant Creek Tuff do not overlap within the margin of analytical uncertainty. Magnetic remanence and trace-element data indicate that the ignimbrite previously recognized as the Conant Creek Tuff actually consists of two units that are similar in age. Unfortunately, the ignimbrites that are representative of most of the Conant Creek Tuff do not have published ages. However, based on similarities in magnetic remanence directions and trace-element geochemistry, most of the Conant Creek Tuff appears to be equivalent to the 4.3-Ma tuff of Kilgore. The published radiometric ages (4.3 Ma and 5.99 Ma) are from the Conant Creek Tuff at Hominy Ridge. Based on similarities in magnetic remanence directions and trace-element contents, the ignimbrite at Hominy Ridge is tentatively correlated with the tuff of Elkhorn Spring, the age of which is between 4.3 and 5.7 Ma. The radiometric ages currently available for the Conant Creek Tuff at Hominy Ridge are not useful in resolving the true age of the unit because of the large age range. More precise dating methods are required to obtain a more accurate age of the Conant Creek Tuff.

REGIONAL STRATIGRAPHIC RELATIONS OF IGNIMBRITES ON THE ESRP

Ignimbrites of the Heise volcanic field

Ignimbrites of the Heise volcanic field are exposed on the margins of the ESRP from the Portneuf Range on the southern margin northeastward to the Teton Range and on the northern margin from the White Knob Mountains to southwestern Montana (Fig. 1). The major ignimbrites generally occur together as a complete stratigraphic package in most exposures. However, on the edges of the field, only one or two of the three major units are exposed.

The Walcott Tuff. The Walcott Tuff was originally recognized and described from exposures near American Falls (Stearns and Isotoff, 1956; Carr and Trimble, 1963) although it has been mapped as far northeast as Higham Park (Fig. 1; Trimble, 1982). Identification of the Walcott Tuff in the field has been difficult because it is typically exposed as isolated, loess-mantled outcrops and because its dense welding and rhyolitic composition are similar to other ignimbrites. Mansfield and Ross (1935) suggested that the Walcott Tuff might correlate with a lithologically similar ignimbrite exposed at Ammon, Idaho. However, Trimble and Carr (1976) noted differences in shard morphology and alteration, heavy mineral content, and the index of refraction of glass shards between the ignimbrite at Ammon and the type Walcott Tuff near American Falls. Despite these differences, they concluded that similar major-element chemistries suggested that these two ignimbrites were contemporaneous and probably shared a similar source. However, the major-element chemistry for most of the voluminous ignimbrites in this area is similar (Morgan and others, 1984) and thus may not be discriminant. Armstrong and others (1975) concluded that the ignimbrite at Ammon (4.1 Ma) was significantly younger than the Walcott Tuff at American Falls (6.5 Ma) (Table 1). The 4.1-Ma age of the ignimbrite at Ammon strongly suggests that this unit is the tuff of Kilgore.

The Conant Creek Tuff. Previous correlations of the Conant Creek Tuff with ignimbrites on the ESRP (Christiansen and Love, 1978) were based primarily on stratigraphic relations and lithologic similarities. Like the three major Heise ignimbrites, the Conant Creek Tuff is stratigraphically below the 2.0-Ma Huckleberry Ridge Tuff in exposures where both units occur, is densely welded, and is rhyolitic. Like the tuffs of Elkhorn Creek and Blue Creek and, locally, the tuff of Kilgore, the Conant Creek Tuff is phenocryst poor and has a nearly aphyric black basal vitrophyre. Prior to this study, the extent of the Conant Creek Tuff had not been determined, its total distribution had not been estimated, and a caldera source had not been located.

PALEOMAGNETIC INVESTIGATIONS

Techniques

An average of 12 samples was collected from each of 60 sites from ignimbrites on the ESRP using a portable, gasoline-powered, rock drill (Morgan, 1988). The units sampled are the tuff of Blacktail Creek (seven sites), the tuff of Blue Creek (seven sites), the tuff of Elkhorn Spring (one site), the tuff of Kilgire (33 sites), the Walcott Tuff (six sites), the Conant Creek Tuff (three sites), and the Huckleberry Ridge Tuff (three sites). Samples were oriented using a sun compass and a magnetic compass. In general, groups of three core samples (each averaging 7.5 cm in length) from four laterally or vertically separated areas about 3 to 4 m apart were collected; an average of eight core samples (two from each group of four) was used in analyzing the remanent magnetization. Depending on the extent of the exposure, samples either were collected from the same welding zone or through a vertical sequence. In two instances, large hand samples were collected after orientation by magnetic compass (sites SRP-227a and SRP-227b), and cores for paleomagnetic analyses were drilled from them in the laboratory.

Samples from the sixty sites were demagnetized using an alternating-current, tumbling-specimen demagnetizer. Remanent magnetization was measured with a spinner magnetometer. At least two cores from each site were subjected to progressive alternating-field demagnetization in four to eight steps to peak fields of 60 to 100 millitesla (mT). Any secondary components of magnetization were typically removed at the 10- to 20-mT level, leaving a stable primary component isolated through progressive alternating-field demagnetization to higher peak fields. Remaining cores from these sites were then demagnetized at peak fields between 20 to 60 mT (Table 2).

Twenty-nine samples from two sites (SRP-165 and SRP-173, Table 2) of the tuff of Kilgore were thermally demagnetized in a noninductively wound furnace. All were heated in air at 100°C intervals from 100 to 400°C and then at smaller intervals up to 600°C. Magnetizations isolated by both alternating-field and thermal demagnetization from specimens from the same core had comparable remanent directions.

Petrographic analyses revealed very fine grained, opaque oxides in nearly all thin sections of the three large-volume ignimbrites from the Heise volcanic field. Large opaque-oxide phenocrysts are rare in ignimbrites from both the Heise and Yellowstone Plateau volcanic fields. Reynolds (1977) inferred that such ubiquitous fine-grained opaques probably carry most of the stable remanence in chemically similar ignimbrites from the Yellowstone Plateau volcanic field.

Site means of magnetic remanence directions and statistics (Fisher, 1953) are summarized in Table 2. As noted, some samples were rejected because they had abnormally high magnetic intensities, probably due to lightning strikes. Others were rejected because the core samples broke during orientation.

Paleomagnetic data

Each ignimbrite exhibits a consistent magnetic remanence direction throughout its vertical section from the basal vitrophyre to the capping vitrophyre (Table 2). This is analogous to the conclusions drawn by Rosenbaum (1986) in his detailed comparison of the Tiva Canyon and Topopah members of the Miocene Paintbrush Tuff in southern Nevada. Relatively thin, densely welded ignimbrites, like the ignimbrites of the Heise volcanic field (typically 10 m to several tens of meters thick) and the Tiva Canyon member (<100 m thick), all yield well-grouped magnetic remanence directions throughout their total thickness. In contrast, thicker, densely welded sections of the Topopah Spring member (up to 300 m thick) had dispersed remanence directions in their basal parts that probably arose from welding that continued after acquisition of magnetization.

Paleomagnetic data of this investigation demonstrate only four distinct magnetic remanence directions for the seven previously defined ignimbrites that were sampled (Fig. 2). However, five, rather than seven, major ignimbrite sheets are probably represented. These are: (1) the tuff of Blacktail Creek, (2) the tuff of Blue Creek correlated with the Walcott Tuff, (3) the Conant Creek Tuff at its type section correlated with the tuff of Elkhorn Spring, (4) the tuff of Kilgore correlated with the rest of the Conant Creek Tuff, and (5) the Huckleberry Ridge Tuff. Magnetic remanence directions of the tuff of Blacktail Creek, the tuff of Blue Creek/Walcott Tuff, and the Huckleberry Ridge Tuff are distinct from the reversely magnetized tuff of Elkhorn Spring tentatively correlated with the Conant Creek Tuff at its type section and the reversely magnetized tuff of Kilgore correlated with most of the Conant Creek Tuff.

TABLE 2. SUMMARY OF PALEOMAGNETIC RESULTS FOR IGNIMBRITES ON THE EASTERN SNAKE RIVER PLAIN

Sample No.	Unit	Locality/Quadrangle	Lat.	Long.	N/N*	H†	MD§	MI**	TD§	TI**	K‡	α95§§
SRP-203	Huckleberry Ridge Tuff	Deadman's Bar	43.75	110.63	8/8	30	206.1	3.7	NA	NA	161.2	4.4
SRP-229	Huckleberry Ridge Tuff	Cliff Lake	44.75	111.58	6/6	40	216.6	-0.1	216.3	-3.8	210.9	4.6
SRP-227b	Huckleberry Ridge Tuff	Juniper Buttes	44.00	111.88	2/2	40	257.5	7.0	NA	NA	...	...
SRP-140	Tuff of Kilgore	Lidy Hot Springs qd	44.25	112.56	5/5	20	178.4	-65.6	168.1	-68.0	616.9	3.1
SRP-142	Tuff of Kilgore	Lidy Hot Springs qd	44.4	112.00	4/6	30	149.4	-58.3	161.3	-52.5	124.6	8.3
SRP-145	Tuff of Kilgore	Higman Peak	43.25	112.20	6/5	20	184.3	-63.2	163.8	-64.8	166.9	5.9
SRP-146	Tuff of Kilgore	Heise cliffs	43.6	112.00	6/4	30	177.9	-63.4	185.8	-54.5	313.8	5.2
SRP-154	Tuff of Kilgore	Goshen qd	43.26	112.03	5/5	20	255.4	-63.9	NA	NA	5.5	35.9
SRP-154	Tuff of Kilgore	Goshen qd	43.26	112.03	5/2	20	176.4	-59.2	NA	NA	443.3	11.9
SRP-164	Tuff of Kilgore	Lincoln Creek	43.25	112.14	8/8	20	179.0	-60.6	199.0	-64.7	254.6	3.5
SRP-169	Tuff of Kilgore	Red Rock Lake qd	44.68	111.95	9/6	20	158.6	-60.3	152.8	-65.6	219.7	4.5
SRP-171	Tuff of Kilgore	Gardner Lake qd	44.35	112.04	6/6	20	194.7	-50.7	200.2	-48.8	134.7	5.8
SRP-173	Tuff of Kilgore	Gardner Lake qd	44.35	112.04	24/20	60	191.9	-59.2	...	...	118.5	3.0
SRP-174	Tuff of Kilgore	Poplar qd	43.37	111.70	10/10	20	169.1	-59.4	186.4	-60.7	263.1	3.0
SRP-175	Tuff of Kilgore	Poplar qd	43.37	111.70	8/8	20	179.9	-55.3	187.0	-54.6	820.8	1.9
SRP-185	Tuff of Kilgore	Rigby SE qd	43.54	111.80	8/8	20	188.8	-62.4	198.5	-62.5	206.7	3.9
SRP-191	Tuff of Kilgore	Ozone qd	43.48	111.83	8/7	20	177.9	-61.0	182.9	-65.3	230.1	4.0
SRP-195	Tuff of Kilgore	Ammon qd	43.4	112.00	8/8	40	173.5	-54.5	167.5	-57.4	267.7	3.4
SRP-196	Tuff of Kilgore	Ammon qd	43.4	111.96	8/8	40	181.1	-56.8	195.0	-57.1	385.6	2.8
SRP-197	Tuff of Kilgore	Wolverine qd	43.38	111.94	6/4	20	199.1	-61.7	181.2	-46.2	13.5	25.9
SRP-198	Tuff of Kilgore	Bone qd	43.32	111.79	6/6	20	176.8	-58.9	183.1	-62.4	387.0	3.4
SRP-199	Tuff of Kilgore	Ozone qd	43.4	111.83	6/6	20	174.0	-61.0	181.6	-58.7	195.6	4.8
SRP-200	Tuff of Elkhorn Spg.	Heise cliffs	43.65	111.70	6/6	30	165.4	-66.1	184.1	-61.9	331.7	3.7
SRP-201	Tuff of Kilgore	Heise cliffs	43.65	111.70	6/6	40	100.3	-64.5	131.1	-62.9	13.3	19.0
SRP-207	Tuff of Kilgore	Shamrock Gulch qd	44.23	112.65	6/6	80	187.9	-51.0	202.9	-51.3	49.6	9.6
SRP-211	Tuff of Kilgore	Lidy Hot Springs qd	44.19	112.62	8/8	40	169.6	-60.2	175.1	-56.5	48.2	8.1
SRP-212	Tuff of Kilgore	Paul Reservoir qd	44.29	112.41	8/8	10	161.8	-61.71	165.2	-71.7	238.0	3.6
SRP-214	Tuff of Kilgore	Indian Butte	44.33	112.38	8/8	60	165.3	-50.2	158.3	-42.7	80.2	6.2
SRP-216	Tuff of Kilgore	Paul Reservoir qd	44.28	112.41	6/4	80	185.7	-60.5	193.4	-58.4	567.6	3.9
SRP-217	Tuff of Kilgore	Small quad	44.23	112.45	7/7	60	165.4	-62.1	160.6	-66.6	277.8	3.6
SRP-218	Tuff of Kilgore	Lidy Hot Springs qd	44.24	112.50	6/5	80	172.4	-54.9	165.9	-57.3	187.5	5.6
SRP-220	Tuff of Kilgore	Small quad	44.22	112.45	8/8	80	150.1	-56.9	156.6	-51.6	223.9	3.7
SRP-221	Tuff of Kilgore	Small quad	44.21	112.40	10/10	20	168.5	-61.3	167.3	-66.3	135.4	4.2
SRP-222	Tuff of Kilgore	Lidy Hot Springs qd	44.2	112.53	6/6	40	173.7	-59.6	182.3	-60.6	1025.9	2.1
SRP-227a	Conant Creek Tuff	Juniper Buttes	44.02	111.88	6/6	40	201.5	-36.5	206.9	-44.3	31.3	12.3
SRP-230	Conant Creek Tuff	Polecat Creek	44.13	110.70	6/6	40	145.6	-51.5	162.9	-61.5	154.0	4.5
SRP-232	Tuff of Kilgore	Heise	43.67	111.64	8/6	40	179.3	-69.1	171.3	-63.1	141.8	5.6
SRP-233	Tuff of Kilgore	Heise	43.65	111.64	8/8	40	201.3	-66.8	186.2	-60.1	102.1	5.5

Figure 3 compares representative magnetic remanent directions for the tuff of Blue Creek (Morgan and others, 1984) and the Walcott Tuff (Stearns and Isotoff, 1956; Carr and Trimble, 1963; Armstrong and others, 1975; Trimble and Carr, 1976) on both margins of the ESRP. Rocks at each locality have a unique normal magnetic remanent direction, suggesting that the tuff of Blue Creek is correlative with most of the Walcott Tuff and that both were erupted and cooled contemporaneously. Unlike many rocks having a normal magnetic remanent direction with a declination to the north, the declination shown by these two units is to the southwest. This correlation is corroborated by the similar radiometric ages (Table 1) and trace-element contents of the two units.

Paleomagnetic data suggest that an ignimbrite previously

TABLE 2. SUMMARY OF PALEOMAGNETIC RESULTS FOR IGNIMBRITES ON THE EASTERN SNAKE RIVER PLAIN (continued)

Sample No.	Unit	Locality/Quadrangle	Lat.	Long.	N/N*	H†	MD‡	MI§	TD**	TI††	K‡	α95§§
SRP-234	Tuff of Kilgore	Hawley Gulch qd	43.65	111.58	8/8	40	158.1	-58.2	183.8	-63.2	276.0	3.3
SRP-236	Tuff of Kilgore	Goshen qd	43.34	112.01	8/6	60	155.6	-60.7	161.4	-54.7	62.3	8.6
3YR-156	Conant Creek Tuff	Hominy Ridge	44.03	110.92	6/6	40	...	...	178.6	-51.2	35.5	11.4
SRP-141	Tuff of Blue Creek	Howe Point	43.83	112.85	3/3	30	229.5	53.9	248.0	56.5	48.1	18.0
SRP-143	Walcott Tuff	Hunt's Party Lndg.	42.75	112.90	6/6	10	234.2	69.2	238.5	74.0	1076.2	2.0
SRP-144	Walcott Tuff	Hunt's Party Lndg.	42.75	112.90	6/6	20	238.3	71.9	244.9	76.6	716.6	2.5
SRP-150	Walcott Tuff	American Falls qd	42.77	112.88	5/4	40	237.1	72.8	251.2	69.2	1041.2	2.8
SRP-162	Walcott Tuff	Steven's Peak	43.14	112.29	6/4	20	249.5	53.5	260.9	49.2	350.2	4.9
SRP-163	Walcott Tuff	Steven's Peak	43.14	112.29	6/4	20	222.0	50.3	239.3	67.8	509.0	4.1
SRP-188	Tuff of Blue Creek	Conant Valley	43.47	111.45	5/5	40	279.0	60.2	259.5	52.4	36.5	12.8
SRP-193	Tuff of Blue Creek	Howe Point	43.83	112.85	6/6	20	208.7	64.4	234.2	70.3	1036.3	2.1
SRP-194	Tuff of Blue Creek	Howe Point	43.83	112.85	15/15	20	217.1	60.2	228.6	72.2	64.4	4.8
SRP-204	Tuff of Blue Creek	Arco Hills	43.73	113.10	6/6	20	229.0	76.8	239.4	70.4	194.3	4.8
SRP-205	Tuff of Blue Creek?	Lidy Hot Spgs. SE qd	44.12	112.57	6/6	40	182.6	65.8	160.0	66.1	31.1	12.2
SRP-237	Tuff of Blue Creek	Richard Butte qd	43.90	112.85	6/4	20	252.4	68.3	248.4	61.6	152.8	7.5
SRP-156	Tuff of Blacktail Crk	Blackfoot	43.14	112.27	6/5	20	4.1	65.8	12.4	56.9	213.3	5.3
SRP-186	Tuff of Blacktail Crk	Blacktail Reservoir	43.51	111.77	6/6	20	12.1	65.4	3.9	62.4	136.9	5.7
SRP-187	Tuff of Blacktail Crk	Conant Valley	43.47	111.45	8/8	20	327.9	61.5	356.2	70.0	038.1	9.1
SRP-192	Tuff of Blacktail Crk	Richard Butte	43.89	112.86	8/8	20	352.2	69.2	16.4	66.7	805.3	2.0
SRP-208	Tuff of Blacktail Crk	Shamrock Gulch qd	44.23	112.72	8/8	40	31.6	73.8	21.8	64.5	2166.8	1.2
SRP-219	Tuff of Blacktail Crk	Edie Ranch qd	...	...	6/4	40	21.0	51.4	...	...	223.7	6.2
SRP-235	Tuff of Blacktail Crk	Paradise Valley qd	44.13	111.90	6/6	40	339.7	70.9	343.1	61.0	17.3	16.6

*Number of samples/number of samples accepted.
†Level of demagnetization, mT.
††Mean declination of remanence.
§Mean inclination of remanence.
**Mean declination of remanence, tectonically corrected.
††Mean inclination of remanence, tectonically corrected.
‡Fisher precision parameter
§§half angle of the cone of 95 percent confidence.

mapped in the Arco Hills of the southern Lost River Range as the tuff of Blacktail Creek (Morgan and others, 1984) is the Walcott Tuff (Fig. 3, Table 2). This unit is a proximal ignimbrite and thus its identification modifies the extent of the Blue Creek caldera, source of the Walcott Tuff, to west of the Arco Hills in the Lost River Range (Figure 4).

Figure 5 compares representative magnetic remanent directions for exposures on both margins of the ESRP that were recognized as the tuff of Kilgore and the Conant Creek Tuff (Christiansen and Love, 1978) and for the Walcott Tuff at a discrete exposure at Higham Peak (Trimble, 1982). Remanence directions of these units are similar and suggest that the tuff of Kilgore is correlative with most of the Conant Creek Tuff and with the "Walcott Tuff" (Trimble, 1982) at Higham Peak. This is corroborated by similar trace-element concentrations, as discussed below.

TRACE-ELEMENT GEOCHEMISTRY

Trace elements are useful in discriminating ignimbrites of the Heise volcanic field and the Walcott and Conant Creek tuffs. As shown in Figure 6, geochemical data for the tuff of Blue Creek and the Walcott Tuff plot in a similar field and are separated from data for the other ignimbrites. Likewise, geochemical data for the tuff of Kilgore and most of the Conant Creek Tuff are similar, but these are different from data of the other ignimbrites (Fig. 6). As shown in the Ba versus Zr and Nb versus Zr plots (Fig. 6), the tuff of Elkhorn Spring and the type section of the Conant Creek Tuff at Hominy Ridge plot in a field distinct from the field for the tuff of Kilgore and the field for the tuff of Blue Creek and Walcott Tuff.

McBroome and others (1981) suggested that the tuff of Blue Creek is correlative with the tuff of Elkhorn Spring because of

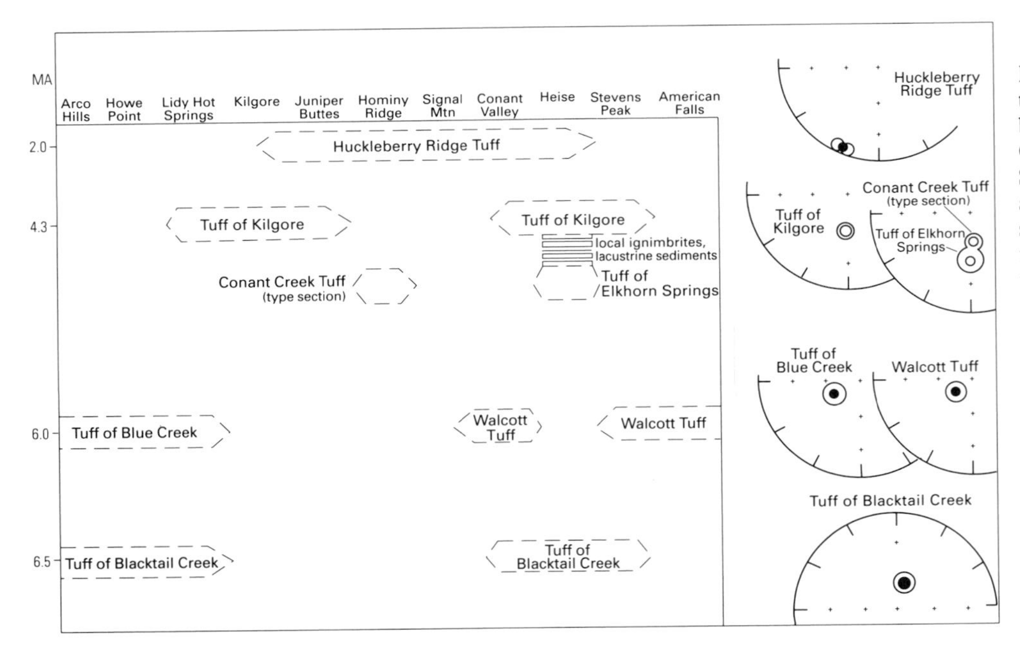

Figure 2. Schematic stratigraphic relationships among the three major ignimbrites of the Heise volcanic field, the Conant Creek Tuff, the tuff of Elkhorn Spring, and the Walcott Tuff. Localities are shown in Figure 1. Representative site mean magnetic remanent directions for each of these units are also depicted; that shown for the Conant Creek Tuff (type section) represents mean magnetic remanent directions for the ignimbrite at the type section of the Conant Creek Tuff at Hominy Ridge and the tuff of Elkhorn Spring exposed at Heise. Table 2 summarizes the paleomagnetic data depicted. Open circles represent upper hemisphere projections, and filled circles represent lower hemisphere projections. Also shown are cones of 95 percent confidence.

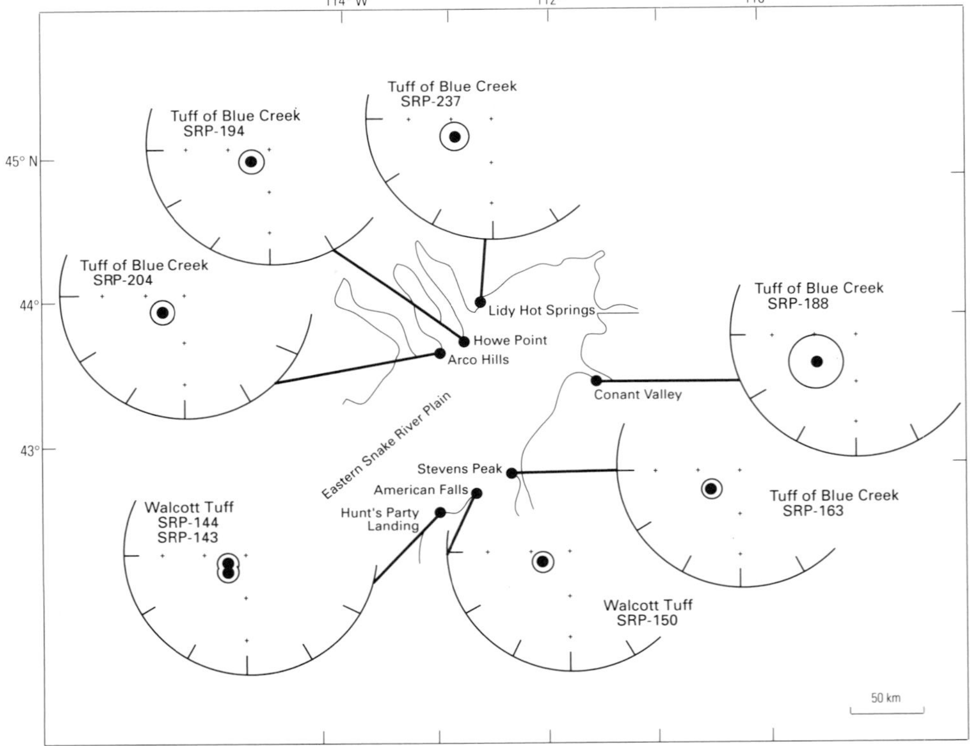

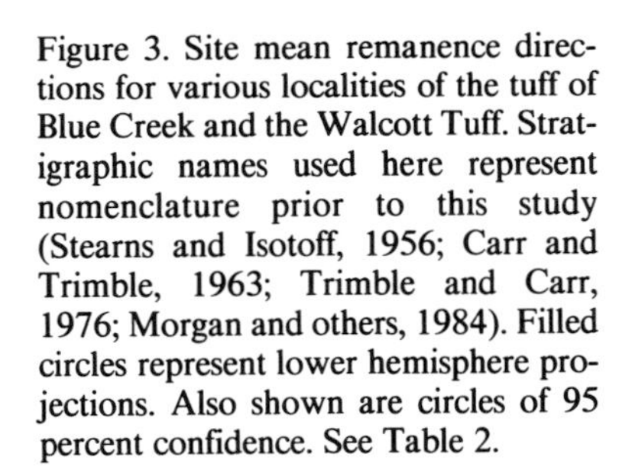
Figure 3. Site mean remanence directions for various localities of the tuff of Blue Creek and the Walcott Tuff. Stratigraphic names used here represent nomenclature prior to this study (Stearns and Isotoff, 1956; Carr and Trimble, 1963; Trimble and Carr, 1976; Morgan and others, 1984). Filled circles represent lower hemisphere projections. Also shown are circles of 95 percent confidence. See Table 2.

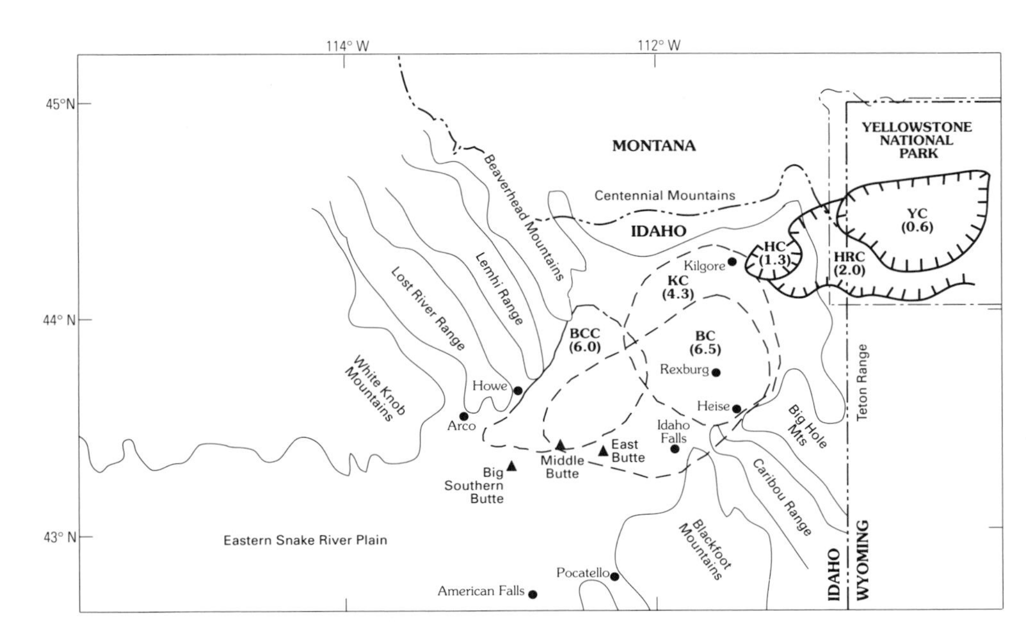

Figure 4. Proposed locations for calderas of the Heise volcanic field (Pierce and Morgan, this volume), as revised from Morgan and others (1984) and based on results from this study. Also depicted are the calderas for the Yellowstone Plateau volcanic field (Christiansen, 1982). Calderas shown are the Kilgore caldera (KC), the Blue Creek caldera (BCC), the Blacktail Creek caldera (BC), the Yellowstone caldera (YC), the Henry's Fork caldera (HC), and the Huckleberry Ridge caldera (HRC). Ages in Ma (given in parentheses) for the calderas are based on data from Christiansen (1982), Morgan and others (1984), and Table 1.

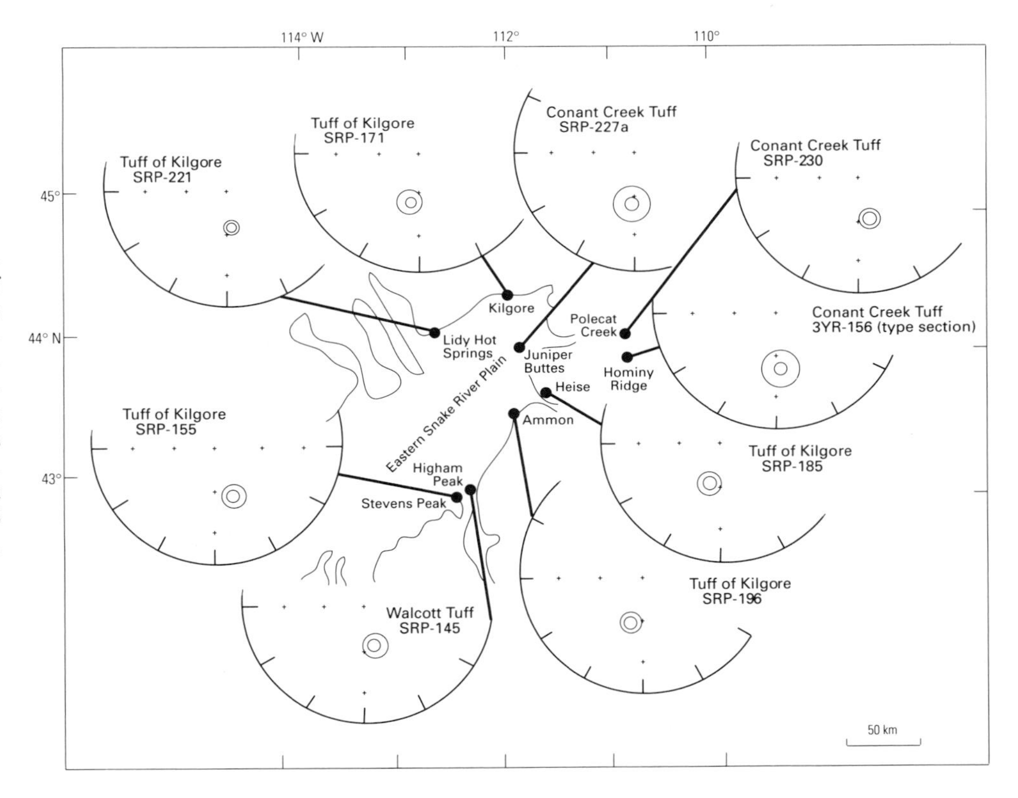

Figure 5. Site mean remanence directions for various localities of the tuff of Kilgore, the Conant Creek Tuff, and the Walcott Tuff at Higham Peak. Stratigraphic names used here represent nomenclature prior to this study (Christiansen and Love, 1978; Trimble, 1982; Morgan and others, 1984). Open circles represent upper hemisphere projections. Also shown are circles of 95 percent confidence. See Table 2. Site 3YR-156 was provided by R. L. Reynolds, U.S. Geological Survey, Denver.

similarities in lithology, phenocryst content, and stratigraphic position. In the southern Beaverhead Mountains, the tuff of Blue Creek is above the 6.5-Ma tuff of Blacktail Creek and below the 4.3-Ma tuff of Kilgore. At Heise, the tuff of Elkhorn Spring is exposed above the tuff of Blacktail Creek and the 5.7-Ma rhyolite of Kelly Mountain (Morgan and others, 1984) and is below the tuff of Kilgore and a sequence of local nonwelded ignimbrites and ash-rich lacustrine sediments (Fig. 2). Magnetic and chemical data suggest that the correlation between the tuff of Blue Creek and the tuff of Elkhorn Spring is erroneous. The tuff of Elkhorn Spring has a reversed magnetic direction that is distinct from that of the tuff of Blue Creek and similar to that of the ignimbrite at the type section of the Conant Creek Tuff and the tuff of Kilgore (Fig. 2). However, the tuff of Elkhorn Spring and the ignimbrite at the type section of the Conant Creek Tuff are chemically distinct from the tuff of Kilgore as well as the tuff of Blue Creek (Fig. 6). Mineral separates from both the crystal-poor Conant Creek Tuff at Hominy Ridge and the tuff of Elkhorn Spring contain a distinct green pyroxene not common in the three major ignimbrites of the Heise volcanic field (D. J. Doherty, written communication, 1988). These characteristics suggest that the tuff of Elkhorn Spring and the ignimbrite at the type section of the Conant Creek Tuff are equivalent units.

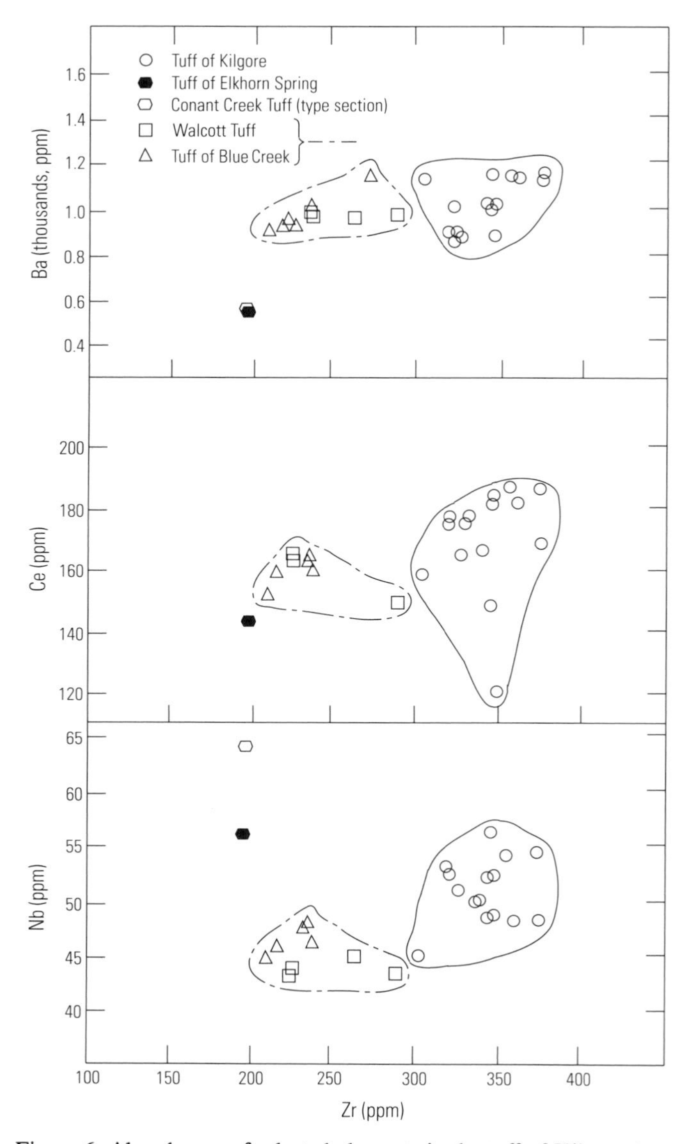

Figure 6. Abundances of selected elements in the tuff of Kilgore (open circles), the tuff of Blue Creek (open triangles), the Walcott Tuff (open squares), the Conant Creek Tuff at its type section Hominy Ridge (open hexagons), and the tuff of Elkhorn Springs (filled hexagons). Chemical analyses obtained by energy dispersive X-ray fluorescence methods by Ross Yeoman, U.S. Geological Survey, Denver. Analyses for sample OYC-590A, from the type section of the Conant Creek Tuff at Hominy Ridge, were provided by R. L. Christiansen, U.S. Geological Survey, Menlo Park, California. A Ce analysis is not available for the Conant Creek Tuff at Hominy Ridge.

OXYGEN ISOTOPE DATA

Oxygen isotope analyses ($\delta^{18/16}O$) of quartz phenocrysts from each of the three ignimbrites of the Heise volcanic field show unique values (Table 3). One oxygen isotope analysis available from the basal vitrophyre of the type section of the Conant Creek Tuff (Hildreth and others, 1984) differs from the value obtained for the tuff of Kilgore. No oxygen isotope data are available for the tuff of Elkhorn Spring.

CONCLUSIONS

Paleomagnetic studies show that most of the major ignimbrites in the Heise volcanic field have distinct and well-defined directions of remanent magnetization. Consequently, paleomagnetic data combined with trace-element chemistry, mapped stratigraphic relations, $\delta^{18/16}O$ analyses, and radiometric ages permit distinction and regional correlation of the units.

Similarities in trace-element concentrations (Fig. 6) and magnetic remanence directions (Fig. 3) and comparable radiometric ages (Table 1) indicate that the tuff of Blue Creek is correlative with the Walcott Tuff. Since the Walcott Tuff is a formally named stratigraphic unit (Stearns and Isotoff, 1956), the informal designation "tuff of Blue Creek" should no longer be used.

The Conant Creek Tuff, as originally described by Christiansen and Love (1978), actually represents two different ignimbrite sheets: (1) a younger ignimbrite equivalent to the tuff of Kilgore and (2) an older ignimbrite tentatively correlated with the tuff of Elkhorn Spring. Although both sheets have similar magnetic remanence directions, they differ in their trace-element

TABLE 3. OXYGEN ISOTOPE DATA FOR IGNIMBRITES OF THE HEISE VOLCANIC FIELD*

Sample No.	Unit	Locality	$\delta^{18}O_{quartz}$	Lithology
SRP-126	Tuff of Kilgore	Lidy Hot Springs	4.3, 3.6	Top vitrophyre
OYC-590A	Conant Creek Tuff	Hominy Ridge	6.3, 6.5	Basal vitrophyre
SRP-127	Tuff of Blue Creek	Lidy Hot Springs	7.0, 7.4	Vapor-phase zone
SRP-41	Tuff of Blue Creek	Lidy Hot Springs	7.39	Basal vitrophyre
SRP-37-79	Tuff of Blue Creek	Lidy Hot Springs	7.52	Basal vitrophyre
SRP-107	Tuff of Blacktail Creek	Heise	6.4	Basal vitrophyre
SRP-109	Tuff of Blacktail Creek	Heise	6.9	Top vitrophyre
SRP-123	Tuff of Blacktail Creek	Lemhi Range	6.4	Basal vitric layer
SRP-122	Tuff of Blacktail Creek	Lemhi Range	7.1, 7.2	Vapor-phase zone

*Hildreth and Morgan, unpublished data.
All analyses are of quartz phenocrysts, except for sample OYC-590A, where the analysis is of glass (Hildreth and others, 1984). Dual values indicate repeat analyses.

chemistry and oxygen isotope values. Furthermore, the stratigraphic relationships at Heise and the lacustrine sediments and ignimbrites between the tuff of Kilgore and the tuff of Elkhorn Spring in the cliffs at Heise support this interpretation. Most of the Conant Creek Tuff correlates with the tuff of Kilgore. A correlation between the Conant Creek Tuff at Hominy Ridge and the tuff of Elkhorn Springs requires further investigation.

Based on data presented in this chapter, the following stratigraphic nomenclature is proposed. The name tuff of Kilgore is to be used for ignimbrites identified as the tuff of Kilgore and most of the Conant Creek Tuff. The name Conant Creek Tuff is to be used for the ignimbrite at the type section at Hominy Ridge, where the Conant Creek Tuff was first formally discribed, and for the tuff of Elkhorn Spring. The name Walcott Tuff should be used for ignimbrites identified as the Walcott Tuff on the southern margin of the ESRP and for the tuff of Blue Creek on the northern margin.

Although there are only four well-defined magnetic remanence directions for the ignimbrites studied, there appear to be five different ignimbrites in this area of the ESRP. In ascending stratigraphic order, these are the 6.5-Ma tuff of Blacktail Creek, the 6.0-Ma Walcott Tuff, the Conant Creek Tuff at Hominy Ridge (tentatively correlated with the tuff of Elkhorn Spring and between 4.3 Ma and 5.7 Ma), the 4.3-Ma tuff of Kilgore (including most of the Conant Creek Tuff), and the 2.0-Ma Huckleberry Ridge Tuff (from the Yellowstone Plateau volcanic field to the northeast).

Correlation of the tuff of Blue Creek with the Walcott Tuff and the tuff of Kilgore with most of the Conant Creek Tuff expands the recognized areal distributions of these major ignimbrites: The Walcott Tuff is estimated to exceed 15,000 km^2, and the tuff of Kilgore is estimated to exceed 18,000 km^2. Furthermore, the source of the Walcott Tuff is inferred to be the Blue Creek caldera, the same as that for the formerly named tuff of Blue Creek. Likewise, the source for most of the Conant Creek Tuff is inferred to be the Kilgore caldera, the same as that for the tuff of Kilgore (Morgan and others, 1984; Morgan, 1988).

ACKNOWLEDGMENTS

I thank H. Z. Albedo, R. P. Hoblitt, J. Morzèl, R. L. Reynolds, and J. G. Rosenbaum for support in the paleomagnetic studies; W. Hildreth for help with the oxygen isotope analyses; and R. Yeoman for help with the chemical analyses. D. E. Trimble kindly provided field notes and thin sections. I thank R. L. Christiansen, G. Demonceaux, G. Desborough, D. J. Doherty, W. R. Hackett, W. Hildreth, R. P. Hoblitt, C. A. Gardner, K. S. Kellogg, W. P. Leeman, J. D. Love, R. L. Reynolds, D. E. Trimble, and G.P.L. Walker for stimulating discussions and for help in various aspects of this study. Special thanks are extended to J. Morzel for help in all aspects of this study. B. Bonnichsen, M. R. Hudson, K. S. Kellogg, M. A. Kuntz, J. G. Rosenbaum, and E. Winget reviewed various drafts of the manuscript. Funds for this study were provided by two Geological Society of America student grants while the author was a Ph.D. candidate at the University of Hawaii, Manoa.

REFERENCES CITED

Anders, M. H., Geissman, J. W., Piety, L. A., and Sullivan, J. T., 1989, Parabolic distribution of circumeastern Snake River Plain seismicity and latest Quaternary faulting: Migratory pattern and association with the Yellowstone hotspot: Journal of Geophysical Research, v. 94, p. 1589–1621.

Armstrong, R. L., Leeman, W. P., and Malde, H. E., 1975, K-Ar dating, Quaternary and Neogene rocks of the Snake River Plain, Idaho: American Journal of Science, v. 275, p. 225–251.

Armstrong, R. L., Harakal, J. E., and Neill, W. M., 1980, K-Ar dating of Snake River Plain (Idaho) volcanic rocks—New results: Isochron/West, no. 27, p. 5–10.

Carr, W. J., and Trimble, D. E., 1963, Geology of the American Falls quadrangle, Idaho: U.S. Geological Survey Bulletin 1121-G, 44 p.

Christiansen, R. L., 1982, Late Cenozoic volcanism of the Island Park area, eastern Idano, *in* Bonnichsen, B., and Breckenridge, R. M., eds., Cenozoic geology of Idaho: Idaho Bureau of Mines and Geology Bulletin 26, p. 345–368.

Christiansen, R. L., and Blank, H. R., Jr., 1972, Volcanic stratigraphy of the Quaternary rhyolite plateau in Yellowstone National Park: U.S. Geological Survey Professional Paper 729-B, 18 p.

Christiansen, R. L., and Love, J. D., 1978, The Pliocene Conant Creek Tuff in the northern park of the Teton Range and Jackson Hole, Wyoming: U.S. Geological Survey Bulletin 1435-C, 9 p.

Dalrymple, G. B., 1979, Critical tables for conversion of K-Ar ages from old to new constants: Geology, v. 7, p. 558–560.

Doherty, D. J., McBroome, L. A., and Kuntz, M. A., 1979, Preliminary geologic interpretation and lithologic log of the exploratory test well (INEL-1), Idaho National Engineering Laboratory, eastern Snake River Plain Idaho: U.S. Geological Survey Open-File Report 79-1248, 10 p.

Embree, G. F., Lovell, M. D., and Doherty, D. J., 1978, Drilling data from the Sugar City geothermal exploration well, Madison County, Idaho: U.S. Geological Survey Open-File Report 78-1095, 1 plate.

Fisher, R. A., 1953, Dispersion on a sphere: Proceedings of the Royal Society of London, v. 217A, p. 295–305.

Grommè, C. S., McKee, E. H., and Blake, M. C., Jr., 1972, Paleomagentic correlations and potassium-argon dating of middle Tertiary ash-flow sheets in the eastern Great Basin, Nevada and Utah: Geological Society of America Bulletin, v. 83, p. 1619–1938.

Hildreth, W., Christiansen, R. L., and O'Neil, J. R., 1984, Catastrophic isotopic modification of rhyolitic magma at times of caldera subsidence, Yellowstone Plateau volcanic field: Journal of Geophysical Research, v. 89, p. 8339–8369.

Kellogg, K. S., and Marvin, R. F., 1988, New potassium-argon ages, geochemistry, and tectonic setting of upper Cenozoic rocks near Blackfoot, Idaho: U.S. Geological Survey Bulletin 1806, 19 p.

Kellogg, K. S., Pierce, K. L., Mehnert, H. H., Hackett, W. R., Rodgers, D. W., and Hladky, F. R., 1989, New ages on biotite-bearing tuffs of the eastern Snake River Plain, Idaho: Stratigraphic and mantle plume implications: Geological Society of America Abstracts with Programs, v. 21, p. 101.

Kuntz, M. A., 1979, Geologic map of the Juniper Buttes area, eastern Snake River Plain, Idaho: U.S. Geological Survey Miscellaneous Field Investigations Map I-1115, scale 1:24,000, 1 plate.

Mansfield, G. R., 1927, Geography, geology, and mineral resources of part of southeastern Idaho: U.S. Geological Survey Professional Paper 152, 453 p.

Mansfield, G. R., and Ross, C. S. 1935, Welded rhyolitic tuffs in southeastern Idaho: EOS Transactions of the American Geophysical Union, p. 308–321.

Marvin, R. F., Mehnert, H. H., and Noble, D. C., 1970, Use of Ar^{26} to evaluate the incorporation of air by ash flows: Geological Society of America Bulletin, v. 81, p. 3385–3392.

McBroome, L. A., 1981, Stratigraphy and origin of Neogene ash-flow tuffs on the north-central margin of the eastern Snake River Plain, Idaho [M.S. thesis]: Boulder, University of Colorado, 74 p.

McBroome, L. A., Doherty, D. J., and Embree, G. F., 1981, Correlation of major Pliocene and Miocene rhyolites, ESRP, Idaho: Geological Society of Montana, Southwest Montana Field Conference, Guidebook, p. 323–330.

Morgan, L. A., 1988, Explosive rhyolitic volcanism of the eastern Snake River Plain [Ph.D. thesis]: Manoa, University of Hawaii, 191 p.

Morgan, L. A., Doherty, D. J., and Leeman, W. P., 1984, Ignimbrites of the ESRP: Evidence for major caldera-forming eruptions: Journal of Geophysical Research, v. 89, p. 8665–8678.

Naeser, C. W., Izett, G. A., and Obradovich, J. D., 1980, Fission track and K-Ar ages of natural glasses: U.S. Geological Survey Bulletin 1489, 31 p.

Oriel, S. S., and Moore, D. W., 1985, Geologic map of the west and east Palisades roadless areas, Idaho and Wyoming: U.S. Geological Survey Miscellaneous Field Studies Map MF-1619-B, scale 1:50,000, 2 plates.

Reynolds, R. L., 1977, Paleomagnetism of welded tuffs of the Yellowstone Group: Journal of Geophysical Research, v. 82, p. 3677–3693.

Rosenbaum, J. G., 1986, Paleomagnetic directional dispersion produced by plastic deformation in a thick Miocene welded tuff, southern Nevada: Implications for welding temperatures: Journal of Geophysical Research, v. 91, p. 12817–12834.

Skipp, B. A., Prostka, H. J., and Schleicher, D. L., 1979, Preliminary geologic map of the Edie Ranch quadrangle, Clark County, Idaho and Beaverhead County, Montana: U.S. Geological Survey Open-File Report 79-845, scale 1:62,500.

Smith, R. L., 1960, Zones and zonal variations in welded ash-flow tuffs: U.S. Geological Survey Professional Paper 354F, p. 148–159.

Stearns, H. T., and Isotoff, A., 1956, Stratigraphic sequence in the Eagle Rock volcanic area near American Falls, Idaho: Bulletin of the Geological Society of America, v. 67, p. 19–34.

Steiger, R. H., and Jager, E., 1977, Subcommittee on geochronology: Convention on the use of decay constants in geo- and cosmochronology: Earth and Planetary Science Letters, v. 365, p. 359–362.

Trimble, D. E., 1982, Geologic map of the Yandell Springs quadrangle, Bannock and Bingham Counties, Idaho: U.S. Geological Survey Geologic Quadrangle Map GQ-1553, scale 1:24,000.

Trible, D. E., and Carr, W. J., 1976, Geology of the Rockland and Arbon quadrangles, Power County, Idaho: U.S. Geological Survey Bulletin 1399, 115 p.

Manuscript Accepted by the Society July 19, 1991

Printed in U.S.A.

Geological Society of America
Memoir 179
1992

Chapter 12

An overview of basaltic volcanism of the eastern Snake River Plain, Idaho

Mel A. Kuntz
U.S. Geological Survey, MS 913, Box 25046, Federal Center, Denver, Colorado 80225
Harry R. Covington
U.S. Geological Survey, Box 327, Mercury, Nevada 89023
Linda J. Schorr
Department of Earth Resources, Colorado State University, Fort Collins, Colorado 80523

ABSTRACT

More than 95% of the eastern Snake River Plain (ESRP) is covered by basaltic lava flows erupted in the Brunhes Normal-Polarity Chron; thus they are younger than 730 ka. About 13% of the area of the ESRP is covered by lava fields of latest Pleistocene and Holocene age <15 ka. More than 90% of the basalt volume of the ESRP is included in coalesced shield and lava-cone volcanoes made up dominantly of tube- and surface-fed pahoehoe flows. Deposits of fissure-type, tephra-cone, and hydrovolcanic eruptions constitute a minor part of the basalt volume of the ESRP.

Eight latest Pleistocene and Holocene lava fields serve as models of volcanic processes that characterize the basaltic volcanism of the ESRP. The North Robbers, South Robbers, and Kings Bowl lava fields formed in short-duration (a few days), low-volume (each <0.1 km^3), fissure-controlled eruptions. The Hells Half Acre, Cerro Grande, Wapi, and Shoshone lava fields formed in long-duration (several months), high-volume (1 to 6 km^3), lava cone- and shield-forming eruptions. Each of these seven lava fields represents monogenetic eruptions that were neither preceded nor followed by eruptions at the same or nearby vents. The Craters of the Moon lava field is polygenetic; about 60 flows were erupted from closely spaced vents over a period of 15,000 years.

Most of the basaltic volcanism of the ESRP is localized in volcanic rift zones, which are long, narrow belts of volcanic landforms and structures. Most volcanic rift zones are collinear continuations of basin-and-range-type, range-front faults bordering mountains that adjoin the ESRP. It is not clear whether the faults extend into the ESRP in bedrock beneath the basaltic lava flows.

The great bulk of basaltic flows in the ESRP are olivine basalts of tholeiitic and alkaline affinities. The olivine basalts are remarkably similar in chemical, mineralogical, and textural characteristics. They were derived by partial melting of the lithospheric mantle at 45 to 60 km, and they have been little affected by fractionation or contamination. Evolved magmas having SiO_2 contents as high as 65% occur locally in and near the ESRP. The chemical and mineralogical variability of the evolved rocks is due to crystal fractionation in the crust and to contamination by crustal minerals and partial melts of crustal rocks. The trace-element compositions of the olivine basalts and the most primitive evolved basalts do not overlap, suggesting that the evolved rocks were derived from parent magmas that are fundamentally different from the parent magmas of the olivine basalts.

Kuntz, M. A., Covington, H. R., and Schorr, L. J., 1992, An overview of basaltic volcanism of the eastern Snake River Plain, Idaho, *in* Link, P. K., Kuntz, M. A., and Platt, L. B., eds., Regional Geology of Eastern Idaho and Western Wyoming: Geological Society of America Memoir 179.

The distribution and character of volcanic rift zones in the ESRP are partly controlled by underlying Neogene rhyolite calderas. Areas that lack basalt vents and have only poorly developed volcanic rift zones overlie calderas or parts of calderas filled by thick, low-density sediments and rocks, which served as density barriers to the buoyant rise of basaltic magma. Volcanic rift zones are locations of concentrated extensional strain; they define regional stress patterns in the ESRP.

INTRODUCTION

We studied the Quaternary basaltic lava fields of the eastern Snake River Plain (ESRP), Idaho, in the period 1975 through 1981 and at various times since then. The study area covers 20,000 km^2 within nearly 250 7½-minute quadrangles and extends from Twin Falls to Ashton (Fig. 1). Our studies included interpretation of volcanic features on aerial photos; field mapping of vents, flow features, and contacts; and sampling of flows and inclusions for petrographic, petrochemical, and dating investigations.

This chapter characterizes the basaltic volcanism of the ESRP based on our field and laboratory studies and the studies of coworkers. We emphasize the latest Pleistocene and Holocene

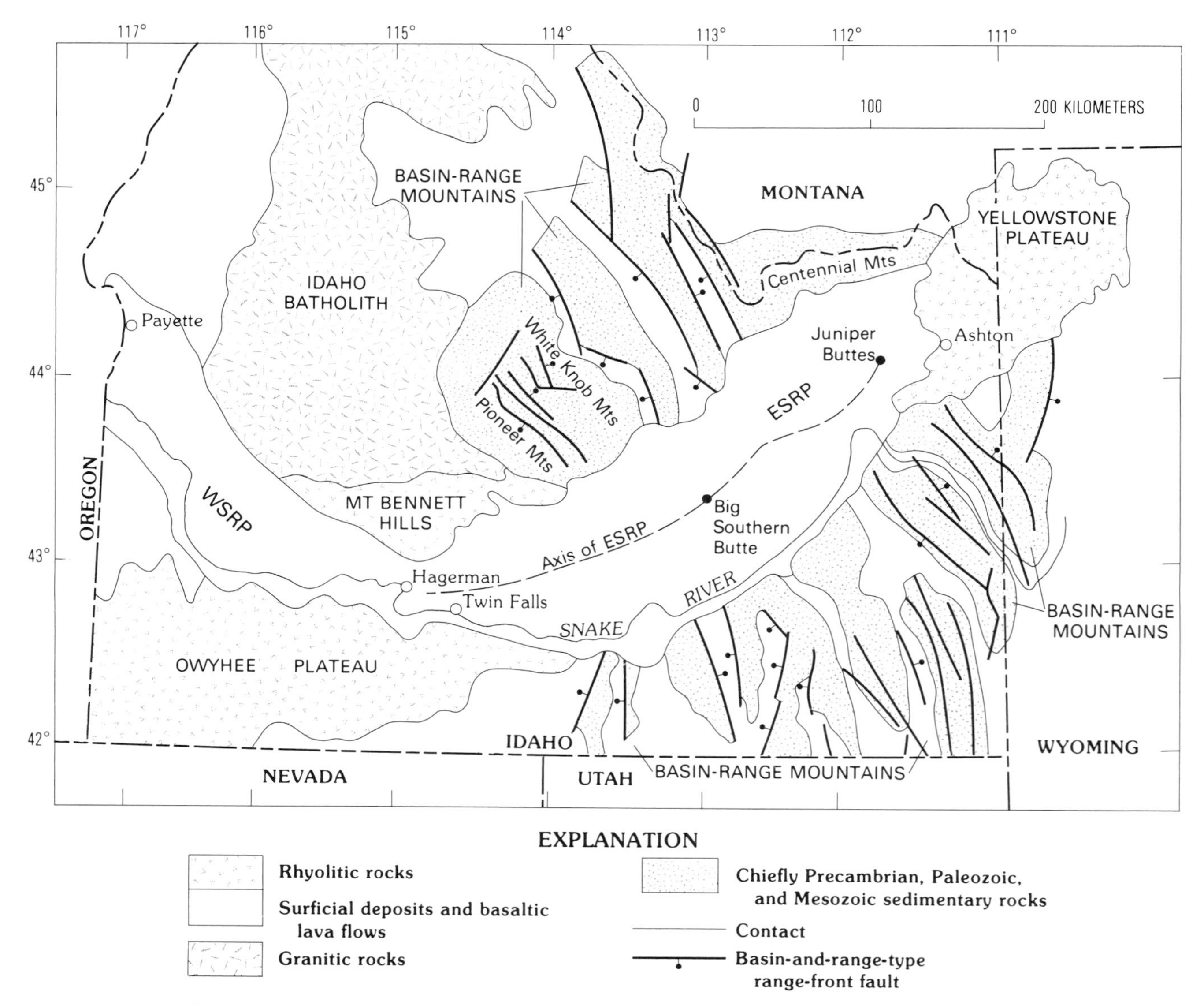

Figure 1. Index map of southern Idaho showing generalized geology, latest Pleistocene-Holocene lava fields, and localities referred to in the text.

lava fields, volcanic rift zones, and the interrelationships between Quaternary basaltic volcanism and Neogene rhyolite calderas. In addition, we describe the volcanic morphology, petrology, petrochemistry, and stratigraphy of the basaltic rocks, the structural control of vents, and magma output rates. We treat only those topics we feel are most important in characterizing basaltic volcanism of the ESRP. The reference section will guide those who wish to investigate these and other topics more completely.

Notes on terminology

Several often-used terms in this chapter are defined here to avoid possible confusion in their meaning. Many of the terms are described by Holcomb (1987) and illustrated in photographs by Takahashi and Griggs (1987). *Lava flow* denotes thin sheets of lava erupted from fissure or pipelike vents. Typical basaltic lava flows of the ESRP are pahoehoe, *slab pahoehoe* (jumbled plates or slabs of broken pahoehoe crust, typically <10 m across), and *shelly pahoehoe* (flows near vents having small open tubes, blisters, and thin crusts that formed from gas-charged lava; Swanson, 1973). *Tube-fed* flows are large-volume pahoehoe flows that advance chiefly by supply of lava through lava tubes; *surface-fed* flows advance by supply of lava in surface conduits such as channels. A'a and block-lava flows are rare; the few that are known in this area are several kilometers wide and several tens of kilometers long and are typically 5 to 15 m thick but as thick as 50 m where ponded.

Flow lobes are nearly contemporaneous discharges of lava that formed during a single eruptive episode. They range from several centimeters to several meters thick. The length of a flow lobe is variable, ranging from local outpourings that moved only a few meters over slightly older surfaces of the same flow to sheets that flowed several kilometers partly or entirely over earlier flow lobes of the same eruption (Wentworth and Macdonald, 1953). In vertical sections and cores in lava flows of the ESRP, it is difficult to determine whether successive flows of similar lava that lack an intervening soil or alluvial horizon are merely different flow lobes of the same major flow or independent flows extruded during different eruptions. *Lava field* refers to a sequence of flows, flow lobes, and associated pyroclastic deposits that accumulated during synchronous eruptions from single or multiple vents along an eruptive fissure. *Volcanic rift zones* are linear arrays of volcanic landforms and structures. The volcanic landforms include eruptive fissures, spatter ramparts, tephra cones, lava cones, shield volcanoes, and dikes at depth. The structures include noneruptive fissures, faults, and grabens.

REGIONAL GEOLOGY AND STRUCTURE

The Snake River Plain

The Snake River Plain is a relatively flat depression, 50 to 100 km wide, that cuts an arcuate swath across southern Idaho. It extends from Payette on the northwest 250 km southeast to Twin Falls and thence 300 km northeast to Ashton (Fig. 1). Elevation of the Snake River Plain rises gradually from 700 m on the west to 2,000 m on the east. The Snake River Plain is bounded on the southeast by folded, thrust-faulted mountains of the Basin and Range province, consisting of upper Precambrian through lower Mesozoic sedimentary rocks that were uplifted along normal faults during Neogene and Quaternary tectonism. The Snake River Plain is bounded on the north by similar basin-and-range type mountains and also by mountains consisting of Mesozoic and lower Tertiary granitic rocks of the Idaho batholith. Southwest of the Snake River Plain are the mid-Miocene rhyolitic and basaltic rocks of the Owyhee Plateau. At the northeast end of the Snake River Plain are the Quaternary rhyolitic and basaltic rocks of the Yellowstone Plateau.

The Snake River Plain is divided into western and eastern parts at about 114° W. longitude. Mabey (1982) defined a central part based chiefly on geophysical characteristics. The eastern and western parts have contrasting structural, geophysical, and geological characteristics, suggesting different modes of origin.

The western Snake River Plain (WSRP) is a late Neogene, fault-bounded graben containing rhyolite and basalt lava flows and thick, interbedded detrital sediments (Malde and Powers, 1962; Malde, 1991). Considerable relief and extensive vertical exposures along the canyons of the Snake River and its tributaries have been important in the geologic study and interpretation of the tectonic history of the western Snake River Plain.

The eastern Snake River Plain (ESRP) is a broad lava plain consisting, at the surface, of mostly Pleistocene and Holocene basaltic lava flows, pyroclastic deposits, and thin, discontinuous deposits of loess, sand dunes, and alluvial sediments. Magnetic polarity determinations and recent radiometric studies (Champion and others, 1988; Kuntz and others, 1992) indicate that most of the surface flows were erupted during the Brunhes Normal-Polarity Chron and thus are younger than 730 ka. Data from wells that penetrate the ESRP to depths as great as 3,500 m (summarized in Embree and others, 1982) show that the lava-flow and sediment sequence is 1 to 2 km thick throughout most of the ESRP. These data also indicate that the thickness of the basaltic volcanic rocks greatly exceeds that of the sedimentary deposits near the center of the ESRP and that the sedimentary deposits are more abundant along the margins.

Drilling and recent field studies (Doherty and others, 1979; Embree and others, 1982; Morgan and others, 1984) show that the basalt-sediment sequence is underlain by rhyolitic lava flows, ignimbrites, and pyroclastic deposits. Volcanologic studies of ignimbrites at localities around the margins of the ESRP indicate the presence of large source calderas beneath the basaltic lava flow and sediment sequence of the ESRP (Pierce and Morgan, this volume; Morgan, this volume).

Available geological and geophysical data suggest that the Snake River Plain has been the site of a northeasterly propagating system of rhyolitic volcanic centers and associated basaltic volcanism (see, for example, Armstrong and others, 1975; Brott and others, 1978, 1981; Mabey, 1982; Malde, 1991; Leeman, 1989;

Pierce and Morgan, this volume). Rhyolitic volcanism began in mid-Miocene time near the southwest corner of Idaho and has migrated northeastward to the site of the most recent volcanism in the Yellowstone Plateau. The volcanic cycle at any one locality is believed to be initiated by generation of basalt magma in the lithospheric mantle. The rise of basalt magma from the mantle produces sufficient heat to cause anatexis of the lower crust and subsequent formation of rhyolitic calderas. Rhyolitic volcanism is generally preceded and followed by basaltic volcanism, thus forming the bimodal basalt-rhyolite volcanic association of the ESRP. Such a volcanic system produces high heat flow, thinning and decreased rigidity of the crust, and progressive formation of a topographic trough along the chain of calderas. The location of the caldera chain in the ESRP may have been governed by a complex interaction between a zone of structural weakness (a crustal "flaw" of Precambrian ancestry) in the crust, an underlying "hot spot" in the mantle, the southwestward drift of the North American plate, and perhaps the major structures of the Basin and Range province.

Major, northeast-southwest–trending faults parallel to the margins of the ESRP have not been identified; thus there is little evidence to support a fault-bounded, rift-valley origin for the ESRP.

Physiography

Mountains north and south of the ESRP are long, narrow, basin-and-range–type mountains, separated by narrow intermontane valleys. Crests of mountain ranges and axes of intermontane valleys are spaced about 35 km apart. Both mountains and valleys trend mostly north or northwest, perpendicular to the long axis of the ESRP, but the Centennial Mountains trend east-west, parallel to the northeast margin of the ESRP (Fig. 1). Most of these ranges are bounded by a single range-front normal-fault system, but others have faults along both margins. The fault-block mountains end abruptly at the margins of the ESRP. The abrupt terminations may reflect their proximity to the walls of calderas that lie within the ESRP.

The Pioneer and White Knob mountains in central Idaho, along the northwest border of the ESRP (Fig. 1), form a broad highland mass consisting of Precambrian rocks of a core complex and Cretaceous and early Cenozoic plutonic rocks of the Idaho batholith. This range has not been broken into distinct basin-and-range–type mountains, even though northwest-trending normal faults in the Big Wood River and Sawtooth valleys suggest incipient basin-and-range deformation (Link and others, 1988).

Drainage in the ESRP is dominated by the Snake River, which flows along the southeast margin of the plain. A topographic ridge of volcanic construction extends from near Hagerman in the western Snake River Plain through Big Southern Butte northeastward to Juniper Buttes near Ashton in the ESRP. The ridge coincides approximately with the long axis of the ESRP (Fig. 1). Rivers entering the ESRP from the north are deflected by the ridge and discharge in closed basins. Because river canyons are absent, little is known of the underlying rocks of the ESRP except from drilling and geophysical data and study of rhyolitic ignimbrites and flows along its margins.

THE STAGES OF BASALTIC ERUPTIONS

Field characteristics of vent areas and lava flows in the ESRP are remarkably similar to those of well-studied volcanoes and volcanic rift zones in Hawaii. By analogy, volcanic processes in the ESRP are assumed to have been similar to Hawaiian volcanic processes, though they may not have occurred at the same rates. Thus, our understanding of ESRP basaltic activity is based partly on Hawaiian studies (for example, Wentworth and Macdonald, 1953; Macdonald and Abbott, 1970; Richter and others, 1970; Swanson and others, 1979; Holcomb, 1987; Peterson and Moore, 1987; Tilling and others, 1987; and Wolfe and others, 1987). From these studies, it is recognized that Hawaiian fissure-fed eruptions generally proceed through four distinct stages, as summarized below.

The *first stage* is characterized by harmonic tremor, ground cracking, and local steam and fume activity. Fissuring is followed by lava fountains, producing a "curtain of fire" that extends for several hundred meters and erupts to heights of as much as 500 m. The early lava is extremely fluid and gas charged. Spatter ramparts typically form along both sides of erupting fissures, and fine tephra is deposited at distances as great as 10 km downwind, depending on lava-fountain height and wind speed. Flows are typically surface-fed, shelly pahoehoe; they accumulate in volumes less than 0.1 km^3 and cover areas less than 5 km^2.

After several days, the *second stage* begins. The erupting fissure system typically diminishes in length, and lava fountains become localized along short segments of a fissure. High fountains form tephra cones over earlier spatter ramparts. Diminished fountaining is typically followed by quiet, more voluminous outpouring of lava over and through tephra-cone walls. Tephra cones may be succeeded by small lava cones or small shields during latter parts of this stage. Flows of the second stage are typically surface-fed pahoehoe, although tube-fed flows form if eruption rates are fairly high.

A *third stage* occurs during long-lived eruptions: Lava fountaining diminishes and is followed by quiet, prolonged, voluminous outpouring of lava for periods of months or a few years, producing a large lava cone or shield volcano. Lava cone or shield summits typically have a crater elongated parallel to the underlying feeder fissure. Large craters may contain a lava lake characterized by pistonlike filling and draining. Proximal flows are largely surface-fed sheets of pahoehoe and slab pahoehoe; medial and distal flows are mostly tube-fed pahoehoe.

The enormous Hawaiian shield volcanoes, such as Mauna Loa and Mauna Kea, represent a *fourth stage* that results from eruptions that take place over thousands or a few million years from a central vent area and from volcanic rift zones. No lava fields of the ESRP correspond to the sustained eruptions of the fourth stage.

During Hawaiian volcanism, long-term eruptions tend to proceed through the four stages, though any one eruptive episode may terminate within the sequence. By inference, some prehistoric volcanic eruptions in the ESRP represent only the first stage of the eruptive sequence, and others proceeded to the second or third stages. Discussions of various lava fields of the ESRP that follow are described in terms of the three-stage eruption sequence.

VOLCANIC LANDFORMS AND LAVA TYPES OF THE ESRP

Shield volcanoes of the ESRP probably began as fissure eruptions, passed through the lava-cone stage, and formed by sustained eruptions lasting several months at a central vent or vent complex. Later-stage eruptive processes typically involved formation of a summit lava lake, pistonlike filling and draining of magma in the vent, collapse of vent walls, and formation of a lava field consisting mainly of tube-fed pahoehoe flows. Shelly-pahoehoe and slab-pahoehoe flows near vents formed from degassed, relatively high viscosity lava by overflow from lava lakes or discharge from rootless vents on lava tubes. Shield volcanoes are distributed widely in the ESRP, although the largest shields are typically located near the plain axis, suggesting that they overlie the main region of magma generation.

Lava flows of shield volcanoes in the ESRP extend as much as 30 km from their vents. Shields cover areas of 100 to 400 km^2 and have volumes of 1 to 7 km^3, the average being about 5 km^3. Dividing this figure into the total volume of basalt in the ESRP ($\sim 4.10^4$ km^3) suggests there may be as many as 8,000 shields in the ESRP. Considering that the ESRP began to form about 10 Ma and assuming a constant eruption rate, the recurrence interval for shield-forming eruptions of the ESRP is about one every 1,000 years.

Lava cones have heights of 10 to 30 m, cover areas of 1 to 50 km^2, and have volumes of 0.05 to 2 km^3. They consist mainly of surface- or channel-fed pahoehoe but have minor amounts of shelly pahoehoe and slab pahoehoe near vents. Tube-fed pahoehoe is present only in some of the larger lava cones. Evidently, sustained eruptions lasting more than several weeks and eruption rates in the range 10^3 to 10^4 kg/sec per meter of eruptive fissure (Kuntz, this volume) were necessary for pahoehoe lava to be distributed by lava tubes.

True a'a flows in the ESRP are found only in lava flows having SiO_2 contents of $>50\%$, but exceptions occur in some flows in the late Pleistocene and Holocene Craters of the Moon (COM) lava field that descended steep slopes, such as the Lava Creek flows (Kuntz and others, 1988).

Although deposits of fissure-type eruptions are widely distributed throughout the ESRP, their contribution to the total volume of basalt lava flows of the ESRP is minimal. Deposits of fissure eruptions are described by Kuntz and others (1988) and are not discussed further here.

Tephra cones in the ESRP occur in two geologic settings: (1) those in the COM lava field formed from lava flows having SiO_2 contents of $>50\%$ and (2) cones formed in low areas where surface- or near-surface water played a part in their formation. Tephra cones in the COM lava field are some of the most imposing volcanic features in the ESRP. The largest cone, Big Cinder Butte (refer to Fig. 3), is 240 m high and covers an area of 3 km^2. Most other cones of the COM lava field are ~ 100 m high and cover areas of ~ 1 km^2 (Kuntz and others, 1988). Six small cones, each <20 m high and <200 m in diameter, occur within a 15-km^2 area of low elevation near Atomic City (refer to Fig. 4). About a dozen small cones occur at the east end of the Spencer–High Point volcanic rift zone (refer to Fig. 9). At this locality, basalt magma was emplaced in water-bearing sediments ponded between the southern margin of the Centennial Mountains and the Spencer–High Point zone. Tephra cones, like eruptive-fissure deposits, constitute a very small part of the total volume of basalt in the ESRP.

Deposits of hydrovolcanic eruptions are widely scattered in the ESRP. Sand Butte and Split Butte (refer to Fig. 7) consist of deposits of palagonitized sideromelane that surround inner lava lakes, which overlie single, pipelike sources of magma discharge (Womer and others, 1982). The Menan Buttes complex of tuff cones, described by Hackett and Morgan (1988), includes several late Pleistocene tuff cones and tuff rings that form the major part of the northwest-trending, 30-km-long Menan volcanic rift zone (refer to Figs. 7 and 18). The largest tuff cones are North Menan Butte (0.7 km^3) and South Menan Butte (0.3 km^3). Hydrovolcanic deposits, like the eruptive fissure and tephra-cone deposits, are also insignificant in the total volume of basalt in the ESRP.

In a series of three similar papers, Greeley (1977, 1982a, 1982b) described the Snake River Plain as being representative of a distinct type of basaltic volcanism, which he termed "plains volcanism." He suggested that this type is gradational between eruptions of flood lava flows and flows that form shield volcanoes. He also suggested that ESRP flows are emplaced primarily in one of three forms: (1) low shield volcanoes having slopes of less than 1°, (2) fissure flows (eruptive-fissure flows in our terminology), and (3) tube-fed flows. Although we do not disagree with Greeley's conclusions, our mapping suggests a different emphasis on the characteristic volcanic morphologies that make up the bulk of basaltic volcanism of the ESRP. We do not draw a distinction between shield volcanoes and tube-fed flows as does Greeley; we find that tube-fed flows constitute the largest volume of the shield volcanoes and parts of some large lava cones on the ESRP.

We conclude that coalesced shield volcanoes and lava cones, the flows of which are mainly of the tube-fed type, constitute more than 95% of the total volume of basalt in the ESRP. The remainder is made up of eruptive-fissure deposits, tephra cones, and deposits of hydrovolcanic eruptions, in that order.

LATEST PLEISTOCENE AND HOLOCENE LAVA FIELDS OF THE ESRP

Eight basalt or basalt-dominated lava fields in the ESRP appear as dark areas of low reflectance on Landsat images

(Lefebvre, 1975; Champion and Greeley, 1977). Because these fields are relatively young, their original surface morphology is well preserved, except where some are locally covered by thin eolian deposits. Radiocarbon studies (Kuntz, Spiker, and others, 1986) show that these fields are very latest Pleistocene or Holocene in age. These eight lava fields are, from west to east, the Shoshone, Craters of the Moon, Wapi, Kings Bowl, North Robbers, South Robbers, Cerro Grande, and Hells Half Acre fields (Fig. 2). Descriptive data for these fields—length and width of fissures, types of flows, areas and volumes of flows and fields, and stages of basaltic volcanism represented—are summarized in Table 1. The best-studied of these lava fields is the Craters of the Moon field.

Craters of the Moon (COM) lava field

The COM lava field, the largest basaltic, dominantly Holocene lava field in the conterminous United States, has been described by Kuntz and others (1982); Kuntz, Champion, and others (1986); Kuntz, Champion, and Lefebvre (1989); Kuntz, Lefebvre, and Champion (1989a, 1989b); and Champion and others (1989). It is therefore reviewed here only briefly. It contrasts with most other lava fields of the ESRP because it is a composite, polygenetic field composed of at least 60 flows erupted from 25 tephra cones and at least eight eruptive fissure systems aligned along the northern part of the Great Rift volcanic rift zone (Fig. 3). Most other lava fields at the ESRP represent single, monogenetic eruptions from vents that are widely scattered in both space and time.

The COM field covers about 1,600 km^2 and contains about 30 km^3 of lava flows and associated vent and pyroclastic deposits. Stratigraphic relationships, paleomagnetic studies, and radiocarbon ages (Kuntz, Spiker, and others, 1986) show that the COM field formed during eight eruptive periods, designated as H, oldest, to A, youngest. Each eruptive period was several hundred years or less in duration and separated from other eruptive periods by recurrence intervals of several hundred to about 3,000 years. The first eruptive period began about 15,000 years ago, and the latest eruptive period ended about 2,100 years ago. Because the present interval of dormancy has lasted ~2,100 years, it seems likely that another eruptive period will occur within the next 1,000 years.

There is a broad range in the chemical composition of lava flows of the Craters of the Moon, Kings Bowl, and Wapi lava fields along the Great Rift. Kuntz, Champion, and others (1986) defined three magma types based on chemical parameters and petrologic characteristics. The types are (1) a contaminated type that has an SiO_2 range of ~49 to 64% and commonly shows petrographic evidence of crustal contamination, (2) a fractionated type that has an SiO_2 range of ~44 to 54% and chemical and mineralogical variation that can be accounted for mainly by crystal fractionation, and (3) a Snake River Plain type that has an SiO_2 range of ~45 to 48% and shows little evidence of fractionation. The contaminated and fractionated magma types are represented by flows of the COM lava field. The Snake River Plain magma type is represented by flows of the Kings Bowl and Wapi lava fields and by the majority of other lava fields of the ESRP.

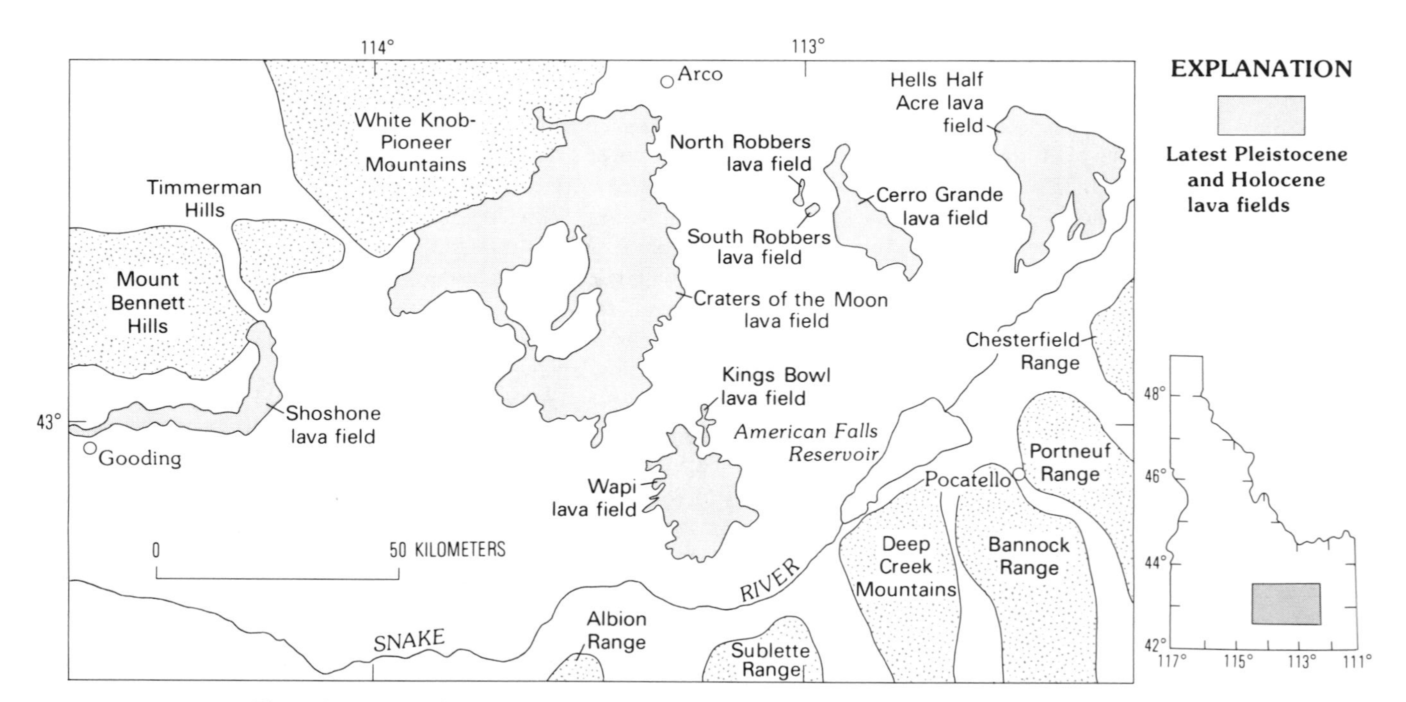

Figure 2. Map showing latest Pleistocene and Holocene lava fields of the eastern Snake River Plain.

Hells Half Acre lava field

The second-largest Holocene lava field in the ESRP is the Hells Half Acre field, a shield volcano about 20 km east of Atomic City (Figs. 2 and 4). The age of the Hells Half Acre lava field is 5,200 ± 150 B.P. (Kuntz, Spiker, and others, 1986). Eruption models (Kuntz, this volume) suggest that the Hells Half Acre field formed over a period of several months. The major study of the field is by Karlo (1977), and most of the field is mapped by Kuntz and others (1992).

The vent area for the Hells Half Acre lava field consists of a flat-floored crater, 0.8 km long and 0.3 km wide, and several spatter cones. A lava lake in the crater floor contains about 10 roughly circular pit craters that are typically <100 m wide and ~5 m deep. Pistonlike draining and filling of the main crater occurred during the last stages of eruptive activity in the crater.

Several small spatter cones and short spatter ramparts lie within the main crater. Five small spatter cones are distributed over a distance of 2 km southwest of the main crater, and two small spatter cones occur about 0.5 km northwest of it. The northwest-southeast–trending eruptive fissure defined by the main crater and spatter cones is about 3 km long.

About 10 km^2 of shelly pahoehoe partly surround and extend out from the main crater for as much as 1.5 km. These flows formed as overflow from the lava lake and from eruptions from spatter cones. As in other latest Pleistocene and Holocene lava fields of the ESRP, lava-tube systems are limited to the proximal and medial parts of the Hells Half Acre lava field; they have not been recognized in distal parts. Two prominent lava-tube systems occur in the proximal parts of the Hells Half Acre lava field. Rootless vents in the tube systems were the sources of some of the youngest pahoehoe flows of the field (Fig. 4).

Slab-pahoehoe flows also partly surround the vent area for distances of several kilometers. In the Hells Half Acre field and other fields in the ESRP, slab pahoehoe formed where partly congealed crust was broken and disturbed as pahoehoe moved over steep, rough topography near vents and also where more viscous degassed lava erupted from lava lakes or rootless vents along lava tubes.

The distal part of the field consists of two major flow lobes, each about 5 km wide and 10 km long. The flows of these lobes moved down the southeast flank of the axis of the ESRP and then extended south and southwest on reaching the flood plain of the Snake River.

Two sets of parallel noneruptive fissures, 1.5 km apart, extend N.50°W. from the Hells Half Acre field (Fig. 4). Each set is about 3.5 km long and consists of en echelon cracks that are 300 m to 1.2 km long. A southeast projection of the noneruptive fissures into the Hells Half Acre lava field forms a three-part fissure system: The eruptive fissure is flanked by noneruptive fissures. This geometry is also present in the Kings Bowl lava field.

Cerro Grande lava field

The Cerro Grande lava field is a small shield volcano fed from a vent that occurs on an extension of a normal fault (down to the east) on the southern flank of Cedar Butte (Fig. 4). The present vent is a shallow depression about 750 m in diameter. An earlier vent was considerably larger, but it was filled by lava that overflowed a late-stage lava lake. Lava tubes occur several kilometers from the vent area, suggesting that near-vent parts of the tubes are filled by lava overflow from the lake. Tephra cones and eruptive fissures characteristic of vent areas of other small shield volcanoes in the ESRP are absent in the Cerro Grande vent area. If originally present, they may also have been covered by overflow from the lava lake. The field was emplaced 13,380 ± 350 B.P. (Kuntz, Spiker, and others, 1986).

The western margin of the southern part of the field is locally indistinct owing to partial cover by eolian sediment. In this respect, the Cerro Grande field is like the North Robbers and South Robbers lava fields, whereas the other latest Pleistocene and Holocene lava fields of the ESRP are essentially free of sediment cover. Porter and others (1983) suggest that Pinedale glacial conditions ended in the ESRP about 11 to 10 ka; thus much of the eolian sediment could have been deposited soon after emplacement.

North and South Robbers lava fields

Two of the smallest late Pleistocene and Holocene lava fields of the ESRP, the North Robbers and South Robbers fields, are a few kilometers west of the Cerro Grande lava field (Fig. 4). The eruptive fissure system for the North Robbers field consists of two segments: The 1-km-long northern segment, which includes both noneruptive and eruptive parts, is offset in alignment about 400 m west from the southern segment, which is 1.2 km long and consists entirely of eruptive fissures. A collinear extension of the northern eruptive fissure, consisting of noneruptive cracks in sediments, extends across the southeast flank of Big Southern Butte for a distance of 700 m. Flows of the North Robbers field were fed mainly by a 2-km-long lava channel that extends from the southern segment of the eruptive fissure system (Fig. 4).

Vents for the South Robbers lava field are aligned along a 1.8-km-long eruptive fissure system. A 150-m-long fissure flanked by spatter ramparts occurs at the south end, and four small tephra cones occur along the north end of the eruptive fissure. A 600-m-long lineation visible on air photos, probably a noneruptive set of cracks, is collinear with and extends northwest from the eruptive fissure system. The South Robbers fissure system lies 1.2 km south and is offset in alignment 600 m west of the eruptive fissures for the North Robbers field. A lava channel that originated at the largest tephra cone was the main distributary for surface-fed flows in the distal part of the South Robbers lava field (Fig. 4).

TABLE 1. CHARACTERISTICS OF FISSURE VENTS AND FLOWS FOR SELECTED ERUPTIONS IN LATEST PLEISTOCENE AND HOLOCENE LAVA FIELDS OF THE EASTERN SNAKE RIVER PLAIN, IDAHO*

LAVA FIELD Flow Name	Length of Eruptive Fissures Single Fissures (km^2)	Whole Zone (km^2)	Estimated Width of Fissures (m)	Area of Lava (km^2)	Estimated Flow/Field Volume (km^3)	Flow Type	Comments
CRATERS OF THE MOON							
Broken Top	Unknown	<0.3	Unknown	11	0.1	Surface- and tube(?)-fed pahoehoe flows.	Intermediate-volume stage-2 eruption. Flow issued from two obscure vents on east and south sides of Broken Top cinder cone. Vents now largely obscured by colluvium
Blue Dragon	0.6	1	1 to 2	280	3.4	Tube-fed pahoehoe flows.	Large-volume sustained lava-cone eruption of stage 3. Eruptions from southernmost cinder cone in Big Crater complex and from 0.5-km-long fissure south of Big Craters. Spatter cones along fissure have internal diameters of 1 to 3 m. Fissure width estimated from smallest diameters.
Trench Mortar Flat	0.3 to 1.3	6	≤2	6	0.03	Shelly, thin pahoehoe flows.	Small-volume stage-1 fissure eruption. Flows erupted from 0.3- to 1.3-km en echelon segments in 6-km zone of eruptive fissures. Fissure widths estimated where enlargement by erosion seemed minimal.
North Crater	Unknown	Unknown	Unknown	1.5	0.01	Surface-fed pahoehoe.	Small-volume eruption from pipelike vent, which is now covered by colluvial cinders from inner wall of North Crater.
Big Craters	≤0.2	0.9	1 to 2	9	0.05	Surface-fed pahoehoe and slab pahoehoe flows.	Small-volume stage-2 eruption. Two source fissures north of Big Craters cinder cone complex are short (≤0.2 km), but Big Craters cinder cone complex is aligned along fissure system 0.9 km long. Fissure width estimated from deepest part of eastern eruptive fissure.
Serrate, Devils Orchard, and Highway	Unknown	...	Unknown	27†	0.5†	Bulbous, block flows.	Vent area near North Crater largely destroyed by eruptions, also covered by younger North Crater flow. Eruption was explosive because remnants of crater walls are contained in flow. Eruptions may have been from central pipelike vent in North Crater.
Vermillion Chasm	0.6 to 1	2.9	1 to 2	20	0.1	Shelly, thin pahoehoe flows.	Small-volume stage-1 fissure eruption. Eruptive fissures enlarged at most localities by explosive venting. Fissure width estimated at deepest, narrowest part of fissure system. Fissure system is 2.9 km long; consists of 3 fissures that range from 0.6 to 1 km in length.
Deadhorse	0.1 to 0.6	10	≤1.5	8	0.04	Shelly pahoehoe and thin pahoehoe flows.	Small-volume stage-1 fissure eruption. Many thin flows were erupted from numerous (>13) en echelon, right-stepping eruptive fissures that range from 0.1 to 0.6 km in length. The Deadhorse fissure system is the longest known fissure system that was active along the Great Rift volcanic rift zone during a single eruptive pulse.
Devils Cauldron	(>0.3)	...	Unknown	90	0.9	Tube- and surface-fed pahoehoe flows.	Intermediate-volume stage-2 eruption from central vent on lava cone. Lava lakes perched above vents. Fissure and vent are obscured by flows and lava lakes.
Minidoka	Unknown	...	Unknown	250	3.0	Chiefly tube-fed pahoehoe flows.	Large-volume, sustained stage-3 eruption from central vent complex that is now covered by flows.

TABLE 1. CHARACTERISTICS OF FISSURE VENTS AND FLOWS FOR SELECTED ERUPTIONS IN LATEST PLEISTOCENE AND HOLOCENE LAVA FIELDS OF THE EASTERN SNAKE RIVER PLAIN, IDAHO (continued)

LAVA FIELD Flow Name	Length of Eruptive Fissures: Single Fissures (km^2)	Length of Eruptive Fissures: Whole Zone (km^2)	Estimated Width of Fissures (m)	Area of Lava (km^2)	Estimated Flow/Field Volume (km^3)	Flow Type	Comments
HELLS HALF ACRE	Unknown	2	Unknown	400	6	Chiefly tube-fed pahoehoe flows with minor surface-fed shelly, and slab pahoehoe flows near vent area.	Large-volume, sustained stage-3 eruption from central vent complex that contained a large lava lake. Collapse pits, spatter cones, and the main depression define the length of main eruptive fissure.
NORTH ROBBERS	<0.5	2.9	Unknown	5	0.05	Chiefly surface-fed pahoehoe and minor shelly pahoehoe flows near vents.	Small-volume stage-1 fissure eruption. Eruptive features defined by spatter ramparts and small cinder cones. Noneruptive fissure 0.7 km long extends north of eruptive fissures.
SOUTH ROBBERS	1.1	1.7	Unknown	3	0.03	Chiefly surface-fed pahoehoe and minor shelly pahoehoe flows near vents.	Small-volume stage-1 fissure eruption. Eruptive fissures defined by spatter ramparts and small cinder cones. A 0.6-km-long noneruptive fissure extends north of eruptive fissures.
CERRO GRANDE	Unknown	Unknown	Unknown	175	2.3	Chiefly tube- and surface-fed pahoehoe flows.	Relatively large-volume stage-3 eruption. Poorly defined vent area filled by a lava lake.
KINGS BOWL	0.1 to 0.5	6.2	0.5 to 1[§]	3.3	0.005	Shelly pahoehoe and thin pahoehoe flows.	Small-volume stage-1 fissure eruption. Eruptions from about a dozen fissures in a zone about 6.5 km long. Dikes ≤1.5 m thick exposed in fissure at Kings Bowl (see Greeley and others, 1977, Figs. 11-14, 11-16; Greeley, 1977, Fig. 3-19).
WAPI	...	>0.6	Unknown	325	6	Surface- and tube-fed pahoehoe flows.	Large-volume stage-3 eruption. Vent complex consists of 11 eruptive centers aligned over a buried eruptive fissure (Champion and Greeley, 1977).
SHOSHONE	Unknown	Unknown	Unknown	190	1.5	Chiefly tube- and surface-fed pahoehoe flows. Shelly pahoehoe and slab pahoehoe flows near vent.	Relatively large-volume stage-3 eruption. Vent area modified by late-stage lava lake. Lava tubes recognized only in proximal parts of lava field. Lava lake activity with pistonlike draining and filling of vent depression.

*Data from Kuntz, Champion, and others (1986, 1992); Kuntz, Lefebvre, and Champion (1989a, b); Kuntz, Champion, and Lefebvre (1989); and this chapter.
†Total for all three flows.
§Measured width.

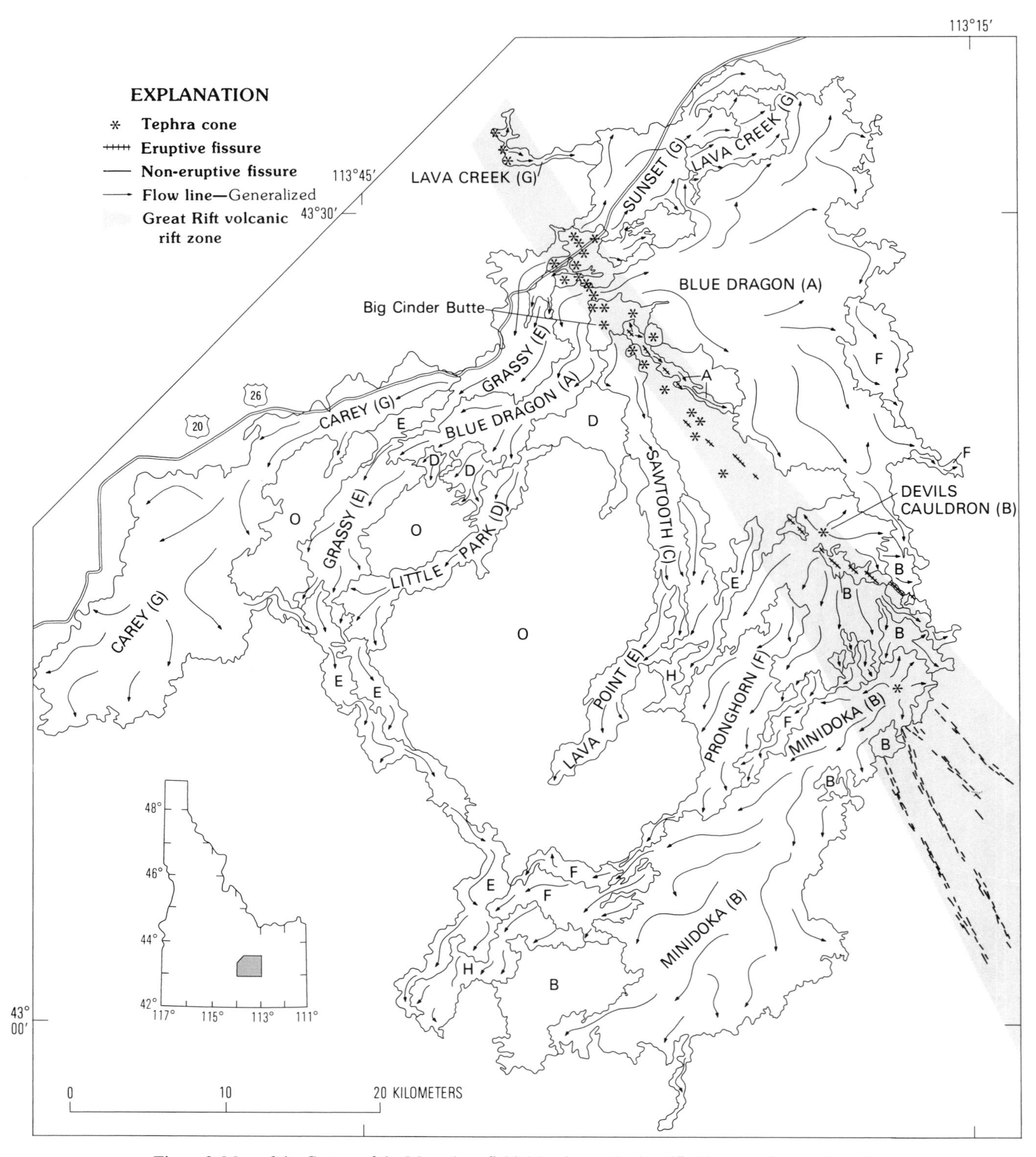

Figure 3. Map of the Craters of the Moon lava field. Map is greatly simplified in area of vents along the Great Rift volcanic rift zone. Some lava flows are shown by name, but most flows are not. However, flow ages are shown by letters that designate eruptive periods from H, oldest, to A, youngest. Several large kipukas within the COM lava field, consisting of flows older than flows of the COM lava field, are designated by the letter O. Adapted from Kuntz and others (1988).

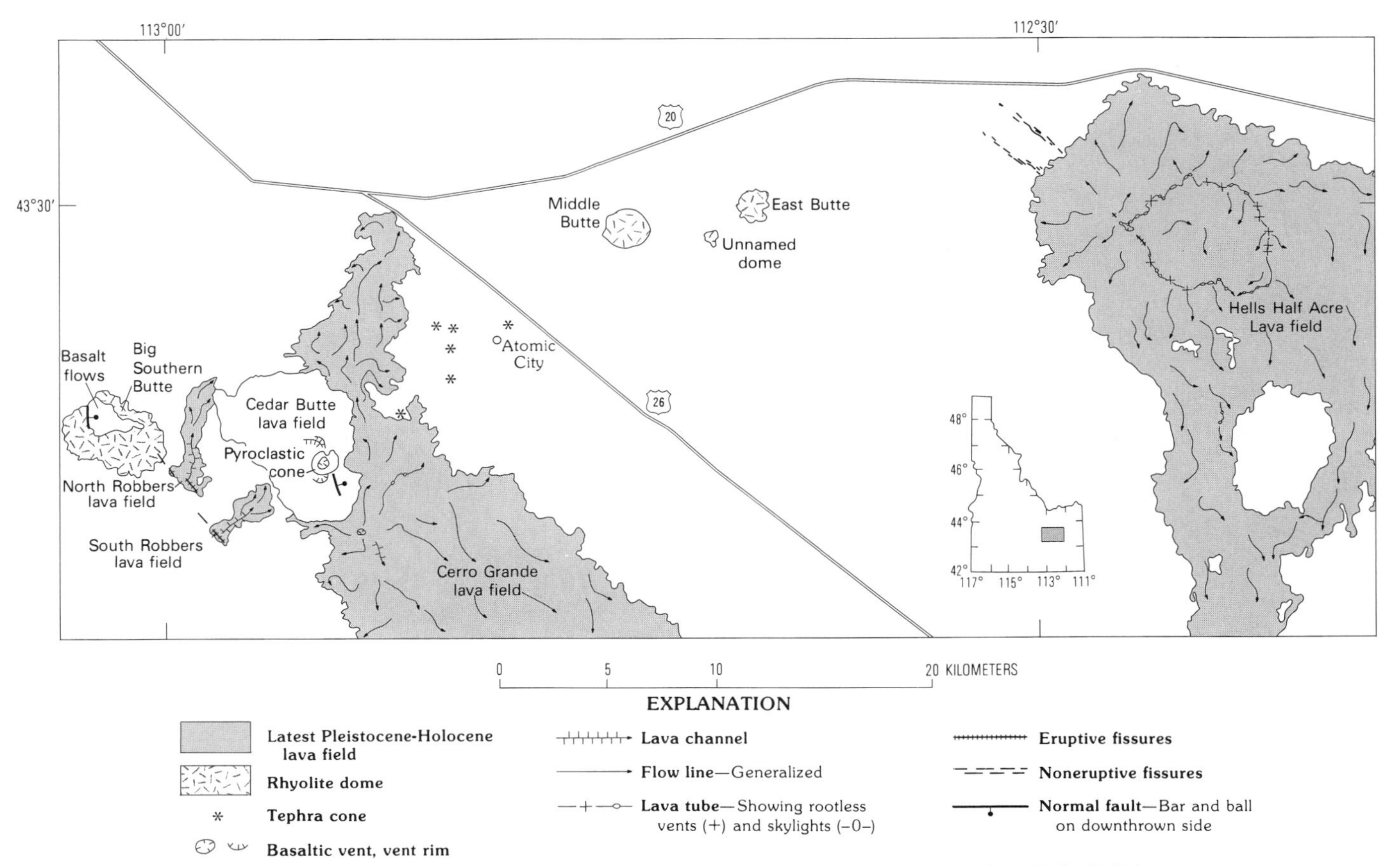

Figure 4. Map of the central part of the eastern Snake River Plain showing parts of the Hells Half Acre and Cerro Grande lava fields and the North Robbers and South Robbers lava fields. Also shown are rhyolite domes (Big Southern Butte, Middle Butte, an unnamed dome, and East Butte) and the Cedar Butte lava field. Adapted from Kuntz and others (1992).

Slab pahoehoe in distal flows and steep flow fronts, as high as 15 m, suggest that flows of the South Robbers field were slightly more viscous and/or were erupted at slower rates than other ESRP basaltic flows.

The North Robbers and South Robbers fields probably formed simultaneously because they are parts of the same eruptive fissure system and have similar paleomagnetic field directions (Champion, 1980). Radiocarbon samples were not obtained for the South Robbers field; the age for the two fields is based on two radiocdarbon ages for the North Robbers field, 11,940 ± 300 and 11,980 ± 300 B.P. (Kuntz, Spiker, and others, 1986).

Flows of the two fields are partly covered by eolian sediment only on their western edges, those closest to sediment sources upwind to the west and southwest. The North and South Robbers fields are mapped by Kuntz and others (1992).

Kings Bowl lava field

The Holocene Kings Bowl lava field lies along the southern part of the Great Rift volcanic rift zone, 12 km southeast of the Craters of the Moon lava field and 1 km north of the Wapi lava field (Figs. 2 and 5; refer also to Fig. 8). The Kings Bowl field has received considerable study (King, 1977; Greeley and others, 1977; Covington, 1977; and Kuntz and others, 1983) because of its small size and easy access. The fissure system in the Kings Bowl area consists of a central eruptive fissure set, about 6.2 km long, trending N.10°W., which is flanked by two subparallel sets of noneruptive fissures that are from 600 to as much as 1,100 m from the main eruptive fissure (Fig. 5). The western noneruptive fissures have a fairly uniform width of about 200 m; the eastern noneruptive fissures have a variable width, ranging from a few meters to 400 m.

The central eruptive fissures consist of discontinuous *en echelon* cracks, 2 to 3 m wide, that are locally filled by breccia and feeder dikes. These short segments, a few meters to 500 m long, are separated by open, noneruptive cracks or by lava that covers eruptive and noneruptive fissures. Feeder dikes, 0.5 to 1 m wide (Figs. 11–14 and 11–16 of Greeley and others, 1977; Fig. 3–19 of Greeley, 1977), are exposed in the central eruptive fissure set, one of the few places where such dikes are exposed in the ESRP.

Small tephra cones, <300 m long and <10 m high, formed

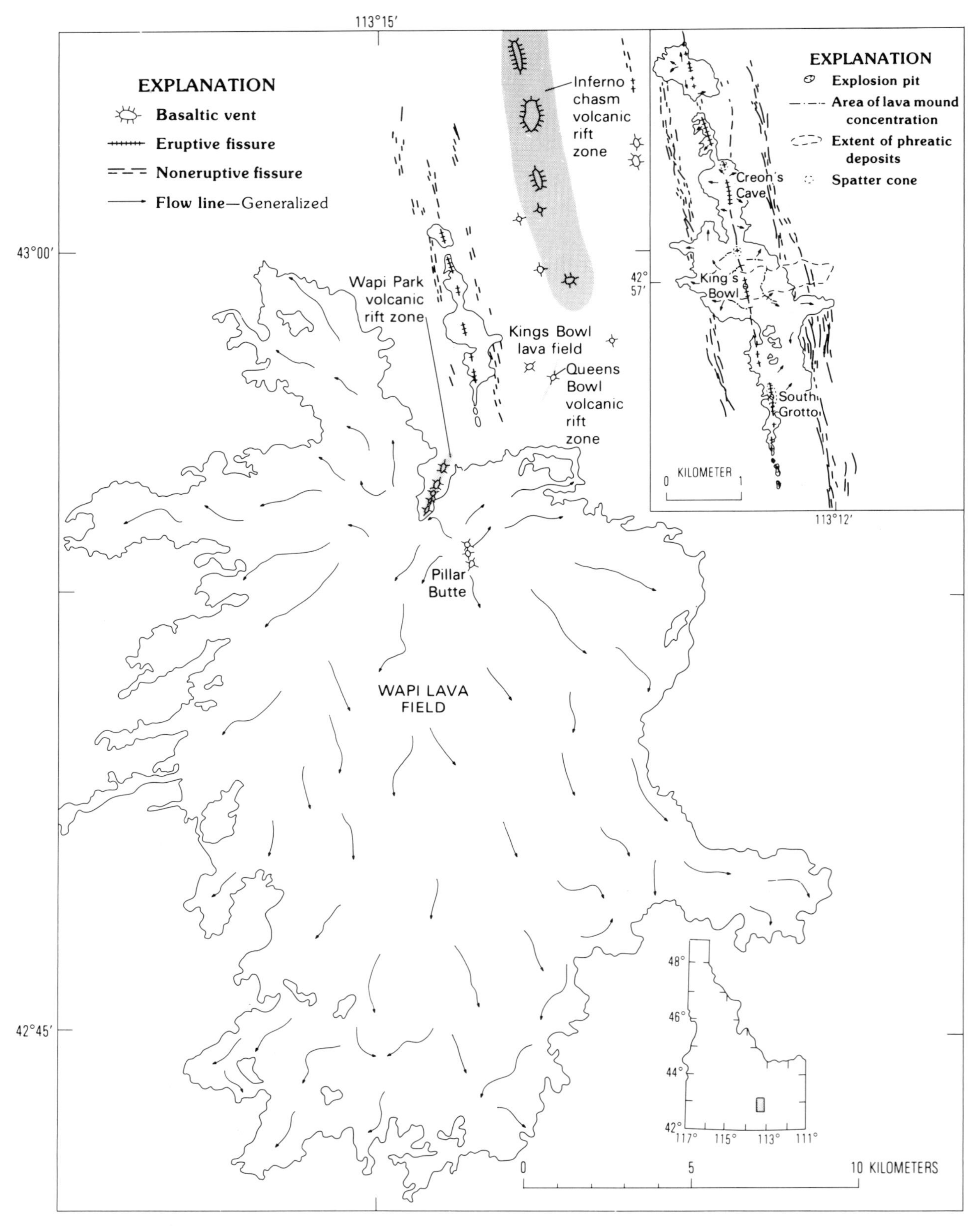

Figure 5. Map of the Wapi and Kings Bowl lava fields. Inset shows greater detail of the Kings Bowl lava field. Also shown are volcanic rift zones discussed in the text. Adapted from King (1977, 1982), Covington (1977), and Kuntz and others (1983, 1988).

along short eruptive fissures at South Grotto and Creons Cave (Fig. 5). Low spatter ramparts occur along a few of the eruptive fissure segments.

The small Kings Bowl lava field possesses excellent examples of certain volcanic landforms that are either poorly developed or simply not present elsewhere in the ESRP. These include explosion pits, lava lakes, squeeze-ups, basalt mounds, and a tephra blanket. The largest explosion pit, Kings Bowl, is 85 m long, 30 m wide, and 30 m deep. A roughly circular lava lake, about 1 km in diameter, surrounded Kings Bowl prior to the explosive eruption. The lake is flat and broken into large plates as a result of subsidence of underlying lava. Basalt mounds several meters wide and <1 m high may be remnants of levees that surrounded the lava lake (King, 1977). A blanket of lapilli tephra east (downwind) of the explosion pit at Kings Bowl is <0.5 m thick, 1.2 km long, and 400 m wide (Fig. 5, inset). King (1977) showed that the volume of the tephra blanket falls short of that needed to occupy the cavity of Kings Bowl, indicating collapse in the vent area subsequent to the explosive event.

The Kings Bowl field represents a single burst of eruptive activity that may have lasted as little as six hours (Kuntz, this volume). A radiocarbon age of 2,222 ± 100 B.P. has been obtained for the Kings Bowl lava field (discussed by Kuntz, Spiker, and others, 1986).

Wapi lava field

Marked contrasts between the Kings Bowl and Wapi fields reflect different styles of eruption. The eruptions that formed the Kings Bowl and Wapi fields probably started simultaneously, but eruption terminated at Kings Bowl after only a few hours and continued at the Wapi site for at least a few months (Kuntz, this volume). The Wapi lava field probably began as a fissure eruption and, with more prolonged activity, progressed to a sustained eruption from a central vent complex, forming a low shield volcano typical of a third-stage eruption.

The vent area for the Wapi field is 4 km south of the southernmost vents of the Kings Bowl lava field (Fig. 5). The Wapi vents are part of the Great Rift volcanic rift zone, but they do not lie on an extension of the fissures of the Kings Bowl rift system. A complex of five major vents and six smaller vents, covering an area of ~0.5 km^2, constitutes the vent area. The vents are steep-sided, roughly circular depressions typically 100 m in diameter and 10 m deep. The largest depression contains several pit craters that display the ledges of lava lakes that filled the crater. Lava tubes are exposed in walls of some of the depressions. Flows of the summit area are chiefly shelly pahoehoe and minor slab pahoehoe. Pillar Butte, a mass of agglutinate and layered flows, rises 18 m above the south side of the largest vent. The origin of Pillar Butte is unclear; its relief is due either to inflation by injection of dikes or to deflation of the area around it (Champion, 1973).

Most flows in the medial and distal parts of the field are tube-fed pahoehoe composed of numerous flow units piled side by side and atop one another, forming a type of lava flow described by Walker (1972) as compound. Exposures in many kipukas along the south and west sides of the lava field show that the flows there have an aggregate thickness of 5 to 10 m. Flow thicknesses near the center of the field are 15 to 25 m except at the vent area, where the total thickness may be 100 m (Champion, 1973). Near the margins of the field, the flow units are larger, tend to have greater local relief (as much as 10 m), and are characterized by large pressure plateaus, flow ridges, and collapse depressions. The transition in size of the flows from the periphery to the interior of the field is apparently a function of proximity to the vent area; thus, closer to the vent, many small pahoehoe flows have filled depressions in earlier flow units and generally leveled the local relief.

A radiocarbon age of 2,270 ± 50 B.P. was obtained for charcoal from a tunnel excavated beneath the Wapi lava field (Champion and Greeley, 1977; Kuntz, Spiker, and others, 1986). Within limits of analytical error, the ages of the Kings Bowl and Wapi lava fields are identical.

Shoshone lava field

The Shoshone lava field is the westernmost of the latest Pleistocene and Holocene lava fields of the ESRP (Fig. 6). Its vent area, Black Butte Crater, is on the northern margin of the ESRP about 20 km northwest of Richfield at the eastern end of the Mt. Bennett Hills. Flows from Black Butte Crater filled parts of the alluvial valleys of the Big Wood and Little Wood rivers, thus forming a lava field that extends 60 km south and west and ranges in width from 2 to 5 km. The Big Wood and Little Wood rivers joined about 10 km southwest of Richfield prior to eruption of the Shoshone flow. At present, the Big Wood River follows the north margin and the Little Wood River follows the south margin of the Shoshone field. These two rivers now join 10 km west of Gooding, 40 km west of their original junction (Fig. 6).

The summit vent area is a complex lava lake that forms a six-petal, flower-shaped depression. The lake has an area of 2 km^2, and its steep walls are as high as 30 m. The lava lake shows evidence of pistonlike filling and draining and extensive overflow by degassed lava. The overflow formed broken shelly-pahoehoe and slab-pahoehoe flows that are distributed in a radial pattern around the lava-lake vents.

There are no obvious features in the vent area, such as eruptive fissures, spatter ramparts, or small cones, that might indicate the length of the original eruptive fissure system. Evidence is also lacking for any structural control for the orientation of the original fissure system, but faults in the Mt. Bennett Hills and the elongation of vents for nearby older volcanoes trend N.35–50°W. (Fig. 6).

A 5-km-long lava tube-channel system extends southeast from the vent area (Fig. 6). The tube system had both roofed and unroofed sections during formation of the field. Many of the formerly roofed parts have collapsed, leaving open trenches that

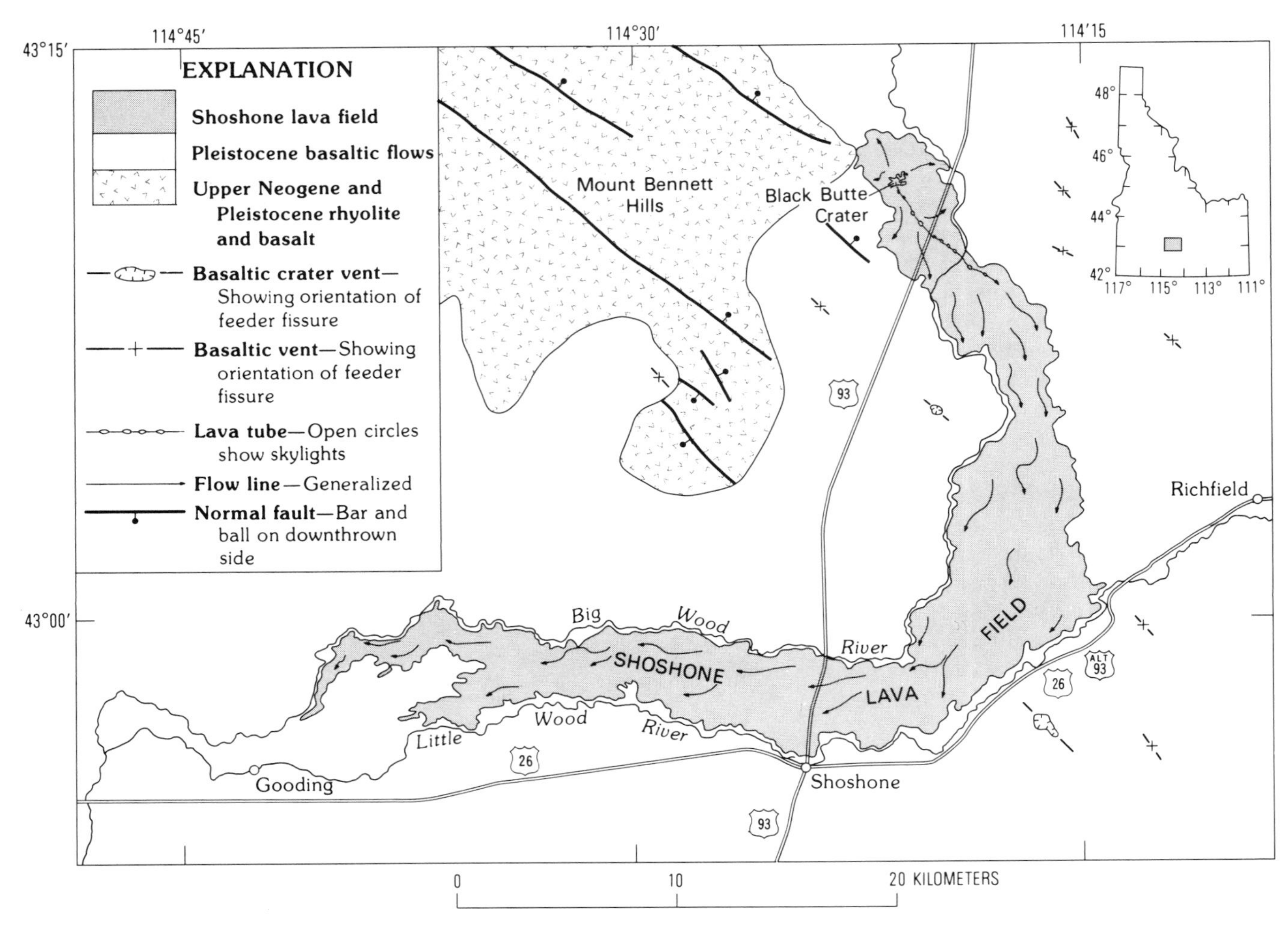

Figure 6. Map of the Shoshone lava field. Based on LaPoint (1977) and unpublished mapping by M. A. Kuntz.

are continuous with the unroofed sections. Smaller distributary tubes that branch from the main tube are still preserved (Greeley, 1977).

The medial and distal parts of the Shoshone lava field consist of pahoehoe flows and flow units that have an average thickness of 6 m. Whether these flows are tube- or surface-fed is not clear. Lava tube systems have not been recognized in medial and distal parts of the lava field, but fields of this size and volume are typically formed of tube-fed pahoehoe. The age of the Shoshone field is 10,130 ± 350 B.P. (Kuntz, Spiker, and others, 1986).

Summary of Holocene and latest Pleistocene basaltic lava fields

The eight latest Pleistocene and Holocene lava fields of the ESRP represent the entire span of lava-field types, eruption processes, and eruption durations described in the section on stages of basaltic eruptions. The fields include short-duration, low-volume eruptions from a long eruptive-fissure system (Kings Bowl, North Robbers, South Robbers fields); intermediate-volume and intermediate-duration eruptions that formed lava cones or small shields (for example, the Holocene Devils Cauldron flow of the COM lava field; see Kuntz, Champion, and Lefebvre, 1989; Fig. 3); and large-volume, relatively long duration eruptions from central vents (Hells Half Acre and Wapi fields).

Individual eruptive fissures range in length from 0.1 km to slightly more than 1 km and average about 0.5 km. The data are highly skewed toward small-volume, eruptive-fissure-dominated eruptions, because those eruptions tend to leave eruptive fissures unburied, whereas large-volume eruptions do not.

Eruptive fissure zones vary in length from <0.3 km to about 10 km; the average is about 3 km. The shorter zones produced small-volume eruptive-fissure eruptions, and the longer ones generated large-volume lava-cone or shield-building eruptions. Sustained eruptions occurred from central vents overlying pipe-like conduits along eruptive fissures. However, the elongation of most vents in lava cones and shield volcanoes shows that eruptive fissures still influenced the location and shape of these central vents.

The areas, volumes, and types of lava fields of the ESRP are functions chiefly of the eruptive stage at which the eruption terminated. Stage-1 fissure eruptions are characterized by flow areas of 2 to 20 km^2, flow volumes of 0.005 to 0.1 km^3, and surface-fed pahoehoe, much of which is shelly pahoehoe near vents. In stage-2 eruptions, lava cones cover areas of 10 to 100 km^2, have volumes of 0.05 to 2 km^3, and consist mostly of surface- and channel-fed pahoehoe. Some large lava cones consist chiefly of tube-fed pahoehoe. Shield volcanoes of stage-3 eruptions cover areas of 100 to 400 km^2, have volumes of 1 to 7 km^3, and consist mainly of tube-fed pahoehoe.

Among these latest Pleistocene and Holocene lava fields (Table 1), the aggregate volume from the stage-1 fissure eruptions (North Robbers, South Robbers, and Kings Bowl) is only 0.085 km^3, whereas the stage-3 shield and lava-cone eruptions (Craters of the Moon, Hells Half Acre, Cerro Grande, Wapi, and Shoshone) produced about 45 km^3. The contrast between these figures illustrates the predominance of the latter type of volcanism in contributing to the total basalt volume of the ESRP (Table 1). The latest Pleistocene and Holocene lava fields cover about 2,600 km^2 (13%) of the 20,000 km^2 area of the ESRP.

VOLCANIC RIFT ZONES

Basaltic volcanism occurs as fissure eruptions within volcanic rift zones in many parts of the world, including Hawaii (Macdonald and Abbott, 1970), Iceland (Macdonald, 1972), and the ESRP (Kuntz, 1977a). The volcanism of the ESRP is concentrated in volcanic rift zones, though it is not confined to them.

In the ESRP, most volcanic rift zones are roughly perpendicular to the axis of the ESRP and roughly parallel to the major basin-and-range structures in mountains that bound the plain. Kuntz (1977a) noted that most of the volcanic rift zones appear to be continuations onto the ESRP of basin-and-range-type, range-front faults and older structures, such as thrust faults, which occur in the bedrock of the mountains outside the ESRP (Figs. 7 and 11). It is important to note that the term *volcanic rift zone,* as applied here, merely designates linear arrays of volcanic landforms and structures; it does not imply that these features have origins, histories, or structures like those of rift valleys.

Some volcanic rift zones of the ESRP are well-defined, narrow, linear to curvilinear belts that contain numerous, geologically young volcanic structures and landforms. These zones illustrate vigorous, relatively concentrated eruptive activity. Examples are the Great Rift and Spencer–High Point volcanic rift zones (Fig. 7; also refer to Figs. 8 and 9). Other volcanic rift zones are less well defined; their presence is indicated by fissure-controlled elongated vents and noneruptive fissures aligned in belts. These zones illustrate sporadic, dispersed eruptive activity. For example, the Lava Ridge–Hells Half Acre zone and other volcanic rift zones in the central part of the ESRP are belts that consist of fewer, relatively older structures and volcanic landforms (Fig. 7).

Radiometric data currently available for ESRP lava flows (Kuntz and others, 1992) suggest that the volcanic activity in the well-defined zones occurred over a period of a few tens or hundreds of thousands of years and that repose intervals between eruptions are on the order of 10^4 to 10^5 years. The radiometric data further suggest that the less well defined volcanic rift zones formed from eruptions that occurred over periods of several hundred thousand to a few million years and that repose intervals between eruptions are on the order of 10^5 to 10^6 years.

Volcanic rift zones of the ESRP typically are 5 to 20 km wide and 50 to 100 km long, are spaced roughly 35 km (range of 20 to 40 km) apart, and trend northwest (Fig. 7). The spacing between rift zones is similar to the 35-km spacing between Basin and Range mountains and their intervening valleys north and south of the ESRP. Two exceptions to the general northwest trend are the Rock Corral Butte and Spencer–High Point volcanic rift zones (Fig. 7; also refer to Figs. 8 and 9).

Some of the major volcanic rift zones of the ESRP are described below in order from west to east.

The Great Rift volcanic rift zone

The Great Rift (Russell, 1902; Stearns, 1928; Murtaugh, 1961; Prinz, 1970; Kuntz and others, 1982, 1983, 1988; Kuntz, Champion, and others, 1986) is the most important volcanic rift zone in the Snake River Plain. It is an 85-km-long, 2- to 15-km-wide system of fractures that extends nearly across the width of the Snake River Plain (Figs. 7 and 8). The Great Rift is defined by an array of tephra cones, shield volcanoes, lava cones, eruptive fissures, and noneruptive fissures. The latest Pleistocene and Holocene Craters of the Moon, Kings Bowl, and Wapi lava fields are aligned along the Great Rift volcanic rift zone (Kuntz and others, 1988; Fig. 8).

The vents of the COM field define the northern 45 km of this volcanic rift zone. These vents include about 25 tephra cones and more than eight eruptive-fissure systems (Figs. 3 and 8). The Great Rift volcanic rift zone extends northwest into the Pioneer Mountains; the tephra-cone vents of the Lava Creek flows of the COM field occur about 10 km beyond the topographic margins of the ESRP (Fig. 3).

The Great Rift continues from the southeastern margin of the Craters of the Moon lava field as two groups of noneruptive fissures (Figs. 3 and 8). The northeastern group consists of individual fissures 150 to 600 m long. This group of fissures is oriented N.45°W. and extends about 8 km beyond the southeast margin of the COM lava field. The southwestern group consists of fissures 100 to 1,000 m long, oriented N.25°W., that extend for about 15 km beyond the southeast margin of the COM field. The eastern and western groups of fissures constitute the open crack rift set of Prinz (1970). The Kings Bowl eruptive and noneruptive fissures, described previously in the discussion of the Kings Bowl lava field, constitute the main part of the Great Rift volcanic rift zone just south of the open crack rift set (Figs. 5 and 8).

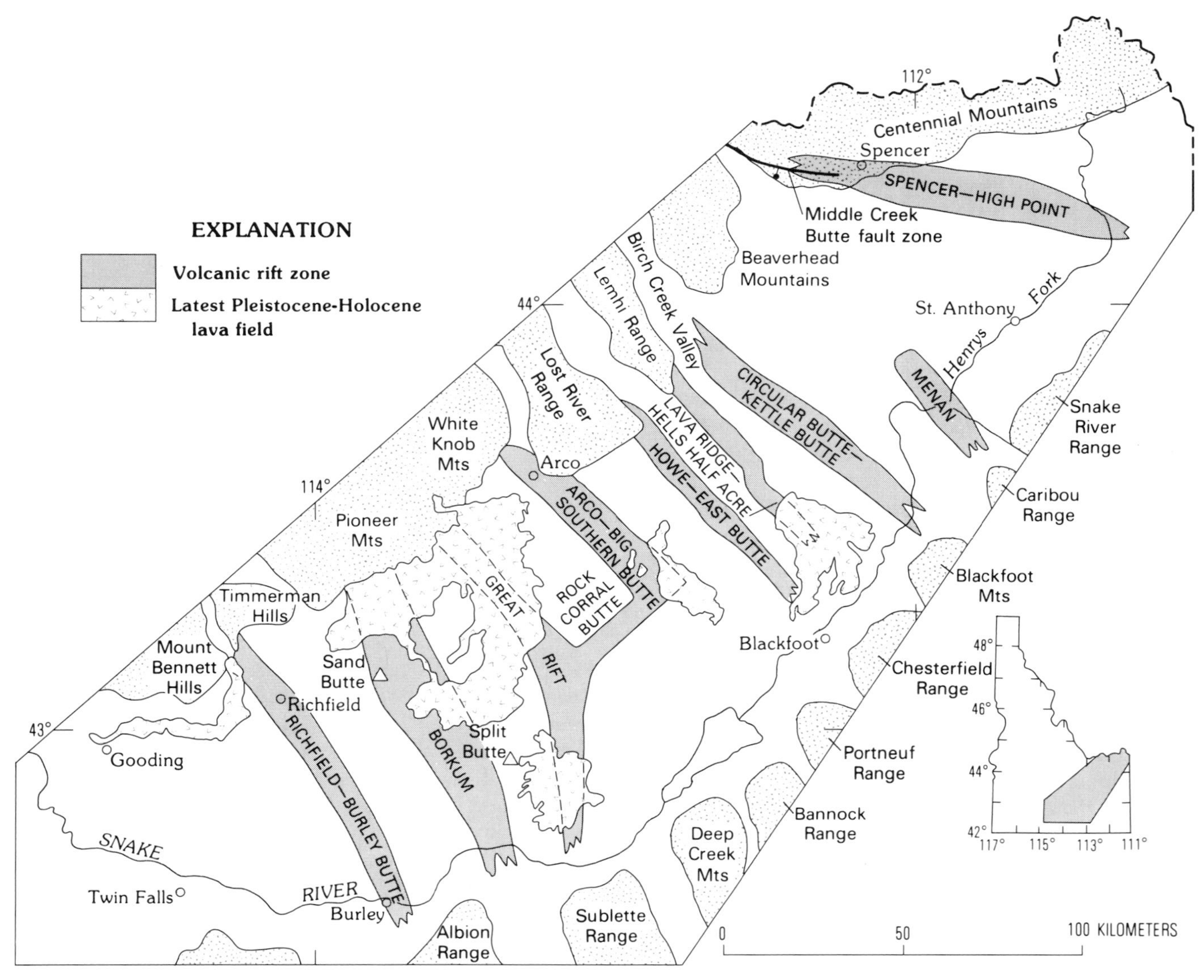

Figure 7. Map of the eastern Snake River Plain showing volcanic rift zones and other geologic and geographic features discussed in the text. Also shown are two hydrovolcanic features—Sand Butte and Split Butte—and latest Pleistocene and Holocene lava fields.

The Inferno Chasm volcanic rift zone (Greeley and King, 1975; King, 1982) is a narrow, 10-km-long, N.10°W.-trending belt that is parallel to and lies 3 km east of the Kings Bowl volcanic rift zone (Fig. 5). The Inferno Chasm zone is marked by six aligned lava cones and shield volcanoes. Flows from vents on the Inferno chasm zone are covered by significantly greater amounts of eolian sediment and vegetation than flows from other latest Pleistocene and Holocene vents on the Great Rift, showing that the Inferno Chasm segment is older than other nearby segments of the Great Rift volcanic rift zone. Greeley and King (1975) and King (1982) identify two other short (<2 km), N.10°E.-trending volcanic rift zones in the Kings Bowl area: the Wapi Park and Queens Bowl zones (Fig. 5). They suggest that both of these north-northeast–trending volcanic rift zones formed before the northwest-trending eruptive and noneruptive fissures of the Great Rift volcanic rift zone.

The Arco–Big Southern Butte volcanic rift zone

The Arco–Big Southern Butte volcanic rift zone extends 50 km southeast from the northwest margin of the ESRP at Arco to the long axis of the ESRP southeast of the Cedar Butte lava field (Figs. 7 and 8). This zone is a locus of extensional faults, grabens, eruptive fissures, and noneruptive fissures. Volcanic features of the volcanic rift zone include fissure-fed basalt flows, the rhyolite dome complex of Big Southern Butte, the andesite volcano at Cedar Butte, and the cones and eruptive-fissure vents of the North

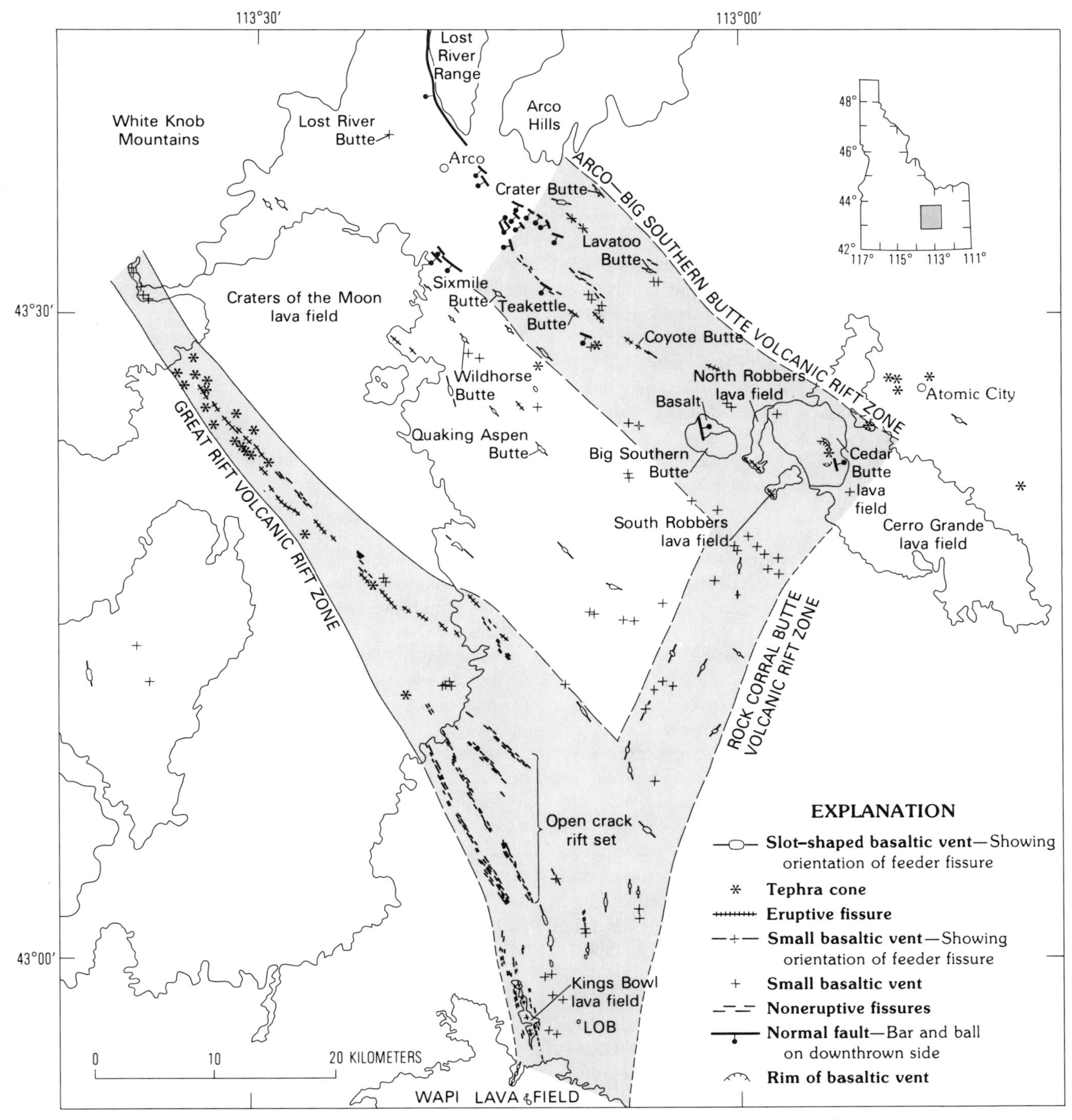

Figure 8. Simplified map showing volcanic and structural features of the Great Rift, Arco–Big Southern Butte, and Rock Corral Butte volcanic rift zones. Adapted from Kuntz and others (1988, 1992).

Robbers, South Robbers, and Cerro Grande lava fields (Spear, 1977; Spear and King, 1982; Kuntz, 1977b; Kuntz and others, 1992). The Arco–Big Southern Butte zone terminates at the vent for the Cerro Grande field, where it joins the northeast-trending Rock Corral Butte volcanic rift zone (Figs. 7 and 8).

Faults, noneruptive fissures, and air-photo lineaments are abundant and well developed along the Big Lost River 3 to 10 km southeast of Arco (Smith and others, 1989; Kuntz and others, 1992; Fig. 8). At that locality, extensional faults having as much as 10 m vertical displacement formed a graben that controls the course of the Big Lost River. Most of these faults show good scarps. At the ends of some faults, however, the surface basalt is unbroken but shows substantial drag across the linear extensions of surface ruptures. The faults range in length from several tens of meters to as much as 4 km. Most scarps are relatively fresh, and the few slickensides observed suggest a purely dip-slip component of movement. Displacements on faults generally decrease to the southeast, and 15 km southeast of Arco, the faults are replaced by open fissures that have purely tensional displacement. Between Arco and Big Southern Butte, northeast-trending lineaments, apparent on aerial photographs, reflect eolian features and prairie-fire scars. Some lineaments having northwest trend, however, are obvious continuations of fissures and faults. Other northwest-trending lineaments are elongated depressions in surficial sediments that are parallel to nearby fissures and faults. Still other northwest-trending features are defined only by luxuriant growth of grass or sagebrush and lack a surface depression or other obvious relation to bedrock deformation. However, all lineaments of the latter type parallel known faults and fissures, suggesting that they too are the surface expressions in sediments of underlying faults and fissures.

Volcanoes in the Arco–Big Southern Butte volcanic rift zone represent all stages of basaltic volcanism. Fissure-fed lava flows occur southeast of Big Southern Butte in the North Robbers and South Robbers lava fields and at Coyote Butte northwest of Big Southern Butte. Small lava cones (Lavatoo Butte, Teakettle Butte, and Lost River Butte) occur at the northwest end of the volcanic rift zone (Fig. 8). These cones are 1 to 5 km in diameter and 50 to 350 m high. Their summits are indented by craters elongated parallel to the long axis of the volcanic rift zone. The largest volcanoes in the Arco–Big Southern Butte volcanic rift zone are shield volcanoes along the margins of the zone. These are the Quaking Aspen Butte, Crater Butte, Sixmile Butte, Wildhorse Butte, and Cerro Grande shield volcanoes (Fig. 8).

Although rhyolitic volcanism and rhyolitic rocks are not the focus of this paper, the Big Southern Butte dome complex must be mentioned because it is an integral part of the Arco–Big Southern Butte volcanic rift zone (Fig. 8). Big Southern Butte consists of two coalesced cumulo domes, both about 300 ka (Spear and King, 1982; Kuntz and others, 1992). The dome complex is 6.5 km in diameter, rises 760 m above the relatively flat surface of the surrounding ESRP, and has an exposed volume of about 8 km^3. Spear and King (1982) suggest that Big Southern Butte formed in several stages, including initial sill and laccolithic stages at depth, followed by extrusion and growth of two endogenous domes on the surface. The southeastern dome consists of spherulitic, flow-banded rhyolite in the core and autoclastic breccias and sugary rhyolite that represent deformed crust above the core. The northwestern dome consists of massive, aphyric rhyolite that formed along the northwest margin of the earlier dome.

A 350-m-thick section of basalt that dips N.45°E. is exposed on the north flank of Big Southern Butte (Fig. 8). Spear and King (1982) suggest that this section is an uplifted and tilted flap of basalt flows of the pre-dome surface of the ESRP. Fifteen to 20 individual flows and flow units are present in the flap. Most flows are olivine basalts and evolved olivine basalts; the uppermost flow is ferrolatite from Cedar Butte volcano.

Cedar Butte, a steep-sided shield volcano, is between Big Southern Butte and the vent area for the Cerro Grande lava field (Figs. 4 and 8). The summit of the shield is defined by several arcuate ridges that outline a vent area 3 km long and 2 km wide. The elongated vent trends N.40°W. A younger pyroclastic cone consisting of rhyolite and obsidian blocks occupies the center of the vent complex. A 2-km-long, northeast-facing scarp of a normal fault extends across the southeast flank of Cedar Butte toward the vent area of the Cerro Grande lava field (Figs. 4 and 8).

Volcanic rift zones in the central part of the ESRP

Volcanic rift zones in the central part are less well defined than zones elsewhere in the ESRP. Because most lava fields in this area are older (300 to 600 ka; Kuntz and others, 1992) than other ESRP fields, many features of these volcanic rift zones have been covered by lava flows and eolian deposits (Pierce and others, 1982).

The Lava Ridge–Hells Half Acre zone is a good example of volcanic rift zones in the central part of the ESRP. This zone extends 50 km southeast across the ESRP from the southern end of the Lemhi Range to the Hells Half Acre lava field (Fig. 7). The southeast end of the zone is well defined by open fissures and the vent complex of the Hells Half Acre lava field. The central part is defined by several small- to medium-sized shield volcanoes near the axis of the ESRP. Vents for these volcanoes are elongated north-south; thus the rift zone is offset in a right-lateral sense in its central part (refer to Fig. 17). The northwest end of the zone is ill defined: Poorly exposed flows of reversed magnetic polarity (older than 730 ka) are mantled by relatively thick fluvial and lacustrine deposits and eolian sediment. Volcanic activity on the Lava Ridge–Hells Half Acre volcanic rift zone is oldest (mostly more than 730 ka) at the northwest end, of intermediate age in the central part (200 to 500 ka), and youngest (about 5 ka) at the southeast end of the zone at the Hells Half Acre lava field (Kuntz and others, 1992).

Spencer–High Point volcanic rift zone

The Spencer–High Point (S-HP) volcanic rift zone occupies the northeastern end of the ESRP (Figs. 7 and 9). The zone extends 70 km from Indian Creek Buttes to High Point, a basaltic vent on Big Bend Ridge. Indian Creek Buttes are rhyolite domes (4.14 ± 0.085 Ma; G. B. Dalrymple, U.S. Geological Survey, written communication, 1979), each about 2 km in diameter, along the Middle Creek Butte fault zone at the western end of the S-HP zone. A few small basalt vents lie along the fault zone near the rhyolite domes (Fig. 9). The main part of the S-HP volcanic rift zone begins where the Middle Creek Butte fault zone enters the ESRP near Spencer (Figs. 7 and 9). There, faults, grabens, a few open fissures, and the feeder fissures for tephra cones, spatter

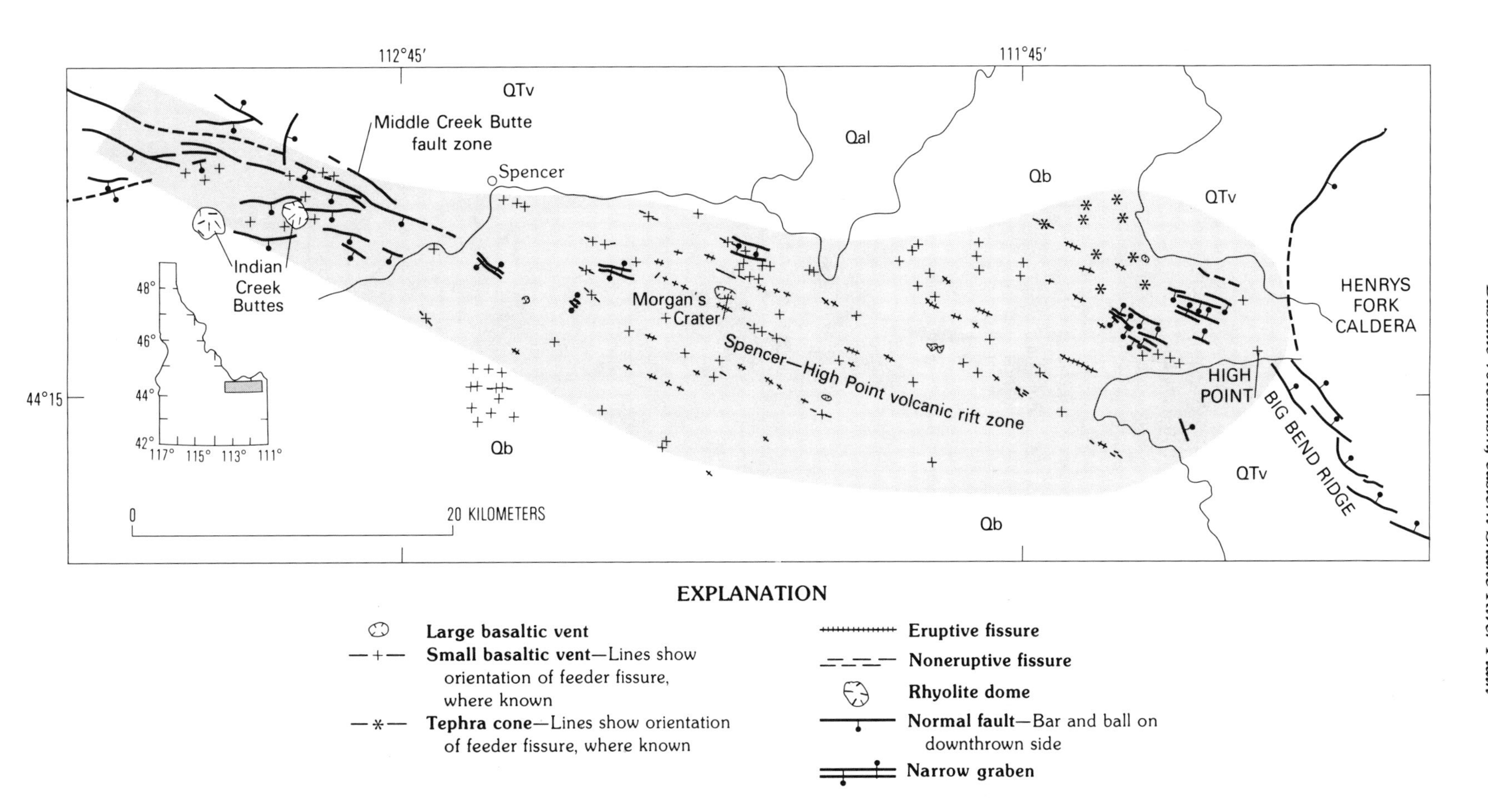

Figure 9. Simplified map showing volcanic and structural features of the Spencer–High Point volcanic rift zone. Abbreviations: QTv, Quaternary and Tertiary rhyolitic and basaltic volcanic rocks; Qb, Quaternary basalt; Qal, Quaternary alluvium. Adapted from Christiansen (1982) and unpublished mapping by M. A. Kuntz, G. F. Embree, and H. J. Prostka.

ramparts, and small shield volcanoes all trend about N.60°W. These structures occur in an east-trending zone about 15 km wide and 55 km long.

The vents and lava fields of the S-HP zone are smaller than other vents and fields in the ESRP. Lava fields in the S-HP zone are mainly of eruptive-fissure and lava-cone types. The flows are typically long and narrow because they moved chiefly south down relatively steep slopes (1 to 2°). Most lava fields cover areas of 10 to 100 km^2 and have volumes of 0.05 to 1 km^3. The larger areas and the higher volumes represent a few lava cones and small shields at the eastern end of the zone. Some of the youngest lava fields of the S-HP zone are believed to be latest Pleistocene or Holocene (<15 ka) in age based on their youthful surface morphology. Suitable material for radiocarbon dating of these fields has been sought but not obtained.

About a half-dozen tephra cones occur near the eastern end of the zone between Morgan's Crater and High Point (Fig. 9). Two factors may explain the restriction of the cones to this locality: (1) Magmas that fed the tephra cones are slightly more siliceous and probably were more viscous than other basalts of the S-HP volcanic rift zone, and (2) magmas in this area passed through alluvial, water-bearing sediments deposited by streams that ponded north of the topographic barrier created by the volcanic rift zone.

The S-HP volcanic rift zone is unique in several respects: (1) It contains the highest concentration of volcanic vents in the ESRP (Fig. 10); (2) it is a wide volcanic rift zone when compared to its length (15 km wide in the 55-km-long stretch between Spencer and High Point); (3) its east-west trend is unlike that of the other volcanic rift zones in the ESRP, most of which trend northwest-southeast; and (4) the vents and structures within the zone have a trend (northwest-southeast) different from that of the zone itself (Fig. 9).

The northwest trend of en echelon volcanic structures in the

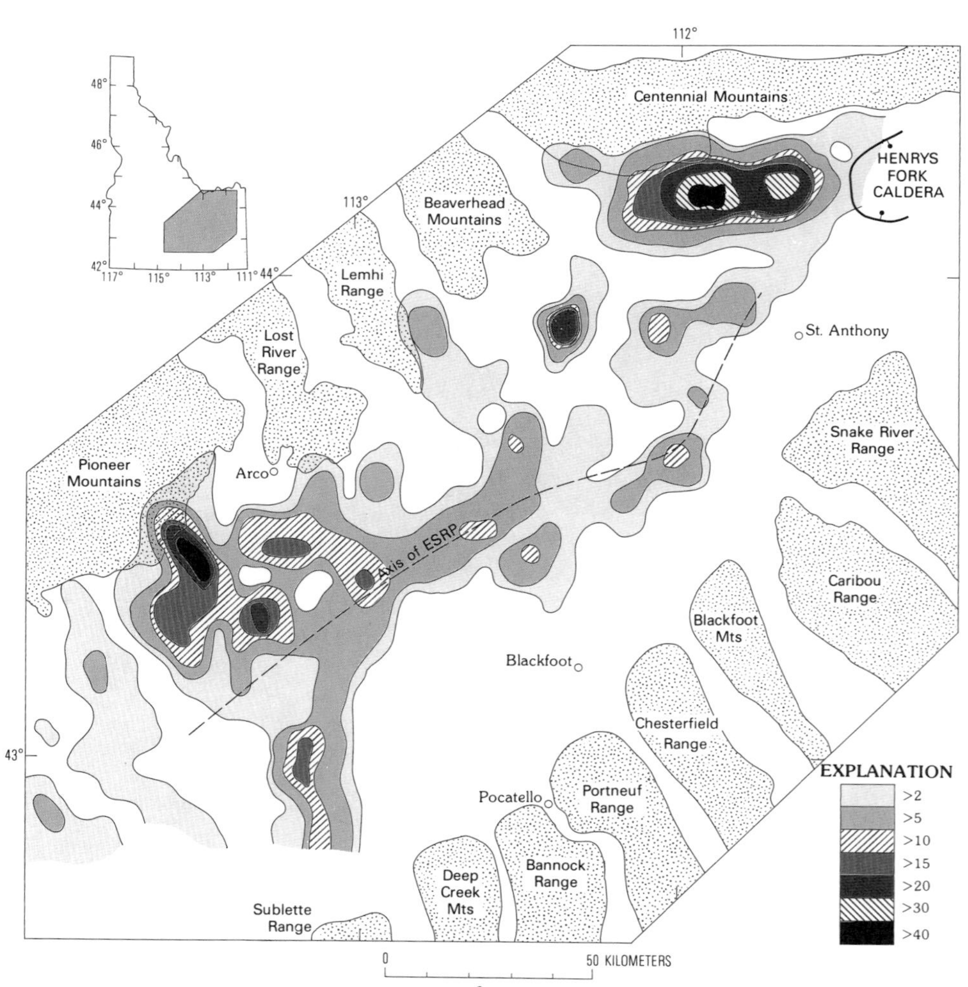

Figure 10. Number of volcanic vents per 140 km^2 (approximate area of a 7½-minute quadrangle) in eastern Snake River Plain.

S-HP volcanic rift zone may result from a component of right-lateral movement on the Middle Creek Butte fault, which produced extension fractures in a broad zone in the overlying flows.

Axis of the ESRP

The axis of the ESRP, from Big Southern Butte to Juniper Buttes, is a broad topographic ridge formed by the accumulation of basaltic lava flows from vents concentrated along it and from local uplift that accompanied emplacement of rhyolite domes. However, the axis of the ESRP is not a volcanic rift zone because the orientations of fissures that control the location and trend of the fissure-dominated vents are perpendicular, rather than parallel, to the axis. Local concentrations of basaltic vents occur along the axis where it crosses volcanic rift zones.

The axis is a belt of basaltic volcanoes and several rhyolite domes in an area about 10 km wide and 100 km long (Fig. 10). The domes and their ages, from west to east, are: Big Southern Butte (300 ka), Middle Butte (<1 Ma), an unnamed dome (1.4 Ma), and East Butte (580 ka) (Kuntz and others, 1992; refer to Fig. 17).

Magnetotelluric soundings (Stanley and others, 1977) and Curie isotherm maps for the ESRP (Bhattacharya and Mabey, 1980) indicate that a ridge of high temperature at shallow depth is roughly parallel to the axis of the ESRP (D. Mabey, oral communication, 1979). Evidently, excess heat resulting from the formation and upward transfer of basalt magma is still present in the deep crust beneath the axis of the ESRP. However, the movement to the surface generally follows fractures oriented perpendicular to the axis. The vertical transfer of heat may also be responsible for melting the deep crust and producing rhyolitic magmas and the rhyolite domes (e.g., Leeman, 1982d).

REGIONAL RELATIONSHIPS OF VOLCANIC RIFT ZONES

Relationships between volcanic rift zones and structures beyond the margins of the ESRP

Based on ideas developed by Stearns (1928), Stearns and others (1938), Prinz (1970), LaPoint (1977), and H. J. Prostka (unpublished) and on his own fieldwork, Kuntz (1977a, 1977b, 1978) suggested that most volcanic rift zones of the ESRP are collinear continuations onto the ESRP of range-front faults in mountains beyond the margins of the plain (Fig. 11). The outlying range-front faults are normal and listric faults that produce fault scarps and steep mountain fronts. Within the ESRP, the continuations of these faults are represented by the structures and volcanic landforms of volcanic rift zones, including normal faults, eruptive and noneruptive fissures, grabens, volcanic vents at the surface, and dikes at depth (Figs. 8 and 9). It is not clear that the range-front faults continue onto the ESRP in bedrock beneath the basaltic lava flows. The three-dimensional relationship between the range-front faults and the structures of the volcanic rift zones can only be inferred from the surface expression of these structures and their inferred arrangement and attitudes at depth. What seems clear at present is that regional extensional strain is manifested in basin-and-range faulting and topography in mountains north and south of the ESRP, but within the ESRP it is expressed only in minor normal faulting, emplacement of dikes, formation of fissure-controlled basaltic eruptions, and development of structures related to eruptions (e.g., faults, grabens, noneruptive fissures) in volcanic rift zones.

Examples of volcanic rift zones as collinear continuations of range-front faults are the Arco–Big Southern Butte, Lava Ridge–Hells Half Acre, and Menan volcanic rift zones, which are collinear with range-front faults of the Lost River, Lemhi, and Caribou Ranges, respectively (Fig. 11). The Spencer–High Point volcanic rift zone is collinear with the Middle Creek Butte fault zone along the southwest margin of the Centennial Mountains (Figs. 9 and 11).

Relations between regional extension, volcanic rift zones, and magma production

Kuntz (this volume) proposes that strain release in the form of basin-and-range faulting and concomitant concentrated extension in volcanic rift zones leads to greater amounts of magma production in the lithospheric mantle, perhaps by decompression melting, which in turn increases the number of suitable fractures for the rise of magma to the surface. The mechanical-thermal feedback nature of this mechanism suggests that volcanic rift zones of the ESRP are long-lived, self-perpetuating zones of volcanism that result from regional extensional tectonics.

We suspect that major periods of faulting near the margins of the ESRP correlate with major periods of volcanism on collinear volcanic rift zones. Two general conclusions support this relationship. (1) The youngest sites of volcanism on the northern margin of the ESRP (Great Rift and Spencer–High Point volcanic rift zones) are near sites of major Quaternary faulting (Pierce and Morgan, this volume; Fig. 11). (2) The area that has only a few, very old volcanic vents and poorly defined volcanic rift zones, along the southeast margin of the ESRP, is adjacent to Belt IV of Pierce and Morgan (this volume), defined by them as an area having range-front faults that were active in late Tertiary time but that show little or no evidence of Quaternary activity (Fig. 11). We conclude that strain release at various times along various basin-and-range faults near the margins of the ESRP also contributed to approximately contemporaneous formation of magma by decompression melting and basaltic volcanism in adjacent volcanic rift zones. Thus, volcanic rift zones may serve as strain recorders that mark the approximate times of basin-and-range–type faulting along the margins of the ESRP and of subsequent volcanism concentrated within the zones. Unfortunately, currently available ages of faulting and volcanism are not yet sufficiently accurate to allow us to investigate this relationship more fully.

Volcanic rift zones and regional tectonic stress

Basin-and-range faults beyond the margins of the ESRP and volcanic rift zones within the ESRP formed in the same regional

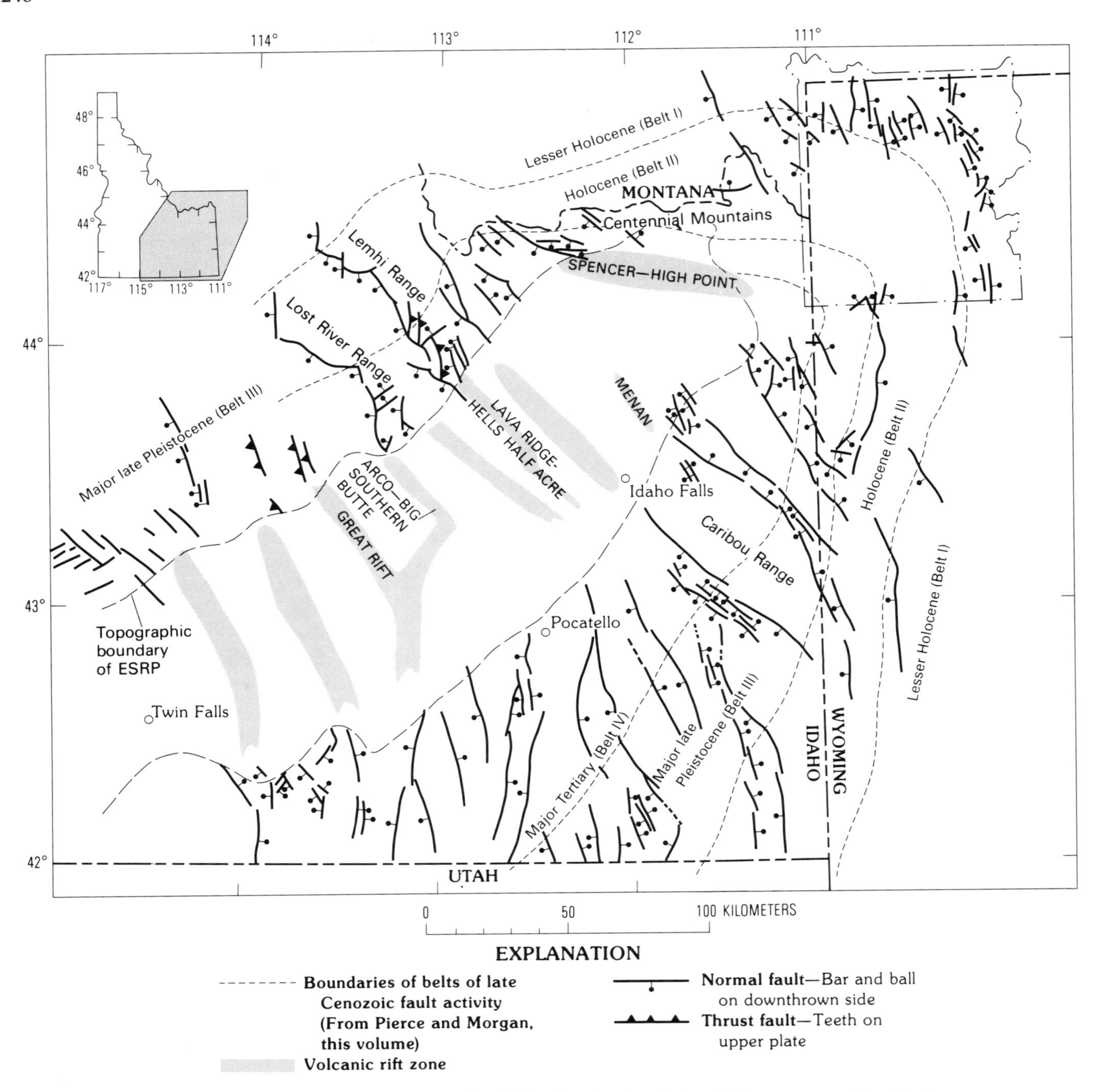

Figure 11. Map of the eastern Snake River Plain, showing the relationship between range-front faults and other structures beyond the margins of the ESRP and volcanic rift zones within the ESRP. Compiled from many sources.

stress system. The axis of least principal (extensional) stress in the region is presently oriented S.45°W. (Stickney and Bartholomew, 1987), which is consistent with regional stress directions identified by Zoback and Zoback (1980). Volcanic rift zones have acted as extensional strain concentrators in the ESRP, just as basin-and-range faults are the loci of regional strain in adjacent regions.

An important feature of the distribution of volcanic vents and volcanic rift zones in the ESRP is the absence of fissure-controlled, elongated vents and volcanic rift zones aligned parallel to and near the margins of the ESRP. If the ESRP were a fault-bounded rift valley, as suggested by some (see, for example, Corbett, 1975; Hamilton, 1987), basaltic vents and volcanic rift zones could be expected to be concentrated along and elongated parallel to boundary faults. On the contrary, volcanic rift zones of the ESRP cut at right angles to the postulated boundary-rift faults (Figs. 7 and 11). Although this observation does not preclude the existence of boundary faults along the margins of the ESRP, it makes their existence unlikely. Faults parallel to the margins of the ESRP have been described from several localities (see, for example, Rodgers and Zentner, 1988). These faults lack regional continuity and basaltic vents are not aligned along them, thus they do not qualify as boundary faults.

AGE OF BASALTIC VOLCANISM OF THE ESRP

Radiometric studies of basaltic lava flows of the ESRP have been undertaken only recently. Radiocarbon study of young lava fields has been an arduous, time-consuming task fraught with many difficulties (see Kuntz, Spiker, and others, 1986). K-Ar dating has been hampered by the low K_2O contents of the basaltic flows (typically less than 0.8%), the relative youth of the flows, and until recently, mass spectrometric and electronic limitations. In spite of these difficulties, about 30 reliable radiocarbon ages and 70 reliable K-Ar ages are presently available for surface and near-surface flows of the ESRP (summarized in Kuntz, Spiker, and others, 1986; Champion and others, 1988; Kuntz and others, 1992).

Field- and laboratory-determined paleomagnetic directions, combined with available radiometric ages, indicate that about 95% of the ESRP is covered by basaltic lava flows that are younger than 730 ka. Most flows have normal polarity and, therefore, were erupted in the Brunhes Normal-Polarity Chron. Surface flows known to be older than 730 ka occur near the mouth of Birch Creek Valley on the northwestern margin of the ESRP and along the northern part of the Lava Ridge–Hells Half Acre volcanic rift zone (refer to Fig. 17). Most flows in the central part of the ESRP are between 200 and 700 ka (Kuntz and others, 1992).

In general, the radiometric data suggest that eruptive activity was earlier and more sporadic near the margins and was later and more persistent near the center of the ESRP. Vents for the latest Pleistocene and Holocene basaltic volcanism are scattered along the entire length of the ESRP and most of them are concentrated near the plain axis. However, vents for the mostly Holocene COM lava field are concentrated near the intersection of the Great Rift volcanic rift zone and the northern margin of the ESRP.

Because vertical exposures are scarce and few core samples of flows have been dated radiometrically, little is known about the age of basalt flows at depth or about the rates at which flows accumulate in various parts of the ESRP. At each locality in the ESRP, all major basalt flows are considered to be younger than the rhyolitic rocks and their associated calderas. This relationship and the known ages of surface and near-surface basaltic flows indicate that most of the basaltic volcanism of the ESRP is Pliocene and younger.

PETROLOGY, PETROCHEMISTRY, AND PETROGENESIS OF BASALTIC LAVA FLOWS OF THE ESRP

Petrologic variability

Stone (1967, 1970) was the first to accurately assess the mineralogical and chemical variability of basaltic rocks of the Snake River Plain. He concluded that (1) they are highly uniform in field, textural, mineralogical, and chemical characteristics; (2) they are mainly olivine tholeiites in terms of mineralogical and chemical character; (3) they constitute a primary magma suite because of their great volume and comparatively uniform composition; (4) they were only slightly affected by magmatic evolution; (5) their minor diversity may reflect small differences in composition of parental magmas as a result of variation in depths of magma genesis; and (6) local compositional variations were produced by contamination by crustal rocks. Subsequent studies have modified but not changed Stone's basic conclusions.

At present, two major types of basaltic rocks are recognized in the Snake River Plain. The majority of lava flows of the Snake River Plain are *olivine basalts* that show little evidence of differentiation or contamination. Most consist of normative olivine and hypersthene (Fig. 12) and are olivine tholeiites in the classification of Yoder and Tilley (1962), but some are alkali olivine basalts based on mineralogical and thermodynamic classifications (Stout, 1975; Stout and Nicholls, 1977). Most basalts shown in Tables 3 and 4 and Figure 12 are olivine tholeiites (normative ol and hy); however, three of the most mafic COM rocks (Table 2), two of 22 analyzed samples from latest Pleistocene and Holocene lava fields (Table 3), and four of 24 olivine basalts of Pleistocene lava fields (Table 4) are slightly nepheline-normative (<3 molecular percent Ne) alkali olivine basalts (Fig. 12).

The second major type of rock is *evolved basaltic* rocks that have SiO_2 contents as high as 65%. Evolved basaltic rocks are now known from at least nine areas on or near the Snake River Plain (Leeman, 1982c). The major areas in the ESRP are the Craters of the Moon lava field, the Cedar Butte field, and the Spencer–High Point area. The most evolved rocks of the COM lava field are quartz normative (<10 molecular percent Q; Table 2 and Fig. 12).

Kuntz, Champion, and others (1986) distinguish three types of basaltic rocks in the COM, Kings Bowl, and Wapi lava fields on the Great Rift volcanic rift zone. The Snake River Plain type (SRPT), represented by rocks of the Kings Bowl and Wapi lava fields, comprises the undifferentiated, uncontaminated olivine basalts discussed above. A fractionated type is represented by COM rocks having an SiO_2 range of 44 to 54%. These rocks are fractionated with respect to SRPT olivine basalts, they mostly lack disequilibrium minerals and textures, and their chemical and

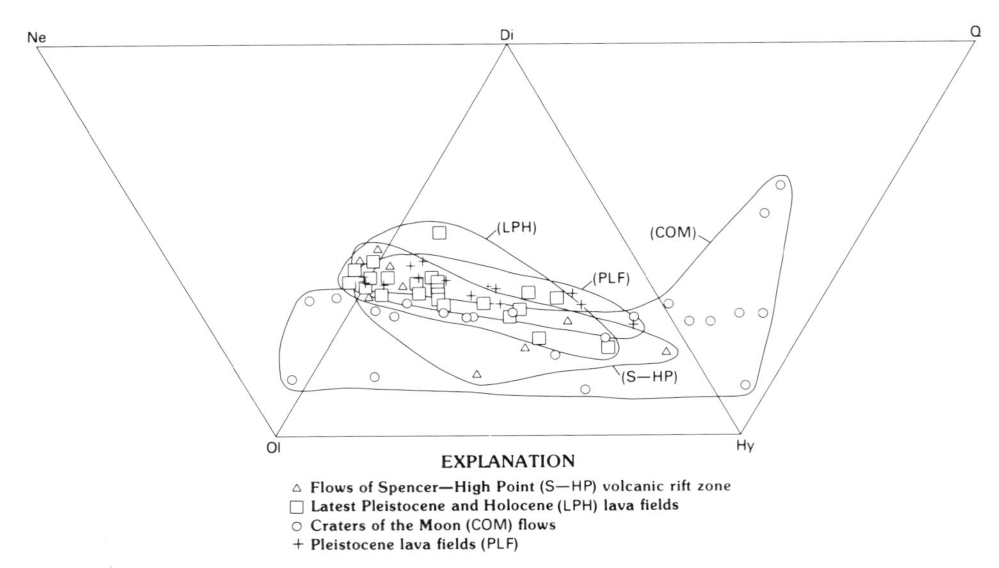

Figure 12. Normative Q-Hy-Di-Ol-Ne for the 70 analyses listed in Tables 2, 3, and 4.

mineralogical variation can be accounted for mainly by fractionation of observed mineral phases. A contaminated type is represented by COM rocks that have an SiO_2 range of 49 to 64%. These rocks commonly contain rhyolitic pumice inclusions that range from microscopic to several meters across, xenoliths of gneissic granulite, and xenocrysts of disequilibrium minerals.

Our field studies indicate that mantle xenoliths are absent in olivine basalts of the ESRP. Karlo (1977) found olivine gabbro xenoliths in rocks of the Hells Half Acre lava field, but these are probably cumulates of phenocryst phases common in Hells Half Acre rocks. Xenoliths and xenocrysts derived from Precambrian gneiss beneath the ESRP are present locally in olivine basalts (Leeman and Manton, 1971; Leeman and others, 1976; Menzies and others, 1984; Leeman and others, 1985; Futa and others, 1988) but are uncommon. The abundant inclusions of rhyolitic rocks in evolved rocks of the COM lava field are absent or extremely rare in olivine basalts of the ESRP. The absence of xenoliths of Upper Precambrian, Paleozoic, and Mesozoic sedimentary rocks (or their metamorphic equivalents) in all basaltic rocks of the ESRP suggests, but does not prove, that these sedimentary and metamorphic rocks do not underlie large tracts of the ESRP.

Petrography

The petrography of the varieties of basaltic rocks of ESRP has been described in detail elsewhere (e.g., Stone, 1967; Leeman, 1974, 1982b, 1982c; Stout and Nicholls, 1977; Kuntz, Champion, and others, 1986) and is reviewed here only briefly. Olivine basalts consist of olivine (Fo_{80-90}), a single clinopyroxene (augite to ferroaugite), plagioclase (An_{70-50}), titanomagnetite, ilmenite, and brown glass. Most samples have a coarse, intergranular texture: Crystals are typically 0.5 to 3 mm long. Other textures range from ophitic and subophitic to intergranular and diktytaxitic. Some samples are inequigranular and consist of 0.5- to 5-mm-long, euhedral and subhedral olivine and plagioclase crystals in a diktytaxitic, subophitic matrix of olivine, plagioclase, clinopyroxene, and opaque minerals $<$0.2 mm in largest dimension.

Evolved rocks of the contaminated type are typically hypocrystalline, and most of their crystals are $<$0.5 mm long. They generally contain crystals of olivine (Fo_{60-25}), plagioclase ($An_{\sim 25-60}$), clinopyroxene, titanomagnetite, and ilmenite in a matrix of brown glass. The rocks typically show some evidence of contamination in the form of disequilibrium textures (resorbed plagioclase, clinopyroxene, and anorthoclase crystals), disequilibrium minerals (hypersthene, anorthoclase), inclusions of rhyolitic pumice and gneissic granulite, and xenocrysts of clinopyroxene, zircon, anorthoclase, and apatite. Xenoliths and xenocrysts are far more abundant in evolved rocks than in olivine basalts, suggesting that the inclusions may be important factors in the chemical and mineralogical diversity of the evolved rocks.

Petrochemistry and petrogenesis

Olivine basalts. Olivine basalts of the ESRP have a fairly narrow range in major-element and minor-element chemistry, as shown in Tables 3 and 4. Covariation between various oxides and Mg value [MgO/(MgO+FeO*)] and between total alkalies and SiO_2 illustrates the chemical diversity of olivine basalts of the ESRP (Figs. 13 and 14). The diagrams also aid in recognizing mechanisms that may have played a role in creating the limited chemical diversity. Note that a coherent covariation trend characterizes the olivine basalts in Figure 13 and that olivine basalts form a tight cluster in Figure 14.

Petrochemical and experimental studies suggest that SRPT olivine basalts were derived by partial melting of spinel lherzolite mantle at depths of 40 to 60 km and that little fractionational crystallization occurred prior to or during eruption. The minor chemical variation in the olivine basalts (Figs. 13 and 14) can be accounted for by fractionation of olivine and plagioclase, which are phenocryst phases in these rocks. Varying degrees of partial melting of peridotite source rocks of the mantle have also been advocated to explain part of the variation in the olivine basalt suite (Stone, 1967). Tilley and Thompson (1970), Thompson (1975), Leeman and Vitaliano (1976), Stout and Nicholls (1977), Leeman (1982b), and Menzies and others (1984) summarize experimental and petrochemical data that support these conclusions.

Rocks from the Spencer–High Point area have affinities with olivine basalts and evolved basalts. The more evolved rocks occur at the low-Mg-value end of the coherent covariation trends for olivine basalts (Fig. 13), suggesting that they may have formed by higher degrees of olivine and plagioclase fractionation than most other SRPT olivine basalts.

Evolved basalts. Evolved basaltic rocks of the ESRP show large variations in major-element and minor-element chemistry (Table 2). The different slopes in the covariation diagrams (Figs. 13 and 14) for the array defined by the three sets of olivine basalts (latest Pleistocene and Holocene lava fields—LPH, Pleistocene lava fields—PLF; and most S-HP) and the array defined by rocks of the COM lava field suggest that fractionation of olivine and plagioclase alone, which accounts for the chemical variation in olivine basalts, is not the same process that accounts for the chemical variation of COM rocks. The different slopes also suggest that significant amounts of crystal fractionation—of olivine, plagioclase, and perhaps other minerals—would be required to derive even the least evolved COM basalts from SRPT olivine basalt.

Studies by Leeman and others (1976), Leeman (1982c), Kuntz, Champion, and others (1986), Futa and others (1988), and Stout and others (1989) identify two mechanisms for the derivation of the evolved basalts from parental SRPT magmas: (1) fractionation of observed phenocryst phases at pressures of ~8 to 10 kb, corresponding to depths of ~30 km, and (2) contamination of primary basalt by mineral components from deep crustal rocks or by partial melts of deep crustal rocks. Either or both of these mechanisms could account for most of the evolved basalts. Stout and others (1989) showed that coherent trends and nonzero intercepts on Pearce element-ratio plots suggest derivation of the evolved COM magmas from basalt by crystal fractionation and by assimilation of pumiceous rhyolite by basalt. Stout and others (1989) also suggested that chemical variations in COM rocks cannot be explained by fractionation of olivine and plagioclase alone but must include augite and Fe-Ti oxide accumulation and/or fractionation as well.

The character of parental magma for the varied COM rocks is problematic. The Kings Bowl and Wapi olivine basalts appear to be likely parental magmas for COM basaltic rocks because they erupted almost simultaneously with the youngest COM rocks along the Great Rift volcanic rift zone. However, Kuntz and others (1985) and Kuntz, Champion, and others (1986) showed that the most mafic basaltic rocks of the COM lava field have a significantly different major-element composition than the Wapi and Kings Bowl basalts. The most mafic COM rocks have rare-earth contents 1.3 to 8 times those of the olivine basalts, represented by latest Pleistocene and Holocene lava rocks (Fig. 15). These geochemical relations suggest that COM basaltic rocks were not derived by fractional crystallization from olivine basalt parental magmas; rather, they were derived from fundamentally different parental magmas.

Petrogenetic schemes to account for the significant enrichment of large-ion-lithophile (LIL) and rare-earth elements (REE) in COM basaltic rocks relative to SRPT olivine basalts might include varying degrees of partial melting of mantle peridotite, crystal fractionation, and interaction of magma with wall rocks in either the crust or the mantle, as suggested by Jamieson and Clarke (1970). In our opinion, the most significant factor that bears on this point is that a 10-km-thick granitic body probably underlies the COM lava field (see Kuntz, this volume). This relatively low density body or the partially melted lower- or middle-crustal rocks beneath it may have inhibited the ascent of magma through the crust and may have promoted enrichment processes such as magma-wall rock interaction and/or crystal fractionation.

Leeman (1982c) and Futa and others (1988) noted that Sr isotopic compositions of COM rocks trend toward typical lower-crustal values with increasing degrees of differentiation and decreasing time, suggesting that progressive assimilation of rhyolitic xenoliths and/or addition of partial melts of lower-crustal granulites may partly account for the compositional variability of the COM suite. Menzies and others (1984) suggest that less than 20% contamination is required to produce the most highly evolved rocks of the COM field.

In summary, the chemical diversity of most SRPT olivine basalts is due to variable degrees of partial melting of mantle source rocks at various depths and minor, subsequent crystal fractionation of olivine and plagioclase. Fractionation of olivine, plagioclase, clinopyroxene, and Fe-Ti oxides and crustal contamination are factors only in formation of more evolved basaltic magmas, such as those of the COM lava field.

MAGMA OUTPUT RATES

Magma output rate (MOR) is the long-term, time-averaged discharge rate of a volcano or volcanic region. Baksi and Watkins (1973) and Nakamura (1974) have shown that the MOR of volcanoes is related to their tectonic environment: It is greatest for intraplate oceanic volcanoes, intermediate for volcanoes at convergent plate margins, and least for intraplate continental volcanoes.

Analyses of volumes and ages of lava flows in the COM lava field show that the MOR was fairly constant at ~1.5 km^3/1,000 yr for the period from 15,000 to 7,000 B.P. and that the rate increased to ~2.8 km^3/1,000 yr from 7,000 to 2,100 B.P.,

TABLE 2. MAJOR-ELEMENT AND TRACE-ELEMENT ANALYSES AND MOLECULAR NORMS OF SELECTED LAVA FLOWS OF THE CRATERS OF THE MOON LAVA FIELD, IDAHO*

Eruptive Period	H	G	G	G	F	F	E	E	D	D	D
Flow	Kimama	Lava Creek	Sunset	Carey	Bottleneck Lake	Pronghorn	Lava Point	Grassy Cone	Little Park	Carey Kipuka	Silent Cone
Age (ka)	15.1	12.8	12.0	12.0	11.0	10.2	7.8	7.3	6.5	6.6	6.5
Sample No.	79K99	79K85	78K184	78K61	78K72	78K73	78K68	78K183	78K65	78K147A	78K150
					Major Oxides (percent)						
SiO_2	45.24	43.50	49.65	47.22	46.58	45.52	45.64	47.25	49.46	50.54	52.78
TiO_2	3.28	3.83	2.44	3.09	3.33	3.20	3.63	3.23	2.55	2.42	1.86
Al_2O_3	14.23	14.20	14.50	13.46	13.42	13.81	13.39	12.49	14.28	14.16	14.06
Fe_2O_3	1.27	2.16	4.53	0.89	1.35	1.20	1.79	8.62	1.14	1.22	0.78
FeO	15.11	14.78	10.47	14.93	14.71	14.96	14.28	7.45	13.55	13.18	11.55
MnO	0.28	0.27	0.24	0.26	0.27	0.24	0.23	0.25	0.24	0.22	0.22
MgO	4.53	5.21	3.14	3.91	4.21	4.10	4.67	3.82	3.08	2.95	1.90
CaO	7.50	8.13	6.47	8.24	8.78	7.48	8.18	8.40	6.49	6.28	5.39
Na_2O	3.50	3.47	4.50	3.47	3.39	4.02	3.55	3.07	4.28	4.28	3.51
K_2O	1.75	1.60	2.35	1.94	1.83	1.92	1.65	1.82	2.30	2.38	2.71
H_2O^+	0.33	0.38	0.12	0.24	0.23	0.32	0.37	0.19	0.43	0.46	0.61
H_2O^-	0.08	0.08	0.16	0.10	0.06	0.09	0.11	0.05	0.14	0.15	0.16
P_2O_5	2.64	2.78	1.56	2.29	2.46	2.18	2.13	2.33	1.60	1.46	1.13
Total	99.74	100.39	100.13	100.04	100.62	99.04	99.62	98.97	99.54	99.70	96.66
					Molecular Norms						
Ap	5.71	5.98	3.34	4.94	5.28	4.73	4.62	5.14	3.45	3.14	2.53
Il	4.73	5.49	3.48	4.44	4.76	4.63	5.24	4.75	3.66	3.47	2.77
Mt	1.95	2.00	1.73	1.89	1.90	1.93	1.91	1.87	1.75	1.71	1.52
Or	10.70	9.73	14.20	11.83	11.10	11.77	10.11	11.35	14.01	14.48	17.11
Ab	32.52	30.91	39.77	32.17	31.26	33.18	33.04	29.10	39.62	39.56	33.69
An	18.57	18.98	12.70	15.92	16.43	14.50	16.31	15.76	13.36	12.76	15.61
Di	1.66	3.06	7.75	8.67	9.44	7.39	9.07	9.73	7.24	7.59	4.06
Hy	6.68	0	0	8.02	5.96	0	1.84	19.97	2.68	6.63	18.42
Ol	17.48	23.16	16.11	12.11	13.85	19.31	17.86	2.33	14.24	10.65	0
Ne	0	0.69	0.92	0	0	2.57	0	0	0	0	0
Q	0	0	0	0	0	0	0	0	0	0	4.29
					Trace Elements (parts per million)						
Ba	1,180	1,100	1,600	1,290	1,160	1,300	944	1,200	1,740	1,490	1,720
Be	3	4	5	3	3	4	2	3	5	4	4
Ce	224	200	250	215	207	230	174	210	274	230	250
Co	35	45	29	35	37	37	40	39	33	28	21
Cr	4	<1	<1	3	4	<1	15	5	4	3	2
Cu	41	29	22	45	50	27	62	46	59	42	38
Eu	nd	8	9	nd	nd	8	nd	nd	nd	nd	nd
Ga	28	28	32	25	26	30	22	26	33	26	29
Ho	nd	5	5	nd	nd	5	nd	nd	nd	nd	nd
La	114	100	130	110	108	110	90	110	143	122	134
Li	21	19	25	20	20	21	16	15	26	23	23
Mo	<2	<2	<2	<2	<2	<2	<2	<2	3	<2	<2
Nb	98	13	<4	107	94	<4	90	108	122	112	126
Nd	nd	120	140	nd	nd	130	nd	nd	nd	nd	nd
Ni	<2	10	<2	<2	<2	3	17	<2	<2	<2	<2
Pb	21	14	17	20	21	15	21	21	30	21	26
Rb	48	50	73	56	52	58	43	48	36	70	84
Sc	28	23	24	29	28	24	30	32	35	30	28
Sr	352	350	330	340	345	353	313	335	373	311	290
V	95	130	45	97	116	98	177	146	84	64	25
Y	114	100	120	109	108	110	92	109	134	113	117
Yb	12	11	13	9	9	11	10	9	15	13	13
Zn	236	230	250	228	224	200	198	225	271	229	229
Zr	994	842	1,228	980	940	965	795	990	1,165	1,170	1,362

TABLE 2. MAJOR-ELEMENT AND TRACE-ELEMENT ANALYSES AND MOLECULAR NORMS OF SELECTED LAVA FLOWS OF THE CRATERS OF THE MOON LAVA FIELD, IDAHO* (continued)

Eruptive Period	C	C	C	C	C	B	B	A	A	A	A	A
Flow	Sentinel	Fissure Butte	Sheep Trail	Sawtooth	Indian Wells N.	Minidoka	Devils Cauldron	Highway	Serrate	Big Craters	Trench Mortar Flat	Blue Dragon
Age (ka)	6.0	6.0	6.0	6.0	6.0	4.5	4.5	2.4	2.4	2.4	2.2	2.1
Sample No.	78K118	78K129	78K92	78K67	78K95A	79K94	78K81	78K151	78K99	78K120	78K126	78K52
Major Oxides (percent)												
SiO_2	44.65	49.58	44.33	55.37	57.20	48.40	48.73	62.88	59.91	52.04	51.08	49.01
TiO_2	3.70	2.41	3.65	1.70	1.29	2.96	2.93	0.67	0.95	2.54	2.57	2.87
Al_2O_3	12.97	14.40	13.77	14.67	15.11	14.06	13.62	14.49	13.94	13.91	13.29	13.41
Fe_2O_3	1.70	16.24	5.06	1.55	1.04	1.09	1.35	0.88	2.94	1.33	1.72	0.77
FeO	14.55	0.02	11.31	11.21	10.23	14.44	13.90	7.58	7.44	13.40	12.16	14.50
MnO	0.27	0.25	0.25	0.20	0.21	0.26	0.22	0.17	0.18	0.24	0.23	0.24
MgO	4.90	2.82	5.57	1.94	1.45	3.83	3.54	0.27	0.52	2.68	3.05	3.34
CaO	9.60	6.36	8.33	4.95	4.12	6.92	7.12	2.94	3.55	6.76	6.82	6.93
Na_2O	3.04	3.86	3.36	4.12	3.97	3.58	4.21	4.20	4.00	3.57	3.55	3.61
K_2O	1.56	2.26	1.50	3.04	3.40	1.96	2.00	4.64	3.91	2.31	2.39	2.03
H_2O^+	0.37	0.32	0.18	0.16	0.43	0.26	0.23	0.14	0.11	0.18	0.13	0.20
H_2O^-	0.06	0.17	0.08	0.04	0.10	0.06	0.08	0.03	0.01	0.03	0.06	0.14
P_2O_5	2.87	1.51	2.77	0.87	0.62	2.12	1.88	0.14	0.29	1.47	1.44	1.77
Total	100.24	100.20	100.16	99.82	99.17	99.94	99.81	99.03	97.75	100.46	98.49	98.82
Molecular Norms												
Ap	6.21	3.29	5.97	1.86	1.34	4.57	4.04	0.30	0.63	3.16	3.15	3.87
Il	5.34	3.50	5.24	2.42	1.85	4.25	4.20	0.96	1.38	3.63	3.74	4.18
Mt	1.93	1.77	1.90	1.50	1.34	1.85	1.81	1.00	1.23	1.74	1.66	1.84
Or	9.54	13.93	9.13	18.39	20.73	11.94	12.15	28.10	24.17	14.02	14.75	12.53
Ab	28.26	36.16	31.09	37.88	36.78	33.16	38.86	38.65	37.57	32.93	33.30	33.85
An	17.74	15.96	18.61	12.86	13.80	17.03	12.71	7.16	8.93	15.52	13.87	15.03
Di	9.72	5.34	4.26	5.17	2.49	3.28	8.78	5.49	6.01	7.25	9.33	7.04
Hy	6.53	12.76	4.00	17.79	16.54	16.53	3.00	8.64	11.30	20.47	19.63	16.97
Ol	14.73	7.28	19.80	0	0	7.39	14.46	0	0	0	0	4.69
Ne	0	0	0	0	0	0	0	0	0	0	0	0
Q	0	0	0	2.12	5.13	0	0	9.72	8.78	1.27	0.57	0
Trace Elements (parts per million)												
Ba	959	1,470	948	1,850	2,180	1,300	1,300	1,900	2,600	1,600	1,500	1,310
Be	3	4	3	5	6	3	4	6	6	5	5	3
Ce	190	244	181	262	276	218	230	270	257	240	240	210
Co	42	24	41	18	16	32	36	22	9	30	32	31
Cr	11	2	19	3	<1	4	<1	<1	<1	<1	8	3
Cu	60	37	64	34	29	54	25	27	26	23	23	42
Eu	nd	nd	nd	nd	nd	nd	8	8	nd	8	8	nd
Ga	26	29	26	31	28	25	30	31	29	31	30	24
Ho	nd	nd	nd	nd	nd	nd	5	5	nd	5	5	nd
La	98	127	95	137	147	112	120	150	142	130	130	109
Li	16	26	17	26	31	20	22	30	30	25	25	20
Mo	<2	<2	<2	<2	<2	<2	<2	3	4	<2	<2	<2
Nb	89	1,284	95	140	136	111	6	22	nd	17	23	110
Nd	nd	nd	nd	nd	nd	nd	130	150	nd	140	130	nd
Ni	8	<2	12	<2	<2	<2	5	<2	<2	<2	10	<2
Pb	21	24	22	25	24	22	18	23	26	19	18	15
Rb	47	50	40	82	77	55	56	90	nd	64	72	53
Sc	31	29	31	28	25	29	25	22	30	25	24	32
Sr	343	338	336	282	256	321	320	270	209	310	290	334
V	172	40	175	15	5	117	110	11	<1	66	99	106
Y	100	122	97	119	124	105	110	120	121	120	110	107
Yb	11	14	9	14	15	12	11	14	15	12	12	12
Zn	209	244	203	231	226	220	210	260	232	250	240	215
Zr	789	1,284	774	1,401	1,344	992	948	1,482	nd	1,113	290	979

See Kuntz and others (1985) for sample localities. nd = no data. USGS analysts: J. Crock, A. Bartel, E. Brandt, J. Carr, M. Elsheimer, L. Espos, L. Jackson, P. Klock, R. Mays, V. McDaniel, R. Moore, J. Rivielle, D. Siems, and K. Wong.

TABLE 3. MAJOR-ELEMENT AND TRACE-ELEMENT ANALYSES AND MOLECULAR NORMS OF LATEST PLEISTOCENE AND HOLOCENE LAVA FIELDS OF THE EASTERN SNAKE RIVER PLAIN, IDAHO*

	Hells Half Acre Lava Field							North Robbers Lava Field		South Robbers Lava Field	
Sample No.	84KS4	6-16-6	6-24-4B	HHA-W	HHA-8D	SRV72-2	SRV72-4	84KS2	75K173	84KS1	84KS3
					Major Oxides (percent)						
SiO_2	47.10	48.40	46.27	46.77	45.72	47.21	47.44	46.40	47.00	44.60	45.20
TiO_2	3.42	2.90	3.02	3.22	3.37	3.44	2.65	2.10	2.30	3.10	3.49
Al_2O_3	14.60	15.30	14.99	13.94	14.87	14.14	14.58	15.40	15.50	14.90	14.70
Fe_2O_3	1.90	2.90	1.06	1.67	1.99	4.12	3.70	1.20	3.70	9.10	2.00
FeO	12.80	12.60	12.66	12.48	13.82	10.53	9.12	11.20	9.20	6.26	12.50
MnO	0.22	0.22	0.20	0.18	0.16	0.22	0.18	0.20	0.16	0.23	0.23
MgO	6.47	6.50	6.54	7.10	5.84	6.40	8.33	9.27	8.80	7.17	6.32
CaO	9.30	8.50	9.12	9.40	9.28	9.60	9.37	10.80	10.10	9.71	9.60
Na_2O	2.60	2.50	2.57	2.40	2.56	2.84	2.83	2.48	2.60	2.93	2.88
K_2O	0.82	0.82	0.83	0.84	0.83	0.87	0.77	0.40	0.43	0.66	0.89
H_2O^+	0.19	0.25	0.17	0.30	0.18	0.10	0.23	0.14	0.23	0.25	0.20
H_2O^-	0.02	0.03	nd	nd	nd	0.12	0.11	0.02	0.06	0.04	0.03
P_2O_5	0.57	0.54	0.68	0.82	0.99	0.11	0.39	0.56	0.67	1.00	1.29
Total	100.01	101.46	98.11	99.12	99.61	99.70	99.70	100.17	100.75	99.95	99.33
					Molecular Norms						
Ap	1.22	1.14	1.48	1.77	2.14	0.24	0.83	1.17	1.40	2.15	2.78
Il	4.88	4.09	4.37	4.63	4.85	4.92	3.74	2.93	3.21	4.43	5.01
Mt	1.73	1.79	1.64	1.68	1.87	1.70	1.47	1.43	1.46	1.72	1.71
Or	4.96	4.90	5.10	5.13	5.07	5.28	4.61	2.37	2.55	4.00	5.42
Ab	23.92	22.72	23.98	22.26	23.75	26.20	25.78	22.31	23.40	25.50	26.64
An	26.38	28.45	27.97	25.60	27.52	23.91	25.17	29.77	29.43	26.22	25.29
Di	13.66	8.53	11.55	13.63	10.70	19.43	15.53	16.21	13.13	13.20	2.07
Hy	13.41	22.18	12.93	16.91	12.47	3.89	5.29	1.84	7.55	0	4.25
Ol	9.84	6.19	10.99	8.40	11.62	14.43	17.59	21.97	17.87	21.88	6.84
Ne	0	0	0	0	0	0	0	0	0	0.89	0
					Trace elements (parts per million)						
Rb	nd	nd	nd	nd	nd	nd	nd	nd	nd	10	15
Sr	341	nd	nd	nd	nd	nd	nd	227	nd	324	336
Zr	243	nd	nd	nd	nd	nd	nd	215	nd	377	428
Nb	21	nd	nd	nd	nd	nd	nd	15	nd	29	44
Y	40	38	nd	nd	nd	nd	nd	32	nd	42	59
La	23	22	nd	nd	nd	nd	nd	20	nd	nd	51
Ce	52	51	nd	nd	nd	nd	nd	44	nd	nd	110
Nd	35	32	nd	nd	nd	nd	nd	27	nd	nd	63
Sm	8.1	7.7	nd	nd	nd	nd	nd	6.10	nd	nd	13
Eu	3.2	2.9	nd	nd	nd	nd	nd	2.4	nd	nd	4.6
Gd	9.9	9.0	nd	nd	nd	nd	nd	7.7	nd	nd	14.3
Tb	1.5	1.4	nd	nd	nd	nd	nd	1.0	nd	nd	2.2
Dy	8.3	3.9	nd	nd	nd	nd	nd	6.5	nd	nd	12.4
Er	4.4	4.0	nd	nd	nd	nd	nd	3.5	nd	nd	6.6
Tm	0.59	0.65	nd	nd	nd	nd	nd	0.49	nd	nd	0.89
Yb	3.89	3.68	ne	nd	nd	nd	nd	3.23	nd	nd	5.89
Lu	0.55	0.51	nd	nd	nd	nd	nd	0.45	nd	nd	0.85

TABLE 3. MAJOR-ELEMENT AND TRACE-ELEMENT ANALYSES AND MOLECULAR NORMS OF LATEST PLEISTOCENE AND HOLOCENE LAVA FIELDS OF THE EASTERN SNAKE RIVER PLAIN, IDAHO* (continued)

	Cerro Grande Lava Field		Kings Bowl Lava Field			Wapi Lava Field			Shoshone Lava Field		
Sample No.	75K166A	SRV72-21	K-KB	IV71-14	IV71-13	K-WAPI	P46HRC	K-SHOS	84KS6	84KS5	SRV72-9
					Major Oxides (percent)						
SiO_2	46.50	45.37	45.80	45.70	45.86	46.20	46.39	46.10	46.40	46.40	46.26
TiO_2	3.20	3.32	2.48	2.29	2.29	2.36	2.17	2.84	2.54	2.68	2.66
Al_2O_3	14.50	14.57	14.70	14.76	14.66	15.00	15.43	14.40	15.00	14.80	14.36
Fe_2O_3	2.40	2.01	1.50	1.78	4.25	1.60	1.05	1.50	1.90	1.70	1.52
FeO	11.60	12.35	11.60	11.03	8.78	11.00	11.03	11.70	11.30	11.70	11.51
MnO	0.17	0.19	0.21	0.18	0.17	0.20	0.19	0.21	0.21	0.21	0.18
MgO	6.80	7.25	9.48	9.86	10.04	8.98	8.34	7.72	8.27	8.14	8.52
CaO	8.90	9.23	10.40	10.13	10.18	10.60	10.77	9.85	10.00	10.10	9.84
Na_2O	2.80	2.92	2.31	2.56	2.56	2.30	2.65	2.52	2.57	2.59	2.88
K_2O	0.89	0.88	0.45	0.41	0.42	0.50	0.61	0.85	0.68	0.69	0.78
H_2O^+	0.45	0.16	0.12	0.10	0.14	0.20	0.14	0.32	0.19	0.25	0.19
H_2O^-	0.13	0.19	nd	0.06	0.10	nd	0.06	nd	0.03	0.03	0.23
P_2O_5	1.40	1.03	0.60	0.55	0.60	0.60	0.67	1.21	0.90	0.95	0.91
Total	99.74	99.47	99.65	99.41	100.05	99.54	99.50	99.22	99.99	100.24	99.84
					Molecular Norms						
Ap	3.01	2.21	1.27	1.16	1.26	1.27	1.42	2.59	1.90	2.01	1.93
Il	4.58	4.74	3.49	3.22	3.21	3.33	3.05	4.05	3.58	3.78	3.75
Mt	1.64	1.69	1.52	1.48	1.47	1.46	1.41	1.55	1.53	1.55	1.51
Or	5.40	5.33	2.69	2.44	2.49	2.99	3.64	5.15	4.07	4.12	4.66
Ab	25.83	26.86	20.96	23.19	23.10	20.92	23.92	23.19	23.35	23.52	26.17
An	25.04	24.64	28.72	27.82	27.41	29.51	28.71	26.10	27.71	27.02	24.24
Di	8.75	12.30	15.58	15.41	15.53	15.84	16.69	12.70	13.23	13.90	15.32
Hy	15.26	1.89	4.88	0.35	0.57	6.33	0	10.47	6.76	6.51	0.17
Ol	10.48	20.35	20.90	24.94	24.96	18.34	21.08	14.19	17.86	17.59	22.26
Ne	0	0	0	0	0	0	0.07	0	0	0	0
					Trace elements (parts per million)						
Rb	25	nd	14	nd	nd	13	nd	21	11	14	nd
Sr	363	nd	264	nd	nd	281	nd	317	304	325	nd
Zr	407	nd	205	nd	nd	211	nd	361	271	288	nd
Nb	40	nd	18	nd	nd	16	nd	36	32	28	nd
Y	nd	nd	35	nd	nd	35	nd	53	40	41	nd
La	nd	nd	22	nd	nd	23	nd	46	33	nd	nd
Ce	nd	nd	47	nd	nd	50	nd	97	71	nd	nd
Nd	nd	nd	28	nd	nd	30	nd	56	41	nd	nd
Sm	nd	nd	6.60	nd	nd	6.3	nd	12	8.6	nd	nd
Eu	nd	nd	2.5	nd	nd	2.5	nd	3.9	3.1	nd	nd
Gd	nd	nd	8.8	nd	nd	9.3	nd	13.1	10.1	nd	nd
Tb	nd	nd	1.4	nd	nd	1.6	nd	1.8	1.4	nd	nd
Dy	nd	nd	6.9	nd	nd	6.9	nd	10.5	8.1	nd	nd
Er	nd	nd	3.9	nd	nd	3.8	nd	5.6	4.3	nd	nd
Tm	nd	nd	0.54	nd	nd	0.51	nd	0.80	0.56	nd	nd
Yb	nd	nd	3.16	nd	nd	3.56	nd	5.19	3.93	nd	nd
Lu	nd	nd	0.50	nd	nd	0.49	nd	0.73	0.54	nd	nd

See Kuntz and others (1985) for sample localities. nd = no data. USGS analysts: J. Crock, B. Anderson, A. J. Bartel, S. Botts, M. Cremer, L. Kapos, Z. A. Hamlin, L. L. Jackson, G. Mason, H. G. Neiman, K. Stewart, C. Stone, and J. Taggart.

TABLE 4. MAJOR-ELEMENT ANALYSES AND MOLECULAR NORMS OF SELECTED PLEISTOCENE LAVA FIELDS, EASTERN SNAKE RIVER PLAIN, IDAHO*

	Central ESRP											
Sample No.	76K50	76K107	75K229A	76K68	76K12	76K11	76K66	76K99	76K176	76K177	76K133	6-14-6
					Major Oxides (percent)							
SiO_2	47.41	46.25	46.50	46.78	45.71	47.10	48.11	46.34	46.24	46.38	47.47	47.50
TiO_2	1.65	3.44	3.70	2.59	2.37	1.26	2.35	2.96	3.48	3.48	3.46	3.20
Al_2O_3	16.42	14.11	14.30	15.59	15.48	16.55	15.59	15.04	14.73	14.78	13.56	15.50
Fe_2O_3	1.54	2.26	3.40	3.30	2.48	1.75	2.22	3.67	3.92	2.94	0.91	2.90
FeO	9.34	12.95	11.90	10.15	10.81	7.38	10.42	10.35	11.63	12.41	13.64	11.60
MnO	0.17	0.20	0.17	0.19	0.21	0.18	0.19	0.21	0.21	0.21	0.23	0.22
MgO	8.78	6.09	6.10	7.79	8.10	9.50	7.67	7.04	6.57	6.61	7.37	4.90
CaO	11.53	9.28	9.40	10.21	11.41	11.00	10.06	10.33	9.38	9.39	10.07	9.50
Na_2O	2.40	2.89	2.50	2.74	2.48	2.36	2.63	2.83	2.73	2.62	2.53	2.70
K_2O	0.32	0.75	0.70	0.49	0.32	0.26	0.66	0.62	0.72	0.73	0.54	0.74
H_2O^+	0.15	0.26	0.48	0.22	0.22	0.44	0.32	0.27	0.30	0.40	0.36	0.35
H_2O^-	0.05	0.10	0.18	0.10	0.12	0.14	0.10	0.10	0.15	0.06	0.06	0.15
P_2O_5	0.33	0.95	0.90	0.58	0.54	0.28	0.57	0.84	0.78	0.81	1.18	1.10
Total	100.09	99.53	100.23	100.73	100.25	98.20	100.89	100.60	100.84	100.82	101.38	100.36
					Molecular Norms							
Ap	0.69	2.05	1.95	1.22	1.14	0.59	1.20	1.78	1.66	1.73	2.50	2.37
Il	2.30	4.95	5.33	3.63	3.34	1.78	3.29	4.18	4.95	4.95	4.88	4.59
Mt	1.25	1.80	1.80	1.53	1.53	1.06	1.45	1.61	1.80	1.79	1.70	1.70
Or	1.89	4.58	4.27	2.92	1.91	1.56	3.92	3.72	4.34	4.40	3.23	4.50
Ab	21.56	26.80	23.20	24.77	21.90	21.53	23.73	25.78	25.02	24.01	22.98	24.97
An	33.11	24.08	26.60	29.00	30.50	34.34	28.93	26.89	26.35	26.96	24.34	28.83
Di	17.58	13.65	12.42	14.56	18.54	15.40	13.99	15.63	12.76	12.15	14.73	9.85
Hy	3.76	9.54	18.21	4.11	0	6.69	10.76	2.80	8.81	11.57	18.65	20.64
Ol	17.86	12.57	6.22	18.25	20.78	17.06	12.73	17.62	14.32	12.45	6.99	2.55
Ne	0	0	0	0	0.37	0	0	0	0	0	0	0

mainly because output of contaminated magma was added to the nearly constant output rate of fractionated magma (Kuntz, Champion, and others, 1986).

The MOR for the entire ESRP for the last 15,000 years, calculated from estimated volumes of the latest Pleistocene and Holocene lava fields, is 3.3 km^3/1,000 yrs. This MOR, when compared to the rate for the COM lava field, indicates that about 75% of the volume of lava erupted in the ESRP in the last 15,000 years has been from the Great Rift volcanic rift zone.

Because there is no reason to assume that the MOR has increased or waned during the history of basaltic volcanism of the ESRP, the latest Pleistocene and Holocene MOR is used to represent the ESRP in Table 5. This value is approximately the same as that for other intraplate-continental, dominantly basaltic lava fields, such as the Springerville and San Francisco volcanic fields in Arizona on the western margin of the Colorado Plateau. These intraplate-continental volcanoes generally have lower MOR values than either intraplate-oceanic or convergent-plate-margin volcanoes, probably because less magma is produced in these regions and/or because they have higher intrusion-extrusion ratios.

RELATIONSHIPS BETWEEN BASALTIC VOLCANISM AND CALDERAS OF THE ESRP

Kuntz and Covington (1979) suggested that the lengths and orientations of volcanic rift zones, the areal density and types of basaltic vents, and subtle topographic features in the ESRP are partly controlled by underlying buried calderas. In addition, they proposed that (1) the orientations of some volcanic rift zones may be controlled by buried ring fractures of underlying calderas, (2) other zones terminate at caldera margins, (3) areas that have few or no vents on the ESRP may overlie areas of low-density rocks within calderas, and (4) subtle arcuate depressions on the ESRP may reflect collapsed parts of underlying calderas. All of these relationships were highly speculative when originally proposed because little was known then about the rhyolite calderas.

Kuntz and Covington used these relationships to predict the size and location of underlying calderas. In hindsight, our ideas about the size and location of the calderas were quite inaccurate when compared to what is presently known or inferred about the calderas. However, we feel that many of the relationships we developed in 1979 remain viable.

TABLE 4. MAJOR-ELEMENT ANALYSES AND MOLECULAR NORMS OF SELECTED PLEISTOCENE LAVA FIELDS, EASTERN SNAKE RIVER PLAIN, IDAHO* (continued)

	Central ESRP			Spencer-High Point Volcanic Rift Zone								
Sample No.	76K119	76K127	76K134	76K28	76K178	76K192	76K205	76K210	76K240	76K196	76K315	77K51
						Major Oxides (percent)						
SiO_2	47.28	46.17	46.35	47.49	46.78	47.78	47.38	47.33	46.17	52.21	50.24	47.34
TiO_2	2.36	3.16	2.58	1.93	2.94	2.87	3.02	1.22	1.74	1.98	1.94	1.21
Al_2O_3	15.21	15.33	15.65	16.25	15.28	15.79	15.24	16.54	16.30	15.89	16.15	16.25
Fe_2O_3	1.70	2.43	3.57	1.86	1.37	3.79	4.40	1.09	1.48	2.74	3.24	0.83
FeO	11.13	12.11	9.89	10.17	12.25	9.84	10.07	9.12	10.37	8.90	8.09	9.52
MnO	0.19	0.20	0.20	0.19	0.19	0.22	0.25	0.17	0.18	0.21	0.17	0.17
MgO	8.10	7.30	7.79	7.90	7.00	4.01	4.05	10.27	9.64	2.88	6.66	9.92
CaO	10.14	9.89	10.03	10.14	9.85	8.44	7.80	11.02	10.28	6.38	8.74	11.10
Na_2O	2.67	2.83	2.92	2.98	3.06	3.63	3.80	2.48	2.48	3.75	3.16	2.72
K_2O	0.67	0.56	0.67	0.54	0.60	1.29	1.51	0.39	0.35	2.71	0.99	0.36
H_2O^+	0.42	0.30	0.46	0.08	0.09	0.27	0.16	0.06	0.08	0.41	0.13	0.27
H_2O^-	0.10	0.08	0.12	0.09	0.08	0.23	0.17	0.11	0.09	0.24	0.11	0.13
P_2O_5	0.58	0.74	0.65	0.60	0.62	1.85	2.12	0.30	0.34	1.06	0.67	0.40
Total	100.55	101.10	101.08	100.22	100.11	100.01	99.97	100.10	99.50	99.36	100.29	100.22
						Molecular Norms						
Ap	1.22	1.56	1.36	1.26	1.31	3.97	4.56	0.62	0.71	2.28	1.41	0.83
Il	3.31	4.44	3.61	2.69	4.15	4.11	4.32	1.68	2.43	2.84	2.72	1.67
Mt	1.48	1.68	1.53	1.38	1.59	1.58	1.68	1.16	1.36	1.36	1.29	1.18
Or	3.99	3.34	3.97	3.20	3.59	7.83	9.17	2.28	2.08	16.48	5.88	2.11
Ab	24.15	25.63	24.80	26.81	27.82	33.49	35.07	21.90	22.35	34.67	28.54	22.23
An	27.74	27.71	27.72	29.43	26.51	23.61	20.63	32.54	32.44	19.07	27.12	30.81
Di	15.29	13.52	14.37	13.64	15.09	5.59	3.93	15.74	13.23	5.11	9.68	16.97
Hy	5.27	3.18	0	1.63	1.33	10.99	8.79	0	1.30	17.11	16.08	0
Ol	17.54	18.94	21.74	19.97	18.62	8.82	11.86	23.98	24.10	1.07	7.29	23.01
Ne	0	0	0.91	0	0	0	0	0.09	0	0	0	1.19

*USGS analysts: S. Botts, L. Espos, Z. A. Hamlin, D. Hopping, P. Klock, S. Morgan, and M. Villareal.

Sample Localities

76K50 = Quaking Aspen Butte
76K107 = Taber Butte
75K229A = Table Legs Butte
76K68 = Crater vent
76K12 = Sixmile Butte
76K11 = Teakettle Butte
76K66 = AEC Butte

76K99 = State Butte
76K176 = Topper Butte
76K177 = Dome Butte
76K133 = Circular Butte
6-14-6 = Kettle Butte
76K119 = Antelope Butte
76K127 = Lava Ridge

76K134 = Teat Butte
76K28 = Morgans Crater
76K178 = Davis Butte
76K192 = Crystal Butte
76K205 = Laird's Butte
76K210 = Big Crater
76K240 = Experimental Crater

76K196 = Red Top Butte
76K315 = Snowshoe Butte
77K51 = Youngest vent, Split Rock Quadrangle

Since that time, much has been learned about the calderas. The Twin Falls and Picabo volcanic fields are described by Pierce and Morgan (this volume); the Taber caldera was originally described and illustrated by Kuntz and Covington (1979); the Blue Creek, Blacktail, and Kilgore calderas are described by Embree and others (1982), Morgan (1988), Morgan and others (1984, 1989), Pierce and Morgan (this volume), and Morgan (this volume); and the Henrys Fork and Yellowstone calderas are described by Christiansen (1982).

The known or inferred location and size of the calderas has been determined mainly from geophysical signatures and also from exposures around the margins of the ESRP. It should be noted that more is known about the calderas in the northeastern part of the ESRP because they are mostly or partly exposed; less is known about calderas farther to the southwest because they are mostly or completely covered by basaltic lava flows. What is presently known or inferred about the location, size, and extent of the rhyolitic calderas and the volcanic rift zones in the ESRP is summarized in Figure 16.

Here, we describe and attempt to evaluate critical field relations of the basaltic volcanism of the ESRP that may be related to underlying calderas. Many of the relationships remain speculative and abstract because of uncertainties about the size, location, and character of the calderas.

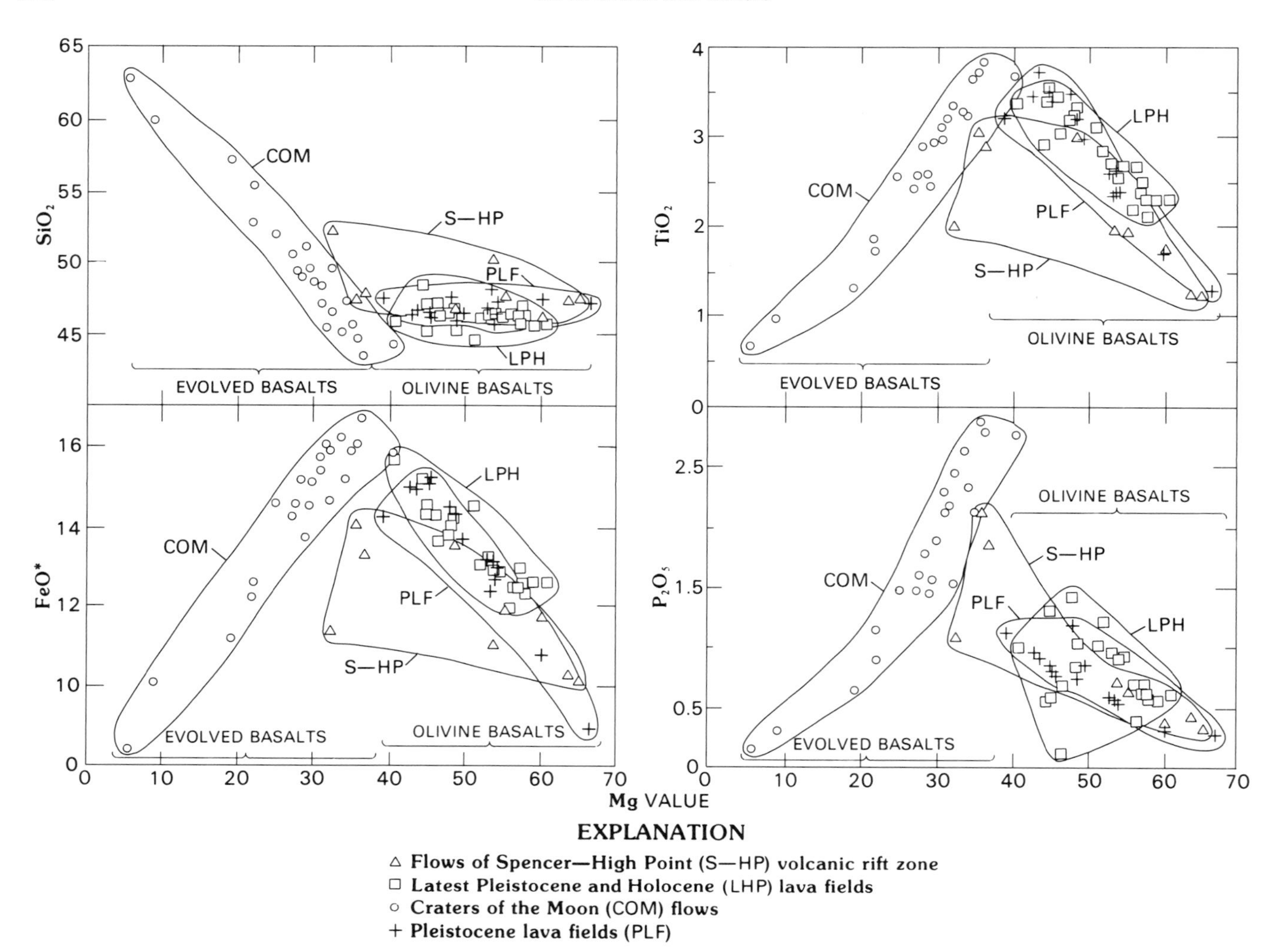

Figure 13. SiO_2, FeO*, TiO_2, and P_2O_5 vs. Mg value [MgO/(MgO+FeO*)] (wt. percent) for the 70 analyses listed in Tables 2, 3, and 4. FeO* is FeO + $0.9Fe_2O_3$.

Relationships between the orientations of volcanic rift zones and underlying calderas

The orientation and distribution of basaltic vents and volcanic rift zones are controlled mainly by regional stress directions and belts of strain concentration, the range-front faults. However, the influence of buried rhyolitic calderas on the orientation and distribution of volcanic vents and volcanic rift zones is important locally within the ESRP.

Several critical observations illustrate the variety and complexity of the influence of buried rhyolitic calderas on the orientation and distribution of basaltic vents and volcanic rift zones. (1) The Rock Corral Butte volcanic rift zone lies on the probable northwest boundary of the Taber caldera (Figs. 16 and 17). (2) Fracture-controlled basaltic vents in the central part of the Lava Ridge–Hells Half Acre volcanic rift zone are elongated north-south, whereas fracture-controlled vents in the northwestern and southeastern parts are elongated northwest-southeast (Fig. 17). (3) Fracture-controlled vents are elongated northeast-southwest between the central parts of the Lava Ridge–Hells Half Acre and Circular Butte–Kettle Butte volcanic rift zones (Fig. 17). (4) The Arco–Big Southern Butte volcanic rift zone terminates at the northern boundary of the Taber caldera (Figs. 16 and 17). (5) Linear concentrations of basaltic vents within calderas in the ESRP occur chiefly where well-developed volcanic rift zones, as collinear continuations of major range-front faults, extend into or cross calderas; major examples are seen where the Spencer–High Point volcanic rift zone crosses the Kilgore caldera and where the Menan volcanic rift zone extends 15 km into but does not completely traverse the Kilgore caldera (Fig. 18). Several concepts may partly explain these field relationships.

1. The regional stress direction in the ESRP (N.45°E.) is evidently deflected and reoriented by the northwest margin of the Taber caldera, producing fracture-controlled basaltic vents within

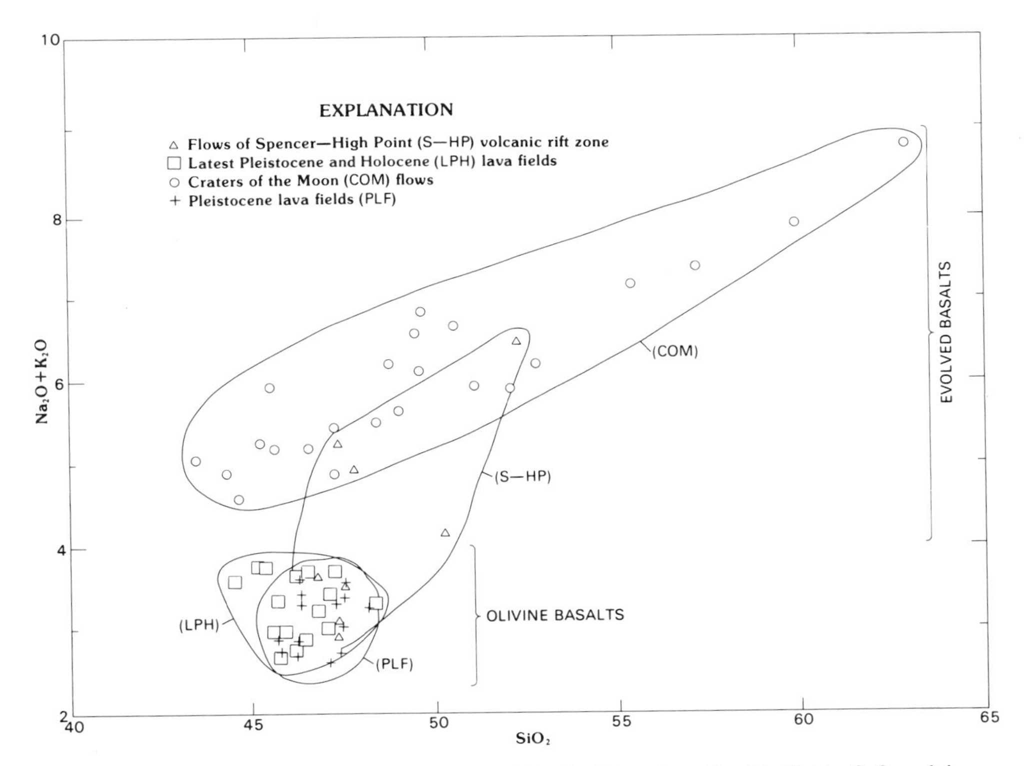

Figure 14. Na_2O + K_2O vs. SiO_2 (wt. percent) for the 70 analyses listed in Tables 2, 3, and 4.

the Rock Corral Butte volcanic rift zone that are oriented nearly parallel to the regional stress direction and, presumably, above and parallel to the caldera margin.

2. The anomalous north-south and northeast-southwest orientation of fracture-controlled basaltic vents occurs at the intersection of the southeast boundary of the Blue Creek caldera and the center of the Lava Ridge–Hells Half Acre volcanic rift zone (Fig. 17). We suggest that regional stress patterns are deflected near the intersection to cause the north-south–trending vents and that the boundary (ring fractures?) of the Blue Creek caldera has controlled the orientation of the northeast-southwest–trending vents.

3. Low-density rocks in calderas may be barriers to the ascent and eruption of basaltic magmas to the surface. As described by Kuntz (this volume), this relationship may explain the termination of the Arco–Big Southern Butte volcanic rift zone at the northwest margin of the Taber caldera (Figs. 16 and 17).

4. Fracture pathways for magma ascent are most abundant in volcanic rift zones that extend into or cross the calderas. The two volcanic rift zones that cross or partly cross the Kilgore caldera are extensions into the caldera of the two major range-front fault systems near its borders (Fig. 19).

5. Calderas may be affected by basaltic eruptions only after subcaldera plutons have cooled and solidified sufficiently that regional stresses can create faults in the plutons (Christiansen, 1982). The time span for the cycle of caldera formation, cooling of subcaldera plutons, subsequent fracturing, and basaltic volcan-

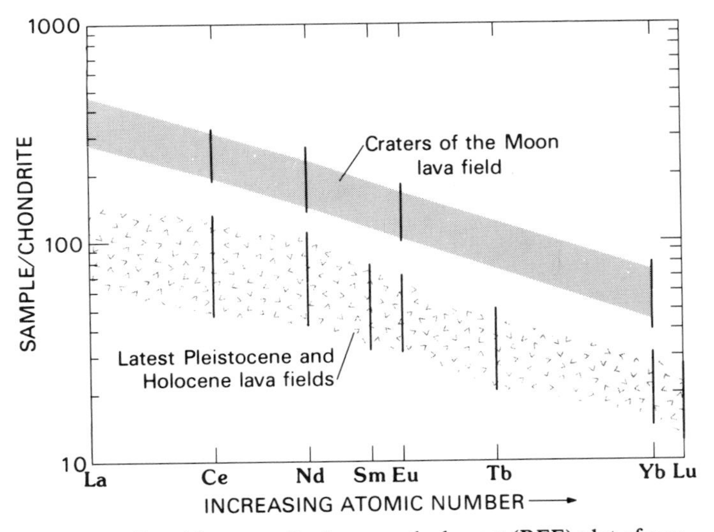

Figure 15. Chondrite-normalized rare-earth element (REE) plot of samples from 23 lava flows from the Craters of the Moon lava field and eight lava flows from latest Pleistocene and Holocene lava fields of the ESRP. Data from Tables 2 and 3. Chondritic abundances used to normalized REE are those of Haskin and others (1969).

TABLE 5. MAGMA OUTPUT RATES FOR VARIOUS VOLCANOES, LAVA FIELDS, AND VOLCANIC REGIONS*

Volcano or volcanic region	Magma Output Rate (km^3/1,000 yr)
CONTINENTAL, DOMINANTLY BASALTIC LAVA FIELDS	
Craters of the Moon lava field, eastern Snake River Plain (15 to 2 ka)	2.4
Entire eastern Snake River Plain (15 to 2 ka)	3.3
San Francisco volcanic field, Arizona (5 Ma to present)	0.2 to 3.3†
Springerville volcanic field, Arizona:	
Peak activity (2 to 1 Ma)	2.5
Average 2 to 0.3 Ma	1.5
CENTRAL VOLCANOES	
Etna, Italy (1759 to 1974)	9.5
Piton de la Fournaise, Reunion Island (1972 to 1981)	9.5
Fuego, Chile	16
Kilauea, Hawaii (1823 to 1975)	22
Mauna Loa, Hawaii (1823 to 1975)	22
LARGE VOLCANIC REGIONS	
Iceland (last 10,000 years)	500
Columbia Plateau, USA	1,000

*Data from Kuntz, Champion, and others, 1986; Tanaka and others, 1986; and Condit and others, 1989.
†Range for various time intervals.

ism can be estimated from radiometric data for rocks in and near the 4.3-Ma Kilgore caldera and the 1.3-Ma Henrys Fork caldera (Figs. 9 and 18). Gerritt Basalt flows that cover the floor of the Henrys Fork caldera were erupted at 200 ka from vents in the southwest part of the caldera (Christiansen, 1982). The Morgans Crater lava field, erupted from one of the oldest vents in the central part of the Spencer–High Point volcanic rift zone, has a K-Ar age of 365 ± 68 ka (G. B. Dalrymple, U.S. Geological Survey, written communication, 1980). These ages suggest that the Spencer–High Point volcanic rift zone had formed and was the locus of eruptions within the Kilgore caldera before 350 ka and extended into Henrys Fork caldera about 200 ka, when the caldera was 1.1 million years old. From this we conclude that roughly one million years was necessary for sufficient cooling to take place before subcaldera plutons were cut by faults.

Relationships between low density or absence of basaltic vents and underlying calderas

Various parts of the ESRP, even within volcanic rift zones, contain few or no basaltic vents (Figs. 10, 17, and 18). Notable examples are: (1) the area within the inferred boundaries of the Taber caldera, (2) the area of the sinks of the Big Lost River, Little Lost River, and Birch Creek along the northwest margin of the ESRP, (3) near the northwest end of the Circular Butte–Kettle Butte volcanic rift zones, (4) the southeast margin of the ESRP between Idaho Falls and Pocatello, (5) an area northwest of Terreton, (6) an area between Idaho Falls and the Menan volcanic rift zone, and (7) the region between the Spencer–High Point volcanic rift zone and Juniper Buttes.

These areas of sparse or no basaltic vents are within the known or inferred boundaries of buried rhyolitic calderas (Figs. 17 and 18). The concept that fill of low-density rocks in calderas created density barriers to the intrusion of basalt magma may partly explain the distribution of sparse or no basaltic vents in some parts of the ESRP. For example, Kuntz (this volume) suggests that the lack of basaltic vents in the area west of Blackfoot may be due to a thick accumulation of low-density rocks throughout the deep, piston-shaped Taber caldera.

The spatial relationship between the sparseness or absence of basaltic vents and low-density rocks in underlying calderas elsewhere in the ESRP is more speculative. Morgan (1988) and Morgan and others (1989) suggest that the Blacktail, Blue Creek, and Kilgore calderas are shallow structures less than 1 km deep and that relatively deep structures, possibly filled by low-density ejecta and/or colluvial materials, occur as separate, grabenlike rhyolitic vents along the margins of these calderas. If the relationship between thick fill of low-density rocks and sparsity of basaltic vents holds, then we propose that elongated, grabenlike rhyolitic vents may occur along the northwest margin of the Blue Creek caldera in the vicinity of the sinks area of the Big Lost River, Little Lost River, and Birch Creek (Fig. 17). A possible caldera-margin rhyolitic vent or vents at this locality may also explain the absence of basaltic vents near the northwest end of the Circular Butte–Kettle Butte volcanic rift zone (Fig. 17).

The absence of basaltic vents along the southeast margin of the ESRP between Idaho Falls and Pocatello (Figs. 10 and 17) may reflect a real lack of basaltic vents or may result from the concealment of older vents by alluvial sediments along the Snake and Henrys Fork rivers. As discussed above, this part ofthe ESRP lies adjacent to Belt IV of Pierce and Morgan (this volume), defined by them as an area having range-front faults that were active in late Tertiary time but show no evidence of Quaternary activity (Fig. 11). By assuming that basaltic volcanism in the ESRP is roughly coeval with basin-and-range deformation along its margins, as discussed above, we conclude that basaltic vents in this region may have formed several million years ago and may be covered by younger alluvial sediments.

Subtle topographic lows generally coincide with areas of sparse or no basaltic vents at four localities in the ESRP: (1) the area northwest of Terreton, (2) the area between Idaho Falls and the Menan volcanic rift zone, (3) the area between the Spencer–High Point volcanic rift zone and Juniper Buttes, and (4) the Taber caldera (Figs. 17 and 18). We suggest that these four regions may be areas of sparse basaltic vents and low elevation because they are adjacent to basaltic vents concentrated in high-

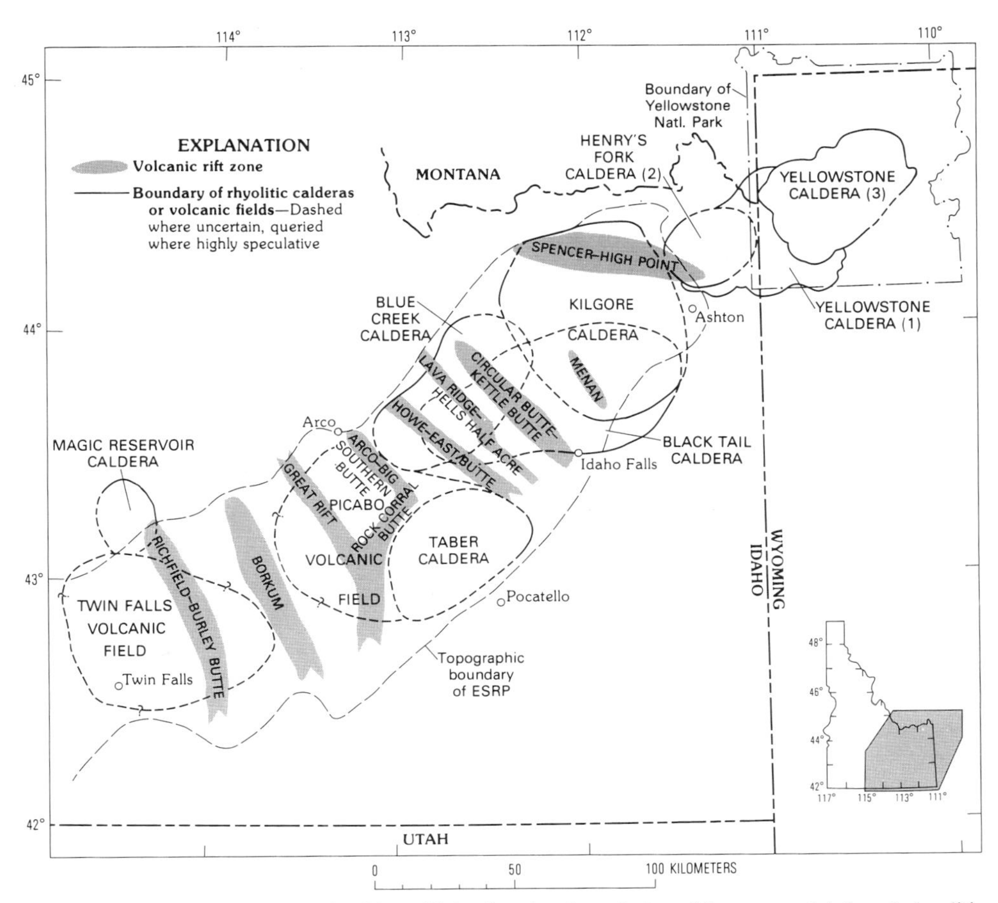

Figure 16. Map of eastern Snake River Plain showing boundaries of known and inferred rhyolitic calderas and volcanic rift zones. (1), (2), and (3) are designations from Christiansen (1982) for the order in which the three calderas formed. Compiled from Embree and others (1982), Christiansen (1982), Morgan (1988), Morgan and others (1984, 1989), Pierce and Morgan (this volume), and Morgan (this volume).

standing volcanic rift zones. Alternatively, these areas may overlie parts of underlying calderas that were once filled with low-density materials and have since compacted, leaving collapsed parts of calderas unrelated to caldera-forming eruptions.

Relationships between high density of basaltic vents, rhyolite domes, and underlying calderas

Within the Arco–Big Southern Butte and Howe–East Butte volcanic rift zones, the highest concentration of basaltic vents occurs near their centers where they intersect the western and southern boundary of the Blue Creek caldera (Fig. 17). This relationship suggests that interaction between fractures of the rift zones and the caldera boundary (ring fractures?) has localized basaltic vents. The rhyolite domes that form Big Southern Butte, Middle Butte, East Butte, and an associated, unnamed dome also are located along this same segment of the Blue Creek caldera boundary (Fig. 17). The Arco–Big Southern Butte and Howe–East Butte volcanic rift zones and the southwestern boundary of the Blue Creek caldera may have concentrated heat for partial melting of deep crustal rocks, which would have led to formation of rhyolite magma and emplacement of the domes (see Leeman, 1982d, for details of this process). This concept is supported by the general correlation between the ages of the rhyolite domes (300 ka to <1.0 Ma) and the ages (200 to 700 ka) of most nearby surface basaltic lava flows (Kuntz and others, 1992).

The rhyolite dome complex at Juniper Buttes (Figs. 18 and 19; Kuntz, 1979) is distinctly different from the rhyolite domes of Big Southern Butte, Middle Butte, East Butte and the unnamed dome in several respects: (1) it is much larger, (2) it is cut by a ditinctive pattern of rectilinear faults, (3) it contains a modified crestal graben, and (4) it does not lie on a volcanic rift zone. With respect to factors 1, 2, and 3, the Juniper Buttes dome complex is similar to resurgent domes of cauldron complexes (compare

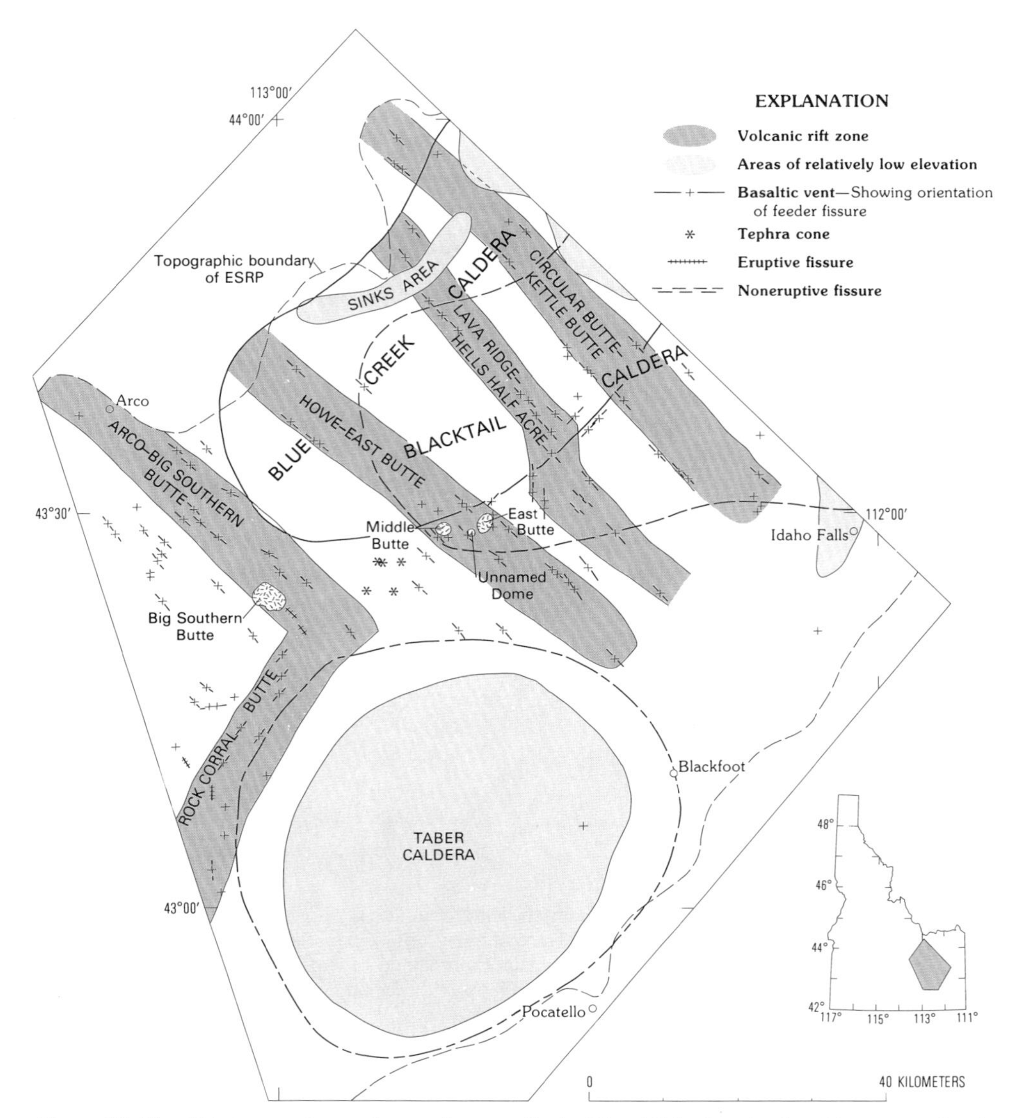

Figure 17. Simplified map of central part of eastern Snake River Plain showing basaltic vents, volcanic rift zones, boundaries of buried calderas, and other features referred to in text. Compiled from Kuntz and others (1992), LaPoint (1977), Pierce and Morgan (this volume), and Morgan (this volume). Map in this figure overlaps southwest boundary of map shown in Fig. 18.

Smith and Bailey, 1968). Morgan (1988) suggested that Juniper Buttes represents a resurgent dome or uplifted fault block in the Kilgore caldera.

Relatively high densities of basaltic vents unrelated to volcanic rift zones occur at several localities within or along the margins of the Kilgore caldera: at Table Butte, Lewisville Knolls, and northeast of Hamer (Fig. 18). The basaltic vents at Table Butte and Lewisville Knolls are not obviously elongated; thus they are probably not related to regional fracture patterns, and they do not lie on volcanic rift zones. The basaltic vents northeast of Hamer are elongated east-west, parallel to one of the fracture sets in the Juniper Buttes dome complex. At present, we have no explanation for the high concentration of basaltic vents at these three localities that might relate to structures or other features of the Kilgore caldera except to note that these vent concentrations occur along inferred caldera margins (Lewisville Knolls, Hamer) or at the intersection of caldera boundaries (Table Butte).

CONCLUDING REMARKS

The formation and petrologic evolution of the ESRP involves extremely complex, large-scale volcano-tectonic processes that have been described in detail only recently (see, for example, Leeman, 1982a, 1989; and Pierce and Morgan, this volume). From these studies, it is clear that the physiographic depression of

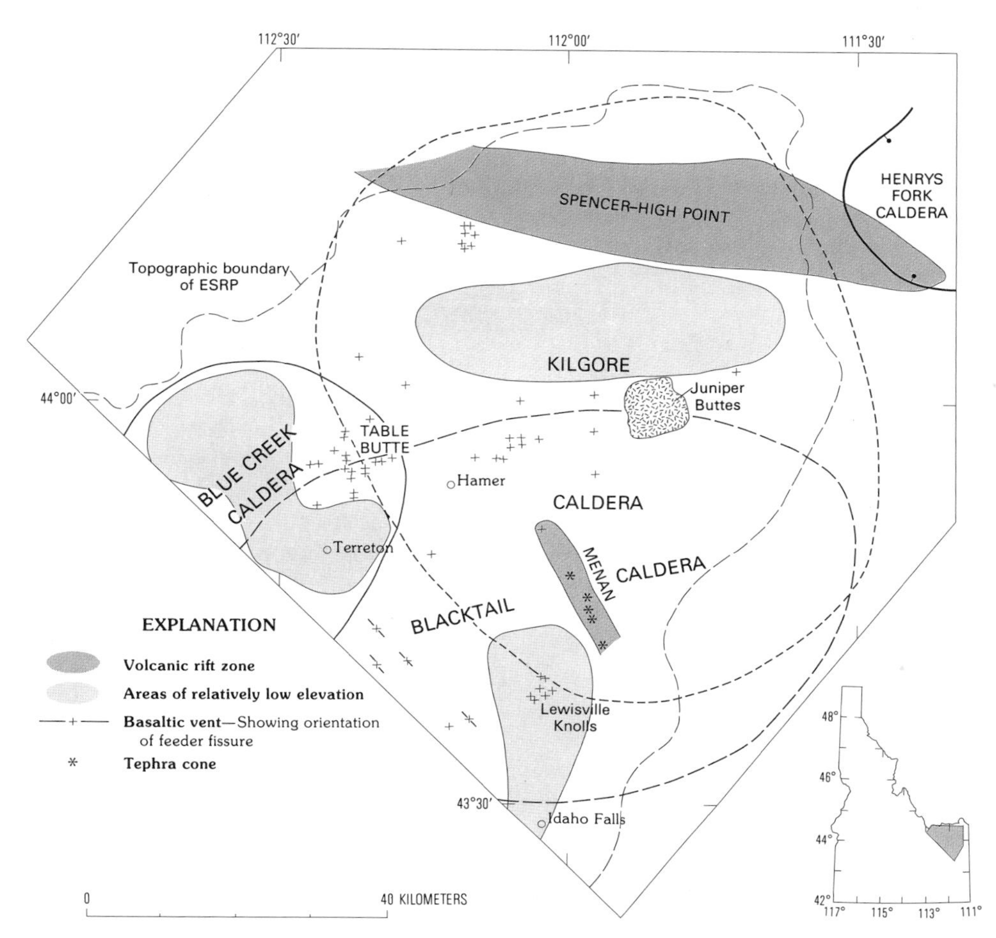

Figure 18. Simplified map of northeastern part of eastern Snake River Plain showing basaltic vents, volcanic rift zones, boundaries of buried calderas, and other features referred to in text. Compiled from unpublished mapping by M. A. Kuntz and G. F. Embree and from LaPoint (1977), Kuntz (1979), Pierce and Morgan (this volume), and Christiansen (1982). For simplicity, basaltic vents and structures in the Spencer–High Point volcanic rift zone are omitted in this figure but are shown in Figure 9. Map in this figure overlaps northeast boundary of map shown in Figure 17.

the ESRP formed initially as a result of earlier rhyolitic volcanism. The rhyolitic volcanism caused high heat flow, thinning and decreased rigidity of the crust, and progressive formation of a topographic trough along the chain of calderas.

The ESRP has been the site of dominantly shield-building basaltic eruptions for the last 4 m.y.; basaltic volcanism followed rhyolitic volcanism after subcaldera plutons first cooled and then fractured in response to regional northeast-southwest extension. Regional extensional strain has been concentrated in basin-and-range faults beyond the margins of the ESRP and in volcanic rift zones within the ESRP. Concentrated extension in volcanic rift zones has resulted in greater amounts of magma production, perhaps by decompression melting. Both of these processes, in turn, increased the number of suitable fractures for intrusion and extrusion of magma. The mechanical-thermal feedback nature of the mechanism created volcanic rift zones that were long-lived and self-perpetuating zones of dike emplacement and surface volcanism.

The ESRP is a unique basalt province in a tectonic sense: Eruptive fissures and volcanic rift zones are oriented essentially perpendicular to the long axis of the physiographic depression of the ESRP, and thus the ESRP is not an intracontinental, fault-bounded, volcanic-rift province, such as the African rift valleys and the Rio Grande rift. Rather, the ESRP is an intracontinental basalt province created through a combination of an earlier history dominated by rhyolitic volcanism that formed the ESRP depression and a more recent history that consists chiefly of basaltic volcanism. The origin of basalt magmas may be due to decompression melting controlled by the timing of strain release, and the location of basaltic vents is controlled by the orientation of contemporary regional stress patterns and resulting fracture pathways.

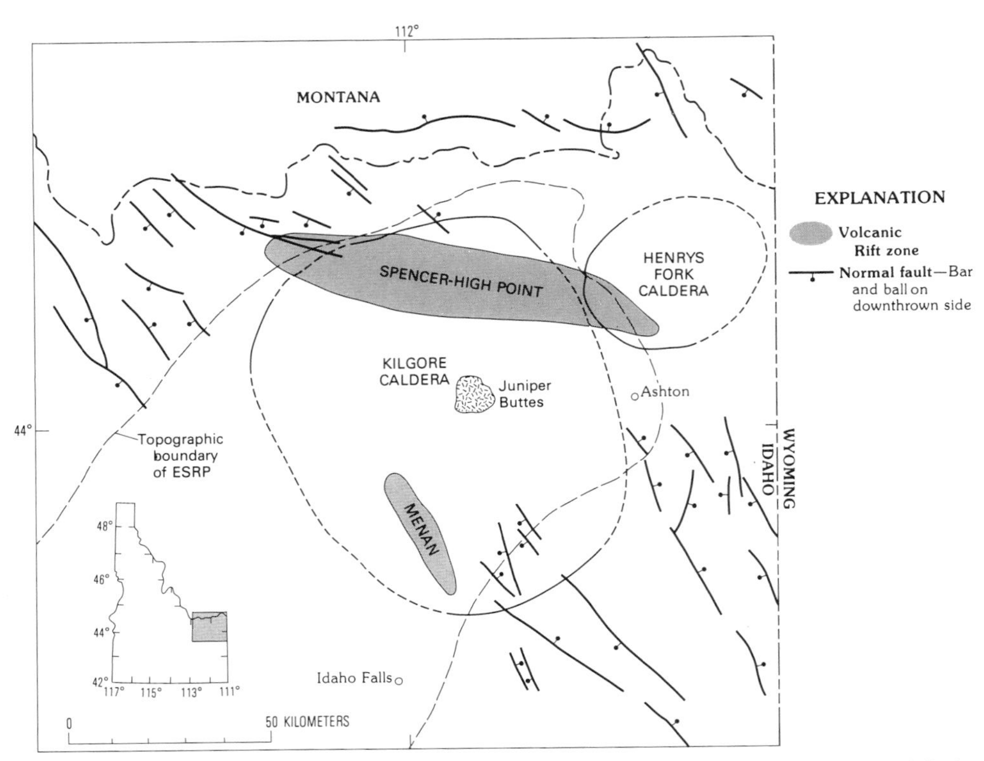

Figure 19. Simplified map of northeastern part of eastern Snake River Plain showing normal faults along margins of the plain, volcanic rift zones, boundaries of calderas, and other features referred to in text. Compiled from unpublished mapping by M. A. Kuntz and G. F. Embree and from Pierce and Morgan (this volume) and Christiansen (1982).

ACKNOWLEDGMENTS

Many people contributed to our studies and interpretations of basaltic volcanism of the ESRP, including U.S.G.S. colleagues D. E. Champion, R. T. Holcomb, S. R. Anderson, and P.J.I. LaPoint; academic colleagues R. H. Lefebvre (Grand Valley State College), W. P. Leeman (Rice University), J. S. King (SUNY-Buffalo), Ronald Greeley (Arizona State University), and W. R. Hackett (Idaho State University); David Clark of Craters of the Moon National Monument; R. P. Smith of EG&G Idaho, Inc., and others. Most of our conclusions and speculations were reached with naive neglect of their views. We thank L. D. Nealey, W. R. Hackett, S. H. Wood, and especially P. L. Martin for helpful reviews of the manuscript and Wayne Hawkins for his skillful preparation of the figures.

Our field studies were conducted under the aegis of the late Steven S. Oriel in his capacity as leader of the U.S. Geological Survey's Snake River Plain project. The guidance, encouragement, and patience that Steve rendered as project leader are deeply appreciated.

REFERENCES CITED

Armstrong, R. L., Leeman, W. P., and Malde, H. E., 1975, K-Ar dating and Quaternary and Neogene volcanic rocks of the Snake River Plain, Idaho: American Journal of Science, v. 275, p. 225–257.

Baksi, A. K., and Watkins, N. D., 1973, Volcanic production rates—Comparison of oceanic ridges, islands, and the Columbia Plateau basalts: Science, v. 180, p. 493–495.

Bhattacharya, B. K., and Mabey, D. R., 1980, Interpretation of magnetic anomalies over southern Idaho using generalized multibody methods: U.S. Geological Survey Open-File Report 80-457, 49 p.

Brott, C. A., Blackwell, D. D., and Mitchell, J. C., 1978, Tectonic implications of heat flow of the western Snake River Plain, Idaho: Geological Society of America Bulletin, v. 89, p. 1697–1707.

Brott, C. A., Blackwell, D. D., and Ziagos, J. P., 1981, Thermal and tectonic implications of heat flow in the eastern Snake River Plain, Idaho: Journal of Geophysical Research, ser. B, v. 86, p. 11709–11734.

Champion, D. E., 1973, The relationship of large-scale surface morphology to lava flow direction, Wapi lava field, southeast Idaho [M.S. thesis]: Buffalo, State University of New York, 44 p.

—— , 1980, Holocene geomagnetic secular variation in the western United States: Implications for global geomagnetic field: U.S. Geological Survey Open-File Report 80-824, 326 p.

Champion, D. E., and Greeley, R., 1977, Geology of the Wapi lava field, Snake River Plain, Idaho, *in* Greeley, R., and King, J. S., eds., Volcanism of the eastern Snake River Plain, Idaho: A comparative planetary geology guidebook: Washington, D.C., National Aeronautics and Space Administration, p. 133–152.

Champion, D. E., Lanphere, M. A., and Kuntz, M. A., 1988, Evidence for a new geomagnetic reversal from lava flows in Idaho: Discussion of short polarity reversals in the Brunhes and late Matuyama polarity chrons: Journal of Geophysical Research, v. 93, p. 11667–11680.

Champion, D. E., Kuntz, M. A., and Lefebvre, R. H., 1989, Geologic map of the North Laidlaw Butte quadrangle, Blaine and Butte Counties, Idaho: U.S. Geological Survey Geologic Quadrangle Map GQ-1634, scale 1:24,000.

Christiansen, R. L., 1982, Late Cenozoic volcanism of the Island Park area, eastern Idaho, *in* Bonnichsen, B., and Breckenridge, R. M., eds., Cenozoic geology of Idaho: Idaho Bureau of Mines and Geology Bulletin 26, p. 345–368.

Condit, C. D., Crumpler, L. S., Aubele, J. C., and Elston, W. E., 1989, Patterns of volcanism along the southern margin of the Colorado Plateau: The Springerville field: Journal of Geophysical Research, v. 94, p. 7975–7986.

Corbett, M. K., 1975, Structural development of the eastern Snake River Plain, Idaho, as an intracontinental rift: Geological Society of America Abstracts with Programs, v. 7, p. 599–600.

Covington, H. R., 1977, Preliminary geologic map of the Pillar Butte, Pillar Butte NE, Pillar Butte SE, and Rattlesnake Buttes quadrangles, Bingham, Blaine, and Power Counties, Idaho: U.S. Geological Survey Open-File Report 77-779, scale 1:48,000.

Doherty, D. J., McBroome, L. A., and Kuntz, M. A., 1979, Preliminary geological interpretation and lithologic log of the exploratory geothermal test well (INEL-1), Idaho National Engineering Laboratory, eastern Snake River Plain, Idaho: U.S. Geological Survey Open-File Report 79-1248, 9 p.

Embree, G. F., McBroome, L. A., and Doherty, D. J., 1982, Preliminary stratigraphic framework of the Pliocene and Miocene rhyolite, eastern Snake River Plain, Idaho, *in* Bonnichsen, B., and Breckenridge, R. M., eds., Cenozoic geology of Idaho: Idaho Bureau of Mines and Geology Bulletin 26, p. 333–343.

Futa, K., Kuntz, M. A., and Stout, M. Z., 1988, Major- and trace-element geochemistry and Sr and Nd isotopic characteristics of flows from Craters of the Moon lava field: Geological Society of America Abstracts with Programs, v. 20, p. 415.

Greeley, R., 1977, Basaltic "plains" volcanism, *in* Greeley, R., and King, J. S., eds., Volcanism of the eastern Snake River Plain, Idaho: A comparative planetary geology guidebook: Washington, D.C., National Aeronautics and Space Administration, p. 23–44.

——, 1982a, The style of basaltic volcanism in the eastern Snake River Plain, Idaho, *in* Bonnichsen, B., and Breckenridge, R. M., eds., Cenozoic geology of Idaho: Idaho Bureau of Mines and Geology Bulletin 26, p. 407–422.

——, 1982b, The Snake River Plain, Idaho: Representative of a new category of volcanism: Journal of Geophysical Research, v. 87, p. 2705–2712.

Greeley, R., and King, J. S., 1975, Geologic field guide to the Quaternary volcanics of the south-central Snake River Plain, Idaho: Idaho Bureau of Mines and Geology Pamphlet 160, 49 p.

Greeley, R., Theiling, E., and King, J. S., 1977, Guide to the geology of Kings Bowl lava field, *in* Greeley, R., and King, J. S., eds., Volcanism of the eastern Snake River Plain, Idaho: A comparative planetary geology guidebook: Washington, D.C., National Aeronautics and Space Administration, p. 171–188.

Hackett, W. R., and Morgan, L. A., 1988, Explosive basaltic and rhyolitic volcanism of the eastern Snake River Plain, *in* Link, P. K., and Hackett, W. R., eds., Guidebook to the geology of central and southern Idaho: Idaho Geological Survey Bulletin 27, p. 283–301.

Hamilton, W., 1987, Plate-tectonic evolution of the western U.S.A.: Episodes, v. 10, p. 271–276.

Haskin, L. A., Haskin, M. A., Frey, F. A., and Wildeman, T. R., 1969, Relative and absolute terrestrial abundances of the rare earth elements, *in* Ahrens, L. H., ed., Origin and distribution of the elements: New York, Pergamon Press, p. 889–912.

Holcomb, R. T., 1987, Eruptive history and long-term behavior of Kilauea Volcano: U.S. Geological Survey Professional Paper 1350, p. 261–350.

Jamieson, B. G., and Clarke, D. B., 1970, Potassium and associated elements in tholeiitic basalts: Journal of Petrology, v. 11, p. 183–204.

Karlo, J. F., 1977, The geology and Bouguer gravity of the Hells Half Acre area and their relation to volcano-tectonic processes within the Snake River Plain rift zone, Idaho [Ph.D. thesis]: Buffalo, State University of New York, 152 p.

King, J. S., 1977, Crystal Ice Cave and Kings bowl Crater, eastern Snake River Plain, Idaho, *in* Greeley, R., and King, J. S., eds., Volcanism of the eastern Snake River Plain, Idaho: A comparative planetary geology guidebook: National Aeronautics and Space Administration, p. 153–163.

——, 1982, Selected volcanic features of the south-central Snake River Plain, Idaho, *in* Bonnichsen, B., and Breckenridge, R. M., eds., Cenozoic geology of Idaho: Idaho Bureau of Mines and Geology Bulletin 26, p. 439–454.

Kuntz, M. A., 1977a, Rift zones of the Snake River Plain, Idaho, as extensions of basin-range and older structures: Geological Society of America Abstracts with Programs, v. 9, p. 1061–1062.

——, 1977b, Extensional faulting and volcanism along the Arco rift zone, eastern Snake River Plain, Idaho: Geological Society of America Abstracts with Programs, v. 9, p. 740–741.

——, 1978, Geologic map of the Arco–Big Southern Butte area, Butte, Blaine, and Bingham Counties, Idaho: U.S. Geological Survey Open-File Report 78-302, scale 1:62,500.

——, 1979, Geologic map of the Juniper Buttes area, eastern Snake River Plain, Idaho: U.S. Geological Survey Miscellaneous Investigations Series Map I-1115, scale 1:48,000.

Kuntz, M. A., and Covington, H. R., 1979, Do basalt structures and topographic features reflect buried calderas in the eastern Snake River Plain (ESRP)?: EOS, Transactions of the American Geophysical Union, v. 60, p. 945.

Kuntz, M. A., Champion, D. E., Spiker, E. C., Lefebvre, R. H., and McBroome, L. A., 1982, The Great Rift and the evolution of the Craters of the Moon lava field, Idaho, *in* Bonnichsen, B., and Breckenridge, R. M., eds., Cenozoic geology of Idaho: Idaho Bureau of Mines and Geology Bulletin 26, p. 423–437.

Kuntz, M. A., Lefebvre, R. H., Champion, D. E., King, J. S., and Covington, H. R., 1983, Holocene basaltic volcanism along the Great Rift, central and eastern Snake River Plain, Idaho: Utah Geological and Mineral Survey Special Studies 61, Guidebook, pt. 3, p. 1–34.

Kuntz, M. A., Elsheimer, N. H., Espos, L. F., and Klock, P. R., 1985, Major-element analyses of latest Pleistocene–Holocene lava fields of the Snake River Plain, Idaho: U.S. Geological Survey Open-File Report 85-593, 64 p.

Kuntz, M. A., Champion, D. E., Spiker, E. C., and Lefebvre, R. H., 1986, Contrasting magma types and steady-state, volume-predictable basaltic volcanism along the Great Rift, Idaho: Geological Society of America Bulletin, v. 97, p. 579–594.

Kuntz, M. A., Spiker, E. C., Rubin, M., Champion, D. E., and Lefebvre, R. H., 1986, Radiocarbon studies of latest Pleistocene and Holocene lava flows of the Snake River Plain, Idaho; data, lessons, interpretations: Quaternary Research, v. 25, p. 163–176.

Kuntz, M. A., Champion, D. E., Lefebvre, R. H., and Covington, H. R., 1988, Geologic map of the Craters of the Moon, Kings Bowl, and Wapi lava fields and the Great Rift volcanic rift zone, south-central Idaho: U.S. Geological Survey Miscellaneous Investigations Series Map I-1632, scale 1:100,000.

Kuntz, M. A., Champion, D. E., and Lefebvre, R. H., 1989, Geologic map of the Fissure Butte quadrangle, Blaine and Butte Counties, Idaho: U.S. Geological Survey Geologic Quadrangle Map GQ-1635, scale 1:24,000.

Kuntz, M. A., Lefebvre, R. H., and Champion, D. E., 1989a, Geologic map of the Inferno Cone quadrangle, Butte County, Idaho: U.S. Geological Survey Geologic Quadrangle Map GQ-1632, scale 1:24,000.

——, 1989b, Geologic map of The Watchman quadrangle, Butte County, Idaho: U.S. Geological Survey Geologic Quadrangle Map GQ-1633, scale 1:24,000.

Kuntz, M. A., and twelve others, 1992, Geologic map of the Idaho National Engineering Laboratory and adjoining areas, eastern Idaho: U.S. Geological Survey Miscellaneous Investigations Series Map I-2330, scale 1:100,000.

LaPoint, P.J.I., 1977, Preliminary photogeologic map of the eastern Snake River Plain, Idaho: U.S. Geological Survey Miscellaneous Field Studies Map MF-850, scale 1:250,000.

Leeman, W. P., 1974, Part 1: Petrology of basalt lavas from the Snake River Plain, Idaho. Part 2: Experimental determination of partioning of divalent cations between olivine and basaltic liquid [Ph.D. thesis]: Eugene, University of Oregon, 337 p.

—— , 1982a, Development of the Snake River Plain–Yellowstone Plateau province, Idaho and Wyoming: An overview and petrologic model, *in* Bonnichsen, B., and Breckenridge, R. M., eds., Cenozoic geology of Idaho: Idaho Bureau of Mines and Geology Bulletin 26, p. 155–177.

—— , 1982b, Olivine tholeiitic basalts of the Snake River Plain, Idaho, *in* Bonnichsen, B., and Breckenridge, R. M., eds., Cenozoic geology of Idaho: Idaho Bureau of Mines and Geology Bulletin 26, p. 181–191.

—— , 1982c, Evolved and hybrid lavas from the Snake River Plain, Idaho, *in* Bonnichsen, B., and Breckenridge, R. M., eds., Cenozoic geology of Idaho: Idaho Bureau of Mines and Geology Bulletin 26, p. 193–202.

—— , 1982d, Rhyolites of the Snake River Plain–Yellowstone Plateau province, Idaho and Wyoming: A summary of petrogenetic models, *in* Bonnichsen, B., and Breckenridge, R. M., eds., Cenozoic geology of Idaho: Idaho Bureau of Mines and Geology Bulletin 26, p. 203–212.

—— , 1989, Origin and development of the Snake River Plain (SRP), Idaho, *in* Ruebelmann, K. L., ed., Snake River Plain–Yellowstone volcanic province: International Geological Congress, 28th, Guidebook T305: Washington, D.C., American Geophysical Union, p. 4–12.

Leeman, W. P., and Manton, W. I., 1971, Strontium isotopic composition of basaltic lavas from the Snake River Plain, southern Idaho: Earth and Planetary Science Letters, v. 11, p. 420–434.

Leeman, W. P., and Vitaliano, C. J., 1976, Petrology of McKinney basalt, Snake River Plain, Idaho: Geological Society of America Bulletin, v. 87, p. 1777–1792.

Leeman, W. P., Vitaliano, C. J., and Prinz, M., 1976, Evolved lavas from the Snake River Plain, Craters of the Moon National Monument, Idaho: Contributions to Mineralogy and Petrology, v. 56, p. 35–60.

Leeman, W. P., Menzies, M. A., Matty, D. J., and Embree, G. F., 1985, Strontium, neodymium, and lead isotopic compositions of deep crustal xenoliths from the Snake River Plain; evidence for Archean basement: Earth and Planetary Science Letters, v. 75, p. 354–368.

Lefebvre, R. H., 1975, Mapping in Craters of the Moon volcanic field, Idaho, with LANDSAT (ERTS) imagery: Proceedings, Ann Arbor, University of Michigan, 10th International Symposium on Remote Sensing of the Environment: v. 2, no. 10, p. 951–957.

Link, P. K., Skipp, B., Hait, M. H., Jr., Janecke, S. U., and Burton, B. R., 1988, Structural and stratigraphic transect of south-central Idaho; a field guide to the Lost River, White Knob, Pioneer, Boulder, and Smoky Mountains, *in* Link, P. K., and Hackett, W. R., eds., Guidebook to the geology of central and southern Idaho: Idaho Geological Survey Bulletin 27, p. 5–42.

Mabey, D. R., 1982, Geophysics and tectonics of the Snake River Plain, Idaho, *in* Bonnichsen, B., and Breckenridge, R. M., eds., Cenozoic geology of Idaho: Idaho Bureau of Mines and Geology Bulletin 26, p. 139–153.

Macdonald, G. A., 1972, Volcanoes: Englewood Cliffs, N.J., Prentice-Hall, 510 p.

Macdonald, G. A., and Abbott, A. T., 1970, Volcanoes in the sea: Honolulu, University of Hawaii Press, 441 p.

Malde, H. E., 1991, Quaternary geology and structural history of the Snake River Plain, Idaho and Oregon, *in* Morrison, R. B., ed., Quaternary nonglacial geology; Conterminous U.S.: Boulder, Colorado, Geological Society of America, The geology of North America, v. K-2, p. 252–281.

Malde, H. E., and Powers, H. A., 1962, Upper Cenozoic stratigraphy of the western Snake River Plain, Idaho: Geological Society of America Bulletin, v. 73, p. 1197–1220.

Menzies, M. A., Leeman, W. P., and Hawkesworth, C. J., 1984, Geochemical and isotopic evidence for the origin of continental flood basalts with particular reference to the Snake River Plain, Idaho, U.S.A.: Transactions of the Royal Society of London, v. A310, p. 643–660.

Morgan, L. A., 1988, Explosive volcanism on the eastern Snake River Plain [Ph.D. thesis]: Manoa, University of Hawaii, 191 p.

Morgan, L. A., Doherty, D. J., and Leeman, W. P., 1984, Ignimbrites of the eastern Snake River Plain: Evidence for major caldera-forming eruptions: Journal of Geophysical Research, v. 89, p. 8665–8678.

Morgan, L. A., Doherty, D. J., and Bonnichsen, B., 1989, Evolution of the Kilgore caldera: A model for caldera formation on the Snake River Plain–Yellowstone Plateau volcanic province; Continental magmatism abstracts: New Mexico Bureau of Mines and Mineral Resources Bulletin 131, p. 195.

Murtaugh, J. G., 1961, Geology of Craters of the Moon National Monument, Idaho [M.S. thesis]: Moscow, University of Idaho, 99 p.

Nakamura, K., 1974, Preliminary estimate of global volcanic production rate, *in* Colp, J. L., ed., The utilization of volcano energy: Albuquerque, N.M., Sandia National Laboratories, p. 273–285.

Peterson, D. W., and Moore, R. B., 1987, Geologic history and evolution of concepts, Island of Hawaii: U.S. Geological Survey Professional Paper 1350, p. 149–190.

Pierce, K. L., Fosberg, M. A., Scott, W. E., Lewis, G. C., and Colman, S. M., 1982, Loess deposits of southeastern Idaho: Age and correlation of the upper two loess units, *in* Bonnichsen, B., and Breckenridge, R. M., eds., Cenozoic geology of Idaho: Idaho Bureau of Mines and Geology Bulletin 26, p. 717–725.

Porter, S. C., Pierce, K. L., and Hamilton, T. D., 1983, Late Pleistocene glaciation in the western United States, *in* Porter, S. C., ed., The late Pleistocene, v. 1 *of* Wright, H. E., Jr., ed., Late Quaternary environments of the United States: Minneapolis, University of Minnesota Press, p. 71–111.

Prinz, M., 1970, Idaho rift system, Snake River Plain, Idaho: Geological Society of America Bulletin, v. 81, p. 941–947.

Richter, D. H., Eaton, J. P., Murata, K. J., Ault, W. U., and Krivoy, H. L., 1970, Chronological narrative of the 1959–1960 eruption of Kilauea Volcano, Hawaii: U.S. Geological Survey Professional Paper 539-E, 73 p.

Rodgers, D. W., and Zentner, N. C., 1988, Fault geometries along the northern margin of the eastern Snake River Plain, Idaho: Geological Society of America Abstracts with Programs, v. 20, p. 465–466.

Russell, I. C., 1902, Geology and groundwater resources of the Snake River Plains of Idaho: U.S. Geological Survey Bulletin 199, 192 p.

Smith, R. L., and Bailey, R. A., 1968, Resurgent cauldrons, *in* Coats, R. R., Hay, R. L., and Anderson, C. A., eds., Studies in volcanology, Geological Society of America Memoir 116, p. 613–662.

Smith, R. P., Hackett, W. R., and Rodgers, D. W., 1989, Surface deformation along the Arco rift zone, eastern Snake River Plain, Idaho: Geological Society of America Abstracts with Programs, v. 21, p. 146.

Spear, D. B., 1977, Big Southern, Middle, and East Buttes, *in* Greeley, R., and King, J. S., eds., Volcanism of the eastern Snake River Plain, Idaho: A comparative planetary geology guidebook: Washington, D.C., National Aeronautics and Space Administration, p. 113–120.

Spear, D. B., and King, J. S., 1982, The geology of Big Southern Butte, Idaho, *in* Bonnichsen, B., and Breckenridge, R. M., eds., Cenozoic geology of Idaho: Idaho Bureau of Mines and Geology Bulletin 26, p. 395–403.

Stanley, W. D., Boehl, J. E., Bostick, F. X., and Smith, H. W., 1977, Geothermal significance of magnetotelluric soundings in the eastern Snake River Plain–Yellowstone region: Journal of Geophysical Research, v. 82, p. 2501–2514.

Stearns, H. T., 1928, Craters of the Moon National Monument, Idaho: Idaho Bureau of Mines and Geology Bulletin 13, 57 p.

Stearns, H. T., Crandall, L., and Steward, W. G., 1938, Geology and groundwater resources of the Snake River Plain in southeastern Idaho: U.S. Geological Survey Water Supply Paper 774, 268 p.

Stickney, M. C., and Bartholomew, M. J., 1987, Seismicity and late Quaternary faulting of the northern Basin and Range province, Montana and Idaho: Bulletin of the Seismological Society of America, v. 77, p. 1602–1625.

Stone, G. T., 1967, Petrology of upper Cenozoic basalts of the Snake River Plain

[Ph.D. thesis]: Boulder, University of Colorado, 392 p.
——, 1970, Highly evolved basaltic lavas in the western Snake River Plain, Idaho: Geological Society of America Abstracts with Programs, v. 2, p. 695–696.
Stout, M. Z., 1975, Mineralogy and petrology of Quaternary lavas, Snake River Plain, Idaho, and the cation distribution in natural titanomagnetites [M.S. thesis]: Calgary, Alberta, University of Calgary, 150 p.
Stout, M. Z., and Nicholls, J., 1977, Mineralogy and petrology of Quaternary lavas from the Snake River Plain, Idaho: Canadian Journal of Earth Sciences, v. 14, p. 2140–2156.
Stout, M. Z., Nicholls, J., and Kuntz, M. A., 1989, Fractionation and contamination processes, Craters of the Moon lava field, Idaho, 2000–2500 years B.P.; Continental magmatism abstracts: New Mexico Bureau of Mines and Mineral Resources Bulletin 131, p. 259.
Swanson, D. A., 1973, Pahoehoe flows from the 1969–1971 Mauna Ulu eruption of Kilauea Volcano, Hawaii: Geological Society of America Bulletin, v. 84, p. 615–626.
Swanson, D. A., Duffield, W. A., Jackson, D. B., and Peterson, D. W., 1979, Chronological narrative of the 1969–71 Mauna Ulu eruption of Kilauea Volcano, Hawaii: U.S. Geological Survey Professional Paper 1056, 55 p.
Takahashi, T. J., and Griggs, J. T., 1987, Hawaiian volcanic features; a photoglossary: U.S. Geological Survey Professional Paper 1350, p. 845–902.
Tanaka, K. L., Shoemaker, E. M., Ulrich, G. E., and Wolfe, E. W., 1986, Migration of volcanism in the San Francisco volcanic field, Arizona: Geological Society of America Bulletin, v. 97, p. 129–141.
Thompson, R. N., 1975, Primary basalts and magma genesis; II, Snake River Plain, Idaho, U.S.A.: Contributions to Mineralogy and Petrology, v. 52, p. 213–232.
Tilley, C. E., and Thompson, R. N., 1970, Melting and crystallization relations of the Snake River basalts of southern Idaho, USA: Earth and Planetary Science Letters, v. 8, p. 79–92.
Tilling, R. I., and seven others, 1987, The 1972–1974 Mauna Ulu eruption, Kilauea Volcano: An example of quasi-steady-state magma transfer: U.S. Geological Survey Professional Paper 1350, p. 405–469.
Walker, G.P.L., 1972, Compound and simple lava flows and flood basalts: Bulletin of Volcanology, v. 36, p. 579–590.
Wentworth, C. K., and Macdonald, G. A., 1953, Structures and forms of basaltic rocks in Hawaii: U.S. Geological Survey Bulletin 994, 98 p.
Wolfe, E. W., Garcia, M. O., Jackson, D. B., Koyanagi, R. Y., Neal, C. A., and Okamura, A. T., 1987, The Puu Oo eruption of Kilauea Volcano, episodes 1–20, January 3, 1983, to June 8, 1984: U.S. Geological Survey Professional Paper 1350, p. 471–508.
Womer, M. B., Greeley, R., and King, J. S., 1982, Phreatic eruptions of the eastern Snake River Plain, Idaho, *in* Bonnichsen, B., and Breckenridge, R. M., eds., Cenozoic geology of Idaho: Idaho Bureau of Mines and Geology Bulletin 26, p. 453–464.
Yoder, H. S., and Tilley, C. E., 1962, The origin of basalt magmas: An experimental study of natural and synthetic rock systems: Journal of Petrology, v. 3, p. 342–532.
Zoback, M. L., and Zoback, M. D., 1980, State of stress in the conterminous United States: Journal of Geophysical Research, v. 85, p. 6113–6156.

MANUSCRIPT ACCEPTED BY THE SOCIETY JULY 19, 1991

Printed in U.S.A.

Geological Society of America
Memoir 179
1992

Chapter 13

Quaternary stratigraphy of an area northeast of American Falls Reservoir, eastern Snake River Plain, Idaho

Brenda B. Houser
U.S. Geological Survey, Tucson Field Office, Gould-Simpson Building #77, University of Arizona, Tucson, Arizona 85721

ABSTRACT

Quaternary rocks northeast of American Falls Reservoir, Idaho, include more than 140 m of poorly indurated sedimentary rocks and intercalated basalt flows. Previous stratigraphic studies in the region include reconnaissance and detailed mapping and topical studies based mostly on surface exposures. The present study extends the Quaternary stratigraphy into the shallow subsurface by using lithologic data from drillers' logs of about 240 water wells.

Pre-Quaternary rocks of the area include Miocene bimodal volcanic and volcaniclastic rocks and Pliocene basalt of Buckskin Basin. The surface developed on the pre-Quaternary rocks slopes gently to the west and northwest except locally, as at Ferry Butte, where doming is inferred to be the result of shallow emplacement of a rhyolite intrusive body.

Quaternary rocks of the study area consist of coarse-grained alluvium deposited by the Snake River, thin distal edges of basalt flows from sources to the northwest, fine-grained alluvial fan deposits derived from highlands to the southeast, two fine-grained fluvial and lacustrine deposits (Raft Formation and American Falls Lake Beds), and deposits of the Bonneville Flood.

The Raft Formation, present only in the subsurface in the area, is correlated with the Raft in exposures along the Snake River southwest of American Falls and at the northern end of the Raft River basin on the basis of stratigraphic position and virtually identical thickness and lithology. The minimum age of the Raft Formation in the study area is constrained by an overlying basalt flow dated at 0.2 Ma.

Driller's logs and surface exposures show that the American Falls Lake Beds are at a higher elevation and are coarser grained northwest of the Snake River than they are southeast of the river. These differences are attributed to slightly different depositional environments caused by the geometry of the lake basin. Dating of a basalt flow that overlies the American Falls Lake Beds north of Springfield will provide valuable information on the history of American Falls Lake.

INTRODUCTION

More than 140 m of sedimentary rocks and intercalated basalt flows have accumulated along the southeastern margin of Idaho's eastern Snake River Plain (Fig. 1) from late Pliocene through Holocene time. Southwest of Blackfoot, the sedimentary rocks consist of fluvial gravel units deposited by the ancestral and modern Snake River and of two finer-grained formations, the Raft Formation and the American Falls Lake Beds, that were deposited in lacustrine as well as fluvial environments. Lacustrine depositional environments have developed from time to time in the Snake River Valley when lava from eruptive centers to the north flowed south and dammed the river. The lava dam at Eagle Rock that formed late Pleistocene American Falls Lake came

Houser, B. B., 1992, Quaternary stratigraphy of an area northeast of American Falls Reservoir, eastern Snake River Plain, Idaho, *in* Link, P. K., Kuntz, M. A., and Platt, L. B., eds., Regional Geology of Eastern Idaho and Western Wyoming: Geological Society of America Memoir 179.

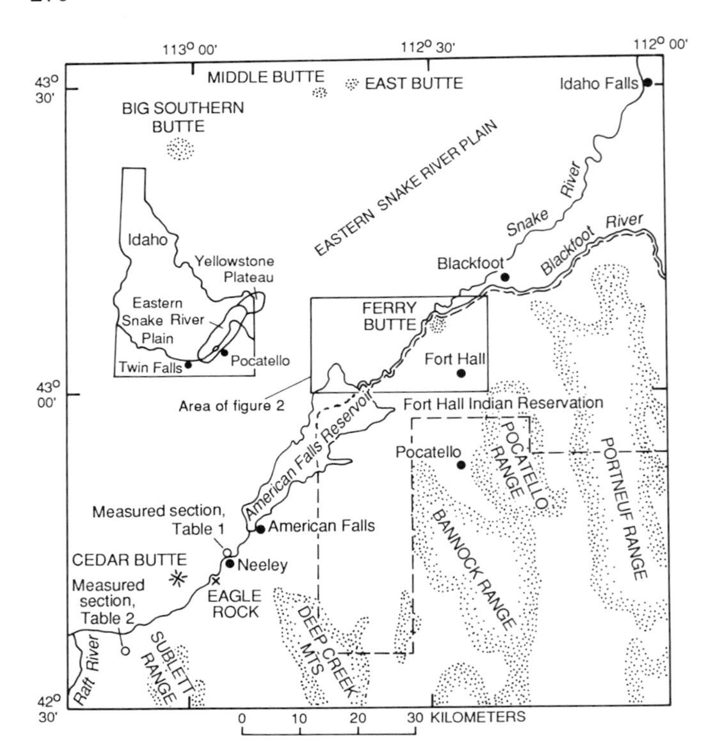

Figure 1. Index map of a part of southeastern Idaho showing major physiographic features adjacent to the study area. Dashed line is approximate boundary of the Fort Hall Indian Reservation.

from Cedar Butte volcano, but the source and location of the lava dam behind which the lacustrine units of the early and middle Pleistocene Raft formation were deposited are not known.

This chapter presents new information on the subsurface distribution and age relationships of the Snake River gravel deposits. Raft Formation, American Falls Lake Beds, and deposits of the Bonneville Flood between the northern end of American Falls Reservoir and the town of Blackfoot, in Bingham and Bannock counties, Idaho (Figs. 1 and 2). Much of this new information was derived from the large data base of drillers' logs of water wells that is available in the farmed areas of the Snake River Plain.

Previous work

Stearns and others (1938) described and named the Raft Formation (originally Raft Lake Beds) and the American Falls Lake Beds in their reconnaissance study of the geology and hydrology of the Snake River Plain. The Raft was named for exposures of light-brown clay, silt, and sand along the Snake River between American Falls and the mouth of the Raft River (Fig. 1). Trimble and Carr (1961) and Carr and Trimble (1963) inferred from well data that the Raft Formation is present in the subsurface 2 to 4 km northwest of the town of American Falls.

Stearns and others (1938) named the American Falls Lake Beds for exposures of chiefly clayey sedimentary rocks overlying the Raft Formation along the Snake River between Eagle Rock and the area around the present-day American Falls Reservoir. They recognized that the sediments of this unit were deposited in a lake upstream from a lava dam at Eagle Rock. The dam was formed by basalt flows dated at 72 ka (Scott and others, 1982) from Cedar Butte, a shield volcano 15 km southwest of American Falls. The distribution of the American Falls Lake Beds mapped by Scott (1982) northwest of the Snake River shows that the lake extended about 65 km northeast of the dam, almost to Blackfoot (Fig. 1).

Based on detailed geologic mapping in the area, Trimble and Carr (1961) and Carr and Trimble (1963) restricted the American Falls Lake Beds to the upper part of the sequence originally assigned to the American Falls by Stearns and others, and they defined a persistent gravel bed as the basal unit of the formation. Ridenour's (1969) sedimentologic study of the American Falls Lake Beds led him to conclude that the fine-grained part of the formation, above the basal fluvial gravel, represents deposition in two environments—fluvial channel and floodplain (lower part) and lacustrine (upper part)—not a lacustrine environment alone as interpreted by Trimble and Carr.

Although evidence is presented in this study that late Pleistocene American Falls Lake had been at least partly drained prior to the Bonneville Flood, the floodwaters undoubtedly completed the breaching of the lava dam at Eagle Rock. Earlier studies of Trimble and Carr (1961) and Malde (1968) indicated that the flood occurred about 30 ka, but Scott and others (1982) presented new K-Ar dates. These dates, in conjunction with studies of the chronology of Lake Bonneville, indicate that the Bonneville Flood (and thus the latest possible time of draining of American Falls Lake) occurred about 15 ka.

Additional aspects of the Bonneville Flood have been investigated by Hearst (1990) and O'Connor (1990). Hearst (1990) discussed the Late Pleistocene Duck Point local fauna and concluded that the gravel body in which the bones are found was deposited by the Bonneville Flood. Ridenour (1969) had correlated the Duck Point gravel with the basal fluvial gravel of the American Falls Lake Beds. O'Connor (1990) studied the hydraulic parameters and sediment transport of the Bonneville floodwaters between Red Rock Pass and Lewiston, Idaho, 1,100 km downstream.

Geologic setting

The eastern Snake River Plain is a northeast-trending, basalt-covered structural downwarp of late Cenozoic age, about 90 km wide and 225 km long, that extends from the Yellowstone Plateau on the northeast to Twin Falls on the southwest (Fig. 1). The downwarp has been interpreted by several authors to be the trace of movement of the North American plate over an upper mantle hot spot, with the Quaternary silicic volcanic eruptive centers of the Yellowstone Plateau marking the northeasten (youngest) end of the inferred hot-spot trace (Morgan, 1972;

Smith and Sbar, 1974; Suppe and others, 1975; Morgan and Pierce, 1990; Pierce and Morgan, this volume).

Along the axis of the plain the downwarp is filled with more than 3,000 m of mafic and silicic volcanic rocks and intercalated volcaniclastic and sedimentary rocks (Whitehead, 1986). Well data indicate that the total thickness of late Miocene to Holocene sedimentary and volcanic rocks at the southeastern edge of the plain beneath the study area is from 185 m to more than 270 m. These rocks record late Miocene and Pliocene basaltic to rhyolitic volcanism; Pliocene and Quaternary basaltic volcanism; Quaternary eolian, fluvial, and lacustrine deposition; and the Bonneville Flood of about 15 ka.

No evidence of faulting along the margin of the plain was found either through stratigraphic analysis of the well data or by examination of surface exposures. The mechanism of downwarping apparent in the upper stratigraphic levels of the eastern Snake River Plain is northwestward tilting amounting to about 0.3° at the surface of the late Pliocene basalt of Buckskin Basin and about 0.2° at the base of the Pleistocene Raft Formation.

Method

The study involved detailed investigation and sampling of surface exposures, reconnaissance mapping, and analysis of drillers' logs of water wells for subsurface control. The initial data base consisted of about 500 well logs, most of which were obtained from the Idaho Department of Water Resources, Boise, Idaho. Logs of four test holes in Fort Hall Bottoms were supplied by George A. Desborough (U.S. Geological Survey) (well nos. 1241, 1242, 1243, and 1244). The locations of the wells were not field checked, although mislocation is a common and acknowledged problem in using water well data. The data base was reduced to about 270 wells by deletion of logs of obviously mislocated wells (those located in the middle of American Falls Reservoir, for example) and questionably located wells, based on stratigraphic criteria. These criteria were derived from the stratigraphic model of the subsurface that was developed and refined by means of an iterative process of cross-section construction using GSLITH, a computer program written by Selner and Taylor (1989). A log that described a sequence consisting entirely of gravel was deleted if, for example, the logs of all the surrounding wells indicated that a 15-m-thick clay layer was interbedded in the gravel sequence.

Locations are exact for about 20 wells drilled for identified public facilities such as schools and churches, and perhaps 50 additional locations, described in terms of street addresses and so forth, are close to exact; most wells are approximately located within 1/16 of a section (40 acres) or 0.4 km, but a few (less than 10%) possibly are erroneous. Because of the low relief of the Snake River Plain, approximate location of wells within 40 acres results in elevation errors of commonly less than ± 3 m.

The geologic cross sections (Figs. 3, 4, and 5) were constructed using wells projected to the lines of section from as far away as 1.4 km on either side of the lines. Thus, the cross sections incorporate strips of well data that are commonly 2.5 km wide. Juxtaposition in the sections of wells that are separated by as much as 2.5 km on the ground is the principal cause in the cross sections of the irregular surfaces and of pinching and swelling of some of the units.

Topographic features such as escarpments and stream valleys are drawn in their real positions on the cross sections; however, between these features, the topographic surface along each line of section is approximated by lines connecting the surface elevations of adjacent projected wells. This approach avoids the problem of having well heads apparently floating above or buried below the ground surface.

STRATIGRAPHY

Pliocene basalt of Buckskin Basin

Southeast of the Snake River, the 2.2- to 3.7-Ma basalt of Buckskin Basin (map unit Tb, Figs. 2, 4, and 5) (Kellogg and Marvin, 1988) is the bedrock upon which the Quaternary sediments were deposited. Thinning of the basalt to the southeast, as shown by data from wells east of the study area and by the work of Kellogg and Marvin (1988, Fig. 3) suggests that the source of the basalt was to the northwest. This source is in apparent contradiction to the general west and southwest slope of the upper surface of the basalt shown on cross sections E-E′, F-F′, G-G′, and H-H′ (Fig. 4). If the source of the basalt of Buckskin Basin was from the northwest, then the west-southwest slope of the basalt surface, amounting to about 0.3°, may be the result of late Pliocene and Quaternary downwarping of the Snake River Plain, although it is possible that some of the slope may have been caused by erosion. No wells in the northwestern part of the area are deep enough to reach the basalt of Buckskin Basin.

Ferry Butte, which is composed of the basalt of Buckskin Basin (Fig. 2), bears a topographic resemblance to a shield volcano but has been inferred to have resulted from doming due to emplacement of rhyolite magma beneath it (Karlo and Jorgenson, 1979; Kellogg and Embree, 1986). No rhyolite is exposed and none was penetrated by wells, but other rhyolite domes, 0.3 to 1.5 million years old, form prominent hills near the center of the eastern Snake River Plain (Big Southern Butte, East Butte; Fig. 1) (Spear and King, 1982).

Upper Pliocene to Holocene alluvium, basalt, and loess

In the Snake River Valley, the Raft Formation and American Falls Lake Beds are enclosed in lithologic envelopes of Snake River alluvial deposits, basalt flows, and loess. Fine-grained alluvial fan deposits are present as a lateral facies on the southeast (Scott, 1982) (Figs. 3, 4, and 5, cross sections A-A′ through J-J′). The following are brief descriptions of these units and of their stratigraphic relationships to the Raft Formation and American Falls Lake Beds.

Snake River gravel deposits. Gravelly alluvial deposits of the ancestral and modern Snake River and its tributaries (unit

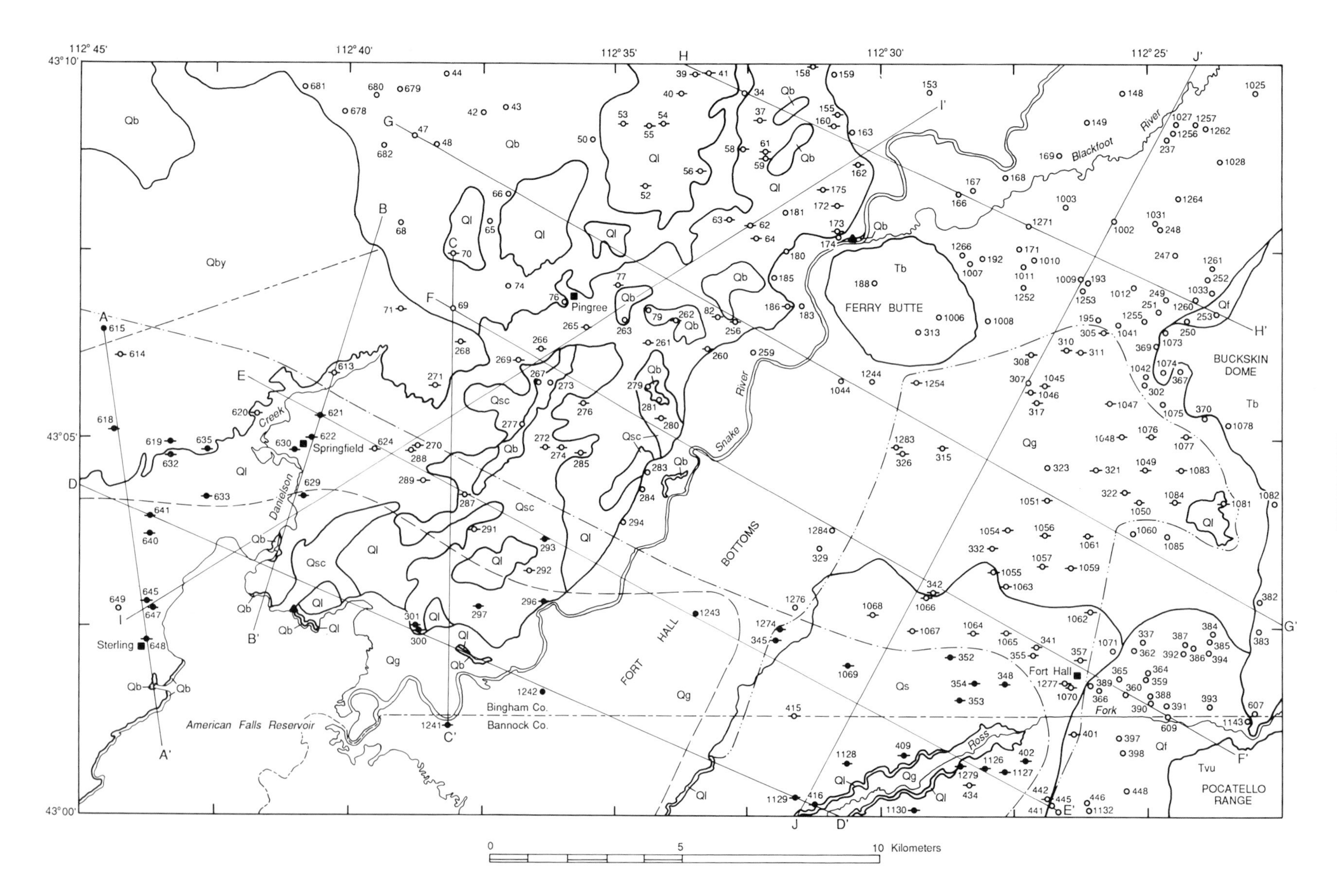

Figure 2. Geologic map of the study area (modified from Scott, 1982).

CORRELATION OF MAP UNITS

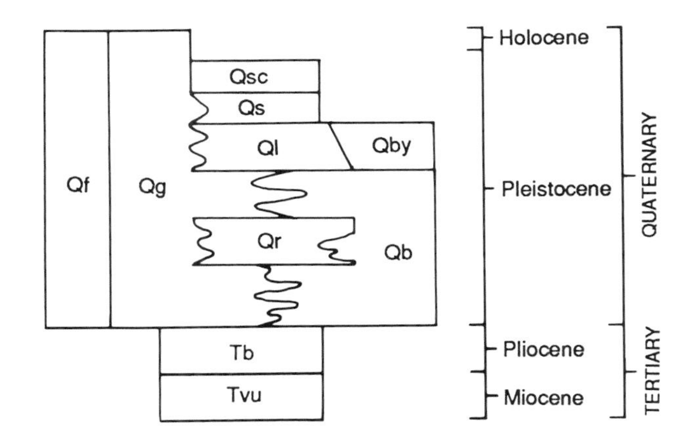

LIST OF MAP UNITS

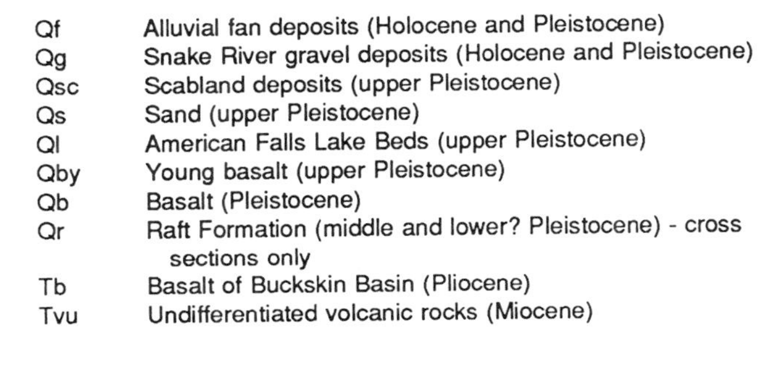

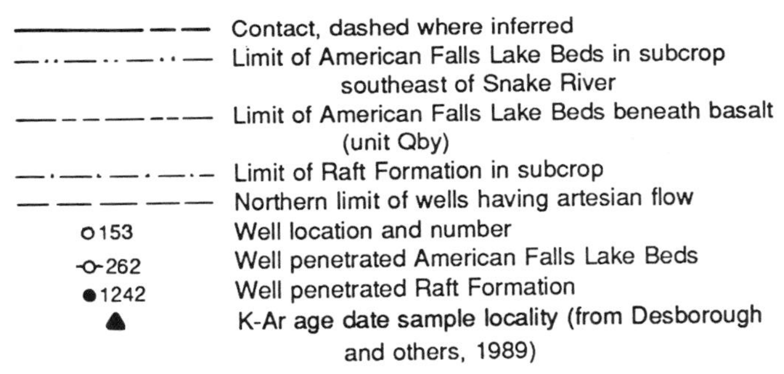

Figure 2 Explanation.

Qg, Fig. 2) constitute the volumetrically most abundant lithologic unit in the study area. Southeast of the Snake River, the gravel directly overlies the basalt of Buckskin Basin and thus may be as old as late Pliocene (Figs. 4 and 5). Northwest of the river beneath the Raft Formation, the gravel is interbedded with lava flows inferred to be early Pleistocene in age (Fig. 3, cross section A-A′). The gravel is as young as Holocene in the present floodplain and channel of the Snake River. Southwest of Ferry Butte, gravel deposition was interrupted during the ponding events represented by the fine-grained Raft Formation and American Falls Lake Beds. Northeast of Ferry Butte, however, gravel deposition apparently was more or less continuous through Pleistocene and Holocene time across much of the Snake River Valley (Figs. 4 and 5, cross sections H-H′ through J-J′). The possibility of erosional events could not be evaluated on the basis of drillers' logs alone.

Drillers' logs and exposures in gravel pits show that alluvium of the Snake River consists chiefly of well-sorted, pebbly to bouldery, sandy gravel and subordinate discrete beds of clay or sand. For example, alluvium of the Snake River below the Raft Formation in well no. 293 (Fig. 4, cross section E-E′) is described as follows: sand and gravel, 6 m; gravel and sand, 3 m; large gravel and clay, 3 m; sand and gravel, 12 m; sandstone, boulders, and clay, 12 m; fine sand and clay, 8 m; gravel and sand, 2 m; large gravel and river rock, 6 m. This log is slightly more detailed than most. A more typical description of this interval might be as follows: gravel, sand, and clay, 36 m; sand and clay, 8 m; gravel and sand, 8 m. Fine-grained intervals within the gravel sequence, such as the 8 m of sand and clay in this log, are shown schematically on the cross sections as lens-shaped bodies.

Distinctive clasts in Snake River alluvial deposits include dark-reddish-brown Late Proterozoic or Paleozoic quartzite pebbles and cobbles and sand-size obsidian grains. Inspection of cuttings by William H. Raymond (U.S. Geological Survey), obtained from drilling done in Fort Hall Bottoms (Fig. 4, cross sections D-D′ and E-E′, wells 1242 and 1243), showed that the clast lithology of at least the upper part of the gravel beneath the Raft Formation is typical of modern Snake River gravel deposits (W. H. Raymond, written communication, 1988).

Alluvial fan deposits. Alluvial fan deposits (unit Qf, Fig. 2) along the base of the highlands in the southeastern corner of the study area consist mostly of interbedded clay, sand, and minor gravel. Drillers' logs indicate that the fan alluvium contains 80 to 90% clay and sand and 10 to 20% sandy gravel. The fine-grained component of the sediment is chiefly loess that was stripped from the adjacent highlands (Scott, 1982). Scott assigned the exposed alluvial fan deposits to the late Pleistocene and Holocene.

Basalt. The surface of the eastern Snake River Plain is made up almost entirely of Quaternary olivine tholeiitic basalt that erupted as flows from low shield volcanoes, as major tube-fed lava flows, and as flows from northwest-oriented fissures (Greeley, 1982; Leeman, 1982a, 1982b; Kuntz and others, 1986, 1988, this volume). The basalt flowed generally south and southeastward and from time to time dammed or diverted the Snake River downstream from the study area.

Quaternary basalt flows (units Qb and Qby, Fig. 2) are confined to the part of the study area northwest of the Snake River, both at the surface and in the subsurface. Drillers' logs and study of surface exposures indicate that the flows become thinner to the south and southeast and are intercalated with Snake River gravel deposits, loess deposits, and both the Raft Formation and American Falls Lake Beds (Figs. 3 and 4, cross sections A-A′ through H-H′). Basalt flows of at least two and probably three different eruptive events are exposed. Desborough and others (1989) reported K-Ar whole rock ages of 0.3 Ma for a basalt just northwest of Ferry Butte and 0.2 Ma for a basalt near the north-

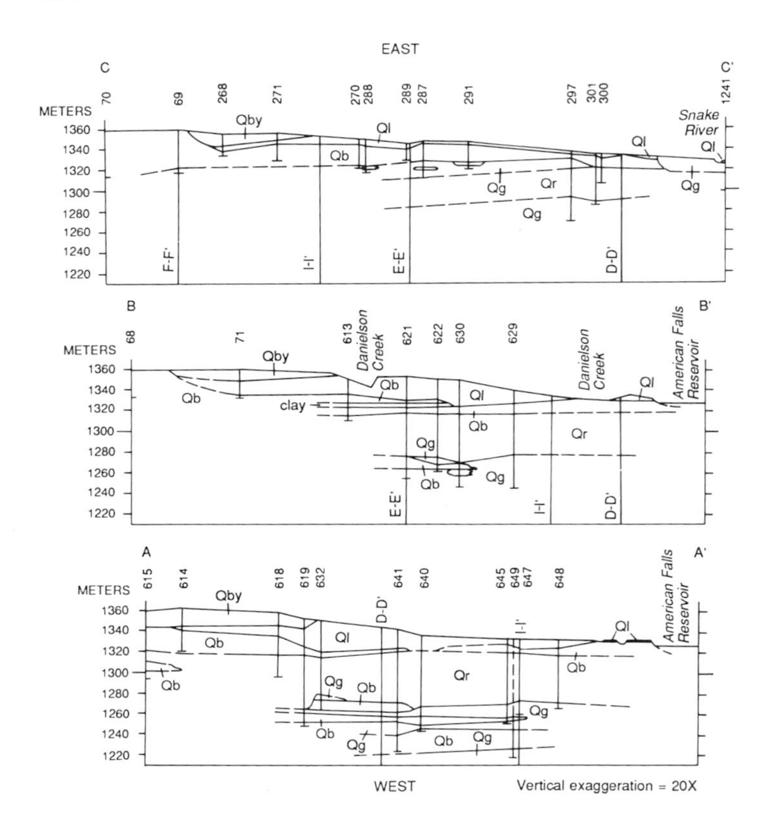

Figure 3. Geologic cross sections A-A′, B-B′, and C-C′. Locations of lines of section and explanation of unit symbols shown on Figure 2. Unlabeled lenticular bodies represent clay or sand lenses within gravel or basalt.

ern end of American Falls Reservoir (Fig. 2). The reconnaissance mapping of the present investigation and the drillers' logs (Figs. 3 and 4, cross sections A-A′ through C-C′ and E-E′) show that the youngest lava flow in the study area (unit Qby) flowed 3 to 4 km out over the American Falls Lake Beds north of Springfield (Fig. 2). Scott (1982) mapped the loess overlying this flow as Upper Pleistocene, implying that the basalt is of latest Pleistocene age. On the basis of his extensive work in the young lava fields of the Great Rift to the northwest, Mel A. Kuntz (U.S. Geological Survey, oral communication, 1990) estimated the age of this basalt to be about 40 ka but probably no younger.

Loess. Loess (silt and silty sand) was deposited in the region chiefly during Pleistocene glacial stages (Scott, 1982) and blankets much of the surface of the area occupied by pre-Holocene units. The age, character, and distribution of loess in southeastern Idaho have been described by Pierce and others (1982) and Lewis and Fosberg (1982). There is general correspondence between increasing age and increasing thickness of loess, as shown on Scott's (1982) surficial geologic map of the region. In the study area loess deposits that overlie latest Pleistocene age units are only 1 to 2 m thick (not shown on Fig. 2 or the cross sections), whereas deposits overlying the Pliocene basalt of Buckskin Basin outside the study area to the southeast are as thick as 35 m (Fig. 4, cross section G-G′ only). Drillers' logs of three wells on Ferry Butte, which may also be underlain by the basalt of Buckskin Basin (Kellogg and Marvin, 1988, p. 16), show that the loess there is 4 to 5 m thick. On the west side of Ferry Butte, exposures in gullies indicate that the loess may be about 9 m thick. An areally extensive sheet of sand as thick as 15 m west and south of Fort Hall (unit Qs, Fig. 2) is part of a larger deposit interpreted by Scott (1982) to be late Pleistocene and Holocene sand dunes.

Thin, discontinuous, widely separated interbeds of brown clay, sandy clay, and sand (shown as lenticular pods on the cross sections) are recorded in the drillers' logs for much of the Pleistocene basalt and gravel in the area. These fine-grained sediments are inferred to most likely be loess intercalated with basalt flows along flow breaks and either loess or overbank deposits in gravel units. Also, the drillers' logs commonly note the presence of 1 to 2 m of clay, inferred to be loess, between the top of the basalt of Buckskin Basin and the base of the ancestral Snake River gravel deposits.

Loess 2.5 to 10 m thick is exposed along the northwestern shore of American Falls Reservoir from Sterling to south of Little Hole, about 15 km to the southwest. This loess bed is not shown on Figure 2 or on Figure 3, cross section A-A′, because of the small scale of the illustrations. The loess overlies the Big Hole Basalt of Carr and Trimble (1963) and unconformably underlies the American Falls Lake Beds. This loess is mostly structureless silt and sandy silt, but some horizons show vague horizontal banding and uncommon thin beds of very fine grained sand with wispy small-scale cross-bedding. Locust burrows are present, and

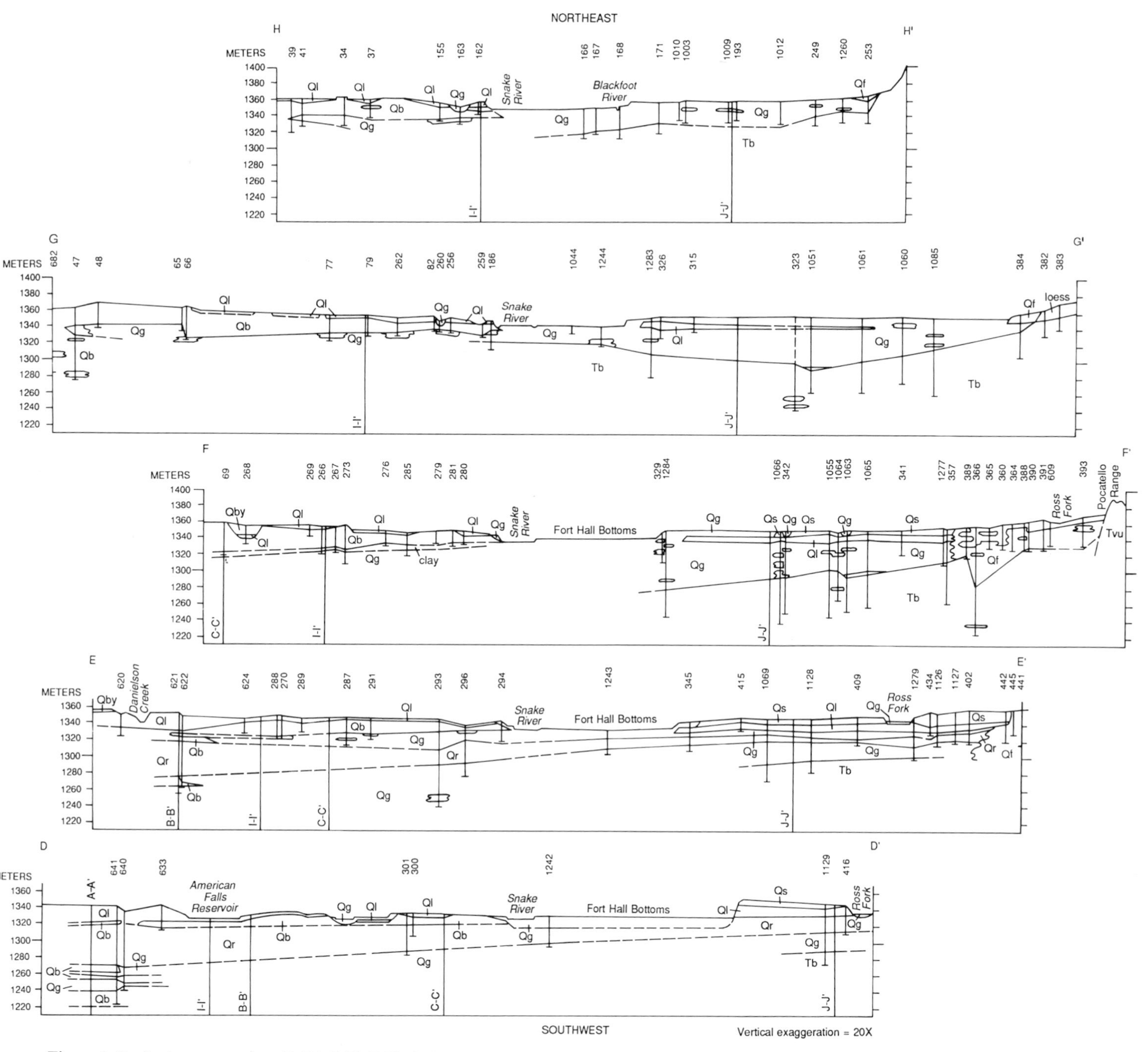

Figure 4. Geologic cross sections D-D′, E-E′, F-F′, G-G′, and H-H′. Locations of lines of section and explanation of unit symbols shown on Figure 2. Unlabeled lenticular bodies represent clay or sand lenses within gravel or basalt.

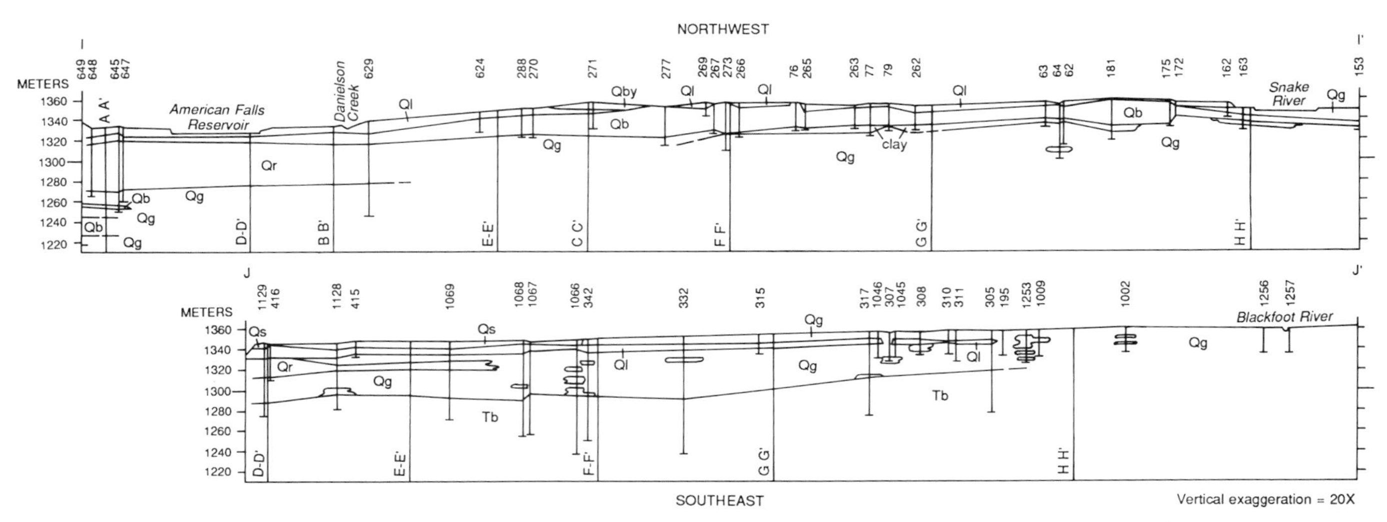

Figure 5. Geologic cross sections I-I' and J-J'. Locations of lines of section and explanation of unit symbols shown on Figure 2. Unlabeled lenticular bodies represent clay or sand lenses within gravel or basalt.

nodular calcite concretions as much as 10 cm across are locally abundant in crudely stratified layers. This loess unit is not present to the northeast between the mouth of Danielson Creek and Crystal Spring. There the American Falls Lake Beds rest directly on the basalt dated at 0.2 Ma that was reported by Desborough and others (1989).

Lower(?) and middle Pleistocene Raft Formation

In the southwestern and south-central part of the study area, drillers' logs show that the gravel deposits of the ancestral Snake River are overlain by a sequence of clay, silt, and sand as thick as 53 m. This fine-grained subsurface sequence is correlated with the Raft Formation, which is exposed southwest of the area along the Snake River and in the Raft River Valley. This correlation is based on the virtually identical thickness, lithology, stratigraphic position, and elevation of the subsurface unit and the Raft Formation.

Surface exposures. The Raft Formation was first described by Stearns and others (1938) for exposures at the northern end of the Raft River Valley and along the Snake River south of American Falls (Fig. 1). Carr and Trimble (1963, p. 21) described the Raft Formation along the Snake River as "light-colored beds of massive silt, and of clay, stratified silt, and sand" and inferred that depositional environments were both fluvial and lacustrine. They noted that near the mouth of the Raft River, the formation is about 60 m thick and includes an upper massive silt unit about 30 m thick; they described the Raft Formation near American Falls, about 35 km southwest of the study area, as "mainly parallel-bedded, calcareous quartzose silt and fine sand with local crossbedding. The presence of nodular concretions in many beds characterizes the formation." (Carr and Trimble, 1963, p. 23). A section of the Raft Formation measured by Carr and Trimble (1963) on the bluffs on the west side of the Snake River near Neeley (Fig. 1) is given in Table 1. Table 2 is a shorter section of the Raft measured during the present study at the northern end of the Raft River basin about 3 km south of the Snake River (Fig. 1).

Scott (1982) described the sediments of the Raft Formation as follows: "Silt, sand, and clay, minor gravel, locally abundant calcareous concretions, includes several thin beds of silicic tephra; moderate to good sorting; parallel bedding and small- and large-scale crossbedding; beds and laminations range from very thin laminae in some silt and clay to very thick beds in massive silt." Scott inferred that the Raft sediments include fine-grained alluvium, lacustrine deposits, and possibly loess.

The Raft Formation unconformably overlies the following Miocene through Pleistocene volcanic units depending on location: the upper tuffaceous unit of the Salt Lake Formation in the Raft River basin (Williams and others, 1982) and the Little Creek Formation, Walcott Tuff, Neeley Formation, or Massacre Volcanics along the Snake River (Carr and Trimble, 1963; Trimble and Carr, 1976). Radiometric dates for some of these units were given by Armstrong and others (1975) and Kellogg and Marvin (1988). The basalt of Radio Relay Butte, shown by Williams and others (1982) to be interbedded with the Raft Formation in the northern part of the Raft River basin, yielded K-Ar whole-rock ages of 0.4 to 0.7 Ma. Scott (1982) estimated that the Raft Formation is of early(?) and middle Pleistocene age based in part on a probable age of less than 1 Ma for an interbedded tephra layer. In the American Falls quadrangle the Raft is unconformably overlain by the Big Hole Basalt and the American Falls Lake Beds (Carr and Trimble, 1963). The Big Hole Basalt may be equivalent to or older than the 0.2-Ma basalt (Desborough and others, 1989) at the northern end of American Falls Reservoir.

TABLE 1. SECTION OF RAFT FORMATION AND AMERICAN FALLS LAKE BEDS*

	Thickness (m)
Top of bluff.	
Dune sand:	
Sand, gray to tan, fine- to medium-grained, loosely indurated	3-4.6
American Falls lake beds:	
Clay, gray-green, noncalcareous; breaks into small conchoidal fragments and forms loose "shaly" slope	2.1
Clay, white to light-tan, diatomaceous, massive, blocky; local platy and prismatic jointing, particularly near top; forms ledge	2.1
Clay, greenish-gray, sandy; pinkish-brown in lower 0.8 m and slightly calcareous; blocky fracture at top; breaks into fine granular or "shaly" rubble; sharp upper contact	2.8
Sand, pinkish-brown, calcareous, fine-grained, loosely indurated; a few pinkish-brown clay layers near the bottom; upper contact gradational	1.5
Clay, pinkish-brown, calcareous, laminated, brittle; in beds 1.3 to 5 m thick; alternates with tan crossbedded silt in layers 0.2 to 0.5 m thick; silt shows wavy oscillatory bedding; contains mollusks; forms ledge	4.3
Sand, gray to tan, fine to very fine grained, loosely indurated; crossbedded on small scale and contains local thin silty layers	2.5
Gravel, coarse, pebbly, moderately indurated; pebbles are well rounded, and consist mainly of quartzite (about 80 percent) but also of basalt and Walcott tuff; layer of yellow clay containing rhyolitic shards and basalt cinders, about 1.3 cm thick, at base; contains mollusks; thickens to about 1.5 m about 9 m downriver from line of section	0.2
Total American Falls lake beds	15.6
Unconformity.	
Raft formation:	
Silt, light-tan, massive; poorly indurated but forms ledge	0.4
Clay and silt, calcareous, laminated; a layer of light-brown loose sand at top of unit. Upper contact gradational	1.8
Sand, tan to light-gray, well-bedded and locally crossbedded; contains several pebbly sand lenses; beds 8 to 10 cm thick; pebbles well-rounded and less than 1.3 cm in diameter; composed of Paleozoic rocks and a few lime concretions	1.5
Silt, tan, calcareous, moderately indurated, massive, with lenses of laminated silt. Forms ledges	0.4
Sand, light-gray to tan, fine-grained, clean, loose; interbedded with tan thin-bedded to laminated silt with platy partings and some crossbedding; pebbly zones about 10 cm thick about 0.6 m above base and at top of unit; lower pebbly zone contains small concretions, platy fragments of Walcott, and rare quartzite	2.0
Silt and fine sand, light-tan to tan, calcareous, moderately well cemented, massive, ledge-forming; fractured case-hardened surface in places and some prismatic jointing; hard irregular concretionary areas in lower and middle parts of unit	4.2
Sand, tan, very fine- to coarse-grained; very loosely cemented by films of calcium carbonate; granular texture in places; contains a few shards and scattered white soft calcareous lumps, 1 to 6 mm across, and lenses of pumice fragments and pebbles of Walcott(?) tuff; upper contact very gradational. Approximate horizon of basaltic ash exposed about 0.4 km up river	2.3
Sand, fine-grained, and tan calcareous silt in beds 10 to 13 cm thick; sand beds contain 0.5 to 3 mm white irregular calcareous particles and very irregular calcareous sandstone concretions, many of which are spindles as much as 0.3 m long that are perpendicular to the bedding	2.1
Silt, tan, calcareous, moderately indurated; rather massive to faintly bedded, with a little very fine sand and several zones of concretions in lower part; forms a ledge together with unit above	4.0
Sand and silt, light tan, calcareous, in alternate beds 2.5 to 10 cm thick, with prominent concretion zone at top	0.6
Sand, tan, calcareous, fine-grained to very fine grained, crossbedded; mostly quartz; contains 3 or 4 discontinuous zones of nearly pure calcareous concretions 5 to 13 cm long and 1.3 to 2.5 cm thick	0.6
Silt and sand, light-tan, calcareous, fine-grained, poorly indurated; contains platy layers and 2.5 cm nodules of calcareous moderately indurated sand	0.9
Total Raft formation measured	23.7
Base of section—4.6 m above river level.	

*Measured on the west bank of the Snake River opposite Neeley in SE¼SW¼Sec.10,T8S,R30E, Rockland Quadrangle, Idaho. Modified from Carr and Trimble (1963).

Subsurface character and distribution. The thickness and distribution of the Raft Formation in the subsurface of the study area are shown in Figure 2 and the cross sections. Appendix Table 1 lists the drillers' lithologic descriptions of the rocks in the stratigraphic interval assigned to the Raft. The drillers' descriptions indicate that the Raft consists of an average of about 45% clay, 15% sandy clay, and 40% sand (Fig. 6). No gravel interbeds were recorded for this interval in any of the logs. The clay cited by most of the drillers is probably actually silt. Horizons inferred to contain nodular calcareous concretions were described in well no. 296 (sand and small sandy stone) and well nos. 1242 and 1243 (indurated lumps). Cross section A-A′ (Fig. 3) indicates that the Raft may be intercalated with lava flows northwest of the study area.

Carr and Trimble (1963) noted that the Raft Formation is about 20 to 25 m thick in exposures along the Snake River downstream from American Falls Dam but thickens in the subsurface to the northwest where 48 m of silt and clay, tentatively assigned by them to the Raft, were penetrated in a well. The same situation persists in the study area, where well data show that the Raft is about 23 m thick beneath the Snake River and about 53 m thick 5 km to the west (Fig. 4, cross sections D-D′ and E-E′).

TABLE 2. SECTION OF THE RAFT FORMATION*

	Thickness (m)
Top of bank	
Loess	1.5
Raft Formation:	
Light-gray sandy clay, consisting of 40 percent irregular calcareous concretions about 20 cm across; upper half is siltier and concretions are larger	0.6
Paleosol, lower part contains burrows and root casts; indistinct bedding may be disrupted laminae; root casts are more abundant in middle part; upper part is a petrocalcic horizon 0.6 m thick	1.5
Laminated silt and silty fine-grained sand in 0.5 to 1.0 cm-thick beds	0.6
Crossbedded fine-grained sand containing rip-up clasts	0.2
Crossbedded silt	0.2
Thin- to thick-bedded silt (2 to 30 cm). Bedding planes are commonly defined by tan clay layers 3 mm thick, by crenulated layers of light yellow clay 3 mm thick, or by concentrations of rusty coloring. Lenses of fine-grained sand 1 to 3 cm thick form load casts in the surrounding silt	1.5
Total Raft Formation measured	4.6

*Measured at the northern end of the Raft River basin 2.5 km south of Cold Water in Sec.36,T9S,R28E, Yale Quadrangle, Idaho.

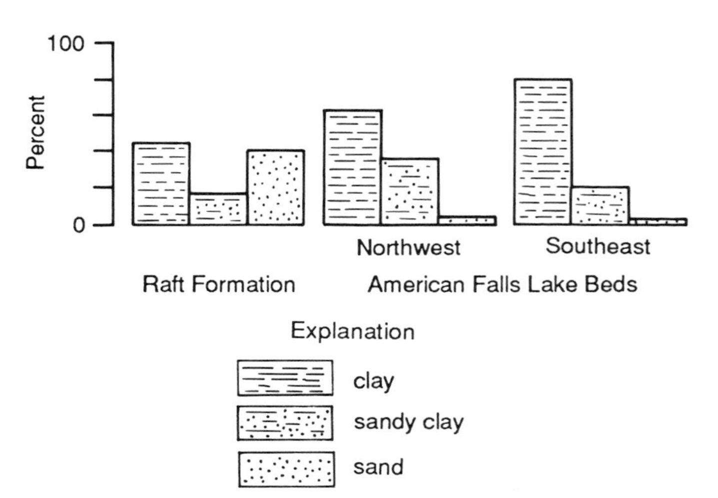

Figure 6. Average grain size of the Raft Formation and American Falls Lake Beds based on rock descriptions given in drillers' logs.

The approximate northern and eastern limit of the Raft subcrop in the study area is shown on Figure 2. Cross sections D-D′ and E-E′ (Fig. 4) show that the Raft thins to the southeast and probably interfingers with alluvial fan sediments about 2.5 km northwest of the foot of the Pocatello Range. Comparison of the thickness of the Raft in cross sections D-D′ and E-E′ (Fig. 4) shows that the Raft thins to the north. Cross section J-J′ (Fig. 5) shows that southeast of the Snake River the Raft Formation probably interfingers with Snake River gravel deposits and pinches out. Northwest of the river, cross section I-I′ (Fig. 5) shows that the Raft probably extends as far northeast as cross section E-E′ (Fig. 4), but its maximum extent is undefined because of the lack of deep wells in this area. Although wells deeper than 30 m are common southeast of the Snake River, on the northwest side of the river deep wells are generally confined to an area of artesian groundwater around the northern end of the reservoir (Fig. 2). The Raft Formation is the aquiclude above this artesian aquifer, so the absence of deep wells to the northeast may indicate that the Raft is not an effective aquiclude in this area; that is, it may be thinner or gravelly or both.

A discontinuous but persistent layer of clay and sand as thick as 8 m is between the basalt and the Snake River gravel deposits north of the inferred northern extent of the Raft Formation subcrop (Fig. 5, cross section I-I′). This fine-grained layer may be loess and may have accumulated contemporaneously with some of the loesslike facies of the Raft.

The subsurface Raft Formation overlies ancestral Snake River gravel deposits in the southeastern part of the area; in the southwestern part of the area, it overlies lava flows inferred to be early Pleistocene in age that are intercalated with ancestral Snake River gravel deposits. It is overlain by the Big Hole Basalt of Carr and Trimble (1963) northwest of American Falls Reservoir, by the basalt dated at 0.2 Ma near Crystal Spring (Desborough and others, 1989), by Snake River gravel deposits, and by the upper Pleistocene American Falls Lake Beds. The subsurface Raft Formation is correlated with Unit B of Desborough and others (1989).

The average elevation of the contact between the Raft Formation and the overlying basalt is 1,319 m, based on the wells shown on cross section A-A′ (Fig. 3). The elevation of the contact between the Raft and the American Falls Lake Beds (Fig. 4, cross section D-D′) is about 1,333 m. Considering the regional dip and the gradient of the Snake River, these values compare well with the elevation of 1,317 m given by Carr and Trimble (1963) for the contact between the Raft and the American Falls Lake Beds along the Snake River in the American Falls quadrangle. Cross sections A-A′ through E-E′ (Figs. 3 and 4) show that both the upper and lower contacts of the Raft Formation dip gently to the north and west. The fact that both contacts are tilted, in conjunction with the fine-grained nature of the Raft, suggests that the dip is tectonic rather than depositional and that it reflects Pleistocene subsidence of the southeastern edge of the eastern Snake River Plain.

The lava flow that presumably dammed the ancestral Snake River and created the depositional basin of the Raft Formation has not been located. The Raft Formation is not exposed west of the mouth of the Raft River, so the dam may not have been very far downstream from this point in the Snake River. The mixture of facies in the Raft indicating fluvial, lacustrine, and perhaps eolian depositional environments suggests that the lava dam may have been leaky or that it may have been emplaced episodically as a number of thin flows. Deposition of the Raft Formation ended prior to about 0.2 Ma, based on the age of the overlying basalt flow at Crystal Spring.

Upper Pleistocene American Falls Lake Beds

A time interval of probably more than 130 k.y. elapsed between the end of deposition of the Raft Formation and the beginning of deposition of the American Falls Lake Beds. Deposition of the American Falls Lake Beds began about 72 ka (Scott and others, 1982) in a lake that formed when basalt flows from a shield volcano at Cedar Butte, 16 km southwest of American Falls, dammed the Snake River (Fig. 1). The geologic map (Fig. 2) shows the surface and subsurface distribution of the American Falls Lake Beds in the study area.

Previous work. In exposures along the Snake River south of American Falls, the American Falls Lake Beds consist of three distinct units: a basal fluvial gravel that locally contains a rich vertebrate fauna (Hopkins, 1951, 1955; Hopkins and others, 1969), a middle fine-grained fluvial unit consisting of channel and floodplain deposits, and an upper, blocky-weathering lacustrine unit consisting of silt, clay, and diatomaceous clay. The section of the American Falls Lake Beds measured by Carr and Trimble (1963) along the Snake River (Fig. 1) is given in Table 1.

Subsequent work has led to reinterpretation of some aspects of stratigraphy and sedimentology of the American Falls Lake Beds as originally described by Carr and Trimble (1963). Ridenour (1969) correlated the Duck Point gravel (exposed near the type area) and its included vertebrate fauna with the basal fluvial gravel. However, Hearst (1990) observed that, because the Duck Point gravel contains rip-up clasts of clay derived from the American Falls Lake Beds, the gravel body is more likely a deposit of the Bonneville Flood. This interpretation has implications for the stratigraphic range of the vertebrate fauna.

Trimble and Carr (1961), Carr and Trimble (1963), and Trimble (1976) assigned all of the fine-grained sediments above the basal gravel to a lacustrine environment, but Ridenour (1969) concluded that only the upper, blocky-weathering unit, consisting mainly of clay, is of lacustrine origin. He assigned the middle unit, which contains considerable cross-bedded sand and silt, to a floodplain environment related to an early lava flow from Cedar Butte that may not have dammed the river completely.

Paleoenvironmental studies by Bright (1982) and Scott and others (1982) of microfossils and of exposures of the middle and upper units of the formation, chiefly along the southeastern shore of American Falls Reservoir, support Ridenour's conclusion. Ostracodes and diatoms in the middle unit indicate a marshy environment of shallow, fresh to saline-alkaline ponds containing abundant aquatic vegetation in a cold temperate, semiarid climate; those in the upper unit are characteristic of a deep, moderately cool, eutrophic to mesotrophic freshwater lake. North of Springfield, Scott (1982) and Scott and others (1982) interpreted interbedded sand and silt as high-energy shoreline deposits; about 3 km north of the study area, they inferred that poorly exposed sand, silt, and clay were topset beds of a delta that presumably formed at the head of American Falls Lake.

Present investigation of American Falls Lake Beds. Field observations, drillers' logs, and mineralogical analysis provide new information that, although in general agreement with the existing stratigraphic framework of the American Falls Lake Beds, suggests some stratigraphic relationships and paleoenvironmental interpretations should be reevaluated.

Basal fluvial gravel. The basal fluvial gravel probably is not present on the northwestern side of the Snake River in the study area, and where it is present in the subsurface on the southeastern side of the river it may be considerably older than the American Falls Lake Beds. Also, although fluvial gravel beds are discontinuously exposed in the bluffs along the Snake River south of American Falls, Hearst (1990) concluded that the gravel at Duck Point, at least, is younger than the American Falls Lake Beds. Ridenour (1969, p. 43) found only one exposure of a basal gravel in the bluffs on the southeastern side of American Falls Reservoir, and he attributed that deposit to a tributary of the Snake River. In exposures on the northern and northwestern sides of the reservoir, either the floodplain or lacustrine facies of the American Falls Lake Beds rests directly on basalt or on the highly irregular surface of a loess deposit. In addition, the drillers' logs do not record an intervening gravel interval where the formation overlies basalt northwest of the river.

Southeast of the river, cross sections D-D′ and J-J′ (Figs. 4 and 5) show that the American Falls directly overlies the Raft Formation at the southern edge of the study area, but cross sections E-E′ and J-J′ (Figs. 4 and 5) show that a little farther to the north a gravel layer 8 to 9 m thick is between the two formations. Because this gravel layer is not exposed, it is not possible to study directly its stratigraphic relationship to the American Falls Lake Beds. The relationships shown on cross section E-E′ (Fig. 4) suggest that the gravel interval may be closer in age to the Raft Formation than to the American Falls Lake Beds. This cross section shows that the gravel interval between the Raft and American Falls Lake Beds southeast of the river possibly can be correlated with a gravel interval between the Raft and basalt flows on the northwestern side of the river. This implies that, in the study area at least, gravel beneath the American Falls Lake Beds may be older than the 0.2- to 0.3-Ma basalt. For this reason, the gravel on the southeastern side of the Snake River was not included in the American Falls Lake Beds in this study.

Fine-grained floodplain and lacustrine units. Exposures of the floodplain and lacustrine units at the northern end of the reservoir and in the bluffs along the southeastern side of Fort Hall Bottoms (Fig. 2) show that these units look about the same in the study area as they do to the southwest where they were described and studied by Carr and Trimble (1963), Ridenour (1969), Trimble (1976), and Scott and others (1982). The floodplain unit is variable, consisting of yellowish-gray, sandy, cross-bedded channel deposits and silty, sandy, massive to thick-bedded overbank deposits. Locust burrows and paleosols are common in the overbank deposits. The overlying lacustrine unit consists of yellowish-gray to light greenish-gray, thin-bedded to laminated, silty clay. Widely separated thin beds of well-sorted, very fine grained sand show ripple cross-bedding. The blocky appearance of the lacustrine unit in some weathered exposures is caused by lack of partings on bedding planes and by vertical joints spaced about 0.3 m apart. Well-preserved fish skeletons have been found in the lacustrine unit (George A. Desborough and Charles A. Repenning, U.S. Geological Survey, oral communication, 1986).

On average, the floodplain unit is 2 to 3 m thick and the lacustrine unit is about 3 to 10 m thick. Exposures in the bluffs around American Falls Reservoir show, however, that the units vary greatly in thickness (for example, see Fig. 5 in Scott and others, 1982). The variation is caused by initial topography on the depositional surface, by local erosion between deposition of the two units, and by erosion of the surface of the lacustrine unit by the Bonneville Flood. Exposures on the northwestern side of the reservoir show that there is as much as 10 m of topographic relief on the surface of the loess that underlies the American Falls Lake Beds. Cross sections A-A′, B-B′, D-D′, and E-E′ (Figs. 3 and 4) indicate that the American Falls Lake Beds (floodplain and lacustrine units combined) are as thick as 30 m in the vicinity of Springfield. Part of this anomalous thickness may possibly include a loess deposit below the American Falls not detected or recorded by the drillers.

Carbonate minerals. Sediments of the American Falls Lake Beds (both floodplain and lacustrine) are composed chiefly of quartz, lithic fragments (volcanic, plutonic, and sedimentary), feldspar, carbonate minerals, and clay minerals (Ridenour, 1969). Identification of dolomite as a major constituent of the carbonate mineral fraction (George A. Desborough, oral communication, 1985) led to an investigation of the source of the dolomite.

In a 6-m-thick section consisting of channel, overbank, and laminated lacustrine facies collected by George Desborough at Bronco Point (Fig. 7), the average amount of carbonate is 20%. X-ray diffraction analysis indicates that the carbonate fraction contains about 60% low-Mg calcite and 40% dolomite. Ridenour's study (1969) showed that, on a lakewide basis, the floodplain unit contains about 10% carbonate and the lacustrine unit about 25%. These abundances are approximately comparable to the average carbonate content of the sediments at Bronco Point; however, Ridenour attributed all the carbonate to calcite. X-ray diffraction patterns of the clay-size fraction (Ridenour, 1969, p. 23) of three samples collected at the type section of the American Falls Lake Beds (Fig. 1) (Carr and Trimble, 1963) show no dolomite peaks for two samples. On the diffraction pattern of the third sample, Ridenour identified a small peak at the position of the dolomite peak as feldspar.

The most likely source for the dolomite and calcite is probably glacial rock flour, although loess is a possible source also. Paleozoic dolostone and limestone are widely exposed in the Snake River drainage basin. American Falls Lake was in existence during most of the Wisconsin glaciation, and because it was the first impoundment of the Snake River downstream from the Yellowstone plateau and the Grand Teton Mountains, it was an

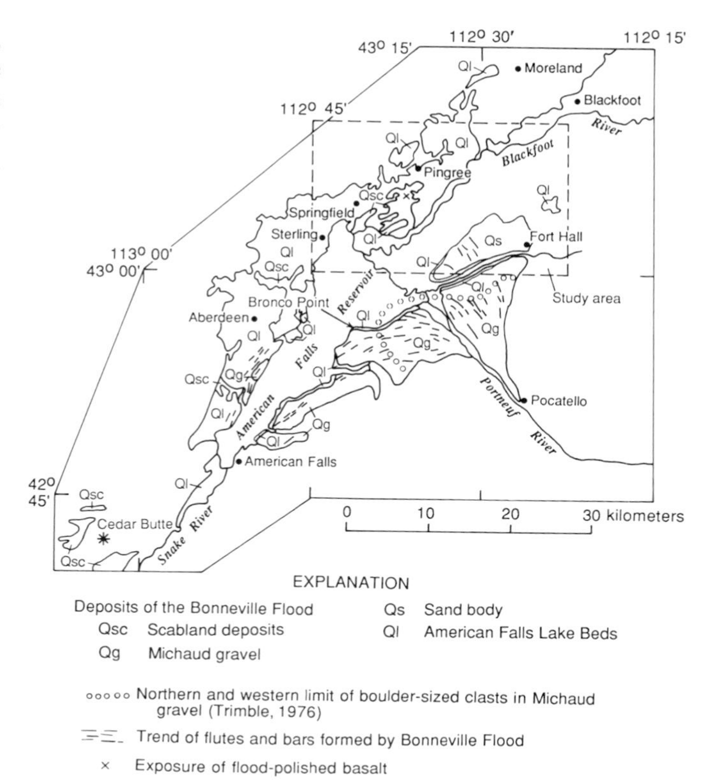

Figure 7. Map showing the areal distribution of the sand body (Qs) and the trend of ridges and swales on its surface in relation to features and deposits of the Bonneville Flood (modified from Scott, 1982).

obvious depositional site for much of the glacially produced sediment carried in suspension by the Snake River and its tributaries.

Analysis of the size distribution of the calcite and dolomite within the silt-size fraction of sediments at Bronco Point is in agreement with the size distribution of these minerals reported by Kennedy and Smith (1977) in the sediment of two glacier-fed lakes. Kennedy and Smith observed that total dolomite decreases and total calcite increases with decreasing grain size (4 to 9 ϕ) in the lakes (Fig. 8). They attributed this distribution to the greater density of dolomite (8% denser than calcite) and to the fact that glacially produced dolomite is coarser grained than calcite. The same relationship of median grain size to the ratio dolomite/(dolomite + calcite) is shown by silt-size calcite and dolomite in the Bronco Point sediments. Although the Bronco Point sediments contain much less total carbonate, the slopes of the curves are quite similar to those shown by Kennedy and Smith. This relationship of size to carbonate mineralogy is not likely to occur if the calcite and dolomite were primary or diagenetic. The observed decrease in dolomite content with decreasing grain size may account for the apparent absence of dolomite in Ridenour's (1969) X-ray diffraction patterns of the clay-size fraction of American Falls Lake sediments.

Geometry of American Falls Lake basin. Northeast of American Falls Reservoir, the following differences between the

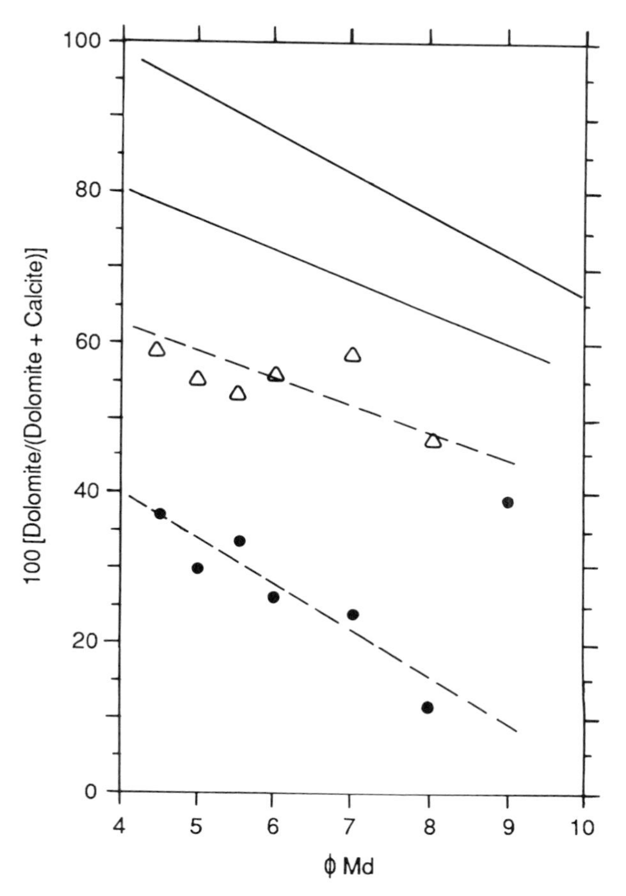

Figure 8. Relative abundance of dolomite and calcite as a function of grain size in sediment from two glacial lakes in Canada (solid lines, Kennedy and Smith, 1977) and from two samples of the floodplain facies of the American Falls Lake Beds at Bronco Point (dashed lines).

American Falls Lake Beds on opposite sides of the Snake River are considered significant and may be explained by the influence of the geometry of the lake basin on deposition. (1) The average maximum elevation of American Falls Lake Beds in exposures northwest of the Snake River is about 1,356 m; southeast of the river, drillers' logs indicate that the average maximum elevation is about 1,349 m, 7 m lower than on the other side of the river (Figs. 3, 4, and 5, cross sections A–A′ through J–J′). (2) In exposures along the northwestern side of American Falls Reservoir, the lacustrine unit of the American Falls Lake Beds is coarser grained and is more irregularly laminated than it is on the southeastern side of the reservoir. (3) Analysis of the rock types recorded by drillers for the American Falls Lake Beds in the study area (Fig. 6; Appendix Tables 2 and 3) shows that northwest of the Snake River the rocks consist of about 60% clay and 40% sandy clay (68 well logs); southeast of the Snake River they consist of about 80% clay and 20% sandy clay (65 well logs).

When the Pleistocene Snake River Valley was inundated by American Falls Lake, the northwestern side of the valley was underlain by coalescing lava flows and was higher than the southeastern side. This provided the setting on the northwest for a broad shoreline zone and a wide, shallow platform above wave base (Figs. 3, 4, and 5, cross sections C–C′ through I–I′). Lacustrine sediments deposited in this environment were likely to be coarser grained and less well bedded than those deposited in the deeper part of American Falls Lake.

The southeastern side of the lake could have been as much as 15 m deeper simply because there were no lava flows here and because this part of the lake floor was the Snake River floodplain and channel (the lowest part of the valley) prior to emplacement of the dam (Figs. 4 and 5, cross sections E–E′ through G–G′ and J–J′). Shoreline facies of the deeper part of the lake would have been deposited in a relatively narrow strip at the southeastern edge of the basin and are probably intercalated with the fine-grained distal alluvial fan deposits (Fig. 4, cross section F–F′).

Deposits of the Bonneville Flood and mininum age of the American Falls Lake Beds

The Bonneville Flood, dated at 14 to 15 ka by Scott and others (1982), entered the Snake River Plain by way of the Portneuf River Valley just south of the study area (Fig. 7) and flowed southwestward along the Snake River Valley. Evidences of the Bonneville Flood in the area are scabland deposits (unit Qsc, Fig. 2) northwest of the Snake River mapped by Scott (1982) and possibly the fluted topographic surface of a body of sand (unit Qs) on the southeastern side of the river across from the scablands (Fig. 7).

The scablands are characterized by a flood-scoured surface on the American Falls Lake Beds and by a thin discontinuous mantle of sand and gravel on the basalt. During reconnaissance studies, an exposure of highly polished basalt was found at the northwestern edge of the flood-scoured strip about 4 km southwest of Pingree (Fig. 7).

The sand deposit southeast of the river is as thick as 15 m and covers an area of about 50 km^2 (Figs. 2 and 7). It was mapped as eolian sand by Scott (1982), but northwest-trending ridges and swales on the surface of the deposit are perpendicular to the prevailing wind direction and to the trend of other sand dunes in the region. It is possible that the ridges on the sand deposit are transverse dunes, but it is also possible that the surface of the sand was reworked by the Bonneville Flood. This second possibility is favored because of the parallel alignment of the sand ridges with giant flutes and bars in cobble and boulder deposits formed by the Bonneville Flood to the south of the study area (Fig. 7). The presence of obsidian grains in the sand indicates that the source of the deposit was the Snake River and not the Bonneville Flood. This implies that if the body of sand was overlying the American Falls Lake Beds prior to the Bonneville Flood, then lacustrine deposition had ceased here some time before the flood. This timing is at variance with the hypothesis of most workers that the lake existed until the Bonneville Flood breached the lava dam at Eagle Rock (Trimble and Carr, 1961; Malde, 1968; Scott and others, 1982), although Hearst (1990) suggested that the lake may have been partially drained or filled with silt prior to the Bonneville Flood.

If the surface of the sand body was reworked by the Bonneville Flood, then an additional implication is that the highest northwest-trending ridge (1,365 m) must have been under water. This is a minimum altitude for the flood crest; it is 9 m higher than Malde's estimate of 1,356 m for the maximum altitude of the flood at the mouth of the Portneuf River (Malde, 1968, Fig. 1) and 19 m lower than O'Connor's maximum estimate of 1,384 m based on the altitude of an eroded scarp 7 km north of Pocatello (O'Connor, 1990, Appendix 2).

Drillers' logs and surface exposures show that north of Springfield a basalt flow overlies the American Falls Lake Beds at an elevation of about 1,353 m (Figs. 3 and 4, cross sections A–A′ through C–C′ and E–E′). Although this unit has not been dated radiometrically, Mel A. Kuntz (oral communication, 1990) estimated that the flow may be at least 40 Ka. Scott (1982) mapped the loess unit overlying the flow as latest Pleistocene. No lacustrine rocks overlie this basalt, so apparently lacustrine deposition had ceased at this elevation by the time the flow was emplaced. A date for this basalt flow may provide additional information as to when the lake was drained (assumed to be about 15 ka at the latest), or it may substantiate the drop in lake level that was suggested by Scott and others (1982) to explain differing thicknesses of loess deposits overlying American Falls Lake Beds southwest of the study area.

DISCUSSION

The conclusions of this study regarding the stratigraphy of the eastern Snake River Plain are similar to, and build upon, those of Carr and Trimble (1963), Ridenour (1969), Scott (1982), and Scott and others (1982). The study demonstrates that Pliocene and Quaternary units exposed at the surface can also be identified and differentiated in drillers' logs of water wells and that subsurface units are similar to or are predictable facies of the surface stratigraphic units.

New information presented in this study concerning the apparent westward continuation and thickening of the Raft Formation in the subsurface indicates that the axis of the Raft depositional basin in the Snake River Plain was west of American Falls Reservoir. Drillers' logs of wells south and west of the study area may contain significant additional information on the lithology and depositional basin of the Raft Formation.

The dip of both the upper and lower surfaces of the Raft to the west and northwest indicates northwestward downwarping of the southeastern edge of the Snake River Plain. The greater dip of the lower surface suggests that downwarping occurred during deposition of the Raft and continued afterward. In the absence of any stratigraphic evidence of faulting, it appears that this part of the Snake River Plain developed through gradual subsidence to the northwest throughout the Pleistocene.

The approximate time of the beginning of deposition of the American Falls Lake Beds is well known because it coincides with the emplacement of the lava dam at Eagle Rock about 72 ka that formed American Falls Lake. Most workers have assumed that the lake persisted until the Bonneville Flood breached the dam at about 15 ka because the surface of the American Falls Lake Beds in exposures around American Falls Reservoir does not show much erosion and was not known to be overlain by any deposits older than Bonneville Flood deposits. Two findings of this study indicate that American Falls Lake may have been drained, in the study area at least, some time before the flood: (1) the presence of a basalt flow overlying the American Falls Lake Beds north of Springfield that is probably older than 15 ka and may be older than 40 ka, and (2) the possibility that the sand body overlying the American Falls Lake Beds in the vicinity of Fort Hall may have been reworked by the Bonneville Flood and thus must be older than the flood. Determination of the minimum age of the American Falls Lake Beds will require additional field work.

ACKNOWLEDGMENTS

The cooperation of the Shoshone-Bannock Tribes in allowing access to the Fort Hall Reservation for field work is gratefully acknowledged. Discussions with K. L. Pierce, W. E. Scott, F. N. Houser, H. E. Malde, R. M. Forester, and G. A. Desborough were helpful to the study. The manuscript was improved by reviews from W. H. Raymond, W. E. Scott, R. N. Breckenridge, and H. T. Ore.

APPENDIX TABLE 1. DRILLERS' DESCRIPTIONS OF SEDIMENTS IN STRATIGRAPHIC INTERVAL ASSIGNED TO RAFT FORMATION

Well No.	Lithology	Total Thickness (m)
293	Brown clay, 18 m	18
296	Sand and heaving(?), 5 m; brown clay, 20 m; sand and small sandy stone(?), 3 m	28
297	Sandy clay, 12 m; sand, 4 m; clay, 6 m; sand, 4 m; brown clay, sand, 1 m	27
300	Brown clay sticky, 5 m; fine brown sand, 8 m; hard black clay, 1 m; fine brown sand, 1 m	15
301	Brown sticky clay, 7 m; brown sand fine, 7 m; hard black clay, 1 m; sandy brown clay, 19 m	34
345	Clay, 5 m; sand, 2 m; clay, 3 m	10
348	Clay, 2 m	2
352	Sand, 3 m; clay yellow, 1 m	4
353	Clay brown, 2 m; sand, 5 m	7
354	Sand, 6 m	6
402	Clay, dark-tan, 4 m; sand brown fine, 6 m	10
409	Clay brown, 3 m	3
416	Brown clay, 5 m; sand, 8 m; brown clay, 8 m	21
613	Sandstone, 4 m	4
615	Gray sandstone, 11 m; lava, 9 m; brown sand and clay, 7 m	27
618	Sandstone, 14 m; clay and sand, 8 m	22
619	Gray clay, 2 m; sand, 11 m; clay, 17 m; sand, 13 m; clay, 5 m; sand, 5 m	53
621	Brown sand, 9 m; blue clay, 5 m; brown sand, 29 m	43
622	Brown sandstone, 6 m; brown sand, 3 m; gray sandstone, 8 m; blue clay-gray sand, 8 m; brown sand, 3 m; gray sand, 11 m; blue clay and sand, 3 m; gray sand, 1 m	43
629	Clay, 10 m; sand, 3 m; blue clay, 5 m; sand, 13 m; clay, 4 m; sand, 2 m	37
630	Brown clay, 9 m; blue sticky clay, 8 m; heaving(?) sand, 18 m; black sand, 14 m	49
632	Clay, 1 m; sand, 13 m; blue clay, 3 m; sand, 16 m; blue clay, 3 m	36
633	Clay, 2 m	2
635	Silty soft clay, 2 m	2
640	Blue clay, 11 m; sand, 3 m; blue clay, 2 m; sand, 18 m; clay, 2 m; sand, 2 m	38
641	White clay, 3 m; brown and blue clay, 9 m; sand, 12 m; clay, 5 m; sand, heaving(?), 12 m; clay, 8 m; sand, 2 m	51
645	Gray, green, and white rock, 2 m; brown clay, 3 m; gray clay, 5 m; gray sand, 3 m; gray and yellow clay, 2 m; gray and black sand, 3 m; gray and blue clay and sand, 2 m; gray sandy clay, 5 m; gray clay, 2 m; gray sand, 12 m; gray clay, 8 m; brown sand, 6 m	53
647	Gray sand, 36 m; sticky brown clay, 9 m; brown sand, 2 m	47
648	Clay, 4 m; brown sand, 8 m; clay, 7 m; black sand, 5 m; blue clay, 6 m; black sand, 3 m; brown clay, 6 m; brown sand, 7 m	46
1069	Sandy brown clay, 9 m	9
1126	Sand and clay, 2 m; brown clay, 7 m	9
1127	Sand and clay, 9 m	9
1128	Brown clay, 5 m	5
1129	Brown sand, 6 m; brown clay and sand, 13 m	19
1130	Gray sand, 4 m; brown sand, 9 m; gray soil, 3 m; brown clay and sand, 2 m; sand, 2 m	20
1241	Brown sand, 2 m; brown clay, 5 m; fine sand, 1 m; medium sand, 4 m	12
1242	Brown clay and fine sand, 1 m; fine sand, 3 m; blue clay, 2 m; brown clay with indurated lumps and caliche, 2 m; fine sand, 6 m; medium sand, 3 m	17
1243	Brown gummy clay, 5 m; fine sand and clay, 2 m; sand, silt, and clay, 2 m; sand and silt with indurated lumps and caliche, 1 m; gummy brown clay, 2 m	12
1274	Clay, 10 m	10
1279	Sand, 2 m; dark-blue clay, 5 m; wet brown sandy clay, 8 m	15

APPENDIX TABLE 2. DRILLERS' DESCRIPTIONS OF SEDIMENTS IN STRATIGRAPHIC INTERVAL ASSIGNED TO AMERICAN FALLS LAKE BEDS NORTHWEST OF THE SNAKE RIVER

Well No.	Lithology			Total Thickness (m)
	American Falls Lake Beds	Overlying Unit	Underlying Unit	
37	Clay	Surface	Lava	4
39	Soil and clay	Surface	Broken lava	1
40	Clay, 9 m; sandy clay, 2 m	Surface	Lava	11
41	Dirt	Surface	Soft black basalt	5
52	Clay	Surface	Gray lava	3
53	Brown silty clay	Surface	Broken black basalt	2
54	Clay	Surface	Broken lava	8
55	Clay	Surface	Lava	6
56	Clay	Surface	Black lava	4
58	Topsoil	Surface	Hard gray basalt	4
59	Clay	Surface	Gray basalt	2
61	Topsoil	Surface	Black basalt	2
62	Clay	Surface	Gray basalt	5
63	Clay	Surface	Lava	5
64	Clay	Surface	Gray basalt	2
70	Clay	Surface	Hard light gray lava	2
71	Clay	Lava	Lava	14
77	Clay	Surface	Gray lava	4
82	Topsoil and basalt	Surface	Solid basalt	6
155	Clay	Surface	Black lava	5
158	Soil	Surface	Broken black lava	3
160	Clay	Surface	Loose lava	14
162	Brown sandy clay	Surface	Gravel and sand	7
172	Clay	Surface	Gravel	5
173	Clay	Surface	Lava	2
175	Clay	Surface	Lava rock	2
186	Sandy brown clay	Surface	Broken lava and clay, 2 m; gray basalt	5
256	Topsoil, 4 m; dry brown clay, 2 m	Surface	Gray basalt	6
260	Brown topsoil and gravel	Surface	Black lava rock	6?
261	Soil, clay	Surface	Gray lava	2
262	Dirt	Surface	Medium hard basalt	7
265	Clay	Surface	Broken basalt and clay, 3 m; gray basalt	7
266	Clay	Surface	Solid gray basalt	4
268	Clay	Surface	Lava	1
269	Clay, 2 m; sandy clay, 2 m; clay, 1 m	Surface	Lava	5
270	Topsoil	Surface	Solid lava	5
271	Clay	Surface	Lava rock	2
272	Topsoil	Surface	Broken basalt	2
274	Brown clay	Surface	Gray lava	1
276	Clay	Surface	Lava	4
280	Sandy clay	Surface	Lava	5
285	Clay	Surface	Lava	2
287	Clay	Surface	Basalt	1
288	Topsoil, 2 m; brown clay, 4 m	Surface	Black lava	6
289	Clay topsoil	Surface	Broken black lava	5
291	Soil	Surface	Gray lava	2
292	Soil	Surface	Solid gray basalt	2
296	Clay and topsoil	Surface	Rock and basalt	2
297	Clay	Surface	Lava	2
300	Very sticky clay	Surface	Gray scoria	14
301	Topsoil	Surface	Gray scoria with crevices	1
613	Sandy clay	Surface	Black lava	13
614	Clay and rock	Soft gray lava	Hard gray lava	4

APPENDIX TABLE 2. DRILLERS' DESCRIPTIONS OF SEDIMENTS IN STRATIGRAPHIC INTERVAL ASSIGNED TO AMERICAN FALLS LAKE BEDS NORTHWEST OF THE SNAKE RIVER (continued)

Well	Lithology			Total Thickness
No.	American Falls Lake Beds	Overlying Unit	Underlying Unit	(m)
618	Light clay, 4 m; dark brown clay, 5 m; sticky clay, 2 m	Hard lava	Lava	11
619	Brown clay, 12 m; blue clay, 5 m; sand, 1 m	Broken lava	Lava	18
620	Light brown clay	Surface	Gray basalt	17
621	Topsoil, 6 m; sandy clay, 17 m	Surface	Lava	23
622	Topsoil, 19 m; sand, 1 m	Surface	Black basalt	20
624	Topsoil clay	Surface	Solid lava rock	6
629	Clay	Surface	Gray basalt	13
630	Brown and blue clay	Surface	Black hard lava	25
632	Topsoil, 4 m; white clay, 4 m; brown clay and sand, 11 m; white sand and clay, 9 m; brown sand, 3 m	Surface	Black basalt	31
633	Clay, 4 m; sand, 15 m	Surface	Broken cracked lava rock	19
635	Soft brown silt and brown clay, 5 m; hard pan blue clay, 2 m; soft brown silt, 10 m	Broken black basalt	Broken black basalt	17
640	Sandy clay, 6 m; brown clay, 15 m; black sand, 9 m	Surface	Blue clay	30
641	Topsoil, 1 m; brown clay, 13 m; blue clay, 5 m	Surface	Broken black lava	19
645	Topsoil	Surface	Black lava	6
647	Topsoil, 1 m; clay, 9 m	Surface	Gray basalt	10
648	Topsoil clay	Surface	Black lava	9

APPENDIX TABLE 3. DRILLERS' DESCRIPTIONS OF SEDIMENTS IN THE STRATIGRAPHIC INTERVAL ASSIGNED TO AMERICAN FALLS LAKE BEDS SOUTHEAST OF THE SNAKE RIVER

Well	Lithology			Total Thickness
No.	American Falls Lake Beds	Overlying Unit	Underlying Unit	(m)
308	Brown clay, 2 m; sand, 2 m	Clean gravel	Sandy gravel	4
303	Clay, 2 m; clay and sand, 4 m	Gravel and sand	Clay	6
310	Wet sandy clay	Sand and gravel	Sand and gravel	2
311	Clay	Gravel	Clay and gravel	3
315	Clay	Sand and gravel	Gravel	5
317	Brown clay	Gravel	Gravel and clay	5
321	Clay and gravel	Sand and gravel	Sand and gravel	3
322	Brown clay	Gravel and sand	Gravel and sand	5
326	Clay, 11 m; sand, 1 m	Gravel	Gravel	12
332	Sandy clay	Sand and gravel	Sand and gravel	5
341	Light brown clay	Sand and dirt	Sand with very small amount of fine gravel	7
342	Clay	Sand and sandy gravel	Sand and fine gravel	6
345	Clay	Sand and gravel	Gravel	6
348	Blue clay	Sand	Gravel and sand	10
352	Blue clay	Sand	Gravel	8
353	Blue clay	Sand, 1 m; white clay, 3 m; sand, 1 m	Gravel	8
354	Sand and blue clay	Sand	Sand and some gravel	3
355	Dry white clay	Black sand	Fine gravel and sand	11
357	Sandy clay, 1 m; clay, 5 m; sandy clay, 2 m	Sand	Gravel	8
401	Tan clay, 7 m; black quick sand, 1 m	Sand	Gravel	8
402	Dry light tan clay, 4 m; blue clay, 8 m	Sand	Fine white sand	12
409	Blue clay	Sand	Gravel	8
415	Brown clay, 7 m; blue clay, 1 m	Blow sand	River bottom gravel	8
416	Brown clay, 8 m; blue clay, 4 m	Topsoil	Brown clay	12
434	Blue clay	Sand and some gravel	Sand and gravel	7
442	Clay	Sand	Gravel and sand	4
1045	Light brown clay	Sand and gravel	Sand and gravel	6
1046	Brown clay	Sand and gravel	Brown clay and gravel	4
1047	Clay	Coarse gravel and clay	Sandy muddy coarse gravel	1
1048	Clay	Large gravel	Gravel	3
1049	Brown clay	Clay, sand, and gravel	Gravel, sand, and clay	5
1050	Clay	Gravel	Gravel	6
1051	Blue clay	Fine gravel	Gravel	4
1054	Green clay	Gravel and sand	Clay and gravel	5
1055	Brown clay, 7 m; blue clay, 5 m	Sand	Gravel	12
1056	Clay	Gravel and clay	Gravel and sand	9
1057	Clay and sand	Gravel	Gravel and clay	10
1059	Clay	Gravel and sand	Gravel and sand	2
1061	Clay	Gravel	Clay, gravel, and sand	2
1062	Clay	Sand	Sand	9
1063	Clay	Sand and gravel	Sand and gravel	6
1064	Sand and clay	Sand	Gravel	6
1065	Clay and sand, 3 m; blue clay, 5 m	Sand	Gravel	8
1066	Clay	Sand	Sand and gravel	4
1067	Clay and sand	Sand	Clay, sand, and gravel	6
1068	White clay	Clay and sand	Sand and gravel	9
1069	Silt, 3 m; white clay, 3 m	Sand	Sand and gravel	6
1070	Sandy clay	Coarse sand	Coarse sand	3
1071	Sand and clay	Sand	Clay and gravel	5

APPENDIX TABLE 3. DRILLERS' DESCRIPTIONS OF SEDIMENTS IN THE STRATIGRAPHIC INTERVAL ASSIGNED TO AMERICAN FALLS LAKE BEDS SOUTHEAST OF THE SNAKE RIVER (continued)

Well No.	Lithology: American Falls Lake Beds	Lithology: Overlying Unit	Lithology: Underlying Unit	Total Thickness (m)
1076	Clay	Sandy gravel	Sand and gravel	4
1077	Brown clay	Sandy gravel	Sandy gravel	4
1081	Clay	Surface	Gravel and clay	5
1083	Brown clay	Gravel	Gravel	5
1084	Brown clay	Dirt and gravel	Clay and gravel	5
1126	Blue clay, 4 m; brown clay, 4 m	Sand	Sand and gravel	8
1127	Blue clay	Sand	Gravel, sand, and clay	8
1128	Blue clay	Sand	Clay and gravel	9
1129	Blue clay	Sand and soil	Brown sand	9
1130	Dark blue clay	Clay and sand	Gray sand	9
1241	Blue clay	Surface	Good gravel	3
1254	Brown clay	Sand and gravel	Sand and gravel	4
1274	Clay with some gravel	Gravel	Sand and gravel	2
1277	Brown clay	Sand	Sand and gravel	11
1279	Brown clay	Sand	Sand and gravel	12
1283	Silty clay, 4 m; sand, 1 m; clay 2 m	Gravel	Gravel	7

REFERENCES CITED

Armstrong, R. L., Leeman, W. P., and Malde, H. E., 1975, K-Ar dating, Quaternary and Neogene volcanic rocks of the Snake River Plain, Idaho: American Journal of Sciences, v. 275, p. 225–251.

Bright, R. C., 1982, Paleontology of the lacustrine member of the American Falls Lake Beds, southeastern Idaho, *in* Bonnichsen, B., and Breckenridge, R. M., eds., Cenozoic geology of Idaho: Idaho Bureau of Mines and Geology Bulletin 26, p. 597–614.

Carr, W. J., and Trimble, D. E., 1963, Geology of the American Falls quadrangle, Idaho: U.S. Geological Survey Bulletin 1121-G, 44 p., scale 1:62,500.

Desborough, G. A., Raymond, W. H., Marvin, R. F., and Kellogg, K. S., 1989, Pleistocene sediments and basalts along the Snake River in the area between Blackfoot and Eagle Rock, southeastern Snake River Plain, southeastern Idaho: U.S. Geological Survey Open-File Report 89-0436, 18 p.

Greeley, R., 1982, The style of basaltic volcanism in the eastern Snake River Plain, Idaho, *in* Bonnichsen, B., and Breckenridge, R. M., eds., Cenozoic geology of Idaho: Idaho Bureau of Mines and Geology Bulletin 26, p. 407–421.

Hearst, J. M., 1990, Paleontology and depositional setting of the Duck Point local fauna (Late Pleistocene; Rancholabrean), Power County, southeastern Idaho [M.S. thesis]: Pocatello, Idaho State University, 275 p.

Hopkins, M. L., 1951, *Bison* [*Gigantobison*] *latifrons* and *Bison* [*Simbobison*] *alleni* in southeastern Idaho: Journal of Mammology, v. 32, p. 192–197.

——, 1955, Skull of fossil camelid from the American Falls Lake Bed area of Idaho: Journal of Mammology, v. 36, p. 278–282.

Hopkins, M. L., Bonnichsen, R., and Fortsch, D., 1969, The stratigraphic position and faunal associates of *Bison* [*Gigantobison*] *latifrons* in southeastern Idaho—A progress report: Tebiwa, v. 12, p. 1–8.

Karlo, J. F., and Jorgenson, D. B., 1979, Fault control of volcanic features southeast of Blackfoot, Snake River Plain, Idaho: Geological Society of America Abstracts with Programs, v. 11, p. 276.

Kellogg, K. S., and Embree, G. F., 1986, Geologic map of the Stevens Peak and Buckskin Basin Areas, Bingham and Bannock Counties, Idaho: U.S. Geological Survey Miscellaneous Field Studies Map MF-1854, scale 1:24,000.

Kellogg, K. S., and Marvin, R. F., 1988, New potassium-argon ages, geochemstry, and tectonic setting of Upper Cenozoic volcanic rocks near Blackfoot, Idaho: U.S. Geological Survey Bulletin 1806, 19 p.

Kennedy, S. K., and Smith, N. D., 1977, The relationship between carbonate mineralogy and grain sizes in two alpine lakes: Journal of Sedimentary Petrology, v. 47, p. 411–418.

Kuntz, M. A., Champion, D. E., Spiker, E. C., and Lefebvre, R. H., 1986, Contrasting magma types and steady-state, volume-predictable, basaltic volcanism along the Great Rift, Idaho: Geological Society of America Bulletin, v. 97, p. 579–594.

Kuntz, M. A., Champion, D. E., Lefebvre, R. H., and Covington, H. R., 1988, Geologic map of the Craters of the Moon, Kings Bowl, and Wapi lava fields, and the Great Rift volcanic rift zone, south-central Idaho: U.S. Geological Survey Miscellaneous Investigations Map I-1632, scale 1:100,000.

Leeman, W. P., 1982a, Development of the Snake River Plain–Yellowstone Plateau province, Idaho and Wyoming—An overview and petrologic model, *in* Bonnichsen, B., and Breckenridge, R. M., eds., Cenozoic geology of Idaho: Idaho Bureau of Mines and Geology Bulletin 26, p. 155–177.

——, 1982b, Olivine tholeiitic basalts of the Snake River Plain, Idaho, *in* Bonnichsen, B., and Breckenridge, R. M., eds., Cenozoic geology of Idaho: Idaho Bureau of Mines and Geology Bulletin 26, p. 181–192.

Lewis, G. C., and Fosberg, M. A., 1982, Distribution and character of loess and loess soils in southeastern Idaho, *in* Bonnichsen, B., and Breckenridge, R. M., eds., Cenozoic geology of Idaho: Idaho Bureau of Mines and Geology Bulletin 26, p. 705–716.

Malde, H. E., 1968, The catastrophic late Pleistocene Bonneville Flood in the Snake River Plain, Idaho: U.S. Geological Survey Professional Paper 596, 52 p.

Morgan, L. A., and Pierce, K. L., 1990, Silicic volcanism along the track of the Yellowstone hot spot: Geological Society of America Abstracts with Programs, v. 22, p. 39.

Morgan, W. J., 1972, Deep mantle convection plumes and plate motions: American Association of Petroleum Geologists Bulletin, v. 56, p. 203–213.

O'Connor, J., 1990, Hydrology, hydraulics, and sediment transport of Pleistocene Lake Bonneville flooding on the Snake River, Idaho [Ph.D. thesis]: Tuscon, University of Arizona, 192 p.

Pierce, K. L., Fosberg, M. A., Scott, W. E., Lewis, G. C., and Colman, S. M., 1982, Loess deposits of southeastern Idaho—age and correlation of the upper two loess units, *in* Bonnichsen, B., and Breckenridge, R. M., eds., Cenozoic geology of Idaho: Idaho Bureau of Mines and Geology Bulletin 26, p. 717–725.

Ridenour, J., 1969, Depositional environments of the late Pleistocene American Falls Formation, southeastern Idaho [M.S. thesis]: Pocatello, Idaho State University, 82 p.

Scott, W. E., 1982, Surficial geologic map of the eastern Snake River Plain and adjacent areas, 111° to 115° W., Idaho and Wyoming: U.S. Geological Survey Miscellaneous Investigations Series Map I-1372, scale 1:250,000.

Scott, W. E., Pierce, K. L., Bradbury, J. P., and Forester, R. M., 1982, Revised Quaternary stratigraphy and chronology in the American Falls area, southeastern Idaho, *in* Bonnichsen, B., and Breckenridge, R. M., eds., Cenozoic geology of Idaho: Idaho Bureau of Mines and Geology Bulletin 26, p. 581–595.

Selner, G. I., and Taylor, R. B., 1989, GSLITH version 2.0: A prototype program to draw cross sections and plot plan views from drill hole data with latitude, longitude coordinates, using an IBM PC (or compatible) microcomputer, digitizer, and plotter: U.S. Geological Survey Open-File Report 89-0114-A, 84 p., and 89-0114-B, 3 p., two 5¼-in. diskettes.

Smith, R. B., and Sbar, M. L., 1974, Contemporary tectonics and seismicity of the western United States with emphasis on the Intermountain seismic belt: Geological Society of America Bulletin, v. 85, p. 1205–1218.

Spear, D. B., and King, J. S., 1982, The geology of Big Southern Butte, *in* Bonnichsen, B., and Breckenridge, R. M., eds., Cenozoic geology of Idaho: Idaho Bureau of Mines and Geology Bulletin 26, p. 395–403.

Stearns, H. T., Crandall, L., and Stewart, W. G., 1938, Geology and ground-water resources of the Snake River Plain in southeastern Idaho: U.S. Geological Survey Water-Supply Paper 774, 268 p.

Suppe, J., Powell, C. and Berry, R., 1975, Regional topography, seismicity, Quaternary volcanism, and the present-day tectonics of the western United States: American Journal of Science, v. 275-A, p. 397–436.

Trimble, D. E., 1976, Geology of the Michaud and Pocatello quadrangles, Bannock and Powers Counties, Idaho: U.S. Geological Survey Bulletin 1400, 88 p., scale 1:48,000.

Trimble, D. E., and Carr, W. J., 1961, Late Quaternary history of the Snake River in the American Falls region, Idaho: Geological Society of America Bulletin, v. 72, p. 1739–1748.

——, 1976, Geology of the Rockland and Arbon quadrangles, Power County, Idaho: U.S. Geological Survey Bulletin 1399, 115 p., scale 1:62,500.

Whitehead, R. L., 1986, Geohydrologic framework of the Snake River Plain, Idaho and eastern Oregon: U.S. Geological Survey Hydrologic Atlas HA-681, scale 1:1,000,000.

Williams, P. L., Covington, H. R., and Pierce, K. L., 1982, Cenozoic stratigraphy and tectonic evolution of the Raft River basin, Idaho, *in* Bonnichsen, B., and Breckenridge, R. M., eds., Cenozoic geology of Idaho: Idaho Bureau of Mines and Geology Bulletin 26, p. 491–504.

Manuscript Accepted by the Society July 19, 1991

Printed in U.S.A.

Geological Society of America
Memoir 179
1992

Chapter 14

A model-based perspective of basaltic volcanism, eastern Snake River Plain, Idaho

Mel A. Kuntz
U.S. Geological Survey, MS 913, Box 25046, Federal Center, Denver, Colorado 80225

ABSTRACT

Models of basaltic volcanism of the eastern Snake River Plain (ESRP), Idaho, provide estimates of crustal basalt magma-driving pressures, magma-supply rates, dike geometry, mass-eruption rates, and durations of eruptions. Magma-driving pressures are 0.1 to 0.2 kb and about 1.2 kb for basalt magma rising from magma reservoirs at 40 and 60 km, respectively. Magma-driving pressures are 0.2 to 0.3 kb and 1.2 kb for basalt and trachyandesite magmas of the Craters of the Moon lava field that rise from a reservoir at 40 km. The lack of vents in several tracts of the ESRP is believed to be related to thick fill of low-density rocks in buried calderas that creates a density barrier that impedes or prevents eruption of basalt magma.

Magma injection by dikes is influenced by regional tectonic extensional stress orientations; eruptive fissures and volcanic rift zones of the ESRP typically are oriented northwest-southeast, perpendicular to minimum regional stresses. Calculated model mass-eruption rates are 9×10^3 kg/sec m for typical ESRP basalt eruptions. Calculated durations of eruptions on the ESRP range from about 25 to 30 hours for the 0.005 km^3 Kings Bowl lava field to about 2 months for the 6 km^3 Wapi lava field.

An inferred magma-plumbing system includes a magma source region in the mantle at depths of 60 km, a magma reservoir at or near the Moho, magma-filled cracks that transfer magma upward from the reservoir, an upper-level reservoir beneath the Craters of the Moon lava field, and dikes that lead from the reservoir to surface vents.

The relatively unfractionated character of ESRP basalts, long repose intervals between eruptions on most volcanic rift zones, and relatively small volumes of lava fields suggest episodic local melting, little or no accumulation of magma in reservoirs, and rise of magma to the surface fairly directly from the source depth.

INTRODUCTION

This paper provides a perspective for the formation, storage, ascent, and eruption of basaltic magmas of the eastern Snake River Plain (ESRP) of Idaho. I incorporate conceptual models and apply mathematical models constrained by field studies to characterize basaltic volcanism of the ESRP. The models account for both the similarities of and differences between basaltic volcanism of the ESRP, the well-monitored volcanism of present-day active volcanoes, and the well-documented basaltic volcanism of the Columbia River Plateau (CRP). However, because the eruptions of the ESRP are prehistoric, there is a lack of data on rates and durations; this restricts both the types of analyses that can be performed and the accuracy of results produced by such analyses.

Estimates for magma-supply rates, magma-driving pressures, dike geometry, magma-rise velocities, mass-eruption rates, and eruption durations constrain a general model perspective of the processes of basaltic volcanism of the ESRP. The general model links magma ascent and eruption to relationships between internal magma pressure, the buoyant rise of magma, rock stresses, and flow paths, all of which are influenced by regional extensional fracture.

The general model involves local and episodic generation of magma in the lithospheric mantle and periodic release of magma

Kuntz, M. A., 1992, A model-based perspective of basaltic volcanism, eastern Snake River Plain, Idaho, *in* Link, P. K., Kuntz, M. A., and Platt, L. B., eds., Regional Geology of Eastern Idaho and Western Wyoming: Geological Society of America Memoir 179.

from upper mantle and/or deep crustal reservoirs by propagation of fractures to the surface. Dike propagation is governed by generation of magma-driving pressure in the reservoir that produces failure by exceeding the tensile strength of the reservoir rocks. Magma rise in fractures is approximated by consideration of buoyant forces generated by the density contrast between magma and overlying rocks in the upper mantle and crust. Increases in tensional stresses due to regional extension lead to higher rates of intrusion frequency.

The general model partly follows that of Shaw and Swanson (1970, p. 271), who obtained a numerical perspective of the volcanism of the CRP region by combining field evidence and estimates for rates of lava eruption and emplacement of flows. They noted that their problem was somewhat stochastic, "one in which each step more or less reflects all preceding steps and affects all subsequent steps, and hence it is not amenable to complete solutions in terms of fixed balances of mass, momentum, and energy. . . .It should be clear that there are no hard numbers here. There are, however, some conspicuous probabilities that agree to some extent with previous intuitions." Similar caution and encouragement applies to the models of basaltic volcanism of the ESRP presented here.

CHARACTERISTICS OF BASALTIC VOLCANISM OF THE EASTERN SNAKE RIVER PLAIN, IDAHO

The surface of the ESRP is covered chiefly by basaltic lava flow rocks and surficial deposits of lacustrine, eolian, and alluvial origin. Drilling and geophysical studies suggest that these flows and sediments also make up a 1- to 2-km-thick sequence that overlies Neogene rhyolitic flow rocks, ignimbrites, and tephra deposits throughout much of the ESRP.

Three stages of basaltic volcanism in the ESRP have been delineated: stage 1—short-term, fissure-type eruptions; stage 2—intermediate-term, lava cone–forming eruptions; and stage 3—long-term, shield-forming eruptions (Kuntz and others, this volume). Lava fields of stage-1, fissure-type eruptions cover areas less than 20 km^2 and have volumes of 0.005 to 0.1 km^3; lava fields of stage-2, lava-cone–type eruptions cover areas of 10 to 100 km^2 and have volumes of 0.05 to 2 km^3; and shield volcanoes of sustained stage-3–type eruptions cover areas of 100 to 400 km^2 and have volumes of 1 to 7 km^3.

The total volume of basaltic lava flows in the ESRP is estimated to be about 4×10^4 km^3. More than 95% of the volume is contained in shield volcanoes and lava cones and the remainder as eruptive-fissure deposits, tephra cones, and deposits of hydromagmatic eruptions. By assuming that the average volume for shield-type and lava-cone–type eruptions is about 3 to 5 km^3 and that the ESRP began to form about 10 Ma, the recurrence interval of shield-forming and lava-cone–forming eruptions is about 1,000 years.

All lava flows issued from eruptive fissures or central conduits confined to eruptive fissures. Eruptive fissures are 0.5 to 2 m wide; the average width is about 1 m. They are 0.1 to 1 km long; the average length is 0.5 km. Eruptive fissures typically are in en echelon arrays that are 0.3 to 10 km long and average about 3 km. Eruptive-fissure arrays are typically less than 1 km wide. Shield volcanoes formed during stage-3 eruptions typically have slot-shaped vents that are 3 km long or less. The shape of these vents and the fact that they overlie and are parallel to eruptive fissures indicate that fissures influenced even the latest parts of the eruptions of these largest volcanic landforms of the ESRP.

Although regional slopes in the ESRP are typically less than 0.5 degrees, flows of shield volcanoes and lava cones are as long as 30 km, and they thicken only slightly toward their vents. These relationships suggest relatively high initial eruption rates and relatively low lava viscosities.

Summary characteristics of selected lava fields in the ESRP are presented in Table 1.

Many eruptive fissures and fissure zones in the ESRP constitute major structures that define *volcanic rift zones.* Volcanic rift zones are linear belts of volcanic landforms and structures related to volcanism. They are 30 to 100 km long and 3 to 15 km wide (Kuntz and others, this volume).

Basaltic lava fields of the ESRP are late Pliocene to Holocene in age, but only a few surface flows are known to be older than 1 Ma. Radiocarbon studies (Kuntz and others, 1986) show that eight lava fields, which cover about 13% of the ESRP, are very latest Pleistocene or Holocene in age. Seven of these eight fields represent single bursts of eruptive activity that were not repeated at the same or nearby vents for long periods of time. Of these seven fields, the North Robbers, South Robbers, and Kings Bowl fields represent stage-1 fissure eruptions. The Hells Half Acre, Cerro Grande, Wapi, and Shoshone fields represent stage-2 and stage-3, lava-cone–forming and shield-forming eruptions. Eruptions that formed the seven fields occurred on volcanic rift zones that have long, but discontinuous, histories; eruptions typically are widely spaced and repose intervals between eruptions are about 10^4 to 10^5 years (Kuntz and others, this volume).

The Craters of the Moon (COM) lava field is a composite of at least 60 individual lava flows that erupted from 25 tephra cones and at least eight eruptive fissure systems. The COM field covers about 1,600 km^2 and contains about 30 km^3 of lava flows. It formed during eight eruptive periods, each several hundreds of years or less in duration, separated from one another by periods of quiescence that lasted from several hundred to about 3,000 years. The average repose interval between eruptions is about 2,000 years. COM magmas have become progressively more silicic, but each eruptive period has produced magma having SiO_2 contents of less than 47% as well as more silicic magmas. The eruptive fissure zone of the COM segment of the Great Rift volcanic rift zone is about 45 km long, but during any one eruptive period, only 0.3- to 10-km-long sections of the zone were active. The COM field began to form about 15 ka, and the latest eruptions occurred about 2 ka. Age and volume data indicate that the eruptive history of the COM field is "volume predictable," that is, the volume of lava erupted in a given eruptive period is proportional to the length of the prior repose interval and the long-term lava output rate (Kuntz and others, 1986).

TABLE 1. SUMMARY OF CHARACTERISTICS OF VENTS AND FLOWS FOR CRATERS OF THE MOON AND OTHER HOLOCENE AND LATEST PLEISTOCENE LAVA FIELDS, EASTERN SNAKE RIVER PLAIN, IDAHO*

Lava Field	Length of single eruptive fissures (Length of zone of fissures in parentheses) (km^3)	Estimated fissure width (m)	Estimated volume of lava fields (km^3)
Craters of the Moon (various selected eruptions in eruptive periods A and B)	0.1–1.3 (0.3–10)	1–2	0.03–3.4
Kings Bowl	0.1–0.5 (4.8)	(0.5–1) (measured)	0.005
Wapi	unknown (0.8)	unknown	6
Hells Half Acre	unknown (1.5)	unknown	6
North Robbers	2.2 (2.9)	1–2	0.05
South Robbers	1.1 (1.8)	1–2	0.03
Cerro Grande	unknown	unknown	2.3
Shoshone	unknown	unknown	1.5

*Data from Kuntz and others, 1988; 1992; this volume, Table 1.

The areal extent and volume of lava fields in the ESRP pale in comparison to the "great flows" of the CRP, which have volumes of 1,000 to 3,000 km^3 and cover areas as large as 100,000 km^2 (Tolan and others, 1989). Significant differences in volumes of erupted basalt between the ESRP and the CRP reflect differences in length of eruptive fissure systems, magma-supply rates, volume of magma reservoirs, and mass-eruption rates but, curiously, not in duration of eruptions. With comparisons of Snake River Plain and Columbia River Plateau volcanism and the constraints provided by the data in Table 1 in mind, I proceed to analyze basaltic volcanism of the eastern Snake River Plain.

ORIGIN OF MAGMA, MAGMA-SUPPLY RATES, AND VOLCANIC RIFT ZONES

Most flow rocks of the ESRP are olivine basalts of olivine tholeiite and alkaline basalt affinities. The olivine basalts have limited mineralogical, chemical, and textural variations that suggest they are the products of partial melting of the lithospheric mantle at depths of 60 km (see, for example, Leeman and Vitaliano, 1976; Stout and Nicholls, 1977; Leeman, 1982a, 1989). Thompson (1975) and Leeman (1982a) speculated that partial crystallization of olivine basalt at about 30 km in the crust formed more fractionated basalts.

A small percentage of flow rocks in the ESRP are more-evolved hawaiite to trachyandesite or latite. The most significant locality for evolved rocks is the Craters of the Moon lava field. Leeman (1982b), Leeman and others (1976), and Stout and others (1989) concluded that the evolved magmas originated from olivine basalt by intermediate pressure fractionation of olivine, plagioclase, augite, and iron-titanium oxide minerals at pressures equivalent to depths of about 30 to 35 km, perhaps in conjunction with contamination of basalt magma by crustal rocks or partial melts derived from crustal rocks.

Upward migration of pockets and films of magma from a source region in the asthenosphere to a storage area at the crust-mantle boundary is driven mainly by buoyant forces, guided by stress conditions in the asthenosphere (see, for example, Waff and Bulau, 1979; Shaw, 1980). Shaw (1980) suggested that shear melting in the asthenosphere and fracture mechanisms for "kneading" melts upward indicate that there is little or no storage of magma at or near the source area. For example, he estimated that the average crustal residence time for basalt at Kilauea Volcano in Hawaii is as short as 10 years; thus, the asthenosphere is sampled directly and quickly with little modification of magma chemistry.

Reservoirs of olivine basalt in the ESRP may occur where upward migration of magma from a primary source area in the mantle is stalled at positions of neutral buoyancy, perhaps at the crust-mantle boundary and/or in the partially melted, lower crustal rocks of the root zone of previously formed rhyolitic calderas, where magma and wall-rock densities may be similar (see, for example, Ryan, 1987a, 1987b; Leeman, 1989).

Production of basalt magma in the ESRP has been ascribed to pressure-release melting in ascending mantle plumes (Morgan, 1972) and to stress relief at the base of the lithosphere due to tectonic extension (Christiansen and McKee, 1978). Christiansen and McKee suggested specifically that (1) greater amounts of extension produce higher percentages of partial melting at shallower depths, and (2) basalts rise buoyantly in areas where favorable structures or large deviatoric stresses can accommodate the rise, such as areas where extensional faulting occurs simultaneously with basaltic volcanism. Lachenbruch and Sass (1978) suggested, however, that rapid extension of the crust causes large amounts of magma to be intruded rather than erupted.

Wright and others (1989) stated the magma-supply rate for the CRP was 75×10^6 m^3/year, and Leeman (1989) estimated the average magma-supply rate for the Snake River Plain for the last 15 Ma to be on the order of 70 to $700 \cdot 10^6$ m^3/year. Fedotov (1981) suggested that magma-supply rates for volcanoes in various tectonic settings are on the order of 2 to $8 \cdot 10^6$ m^3/year. For comparison, the long-term lava-output rate for ESRP as a whole for the last 15 k.y. is approximately $2 \cdot 10^6$ m^3/year (Kuntz and others, 1986). If Leeman's estimated magma-supply rate is reasonable, only 0.3 to 3% of the supplied magma has been erupted at the surface of the ESRP in the last 15 k.y.; however, if Fedotov's value is a better estimate of the magma-supply rate for the ESRP, then almost all magma has been erupted at the surface.

The Great Rift and Spencer–High Point volcanic rift zones of the ESRP are unique in that multiple episodes of volcanism occurred along the same fissure system in relatively short periods of time (Kuntz and others, this volume). These two volcanic rift zones may constitute a mechanical-thermal feedback system in which localized, concentrated extension in volcanic rift zones produced a higher rate of partial melting in the upper mantle, perhaps by localized decompression melting (White and McKenzie, 1989) beneath these linear belts that, in turn, increased the number of suitable fractures for the rise of magma to the surface. These relationships suggest that these volcanic rift zones were and remain long-lived, self-perpetuating zones of volcanism.

Many prominent volcanic rift zones of the ESRP are extensions onto the ESRP of range-front faults for Basin and Range mountains at the margins of the ESRP (Kuntz and others, this volume). The self-perpetuation concept suggests that volcanic rift zones in the ESRP may represent strain recorders that mark the approximate times of regional extension marked by Basin and Range–type faulting along the margins of the ESRP, related decompression melting of the upper mantle, and subsequent volcanism concentrated in volcanic rift zones. Integrated studies in the ESRP of the age of faulting along range-front faults and the age of volcanism along colinear volcanic rift zones may provide evidence that either corroborates or negates this idea. Unfortunately, such data are not available at present.

SHAPE AND VOLUME OF MAGMA RESERVOIRS AND REPOSE INTERVALS BETWEEN ERUPTIONS

The shape of magma reservoirs in the ESRP cannot be determined with certainty in the absence of geophysical data on the events preceding and accompanying eruptions. Concerning the subsurface configuration of magma conduits, storage reservoirs, and the zone of magma generation in the upper mantle for the Yakima Basalt of the Columbia Plateau, Swanson and others (1975, p. 896) suggested that "the simplest possibility is that a vent system directly reflects the geometry of its subsurface plumbing. If it does, then the linear vent systems imply that magma was stored or even produced along narrow zones in the mantle. . . ." Wright and others (1989) thus suggested that magma reservoirs for CPR eruptions were sill-like and located directly beneath eruptive fissure systems and that the minimum volume is represented by the volume of the largest flows. Similar shapes were proposed by Lachenbruch and Sass (1978) for magma bodies in an extending crust that accommodates intrusion of melt from the asthenosphere. Applying Swanson's and Wright's ideas to the ESRP and assuming the largest volume of erupted magma to be about 5 km^3, the minimum dimensions of sill-like magma reservoirs that underlie some of the largest eruptive fissure zones in the ESRP (10 km long and 1 km wide) are 10 km long, 1 km wide, and 0.15 km thick. Assuming magma-supply rates for the ESRP of $5 \cdot 10^6$ m^3/year and $75 \cdot 10^6$ m^3/year, ussed above, magma reservoirs would fill in about 100 to 1,000 years. The actual volumes of magma reservoirs beneath the ESRP are probably considerably larger but cannot be quantified.

The short repose intervals (~0.3 to 75 years) between eruptions for Hawaiian volcanoes and Etna and Fuego volcanoes contrast markedly with the long repose intervals (10^3 to 10^5 years) between eruptions on volcanic rift zones in the ESRP (Kuntz and others, 1986). Thus, eruptions on these volcanic rift zones represent discrete events. In addition, latest Pleistocene and Holocene lava fields are widely scattered over the ESRP. These factors suggest that magma is generated locally and quickly in the ESRP, has short residence time in upper mantle and/or lower crustal reservoirs, and produces relatively low volume eruptions. If melt migrates quickly and is not stored for long periods in reservoirs beneath the ESRP, then the long repose intervals suggest episodic melting, accumulation, and rise of magma to the surface fairly directly from the source depth.

Magma-supply rate and the pressure regime of the magma reservoir are critical factors that affect the type of volcanic eruption (Shaw, 1980; Fig. 1). Two limiting cases of stress adjustment of the reservoir walls affect the release of magma during eruption.

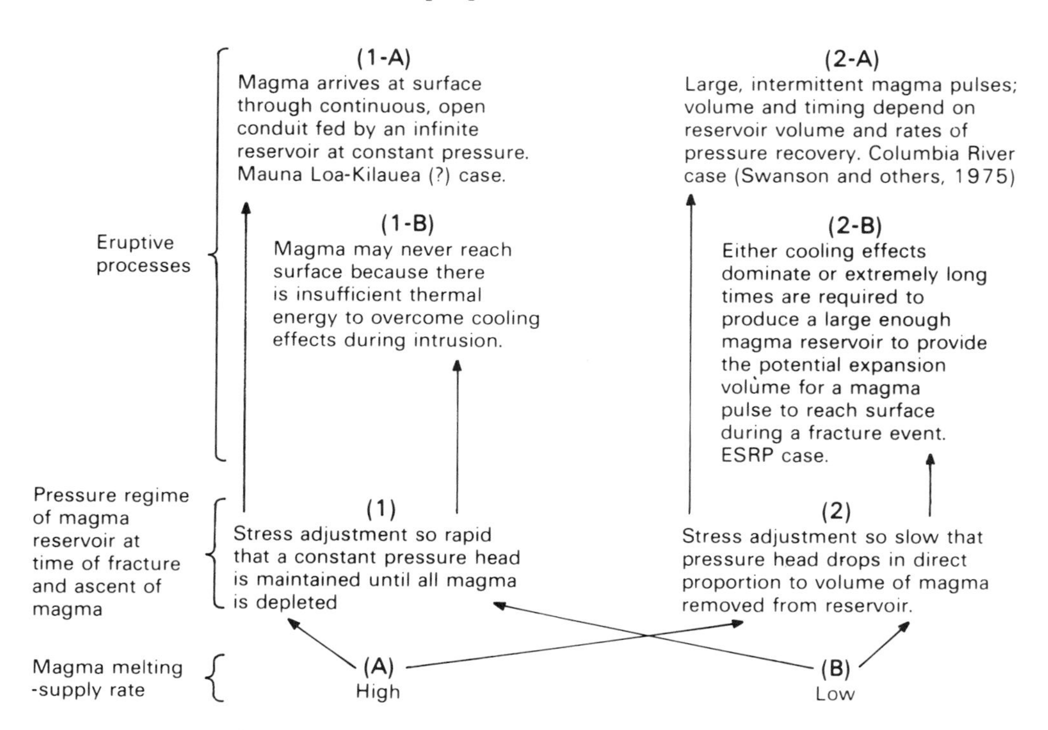

Figure 1. Relationships of magma-supply rate, pressure regime of magma reservoir, and eruptive processes (modified from Shaw, 1980).

Rapid adjustment of reservoir walls as magma is withdrawn from the reservoir produces a constant pressure potential so that almost all magma is expelled from the reservoir, which tends to promote long-lived, large-volume eruptions. Slow adjustment produces a progressive pressure drop in proportion to the volume of expelled magma, and the eruption terminates when magma pressure reaches a critical level; thus, eruptions are short lived and of relatively low volume. Low magma-supply rates and slow stress adjustment of reservoir walls on eruptions are factors that are believed to control volcanism in the ESRP and to produce relatively low volume, short-duration eruptions.

STRUCTURE AND DENSITY OF THE CRUST OF THE ESRP AND MAGMA-DRIVING PRESSURES

The principal source of magma-driving pressure (Johnson and Pollard, 1973) is the density contrast between magma in a reservoir and the rocks that surround and lie above the reservoir. As suggested by Shaw and Swanson (1970) and Wilson and Head (1981), nonhydrostatic processes such as viscous creep of mantle or lower crustal rocks and volume changes on melting and partial crystallization are ignored here. Wilson and Head show that because gas exsolution is common in magmas at depths of less than 2 km, the ascent of bubble-free magma can be treated almost completely by the buoyant-rise model.

Structure and density of the crust and upper mantle of ESRP

The buoyant-rise model is dependent on the density contrast between magma and surrounding rocks, and thus the structure and density of the crust in the ESRP are important factors in constraining the model. Seismic studies (Smith and others, 1982; Braile and others, 1982; Sparlin and others, 1982) provide velocity and density data for the lithosphere of the ESRP (Fig. 2).

In a typical crustal and upper mantle section of the ESRP (Fig. 2A), the uppermost layer, 2 km thick, consists of Pliocene to Holocene basaltic lava flow rocks and interlayered sediments (density = 2,000 kg/m^3). Below the uppermost layer is a layer, 8-km-thick, interpreted as rhyolitic flow rocks and tuffs (density = 2,450 to 2,550 kg/m^3) that formed during late Miocene–Pliocene rhyolitic volcanism in the ESRP (summarized by Pierce and Morgan and by Morgan, this volume). The two volcanic layers lie above a greatly thinned, 10-km-thick, upper crustal layer (density = 2,670 kg/m^3) related to the rhyolitic volcanism. A middle crustal layer, 10-km-thick (density = 2,880 kg/m^3), is interpreted as mafic material intruded into crustal rocks. The 10-km-thick lower crust (density = 3,000 kg/m^3) has a relatively uniform thickness under the ESRP. The Moho is at a depth of about 40 km. The density of the lithospheric mantle is 3,300 kg/m^3 and V_p is 7.95 km/second. These data suggest crustal underplating in the

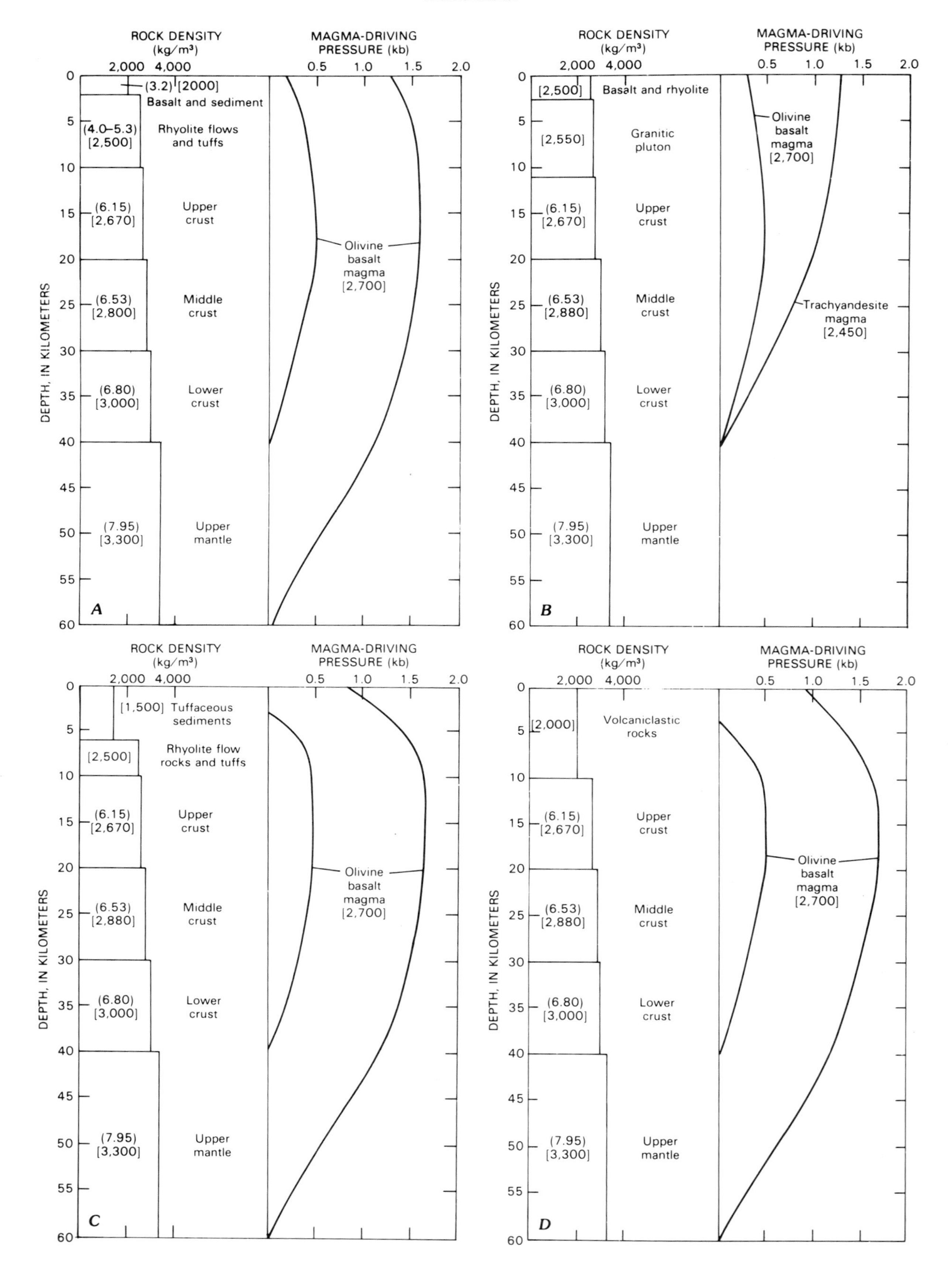
ROCK DENSITY (kg/m³)
2,000 4,000
MAGMA-DRIVING PRESSURE (kb)
0.5 1.0 1.5 2.0
DEPTH, IN KILOMETERS
(3.2) [2000] Basalt and sediment
(4.0–5.3) [2,500] Rhyolite flows and tuffs
(6.15) [2,670] Upper crust
(6.53) [2,800] Middle crust
(6.80) [3,000] Lower crust
(7.95) [3,300] Upper mantle
Olivine basalt magma [2,700]
A
[2,500] Basalt and rhyolite
[2,550] Granitic pluton
(6.15) [2,670] Upper crust
(6.53) [2,880] Middle crust
(6.80) [3,000] Lower crust
(7.95) [3,300] Upper mantle
Olivine basalt magma [2,700]
Trachyandesite magma [2,450]
B
[1,500] Tuffaceous sediments
[2,500] Rhyolite flow rocks and tuffs
(6.15) [2,670] Upper crust
(6.53) [2,880] Middle crust
(6.80) [3,000] Lower crust
(7.95) [3,300] Upper mantle
Olivine basalt magma [2,700]
C
[2,000] Volcaniclastic rocks
(6.15) [2,670] Upper crust
(6.53) [2,880] Middle crust
(6.80) [3,000] Lower crust
(7.95) [3,300] Upper mantle
Olivine basalt magma [2,700]
D

upper mantle by accumulation of dikes and sills of basalt (Furlong and Fountain, 1986).

Greensfelder and Kovach (1982) attributed a decrease in shear-wave velocity at depths of 20 to 40 km in the middle and lower crust to granulites that were depleted of quartz and, presumably, other low-temperature-melting minerals during earlier episodes of partial melting that formed rhyolitic magmas. They discounted the presence of "significant amounts of partial melt" to explain the velocity decrease because they considered the lower crust to have a temperature below the solidus of dry granodiorite. Priestly and Brune (1982) found a low-velocity zone in the upper mantle of the region extending from about 50 to 180 km that they interpreted to be a zone of partial melt in the upper mantle.

Gravity, magnetic, and magnetotelluric investigations by D. R. Mabey and W. D. Stanley (in Kuntz and others, 1983) reveal additional details of the upper crustal structure beneath the Great Rift volcanic rift zone (GRVRZ). A positive magnetic anomaly of about 70 gammas centered along the northern part (Fig. 3A) is approximately coincident with an elongated gravity low (Fig. 3B). The southeast edge of the magnetic anomaly is sharp and coincides with the southeast limit of eruptive vents along the COM segment of the GRVRZ (Kuntz and others, 1988) (Fig. 3A). The magnetic and gravity data suggest that a granitic intrusive body having an area of about 500 km^2 and a density of about 2,550 kg/m^3 (D. R. Mabey, U.S. Geological Survey, oral communication, 1983) underlies the COM lava field (Fig. 3A). Multibody inversion analysis of the regional magnetic field in the area of the GRVRZ (Bhattacharya, 1980; Bhattacharya and Mabey, 1980) indicates that the plutonic body is highly magnetized and has a top surface 1 to 3.5 km below sea level and a base 10.5 to 11.5 km below sea level.

The base of the magnetized crust is about 10 km below sea level along the northern part of the GRVRZ, about 17 km below sea level along the southern part of the zone, and about 12 km below sea level beneath the Wapi lava field (Fig. 3C). Bhattacharya and Mabey (1980) and Stanley and others (1977) suggested that the base of the magnetized crust may coincide with the Curie-point isotherm: thus, temperatures of 500 to 580°C may exist at depths as shallow as 10 km beneath the northern part of the volcanic rift zone and at depths of 16 to 18 km along the zone at the southern end of the COM lava field, as shown by the shaded area in Figure 4. This extremely high heat flow is comparable to that of the Battle Mountain, Nevada, area (Lachenbruch and Sass, 1978). High heat flow in the ESRP is also suggested by the work of Brott and others (1981), and heat-flow values of 2.6 HFU have been determined from the 3-km-deep borehole at the Idaho National Engineering Laboratory (D. A. Blackwell, oral communication, 1983). The higher geothermal gradients beneath the northern part of the Great Rift volcanic rift zone, compared to other parts of the ESRP, probably result from longer magma residence time in that area, which may promote greater amounts of partial melting of reservoir rocks and the production of more siliceous magmas. The data in Figure 4 suggest that the base of the granite pluton may be near conditions of partial melting and may be the site of contamination that forms the more siliceous basaltic magmas of the COM lava field.

A 70 × 30 km area of the ESRP near Blackfoot that contains no basalt vents overlies the buried Taber caldera (see Kuntz and others, this volume, Figs. 10, 16, 17). On the basis of Schlumberger soundings along a traverse across the northeastern part of the caldera, Zohdy and Stanley (1973) inferred that the upper crust in that area consists of several geoelectrical units: The uppermost layer (300 to 3,000 ohm-m), about 1 to 2 km thick, consists of basalt and water-saturated basalt; an intermediate layer (20 to 40 ohm-m), 1 to more than 6 km thick, is made up of sedimentary rocks and/or rhyolitic ash-flow tuff; and the geoelectrical basement below the intermediate layer consists of ~40-ohm-m rocks. A.A.R. Zohdy (U.S. Geological Survey, oral communication, 1989) suggested that materials of the intermediate layer are probably low-density, noncompacted tuffs, breccias, and volcaniclastic sedimentary rocks. Density models that incorporate relatively low density rocks inferred from the electrical sounding data are shown in Figures 2C and D.

A constraint on magma stored in an upper mantle or deep-crustal reservoir for long periods of time is that it must retain enough heat so that its temperature remains above the solidus at that depth. Figure 4 shows solidi and liquidi for the McKinney and King Hill basalts (Tilley and Thompson, 1970; Thompson, 1972), representative of Snake River Plain olivine basalt and trachyandesite flow rocks from the COM lava field, respectively. The experimental data suggest that temperatures are probably 1,100 to 1,250°C at the Moho, corresponding to about 2.2 to 2.5 HFU, or roughly equivalent to the high Basin-Range gradient of Lachenbruch and Sass (1978).

Figure 2. Structure and density distributions of the upper mantle and crust of the eastern Snake River Plain, Idaho, and magma-driving pressures for various upper mantle and crustal density-magma density pairs. Upper mantle and crustal density models mainly from Smith and others (1982), Sparlin and others (1982), Braile and others (1982), and Priestly and Brune (1982). For A, B, C, and D, numbers in parentheses are V_p in km/second, and numbers in brackets are rock density in kg/m^3. A, Typical olivine basalt magma of ESRP; typical upper mantle and crustal density distribution for ESRP. B, Crustal density model for COM segment of Great Rift volcanic rift zone (GRVRZ), adapted from D. R. Mabey and D. B. Stanley (in Kuntz and others, 1983); olivine basalt magma and trachyandesite magma typical of COM lava field. C, Model 1 for Taber caldera, which is made up of 6-km-thick section of low-density tuffaceous sedimentary rocks above a 4-km-thick section of rhyolite flow rocks and tuffs; olivine basalt magma. D, Model 2 for Taber caldera, which is made up of 10-km-thick section of volcaniclastic rocks; olivine basalt magma. Crustal density models in C and D adapted from Zohdy and Stanley (1973).

Magma-driving pressures and their implications

Magma-driving pressures (the "dynamic pressure head" in the terminology of Shaw and Swanson, 1970) were calculated according to the relation (Johnson and Pollard, 1973):

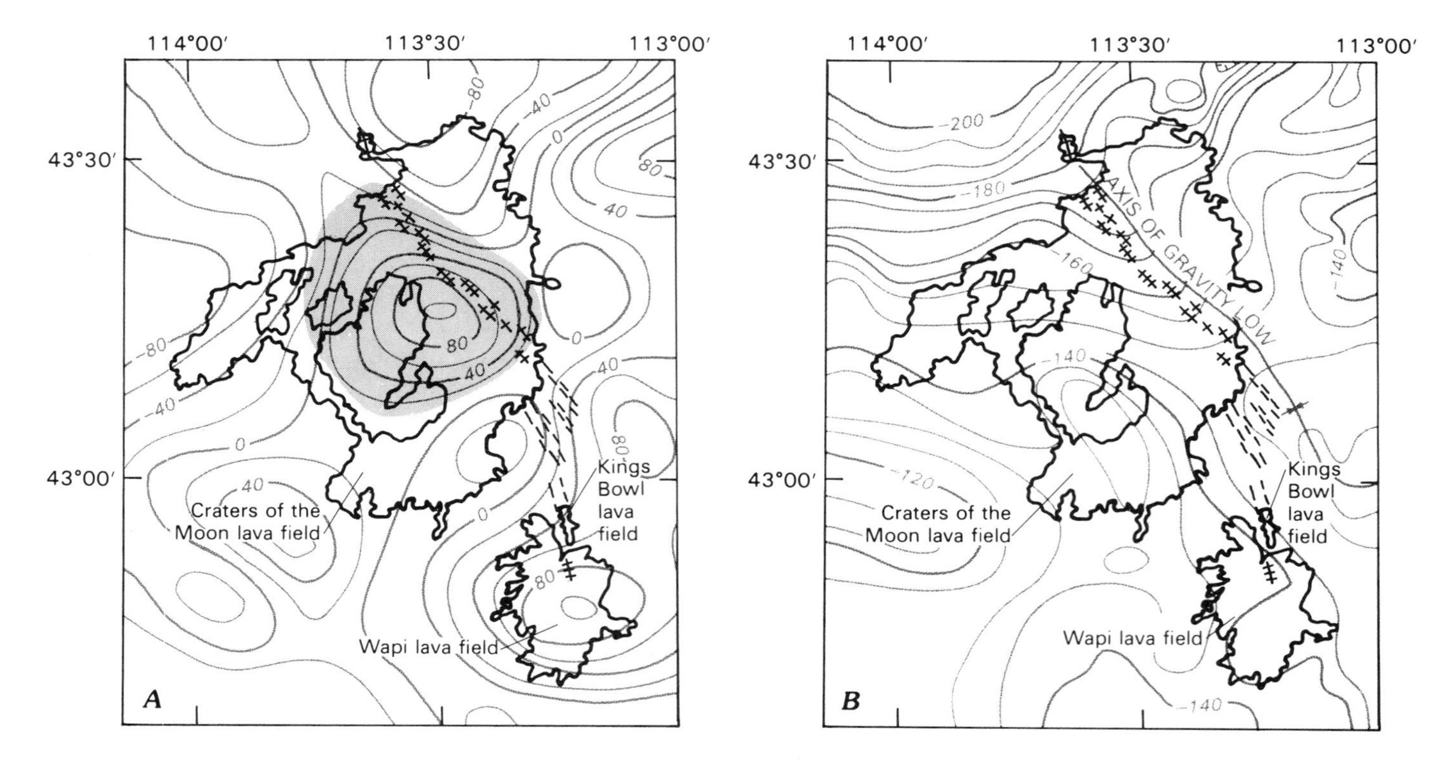

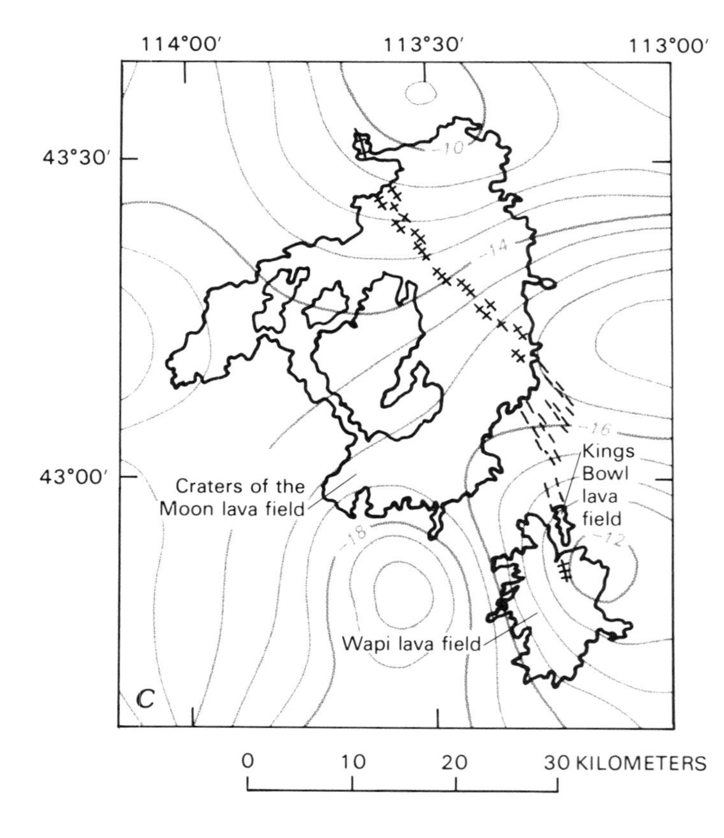

Figure 3. Magnetic and gravity anomalies and inferred thermal characteristics of the crust in the Great Rift area, eastern Snake River Plain, Idaho. For all maps, cross-hatched and en echelon lines show locations of eruptive fissures and noneruptive fissures that constitute the GRVRZ. Outlines of Craters of the Moon, Kings Bowl, and Wapi lava fields also shown. A, Residual magnetic map continued upward to 11.4 km above sea level; contour interval 20 gammas; shaded area shows inferred granitic pluton; modified from Bhattacharyya and Mabey (1980). B, Regional gravity map; contour interval 5 mgal; modified from D. R. Mabey (in Kuntz and others, 1983). C, Map of the depth below sea level of the base of the magnetized crust; contour interval 1 km; modified from Bhattacharyya and Mabey (1980).

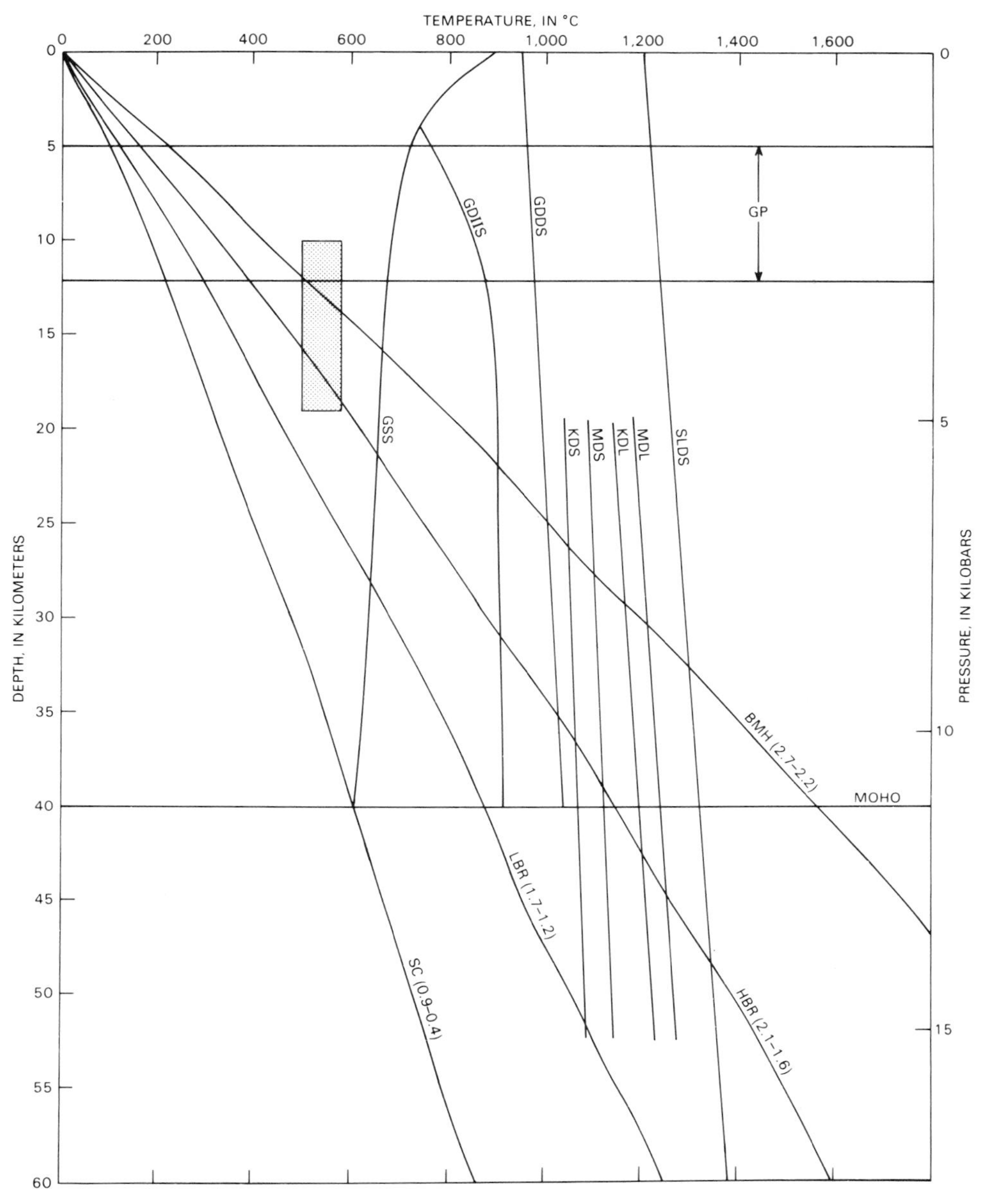

Figure 4. Geothermal gradients and pertinent experimental data related to generation, storage, and migration of basaltic magmas in the eastern Snake River Plain, Idaho. Geothermal gradients from Lachenbruch and Sass (1978): BMH, Battle Mountain high; HBR, high Basin and Range; LBR, low Basin and Range; SC, stable continent. Numbers in parentheses on each gradient are heat flow in HFU. Horizontally ruled area represents inferred granitic pluton beneath Craters of the Moon laval field and temperature-depth relations in crust along the COM segment of the Great Rift volcanic rift zone, as described in text. Experimental data: GSS, granite water-saturated solidus (Merrill and others, 1970); GDIIS, granodiorite type II solidus (Robertson and Wyllie, 1971); GDDS, granodiorite dry solidus (Robertson and Wyllie, 1971); SLDS, spinel lherzolite dry solidus (Kushiro and others, 1968); MDL, McKinney basalt dry liquidus (Thompson, 1972); MDS, McKinney basalt dry solidus (Thompson, 1972); KDL, King Hill basalt dry liquidus (Thompson, 1972); KDS, King Hill basalt dry solidus (Thompson, 1972).

$$P_{m_h} = (P_{l_s} - \rho_m gH) - P_{l_h}$$

where P_{m_h} = magma-driving pressure at depth h above the source, P_{l_s} = lithostatic pressure at the source, ρ_m = magma density, g = 9.81 m/sec^2, H = height above magma reservoir, and P_{l_h} = lithostatic pressure of overlying column of rocks at depth h.

Magma-driving pressures have been determined for a combination of various crustal and upper mantle density models and for magmas having different densities (Fig. 2). The combinations were chosen to correspond to specific crustal and upper mantle density and magma density pairs in the ESRP.

The density of an average ESRP olivine basalt magma at 1,250°C, calculated by the method of Bottinga and Weill (1970), is about 2,700 kg/m^3. The driving pressure at the surface for a magma of this density rising through a typical ESRP crust from a reservoir at the Moho is 0.1 to 0.2 kb and about 1.2 kb for a magma rising from a source region at 60 km (Fig. 2A). The driving-pressure model shows that olivine basalt magmas of density 2,700 kg/m^3 must rise from reservoirs more than 35 km deep in those parts of the ESRP with a typical crustal and upper mantle density distribution. These relationships suggest that most areas of the ESRP are subject to eruption of olivine basalt by buoyant rise of magma, assuming that the following conditions, among others, are met: (1) magma rises from reservoirs near or below the Moho, and (2) cooling effects of wall rocks are compensated by critical magma-rise velocities.

The crustal-density distribution used for calculation of driving pressures for magmas of the COM lava field, rising from a reservoir at the Moho, includes the inferred granitic pluton (Fig. 2B). Densities of magmas at 1,250°C for the COM lava field range from 2,700 kg/m^3 for the Sheep Trail Butte flow to 2,450 kg/m^3 for the Highway trachyandesite flow; the magma-driving pressure curves in Figure 2B were calculated for these density extremes. Magma-driving pressures are 0.2 to 0.3 kb and about 1.2 kb for basalt and trachyandesite magmas, respectively (Fig. 2B). The buoyant rise model indicates that basalt magmas must rise from depths greater than 30 km and that trachyandesite magma could rise from reservoirs at any depth in the crust beneath the COM lava field. However, the heat-flow and experimental data (Fig. 4) suggest that reservoirs of trachyandesite magma probably occur at depths greater than 25 km. These factors, as well as the compositional variety of COM magmas cited earlier, suggest that (1) dikes tap olivine basalt magma near the crust-mantle boundary and more siliceous magma in reservoirs in the lower or middle crust or that (2) magma reservoirs are compositionally zoned dikes. For the second case, each eruptive period in the COM lava field approaches evacuation of the dikelike magma reservoir(s) by removing more silicic magma at the top early in the eruptive period, followed by eruption of progressively more mafic magma at deeper levels. The higher driving pressures for COM magmas that result from less dense, more silicic magmas and/or the presence of the granite pluton may account, in part, for the greater volume of magma erupted along the GRVRZ than from other parts of the ESRP during the latest Pleistocene and Holocene.

Figures 2C and D show crust and upper mantle density models for the Taber caldera: One has a 6-km-thick fill of tuffaceous sedimentary rocks (density = 1,500 kg/m^3) above a 4-km-thick section of rhyolite flows and tuffs (density = 2,500 kg/m^3), and the other model has a 10-km-thick fill of volcaniclastic rocks (density = 2,000 kg/m^3). For both models, driving pressures for magmas of density 2,700 kg/m^3 rising from reservoirs at the Moho become negative at depths of about 4 km. Olivine basalt magma rising from a source region at 60 km generates sufficient driving pressure to penetrate the Taber caldera, as represented in both density models, but the driving pressures are considerably less than the driving pressure for basalt that rises through a typical crust and upper mantle of the ESRP (compare figures 2A, C, D). These decreased and negative pressures suggest that low-density volcanic rocks and/or volcaniclastic sediments that fill calderas form density barriers that prevent, or decrease the likelihood of, surface eruption of basalt magmas.

Pierce and Morgan (this volume) suggest that the width of the hot spot beneath the ESRP is on the order of 1,500 km; theoretically, therefore, basaltic vents should be plentiful beyond the margins of the ESRP. The few sites of Quaternary basaltic volcanism beyond the margins of the ESRP may be related, in part, to the crustal-density structure in those areas: low-density Precambrian, Paleozoic, and Mesozoic rocks in the upper crust beyond the margins of the ESRP may inhibit eruption of basaltic magmas. Basaltic lava fields beyond the margins of the ESRP typically occur in intermontane valleys where the crust has been thinned by extensional and/or thrust faulting.

All calculated magma-driving pressures would be less if the magmas had significant shearing strength (Johnson and Pollard, 1973). Yield strengths of basaltic magmas of the ESRP, determined by a relation given by Wilson and Head (1981, p. 2,976), are about 3 N/m^2. All calculated driving pressures are smaller by only 5% or less if yield strength of magma is also considered.

STRESS ORIENTATION AND FAILURE CRITERIA

The orientation of fissures in basaltic eruptions is related to regional tectonic stress orientations (for example, Nakamura, 1969; Shaw, 1980; Zoback, 1989). Nakamura (1977) showed that formation of volcanic fissures is similar to large-scale, magma-fracturing experiments.

Failure criteria for fracture propagation assumes a triaxial stress condition in which σ_1 and σ_3, the maximum and minimum effective normal stresses, are oriented horizontally, and σ_2, the intermediate effective normal stress, is oriented vertically, following the conventions of Shaw (1980). The vertical orientation of dikes and eruptive fissures indicates that the direction of the least principle stress axis in the crust of the ESRP is horizontal and oriented about northeast-southwest, as predicted in the extensional tectonic regime of the Snake River Plain (see, for example,

Zoback and Zoback, 1980; Stickney and Bartholomew, 1987; Zoback, 1989).

The magma-driving pressure produces a stress σ_p perpendicular to the walls of the magma reservoir. A vertical fracture extending from the reservoir is produced when the value of σ_p reaches τ_0, the tensile strength of the reservoir walls. The criterion for the minimum stress (Wadge, 1977) is:

$$\sigma_{p_{min}} = \tau_0 = 3\sigma_3 - \sigma_1 - P_m$$

where P_m is internal magma pressure. Fracture occurs in a vertical plane along the direction of σ_2 when $P_m > 3\sigma_3 + \tau_0 - \sigma_1$. Gudmundsson (1988) suggested that reduction of σ_3 as a result of tensile stress concentration, a function of regional extension and the aspect ratio of magma reservoirs, is the ultimate cause of dike intrusion in deep-seated reservoirs. Thus, increased intrusion frequencies are promoted by higher crustal spreading rates and/or higher aspect ratios of magma chambers.

Pollard (1973) noted that intrusions propagate along extension fractures when:

$$\sigma_1 - \sigma_3 - P_m + P_p + (P_m - \sigma_3) \cdot 2L/T = \tau_o$$

where L and T are the length and thickness of the fracture and P_p is pore pressure. Propagation is predicted at even lower values of magma pressure as compared to the previous criterion because the term $(P_m - \sigma_3)$ is multiplied by L/T, a large number. As Pollard stated, dikes have a tremendous mechanical advantage over the host rock because of large L/T values and numerous, critically oriented microscopic flaws, and thus they will propagate at very small values of $P_m - \sigma_3$. In practical terms, Wilson and Head (1981) suggested that excess magma pressures need not exceed 0.2 kbar to cause fracture of reservoir walls at all reasonable depths.

DIKE GEOMETRY, MAGMA-RISE VELOCITIES, MASS-ERUPTION RATES, AND ERUPTION DURATIONS

Application of models of the dynamic flow of magma in dikes provides insight into dike geometry, magma-rise velocities, mass-eruption rates, and duration of eruptions for ESRP volcanoes.

By way of a general perspective, Delaney and Pollard (1982) and Bruce and Huppert (1989) determined that for flow to be sustained in dikes, the heat lost to wall rocks must be replenished by heat advected from the magma reservoir. Bruce and Huppert discussed the competing processes of dike blockage due to solidification of magma on dike walls and dike widening due to meltback. They concluded that eruptions must become localized at one or several isolated, probably cylindrical vents for fissure eruptions to continue for more than a few days and that eruptions from localized vents will continue until limited or terminated by other volcanic processes. These conclusions substantiate observations made in Hawaii and in the ESRP that eruptions begin on an eruptive fissure system as long as several kilometers, then proceed to eruptions from tephra cones or lava cones at one or several central vents, and that sustained eruptions occurred at slot-shaped vents, typically 1 km or less in length, at lava cones and shield volcanoes.

Dike geometry

Koide and Battacharji (1975) showed that effective stresses are greatest at the top of magma reservoirs, and because fractures are in areas of largest effective stresses, reservoirs tend to grow or expand as an upward-developing plexus of dikes.

As a first approximation, dike geometry can be estimated by assuming that dike width or conduit radius is the main factor limiting magma rise due to buoyant forces, rather than cooling effects related to magma viscosity or to finite magma yield strength (Wilson and Head, 1981; Delaney and Pollard, 1982). Minimum dike width at depth can be estimated by using the relationship of critical dike half-width that just permits movement of magma due to buoyant forces (Wilson and Head, 1981):

$$r_{crit.} = \left[\frac{ZHA\eta}{8g\,(\rho_m - \rho_h)}\right]^{1/4}$$

where Z is a constant ($1.25 \cdot 10^{-6}$ m^2/sec), H is dike height, A is a constant (24), η is magma viscosity, g is the acceleration of gravity, and $(\rho_m - \rho_h)$ is the density contrast between magma and wall rocks. For conditions in the lower crust, assuming open conduit flow, H is 30 to 40 km, η is 300 Pa · s (Wilson and Head, 1981), and a density contrast of 300 kg/m^3, dike half-width is about 0.4 m. For a near-surface, multifracture model in which H = 5 km, dike half-width is about 0.2 m, comparable to observed dike half-widths at the surface in the ESRP (Table 1). At a depth of about 15 km, where the density contrast decreases to about 100 kg/m^3, dike half-width is about 0.4 m for the open conduit model and about 0.2 to 0.3 m for the multifracture model for crack heights of 1 to 5 km. These calculations suggest that minimum dike widths are about 0.4 to 0.8 m at depth, only slightly less than the observed widths of feeder fissures at the surface (Table 1).

Width/length ratios of fissures at various depths can be estimated according to the relation of Fedotov (1978):

$$\frac{w}{l} = \frac{4\,(1 - \nu^2)\,P_m}{E}$$

where l is fissure half length, ν is Poisson's ratio, P_m is excess magma pressure, and E is Young's modulus. R. P. Denlinger (U.S. Geological Survey, written communication, 1989) suggested Poisson's ratio is 0.4 to 0.5 for deformation of the lithosphere at elevated temperatures beneath the ESRP. Young's modulus is 20 GPa near the surface and about 100 GPa at the base of the crust (Gudmundsson, 1988). Using values for excess magmatic pressure (magma-driving pressure) as shown in Figure 2A for olivine basalt, width/length ratios range from about $2 \cdot 10^{-3}$ near the surface to about $3 \cdot 10^{-4}$ in the lower crust. For an average dike width of 1 m at the surface and P_m of 15 MPa for olivine basalt, fissure lengths of about 1 km are predicted, equal to or slightly less than those observed in the ESRP (Table 1). Fissure widths are wider at depth by as much as a factor of two. For trachyandesite magma from the COM lava field, assuming a dike width of 1 m and P_m of 120 MPa (Fig. 2B), a fissure length of about 100 m is predicted. Such a short fissure length is in accord with the observation that trachyandesite flows, specifically the Serrate, Devils Orchard, and Highway flows of the COM lava field, were erupted from central pipelike vents rather than from eruptive fissures (Table 1).

The heights of magma-filled fissures at various depths can be estimated by the relationship (Wadge, 1982):

$$H \leqslant \left[\frac{K_c}{\sqrt{\pi}\ g\,(\rho_m - \rho_h)}\right]^{2/3}$$

where K_c is fracture toughness, a property of rocks that measures resistance to crack propagation, and $(\rho_m - \rho_h)$ is the density contrast between magma and surrounding rock. Experimental data indicate that K_c values are on the order of 1 to 5 MPa/m^2 for sedimentary rocks, but K_c values of 100 MPa/m^2 were calculated for sedimentary host rocks penetrated by dikes near Ship Rock, New Mexico (Delaney and Pollard, 1981). Using the value $K_c = 100$ MPa/m^2, and density contrasts of 300 kg/m^3 for the lower crust and 150 kg/m^3 for the middle crust (Fig. 2), crack heights of about 1.5 km and 2.3 km, respectively, are predicted.

These data suggest that magma-filled cracks in the overlying lower crust have lengths of about 5 to 10 km, widths of about 0.5 to 2 m, and heights of about 2 km, yielding volumes of $4 \cdot 10^6$ to $3 \cdot 10^7$ m^3. racks of these dimensions represent the magma plumbing system from reservoir to surface, they would contain volumes on the order of 10^8 to 10^9 m^3 of magma.

Magma-rise velocities

An estimate of the rise velocity of magma in cracks at all Reynolds numbers, assuming that the rise is not limited by excessive cooling, may be determined by the relation (Wilson and Head, 1981):

$$V = \frac{A\eta}{4K\rho_m\nu}\left\{\left[1 + \frac{64gr^3\,(\rho_m - \rho_h)\,K\rho_m}{A^2\eta^2}\right]^{1/2} - 1\right\}$$

where K is a constant (0.01), r is the crack half-width, and the other variables are as defined previously. The average crustal density contrast for most magmas of the ESRP is 150 kg/m^3, and magma viscosity is taken as 300 Pa·s (Wilson and Head, 1981). For dike half-widths of 1, 0.75, and 0.5 m, the respective velocities are about 1.6, 0.9, and 0.4 m/second. These velocities correspond to travel times of magma from a reservoir at the Moho to surface of approximately 7, 12, and 28 hours, respectively. Wilson and Head (1981) stated that rise of magmas that possess yield strengths greater than $10^3 N/m^2$ will be limited by yield strength rather than by cooling. However, calculations discussed previously suggest that magmas of the ESRP had shearing strengths of about $3N/m^2$; thus magma rise velocities would be decreased by extremely small amounts. Therefore, it is reasonable to assume that cooling effects and dike geometry are more important factors than yield strength in determining magma-ascent rates.

Intrusion/extrusion ratio and its implications

The intrusion/extrusion ratio is poorly known for even well-monitored, active volcanoes and is probably impossible to estimate for ESRP volcanoes. General conclusions reached by Gudmundsson (1988) are that greater extrusion frequency is favored by (1) higher tensile stress concentration around magma chambers, (2) the narrower width of zones undergoing tensile strain, and (3) higher spreading rates. The second factor is reflected in repose intervals of volcanic rift zones of the ESRP: The narrow (<3 km) northern part of the Great Rift volcanic rift zone has had eruptions that are closely spaced in space and time, whereas wider (<15 km) volcanic rift zones have had eruptions that are more widely spaced in space and time. The third factor suggests that spreading rates are higher on the Great Rift and Spencer–High Point volcanic rift zones than other volcanic rift zones of the ESRP.

Mass-eruption rates

Initial mass-eruption rates of basaltic eruptions for the ESRP can be estimated from the relation (Wilson and Head, 1981):

$$E_0 = \frac{A\eta}{2K}\left\{\left[1 + \frac{64gr^3\,(\rho_m - \rho_g)\,K\rho m}{A^2\eta^2}\right]^{1/2} - 1\right\}$$

where variables and constants are as previously defined. For values of $\eta = 300$ Pa·s, r = 1 m, a density contrast of 150 kg/m^3, and a magma density of 2,700 kg/m^3, the mass-eruption rate is $9 \cdot 10^3$ kg/sec m. For comparison, Wilson and Head (1981) determined that the mass-eruption rate for the Askja, Iceland, eruption of 1961 was $1 \cdot 10^4$ kg/sec m and the rate for the Lakagigar, Iceland, eruption of 1783 was at least $1.5 \cdot 10^3$ kg/sec m. Tolan and others (1989) estimated mass-eruption rates of 3 to $9 \cdot 10^4$ kg/sec m for the "great flows" of the Yakima member of the Columbia Plateau.

Duration of eruptions

To obtain an initial perspective on the duration of eruptions of the ESRP, we assume that (1) the eruption rate was constant (probably unreasonable, as evaluated later), (2) ESRP eruptions occurred along fissures 1 to 5 km long, and (3) most lava flows of the ESRP have an average density of 2,700 kg/m^3. The large Blue Dragon, Minidoka, and Carey-Sunset pahoehoe flows of the COM lava field, which have average volumes of about 3 km^3, may have formed in about two days for a fissure length of 5 km and in about 10 days for a fissure length of 1 km. Minimum eruption durations for the larger, latest Pleistocene and Holocene lava fields of the ESRP, which have volumes of as much as 6 km^3, are about four days for a fissure length of 5 km and about 20 days for a fissure length of 1 km. The calculated mass-eruption rate for the Kings Bowl lava field, assuming a fissure half-width of 0.5 m, a magma density of 2,700 kg/m^3, and a density contrast of 100 kg/m^3, is about 8×10^2 kg/m sec. By assuming a constant eruption rate, an eruptive fissure 5 km long, a lava flow density of 2,700 kg/m^3, and a volume of 0.005 km^3, the Kings Bowl lava field may have formed in as little time as one hour.

A more realistic estimate of the duration of eruptions for the emplacement of these lava fields can be modeled by continuous flow from an eruptive fissure, but with decreasing flow rate with time. This corresponds to the case of decreasing head for a fixed-volume eruption model, as discussed by Shaw and Swanson (1970). Machado (1974) and Scandone (1979) noted that the flow rate for many eruptions decrease exponentially according to the relation:

$$E_{(t)} = E_0 \cdot e^{-At}$$

where E_0 is the initial mass-eruption rate, A is a constant, and t is time. Also,

$$E = \frac{dV}{dT}$$

Substituting the first equation into the second and integrating both sides yields

$$\int_0^{\infty} E_0 \cdot e^{-At} \cdot dt = \int_0^{\infty} dV$$

where

$$A = \frac{E_0}{V_{tot}}$$

where V_{tot} is the total flow volume. For the lava flows of a volume of 3 km^3, where $E_0 = 4.5 \cdot 10^7$ kg/second for a 5-km-long fissure and lava density is 2,500 kg/m^3, $A = 3 \cdot 10^{-6}$ seconds. From the properties of the exponential curve, 63% of the total volume is extruded in $t_{63} = 1/A$ seconds, or about four days, and 95% of the total volume is extruded in $t_{95} = 3/A$ seconds, or about 12 days.

Applying the same model to the Kings Bowl lava field and using $E_0 = 4 \cdot 10^6$ kg/second for a 5-km-long eruptive fissure and a lava field volume of $2.5 \cdot 10^{10}$ kg, t_{63} is about two hours and t_{95} is about 5.5 hours. Assuming that eruptions occurred on a 1-km-long section of the fissure, t_{63} for the Kings Bowl lava field is nine hours and t_{95} is 26 hours.

The Wapi lava field, representative of large, latest Pleistocene and Holocene shield volcanoes of the ESRP, has a volume of 6 km^3. The vent complex of the Wapi field consists of a line of pit craters and vents that extends for a distance of about 800 m. By applying the exponential decreasing head model and assuming a mass-eruption rate of $9 \cdot 10^3$ kg/m and a fissure length of 1 km, t_{63} of the Wapi lava field is about three weeks and t_{95} of the field is about two months.

For comparison, Shaw and Swanson (1970) and Tolan and others (1989) estimated that flows of the Columbia River Plateau, even the great flows (>1,000 km^3), were emplaced in periods of a few weeks or even less. Tolan and others (1989) suggested that because fissure systems were not overly large for the great flows, mass-eruption rates must have been significantly faster for great flows than for other flows of the Columbia River Plateau.

MAGMA PLUMBING SYSTEM AND SUMMARY OF VOLCANIC PROCESSES

Common elements in plumbing systems of basalt volcanoes include (1) a magma source region in the upper mantle, (2) a magma column that leads upward from the source region, (3) a magma reservoir at or near the Moho, (4) magma-filled cracks that transfer magma upward, (5) an upper level, subvolcanic reservoir (not always present), and (6) conduits from the subvolcanic reservoir to volcanic vents. These elements are based on studies, for example, of Kilauea (Ryan and others, 1981), Mt. Etna (Wadge, 1977), and Fuego (Martin and Rose, 1981). Modifications of these elements, based on conceptual and mathematical models presented in this paper, lead to the magmatic plumbing systems for basaltic volcanoes for most areas of the ESRP and for the COM lava field shown in Figure 5.

Most olivine basalts of the ESRP have limited mineralogical, chemical, and textural variations that suggest they are the products of partial melting of the lithospheric mantle at depths of about 60 km and that their magmas had short residence times in upper mantle and/or lower crustal reservoirs. The relatively unfractionated character of ESRP basalts, long repose intervals between eruptions, and relatively small volumes of lava fields suggest episodic local melting, little or no accumulation of magma in reservoirs, and rise of magma to the surface fairly directly from the source depth. Primary magma reservoirs formed locally at the base of or within the lower crust where magma stagnated, owing perhaps to comparable magma and wall-rock densities. Partial crystallization of olivine basalts in lower-crustal reservoirs at about 30 km probably formed more fractionated basalts. Evolved magmas originated from olivine basalt by intermediate pressure fractionation of olivine, plagioclase, augite, and iron-titanium oxide minerals in crustal reservoirs, also at depths of about 30 km, perhaps in conjunction with contamination of basalt magma by crustal rocks or partial melts derived from crustal rocks.

Magma is believed to have been produced and stored in narrow zones directly beneath eruptive fissures. The reservoirs are probably long, narrow sills that have a minimum volume of about 5 km^3 and minimum dimensions of 10 km long, 1 km wide, and 0.15 km thick.

Fracture systems traverse the entire thickness of the crust,

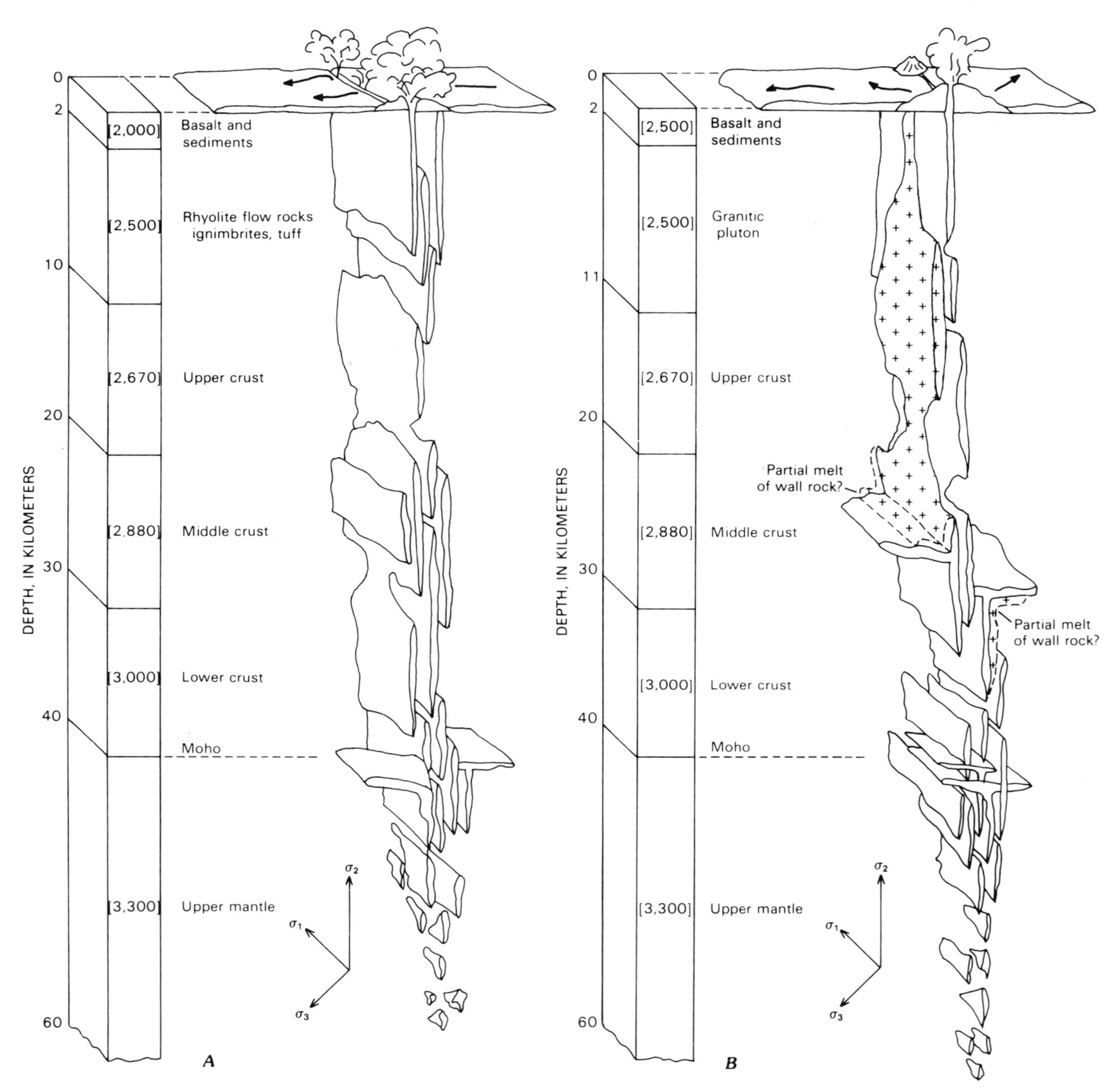

Figure 5. Schematic representation of magmatic plumbing systems in the eastern Snake River Plain, Idaho. A, Plumbing system for most parts of eastern Snake River Plain. B, Plumbing system for Craters of the Moon lava field. See text for discussion. Stress orientations and schematic representation of area of origin of basaltic melt after Shaw (1980).

and they may locally penetrate reservoirs in the lower and middle crust. The vertical orientation of dikes and eruptive fissures indicates that the direction of the least principle stress axis in the crust of the ESRP is horizontal and oriented about northeast-southwest, as predicted in the extensional tectonic regime of the Snake River Plain. The magma-filled cracks have lengths of about 5 to 10 km, widths of about 0.5 to 2 m, and heights of about 2 km, yielding volumes of $4 \cdot 10^6$ to $3 \cdot 10^7$ m^3.

Calculated mass-eruption rates and a decreasing head for a fixed-volume eruption model allow for emplacement of about 6 km^3 of lava in a period of several months. Eruptions end because of an exponential decrease in eruption rate due to the slow adjustment of the magma reservoir on magma withdrawal.

ACKNOWLEDGMENTS

Field studies of basaltic volcanic rocks of the ESRP were carried out under the aegis of the late S. S. Oriel; his guidance and patience are deeply appreciated. H. R. Shaw provided encouragement and a review of a preliminary version of the paper. R. P. Denlinger, P. T. Delaney, and K. H. Wohletz provided helpful reviews of the paper and clarified the author's meaning and intent. D. L. Campbell's help with mathematical problems is appreciated. Wayne Hawkins drafted the figures.

REFERENCES CITED

Bhattacharya, B. K., 1980, A generalized multibody model for inversion of magnetic anomalies: Geophysics, v. 45, p. 255–270.

Bhattacharya, B. K., and Mabey, D. R., 1980, Interpretation of magnetic anomalies over southern Idaho using generalized multibody methods: U.S. Geological Survey Open-File Report 80-457, 49 p.

Bottinga, Y., and Weill, D. F., 1970, Densities of liquid silicate systems calculated from partial molar volumes of oxide components: American Journal of Science, v. 269, p. 169–182.

Braile, L. W., and nine others, 1982, The Yellowstone-Snake River Plain seismic profiling experiment—Crustal structure of the eastern Snake River Plain: Journal of Geophysical Research, v. 87, p. 2597–2609.

Brott, C. A., Blackwell, D. A., and Ziagos, J. P., 1981, Thermal and tectonic implications of heat flow in the eastern Snake River Plain, Idaho: Journal of Geophysical Research, v. 86, p. 11709–11734.

Bruce, P. M., and Huppert, H. E., 1989, Thermal controls of basaltic fissure eruptions: Nature, v. 342, p. 665–667.

Christiansen, R. L., and McKee, E. H., 1978, Late Cenozoic volcanic and tectonic evolution of the Great Basin and Columbia intermontane regions, *in* Smith, R. B., and Eaton, G. P., eds., Cenozoic tectonics and regional geophysics of the Western Cordillera: Geological Society of America Memoir 152, p. 283–312.

Delaney, P. T., and Pollard, D. D., 1981, Deformation of host rocks and flow of magma during growth of minette dikes and breccia-bearing intrusions near Ship Rock, New Mexico: U.S. Geological Survey Professional Paper 1202, 61 p.

——, 1982, Solidification of basaltic magma during flow in a dike: American Journal of Science, v. 282, p. 856–885.

Fedotov, S. A., 1978, Ascent of basaltic magmas in the crust and the mechanism of basaltic fissure eruptions: International Geology Revue, v. 20, p. 33–48.

——, 1981, Magma rates in feeding conduits of different volcanic centres: Journal of Volcanology and Geothermal Research, v. 9, p. 379–394.

Furlong, K. P., and Fountain, D. M., 1986, Continental crustal underplating—Thermal considerations and seismic-petrologic consequences: Journal of Geophysical Research, v. 91, p. 8285–8294.

Greensfelder, R. W., and Kovach, R. L., 1982, Shear wave velocities and crustal structure of the eastern Snake River Plain, Idaho: Journal of Geophysical Research, v. 87, p. 2643–2653.

Gudmundsson, A., 1988, Effect of tensile stress concentration around magma chambers on intrusion and extrusion frequencies: Journal of Volcanology and Geothermal Research, v. 35, p. 179–194.

Johnson, A. M., and Pollard, D. D., 1973, Mechanics of growth of some laccolithic intrusions in the Henry Mountains, Utah, pt. I: Tectonophysics, v. 18, p. 261–309.

Koide, H., and Bhattacharji, S., 1975, Formation of fractures around magmatic intrusions and their role in ore localization: Economic Geology, v. 70, p. 781–799.

Kuntz, M. A., and seven others, 1983, Geologic and geophysical maps of the Great Rift Instant Study area, Blaine, Butte, Minidoka, and Power Counties, Idaho: U.S. Geological Survey Miscellaneous Field Studies Map MF-1462, scale 1:100,000.

Kuntz, M. A., Champion, D. E., Spiker, E. C., and Lefebvre, R. H., 1986, Contrasting magma types and steady-state, volume-predictable basaltic volcanism along the Great Rift, Idaho: Geological Society of America Bulletin, v. 97, p. 579–594.

Kuntz, M. A., Champion, D. E., Lefebvre, R. H., and Covington, H. R., 1988, Geologic map of the Craters of the Moon, Kings Bowl, and Wapi lava fields and the Great Rift volcanic rift zone: U.S. Geological Survey Miscellaneous Investigations Map I-1632, scale 1:100,000.

Kuntz, M. A., and twelve others, 1992, Geologic map of the Idaho National Engineering Laboratory and adjoining areas, eastern Idaho: U.S. Geological Survey Miscellaneous Investigations Series Map I-2330, scale 1:100,000.

Kushiro, I., Syono, Y., and Akimoto, S., 1968, Melting of peridotite at high pressures and high water pressures: Journal of Geophysical Research, v. 73, p. 6023–6029.

Lachenbruch, A. H., and Sass, J. H., 1978, Models of an extending lithosphere and heat flow in the Basin and Range province, *in* Smith, R. B., and Eaton, G. P., eds., Cenozoic tectonics and regional geophysics of the western Cordillera: Geological Society of America Memoir 152, p. 209–250.

Leeman, W. P., 1982a, Olivine tholeiitic basalts of the Snake River Plain, Idaho, *in* Bonnichsen, B., and Breckenridge, R. M., eds., Cenozoic geology of Idaho: Idaho Bureau of Mines and Geology Bulletin 26, p. 181–191.

——, 1982b, Evolved and hybrid lavas from the Snake River Plain, Idaho, *in* Bonnichsen, B., and Breckenridge, R. M., eds., Cenozoic geology of Idaho: Idaho Bureau of Mines and Geology Bulletin 26, p. 193–212.

——, 1989, Origin and development of the Snake River Plain (SRP), Idaho: 28th International Geological Congress, Guidebook T305, p. 4–12.

Leeman, W. P., and Vitaliano, C. J., 1976, Petrology of McKinney basalt, Snake River Plain, Idaho: Geological Society of America Bulletin, v. 87, p. 1777–1792.

Leeman, W. P., Vitaliano, C. J., and Prinz, M., 1976, Evolved lavas from the Snake River Plain, Craters of the Moon National Monument, Idaho: Contributions to Mineralogy and Petrology, v. 56, p. 35–60.

Machado, F., 1974, The search for magmatic reservoirs, *in* Civetta, L., Gasparini, P., Luongo, G., and Rapolla, A., eds., Physical volcanology: Amsterdam, Elsevier, p. 255–273.

Martin, D. P., and Rose, W. I., Jr., 1981, Behavioral patterns of Fuego volcano, Guatemala: Journal of Volcanology and Geothermal Research, v. 10, p. 67–81.

Merrill, R. B., Robertson, J. K., and Wyllie, P. J., 1970, Melting reactions in the system $NaAlSi_3O_8$-$KAlSi_3O_8$-SiO_2-H_2O to 20 kilobars compared with results for other feldspar-quartz-H_2O and rock-H_2O systems: Journal of Geology, v. 78, p. 558–569.

Morgan, W. J., 1972, Convection plumes and plate motions: American Association of Petroleum Geologists Bulletin, v. 56, p. 203–213.

Nakamura, K., 1969, Arrangement of parasitic cones as a possible key to regional stress field: Bulletin of the Volcanological Society of Japan, v. 14, p. 8–20.

—— , 1977, Volcanoes as possible indicators of tectonic stress orientation—Principle and proposal: Journal of Volcanology and Geothermal Research, v. 2, p. 12–16.

Pollard, D. D., 1973, Derivation and evaluation of a mechanical model for sheet intrusions: Tectonophysics, v. 19, p. 233–269.

Priestly, K., and Brune, J., 1982, Shear wave structure of the southern Volcanic Plateau of Oregon and Idaho and the northern Great Basin of Nevada from surface wave dispersion: Journal of Geophysical Research, v. 87, p. 2671–2675.

Robertson, J. B., and Wyllie, P. J., 1971, Rock-water systems with special reference to the water deficient region: American Journal of Science, v. 271, p. 252–277.

Ryan, M. P., 1987a, Elasticity and contractancy of Hawaiian olivine tholeiite and its role in the stability and structural evolution of subcaldera magma reservoirs and rift systems, *in* Decker, R. W., Wright, T. L., and Stauffer, P. H., eds., Volcanism in Hawaii: U.S. Geological Survey Professional Paper 1350, p. 1395–1447.

—— , 1987b, Neutral buoyancy and the mechanical evolution of magmatic systems, *in* Mysen, B. O., ed., Magmatic processes—Physicochemical principles: Geochemical Society Special Publication 1, p. 259–287.

Ryan, M. P., Koyanagi, R. Y., and Fiske, R. S., 1981, Modeling the three-dimensional structure of macroscopic magma transport systems: Application to Kilauea volcano, Hawaii: Journal of Geophysical Research, v. 86, p. 7111–7129.

Scandone, R., 1979, Effusion rate and energy balance of Paricutin eruption (1943–1952), Michoacan, Mexico: Journal of Volcanology and Geothermal Research, v. 6, p. 49–59.

Shaw, H. R., 1980, The fracture mechanics of magma transport from the mantle to the surface, *in* Hargraves, R. B., ed., Physics of magmatic processes: Princeton, N.J., Princeton University Press, p. 202–264.

Shaw, H. R., and Swanson, D. A., 1970, Eruption and flow rates of flood basalts, *in* Gilmour, E. H., and Stradling, D., eds., Proceedings, 2nd Columbia River Basalt Symposium: Cheney, Eastern Washington State College Press, p. 1505–1526.

Smith, R. B., and nine others, 1982, The 1978 Yellowstone–eastern Snake River Plain seismic profiling experiment–Crustal structure of the Yellowstone region and experiment design: Journal of Geophysical Research, v. 87, p. 2583–2596.

Sparlin, M. A., Braile, L. W., and Smith, R. B., 1982, Crustal structure of the eastern Snake River Plain determined from ray trace modeling of seismic refraction data: Journal of Geophysical Research, v. 87, p. 2619–2633.

Stanley, W. D., Boehl, J. E., Bostick, F. X., and Smith, H. W., 1977, Geothermal significance of magnetotelluric soundings in the eastern Snake River Plain–Yellowstone region: Journal of Geophysical Research, v. 82, p. 2501–2514.

Stickney, M. C., and Bartholomew, M. J., 1987, Seismicity and late Quaternary faulting of the northern Basin and Range province, Montana and Idaho: Seismological Society of America Bulletin, v. 77, p. 1602–1625.

Stout, M. Z., and Nicholls, J., 1977, Mineralogy and petrology of Quaternary lavas from the Snake River Plain, Idaho: Canadian Journal of Earth Sciences, v. 14, p. 2140–2156.

Stout, M. Z., Nicholls, J., and Kuntz, M. A., 1989, Fractionation and contamination processes, Craters of the Moon lava field, Idaho, 2,000–2,500 years B.P.: New Mexico Bureau of Mines and Mineral Resources Bulletin 131, p. 259.

Swanson, D. A., Wright, T. L., and Helz, R. T., 1975, Linear vent systems and estimated rates of magma production and eruption for the Yakima basalt on the Columbia Plateau: American Journal of Science, v. 275, p. 877–905.

Thompson, R. N., 1972, Melting behavior of two Snake River lavas at pressures up to 35 kb: Carnegie Institution of Washington, Geophysical Laboratory Yearbook 71, p. 406–410.

—— , 1975, Primary basalts and magma genesis—II. Snake River Plain, Idaho: Contributions to Mineralogy and Petrology, v. 52, p. 213–232.

Tilley, C. E., and Thompson, R. N., 1970, Melting and crystallization relations of the Snake River basalts of southern Idaho, USA: Earth and Planetary Sciences Letters, v. 8, p. 79–92.

Tolan, T. L., Reidel, S. P., Beeson, M. H., Anderson, J. L., Fecht, K. R., and Swanson, D. A., 1989, Revisions to the estimates of the areal extent and volume of the Columbia River Basalt Group, *in* Reidel, S. P., and Hooper, P. R., eds., Volcanism and tectonism in the Columbia River flood-basalt province: Geological Society of America Special Paper 239, p. 1–20.

Wadge, G., 1977, The storage and release of magma on Mount Etna: Journal of Volcanology and Geothermal Research, v. 2, p. 361–384.

—— , 1982, Steady state volcanism: Evidence from eruption histories of polygenetic volcanoes: Journal of Geophysical Research, v. 87, p. 4035–4049.

Waff, H. S., and Bulau, J. R., 1979, Equilibrium fluid distribution in an ultramafic partial melt under hydrostatic stress conditions: Journal of Geophysical Research, v. 84, p. 6109–6114.

White, R., and McKenzie, D., 1989, Magmatism and rift zones—The generation of volcanic continental margins and flood basalts: Journal of Geophysical Research, v. 94, p. 7685–7729.

Wilson, L., and Head, J. W., III, 1981, Ascent and eruption of basaltic magma on the earth and moon: Journal of Geophysical Research, v. 86, p. 2971–3001.

Wright, T. L., Mangan, M., and Swanson, D. A., 1989, Chemical data for flows and feeder dikes of the Yakima Basalt Subgroup, Columbia River Basalt Group, Washington, Oregon, and Idaho, and their bearing on a petrogenetic model: U.S. Geological Survey Bulletin 1821, 71 p.

Zoback, M. L., 1989, State of stress and modern deformation of the northern Basin and Range Province: Journal of Geophysical Research, v. 94, p. 7105–7128.

Zoback, M. L., and Zoback, M. D., 1980, State of stress in the conterminous United States: Journal of Geophysical Research, v. 85, p. 6113–6156.

Zohdy, A.A.R., and Stanley, W. D., 1973, Preliminary interpretation of electrical sounding curves obtained from the Snake River Plain from Blackfoot to Arco, Idaho: U.S. Geological Survey Open-File Report 73-370, 5 p.

MANUSCRIPT ACCEPTED BY THE SOCIETY JULY 19, 1991

Index

[Italic page numbers indicate major references]

Typeset by WESType Publishing Services, Inc., Boulder, Colorado
Printed in U.S.A. by Malloy Lithographing, Inc., Ann Arbor, Michigan